T0183083

Lecture Notes in Artificial Intelligence   9101

Subseries of Lecture Notes in Computer Science

LNAI Series Editors

Randy Goebel
University of Alberta, Edmonton, Canada
Yuzuru Tanaka
Hokkaido University, Sapporo, Japan
Wolfgang Wahlster
DFKI and Saarland University, Saarbrücken, Germany

LNAI Founding Series Editor

Joerg Siekmann
DFKI and Saarland University, Saarbrücken, Germany

More information about this series at http://www.springer.com/series/1244

Association for Consumer Electronics (TACE), and Texas State University – San Marcos. We would also like to thank the keynote speakers who shared their vision on applications of intelligent systems.

April 2015

Moonis Ali
Young Sig Kwon
Chang-Hwan Lee
Juntae Kim
Yongdai Kim

Moonis Ali · Young Sig Kwon
Chang-Hwan Lee · Juntae Kim
Yongdai Kim (Eds.)

Current Approaches in Applied Artificial Intelligence

28th International Conference
on Industrial, Engineering and Other Applications
of Applied Intelligent Systems, IEA/AIE 2015
Seoul, South Korea, June 10–12, 2015
Proceedings

Springer

Editors
Moonis Ali
Texas State University
San Marcos, Texas
USA

Juntae Kim
Dongguk University
Seoul
Korea, Republic of (South Korea)

Young Sig Kwon
Dongguk University
Seoul
Korea, Republic of (South Korea)

Yongdai Kim
Seoul National University
Seoul
Korea, Republic of (South Korea)

Chang-Hwan Lee
Dongguk University
Seoul
Korea, Republic of (South Korea)

ISSN 0302-9743   ISSN 1611-3349 (electronic)
Lecture Notes in Artificial Intelligence
ISBN 978-3-319-19065-5   ISBN 978-3-319-19066-2 (eBook)
DOI 10.1007/978-3-319-19066-2

Library of Congress Control Number: 2015938754

LNCS Sublibrary: SL7 – Artificial Intelligence

Springer Cham Heidelberg New York Dordrecht London
© Springer International Publishing Switzerland 2015

Printed on acid-free paper

Springer International Publishing AG Switzerland is part of Springer Science+Business Media
(www.springer.com)

Preface

The International Society of Applied Intelligence (ISAI), through its annual IEA/AIE conferences, provides a forum for the international scientific and industrial community in the field of applied artificial intelligence to interactively participate in developing intelligent systems that are needed to solve the 21st century's ever-growing problems in almost every field.

The 28th International Conference on Industrial, Engineering & Other Applications of Applied Intelligent Systems (IEA/AIE 2015), held at COEX, Seoul, Korea from June 10 to 12, 2015, followed the IEA/AIE tradition of providing an international scientific forum for researchers in the diverse field of applied artificial intelligence.

IEA/AIE 2015 accepted 72 papers for inclusion in these proceedings out of many papers submitted from 36 different countries. Each paper was reviewed by at least two members of Program Committee, thereby facilitating the selection of high-quality papers. The papers in the proceedings cover a wide range of topics in applied artificial intelligence including reasoning, robotics, cognitive modeling, machine learning, pattern recognition, optimization, text mining, social network analysis, and evolutionary algorithms. These proceedings cover both the theory and applications of applied intelligent systems. Together, these papers highlight new trends and frontiers of applied artificial intelligence and show how new research could lead to innovative applications of considerable practical significance.

It was a great pleasure for us to organize this event. However, we could not have done it without the valuable help of many colleagues around the world. First we would like to thank the Organizing Committee members and Program Committee members for their extremely hard work and the timely return of their comprehensive reports. Without them, it would have been impossible to make decisions and to produce such high-quality proceedings on time. Second, we would like to acknowledge the contributions of all the authors of the papers submitted to the conference.

We also would like to thank our main sponsor, ISAI. The following cooperating organizations deserve our gratitude: the Association for the Advancement of Artificial Intelligence (AAAI), the Association for Computing Machinery (ACM/SIGART), the Austrian Association for Artificial Intelligence (GAI), the Catalan Association for Artificial Intelligence (ACIA), the International Neural Network Society (INNS), the Italian Association for Artificial Intelligence (AI*IA), the Japanese Society for Artificial Intelligence (JSAI), Korea Business Intelligence and Data Mining Society, Korean Information Processing Society (KIPS)., the Lithuanian Computer Society - Artificial Intelligence Section (LIKS-AIS), the Spanish Society for Artificial Intelligence (AEPIA), the Society for the Study of Artificial Intelligence and the Simulation of Behaviour (AISB), the Taiwanese Association for Artificial Intelligence (TAAI), the Taiwanese

Organization

IEA/AIE 2015 Organizing Committee Members

General Chairs

Moonis Ali	Texas State University, San Marcos, TX, USA
Young Sig Kwon	Dongguk University, Seoul, Korea
Chang-Hwan Lee	Dongguk University, Seoul, Korea

Program Chairs

Juntae Kim	Dongguk University Seoul, Korea
Kyoung-Jae Kim	Dongguk University, Seoul, Korea
Jae. C. Oh	Syracuse University, USA

Local Chairs

Byungun Yoon	Dongguk University, Seoul, Korea
Mina Jung	Syracuse University, USA

Finance Chair

Yung-Seop Lee	Dongguk University, Seoul, Korea

Publication Chair

Seoung Bum Kim	Korea University, Seoul, Korea

Special Session Chair

Yongdai Kim	Seoul National University, Seoul, Korea

Web Chair

Chan-Kyoo Park	Dongguk University, Seoul, Korea

Advisory Committee

Sungzoon Cho	Seoul National University, Seoul, Korea
Hyunjoong Kim	Yonsei University, Seoul, Korea

Program Committee Members

Anwar Althari
Hakan Altincay
Gustavo Arroyo
Youngchul Bae
Alexandra Balahur
Suzanne Barber
Edurne Barrenechea
Fevzi Belli
Jamal Bentahar
Leszek Borzemski
Patrick Brezillon
Andres Bustillo
Humberto Bustince
Erik Cambria
Joao Paulo Carvalho
C.W. Chan
Chien-Chung Chan
Darryl Charles
Shyi-Ming Chen
Hyuk Cho
Sung-Bae Cho
Jyh Horng Chou
Paul Chung
Amitava Das
Aida De Haro
J. Valente de Oliveira
John Dolan
Georgios Dounias
Shaheen Fatima
Duco Ferro
Alois Ferscha
Enrique Frias-Martinez
Hamido Fujita
Nicolás García-Pedrajas
Ashok K. Goel
Maciej Grzenda
Rodolfo Haber
Francisco Herrera
Koen Hindriks
Kaoru Hirota
Emily Hou
Chih-Cheng Hung
Bipin Indurkhya
Takayuki Ito

He Jiang
Vicente Julian
Tetsuo Kinoshita
Frank Klawonn
Amruth Kumar
Bora İ Kumova
Kyoungsup Kim
Jean-Charles Lamirel
Jooyoung Lee
Greg Lee
Mark Sh. Levin
Ming Li
Hong Liu
Vincenzo Loia
Manuel Lozano
Paul Lukowicz
Kurosh Madani
Francesco Marcelloni
Jesus Maudes
Kishan Mehrotra
Riichiro Mizoguchi
Yasser Mohammad
Ariel Monteserin
Nadia Nedjah
Ngoc-Thanh Nguyen
Muaz Niazi
Shogo Okada
Hiroshi G. Okuno
Jose Angel Olivas
Gregorio Sainz Palmero
Alexis Arroyo Pena
Jing Peng
Alexander Pokahr
Don Potter
Dilip Pratihar
S. Ramaswamy
Antonio Bahamonde Rionda
Djamel F.H. Sadok
Joao M. Sousa
Kazuhiko Suzuki
Valery Tereshko
Lin-Yu Tseng
Vincent S. Tseng
Marco Valtorta

Sander van Splunter
Zsolt Janos Viharos
Martijn Warnier
Don-Lin Yang
De-Chuan Zhan
Lei Zhang

Zhi Hua Zhou
Sung-Hwa Jung
Sun Mars
Alex Syaekhoni

Reviewers

Jens Allmer
Sudhir Barai
Wenxuan Gao
Ashish Ghosh
Akira Imada
Pabitra Mitra
Jayanta Mukhopadhyay
Mahmuda Rahman

Bhavanandan Rathinasamy
Lenin Rathinasamy
Faisal Riaz
Asimava Roychoudhury
Sithu Sudarsan
Hong Qian
Brian Soeder

Contents

Optimization

Machine Learning

Classification

Intelligent Systems Applications

Theoretical AI

Uncertainty Management in Multi-leveled Risk Assessment: Context of IMS-QSE

Marwa Ben Aissia$^{(\boxtimes)}$, Ahmed Badereddine, and Nahla Ben Amor

LARODEC, Université de Tunis, Institut Supérieur de Gestion de Tunis,
41 Avenue de la liberté, cité Bouchoucha, 2000 Le Bardo, Tunisia
benaissiamarwa@gmail.com, badreddine.ahmed@hotmail.com,
nahla.benamor@gmx.fr

Abstract. Managing uncertainty in risk assessment is a crucial issue for better decision making and especially when it is adapted to the three standards ISO 9001, OHSAS 18001 and ISO 14001. This paper proposes a new risk assessment approach able to manage risk in the context of integrated management system (IMS-QSE) while taking into account the uncertainty characterizing the whole process. The proposed approach is mainly based on fuzzy set theory and Monte Carlo simulation to provide an appropriate risk estimation values and adequate decisions regarding the three management systems Quality, Security and Environment. In order to show the effectiveness of our approach, we have performed simulations on real database in the petroleum field at TOTAL TUNISIA company.

Keywords: Risk assessment · IMS-QSE · Uncertainty · Fuzzy set theory · Monte carlo simulation

1 Introduction

The risk management is defined as a set of principles and practices whose purpose is to identify, analyze, evaluate and treat eventual risk factors seen as events hindering an organization to reach their objectives. The risk management in projects is currently one of the main topics of interest for researchers and practitioners working in the area of project management. In fact, it became an important factor to ensure the integration of the three most known management systems, namely, the Quality Management System (ISO 9001), the Environmental Management System (ISO 14001) and the Occupational Health and Safety Management System (OHSAS 18001). In fact, the same source of hazard can cause several risks relative to Quality, Security and Environment (QSE) management systems. This means that the risk management can be seen as a common factor that can be used to identify and evaluate different risks with a view to achieve different QSE objectives. In this context, we are interested by the process-based approach proposed by Badreddine et al. [1] and particularly to implement its plan phase aiming to identify, analyze and evaluate risks relative to each QSE objectives.

© Springer International Publishing Switzerland 2015
M. Ali et al. (Eds.): IEA/AIE 2015, LNAI 9101, pp. 3–12, 2015.
DOI: 10.1007/978-3-319-19066-2_1

However, implementing a risk management approach in the context of an integrated management system QSE is not an easy task. In fact, the existing qualitative approaches such as *Preliminary Risk Analysis (PRA)* [11], *Failure Mode and Effects Analysis (FMEA)* [9], *Hazard and Operability Study (HAZOP)* [10] and quantitative approaches such as *Bow Tie Diagrams* [12], *Monte Carlo Analysis (MCA)* [6] are not appropriate to deal with several management areas simultaneously and to take into account the uncertainty aspect involved in the QSE Objectives. Thus, we propose in this paper, a new approach to implement the risk management process able to deal jointly with multi-leveled QSE risks and uncertainty issue.

This paper is organized as follows: Section 2 presents a brief recall on the process-based approach IMS-QSE. Section 3 discusses the new proposed approach to manage uncertainty in risk assessment regarding QSE management systems. Finally, Section 4 presents an illustrative example in the petroleum field.

2   Brief Recall on the Process-Based Approach for IMS-QSE

This section presents a brief recall on the process-based approach for implementing an integrated Quality, Security and Environment management systems proposed in [1]. This approach is based on three integrated factors: *Risk management* to increase the compatibility and the correspondence between the three systems, *Process-based approach* to deal with coordination and the interactions between different activities, *Monitoring system* to ensure the monitoring of the global system and the integration as a continuous improvement of the performance. The proposed approach using the PDCA (Plan, Do, Check, Act) is based on three phases detailed as follows:

1. **Plan phase:** ensuring the risk management process aiming to carry out the objectives. It consists in setting up all quality, security and environment objectives in order to provide their deployment in each process. In fact, all the existing risks in relation with these objectives are identified and analyzed in order to select the most critical ones leading to a possible failure to reach up the objectives. Thus, the levels of risk according to each objective are calculated. Finally, an evaluation of each selected risk is proposed in order to assist the decision makers to define the appropriate treatments related to each risk and the protective actions to reduce its severity.
2. **Do phase:** developing strategies and achieving the policies, objectives and targets according to the global management plan QSE.
3. **Check and Act phase:** finalizing the integration process by selecting and implementing the most appropriate management plan to reach the QSE objectives.

Our aim in this paper is to focus on the effective implementation of the plan phase regarding QSE management systems in order to provide multi-values of risks and to take into consideration the uncertainty characterizing the different QSE objectives.

3 New Approach to Manage Uncertainty in Multi-leveled Risk Assessment: Context of IMS-QSE

As mentioned before, our objective is to implement the *plan phase* of the process-based approach of IMS-QSE [1] to deal with multi-objective risk management and uncertainty jointly. In fact, several researches have addressed the uncertainty issue using different theories of uncertainty (e.g. probability, fuzzy set theory) [2–4]. In our case, we propose to extend the recent combined probabilistic-fuzzy approach of Arunraj and al. [4] which was basically proposed by Kentel and Aral [2]. This approach substitutes vertex method by DSW algorithm [8] and uses lognormal PDF of failure in order to capture uncertainty in a better way. It is the more appropriate and recent approach in risk management. However, the major problem is its limitation to a unique management area. For this purpose, our proposed approach named Two-dimensional QSE Fuzzy Monte Carlo Analysis (2D QSE-FMCA for short), depicted in Figure 1, is based on three main phases detailed below.

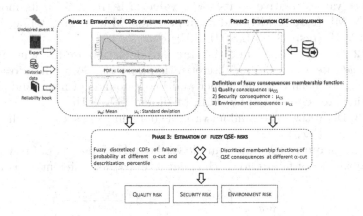

Fig. 1. Proposed uncertainty management approach: 2D QSE-FMCA

3.1 Phase 1: Estimation of Fuzzy CDFs of Failure Probability

The objective of this phase is the estimation of failure probability of an undesired event. For this purpose, we have selected the Cumulative Distribution Function (CDF) because it can be especially informative for illustrating the percentile corresponding to a particular risk level in a fuzzy way. The estimation procedure of fuzzy CDFs of failure probability was initially proposed by Kentel and Aral [3] and recently enhanced by Arunraj and al. [4]. This phase has as input the probability of an undesired event or situation represented by a continuous random variable X. Let M and S be its mean and standard deviation, respectively. M and S are used as arguments of the CDF function. These parameters are presented as fuzzy numbers and they are extracted from historical databases and/or experts opinions. Indeed, the definition of mean and standard deviation is established using triangular Membership Functions (TMF). Using these inputs, this phase aims to find the estimation of probability of failure at different percentiles. Note that Log normal distribution is used for studying the probability of events given its capacity to take only positive argument's value which is wide counseled distribution. Thus, the probability of failure for each value x of X is assumed to follow a log normal distribution expressed by:

$$F(x, M, S^2) = Pr\left(X \leq x\right) = \phi\left(\frac{\ln(x) - M}{S}\right) \tag{1}$$

where $\phi\left(\frac{\ln(x) - M}{S}\right)$ is the cumulative distribution function of a log normal distribution. This phase is composed of five steps:

- **Step 1**: Defines the triangular fuzzy membership function for mean μ_M and standard deviation μ_S of the log normally distributed random variable X. More precisely, the mean M (resp. standard deviation S) will be represented by three values relative to its lower value M.L (resp. S.L), its most likely value M.ML (resp. S.ML) and its upper value M.U (resp. S.U) i.e. M=[M.L M.ML M.U] (resp. S=[S.L S.ML S.U]).
- **Step 2**: Since the mean and the standard deviation are fuzzy numbers, we use α_cut interval $= 0.5$ for discretization purpose. So, we have $\alpha=$ [0: 0.5: 1]. Then, for each α value, upper and lower bounds of mean and standard deviation are represented by [M.L, M.U] and [S.L, S.U].
- **Step 3**: Discretizes the random variable domain (X) as x $= x_i$; where i $= 1$, 2, 3,..., n $\forall$ F(x_n)$= 1$.
- **Step 4**: Computes the upper and lower bounds of interval of fuzzy function for probability of failure for each α_cut level (α) for different x_i as follows:

$$F(x_i)^{\alpha,p} = \begin{cases} F(x_i)^{\alpha}_{lower} = \phi\left(\frac{\ln(x_i) - M.L^{\alpha}}{S.U^{\alpha}}\right) \\ F(x_i)^{\alpha}_{upper} = \phi\left(\frac{\ln(x_i) - M.U^{\alpha}}{S.L^{\alpha}}\right) \end{cases} \tag{2}$$

The outputs of this phase is a set of fuzzy CDFs (One for upper and one for lower bound) for each discretization percentile p and the correspondent triangular membership function.

In fact, two CDFs (one for upper and one for lower bound) for $\alpha = 0$ and $\alpha = 0.5$ are generated contrary to $\alpha = 1$ where we have same CDF curve for upper and lower. In the generation of CDFs for probability of failure, 5 different combinations of mean and standard deviations are used. For each mean and standard deviation combination, a number of Monte Carlo simulations were conducted to generate a CDF. This means that, the total number of CDFs is twice the number of used α_cut minus 1. In our experimental study, we select $\alpha \in [0: 0.5: 1]$. So, the total number of CDFs = 5 (see Figure 2.(a)). The CDF is informative for illustrating the percentile corresponding to a particular level of concern. For this reason, the fuzzy CDFs resulting from this phase are discretized into percentiles $p= 10\%, 20\%, \ldots, 100\%$ in order to obtain the fuzzy membership function for failure probability at each fractile. Each discretization percentile is presented graphically by an horizontal line crossing the CDFs. It consists on finding the least, most likely and highest values of the fuzzy triangular membership function of failure probability (denoted TMF.L, TMF.ML and TMF.U, respectively) for each discretization percentile p as illustrated graphically (see Figure 2.(b)).

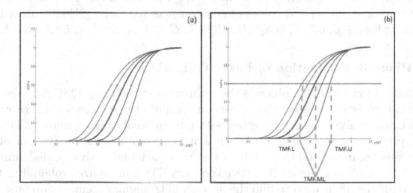

Fig. 2. CDFs of failure probability at different α_cut values (a) and at 60% of discretization percentile (b)

The risk manager will select the adequate percentile of risk estimation depending on the collected amount of information. He may choose a lower percentile of risk estimates (e.g. the 5t%) as the most representative of the risk estimate if a sufficient amount of information has been collected. Conversely, when the risk manager believes that the risk estimates are more likely to underestimate true risks, or if there is considerable uncertainty in the accuracy of the risk estimates, the risk manager may choose a higher percentile of risk estimates (e.g. the 90%).

3.2   Phase 2: Estimation of Fuzzy QSE Consequences

As mentioned before, each risk R has possible consequences on three areas i.e. quality, security and environmental. The main objective of this phase is the estimation of these fuzzy QSE consequences. This process is generally uncertain. For that, the use of fuzzy membership function is the better solution. This phase has as inputs the CQ.U (resp. $CQ.ML, CQ.L$) which is the upper (resp. most likely, lower) value for *quality* consequence, the CS.U (resp. $CS.ML, CS.L$) which is the upper (resp. most likely, lower) value for *security*consequence and the CE.U (resp. $CE.ML$, $CE.L$) which presents the upper (resp. most likely, lower) value for *environment* consequence. This values are extracted from historical data or expert's opinion. Using these inputs, this phase aims to compute the fuzzy triangular membership functions of QSE-consequences. This function is defined as follows for quality consequence μ_{CQ}:

$$\mu_{CQ} = \begin{cases} \frac{C-CQ.L}{CQ.ML-CQ.U}; & CQ.L \leq C \leq CQ.ML \\[2mm] \frac{C-CQ.U}{CQ.ML-CQ.U}; & CQ.ML \leq C \leq CQ.U \end{cases} \tag{3}$$

Note that, the same formulation is available for security and environment to compute μ_{CS} and μ_{CE} simply by replacing CQ by CS and CE In equation 3. As outputs, we find discretized values of consequences for each QSE area according to α_cut values $\in [0, 0.5, 1]$ ($CQ^{\alpha}.U$, $CQ^{\alpha}.L$, $CS^{\alpha}.U$, $CS^{\alpha}.L$, $CE^{\alpha}.U$, $CE^{\alpha}.L$).

3.3   Phase 3: Estimation of Fuzzy QSE Risks

The main objective of this phase is the estimation of the fuzzy QSE risks. Based on the two phases discussed above, we can compute these risks. In this context, Monte Carlo analysis is used together with interval analysis to generate CDFs of failure probability at each percentile p. Furthermore, interval analysis involves discretizing the membership domain of the fuzzy variable into a specified number of α_cut levels [3] in order to discretize the fuzzy CDFs of failure probability. The DSW Algorithm [8] is used to find the product of triangular membership function of failure probability and fuzzy membership function of QSE-consequences of failure μ_{CQ}, μ_{CS} and μ_{CE}.

The principle of computation of fuzzy multi-leveled QSE-risks aims to discretize each membership function of QSE-consequences using α_cut level and then, to compute for each α_cut level α and percentile of discretization p the lower, most likely and upper bounds of fuzzy QSE-risks. The computation of fuzzy membership function of quality-risk is defined as follows:

$$\mu_{RQ^{\alpha,p}} = \begin{cases} RQ.L^{\alpha,p} = min\left((TMF.L^{\alpha,p} \times CQ^{\alpha}.U), (TMF.U^{\alpha,p} \times CQ^{\alpha}.L)\right) \\[2mm] RQ.ML^{\alpha,p} = CQ.ML \times (TMF.ML^{\alpha,p}) \\[2mm] RQ.U^{\alpha,p} = max\left((TMF.L^{\alpha,p} \times CQ^{\alpha}.U), (TMF.U^{\alpha,p} \times CQ^{\alpha}.L)\right) \end{cases} \tag{4}$$

where $RQ.L^{\alpha,p}$, $RQ.ML^{\alpha,p}$ and $RQ.U^{\alpha,p}$ are the upper value where the risk level is assumed as intolerable, the most likely value where the risk occurs frequently and a lower value where the risk level is considered as negligible.

Note that, the same formulation is available for security and environment to compute $\mu_{RS^{\alpha,p}}$ and $\mu_{RE^{\alpha,p}}$ simply by replacing RQ by RS and RE in Equation 4. In total, for each discretization percentile p and each α value, three membership functions of risk are conducted namely μ_{RQ}, μ_{RS} and μ_{RE}. Frequently, an overall consequence of any failure event is handled using fuzzy numbers and combining all the consequence factors in one function. Whereas, the most commonly used consequence estimation is general and uses a simple and imprecise assimilation of losses [4]. This leads to the deterioration of quality of estimated risk value. So, with our approach, the consequence of failure probability of an undesired event is treated separately μ_{CQ}, μ_{CS} and μ_{CE} instead of an overall consequence.

3.4 Decision Making and Conformance to Compliance Guidelines

Once QSE-risks are estimated, the decision maker have to take some decisions. For that, our approach will provide some indices of uncertainty helping him to make better estimation. Then, we will deal with compliance guideline in order to estimate either a given risk is accepted or rejected.

Estimation of Support and Uncertainty Index: Based on fuzzy membership functions of QSE-risks, two meaningful indices of uncertainty can be used to manage uncertainty in order to make adequate decisions. The support (denoted SU) and the uncertainty index (denoted UI) are widely used in fuzzy literature [7]. They provide extra information about the resulting uncertainty. Indeed, the support of a membership function of fuzzy risk R is the interval value between the maximum and minimum estimated values [4]. The uncertainty index is obtained by dividing the support value with the most likely value of risk. Thus $SU = R.U - R.L$ and $UI = SU/R.ML$. The uncertainty associated with a risk that has a small SU and/or UI is respectively smaller than that of a risk which has a larger SU and/or UI. The next step consists in treating the acceptability of the estimated risks. It is computed here with respect to a maximum reference value.

Conformance to Compliance Guideline: To assess the importance of obtaining results, comparison with compliance guidelines should be discussed. The main objective is to identify either the risk is accepted or rejected for a specific given guideline (i.e. considered here as non fuzzy 'crisp' proposition). Guidelines should be set either by regulatory authorities or by expert and especially in industrial safety as the highest acceptable risk in order to avoid catastrophic situation. To this end, we propose to use the possibilistic approach proposed by Guyonnet and al. [5] in order to evaluate the proposition : *'Is the fuzzy risk R equal or lower to the Compliance Guideline CG for each QSE area ?'*. This approach is based on the estimation of two dual measures:

- Possibility measure Π for the proposition $R \leq CG$ being true expressed by:

$$\Pi\left(R \leq CG\right) = \underset{x}{Sup\ min}\left[\mu_R(x), \mu_{CG}(x)\right] \tag{5}$$

where $\mu_R(x)$ is the membership function of risk R for any value x and $\mu_{CG}(x)$ is the membership of compliance guideline CG for any value of x.

- Necessity measure N for this proposition defined by:

$$N\left(R \leq CG\right) = \underset{x}{Inf\ max}\left[1 - \mu_R(x), \mu_{CG}(x)\right] \tag{6}$$

Both of these measures provide two different forms of information used in decision making. In fact, a possibility measure can be considered as the criteria of an optimistic decision maker whereas the necessity measure can be thought as a pessimistic decision maker. Several studies have used these measures in the decision making process related to possibilistic risk assessment [2,4,5]. In our approach, we will provide both of them namely Π and N measures for better comparisons and decision making. In fact, the possibility measure is performed using the left arm of the triangular membership function whereas the necessity measure is computed using the right arm of the triangular membership function [4]. In fact, each QSE multi-leveled risk and total risk μ_R at each α_cut and p are compared to the same compliance guideline CG.

The main issue is to examine the acceptability of the estimated QSE-risks. In fact, the decision maker must be aware either the risk is rejected or accepted according to the compliance guideline in order to take earlier corrective and preventive actions. Several researches have been proposed to deal with the acceptability of a fuzzy risk [2]. Indeed, each fuzzy risk is evaluated with respect to a set of guidelines. Unless, these studies deal only with total-risk estimation. Our aim is to compare QSE-risks and relative total risk regarding the same guidelines. Our method proved that an accepted risk in quality context can be either rejected or accepted according to security or environment area. For these purposes, managing the acceptability of a fuzzy risk regarding QSE management systems is more significant and precise than overall area.

4 Illustrative Example

In order to illustrate our approach, an implementation in the petroleum field in TOTAL TUNISIA company is proposed. This company is certified in quality, security and environment management systems. We will consider three QSE objectives (O_1, O_2, O_3) and three risks (R_1, R_2, R_3)[1]:

- O_1: Gain market share by providing superior all round by decreasing the product of non conformity level service to the customer
- O_2: Increase safety staff by decreasing the number of day off of employees
- O_3: Minimize the environmental waste
- R_1: A major fire and explosion on tanker truck carrying hydrocarbon
- R_2: Dispersal of products following collision between two vehicles
- R_3: Overflow tank during reception

Table 1 proposes a comparison between the three risks R_1, R_2 and R_3 regarding QSE-objectives for $\alpha\_cut =1$ and $p= 50\%$, where nine fuzzy QSE-risks are calculated. It is clear that R_1 has the highest effect on quality and environment areas while R_3 is the maximum contributor in security area with the highest value of risk which reaches 8.716E-02 for 50% percentile. Table 2 presents the corresponding total risks R_1, R_2 and R_3. It is clear that the total risk estimation can not provide a detailed estimation for decision maker since it provides a global vision.

---

[1] For the lack of space we cannot give all the numerical data.

Table 1. R_1, R_2 and R_3 values for $p = 50\%$ and $\alpha\_cut=1$

Quality	R_1: [4.265E-02 5.093E-02 6.037E-02] R_2:[2.756E-02 3.296E-02 3.911E-02] R_3:[3.753E-02 4.494E-02 5.340E-02]
Security	R_1:[5.761E-02 6.891E-02 8.180E-02] R_2:[5.636E-02 6.741E-02 8.001E-02] R_3:[6.135E-02 7.340E-02 8.716E-02]
Environment	R_1:[8.79E-02 1.0486E-01 1.2431E-01] R_2:[3.379E-02 4.045E-02 4.804E-02] R_3:[7.145E-02 8.539E-02 1.0127E-01]

Table 2. Fuzzy total risks R_1, R_2 and R_3 values for $p = 50\%$ and $\alpha\_cut=1$

Identified risks / TMF values	R_1	R_2	R_3
Lower	6.268253E-02	3.923893E-02	5.678007E-02
Most-likely	7.490000E-02	4.693733E-02	6.790933E-02
Upper	8.882373E-02	5.572320E-02	8.060813E-02

Comparison with compliance guidelines should be discussed to prove the efficiency of our approach. Based on the expert's judgments of the petroleum field in TOTAL TUNISIA company, three guidelines have been identified namely CG_1, CG_2 and CG_3. Our main objective is to identify if the risk is accepted or rejected according to these guidelines. For instance if we focus on the risk R_1, the three guidelines can be defined as follows:

- CG_1: QSE risks at $\alpha\_cut=0$ and $p=50\% \leq 9.1$E-02 for $\Pi= 0.34$.
- CG_2: QSE risks at $\alpha\_cut=0.5$ and $p=10\% \leq 4.2$E-02 for $N= 0.53$.
- CG_3: QSE risks at $\alpha\_cut=0.5$ and $p=60\% \leq 8$E-02 for $\Pi= 0.82$.

For the sake of simplicity, we will present only the acceptability of R_1 at $\alpha\_cut=0$ and $p=50\%$ with different guidelines. Table 3 shows the acceptability of the QSE fuzzy risks and the total risk R_1 for different compliance guidelines.

Table 3. Comparison of fuzzy R_1 QSE-risks to various compliance guidelines

Guidelines / QSE and total risks	CG_1	CG_2	CG_3
Quality	Accepted	Rejected	Accepted
Security	Accepted	Rejected	Rejected
Environment risk	Rejected	Rejected	Rejected
Total risk	Accepted	Rejected	Rejected

Clearly, the quality and security risks at $\alpha\_cut=0$ and $p=50\%$ have not exceed 9.1E-02 for a possibility measure Π of 0.34 according to CG_1. However, they have greater values then 4.2E-02 at $\alpha\_cut=0.5$ and $p=10\%$ for a necessity measure N of 0.53 which according to CG_2. For example, according to CG_1, the risk R_1 is accepted on quality and security areas. However, it is rejected on environment area. To conclude, conducting QSE-risks estimation instead of total estimation seems more adequate for better decision making. This estimation can help decision makers to select preventive and corrective decisions according to the acceptability of the risk regarding specific area.

5   Conclusion

This paper proposes a new approach to implement the plan phase of the process-based approach for integrating Quality, Security and Environment management systems that was proposed in [1]. The proposed approach is able to manage multi-leveled risk estimation in an uncertain framework regarding QSE areas. It is based on a combined risk assessment approach for reasoning under uncertainty. This choice was motivated by the fact that this kind of approaches have proven its capacities in capturing sparse and vague data. Indeed, probability theory and fuzzy set theory have been combined to integrate both variability and uncertainty in risk estimation. Comparison with compliance guidelines were also established to discuss the acceptability of given estimated risks. An interesting future work is to take into account the dependence between different risks which will give a better fit to real problems.

References

1. Badreddine, A., Ben Romthan, T., Ben Amor, N.: A Multi-Objective Approach to Generate an Optimal Management Plan in an IMS-QSE. Electronic Engineering and Computing Technology **60**, 335–347 (2010)
2. Kentel, E., Aral, M.M.: 2D Monte Carlo versus 2D fuzzy Monte Carlo health risk assessment. Stochastic Environmental Research and Risk Assessment **19**(1), 86–96 (2005)
3. Kentel, E., Aral, M.M.: Probabilistic-fuzzy health risk modeling. Stochastic Environmental Research and Risk Assessment **18**(5), 324–338 (2004)
4. Arunraj, N.S., Mandal, S., Maiti, J.: Modeling uncertainty in risk assessment: An integrated approach with fuzzy set theory and Monte Carlo simulation. Accident Analysis Prevention **55**, 242–255 (2013)
5. Guyonnet, D., Bourgine, B., Dubois, D., Fargier, H., Côme, B., Chilès, J.P.: Hybrid approach for addressing uncertainty in risk assessments. Journal of Environmental Engineering **129**(1), 68–78 (2003)
6. Schuhmacher, M., Meneses, M., Xifro, A., Domingo, J.L.: The use of Monte-Carlo simulation techniques for risk assessment: study of a municipal waste incinerator. Chemosphere **43**(4), 787–799 (2001)
7. Ross, T.J.: Fuzzy logic with engineering applications. John Wiley & Sons (2009)
8. Dong, W.M., Shah, H.C., Wongt, F.S.: Fuzzy computations in risk and decision analysis. Civil Engineering Systems **2**(4), 201–208 (1985)
9. Do, D.: Procedure for Performing a Failure Mode, Effects and Criticality Analysis (1980)
10. Kletz, T.A.: The origins and history of loss prevention. Process Safety and Environmental Protection **77**(3), 109–116 (1999)
11. Fullwood, R.R., Hall, R.E.: Probabilistic risk assessment in the nuclear power industry (1988)
12. Couronneau, J.C., Tripathi, A.: Implementation of the new approach of risk analysis in france. In: 41st International Petroleum Conference, Bratislava (2003)

Implementation of Theorem Prover
of Relevant Logic

Noriaki Yoshiura[✉]

Department of Information and Computer Sciences, Saitama University,
255, Shimo-ookubo, Sakura-ku, Saitama, Japan
yoshiura@fmx.ics.saitama-u.ac.jp

Abstract. The formalization of human deductive reasoning is a main issue in artificial intelligence. Although classical logic is one of the most useful ways for the formalization, the material implication of classical logic has some fallacies. Relevant logic has been studied for the removal of material implication fallacies in classical logic for the formalization of human deductive reasoning. Relevant logic ER is free from fallacies of material implication and has more provability than typical relevant logic R. Moreover, ER has a decision procedure, even though almost all relevant logics are undecidable. This is one of the reasons why ER is an appropriate logic for the formalization of human deductive reasoning. This paper implements automated theorem prover of Relevant logic ER and experiments on this theorem prover to check the processing time of implemented automated theorem prover.

1 Introduction

In classical logic, $A \to B$ is inferred from B even if A is an arbitrary formula and has no relation with B. However, from the viewpoint of the meaning of implication in natural languages, the inference of $A \to B$ requires some relation between A and B. Thus, it is not reasonable to infer $A \to B$ from B for any arbitrary formula A. In this sense, the implication in classical logic has fallacies.

Relevant logic has been studied from the viewpoint of philosophy[2]. The aim of relevant logic is to remove the fallacies of implication from classical logic. The formalization of human reasoning is a main issue in artificial intelligence. As a method of formalization of knowledge reasoning, relevant logic is more suitable than classical logic in the sense that the fallacies of implication are removed[12]. Several relevant logics have been proposed. Church[3] and Moh[6] proposed the implication fragment logic $R_\to$. Ackermann proposed the concept of entailment, which is the most strict implication[1]. Anderson and Belnap proposed full logic R, which includes $R_\to$. They also proposed E that includes entailment. Moreover, many other logics have been proposed[8].

R is a typical relevant logic but undecidable[9]. E is also undecidable[2]. Thus, decidable relevant logics have been researched. The contraction rule in sequent calculus is a problem for constructing decidable relevant logics[5]. In logics with this rule, it is difficult to examine whether there is a proof of a formula to

© Springer International Publishing Switzerland 2015
M. Ali et al. (Eds.): IEA/AIE 2015, LNAI 9101, pp. 13–22, 2015.
DOI: 10.1007/978-3-319-19066-2_2

decide that the formula is a theorem, because the maximum number of uses of the contraction rule in a proof is difficult to decide. Some relevant logics without contraction rule are proposed [7]. Fallacies of implication are removed from these logics, but they are weaker logics than R.

In [10,11], one of the causes of fallacies is redundant inference. Decidable relevant logic ER is proposed by prohibiting such inferences [10,11]. Relevant logic ER is free from fallacies of implication and it is suitable for formalizing knowledge reasoning[10,11]. ER is a labeled deductive system[4]. ER uses an attribute value to remove fallacies of implication. The decidability of ER is proved by constructing decision procedure[11]. There are a few decidable relevant logics and there are no theorem provers of decidable relevant logics. Thus, this paper implements a theorem prover of Relevant logic ER and evaluates the implemented theorem prover by experiment.

This paper is organized as follows; section 2 explains relevant logic ER. Section 3 gives decision procedure of ER. Section 4 implements the theorem prover based on the decision procedure. Section 5 shows the experiments of the theorem prover. Section 6 concludes this paper.

2   Relevant Logic ER

Relevant logic ER is defined as a sequent style natural deduction system [11]. The difference between ER and usual natural deduction is that ER is a kind of labeled deduction system[4]. A formula has an attribute value, which shows how the formula is inferred and which kind of rule can be applied to the formula. Using attribute values restricts the applicability of inference rules to remove the fallacies of implication from ER.

Definition 1. *Atomic propositions are formulas of ER. If A and B are formulas of ER, ¬A, A ∧ B, A ∨ B and A → B are formulas of ER.*
"e", "i" and "r" are defined to be attribute values of formula. "e" indicates that a formula with such an attribute value can be major premise of elimination rules. "i" indicates that a formula with such an attribute value can not be major premise of elimination rules. "r" indicates that a formula with such an attribute value can be discharged by the rule RAA.

In the following, φ, φ_1, $\cdots$, ϕ,ϕ_1, $\cdots$ are used as meta variables of attribute value. $A : \varphi$ is defined to be *attribute formula* where A is a formula and φ is an attribute value.

Definition 2. $\Gamma \vdash A$ *is defined to be sequent where Γ is a multiset of attribute formulas and A is an attribute formula or an empty formula. A is a theorem of ER if and only if there exists a proof of $\vdash A : \varphi$ by using the natural deduction system in Fig.1.*

This paper uses the natural deduction system ERM in Fig.1 for ER because ERM is used in decision procedure[11]. ERM is a modified version of the original

$$A : e \vdash A : e \qquad \neg A : r \vdash \neg A : r \qquad \text{Axiom}$$

$$\frac{\Gamma, \neg A : r \vdash}{\Gamma \vdash A : e} \; RAA \qquad \frac{\Gamma, \Pi_1 \vdash A : \varphi_1 \quad \Delta, \Pi_2 \vdash B : \varphi_2}{\Gamma, \Delta, \Pi \vdash A \wedge B : i} \; \wedge I$$

$$\frac{\Gamma \vdash A \wedge B : e}{\Gamma \vdash A : e} \; \wedge E1 \qquad \frac{\Gamma \vdash A \wedge B : e}{\Gamma \vdash B : e} \; \wedge E2 \qquad \frac{\Gamma \vdash A : \varphi}{\Gamma \vdash A \vee B : i} \; \vee I1 \qquad \frac{\Gamma \vdash B : \varphi}{\Gamma \vdash A \vee B : i} \; \vee I2$$

$$\frac{\Gamma, \Pi_3 \vdash A \vee B : e \quad \Delta_1, \Pi_1, A : e \vdash C : e \quad \Delta_2, \Pi_2, B : e \vdash C : e}{\Gamma, \Delta_1, \Delta_2, \Pi \vdash C : e} \; \vee E1$$

$$\frac{\Gamma, \Pi_3 \vdash A \vee B : e \quad \Delta_1, \Pi_1, A : e \vdash C : i \quad \Delta_2, \Pi_2, B : e \vdash C : \varphi}{\Gamma, \Delta_1, \Delta_2, \Pi \vdash C : i} \; \vee E2$$

$$\frac{\Gamma, \Pi_3 \vdash A \vee B : e \quad \Delta_1, \Pi_1, A : e \vdash C : \varphi \quad \Delta_2, \Pi_2, B : e \vdash C : i}{\Gamma, \Delta_1, \Delta_2, \Pi \vdash C : i} \; \vee E3$$

$$\frac{\Gamma, \Pi_3 \vdash A \vee B : e \quad \Delta_1, \Pi_1, A : e \vdash \quad \Delta_2, \Pi_2, B : e \vdash}{\Gamma, \Delta_1, \Delta_2, \Pi \vdash} \; \vee E4$$

$$\frac{\Gamma, A : e \vdash B : \varphi}{\Gamma \vdash A \to B : i} \to I \qquad \frac{\Gamma, \Pi_1 \vdash A \to B : e \quad \Delta, \Pi_2 \vdash A : \varphi}{\Gamma, \Delta, \Pi \vdash B : e} \to E$$

$$\frac{\Gamma, \Pi_1 \vdash A : \varphi \quad \Delta, \Pi_2 \vdash \neg A : e}{\Gamma, \Delta, \Pi \vdash} \; \neg E1 \qquad \frac{\Gamma, \Lambda \vdash A : e \quad \neg A : r \vdash \neg A : r}{\Gamma, \neg A : r \vdash} \; \neg E2 \qquad \frac{\Gamma, A : e \vdash}{\Gamma \vdash \neg A : i} \; \neg I$$

$$\frac{\Gamma, \neg A : e \vdash B : \varphi}{\Gamma \vdash A \vee B : i} \; EM1 \qquad \frac{\Gamma, \neg B : e \vdash A : \varphi}{\Gamma \vdash A \vee B : i} \; EM2$$

$$\frac{\Gamma, \Pi_1 \vdash A \vee B : e \quad \Delta, \Pi_2, A : e \vdash}{\Gamma, \Delta, \Pi \vdash B : e} \; DS1 \qquad \frac{\Gamma, \Pi_1 \vdash A \vee B : e \quad \Delta, \Pi_2, B : e \vdash}{\Gamma, \Delta, \Pi \vdash A : e} \; DS2$$

(*1)

Π_1, Π_2, Π_3 and Π satisfy the following conditions in $\wedge I$, $\to E$, $\neg E1$, $DS1, DS2, \vee E1, \vee E2, \vee E3$ and $\vee E4$.
$\Pi_1 \subseteq \Pi \cup C(\Pi)$, $\Pi_2 \subseteq \Pi \cup C(\Pi)$, $\Pi_3 \subseteq \Pi \cup C(\Pi)$, $\Pi \subseteq \Pi_1 \cup \Pi_2 \cup \Pi_3$

(*2)

In $\neg E2$, Λ satisfy the following conditions.

 – If $\neg A : e \in \Gamma$, $\Lambda = \emptyset$
 – If $\neg A : e \notin \Gamma$, Λ is $\{\neg A : e\}$ or $\emptyset$.

We define $C(\Pi)$ to be a multiset which does not include the same plural elements as follows: $C(\Pi) = \{\neg P : e \mid \neg P : r \in \Pi \text{ and } \neg P : e \notin \Pi\}$

Fig. 1. Inference rules of ERM

natural deduction system, which is presented in [11]. The difference between ERM and the original natural deduction is that ERM has no rule corresponding to contraction rules of the original natural deduction. Each rule of ERM has the function of contraction rules.

ER has the the following properties [10, 11].

1. Removal of fallacy of relevance
   $A \to (B \to A)$ is not a theorem in ER.
2. Removal of fallacy of validity
   $A \to (B \vee \neg B)$ and $(A \wedge \neg A) \to B$ are not theorems in ER.
3. Variable-sharing
   If $A \to B$ is a theorem, A and B contain an atomic proposition in common.
4. Disjunctive Syllogism

In ER, Disjunctive Syllogism(DS) holds. In almost all relevant logics, this rule does not hold.

$$\frac{A \vee B \quad \neg A}{B} \; DS$$

5. Provability
   ER is stronger than relevant logic R.
6. Decidability
   ER is decidable, while almost all relevant logics are undecidable.

Variable-sharing is a necessary property of relevant logic and fallacies of implication are removed from ER as much as from R. Thus, ER is more suitable for formalization of knowledge base reasoning than R.

$$\frac{\dfrac{\dfrac{A : e \vdash A : e \quad B : e \vdash B : e}{A : e, B : e \vdash A \wedge B : i} \wedge I}{A : e \vdash B \to A \wedge B : i} \to I}{\vdash A \to (B \to A \wedge B) : i} \to I$$

Fig. 2. A proof tree of $A \to (B \to A \wedge B)$

3   Decision Procedure of ER

To decide whether a formula is a theorem of ER requires to construct a proof of the formula in ERM. This section explains the decision procedure of ER.

Definition 3 (Proof tree). *Proof tree is defined to be a tree which satisfies the following conditions.*

- *A node is sequent.*
- *Child node and parent node have a relation of upper side and lower side of one of the rules of ERM.*
- *Axiom of ERM occurs only as a leaf of a proof tree.*

A proof tree of sequent $\Gamma \vdash A : \varphi$ is defined as a proof tree whose root node is $\Gamma \vdash A : \varphi$. A proof tree of formula A is defined to be a proof tree of sequent $\vdash A : i$. A proof tree is closed if and only if all leaves of it are axioms.

Fig.2 is one of the proof trees of $A \to (B \to A \wedge B)$ (there exist many proof trees of this formula). The bottom sequent of Fig.2 is the root of this proof tree and $A : e \vdash A : e$ and $B : e \vdash B : e$ are leaves. The child nodes of $A : e, B : e \vdash A \wedge B : i$ are $A : e \vdash A : e$ and $B : e \vdash B : e$. This proof tree is closed because all leaves of it are axioms.

Definition 4 (Expansible node). *A node $\Gamma \vdash A : \varphi$ of a proof tree is an expansible node if and only if it is not an axiom and has no ancestor $\Delta \vdash A : \varphi$ satisfying the following conditions.*

- $\Delta \subseteq \Gamma$
- If $B : \phi \in \Gamma - \Delta$, $B : \phi \in \Delta$.

A leaf of a proof tree is an expansible leaf if and only if it is expansible node.

For example, a node $\Gamma, A : e, A : e \vdash$ is not expansible if $\Gamma, A : e \vdash$ is an ancestor of it.

Definition 5 (Subformula Condition). *Subformula Condition of $\Gamma \vdash A : \varphi$ is that if φ is e, there exists $P : \phi \in \Gamma$ such that $A \sqsubseteq P^1$. A proof tree satisfies Subformula Condition if and only if all nodes of a proof tree satisfy Subformula Condition.*

Definition 6 (Proof Tree Construction Procedure). *Proof Tree Construction Procedure constructs a set of proof trees of A as follows:*

Step1 *Let S be a set of proof trees which includes only a proof tree that consists of the node $\vdash A : i$.*

Step2 *If S includes no proof tree which has an expansible leaf, this procedure terminates.*

Step3 *If S includes a proof tree T which has an expansible leaf, T is removed from S. By applying ERM rules to expand leaves of T, this procedure constructs possible proof trees satisfying Subformula Condition as much as possible. All the constructed proof trees are added to S. Go to Step 2.*

The following briefly explains Proof Tree Construction Procedure; this procedure constructs proof trees from $\vdash A : i$ by repeatedly applying ERM rules to expansible leaves and creating new leaf nodes. Applying the ERM rules which have one sequent in upper side obtains one proof tree, but applying the ERM rules which have more than one sequents in upper side obtains more than one proof trees. The number of obtained proof trees is equal to the number of ways of dividing a multiset Γ where $\Gamma \vdash A : \varphi$ is an expansible leaf. In the expansion of proof trees, Subformula Condition must be satisfied to terminate this procedure.

Definition 7 (Decision Procedure). *Decision Procedure decides that A is a theorem of ER if and only if Proof Tree Construction Procedure obtains a closed proof tree of formula A.*

This decision procedure is complete and sound[11]. This decision procedure terminates for any formula[11].

4   Implementation of Theorem Prover of *ER*

This section implements the theorem prover of *ER* based on the decision procedure. The implemented theorem prover constructs proof trees of a formula. If a

---

[1] $A \sqsubseteq B$ if and only if A is a subformula of B and B is not $\neg A$. $A \sqsubset B$ if and only if $A \sqsubseteq B$ and A is not B.

closed proof tree is constructed, the theorem prover decides that the formula is a theorem and otherwise, the theorem prover decides that the formula is not a theorem. Improving the theorem prover must solve the following problems.

– **First problem**
  A point of the improvement is repetition of constructing the same proof tree; in the theorem prover, inference rules are applied to a sequent in a proof tree to construct proof trees. However, the same sequent occurs repeatedly in the construction of proof trees. For example, to apply $\vee E1$ to sequent $A \to B : e, A : e, B : e \vdash B : e$ has 48 patterns of the child nodes. Moreover, many kinds of applications of inference rules may obtain the same sequent as leaf nodes. The inference rules $\vee E1$, $\vee E2$ and $\vee E3$ obtain child nodes that are the same as the parent nodes. To omit constructing the same proof trees is important to decrease processing time of the theorem prover.
– **Second Problem**
  Second problem is that several proof trees can be constructed from one sequent. Several inference rules can be applied to one sequent to construct proof trees of the sequent. For example, $\vee E2$, $\vee E3$, $\to E$ and $\to I$ can be applied to $A \to (A \to B) : e, A \vee B : e \vdash A \to B : i$. As described in the first problem, there are many patterns of applying inference rules. To check whether a formula is a theorem requires to construct many proof trees. To decrease constructed proof trees is important to decrease processing time of the theorem prover.

This paper uses the following speed-up techniques to implement the theorem prover.

1. Removing the same sequent in the construction of proof trees
   In construction of proof trees, if the sequent that has already occurred occurs again, the theorem prover decides whether closed proof trees can be obtained from the sequent without constructing proof trees from the sequent.
2. Parallel processing
   Multi threads are used to improve the theorem prover. In applying inference rules, which have several premises, to a sequent $\Gamma \vdash A : \varphi$, the theorem prover generates threads for each premise of the inference rules. Each thread checks whether there is a closed proof tree whose root is each premise. If one of threads concludes that there is no closed proof tree, applying the inference rules to the sequent $\Gamma \vdash A : \varphi$ obtains no closed proof tree. Thus, the theorem prover stops all the threads and applies another inference rule to the sequent to obtain a closed proof tree. To use multi threads, the theorem prover stops constructing unclosed proof trees.
3. SAT solver
   A formula that is not a theorem of classical logic is not a theorem of relevant logic and theorem proving in classical logic is easier than that in relevant logic. SAT solver is a theorem prover for classical logic. The theorem prover uses MiniSat[13], which is one of the best SAT solver, to check whether a formula is a theorem. If the formula is not a theorem of classical logic, the theorem prover decides that the formula is not a theorem.

4. Restriction of application of inference rules

This paper restricts the following application of inference rules to improve the theorem prover.

- The inference rule RAA is not applied to the sequent whose right side is of the form $A \wedge B$, $A \rightarrow B$ or $\neg A$.
- The inference rule $\vee E$ is not applied to the sequent whose right side is of the form $A \rightarrow B$ or $\neg A$.

This restriction does not decrease provability of the theorem prover because of the following reason; if the application of the restricted inference rules to the sequent obtains a closed proof tree, a closed proof tree is obtained without the application of the restricted inference rules. Fig.3 and Fig.4 are closed proof trees of $\Gamma \vdash C \rightarrow (A \rightarrow B) : i$. In Fig.3, RAA is applied to the sequent whose right side is $A \rightarrow B$. To modify the proof in Fig.3 obtains the proof in Fig.4, where RAA is not applied to the sequent. Therefore, the restriction does not decrease provability of the theorem prover.

$$\frac{\neg(A \rightarrow B) : r \vdash \neg(A \rightarrow B) : r \quad \Delta, C : e \vdash A \rightarrow B : e}{\Delta, C : e, \neg(A \rightarrow B) : r \vdash} \neg E2$$

$$\frac{\Gamma, C : e, \neg(A \rightarrow B) : r \vdash}{\dfrac{\Gamma, C : e \vdash A \rightarrow B : e}{\Gamma \vdash C \rightarrow (A \rightarrow B) : i} \rightarrow I} RAA$$

Fig. 3. Proof tree 1

$$\frac{\neg B : r \vdash \neg B : r \quad \dfrac{\Delta, C : e \vdash A \rightarrow B : e \quad A : e \vdash A : e}{\Delta, A : e, C : e \vdash B : e} \neg E2}{\Delta, C : e, A : e \neg B : r \vdash} \neg E2$$

$$\frac{\Gamma, C : e, \neg B : r \vdash}{\dfrac{\dfrac{\Gamma, C : e, A : e \vdash B : e}{\Gamma, C : e \vdash A \rightarrow B : i} \rightarrow I}{\Gamma \vdash C \rightarrow (A \rightarrow B) : i} \rightarrow I} RAA$$

Fig. 4. Proof tree 2

5. Usage of subformula condition

In ER, if $\Gamma \vdash A : e$ is proved, Γ includes $P : \varphi$ such that $A \sqsubseteq P$[11]. Moreover, if $\Gamma \vdash A : e$ is proved and Γ includes $\neg B : r$, Γ includes $Q : \varphi$ such that $B \sqsubseteq Q$. Thus, in constructing proof trees, the inference rules that can be applied to an expansible leaf whose right side has attribute value e are restricted. The inference rules that can be applied to an expansible leaf whose left side includes $\neg B : r$ are also restricted. Therefore, if several sequents are generated in construction of proof trees, among the sequents the theorem prover selects the sequent whose right side has attribute value e and searches a closed proof tree of the sequent first. Similarly, the theorem prover selects the sequent whose left side includes $\neg B$ and and searches a closed proof tree of the sequent first. This strategy of the theorem prover enables to decrease the time that is taken by the theorem prover.

6. Reusage of construction of proof trees

In the construction of proof tree, the theorem prover tries to construct closed proof trees. However, the theorem prover does not always construct closed proof trees and obtains failure examples of constructing closed proof trees. The failure examples are useful to avoid constructing proof trees that are unuseful to construct closed proof trees; if the theorem prover tries to construct a closed proof tree from a sequent whose closed proof tree is not proved to be constructed because of the failure examples, the theorem prover omits constructing a closed proof tree from a sequent.

The theorem prover also obtains the successful examples of constructing closed proof trees. If the theorem prover tries to construct a closed proof tree from a sequent whose closed proof tree is proved to be constructed because of the successful examples, the theorem prover omits constructing a closed proof tree from a sequent and decides that there is a closed proof tree from the sequent.

For example, only the proof tree in Fig.5 is obtained from the following sequent $P : e, P : e \vdash P : e$. Closed proof trees are not obtained from $P : e, P : e \vdash P : e$ because no inference rules can applied to $P : e, P : e, \neg P : e, \neg P : e \vdash$. Therefore, the theorem prover decides that the sequent $P : e, P : e \vdash P : e$ has no closed proof trees without trying to construct a closed proof tree.

$$
\cfrac{\cfrac{\cfrac{\cfrac{P : e, P : e, \neg P : e \vdash P : e \quad \neg P : r \vdash \neg P : r}{P : e, P : e, \neg P : e, \neg P : r \vdash} \neg E2}{P : e, P : e, \neg P : e \vdash \neg P : e} RAA \quad \neg P : r \vdash P : r}{P : e, P : e, \neg P : r \vdash} \neg E2}{P : e, P : e \vdash P : e} RAA
$$

Fig. 5. Proof tree of $P : e, P : e \vdash P : e$

5   Experiment

The experiment of this paper compares the naive implementation and the revised implementation of the theorem prover. The environment of the experiment is as follows; OS is Fedora 16 Linux, CPU is Core i7 and memory size is 24GB. Table 1 shows the result of the experiment. In the table, "o" represents that the implemented theorem prover decides that a formula is a theorem and "×" represents

Table 1. The result of experiment

Input		S	A	B	C	D	E	F	G
		Time(second)							
$(A \to B) \to ((B \to C) \to (A \to C))$	○	0.01	0.02	0.02	0.04	0.01	0.01	0.01	1.01
$(A \to (A \to B)) \to (A \to B)$	○	0.00	0.00	0.01	0.82	0.00	0.00	0.00	0.81
$((A \to C) \wedge (B \to C)) \to ((A \vee B) \to C))$	○	-	-	-	-	-	-	-	1.63
$(A \wedge (B \vee C)) \to ((A \wedge B) \vee C)$	○	-	0.08	-	-	-	-	-	13.31
$(A \to \neg A) \to \neg A$	○	0.00	0.00	0.01	0.71	0.00	0.00	0.00	0.82
$(A \to \neg B) \to (B \to \neg A)$	○	0.00	0.00	0.00	0.40	0.00	0.00	0.00	0.48
$A \to (B \to A)$	×	0.00	0.00	0.00	0.10	0.00	0.00	0.00	0.14
$A \to (B \vee \neg B)$	×	-	0.44	0.00	1.18	-	-	-	1.36
$(A \to B) \to ((A \to B) \to (A \to B))$	×	-	-	-	-	-	-	-	214.19
$(A \vee B) \to ((A \vee B) \to (A \vee B))$	-	-	-	-	-	-	-	-	-

the implemented theorem prover decides that a formula is not a theorem. "−" represents that the theorem prover does not obtain decision after the theorem prover runs for two days. The paper implements the theorem prover with the several techniques and obtains 8 kinds of implementation as follows.

S. Naive implementation without any technique
A. Removing the same sequent in the construction of proof trees
B. Parallel processing
C. SAT solver
D. Restriction of application of inference rules
E. Usage of subformula condition
F. Reusage of construction of proof trees
G. Using all techniques from A. to F.

Table 1 shows that the implementations of D. E. and F. do not change the processing time. One of the reasons is that the techniques of D. E. and F. are available to several kinds of sequents and are not used in the process of checking the formulas in the experiment. The implementation of C. requires more time than that of S. One of the reasons is that SAT solver requires processing time. The techniques of A. and B. enables the theorem prover to handle the formulas that cannot be handled by the naive implementation of the theorem prover. The implementations from A. to F. do not decide that $(A \to B) \to ((A \to B) \to (A \to B))$ is a theorem or not, but the implementation of G decides that $(A \to B) \to ((A \to B) \to (A \to B))$ is a theorem. Thus, the techniques from A. to F. are efficient to implement the theorem prover. Whether $(A \vee B) \to ((A \vee B) \to (A \vee B))$ is a theorem cannot be decided by all the implementations for two days. The result of the experiment shows that the formulas with logical operator "∨" is difficult to handle in the theorem prover. Thus, the inference rules $\vee E_1$, $\vee E_2$, $\vee E_3$, $\vee E_4$, $\vee I_1$ and $\vee I_2$ are important to improve the theorem prover.

6   Conclusion

This paper implements the theorem prover of relevant logic ER. The naive implementation of the theorem prover is not powerful because it cannot handle many formulas. This paper revises the implementation by several techniques. The experiment shows that the revised implementations can handle several formulas that cannot be handled by the naive implementation. One of the future works is to speed up the theorem prover.

References

1. Ackermann, W.: Begründung einer strenger Implikation. Journal of Symbolic Logic **21**, 113–128 (1956)
2. Anderson, A.R, Belnap, N.D.: Entailment: The logic of relevance and necessity, vol. 1. Princeton University Press (1975)
3. Alonzo, C.: The weak positive implicational propositional calculus (abstract). The Journal of Symbolic Logic **16**, 238 (1951)
4. Gabbay, D.M.: Labelled Deductive Systems, Oxford Logic Guides 33, vol. 1 (1996)
5. Gabbay, D., Guenthner, F. (eds.): Handbook of philosophical logic, vol. 3, pp. 150–180 (1986)
6. Moh, S.-K.: The deduction theorems and two new logical systems. Methodos **2**, 6–75 (1950)
7. Brady, R.T.: Gentzenization and Decidability of Some Contraction-Less Relevant Logic. Journal of Philosophical Logic **20**, 97–118 (1991)
8. Ross, T.B.: Relevant Implication and the Case for a Weaker Logic. Journal of Philosophical Logic **25**, 151–183 (1996)
9. Urquhart, A.: The Undecidability of Entailment and Relevant Implication. Journal of Philosophical Logic **49**, 1059–1073 (1984)
10. Yoshiura, N., Yonezaki, N.: Provability of relevant logic ER. In: Information Modeling and Knowledge Bases XIII, pp. 100–114. IOS Press (2001)
11. Yoshiura, N., Yonezaki, N.: A Decision Procedure for the Relevant Logic ER, Tableaux 2000, published in University of St. Andrews Research Report CS/00/01, pp. 109–123 (2000)
12. Cheng, J.: The Fundamental Role of Entailment in Knowledge Representation and Reasoning. Journal of Computing and Information **2**(1), 853–873 (1996)
13. The MiniSat Page. http://minisat.se/

Finding Longest Paths in Hypercubes: 11 New Lower Bounds for Snakes, Coils, and Symmetrical Coils

Seth J. Meyerson[1(✉)], Thomas E. Drapela[2], William E. Whiteside[1], and Walter D. Potter[1,2]

[1] Computer Science Department, University of Georgia, Athens, USA
Seth.Meyerson@gmail.com
[2] Institute for Artificial Intelligence, University of Georgia, Athens, Georgia, USA

Abstract. Since the problem's formulation by Kautz in 1958 as an error detection tool, diverse applications for long snakes and coils have been found. These include coding theory, electrical engineering, and genetics. Over the years, the problem has been explored by many researchers in different fields using varied approaches, and has taken on additional meaning. The problem has become a benchmark for evaluating search techniques in combinatorially expansive search spaces (NP-complete Optimizations).

We build on our previous work and present improved heuristics for Stochastic Beam Search sub-solution selection in searching for longest induced paths: open (snakes), closed (coils), and symmetric closed (symmetric coils); in n dimensional hypercube graphs. Stochastic Beam Search, a non-deterministic variant of Beam Search, provides the overall structure for our search. We present eleven new lower bounds for the Snake-in-the-Box problem for snakes in dimensions 11, 12, and 13; coils in dimensions 10, 11, and 12; and symmetric coils in dimensions 9, 10, 11, 12, and 13. The best known solutions of the unsolved dimensions of this problem have improved over the years and we are proud to make a contribution to this problem as well as the continued progress in combinatorial search techniques.

Keywords: Stochastic beam search · Snake-in-the-box · Combinatorial optimization · Graph search · Hypercube · Heuristic search

1   Introduction

The Snake-in-the-Box problem (SIB) is a computationally hard problem concerned with finding the longest induced path through an n-dimensional hypercube (or n-cube) graph. Finding longest induced paths is well known to be NP-complete. A hypercube graph consists of a set of binary numbered nodes $\{0 \dots 2^{n-1}\}$ in which two nodes are adjacent if and only if their binary values differ by exactly one bit. An induced path (achordal path) is a path in which each node in the path is adjacent only to its immediate predecessor and successor nodes within the path. In the SIB problem, open achordal paths are called snakes, and closed achordal paths are called coils. For snakes, any path which cannot be extended is called a maximal path, with the *longest*

© Springer International Publishing Switzerland 2015
M. Ali et al. (Eds.): IEA/AIE 2015, LNAI 9101, pp. 23–32, 2015.
DOI: 10.1007/978-3-319-19066-2_3

of these being the longest maximal path. The longest maximal snake for any n-cube is the snake solution (absolute bound) for SIB in dimension-n. Similarly, the longest possible coil for any n-cube is the coil solution (absolute bound) for dimension-n. Symmetric coils are a special case of coil, where the transition sequence for the first half of the path is equal to the transition sequence for the second half of the path. Transition sequence notation describes paths using the bit positions that change between two nodes. A symmetrical coil described by the node sequence {0, 1, 3, 7, 6, 4, 0} and has the transition sequence 012012. For these and other SIB terms and definitions, please visit the UGA Institute for Artificial Intelligence's webpage [1].

Introduced by Kautz in 1958, the Snake-in-the-Box problem has been extensively studied for over 50 years. The problem has found relevance in many fields including coding theory [2], electrical engineering [3], analog to digital conversion [4], precision high speed rotational measurement devices [5], disk sector encoding [6], rank modulation schemes for charge distribution in multi-level flash memories [7], and genetics related to embryonic development [8].

For dimensions n > 7, exhaustive search remains infeasible for finding solutions due to the exponential growth of the problem as dimensionality increases. It is estimated that it would take many years of computing time to exhaustively search dimension 8 [12]. In lieu of this, non-exhaustive techniques, both computational search and mathematical construction based approaches have been applied to improving the known lower bounds of higher dimensions. These include: genetic algorithms [10], hybrid genetic algorithms [13], nested Monte Carlo Search schemes [14], particle swarm optimization [16], population-based stochastic hill-climbers [16], distributed heuristic computing [17], neural networks [18], hybrid evolutionary algorithms [16] and mathematical constructive techniques [19, 20, 21]. With each successive technical improvement, enhanced by increases in computing power, the lower bounds for these higher dimensions continue to be improved. For solution lengths of n-snakes, n-coils, and n-symmetrical coils for n ≤ 7, as well as current lower bounds non-exhaustively searched for 8 ≥ n ≤ 13 see [22].

2   Overview of Our Previous Work

Building on our previous work [23], we again implement our non-deterministic variant of Beam Search, which we refer to as Stochastic Beam Search, which combines elements of graph search with methods common to evolutionary computing. Our Stochastic Beam Search (SBS) differs from BS solely in how the decision to prune is made. Whereas BS retains only the absolute best sub-solutions at each step, SBS provides for randomness similar to the tournament selection process commonly used in Genetic Algorithms (GA) to determine which sub-solutions are retained. Basic tournament selection retains only the single best sub-solution from among a randomly selected subset of a population. SBS tournament selection, on the other hand, is a reverse tournament selection that discards only the single worst sub-solution found in each tournament. Tournaments continue to be run while the population of sub-solutions exceeds the beam width.

In our previous work, we extended the excellent "available nodes" metric [14] [25] by determining which available nodes were actually reachable from a sub-solution's head. We called these "reachable available nodes". Our base implementation in our previous work searched only for snakes. To search for coils we added an additional check to our base fitness function which allowed us to discard any sub-solution that did not retain the possibility of becoming a coil based on end point existence. For more information on our previous path exploration optimization and fitness function see [23].

3   Improvements to Representation and Selection

Our previous implementation was constrained by RAM for how large a beam width we could practically use. In order to use a wider beam that considers many more sub-solutions, we needed to reduce our application's memory footprint. We improved our sub-solution representation from a linked list to a bitmap in which a sub-solution is stored as an array of bits rather than a list of integers or shorts. This new representation offered a huge improvement over the previous linked list. Using this representation netted a memory savings of greater than 96%. A representation like this was first suggested by [26]. This improvement allowed us to dramatically increase our beam size, and moved the bottleneck of our process from memory to processing.

Because of this improvement in our representation we were able to increase the beam width from 50,000 to 1,000,000 sub-solutions for searches in dimension 12. Correspondingly, we improved on our previously reported new lower bound in dimension 12 [23] by 8 nodes.

4   Improvements to Fitness for Coil Search

Here we present our latest fitness improvements for searching for coils. Our symmetrical coil search is addressed in Section 5.

4.1   Removing Dead End Nodes

In our previous work, we introduced connectivity values for reachable available nodes [23]. The connectivity value of a reachable available node is equal to the number of reachable available nodes that are adjacent to it. *Dead End Nodes* are reachable available nodes with only one reachable available node neighbor, and thus have a connectivity value of 1. Dead end nodes may only become: (1) the terminating node of a snake, (2) a skin node to its sole reachable available node neighbor, or (3) an isolated node [23].

Dead ends correlate to exclusive OR forks in the pool of reachable available nodes. These are points where large numbers of nodes that were previously additive to the fitness value as reachable nodes, become isolated once the path of the snake or coil passes this fork. Being able to efficiently detect and measure the impact of an exclusive OR fork as soon as it appears in a sub-solution's hypercube is of great predictive value in determining which sub-solutions should be carried forward for further consideration (fitness and selection).

In coil search, the end points of our search process are adjacent to our start node (node 0). We call these nodes coil-forming nodes [23]. For coils, any dead end node that is not a coil-forming node is an unusable dead end and cannot be part of a valid coil solution. We remove these nodes from the total of reachable available nodes that serves as our primary fitness value. Removing dead end nodes from reachable available nodes in our fitness calculation, with a beam width of 100,000, yielded length 92 coils in dimension 8.

4.2   Extension of Dead End Concept - Blind Alleys

A natural next step to identifying and removing dead end nodes is to look for subpaths which lead without alternatives to dead ends. We call such paths blind alleys. See Figure. 1. Any reachable available node that is adjacent to a dead end node and only one other reachable available node is a blind alley node. Furthermore, any node that would become a dead end upon the removal of a dead end node is also a blind alley node. In coil search, we have known end points, so we can remove all dead ends that are not coil nodes without concern.

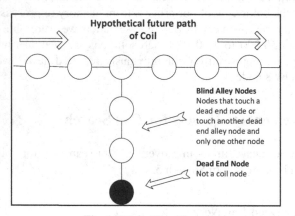

Fig. 1. Blind Alley Nodes

Reducing our fitness value by the number of blind alley nodes can be done efficiently. For each dead end node removed, we need only determine if its reachable available node neighbor(s) has become a dead end node. If so, we remove this node from the sub-solution reachable available node set. We repeat this process until no more dead end nodes are found in the sub-solution graph.

Once we have completed the removal of unreachable available, dead end, and blind alley nodes, the remaining nodes form a much better basis for fitness than available nodes alone. When we use a fitness value refined by removing dead end and blind alley nodes from the count of the reachable available nodes set, we are able to routinely match the record of length 96 coils in dimension 8 with a beam width of only 100,000 sub solutions.

4.3 Concurrent Stochastic Selection

Because our improved representation enables us to increase the beam width of our search, our previous serial selection process becomes a pressing core utilization bottleneck. To address this constraint, we implement a concurrent selection process. We first randomize the order of our population of sub-solutions, then partition the population and conduct separate concurrent tournaments within each partition; reducing the population of each partition until the total population of all partitions is less than or equal to the beam width. We first randomize the order of the population of sub-solutions because, without this step, the more closely related sub-solutions are to each other, the closer in the population list they appear. This would create a positional bias that has nothing to do with our mission of finding the best possible solution. We found performance to be better when we incurred the small computational expense of randomizing the order of the population list before partitioning the list and performing selection.

5   Specific to Symmetric Coil Search

Our long symmetric coil finding implementation is based on our long coil finding approach, and differs from previous approaches that either grow individual n-dimensional snakes and then attempt to "double" them to form coils / symmetrical coils in n+1 [21], or search for snakes and occasionally find coils.

In our process we specifically set out to find / build long symmetrical coils. Within a single sub-solution, we simultaneously "grow" two valid, bit transition identical, sub-snakes in a single hypercube. Within this sub-solution, we refer to these sub-snakes as snake1 and snake2. When we refer to the sub-solution, we are referring to both snake1 and snake2 together. The objective is to link the head of snake1 to the tail of snake2, and the head of snake2 to the tail of snake1 in order to create a long valid symmetrical coil.

To remain valid, neither snake1 nor snake2 may violate the others or its own achordal path constraints. In addition sub-solutions that are retained after each construction step retain the potential to be linked; one or more of snake1's available coil-forming nodes remains reachable by snake2's head node, and one or more of snake2's available coil-forming nodes remains reachable by snake1's head node. There is one last constraint we impose on the growing sub-solution. Snake1 must conform to our previous implementation of Kochut's symmetry reducing constraint [11] [23].

During each construction step, new valid sub-solutions are constructed from a previous sub-solution by adding a single node to the head of snake1. Then a new valid node is added to the head of snake2. This new addition to snake2 must share the bit shift position of the node added to snake1 (relative to the previous node in snake1). We call this relationship of the addition of a new head to snake1 and the next addition to snake2 a symmetrical move. If no valid symmetric move is available for snake2 then this partially constructed and invalid child is discarded.

In contrast to our coil and snake process, we initialize our process with more than a single sub-solution. We create an initial population containing every valid sub-solution in which snake1 starts at node 0 and snake2 starts at some other valid node in the hypercube. Sub-solutions are not created where snake1 and snake2 are adjacent. If the two nodes are adjacent, then this would represent an achordal constraint violation.

5.1   Construction Phase

For every sub-solution in the current population, and for each available node that can be validly added to snake1 as a new head, we perform the steps below.

1. For each reachable available node adjacent to the current head of snake1: If snake2 has a valid symmetrical move, we add the corresponding node to be the new head of snake2; otherwise, the new sub-solution is discarded, as it can never form a symmetrical coil.
2. If the node newly added to snake1 is one of snake2's coil-forming nodes, the sub-solution has terminated in a symmetrical coil solution. Symmetrical coil solutions long enough to be of interest are reported. If the sub-solution (now two nodes longer than its parent) has not terminated, it is added to the successor population.
3. After all members of the current population have been processed to this point, the successor population replaces the current population.

5.2   Fitness Evaluation Phase

For every sub-solution in the successor population, we perform the following steps in order to compute its fitness value:

1. Find the set of available nodes reachable from the head of snake1, and separately the set of available nodes reachable from the head of snake2. As we do for regular coils, we allow coil-forming nodes to be added to the reachable available node sets; however, reachability exploration is not continued from these nodes. As a coil-forming node is the last node to be added to any coil solution, any nodes reachable only through adjacency to it are effectively unreachable in the context of coils.
2. If none of snake2's coil-forming nodes are in the reachable available node set of snake1, we discard the sub-solution; as it can never form a symmetrical coil. This one constant time check drives a large reduction in the number of solutions considered.
3. Ascertain if the two sets of reachable available nodes are equal or disjoint. Because of the symmetry of the hypercube and of our search, the reachable available node sets of snake1 and snake2 will either be equal or disjoint. If equal, all reachable available nodes are reachable from the heads of snake1 and snake2, and our fitness evaluation will not require adjustment. If disjoint, the set of all reachable available nodes is split into two distinct subsets, which (again because of symmetry) are functionally identical. To account for this, we will double the fitness value at the end of the fitness evaluation phase. In either case (disjoint or equal), because of symmetry, we only need to work with Snake1's reachable available node set from this point forward.
4. Remove blind alley nodes from snake1's reachable available node set using snake2's coil-forming nodes as our end points; just as we do for standard coils.

5. Calculate a connectivity value as in [23] from the remaining nodes in snake1's reachable available node set.
6. Lastly, double the fitness value if the reachable available node sets were determined to be disjoint in step 3.

5.3   Termination

As previously described in all of our approaches, the construction and fitness evaluation processes continue until there are no longer any sub-solutions in the population.

6   Results

We found new lower bounds for snakes in dimensions 11, 12, and 13; coils in dimensions 10, 11, and 12; and symmetrical coils in dimensions 9, 10, 11, 12, and 13. Table 1 shows current lower bounds for dimensions 8–13. Parenthetical values indicate previously known lower bounds. Asterisks indicate where we have improved on our previously published [23] records. An up to date list of snake and coil, absolute and lower bounds, is maintained at The University of Georgia SIB Records Page [22]. Please find examples for new record lower bounds for snakes, coils, and symmetrical coils at http://www.cs.uga.edu/~potter/CompIntell/IEAAIE-snakes.pdf.

Table 1. Current SIB Absolute Bounds

Dimension	Length		
	Snakes	*Coils*	*Symmetric Coils*
8	98	96	
9	190	188	(180) 186
10	370	(*358) 362	(340) 362
11	(*705) 707	(*666) 668	(640) 662
12	(1280) 1302	(*1268) 1276	(1128) 1222
13	(2466) 2520	2468	(1898) 2354

7   Conclusions

Our approach was largely inspired by efforts in [13], [16], [25], [17], and [10] to use a GA to improve known lower bounds on this difficult problem. The GA will usually yield good solutions to the SIB problems, and by applying more computation effort, the solutions can be improved to a point; however, the GA applied to the SIB problem runs into diminishing returns.

Our use of fitness relentlessly selects dense solutions that are highly likely to dead end, but are the best prospects for an excellent result. We expand upon the small fraction we are able to practically process that we judge through fitness as good prospects.

Improvements in the predictive value of our fitness function resulted in new lower bounds in several dimensions and solution classes (snakes / coils / symmetrical coils) while increasing the beam width process by 50–100 times yielded only small improvements in results.

Exploiting the inherent constraints of specific solution classes (specifically coils and symmetrical coils) in the Snake-in-the-Box problem enabled us to dramatically reduce the effective search space, and to impose powerful checks on the validity of sub-solutions. For example, the symmetrical coil aspect of the SIB problem is the most constraining. We checked that sub-solutions could reach coil-forming end points and thus retain coil forming potential. We also checked sub-solutions conformance to the symmetrical property of symmetric coils. If the sub-solution did not pass these checks, it was discarded. This left us with a small fraction of the search space that we would have had to traverse without the intelligent application of known constraints. This approach explains the particular strengths of our process when applied to coils and symmetrical coils. We are particularly pleased with the improvements achieved in symmetrical coil lower bounds in dimensions 9–13, which saw lower bounds improvements of 6, 12, 22, 94, and 456 nodes respectively.

Beam search is an excellent platform to experiment with approaches to assess the state of the hypercube (fitness functions). Once the beam width is sufficiently large, small changes in the efficacy of the fitness function led to large performance changes. We expect that more algorithms common to graph search may merit application to this problem.

8   Future Work

In future work we plan to explore more graph theory based improvements to our fitness function. One of the most promising we are testing is articulation point / biconnected component detection algorithms. The techniques we have described including potential future work with articulation points may lend themselves to searching for snakes using defined end points; and could include techniques like our symmetric coil search, where multiple sub-solutions with many different end points are considered within the same search; much like considering multiple start points for complimentary symmetrical snakes.

Acknowledgments. We would thank Dustin Cline and Thomas Horton for their important contributions to this project, Snake-in-the-Box research, and the Institute for Artificial Intelligence. This study was partially supported by resources and technical expertise from the University of Georgia Computer Science Department and the Georgia Advanced Computing Resource Center (GACRC).

References

1. Potter W.: Snake-In-The-Box Dictionary (2014). http://ai1.ai.uga.edu/sib/sibwiki/doku. php/dictionary (accessed June 5, 2014)
2. Kautz, W.: Unit-distance Error-Checking Codes. IRE Trans. Electron. Comput. **EC-7**(2), 179–180 (1958)
3. Klee, V.: What is the Maximum Length of a d-Dimensional Snake? The Amer. Math. Monthly **77**(1), 63–65 (1970)
4. Hiltgen, A., Paterson, K.: Single Track Circuit Codes. IEEE Trans. on Inform. Theory **47**, 2587–2595 (2000)
5. Zhang, F., Zhu, H.: Determination of optimal period of absolute encoders with single track cyclic gray code. J. of Central South University of Technology **15**(2 suppl.), 362–366 (2008)
6. Blaum, M., Etzion, T.: "Use of snake-in-the-box codes for reliable identification of tracks in servo fields of a disk drive," U.S. Patent 6 496 312, December 17, 2002
7. Yehezkeally, Y., Schwartz, M.: Snake-in-the-Box Codes for Rank Modulation. IEEE Trans. Information Theory **58**(8), 5471–5483 (2012)
8. Zinovik, I., Chebiryak, Y., Kroening, D.: Periodic Orbits and Equilibria in Glass Models for Gene Regulatory Networks. IEEE Trans. Information Theory **56**(2), 805–820 (2010)
9. Davies, D.: Longest 'separated' paths and loops in an N cube. IEEE Trans. Electron. Comput. **14**, 261 (1965)
10. Potter, W., Robinson, J., Miller J., Kochut, K.: Using the genetic algorithm to find snake-in-the-box codes. In: Proc. 7th Int. Conf. Industrial & Engineering Applications of Artificial Intelligence and Expert Systems, Austin, TX, pp. 421–426 (1994)
11. Kochut, K.: Snake-In-The-Box Codes for Dimension 7. J. Combinatorial Math. and Combinatorial Computing **20**, 175 185 (1996)
12. Kinny, D.: A new approach to the snake-in-the-box problem. In: Proc. 20th European Conf. Artificial Intelligence, ECAI 2012, Montpellier, France, 2012 © The Author. doi:10.3233/978-1-61499-098-7-462
13. Carlson, B., Hougen, D.: Phenotype feedback genetic algorithm operators for heuristic encoding of snakes within hypercubes. In: Proc. 12th Annu. Genetic and Evolutionary Computation Conf., GECCO 2010, Portland, Oregon, pp. 791–798 (2010)
14. Kinny, D.: Monte-carlo search for snakes and coils. In: Sombattheera, C., Loi, N.K., Wankar, R., Quan, T. (eds.) MIWAI 2012. LNCS, vol. 7694, pp. 271–283. Springer, Heidelberg (2012)
15. Brooks, P.: Particle Swarm Optimization and Priority Representation: M.S. thesis, Artificial Intelligence, Univ. Georgia, Athens (2012)
16. Casella, D., Potter, W.: Using evolutionary techniques to hunt for snakes and coils. In: Proc. IEEE Congress on Evolutionary Computation, CEC 2005, Edinburgh, Scotland, pp. 2499–2505 (2005)
17. Juric, M., Potter, W., Plaskin, M.: Using the parallel virtual machine for hunting snake-in-the-box codes. In: Proc. 7th Conf. North American Transputer Research and Applications Conf., NATUG-7, Athens, GA, pp. 97–102 (1994)
18. Bishopm, J.: Investigating the snake-in-the-box problem with neuroevolution. Dept. of Computer Science, Univ. Texas, Austin (2006)
19. Adelson, L., Alter, R., Curtz, T.: Long snakes and a characterization of maximal snakes on the d-cube. In: Proc. 4th SouthEastern Conf. Combinatorics, Graph Theory, and Computing, Congr. No. 8, Boca Raton, FL, pp. 111–124 (1973)

20. Abbott, H., Katchalski, M.: On the Construction of Snake in the Box Codes. Utilitas Mathematica **40**, 97–116 (1991)
21. Wynn, E.: Constructing circuit codes by permuting initial sequences. In: Proc. 20th European Conf. Artificial Intelligence, ECAI 2012, Montpellier, France, 2012 © The Author. doi:10.3233/978-1-61499-098-7-468 http://arxiv.org/abs/1201.1647 (accessed June 5, 2014)
22. Potter, W.: SIB Records (2014). http://ai.uga.edu/sib/sibwiki/doku.php/records (accessed June 5, 2014)
23. Meyerson, S., Drapela, T., Whiteside, W. Potter, W.: Finding longest paths in hypercubes, snakes and coils. In: Proc. 2014 IEEE Symp. Computational Intelligence for Engineering Solutions, CIES 2014, Orlando, FL, pp. 103–109 (2014)
24. Pinedo, M.: Scheduling: Theory, Algorithms, and Systems. Prentice-Hall, New York (1995)
25. Tuohy, D., Potter, W., Casella, D.: "Searching for snake-in-the-box codes with evolved pruning models. In: Proc. 2007 Int. Conf. Genetic and Evolutionary Methods, GEM 2007, Las Vegas, NV, pp. 3–9 (2007)
26. Diaz-Gomez, P., Hougen, D.: Genetic algorithms for hunting snakes in hypercubes: fitness function analysis and open questions. In: 7th ACIS Int. Conf. Software Engineering, Artificial Intelligence, Networking and Parallel/Distributed Computing, SNPD 2006, Las Vegas, NV, pp. 389–394 (2006)

Novel AI Strategies for *Multi*-Player Games at Intermediate Board States

Spencer Polk[✉] and B. John Oommen

School of Computer Science, Carleton University, Ottawa, Canada
andrewpolk@cmail.carleton.ca, oommen@scs.carleton.ca

Abstract. This paper considers the problem of designing efficient AI strategies for playing games at intermediate board states. While general heuristic-based methods are applicable for *all* boards states, the search required in an alpha-beta scheme depends heavily on the move ordering. Determining the best move ordering to be used in the search is particularly interesting and complex in an intermediate board state, compared to the situation where the game starts with an initial board state, as we do not assume the availability of "Opening book" moves. Furthermore, unlike the *two*-player scenario that is traditionally analyzed, we investigate the more complex scenario when the game is a *multi*-player game, like Chinese Checkers. One recent approach, the Best-Reply Search (BRS), resolves this by a process of grouping opponents, which although successful, incurs a very large branching factor. To address this, the authors of this work earlier proposed the Threat-ADS move ordering heuristic, by augmenting the BRS by invoking techniques from the field of Adaptive Data Structures (ADSs) to order the moves. Indeed, the Threat-ADS performs well under a variety of parameters when the game was analyzed at or near the game's initial state. This work demonstrates that the Threat-ADS also serves as a solution to the unresolved question of finding a viable solution in the far-more variable, intermediate game states. Our present results confirm that the Threat-ADS performs well in these intermediate states for various games. Surprisingly, it, in fact, performs better in some cases, when compared to the start of the game.

1   Introduction

AI techniques have been used extensively to play games, and well-known approaches such as alpha-beta search are known to gain improved efficiency and performance through strong move ordering, or attempts to search the best move first [1]. While these techniques are generally applicable for *all* board positions, their applicability for intermediate positions is far from obvious. This is because, in investigating the performance of move ordering techniques, rather than considering random game positions or the initial board state, analysis at *intermediate*

Chancellor's Professor; *Fellow: IEEE* and *Fellow: IAPR*. The second author is also an *Adjunct Professor* with the Dept. of ICT, University of Agder, Grimstad, Norway.

M. Ali et al. (Eds.): IEA/AIE 2015, LNAI 9101, pp. 33–42, 2015.
DOI: 10.1007/978-3-319-19066-2_4

board states is particularly fascinating. This is because the number of possible positions and the consequent potential moves, in an intermediate position is far more than the number encountered at the starting position. This is the focus of this paper. However, rather than considering move ordering-based AI search strategies for the classical family of *two*-person games, we consider here the scenario when the game is a *multi*-player game. It is well-established that the research domain associated with Multi-Player Game Playing (MPGP) of adversarial games is relatively unexplored[1].

The majority of techniques utilized by MPGP, such as the Paranoid and Max-N algorithms, are extensions of well-understood, powerful two-player techniques, such as Alpha-Beta Search, to a multi-player environment [5,6]. However, while these techniques are natural and make intuitive sense, for a variety of issues, they have trouble performing at a level comparable with their two-player counterparts, necessitating the development of new techniques, and improvements to existing ones [4,7]. One very recent strategy, attempting to overcome weaknesses in tree pruning and processing time in multi-player environments, is the Best-Reply Search (BRS), which has been found to outperform its competitors in a variety of environments [8]. Despite its power, however, the nature of the BRS leads to a large branching factor in the game tree, thus making tree pruning an important area of consideration in its use [8].

The authors of this work proposed in [9] the Threat-ADS heuristic, which makes use of techniques from the formerly unrelated field of Adaptive Data Structures (ADS) to improve move ordering in the context of the BRS. The Threat-ADS heuristic was found to produce meaningful improvements in a variety of cases [9,10] when the game started from its initial board position. This paper demonstrates that the Threat-ADS is also a viable solution to efficiently resolve the search problem for MPGP analyzed from intermediate states.

The remainder of the paper is laid out as follows. After a motivating section, Section 3 briefly discusses existing techniques for MPGP including the Threat-ADS. Section 4 describes the game models that we employ and specifies the experiments we have performed. Section 5 presents our results. Finally, Sections 6 and 7 contain our analysis and conclusions, and open avenues of further research.

2   Motivation

Investigation into AI-based game playing from intermediate board states is an interesting topic, as it involves a much richer area of inquiry than the starting position of the game. This is because:

---

[1] Recent years have seen a rising focus in the literature on this domain, accelerated, in part, by the growing popularity of multi-player board games, such as the highly-regarded Settlers of Catan, and the multi-player nature of electronic games [2,3]. Despite increasing interest in both the academic and public domains, MPGP remains a relatively unexplored field, and significant work remains to be done before knowledge of MPGP approaches the vast body of work dedicated to adversarial two-player game playing [1,2,4].

1. "Opening book" moves are not available at an intermediate board state [11].
2. In all non-trivial games, the variability in possible intermediate board states is almost always exponentially large.
3. Unlike two-player games, in multi-player games, the number of opponents who pose an adversarial threat in an intermediate board state could be larger than the number that the perspective player faces at the start of the game.
4. In some cases, at an intermediate board state, the number of adversarial opponents could have reduced, due to player eliminations.
5. Apart from the number of opponents, the identity of the opponents that could potentially threaten the perspective player could change in the intermediate board sate.

To explain the above issues, it is clear that if a game is well known, a game playing engine will, generally, make use of an opening book of strong, known moves that are derived from either expert knowledge, or trained from previous experience with the game [11]. In the case of intermediate game states, however, such well-known opening moves are unavailable, and thus the efficiency of the game search is more important. Furthermore, the opening stages of the game have a relatively small degree of variability, compared to the vast number of possible intermediate board states. Thus, when analyzing the performance of move ordering, addressing intermediate board states provides much more powerful results related to its performance.

The problem is accentuated in the case of MPGP. In the case of a multi-player game, there are other considerations at work aside from the variability of the intermediate board states. It is typical that in a multi-player game, the winner is determined by the last player remaining in the game, thus over time it is possible that players will be eliminated from play. Considering that the Threat-ADS makes gains by prioritizing the most threatening player, it is possible that the removal of players from contention will have an impact on its performance. Furthermore, depending on the qualities of the game, different opponents may be more threatening at different stages of play. For example, if one opponent has performed very well thus far, he may be more capable of minimizing the perspective player's score and pose a greater threat than he posed earlier in the game. Figure 1 shows an example of these effects in the context of the Virus Game [9,10], explained below. An alternate example when the number of adversarial opponents increases, e.g., in Chinese Checkers, is easily derived.

Our goal is to find any expedient solution to the search problem when one starts from these intermediate positions. Of course, while one can use the established BRS to tackle the problem, we unarguably demonstrate that by augmenting it with an ADS, the previously-proposed Threat-ADS is capable of adapting to these changing factors, and that it will still perform, well deeper in the tree.

3   Previous Work

The dominant search techniques for MPGP can be broadly divided into stochastic methods, such as the popular UCT algorithm [12], and deterministic methods,

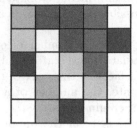 

Fig. 1. The Virus Game at its initial state, and ten turns into the game. Observe that two players have been eliminated, and the pieces are more closely grouped together.

generally derived from the well-known Mini-Max approach for two-player games [1,2]. In the case of a multi-player environment, there does not exist the simple paradigm of one player maximizing his score, while the others attempt to minimize it, as one player's gain does not necessarily equally impact all opponents. This must be addressed when extending Mini-Max to a multi-player environment. The Paranoid algorithm seeks to accomplish this by considering all opponents to minimize the perspective player's score, whereas the Max-N algorithm assumes each opponent will seek to minimize his own score [5,6]. The more recent BRS attempts to overcome the weaknesses of assuming all opponents are a coalition, as in the Paranoid approach, and the tree pruning difficulties that the Max-N algorithm encounters [8]. It does this by assuming all opponents are a single entity, who may take only one action between the perspective player's turns. While this implies it considers invalid board states, it was found to perform better than the Paranoid and Max-N in a variety of environments [8].

As the BRS groups opponents together into a single minimizing "super opponent", the Min nodes will naturally have a larger branching factor when searching the game tree than any single player, as all possible moves for all opponents are grouped together. As this grouping is a unique feature of the BRS, it was recognized that using some form of ranking of the opponents could potentially lead to a novel move ordering strategy, thus implying that one could apply methods previously unused in the context of move ordering. Using this, the present authors proposed, in [9], the Threat-ADS heuristic, which gathers moves from each opponent in the order of their relative threats. The Threat-ADS heuristic employs a list-based ADS, which is a list designed to quickly and efficiently update its structure to improve query efficiency over time [13,14]. In this context, the list is "queried" when an opponent provides the most minimizing move at a Min node of the tree (which is discovered naturally in the execution of the BRS), and the resultant structure is interpreted as a ranking of opponent threats. Figure 2 shows a comparison of a Min level of the BRS using the Threat-ADS, and not, showing how the ADS can assist in finding the most minimizing move quickly.

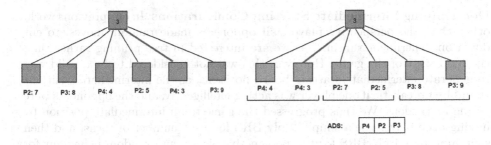

Fig. 2. The BRS not using Threat-ADS on the left, and using it, on the right. Note how the ranking of opponents by their threat levels improves move ordering.

In its introductory paper, the Threat-ADS was shown to provide statistically significant gains in terms of Node Count (NC), using a standard T-Test, over several turns of a variety of game models, using a "Move-to-Front" update mechanism for the ADS [9]. It was later shown to retain these improvements under a variety of alternative update mechanisms, although the "best" one to use can vary depending on the game in question, and to retain its performance in trees of deeper ply depth [10].

Our previous work demonstrated the benefits of Threat-ADS under a variety of game models, and using different ADS update mechanisms. However, all measurements were taken beginning at the initial board state of the game, and a few turns into the future. The question now is whether the Threat-ADS can also serve as a feasible solution to the intermediate board positions, and we shall demonstrate that it is.

4   Game Models and Experimental Setup

For the purposes of consistency with our previous work, we have elected to use the same game models that we made use of in [9] and [10]. Specifically, we make use of the well-known multi-player games Focus and Chinese Checkers, as well as a game akin to the well-known "Game of Life" called the Virus Game. Focus and Chinese Checkers were chosen due to their use in the work that introduced the BRS, to demonstrate its efficacy under a variety of game models [8]. The rules and evaluation functions for all the three games employed are the same as in the previous work in [9] and [10].

The Virus Game is a territory control game played on an N-by-N sized board, where any number of k players, where $k \geq 2$, have a configured number of pieces placed randomly on the board. During their turn, a player may "infect" a square occupied by or adjacent to (interpreted as horizontally or vertically adjacent, but not diagonally) one of their pieces. The player claims the infected square, and gains pieces on that square, and all adjacent squares. Further details of the games are omitted here in the interest of space, but can be found in [9].

Determining Intermediate Starting Configurations: In our previous work, other than the perspective player, all opponents made random moves, to cut down on experiment runtime, as we are interested in tree pruning rather than the final state of the game. However, this was not considered to be a valid way to generate intermediate starting board positions, as the intermediate positions would be unrealistic if one player was acting intelligently, and the opponents were acting at random. We thus progressed the game to an intermediate position by having each player use a simple 2-ply BRS for a set number of turns, and then switching to a 4-ply BRS for the perspective player, and random behaviour for the opponents (as in the previous work), while measurements were taken. The number of turns we advanced into the game in this way was fifteen for the Virus Game, ten for Chinese Checkers, and, given its short duration, three for Focus.

The games were run for the same number of turns as in [9] and [10], and the sum of the Node Count (NC) over those turns was taken as our metric. This was ten turns for the Virus Game, five for Chinese Checkers, and three for Focus.

In the case of the Virus Game, we set the number of players to be equal to five, and four players for Focus. In the case of Chinese Checkers, unlike in previous work, we present our results for both the four-player and the six-player case. Each of these trials was repeated fifty times, for each of the adaptive list mechanisms from [10], with the exception of the Stochastic Move-to-Rear absorbing rule, which was found to not produce any effect, and for the unaugmented BRS. These mechanisms were, specifically, the Move-to-Front, Transposition, Move-Ahead-k, and POS(k) rules (exact specification of the implementation of these update mechanisms is omitted from this work, and the reader is referred to [10]). As in previous work, the value for k was chosen to be two, to insure Move-Ahead-k and POS(k) did not perform identically to the Move-to-Front or Transposition.

In our previous work, we performed a total of two hundred trials. This was subsequently found to be very time consuming, and more than necessary to demonstrate a clear improvement from the use of the Threat-ADS heuristic. Because of the large sample, the results in [9] and [10] pass the test for normalcy, although they do not pass it in this work. Thus, rather than make use of the one-tailed T-Test, we have instead elected to employ the non-parametric Mann-Whitney test for statistical significance. We have furthermore included the Effect Size measure, which is the standardized mean difference between the two data sets, to make the degree of impact from the Threat-ADS more intuitive [15].

Our results and statistical analysis are presented in the next section.

5 Results

Consider Table 1, where we present our results for the Virus Game. We observe that, as was the case in [10], all ergodic ADSs produced a statistically significant improvement in pruning, ranging from a 5% – 10% reduction in NC. The Move-to-Front and Transposition performed the best, and equally well, in this case. The Effect Sizes ranged between approximately 0.5 and 0.9, with the majority near 0.8.

Table 1. Results for the Virus Game

Update Mechanism	Avg. NC	Std. Dev	P-Value	Effect Size
None	303,000	35,000	-	-
Move-to-Front	275,000	40,000	2.5×10^{-4}	0.82
Transposition	275,000	37,000	6.0×10^{-5}	0.87
Move-Ahead-k	280,000	30,000	1.4×10^{-4}	0.72
POS(k)	286,000	40,000	5.7×10^{-3}	0.56

Table 2 shows our results for Focus. In this case, again, all adaptive update mechanisms produced a statistically significant improvement in tree pruning. In this case it was a very large reduction of roughly 20% for all cases. However, here the Effect Size lingered around 0.45, given the much larger variance of the datasets. As there are a total of three opponents, the Move-Ahead-k and POS(k) rules are identical to the Move-to-Front and Transposition strategies, respectively, and thus the results are duplicated in the table, and marked '*'.

Table 2. Results for Focus

Update Mechanism	Avg. NC	Std. Dev	P-Value	Effect Size
None	16,767,000	7,161,000	-	-
Move-to-Front	13,592,000	5,240,000	0.01	0.44
Transposition	13,671,000	4,240,000	0.03	0.43
Move-Ahead-k*	13,592,000*	5,240,000	0.01	0.44
POS(k)*	13,671,000*	4,240,000	0.03	0.43

Our results for four-player Chinese Checkers are shown in Table 3. Again, the Move-to-Front and Transposition are identical to Move-Ahead-k and POS(k), respectively. Here we saw a statistically significant improvement from the Move-to-Front/Move-Ahead-k rule (a 13% reduction in tree size), however, while the average NC was reduced, the Transposition/POS(k) fell outside 95% certainty.

Table 4 shows our results for six-player Chinese Checkers. In this case, Threat-ADS had a smaller impact than the others. However, we do observe that the average NC is lower when Threat-ADS is employed in all cases. The best performance was obtained from the Transposition rule, with an 8% reduction in tree size, which fell within 90% certainty, although outside 95% certainty.

Table 3. Results for four-player Chinese Checkers

Update Mechanism	Avg. NC	Std. Dev	P-Value	Effect Size
None	3,458,000	905,000	-	-
Move-to-Front	3,024,000	816,000	0.01	0.46
Transposition	3,206,000	798,000	0.12	0.26
Move-Ahead-k*	3,024,000*	816,000	0.01	0.46
POS(k)*	3,206,000*	798,000	0.12	0.26

Table 4. Results for six-player Chinese Checkers

Update Mechanism	Avg. NC	Std. Dev	P-Value	Effect Size
None	8,168,000	2,560,000	-	-
Move-to-Front	7,677,000	2,280,000	0.18	0.30
Transposition	7,494,000	1,670,000	0.09	0.26
Move-Ahead-k	7,644,000	1,830,000	0.21	0.20
POS(k)	7,975,000	2,190,000	0.44	0.08

6   Discussion

Our results in this work further reinforce the conclusions from [9] and elaborated upon in [10]. In the case of each of our game models, we see that the use of Threat-ADS resulted in a decrease in the average NC, over the course of several turns. In the case of the Virus Game and Focus, the results were always statistically significant within a 95%, or greater, certainty threshold. While we always see an improvement in the case of Chinese Checkers, in a number of cases this improvement falls outside 95% certainty. *These results demonstrate that the Threat-ADS is able to produce meaningful savings in tree pruning in a variety of intermediate game states, for the games we have analyzed.* Given that the games vary substantially from each other, being based on piece capturing, racing, and territory control for Focus, Chinese Checkers, and the Virus Game, respectively, we hypothesize that the Threat-ADS will continue to perform well in other multi-player games too.

We observe that in our previous experiments, the improvement in tree pruning would range between a 5% and 10% reduction in average NC [10]. Our results for the Virus Game in this work is consistent with that. However, we note that in the context of Focus, we observed a much stronger 20% reduction in average NC. We suspect the reason for this is that, after some time is passed, it is likely that one or more of the opponents will have performed well in the opening moves of the game, and thus pose a greater threat to the perspective player. By prioritizing that opponent, the Threat-ADS is therefore able to make substantial savings, whereas at the beginning of the game, all players start at equal footing. This demonstrates that in fact the benefits of the Threat-ADS can improve over the course of the game, generating greater savings as time progresses.

In the case of Chinese Checkers, our best results for four-player Chinese Checkers resulted in a 13% saving in tree pruning, and the Transposition rule generated an 8% reduction in NC in the six-player case, which is a greater reduction than was observed in [10]. However, the best result for six-player Chinese Checkers fell outside 95% statistical certainty, regardless of the greater reduction in tree size. We suspect this is due to the large variance, which is to be expected, given that we are beginning measurements from a different starting position each time. Given that the use of Threat-ADS produces a reduction in tree size at all times, we suspect that it is not the case that the Threat-ADS is performing worse in the context of intermediate board states, but simply that the increased variance makes each pairwise comparison less likely to be statistically significant within 95% certainty. This also leads to a reduced Effect Size metric.

The Effect Size provides an easily-understood metric for demonstrating the degree of impact that a new technique has, in a way that is immediately obvious to the reader. While these rules are not universal, an Effect Size of 0.2 is considered to be small, 0.5 to be medium, and 0.8 to be large [15]. Given that our best-performing strategies are normally between 0.5 and 0.8 or larger, we conclude that Threat-ADS' contribution to tree pruning is not coincidental.

7 Conclusions and Future Work

In this paper, we have considered the task of designing an efficient AI scheme for multi-player games analyzed from intermediate (as opposed to starting) board states. Our results clearly demonstrate that the scheme presented in [9] and [10], i.e., the Threat-ADS, is an expedient strategy. Indeed, it maintains its performance when applied to intermediate board states, and does not function disproportionately well at the start of the game. In fact, our results confirm that the Threat-ADS can obtain greater performance in some cases later in the game, as we observed with Focus.

The Virus Game, Focus, and Chinese Checkers are different enough from each other that one can reasonably hypothesize that the Threat-ADS will remain viable under other game models. However, we have currently not explored its function in a very wide set of games, and thus exploration of a larger set of games is a potential future research direction. We have also shown here that Threat-ADS' performance within Chinese Checkers varies depending on whether the game is played with four or six players, showing that examination of different numbers of players may lead to interesting results as well.

As in our previous work, however, what the Threat-ADS' consistent results, with its inexpensive cost, demonstrates is the potential benefits ADSs hold for game playing. Hopefully, this work will inspire others and provide a basis for further exploration of ADS-based techniques within area of game playing.

References

1. Russell, S.J., Norvig, P.: Artificial Intelligence: A Modern Approach, 3rd edn, pp. 161–201. Prentice-Hall Inc, Upper Saddle River (2009)
2. Sturtevant, N.: Multi-Player Games: Algorithms and Approaches. PhD thesis, University of California (2003)
3. Szita, I., Chaslot, G., Spronck, P.: Monte-carlo tree search in settlers of catan. In: van den Herik, H.J., Spronck, P. (eds.) ACG 2009. LNCS, vol. 6048, pp. 21–32. Springer, Heidelberg (2010)
4. Sturtevant, N., Bowling, M.: Robust game play against unknown opponents. In: Proceedings of AAMAS 2006, the 2006 International Joint Conference on Autonomous Agents and Multiagent Systems, pp. 713–719 (2006)
5. Luckhardt, C., Irani, K.: An algorithmic solution of n-person games. In: Proceedings of the AAAI 1986, pp. 158–162 (1986)
6. Sturtevant, N.: A comparison of algorithms for multi-player games. In: Schaeffer, J., Müller, M., Björnsson, Y. (eds.) CG 2002. LNCS, vol. 2883, pp. 108–122. Springer, Heidelberg (2003)
7. Sturtevant, N., Zinkevich, M., Bowling, M.: Prob-maxn: playing n-player games with opponent models. In: Proceedings of AAAI 2006, the 2006 National Conference on Artificial Intelligence, pp. 1057–1063 (2006)
8. Schadd, M.P.D., Winands, M.H.M.: Best Reply Search for multiplayer games. IEEE Transactions on Computational Intelligence and AI in Games 3, 57–66 (2011)
9. Polk, S., Oommen, B. J.: On applying adaptive data structures to multi-player game playing. In: Proceedings of AI 2013, the Thirty-Third SGAI Conference on Artificial Intelligence, pp. 125–138 (2013)
10. Polk, S., Oommen, B. J.: On enhancing recent multi-player game playing strategies using a spectrum of adaptive data structures. In: Proceedings of TAAI 2013, the 2013 Conference on Technologies and Applications of Artificial Intelligence (2013)
11. Levene, M., Bar-Ilan, J.: Comparing typical opening move choices made by humans and chess engines. Computing Research Repository (2006)
12. Gelly, S., Wang, Y.: Exploration exploitation in go: UCT for monte-carlo go. In: Proceedings of NIPS 2006, the 2006 Annual Conference on Neural Information Processing Systems (2006)
13. Corman, T.H., Leiserson, C.E., Rivest, R.L., Stein, C.: Introduction to Algorithms, 3rd edn, pp. 302–320. MIT Press, Upper Saddle River (2009)
14. Hester, J.H., Hirschberg, D.S.: Self-organizing linear search. ACM Computing Surveys 17, 285–311 (1985)
15. Coe, R.: It's the effect size, stupid: what effect size is and why it is important. In: Annual Conference of the British Educational Research Association. University of Exeter, Exeter, Devon (2002)

AIs for Dominion Using Monte-Carlo Tree Search

Robin Tollisen, Jon Vegard Jansen$^{(\boxtimes)}$, Morten Goodwin,
and Sondre Glimsdal

Department of ICT, University of Agder, Grimstad, Norway
park533@gmail.com

Abstract. Dominion is a complex game, with hidden information and stochastic elements. This makes creating any artificial intelligence (AI) challenging. To this date, there is little work in the literature on AI for Dominion, and existing solutions rely upon carefully tuned finite-state solutions.

This paper presents two novel AIs for Dominion based on Monte-Carlo Tree Search (MCTS) methods. This is achieved by employing Upper Confidence Bounds (UCB) and Upper Confidence Bounds applied to Trees (UCT). The proposed solutions are notably better than existing work. The strongest proposal is able to win 67% of games played against a known, good finite-state solution, even when the finite-state solution has the unfair advantage of starting the game.

Keywords: Dominion · MCTS · UCB · UCT

1 Introduction

Dominion is a deck-building card game for two to four players. The game is played with a subset of 10 cards from the game, chosen by the players. The original game contains 25 different cards to choose from, and there are numerous expansions which contain more cards. Each player attempts to optimize their deck depending on the subset, and what the opposing players decide.

This makes any successful, automated strategy for Dominion extremely complex. To date, to the best of our knowledge, there is no good AI available for Dominion. The complexity is caused by the stochastic environment, the hidden game elements and the large number of possible game states. Brute force and state searching approaches are in practice impossible. Furthermore, since a subset of 10 cards out of 25, only including the original Dominion game, yields more than three million game variants, it is difficult to create a successful AI applicable to the entire set of cards.

This article explores using a Monte Carlo Tree Search (MCTS) based AI to play the card game Dominion [1],[2]. The paper presents two novel variants: (1) **AI-UCB** which uses and Upper Confidence Bounds without trees, and (2) **AI-UCT** which applies Upper Confidence Bounds applied to Trees.

---

[1] http://en.wikipedia.org/wiki/Dominion_(card_game)
[2] Part of this work was carried out in a Master's Thesis at University of Agder [1]

© Springer International Publishing Switzerland 2015
M. Ali et al. (Eds.): IEA/AIE 2015, LNAI 9101, pp. 43–52, 2015.
DOI: 10.1007/978-3-319-19066-2_5

2   Dominion

The basic gameplay consists of each player having five cards on hand. When it is a player's turn, They start with what is called one action and one buy. Then, the player goes through three phases: (1) The action phase, (2) the buy phase and (3) the cleanup phase. During the action phase, the player can play action cards from hand, which give a variety of effects beneficial to the player. Initially, due to having only one action, they cannot play more than one action card from hand. However, cards can give more actions, allowing a player to play more action cards afterwards. This opens up for creating interesting combinations when the players carefully choose which cards to use in their decks.

2.1   AI Challenges for Dominion

Creating an AI for Dominion is particularly challenging because of the following:

- The game operates in a **stochastic environment**, as the players' decks are shuffled.
- Dominion is in practice **many different games**. At the start of a game, the players choose 10 out of 25 card types with different behavior. This gives $\binom{25}{10}$ variants of the original game, which in turn makes it difficult to learn one good prior strategy for all variants.
- The players interact in such a way that one players action limits or enables other possible actions. This means that the **number of possible game states is extremely high**.
- Every game has **two possible endings**: Either all the province cards are emptied from supply, or three other card stacks are empty. This means that a strategy needs to optimize for two possible ways of ending the game.
- Part of the **game state is unknown**, as players keep their cards in hand hidden.

This paper presents solutions to these challenges, as well as performance statistics for different MCTS algorithms applied to Dominion.

2.2   Existing Dominion AIs

To the best of our knowledge, very little academic work has been done in the field of Dominion AIs. Fynbo and Nelleman created an AI for Dominion using Artificial Neural Networks, with limited success [2]. However, there exists at least two other AIs of interest for Dominion without academic papers. (1) The official set of AI at playdominion.com from the publishers of the Dominion card game, and (2) the Provincial Dominion AI found at Matt's Webcorner [3].

As far as we can determine, the official AI is made up of different state machines, which employ pre-set rules to the current game state to decide which action to take. We employ similar rule based AIs to measure the strength of our solutions in the experiments section. The Provincial AI utilizes generational co-evolution to evolve tactics based on an initial set of candidates. Neither the official AI nor Provincial provide any results for us to directly compare against. Fynbo and Nelleman provide some results, which are mentioned in the conclusion.

A brute force tree search is not applicable for a Dominion AI. On the first turn, there will only be 21 possible choices and 99 during the second turn. But after six turns, the number of possible choices reach over 15 million, with a branching factor of 22.6 on that turn. A brute force approach would require to exhaustively search to the end of the game — which is usually at least 20 turns.

2.3 Monte Carlo Tree Search Variants

MCTS appeared in different versions in 2006 [4], where the UCT variant was proposed by Kocsis and Szepesvári [5]. UCT is MCTS using any UCB selection formula.

The original formula proposed UCB formula is shown in Equation 1 [6].

$$v_i + \sqrt{\frac{2 \ln n_p}{n_i}} \tag{1}$$

UCB1 is a modification of UCB used for UCT [5] shown in Equation 2:

$$v_i + C \times \sqrt{\frac{\ln n_p}{n_i}} \tag{2}$$

where v_i is the value of the currently evaluated node, C is the exploration constant, n_i is the number of times node i has been visited and n_p is how many times the parent of node i has been visited.

The difference between the two formulas is the inclusion of the exploration and exploitation term, C. The original formula in Equation 1 use $\sqrt{2}$ as value for C, while the formula in Equation 2 is more flexible, allowing this value to be manually set depending on the application.

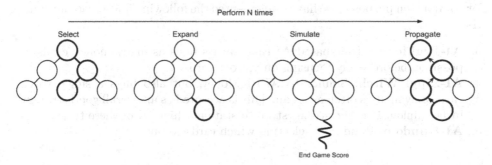

Fig. 1. The flow of the UCT algorithm [4]

After the original was proposed, the UCB1 formula has been subject to changes and improvements, such as UCB1-Tuned and UCB Improved [7,8]. Although UCT is an extension of flat UCB, UCT was shown to sometimes be overly optimistic compared to flat UCB [9].

2.4   UCT for Games

After UCT was introduced by Kocsis and Szepesvári it has been applied to AIs in many games. The most notable research done for the game Go [7,10]. Even though Go is a difficult game with many game states, it is notably different than Dominion, because Go has neither hidden, nor stochastic game elements. However, there is some work on games similar to Dominion, most notably *Magic: The Gathering* (M:TG) [11] and *The Settlers of Catan* (Settlers) [12].

M:TG is a card game quite similar to Dominion in nature, containing the same stochastic elements relating to card draws [13]. The work on the M:TG AI touched upon utilizing UCT, but ended in a flat UCB due to the difficulty in the stochastic nature making tree creation. Although in a later paper, they apply MCTS with Ensemble Determinization with some success [14].

Szita et al. apply MCTS to Settlers [15]. However they are unable to achieve high performance against human players. Still, their work shows that MCTS is both applicable for Settlers, and they also claim that "MCTS is a suitable tool for achieving a strong Settlers of Catan player" [15].

There is, to the best of our knowledge, no existing work using MCTS for Dominion.

3   Approach

This section presents our approach to creating an AI for Dominion. This includes two AI variants:

1. **AI-UCB**: Using MCTS with a flat UCB.
2. **AI-UCT**: Using MCTS with UCT.

For comparison purposes, we have implemented the following finite-state machine AIs:

3. **AI-BigMoney**: Rule-based AI that always buys as many money cards as possible, before solely focusing on victory cards.
4. **AI-SingleWitch**: Similar to AI-BigMoney, but also buys a single witch card, playing it as often as it can. This is accepted as an overall good strategy in Dominion. It is the strongest finite-state machine AI of these three.
5. **AI-Random**: Randomly selecting which cards to buy.

3.1 Stochastic Card Draws

A game turn in Dominion consists of three phases (see section 2). In the third phase (clean-up), the player will have to draw new cards from deck. The new hand will consist of five random cards. This is a variable that makes creating an AI more challenging. It should be noted that even though the cards are randomly drawn, the probability of each card is known as a strong player can have full knowledge of what is left in deck. Hence, it is possible to calculate the probability of drawing each card.

We explored two different approaches to deal with this for both AIs.

1. Create a search node for each possible combination of drawn cards, and associate a probability with each node.
2. Create only one node with a proportionally sampled hand of cards, and create multiple trees to increase reliability.

A search node for each combination. The first approach (see an example in Fig. 2) requires the AIs to create many nodes. This is not very far from a brute force solution, since the number of cards drawn is the biggest factor when the size of the tree increases. For example, in a simple deck with four types of cards and five cards of each type (20 cards in total), there are a total of 56 card combinations. Towards the end of a game of Dominion it is not unusual to have a deck of 40 cards, with possibly 17 different types.

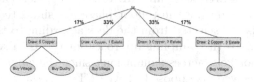

Fig. 2. Example of creating a node for each possible card draw, and assigning the edges of each card draw an associated probability

In order to create all nodes associated with card draws, all possibilities must first be found along with their associated probabilities. Hence, instead of expanding one node at a time, when encountering a node that has card draws beneath, all card draws are created at once, and then one of them is selected afterwards using the associated probability multiplied with the score from the UCB1 formula, to let likely nodes be explored more. However, since each node should be visited at least once, this approach soaks up a lot of simulations, exploring unlikely combinations of cards. One solution could be to set the counter for times visited to one for each draw upon creation, so that all do not need to be explored. Still, the number of nodes needed for this approach to be effective will likely be enormously high.

Proportional Sampling. The complexity of creating a search node for each combination, makes it more attractive to proportionally sample single cards. In this approach, whenever there is need to draw cards in the search tree, the probabilities for drawing card combinations are used, resulting in a single card draw. This card draw is then used for all further simulations, even though the game may never progress to that specific card draw. This works because the card draws are made using proportional sampling, and can be further increased using parallelization to create more realistic card draws. Thus, instead of only having one set of card draws, multiple search trees will view different possibilities of card draws, resulting in better game estimation. Fig. 3 shows how multiple search trees increases card draw sampling.

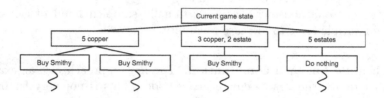

Fig. 3. Different card draws will result in different trees. Three parallell trees could yield the three different draw outcomes.

For the other AI-variant, AI-UCB, card draws are not such a big challenge. Due to the lack of a search tree, all card draws are taken care of during rollout phase. Card draws during the rollout phase are simulated as a real game, where random cards are selected from the deck according to their proportional probabilities, similar to the sampling approach for the UCT variant. However, these samples are not kept across simulations, since the game state beyond the current node is not saved, as opposed to AI-UCT.

3.2 Player Interaction

For AI-UCT, the interaction between players is taken care of at three places:

1. **UCT tree nodes:** The UCT search tree creates nodes for both players.
2. **Propagation phase:** UCT only propagates score according to the same player to corresponding nodes, effectively skipping the other player every other turn. Instead of having a behavior where the AI selects nodes where the opponent performs badly, this ensures that it selects leaf nodes where the opponent will perform their best moves as well. This resembles a Minimax behavior, since the AI will select nodes where both players perform well.
3. **Rollout phase:** The rollout policy is the same for both players, but could have been modified to resemble the opponent player. For this AI however, one assumes that both players use the epsilon heuristic greedy policy.[3]

---

[3] Full details about the rollout policy can be found in the complete thesis [1].

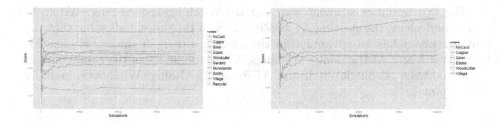

(a) Relative choice likelihood for AI-UCB   (b) Relative choice likelihood for AI-UCT

Fig. 4. Score for each possible decision node at 100 000 simulations after turn 1

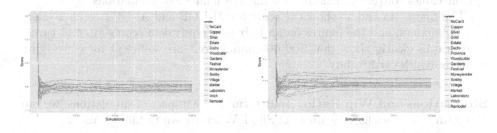

(a) Score for each possible choice for AI-UCB   (b) Score for each possible choice for AI-UCT

Fig. 5. Score for each possible decision node at 100 000 simulations after turn 11

For AI-UCB, only the Rollout policy is included. This is because AI-UCB does not include trees, which makes propagation and tree nodes not applicable.

For both AI-UCT and AI-UCB, we note that complete information about the players' discard piles is known to both players. Since drawing cards from the deck is random, information regarding which cards are left can give a slight, but neglectable advantage. In this implementation, information about the opponent's next draw is available during rollout. This is neglectable, due to the subset of cards.

4   Experiments

This section presents some of the most relevant experiments and statistics which display the performance of the Dominion AIs.[4]

The experiments are carried out with the five AI alternatives: AI-UCB, AI-UCT, AI-BigMoney, AI-SingleWitch and AI-Random (see section 3).

---

[4] Additional experiments have been carried out. These are available in the complete thesis [1].

The first initial experiment was AI-SingleWitch against AI-SingleWitch. In this experiement, the starting AI won in 54.1% of the times, indicating that the starting player has an advantage in the game. Further, in games between AI-SingleWitch and AI-BigMoney, AI-SingleWitch won 98.8% of the games, and in games between AI-SingleWitch and AI-Random, AI-SingleWitch won in 100% of the games.

4.1 Converging on an Action

To determine the appropriate amount of simulations to run for each turn, we used a detailed plot of the score for the possible options throughout a game for AI-UCT and AI-UCB. By plotting these values it is possible to see when the option values converge, thus find a sufficient number of simulations.

The figures 4a and 4b show the score for AI-UCB and AI-UCT as an average of 1 000 games, and the figures 5a and 5b show equivalent scores after 11 turns. The graphs clearly show that 100 000 simulations are sufficient for converging towards which card to buy.

Simulations and Win Ratio. To determine the impact of simulations, we ran 100 games with each AI against AI-SingleWitch shown in table 1 and 2.

Table 1. Win ratio from AI-UCB/AI-UCT vs. AI-SingleWitch exploration constant to 5.0

Simulations	AI-UCB	AI-UCT
1 000	15.00%	2.00%
10 000	67.00%	20.00%
100 000	66.00%	56.57% (99 games)

Table 2. Win ratio from AI-UCB/AI-UCT at 100 000 simulations versus AI-SingleWitch, varying the exploration constant

C	AI-UCB	AI-UCT
1.0	60.00%	34.00%
5.0	66.00%	56.57% (99 games)
10.0	66.00%	44.00%
50.0	60.00%	21.00%

Table 1 shows that AI-UCB is far stronger at 10 000 simulations, however at 100 000 the difference is almost gone. For AI-UCB, we see that 10 000 simulations are almost as good as 100 000 simulations, while for AI-UCT in table 2, 10 000 are not nearly as good as 100 000.

Winning more than 50% of the games, when the opponent is starting, is quite good when determining the total strength of the AIs.

Different Exploration Values. To test the effectiveness of changing the exploration constant, we tested a few different values to see if changing it in any direction could affect the win ratio, shown in table 2.

We notice that the exploration constant does not seem to have to much effect for AI-UCB. For AI-UCT however, the exploration constant appears to be vital for the performance of the algorithm. The reason for this could be that our rollout function is extremely strong in this partical set of cards. When AI-UCB performs rollout from the initial game state repeatedly, the score will mostly be the same, thus the results will not vary much regardless of the number of visits to each node. For AI-UCT however, which will explore options that AI-UCB would never reach, will score worse with too much or too little exploration. AI-UCB performs based on the expert knowledge of the rollout function for this particular set of cards.

Statistics for AI-UCB vs. AI-UCT. We played 200 games of AI-UCT vs. AI-UCB, each algorithm taking the starting position in 100 of them. Both algorithms run with 100 000 simulations.

Table 3. Results from AI-UCB versus AI-UCT

Starting Player	AI-UCT Winrate	AI-UCB Winrate
AI-UCT	52.00%	48.00%
AI-UCB	24.00%	76.00%

As seen here, AI-UCB performs stronger than AI-UCT, as expected from the results in the previous tests.

5 Conclusion

This paper presents completely new AIs for the game Dominion. The novel approaches are presented as: AI-UCB and AI-UCT using UCB and UCT respectively.

AI-UCB is capable of winning 67.00% games against the AI-SingleWitch strategy (a possibly sub-optimal, finite-state machine based solution), even when AI-SingleWitch has the unfair advantage of starting. In the same scenario, AI-UCT is able to win 56.57% of the games. In this particular subset, the AI-SingleWitch appears to be extremely strong and may possibly be close to a sub-optimal tactic.

There is no directly comparable results in the literature. The closest related work is by Fynbo and Nellman who carried out several experiments including a scenario with three opponents. They achieved win rates from 48%, and in the best scenario a win rate of 88%.

6 Future Work

For future work, we would like to test different subsets of cards, which add more complex mechanics to the game, where AI-UCT may be stronger.

We would like to test some improvements, such as changing the scoring system in the propagation phase, the rollout policies and values for C.

Using UCT and UCB for a Dominion AI yields promising results, and can result in extremely strong Dominion players.

References

1. Jansen, J.V., Tollisen, R.: An ai for dominion based on monte-carlo methods (2014)
2. Fynbo, R.B., Nellemann, C.: Developing an agent for dominion using modern ai-approaches. Master's thesis, IT University of Copenhagen (2010)
3. Fisher, M.: Provincial: A kingdom-adaptive ai for dominion (2014)
4. Chaslot, G.: Monte-Carlo Tree Search. PhD thesis, Maastricht University (2010)
5. Kocsis, L., Szepesvári, C.: Bandit based monte-carlo planning. In: Fürnkranz, J., Scheffer, T., Spiliopoulou, M. (eds.) ECML 2006. LNCS (LNAI), vol. 4212, pp. 282–293. Springer, Heidelberg (2006)
6. Auer, P., Cesa-Bianchi, N., Fischer, P.: Finite-time analysis of the multiarmed bandit problem. Machine learning **47**(2–3), 235–256 (2002)
7. Gelly, S., Wang, Y.: Exploration exploitation in go: Uct for monte-carlo go (2006)
8. Auer, P., Ortner, R.: Ucb revisited: Improved regret bounds for the stochastic multi-armed bandit problem. Periodica Mathematica Hungarica **61**(1–2), 55–65 (2010)
9. Coquelin, P.A., Munos, R.: Bandit algorithms for tree search (2007). arXiv preprint cs/0703062
10. Gelly, S., Silver, D.: Monte-carlo tree search and rapid action value estimation in computer go. Artificial Intelligence **175**(11), 1856–1875 (2011)
11. Wizards of the Coast. Magic: The Gathering (2013) http://www.wizards.com/Magic/Summoner/ (visited on 05/30/2014)
12. GmbH, C.: The official website for the settlers of catan (2014)
13. Ward, C.D., Cowling, P.I.: Monte carlo search applied to card selection in magic: the gathering. In: IEEE Symposium on Computational Intelligence and Games, CIG 2009, pp. 9–16. IEEE (2009)
14. Cowling, P.I., Ward, C.D., Powley, E.J.: Ensemble determinization in monte carlo tree search for the imperfect information card game magic: The gathering. IEEE Transactions on Computational Intelligence and AI in Games **4**(4), 241–257 (2012)
15. Szita, I., Chaslot, G., Spronck, P.: Monte-carlo tree search in settlers of catan. In: van den Herik, H.J., Spronck, P. (eds.) ACG 2009. LNCS, vol. 6048, pp. 21–32. Springer, Heidelberg (2010)

Evolutionary Dynamic Scripting: Adaptation of Expert Rule Bases for Serious Games

Reinier Kop[1], Armon Toubman[2(✉)], Mark Hoogendoorn[1], and Jan Joris Roessingh[2]

[1] Department of Computer Science, VU University Amsterdam,
De Boelelaan 1105, 1081HV, Amsterdam, The Netherlands
{r.kop,m.hoogendoorn}@vu.nl
[2] National Aerospace Laboratory NLR, Anthony Fokkerweg 2, 1059CM,
Amsterdam, The Netherlands
{Armon.Toubman,Jan.Joris.Roessingh}@nlr.nl

Abstract. Automatically generating behavior for Non-Player Characters (NPCs) in serious games can be problematic as the specification of their behavior heavily relies on the availability of domain expertise. This expertise can be difficult and costly to extract, and the specified behavior usually does not allow for generalization to new scenarios or users. Alternatively, behavior can be generated using a pure machine learning approach. However, such NPCs may quickly develop static, non-adaptive behavior by exploiting the environment without proper constraints. In this paper, an approach called Evolutionary Dynamic Scripting (EDS) is presented to effectively cope with the disadvantages of the two extremes sketched above. This technique combines the generative characteristics of an evolutionary approach with an adaptive reinforcement learning method called Dynamic Scripting. Dynamic Scripting essentially learns how to prioritize rules from a fixed rule-base specified by domain experts. EDS was tested in an air combat simulation in which agents co-evolve their tactics using EDS. EDS was able to generate improved behavioral rules over the original Dynamic Scripting approach, given the same initial rule-bases. Both generalization to new situations and specialization into roles for the agents were observed.

Keywords: Serious gaming · Dynamic scripting · Evolutionary algorithms

1 Introduction

In the last decade, serious gaming approaches to (military) training applications have gained widespread popularity. Serious games have a number of benefits over 'classical' training, such as cost reduction, the possibility of training with more individualized scenarios, and the ability to produce events that cannot easily be staged in real life.

In many serious games, the behavior of the Non-Player Characters (NPCs) is of crucial importance. Whether filling the role of teammates, tutors, or adversaries, NPCs should exhibit dynamic, adaptive behavior. However, generating such behavior

M. Ali et al. (Eds.): IEA/AIE 2015, LNAI 9101, pp. 53–62, 2015.
DOI: 10.1007/978-3-319-19066-2_6

often requires complex models based on expert knowledge (see e.g. [1]). Creating expert models for a sufficiently large number of training scenarios can be cumbersome. As a result, these models often do not exhibit a sufficiently rich palette of behaviors, and hence result in a highly predictive learning experience. An alternative is to deploy machine learning to generate appropriate behavior models from scratch (see e.g. [2]). Still, guaranteeing dynamic behavior remains problematic.

Several approaches that try to combine the best of both worlds have been proposed, among which the Dynamic Scripting (DS) technique [3]. DS is a reinforcement learning (RL) technique that starts with a set of generally applicable behavior rules provided by domain experts, and lets agents learn how to prioritize these rules for a specific scenario in a series of RL trials. Favoring adaptation speed over optimal performance, this method is not like traditional RL methods, which attempt to create a policy (a mapping of optimal actions to take given a situation). DS heavily relies on the assumption that the initial rule set provides 'good-enough' rules for all possible scenarios, which might not always be the case as the need for additional rules might occur in novel scenarios.

In this paper, a technique is presented which is called Evolutionary Dynamic Scripting (EDS). In EDS, the DS learning process is embedded in an evolutionary method which enables the discovery of new rules. In essence, DS serves as the fitness evaluation function for the evolutionary method. This combination allows for more flexibility in the generation of specific behavior for novel scenarios, while using more general rules designed by a domain expert as a starting point. This way, appropriate domain knowledge is utilized and new behavior is generated where it is shown to be effective, while reducing the workload of the domain expert. The approach is evaluated in an air combat simulation [4], as several DS case studies have already been performed in this domain and can be used as a benchmark.

This paper is organized as follows. Section 2 sketches the relevant background of this work. The EDS approach is described in Section 3 and the experimental setup is described in Section 4. Section 5 presents the results, and finally, Section 6 concludes the paper.

2   Background

A variety of techniques have been proposed that enable the generation of NPC behavior based on domain expertise. Many of these techniques come in the form of cognitive models that stem from human decision-making processes. These techniques mainly focus on the realism of the generated behavior. Swartout et al. [1] for example introduce an architecture for virtual reality-based training in which the virtual agents use task models to reason about causality and the distribution of tasks between human and virtual agents. In [5], Merk presents a number of cognitive models, addressing various aspects of fighter pilot behavior. These models were validated in evaluations in which fighter pilots received training in simulators using enemies driven by the cognitive models. Methods that are still cognitive-based but do add some anticipation elements include theory of mind based approaches (see e.g. [6] and [7]). However, these approaches all heavily rely on domain expertise.

On the other side of the spectrum, various approaches are based on pure machine learning techniques to establish adaptive NPC behavior. A complete overview of this domain is beyond the scope of this paper, but examples for the domain of fighter pilot behavior can be found in [2]. Bellotti *et al.* [8] introduce an agent based on machine learning that is able to adapt the flow of a serious game during play.

The current research is not the first attempt to combine expert knowledge with machine learning techniques. As said, the DS technique is based on the prioritization of rules depending on their appropriateness for the situation at hand. An alternative approach that has been proposed is to tailor a cognitive model to the situation at hand by means of adapting the parameters of the model, also allowing for a form of adaptation [9]. Although these techniques provide ways to adapt existing behavior to new scenarios, they do so in a relatively limited way as they are based on existing rules. Generating new rules is the focus of the work presented here, thereby still taking advantage of information obtained from domain experts. An additional advantage of using and adapting behavior rules is that all NPC behavior is defined in a human-readable way. After the machine learning process, a domain expert or training instructor can always review the learned behavior and make manual changes, if needed.

3   Method

EDS attempts to improve NPC behavior by repeatedly evaluating the performance of an agent's rules using DS, and evolving the evaluated rules using an evolutionary method based on genetic programming (GP). Implementing such a system has two major advantages. First, as with regular DS, the use of behavior rules provides transparency throughout the learning process, as the behavior model is always human-readable. Second, it decreases the need for a domain expert, since the evolutionary method is able to optimize existing rules, or even discover completely new rules.

EDS can be classified as a specialized Learning Classifier System (LCS). LCSs attempt "to evolve a system that will respond to the current state of its environment" [13]. An LCS is usually based on optimizing a policy, using a genetic algorithm (GA) and a reinforcement component. Instead of a GA, EDS uses GP and instead of traditional RL for the reinforcement component, EDS uses DS. EDS is thus able to use rule structures, rather than binary sequences. Moreover, it doesn't necessarily have to create rules from scratch (such as the LCS used in [2]), but is able to optimize existing rules, which it can evaluate rapidly.

The main EDS loop is shown in Algorithm 1. The algorithm is initialized with an initial rule base that is either predefined or generated from scratch. Then, the evolutionary loop starts (lines 3). First, the fitness (or expected effectiveness) of each individual rule is evaluated in the *DS component* (starting in line 4). In the DS component, the rule base is optimized by DS, which results in a fitness value for each rule. The fitness may be revaluated multiple times at each generation (lines 4-6), allowing (weighted) averaging of a rule's fitness values across multiple DS *learning episodes*, thereby increasing the robustness of EDS. Once the fitness is known and stored (line 6), the *GP component* alters various rules in the rule base based on their fitness (shown in line 7). This constitutes one *generation*, after which the (new) rule base can again be evaluated using the DS component. This loop terminates when some termination condition is met (such as a maximum number of generations).

Algorithm 1. EDS
1: rule_base ← initial_rule_base
2: results ← array[max_episodes]
3: **for** generation ← 1 **to** max_generation **do** // EDS loop
4:      **for** episode ← 1 **to** max_episode **do** // DS loop
5:          results[episode] ← perform_DS(rule_base)
6:      fitness ← evaluate_fitness(result)
7:      rule_base ← evolve(rule_base, fitness) // GP component

3.1 Dynamic Scripting Component

DS is a form RL which allows an autonomous agent to dynamically adjust its behavior based on feedback from the environment [3]. DS was originally designed for computer role-playing games, but it has since been adapted for usage in other genres such as real-time strategy games [10, 11], first-person shooters [12], and air combat simulators [4]. DS is not like other RL methods because it requires predefined behavioral rules which are not modified during the learning process. While agents using DS are unable to find new behavior, it also guarantees that agents cannot learn behavior that is worse than the predefined rules. Another difference with DS and traditional RL is the fact that DS does not attempt to make a mapping between observed states and desired actions. This makes it easier for DS to (dynamically) adapt to new situations.

Each agent using DS maintains a set of predefined rules in its rule base, together with a weight value for each rule. For every encounter, a subset of rules is stochastically selected from the rule base, directly proportional to the weight value for each rule.

The selected rules form a script that is used to control the agent during an encounter. After each encounter, the DS algorithm adjusts each rule's weight value based on the outcome of the encounter. If the rules in the script performed well during the encounter, their weights are increased; conversely, if the rules in the script performed badly during the encounter, their weights are decreased. The weight value of each rule therefore comes to represent the expected effectiveness of the rule against the current opponent(s). This redistribution of weights leads to reselection of favorable rules.

After a sufficient number of encounters, we treat the weight of each rule as its fitness value for use in the GP component. When evolving a large set of rules (as we are trying to accomplish with EDS), DS has the advantage of being able to evaluate subsets of the rules in various specific circumstances. Which subsets are applicable under what circumstances is detected automatically by the DS algorithm. At the same time, because only one large set of rules is being evolved, it is possible to maintain a level of general applicability.

3.2 Genetic Programming Component

In the GP component (line 7 in Algorithm 1), the rule adaptation takes place. Once the rules' fitness values are known, we adapt the rules using evolutionary principles, i.e. offspring is created and survivors are selected. The steps of the GP component are shown in Algorithm 2.

Algorithm 2. evolve(rule_base, fitness)

1:	child_num ← 0
2:	**while** child_num < max_children **do**
3:	parents ← select_parents(rule_base, fitness)
4:	**if** random() < prob_crossover **then**
5:	children ← crossover(parents)
6:	**else**
7:	children ← mutate(parents)
8:	**for** child in children **do**
9:	rule_base ← survivors(rule_base + child)
10:	child_num ← child_num + 2
11:	**return** rule_base

Parent selection (line 3) is done through fitness proportionate selection. Two parents are selected per cycle. After selection, crossover is applied with a probability *prob_crossover* (lines 4-5), or else they are mutated (lines 6-7). Applying *either* mutation *or* crossover (as opposed to both) is a common procedure for GP [13].

In lines 8-9, the newly generated children are inserted in the rule base. After adding a child, *survivors* are immediately *selected*. The rule with the lowest fitness value is deterministically removed.

Genetic Operators

To facilitate the use of genetic operators, the rules in the rule base are represented as tree structures. The crossover operator used is subtree crossover, which creates new children by randomly exchanging subtrees of two parents. Each of the subtrees is selected with a probability $p = 1/n$, where n is the number of expressions in the rule.

The mutation operator is one of three methods, chosen randomly (with equal probabilities): point, subtraction, and addition mutation. In point mutation, each expression in a rule has a probability $p = 1/n$ of changing to another random expression, where n is the number of expressions in the rule. Subtraction mutation randomly removes an entire subtree from the rule. This subtree is selected identical to the aforementioned crossover method. With equal probabilities, addition mutation either takes a subtree subtracted by a previously applied subtraction mutation, or randomly generates and adds a subtree to the rule. This random generation is done using a simple grammar in which all valid tree structures are expressed. This grammar prevents invalid rules and erroneous combinations from being created, such as 'fire a missile at an ally'. The choices made in the grammar are always made with equal probabilities.

After performing these steps, the contents of the rule base will have changed, meaning each rule's fitness can be evaluated again using the DS component. This evaluation also includes existing rules that survived the previous generations.

4   Experimental Setup

To investigate the proposed method from various angles, we split the experiments in three stages: rule base generation, validation, and generalization. These stages and the scenario in which they are applied are explained below.

Scenario

A simulated scenario was used (identical to the scenario in [4]) in which two F-16 aircraft (the blues), a 'flight lead' and a 'wingman', engage an enemy F-16 aircraft (the red). The latter is performing a so called Combat Air Patrol (CAP), i.e. repeatedly flying a circular pattern in the airspace that it needs to defend (see Figure 1). Each simulated encounter, the team that first eliminates an aircraft from the other team wins. The blue aircraft are controlled by agents whose rule bases are generated using the EDS approach. The red aircraft is controlled by an agent that uses one of six tactics:

- $t_{default}$ represents the default opponent's tactic. Red flies a counter-clockwise CAP, and fires at enemies upon detection.
- $t_{evading}$ is as $t_{default}$, but includes evasive maneuvers.
- t_{close_range} is as $t_{default}$, but fires missiles from a shorter range.
- $t_{default\_alt}$, $t_{evading\_alt}$, $t_{close\_range\_alt}$ are as $t_{default}$, $t_{evading}$, t_{close_range}, respectively, but flies the CAP in a clockwise fashion.

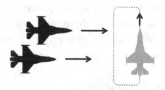

Fig. 1. The two blues (left) try to intercept red (right), who is flying a Combat Air Patrol.

Rule Base Generation

In this experimental stage, EDS is applied to the rule bases of both blues, in an attempt to generate new, improved rule bases. This experiment is meant to show that EDS can find rules that are competitive to rules provided by experts, provided that a reasonably rich initial rule base is present. For the used virtual environment, an Expert Rule Base (ERB) for each blue agent is available and known to be working well with respect to the described scenario. However, we cannot be sure whether these ERBs can be further improved by EDS (it may be the ERBs are already the best possible set of rules for this scenario, given that an expert has specified it). To be certain the initial rule bases are non-optimal, we weaken the ERBs with respect to their originals. This provides us with a Degenerated Rule Base (DRB) for each blue agent.

Both agents' DRBs are evolved concurrently against each of the opponent's tactics. Each application of EDS is performed thirty generations, with ten DS learning episodes per generation, and fifty encounters per learning episode. Ten EDS applications are run per tactic of red. These parameters are based on prior tests.

Each generation, each blue agent generates new rules based on old ones using the mutation operator only. This was empirically found to outperform the use of both crossover and mutation. This predominantly results in exploitation of the rule base (associated with mutation), as opposed to exploration (associated with crossover). When mutation is applied, one of the three mutation operators is randomly chosen.

The performance of the blues in this phase is the ratio of blue wins with respect to the total number of performed encounters.

Validation

The output of the rule generation phase is six sets of rule bases per blue agent (i.e. one per enemy tactic). Ideally, we would have a single rule base which is able to adapt to multiple tactics, the big advantage of regular DS. To achieve this, rules of the rule bases obtained from the previous section are combined so that we end up with a Combined Rule Base (CRB) for each blue agent, equal in size to the ERB and the DRB. The CRB is made by taking the inclusive disjunction of the each of the tactic's best performing rule base from the previous section, ordered by fitness. This set is cut off at the correct rule base size mark (31 for the flight lead, 32 for the wingman).

To validate whether EDS has had any significant impact on the blue's rule bases, the performance of the DRB, ERB, and CRB against each of the six opponent's tactics are all measured using regular DS (since we just want to evaluate the performance of rule bases resulting from EDS). The significance of this comparison is as follows: if the CRB outperforms both other rule bases, we can conclude that the rules evolved by EDS in some way improved the behavior of the ERB, which was one of the goals of this research. We hypothesize that the CRB will outperform the DRB (hypothesis 1), and perform at least as well as the ERB (hypothesis 2) when applying regular DS on either. Hypothesis 2 is a bit more conservative as ERB might already be (near) optimal. The blue agents are trained during 100 learning episodes against each tactic. Each learning episode simulates 250 encounters. During this stage, performance is measured as the running average of the current win ratio (window size 20), averaged over the 100 learning episodes. The large number of learning episodes is chosen to increase the likelihood of the hypothesized effect, given the stochastic nature of both the simulation environment and the DS/EDS techniques.

Generalization

To study the generated rules' ability to generalize, we combine CRBs using five out of the six evolved rule bases. This is done for each combination of five rule bases against its respective previously unseen tactic. For example, we can measure $CRB_{\sim default}$'s performance (with the $\sim$ representing 'not'), consisting of all evolved rule bases except for the one evolved against $t_{default}$, against tactic $t_{default}$. As such, six partially combined rule bases $CRB_{\sim n}$ are generated per blue agent, where n indicates an adversary's tactic. Generating a partially combined rule base is done in a fashion identical to how the CRB was generated. Similar to the previous experimental phase, each of these partially combined rule bases is then used for running DS against each tactic. Again, we apply 100 learning episodes, and 250 encounters per learning episode, where the performance is the running average of window size 20. We hypothesize that the regular CRB will outperform each $CRB_{\sim n}$ using regular DS for both cases (hypothesis 3), because it has complete knowledge of the current opponent's tactic.

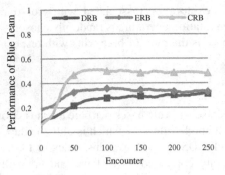

Fig. 2. Results of DS with various initial rule bases, averaged over all tactics, (100 learning episodes per tactic)

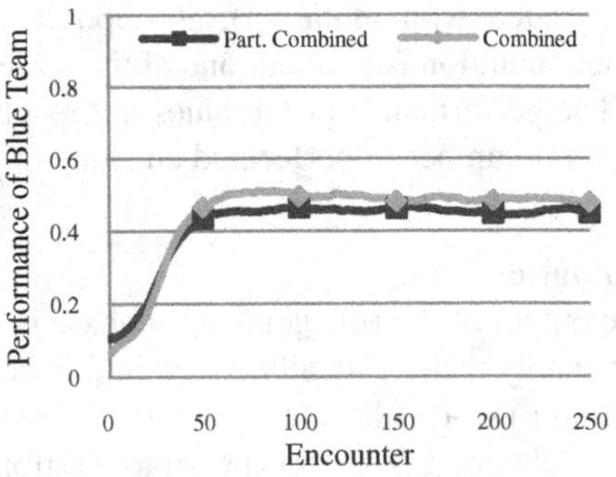

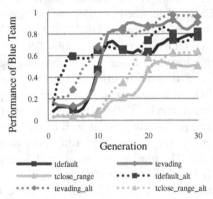

Fig. 3. Results of DS with fully and partially combined rule bases, averaged over all tactics (100 learning episodes per tactic)

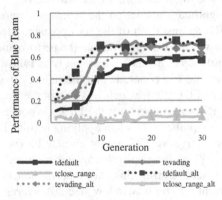

Fig. 4. Average of 10 EDS runs for each tactic played against

Fig. 5. Best of 10 EDS runs for each tactic played against

5   Results

Rule Base Generation

Figure 2 shows the average over ten EDS runs against each tactic. The runs against t_{close_range} and $t_{close\_range\_alt}$ evidently perform much worse relative to the runs against the other four tactics. The average final performance per tactic is highest against $t_{default\_alt}$ (0.728), followed by $t_{evading}$ (0.682), $t_{evading\_alt}$ (0.629), $t_{default}$ (0.573), $t_{close\_range\_alt}$ (0.116), and t_{close_range} (0.052).

Figure 3 shows the best EDS run out of 10. The lines in the figure are smoothed using a running average with a window size of 3. t_{close_range} and $t_{close\_range\_alt}$ are in fact able to reach a relatively high performance in a few cases. The best performance per tactic in the final generation is greatest against $t_{evading\_alt}$ (0.956), followed by $t_{evading}$ (0.905), $t_{default\_alt}$ (0.819), $t_{default}$ (0.771), $t_{close\_range\_alt}$ (0.637), and t_{close_range} (0.506).

The results show that EDS is able to generate rules of increasing quality against four out of six tactics. This is in line with the research reported in [4], where the blues against t_{close_range} and $t_{close\_range\_alt}$ were found to perform worse when applying DS. This suggests that the close range tactics are simply more difficult to win against.

When exploring the behavior exhibited by the newly generated tactics, a limited form of role specialization was observed. For example, one agent would behave as an engaging agent, with rules describing how to engage the opponent. Another agent behaved as an evading agent, with mostly rules of evasion. After the experiment, the engaging agent had many near-identical rules, e.g. a rule for firing missiles when no other missile is flying towards the target, and another for firing missiles only if the agent's wingman has no missiles left. A similar situation held for the evading agent. This indicates a certain degree of convergence has taken place for these rule bases.

Validation
The performances of DRB, ERB, and CRB have been averaged over the different opponent tactics. An independent two-tailed t-test assuming unequal variances ($\alpha = 0.05$) shows that the average performance after the final encounter of the CRB is significantly higher than both DRB ($p \ll 0.05$) and ERB ($p \ll 0.05$) performances, as suggested in Figure 4. This confirms hypothesis 1 and 2, suggesting the CRB (and thus EDS) make a significant contribution to the scenario. Additionally, the ERB and DRB performances do not differ significantly from each other ($p = 0.513$).

Generalization
To investigate generalizability, again the performances for all tactics are averaged, providing us with results for which the relevant adversary tactic was used during the learning process to form the rule set (fully combined), and results for which it was not (partially combined). A two-tailed t-test assuming unequal variances ($\alpha = 0.05$) shows there is no difference in performance ($p = 0.086$), as Figure 5 suggests. Thus, hypothesis 3 is rejected, suggesting a certain measure of redundancy was introduced when evolving rule bases against different tactics. Finally, the graphs in Figure 4 and Figure 5 stabilize at the 0.5-mark. This is likely due to chance; the red and blue team setups are different from one another. Slight changes would quickly shift this equilibrium.

6   Conclusion

Though NPCs are rapidly becoming more intelligent, there is still a long way to go before they can match human intelligence. EDS attempts to partially bridge this gap by focusing on the adaptivity by evolving rule bases, while at the same time reducing the required domain knowledge. We investigated whether EDS could be an improvement relative to DS in terms of behavior, domain expertise, and generalization.

We showed that EDS is able to generate improved behavior. Simpler and sometimes more specialized rules were found. This suggests that EDS may be able to reduce the workload of domain experts, since they no longer have to focus on creating different roles for NPCs; they can simply design a generic set of rules, after which EDS automatically assigns certain roles to certain NPCs. Whether the observed generalizability of the rules also holds in other scenarios should be investigated.

Regarding future work, there are a number of improvements and interesting possibilities left to explore. First, the realism of the behavior in the system needs to be investigated. Second, generating rules from scratch as opposed to using a predefined rule base may create more diverse behavior. Finally, since a rule's fitness is evaluated in the context of script and not in isolation, it is worthwhile to investigate how to incorporate rule's interactions with each other when assigning fitness values.

References

1. Swartout, W., Gratch, J., Hill, R., Hovy, E., Marsella, S., Rickel, J., Traum, D.: Towards Virtual Humans. AI Magazine **27**, 96–108 (2006)
2. Smith, R., El-Fallah, A., Ravichandran, B., Mehra, R., Dike, B.: The fighter aircraft LCS: a real-world, machine innovation application. In: Applications of Learning Classifier Systems, pp. 113–142 (2004)
3. Spronck, P., Ponsen, M., Sprinkhuizen-Kuyper, I., Postma, E.: Adaptive game AI with dynamic scripting. Machine Learning **63**(3), 217–248 (2006)
4. Toubman, A., Roessingh, J.J., Spronck, P., Plaat, A., van den Herik, J.: Dynamic scripting with team coordination in air combat simulation. In: Ali, M., Pan, J.-S., Chen, S.-M., Horng, M.-F. (eds.) IEA/AIE 2014, Part I. LNCS, vol. 8481, pp. 440–449. Springer, Heidelberg (2014)
5. Merk, R.-J.: Cognitive Modelling for Opponent Agents in Fighter Pilot Simulators, Agent Systems Research Group, VU University, Amsterdam: Ph.D. Thesis (2013)
6. Harbers, M., Van den Bosch, K., Meyer, J.: Modeling agent with a theory of mind. In: Proceedings of the 2009 IEEE/WIC/ACM International Joint Conference on Web Intelligence and Intelligent Agent Technology (2009)
7. Hoogendoorn, M., Soumokil, J.: Evaluation of virtual agents attributed with theory of mind in a real time action game. In: van der Hoek, Kaminka, Lesperance, Luck, Send (eds.) Proceedings of the Ninth International Conference on Autonomous Agents and Multiagent Systems (2010)
8. Bellotti, F., Berta, R., de Gloria, A., Primavera, L.: Adaptive Experience Engine for Serious Games. IEEE Transactions on Computation Intelligence and AI in Games **1**(4), 264–280 (2009)
9. Koopmanschap, R., Hoogendoorn, M., Roessingh, J.: Tailoring a Cognitive Model for Situation Awareness using Machine Learning. Applied Intelligence **42**(1), 36–48 (2015)
10. Dahlbom, A., Niklasson, L.: Goal-directed hierarchical dynamic scripting for RTS games. In: AIIDE (2006)
11. Ponsen, M.: Improving adaptive game AI with evolutionary learning, TU Delft: PhD Thesis (2004)
12. Policarpo, D., Urbano, P., Loureiro, T.: Dynamic scripting applied to a first-person shooter. In: 5th Iberian Conference on Information Systems and Technologies (CISTI) (2010)
13. Eiben, A.E., Smith, J.E.: Introduction to evolutionary computing. Springer, Berlin (2010)

Particle Filter-Based Model Fusion for Prognostics

Claudia Maria García[1], Yanni Zou[2], and Chunsheng Yang[3](✉)

[1] Grup Rubí Social, 08191 Rubí (Barcelona), Spain
[2] Jiujiang University, Jiujiang, Jiangxi, China
[3] National Research Council Canada, Ottawa, ON, Canada
Chunsheng.Yang@nrc.gc.ca

Abstract. Predictive maintenance is an emerging technology which aims at increasing availability of systems, reducing maintenance cost, and ensuring the safety of systems. There exist two main issues in predictive maintenance. The first challenge is the system operation region definition, detection and modelling; and another one is estimation of the remaining useful life (RUL). To address these issues, this paper proposes a particle filter (PF)-based model fusion approach for estimating RUL by classifying the system states into different operation regions in which a data-driven model is developed to estimate RUL corresponding to each region, and combined with PF-based fusion algorithm. This paper reports the proposed approach along with some preliminary results obtained from a case study.

Keywords: Particle filter (PF) · Predictive maintenance · Remaining useful life /cycle (RUL/RUC) · Time to failure (TTF) · Operation region · Classification

1   Introduction

Predictive maintenance based on fault prognosis can reduce maintenance and production costs, and increase the availability of systems. Hence, prognosis such as data-driven prognostics has become an active research area [1] [2]. The main goal of a prognostic system is to predict failures before accruing and estimate the Remaining Useful Life or Cycles (RUL/ RUC) or time to failure (TTF). The operation regions covered from normal-operation to failure, i.e. the number of cycles remaining between the present and the instance when a system can no longer perform a complete cycle. State prediction can help to detect the faults as early as possible and to deal with them when the current measurements are still within the normal phase of operation. However, how to define the normal operation region remains an issue. Due to the increment of systems complexity with large number of components, model decomposition is attracting a lot of attention from research community [3]. In other words, operation region is decomposed into independent subspaces by using measured signals as local inputs. Each local estimator corresponding to operation region operates independently, and the damage estimation becomes naturally distributed [4] [5].

In general, prognostics can be performed using either data-driven methods or physics-based approaches. Data-driven prognostic methods use pattern recognition and machine learning techniques to detect changes in system states [7, 8]. Data-driven

© Her Majesty the Queen in Right of Canada 2015
M. Ali et al. (Eds.): IEA/AIE 2015, LNAI 9101, pp. 63–73, 2015.
DOI: 10.1007/978-3-319-19066-2_7

prognostic methods rely on past patterns of the degradation of similar systems to project future system states; their forecasting accuracy depends on not only the quantity but also the quality of system history data, which could be a challenging task in many real applications [6, 19]. Another principal disadvantage of data-driven methods is that the prognostic reasoning process is usually opaque to users [20]; consequently, they sometimes are not suitable for some applications where forecast reasoning transparency is required. Physics-based approaches typically involve building models (or mathematical functions) to describe the physics of the system states and failure modes; they incorporate physical understanding of the system into the estimation of system state and/or RUL [21-23]. Physics-based approaches, however, may not be suitable for some applications where the physical parameters and fault modes may vary under different operation conditions [24]. On one hand, it is usually difficult to tune the derived models *in situ* to accommodate time-varying system dynamics. Recently the particle filter(PF)-based approaches have been widely used for prognostic applications [25-29], in which the PF is employed to update the nonlinear prediction model and the identified model is applied for forecasting system states. It is proven that FP-based approach, as a Sequential Monte Carlo (SMC) statistic method [30, 31], is affective for addressing the issues that data-driven and physic-based approach face by fusing different models to improve the performance of models.

This paper presents a general PF-based model fusion methodology for constructing RUL monitoring/predicting systems. The proposed method is formulated and illustrated through a real Auxiliary Power Unit (APU) prognostic application. This paper also presents the the preliminary experimental results from the case study. The organization of the paper is as follows. Section 2 presents the proposed PF-based model fusion methods; Section 3 introduces the case study to demonstrate the implementation of the proposed methods along with some experimental results; Section 4 discusses the results and future work; and the final section concludes paper.

2   FP-Based Model Fusion Method

The stochastic approximation is used when functions cannot be computed directly and the system states may be estimated via acquired real data (noisy observations). Stochastic approximation uses probability theory to estimate the monitoring system state. Let's get started by Monte Carlo equation, concerning to the approximation of some probability distribution $p(x)$, $x \in \mathcal{R}^l$, being l the operation region, the approximation yields to

$$E\{X\} = \int f(x) p(x) dx \tag{1}$$

where $f(\cdot) \in \mathcal{R}^n$ is some useful function for estimation, in the n domain of interest. In cases where this cannot be achieved analytically the approximation problem can be tackled indirectly by drawing random samples σ_i^l, in l region, from a distribution $q(x)$ instead of $p(x)$, like:

$$E\{X\} = \int f(x)p(x)dx = \int f(x)\frac{q(x)p(x)}{q(x)}dx$$

$$\approx \int f(x)\frac{p(x)}{q(x)}\frac{1}{N}\sum_{i=1}^{N}\delta_{\sigma_i}(x)dx \tag{2}$$

$$= \frac{1}{N}\sum_{i=1}^{N}\frac{p(\sigma_{i^l})}{q(\sigma_{i^l})}f(\sigma_{i^l})$$

with, $\dfrac{p(\sigma_{i^l})}{q(\sigma_{i^l})}$ as normalizing term, since now weight parameter Θ, then (1) results

in the function approximator

$$\hat{X}(x:\theta,\sigma) = \sum_{i=1}^{N}\theta_i f(x:\sigma_{i^l}) \tag{3}$$

where the notation $\hat{X}(x:\theta,\sigma)$ means the value of $\hat{X}(x:\theta,\sigma)$ evaluated at $x \in \mathcal{R}^l$ given $\theta \in \mathcal{R}^N$ and the parameter $\sigma \in \mathcal{R}^l$ in which dimension depends on the approximator. The approximator has a linear dependence on θ but a nonlinear dependence on σ. In case of Non-Linearly Parametrized Approximators (NLIP), the parameters denoted by σ and the adaptable weights θ are updated. Therefore, we can think informally of the nonlinear estimation structure,

$$y = g^l(\hat{x},\alpha^l) \tag{4}$$

$$\hat{x} = \sum_{i=1}^{N}\theta_i f(x:\sigma_i) \tag{5}$$

$$\theta = \eta(g^l,q,\sigma) \tag{6}$$

$$\sigma = \zeta(y,l) \tag{7}$$

where (4) is the observation equation, (5) is the nonlinear estimation equation, (6) is the importance on-line adjustable weight correction and (7), Particle selection, is the particles evolution depending on the noisy observations y and the operation region l. Next we proceed to the design of η, ζ, f and l, the operation-region partition as shwon in Figure 1.

Figure 1 shows the scheme of PF-based model fusion proposed for TTF estimation for prognostics. There are three main components: region selection, regional models for TTF (RUL) estimation, PF-based model fusion algorithm named as On-line Approximate Classifier (OLAC). Following is description of these main components

2.1 Regional Selection (ζ)

In prognostics systems, to detect and predict the risk to complete a planned mission, the monitored data variance may be sufficient to differentiate a normal-operation and failure; the variance is a measure of how far a set of input data is from the normal-operation region. Signs of aging or degradation are detectable prior failure when the monitored model incorporate variables with high variance.

Principal Component Analysis (PCA) is a multivariable statistical approach used to study the variance of a data set, it is widely applied in industry for process monitoring. We developed a PCA search algorithm aiming to sort the input data vector $\vec{y}_j, j = \{1..N\}$ with N the number of input measurements and find indices $v_j, j = \{1..N\}$ that describe the variation in the model in order of their importance. The ordering is such that $Var(v_1) \geq Var(v_2) \geq ... \geq Var(v_j)$, where $Var(v_j)$ denotes the variance of $\vec{y}_j$; being the variance a parameter describing the actual, normal-operation to failure, probability distribution of $\vec{y}_j$. This algorithm computes variable variance corresponding to a data-model $M = [Var(\vec{y}_j), j = \{1..a\}]$. Given M the mean μ and standard deviation λ for each $\vec{y}_j$ computes standardized $\bar{\chi}$ model. The covariance matrix, E will be computed for each $\vec{y}_j$. So, the eigenvalues, EIG is evaluated with a design parameter, Γ; this process permits extract lines that characterize the data. A result table, $\{v,\omega\}$ giving up with the more representative variable, $Var(v_1) \geq Var(v_2) \geq ... \geq Var(v_j)$. This information is used to replace the high variance attributes $Var(v_1), Var(v_2), ..., Var(v_n)$ by particles $\sigma_i^l \in \mathcal{R}^l$. The particles, σ_i^l of the PF model are selected among the centers, a_j of the fuzzy subspaces, according to the minimum distance criterion:

$$\sigma_i^l = 1 + round\left(\frac{x_n}{d\rho_n}\right) \tag{8}$$

with $d\rho_n = dx_n$ to $dx_n / 2$ neglecting evaluation centers within $dx_n / 2$ of the edges of $a_j \in \mathcal{R}^n$. This criterion was chosen since it extends the fundamental notion of fuzzy membership degree in the multidimensional input space and can be used to

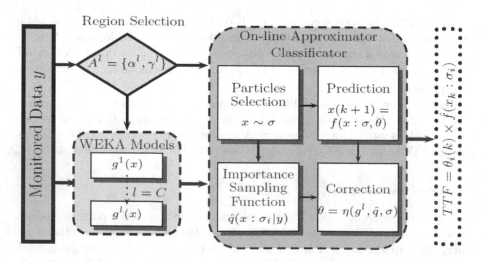

Fig. 1. The PF-based model fusion scheme for TTF estimation

activate the closest particles, σ_i in a certain domain, l, dealing in such a way with the particles degeneracy problem. The method considers all the centers as candidates for locating the particles, however, only a subset of the fuzzy centers is selected as the algorithm proceeds, the ones that are close to the observation data according to the Euclidean minimum distance criterion. At each time instant the number of selected fuzzy functions is equal to the number of particles. And if, the measurement get to a different operation-region, A^l, the algorithm selects other particles and memberships, and PF estimator never gets to zeros.

2.2   TTF Estimation

In the estimation step, the expectation of the state given by $f(x)$ is approximated by choosing an importance distribution $q(x)$ that is similar to $f(x)$

$$
\begin{aligned}
E\{f(x)\} &= \int_x f(x) \times p(x)dx \\
&= \int_x f(x)\left(\frac{p(x)}{q(x)}\right) \times q(x)dx
\end{aligned}
\tag{9}
$$

by using particles σ_i^l, $i=\{1..M\}$ from the importance distribution $q(x)$, the proposal sample distribution lead to $q(x_k | y_k)$,

$$
\begin{aligned}
\hat{f}(k): &= E\{f(x)\} \\
&= \int_x f(x_k : \sigma_i^l)\left(\frac{p(x_k \mid y_k)}{q(x_k \mid y_k)}\right) \times q(x_k \mid y_k)dx
\end{aligned}
\tag{10}
$$

If, applying *Bayes rule* and defining a weight law as:

$$
\begin{aligned}
\Theta(k) &\propto \frac{p(x_k \mid y_k)}{q(x_k \mid y_k)} = \\
&= \frac{g^l(y_k \mid x_k, \alpha^l) \times p(x_k)}{q(x_k \mid y_k)}
\end{aligned}
\tag{11}
$$

and, replacing (10) in (11)

$$
\begin{aligned}
\hat{f}(k) &= \frac{1}{p(y_k)}\int f(x_k : \sigma_i^l) \\
&\left[\frac{g^l(y_k \mid x_k, \alpha^l) \times p(x_k)}{q(x_k \mid y_k)}\right] \times q(x_k \mid y_k)dx_k
\end{aligned}
\tag{12}
$$

which is the expectation of the weighted function $\hat{f}(k : \theta, \sigma) = \Theta(k) \times f(x_k : \sigma_i^l)$ scaled by a normalizing term.

2.3   Correction (η)

In the weight calculation phase, you get the weight Θ_i of each particle $f(x : \sigma_i^l)$ from the observation model $g^l(y_k \mid x_k, \alpha^l)$. This probability is calculated as a multi-variate distribution $q(x_k|y_k)$. This importance function is evaluated for each particle $x:\sigma$ sampled in the prediction step. Finally, the weights of all the particles are normalized, θ, always ensuring sum of one. Arriving to the final estimate, $\hat{f}(k) \approx \theta_i(k) \times f(x_k : \sigma_i^l)$.

Generalizing the on-line weights law, using (11), $\Theta(k) \cdot q(x_k \mid y_k) = g^l(y_k \mid x_k, \alpha^l) \times p(x_k)$, thus:

$$\hat{q}(x_k : \sigma_i^l \mid y_k) \approx \frac{1}{N} \sum_{i=1}^{N} \delta(x_k - x_k : \sigma_i^l) \tag{13}$$

Now, introducing the particles from the proposal $\sigma_i^l \sim q(x_k \mid y_k)$ and using the Monte Carlo approach leads to the desired result. That is, the *Importance Sampling Function*:

$$\hat{q}(x_k : \sigma_i^l \mid y_k) \approx \frac{1}{N} \sum_{i=1}^{N} \delta(x_k - x_k : \sigma_i^l) \tag{14}$$

and therefore substituting, applying the sifting property of the Dirac delta function and defining the normalized weights as

$$\theta(k) = \frac{\Theta^i(k)}{\sum_{i=1}^{N} \Theta_i(k)} \tag{15}$$

$$\Theta(k) = \frac{g^l(y_k \mid x_k : \sigma_i^l) \times p(x_k : \sigma_i^l)}{\hat{q}(x_k : \sigma_i^l \mid y_k)} \tag{16}$$

3   Implementation: A Case Study of APU RUC Estimation

3.1   APU and APU Data

The APU engines on commercial aircrafts are mostly used at the gates. They provide electrical power and air conditioning in the cabin prior to the starting of the main engines and also supply the compressed air required to start the main engines when the aircraft is ready to leave the gate. APU is highly reliable but they occasionally fail to start due to failures of components such as the Starter Motor. APU starter is one of the most crucial components of APU. During the starting process, the starter accelerates APU to a high rotational speed to provide sufficient air compression for self-sustaining operation. When the starter performance gradually degrades and its output power decreases, either the APU combustion temperature or the surge risk will increase significantly. These consequences will then greatly shorten the whole APU life and even result in an immediate thermal damage. Thus the APU starter degradation can result in unnecessary economic losses and impair the safety of airline operation. When Starter fails, additional equipment such as generators and compressors must be used to deliver the functionalities that are otherwise provided by the APU. The uses of such external devices incur significant costs and may even lead to a delay or a flight cancellation. Accordingly, airlines are very much interested in monitoring the health of the APU and improving the maintenance.

For this study, we considered the data produced by a fleet of over 100 commercial aircraft over a period of 10 years. Only ACARS (Aircraft Communications Addressing and Reporting System) APU starting reports were made available. The data consists of operational data (sensor data) and maintenance data. The maintenance data contains reports on the replacements of many components which contributed the different failure modes. Operational data are collected from sensors installed at strategic locations in the APU which collect data at various phases of operation (e.g., starting of the APU, enabling of the air-conditioning, and starting of the main engines). The collected data for each APU starting cycle, there are six main variables related to APU performance: ambient air temperature (T_1), ambient air pressure (P_1), peak value of exhaust gas temperature in starting process (EGT_{peak}), rotational speed at the moment of EGT_{peak} occurrence (N_{peak}), time duration of starting process (t_{start}), exhaust gas temperature when air conditioning is enable after starting with 100% N (EGT_{stable}). There are 3 parameters related to starting cycles: APU serial number (S_n), cumulative count of APU operating hours (h_{op}) , and cumulative count of starting cycles(cyc). In this work, in order to find out remaining useful cycle, we define a remaining useful cycle (RUC) as the difference of cyc_0 and cyc. cyc_0 is the cycle count when a failure happened and a repair was token. When RUC is equal to zero (0), it means that APU failed and repair is needed. RUC will be used in PF prognostic implementation in the following.

3.2   Implementation for APU RUC Estimation

This section presents an implementation of PF-based prognostics for APU starter. As we mentioned in Section 2, we first partition the operation state into 16 regions by using our fuzzy-based classification techniques (will be reported in other paper). Then for each region, we build a data-driven model to estimate the TTF given the operation condition. In this work, the region models are trained with SMO (Sequential Minimal Optimization) SVM(support Vector Machine) algorithm. The models noted SMO0, SMO1, SMO2, ... SMO 16. The detail on data-driven methodology is refereed in [32]. Finally we implement the OLAC algorithm following the proposed PF-based model fusion schema. The OLAC algorithm combines 16 regional models to generate a relatively precise TTF (RUC) estimation.

3.3   Experimental Results

We conducted the experiments based on implementation of the proposed PF-based model fusion algorithm (OLAC) by using the data-driven models developed for each region. The data-driven models are developed following a data-mining-based methodology [32] from 10-year operational data and maintenance data from operator. We run the testing data on all models by using region selection, PF-based correction to verify if the algorithm improved the precision of the RUL estimations. To evaluate the performance of the model fusion algorithm, several criteria are deployed from statistics. They are the mean absolute error (MAE), the mean squared error (MSE), and the mean absolute percentage error ($MAPE$) defined as follows.

$$MAE = \frac{1}{N}\sum_{i=1}^{N} |e_i|$$

(17)

$$MSE = \sum_{i=1}^{N} \sqrt{e_i} \qquad (18)$$

$$MAPE = \frac{1}{N} \sum_{i=1}^{N} |\frac{e_i}{y_i}| \times 100 \qquad (19)$$

We computed these criteria for each region based on deployed models selected by OLAC algorism, and the results are shown in Table 1. The estimation results are plotted in Figure 2

4   Discussion

As mentioned above, Table 1 shows the statistical analysis results of RUL estimation from OLAC model combination. Figure 2 shows the plotting figures of the experimental results for 16 regions noted as *test0* to *test15*. In these Figures, the x-axis is the predicted RUC and y-axis is the actual RUC. When the dots form a direct line with slop ~1.0, the estimation is the close to actual value. This is the ultimate goal of our PF-based model fusion techniques. From these preliminary results, it is obvious that Region 0, 1, 2, 4, 7, 8, 9, 10, 11, and 15 demonstrated a high precision of the RUC estimations, and rest are needed to be further investigated. The experimental results above demonstrated that PF-based model fusion methods are useful and effective for combining multiple region models to improve the precision of RUC.   Since there is existing a large variance in the different failure models, the precise RUC prediction for a particular APU Starter is really challenged.

Table 1. The evaluation results for each region

Region Name	MAE	MSE	MAPE	MODEL ACTIVATED
Test 0	0.8689	1,2918	893,7902	SMO0
Test 1	6.1142	37,7908	7.05e+03	SMO1, SMO6
Test 2	16,7287	347,7456	4.08e+03	SMO2, SMO6
Test 3	52,6930	4.42e+03	4.79e+04	SMO 3, SMO6, SMO9, SMO14
Test 4	27,1294	768,7272	5.57e+04	SMO4, SMO13
Test 5	8,2586	89,6823	2.63e+03	SMO5, SMO6
Test 6	50,9350	3.78e+03	1.01e+05	SMO6
Test 7	10,0948	117,3304	1.11e+04	SMO6,SMO7
Test 8	15,7412	350,4517	2.43e+04	SMO8
Test 9	74,1271	5.99e+03	9.17e+04	SMO6, SMO9, SMO14
Test 10	12,0190	209,1858	1.88e+04	SMO6, SMO10
Test 11	23,4310	834,0634	2.51e+03	SMO11, SMO12
Test 12	51,7319	4.87e+03	1.40e+05	SMO12, SMO14
Test 13	15,9962	393,2352	6.49e+04	SMO4, SMO13
Test 14	50,5856	5.14e+03	1.35e+05	SMO0, SMO6, SMO14
Test 15	31,2127	1.35e+03	1.167e+03	SMO6, SMO15

In this paper, we only report the preliminary results. It is worth to note the number of regions is empirical and need more investigation to find out the right number such that the precision of RUC estimation will be largely improved. At moment, we developed a fuzzy- based operation region definition and modeling techniques. Its details will be reported in other papers. It is possible to investigate the other method to improve the performance of region classification models. One potential way is to classify the state region based on failure modes by integrating with FMEA (failure mode effective analysis). This will be our future work.

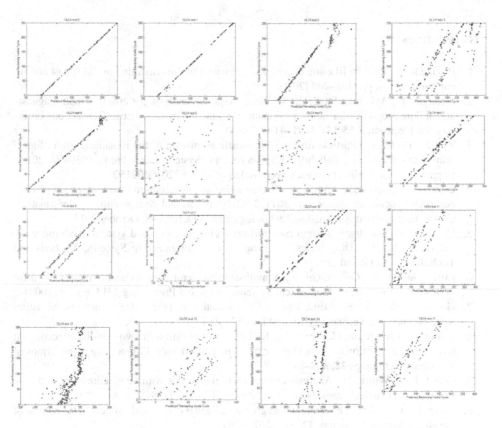

Fig. 2. The RUC estimation results from each regional estimator (16 regions)

5   Conclusions

In this paper we developed a PF-based model fusion method to estimate the RUL or TTF for predictive maintenance and applied it to APU Starter prognostics as a case study. We implemented the PF-based OLAC algorithm by using sequential importance sampling, and conducted the experiments with 10 years historic operational data provided by an airline operator. From the experimental results, it is obvious that the developed PF-based model fusion technique is useful for estimating relative precise remaining useful life for the monitored components or machinery systems in predictive maintenance.

Acknowledgment. Many people at the National Research Council Canada have contributed to this work. Special thanks go to Jeff Bird and Craig Davison. We also thank for Air Canada to provide us APU historic data. This work is supported by the Natural Science Foundation of China (Grant No. 61463031).

References

1. P. G. Zio, E.: Particle filtering prognostic estimation of the remaining useful life of nonlinear components, pp. 403–409 (2010)
2. Strangas, E., Aviyente, S., Zaidi, S.: Time-frequency analysis for efficient fault diagnosis and failure prognosis for interior permanent-magnet ac motors. IEEE Transactions on Industrial Electronics, **55**(12), 4191–4199 (2008)
3. Zio, E., Peloni, G.: Particle filtering prognostic estimation of the remaining useful life of nonlinear components. Reliability Engineering and System Safety **96**(3), 403–409 (2011). http://www.sciencedirect.com/science/article/pii/S0951832010002152
4. Garcia, C., Quevedo, J.: Agile accelerated aging and degradation characterization of a v/f controlled induction motor. In: 2011 XXIII International Symposium on Information, Communication and Automation Technologies (ICAT), pp. 1–8, October 2011
5. Garca, C.M.: Intelligent system monitoring: on-line learning and system condition state. In: Kadry, S. (ed.) Diagnostics and Prognostics of Engineering Systems: Methods and Techniques. IGI Global (2013)
6. Liu, J., Wang, W., Golnaraghi, F.: A multi-step predictor with a variable input pattern for system state forecasting. Mechanical Systems and Signal Processing **2315**, 86–99 (2009)
7. Gupta, S., Ray, A.: Real-time fatigue life estimation in mechanical structures. Measurement Science and Technology **18**, 1947–1957 (2007)
8. Yagiz, S., Gokceoglu, C., Sezer, E., Iplikci, S.: Application of two non-linear prediction tools to the estimation of tunnel boring machine performance. Engineering Applications of Artificial Intelligence **22**, 808–814 (2009)
9. Groot, C.D., Wurtz, D.: Analysis of univariate time series with connectionist nets: a case study of two classical cases. Neurocomputing **3**, 177–192 (1991)
10. Tong, H., Lim, K.S.: Threshold autoregression, limited cycles and cyclical data. Journal of the Royal Statistical Society **42**, 245–292 (1991)
11. Subba, R.T.: On the theory of bilinear time series models. Journal of the Royal Statistical Society **43**, 244–255 (1981)
12. Friedman, J.H., Stuetzle, W.: Projection pursuit regression. Journal of the American Statistical Association **76**, 817–823 (1981)
13. Friedman, J.H.: Multivariate adaptive regression splines. Annals of Statistics **19**, 1–141 (1981)
14. Brillinger, D.R.: The identification of polynomial systems by means of higher order spectra. Journal of Sound and Vibration **12**, 301–313 (1970)
15. Atiya, A., El-Shoura, S., Shaheen, S., El-Sherif, M.: A comparison between neural-network forecasting techniques-case study: river flow forecasting. IEEE Transactions on Neural Networks **10**, 402–409 (1999)

16. Liang, Y., Liang, X.: Improving signal prediction performance of neural networks through multi-resolution learning approach. IEEE Transactions on Systems, Man, and Cybernetics-Part B: Cybernetics **36**, 341–352 (2006)
17. Husmeier, D.: Neural networks for conditional probability estimation: forecasting beyond point prediction. Springer, London (1999)
18. Korbicz, J.: Fault Diagnosis: Models, Artificial Intelligence, Applications. Springer, Berlin (2004)
19. Wang, W., Vrbanek, J.: An evolving fuzzy predictor for industrial applications. IEEE Transactions on Fuzzy Systems **16**, 1439–1449 (2008)
20. Tse, P., Atherton, D.: Prediction of machine deterioration using vibration based fault trends and recurrent neural networks. Journal of Vibration and Acoustics **121**, 355–362 (1999)
21. Adams, D.E.: Nonlinear damage models for diagnosis and prognosis in structural dynamic systems. Proc. SPIE **4733**, 180–191 (2002)
22. Luo, J., et al.: An interacting multiple model approach to model-based prognostics. System Security and Assurance **1**, 189–194 (2003)
23. Chelidze, D., Cusumano, J.P.: A dynamical systems approach to failure prognosis. Journal of Vibration and Acoustics **126**, 2–8 (2004)
24. Pecht, M., Jaai, R.: A prognostics and health management roadmap for information and electronics-rich systems. Microelectronics Reliability **50**, 317–323 (2010)
25. Saha, B., Goebel, K., Poll, S., Christophersen, J.: Prognostics methods for battery health monitoring using a Bayesian framework. IEEE Transactions on Instrumentation and Measurement **58**, 291–296 (2009)
26. Liu, J., Wang, W., Golnaraghi, F., Liu, K.: Wavelet spectrum analysis for bearing fault diagnostics. Measurement Science and Technology **19**, 1–9 (2008)
27. Liu, J., Wang, W., Golnaraghi, F.: An extended wavelet spectrum for baring fault diagnostics. IEEE Transactions on Instrumentation and Measurement **57**, 2801–2812 (2008)
28. Liu, J., Wang, W., Ma, F., Yang, Y.B., Yang, C.: A Data-Model-Fusion Prognostic Framework for Dynamic System State Forecasting. Engineering Applications of Artificial Intelligence **25**(4), 814–823 (2012)
29. García, C.M., Chalmers, J., Yang, C.: Particle filter based prognosis with application to auxiliary power unit. In: The proceedings of the Intelligent Monitoring, Control and Security of Critical Infrastructure Systems, September 2012
30. G. S. A. C. Doucet: On sequential Monte Carlo sampling methods for Bayesian filtering. Statistics and Computing, 197–208 (2000)
31. Arulampalam, M., Maskell, S., Gordon, N., Clapp, T.: A tutorial on particle filters for online nonlinear/non-gaussian bayesian tracking. Trans. Sig. Proc. **50**(2), 174–188 (2002). http://dx.doi.org/10.1109/78.978374
32. Yang, C., Létourneau, S.: Learning to predict train wheel failures. In: Proceedings of the 11th ACM SIGKDD International Conference on Knowledge Discovery and Data Mining (KDD2005), Chicago, USA, pp. 516–525, August 2005

Fuzzy Q-Learning Approach to QoS Provisioning in Two-Tier Cognitive Femtocell Networks

Jerzy Martyna[✉]

Institute of Computer Science, Faculty of Mathematics and Computer Science,
Jagiellonian University, ul. Prof. S. Lojasiewicza 6, 30-348 Cracow, Poland
jerzy.martyna@ii.uj.edu.pl

Abstract. This paper addresses the problem of effective quality-of-service (QoS) provisioning in two-tier cognitive radio femtocell networks. The incorporation of primary user activity in the design of radio resource allocation technique is provided for this. Considering the spectrum environment as time-varying and that each femtocell network is able to use an adaptive strategy, the QoS provisioning provisioning guarantees are identified by finding the effective capacity of the femtocell and the femtouser. An integrated method based on fuzzy reinforcement learning algorithm is proposed to solve the formulated problem. It is confirmed by the simulation study that the effective solution of QoS provisioning in these networks is achieved.

1 Introduction

Artificial intelligence can be considered as the ability to act appropriately in an uncertain environment where success is the achievement of behavioural sub-goals that support the system's ultimate goal. Cognitive communications, including both cognitive radio and cognitive networks, represent the tip of the iceberg, when it comes to how artificial intelligence can be used in wireless communication systems. This paper is mainly focussed on fuzzy reinforcement-based learning actively applied to heteregenous cognitive radio-based systems.

A femtocell-based cognitive radio architecture was defined [7] as the conventional femtocell system with an infrastructure-based overlay cognitive network paradigm. The cognitive femtocell concept leads to the simpler and easier proliferation of cognitive radio into practical systems. Examples of cognitive femtocells include home LAN access points or urban hot-spot data access points incorporated in the architecture of a cognitive radio network. Thus, a radio resource management scheme for each cognitive femtocell should be able to "autonomously" utilise cognitive radio resources.

Two-tier femtocell networks consist of a small range data access point situated around hot spots with high user density serving stationary or low-mobility users. These networks have been studied in a number of papers. Among others, interference control and QoS awareness are the main contribution of the paper by [8]. Effective QoS provisioning and fair admission control appear as two crucial challenges in OFDMA hybrid small cells was proposed in the paper by

© Springer International Publishing Switzerland 2015
M. Ali et al. (Eds.): IEA/AIE 2015, LNAI 9101, pp. 74–83, 2015.
DOI: 10.1007/978-3-319-19066-2_8

Balakrishnan and Canberle [2]. While these approaches have achieved promising results, they cannot be directly used in context-aware multimedia applications since they are devoted to data communications, but do not explicitly consider the characteristics of video content and the resulting impact on the QoS parameters.

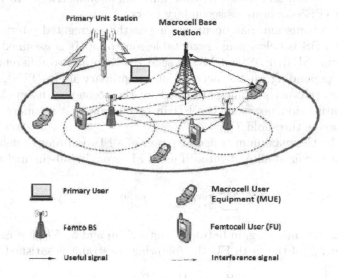

Fig. 1. An example of two-tier cognitive femtocell network

In this paper, the problem of fuzzy Q-learning algorithms obtaining QoS aware traffic in two-tier cognitive femtocell networks is considered. Additionally, the concept of effective capacity theory was used to define the maximum arrival rate that a cognitive radio system can support the given QoS requirements. The presented fuzzy Q-learning algorithm determines radio parameters for all given radio channel states with optimisation goals. The obtained simulation results show that the fuzzy Q-learning algorithm achieves good performance in accuracy.

The remainder of this paper is organised as follows. Section 2 describes the network and cognitive channel models. Section 3 provides statistical QoS guarantees. Section 4 proposes the fuzzy Q-learning algorithm to QoS provisioning in two-tier cognitive femtocell networks. Simulation results are presented in Section 5. Finally, concluding remarks are given in Section 6.

2   Network and Cognitive Channel Modelling

2.1   Network Model

We consider a Macrocell Base Station (MBS) that orchestrates the Macrocell User Equipments (MUEs) in a given coverage area, as shown in Fig. 1. There are some hybrid access OFDMA femtocell networks in the network. We suppose that there are K MUEs and M Femtocell Users (FUs), which are randomly located inside the macrocell coverage area. Each femtocell has only one FU. We assume that the total bandwidth is divided into N subcarriers with two of them being grouped into one subchannel. All femtocells and macrocells operate in the same frequency band and have the same numbers of subcarriers. All Femtocell Base Stations (FBS) operate as secondary users.

Two-tier transmission may occur as long as the aggregated interference incurred by the FBS is below some acceptable constraint. It is assumed that the secondary users (SUs), FBS and MUEs send the data at two different average power levels, depending on the activity of the primary users (PUs), which is determined by channel sensing performed by the secondary users. Moreover, the total transmission power of the FU in each channel is no more than the interference power threshold P_{int}^{max}.

Let P_{max} be the maximum transmit power of a FU. In order to define maximum transmit power of all M femtocell users FU over the subchannel n, we can write:

$$\sum_{m=1}^{M} P_m^n G_{m,k}^n \leq P_{int}^{max} \qquad (1)$$

where $G_{m,k}^n$ is the channel gain between femtocell m and MU k. For each transmission power P_m^n of the m-th FU the following constraint is satisfied, namely

$$P_{min} \leq P_m^n \leq P_{max} \qquad (2)$$

where P_{min} is the minimum of transmit power of femtocell.

The downlink signal-to-interference-plus-noise-ratio (SINR) from femtocell m to FU m over the subchannel n is defined as

$$SINR_{m,n} = \frac{P_m^n H_{f,m}^n}{N_0 W^n + I_m^n} \qquad (3)$$

where $H_{f,m}^n$ is the channel gain between m-th FU and m-th femtocell on subchannel n. N_0 is the power spectral density of the white Gaussian noise, W^n is the bandwidth of subchannel n. I_m^n is the co-channel interference (CCI) on the subchannel n of m-th FU in the coverage area of femtocell m and is given by

$$I_m^n = \sum_{k \in K} P_k^n G_{k,m}^n * H_{k,m}^n + \sum_{m' \in M'} P_{m'}^n G_{m',m}^n \qquad (4)$$

where $P_k^n G_{k,m}^n$ is the interference between BS and the m-th FU, $H_{k,m}^n$ is the frequency selective fading on the subchannel n between BS and m-th FU, $P_{m'}^n G_{m',m}^n$ is the interference from other femtocells.

2.2 Cognitive Channel Model

The cognitive radio channel model presents transmission details in the presence of primary users. We assume that the cognitive radio will be tested by secondary

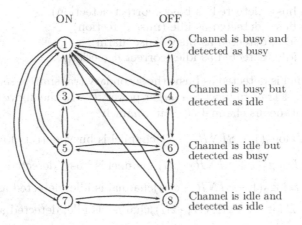

Fig. 2. State transition model for the cognitive radio channel

users. If the secondary transmitter selects its transmission when the channel is busy, the average power is $\overline{P}_1$ and the rate is r_1. When the channel is idle, the average power is $\overline{P}_2$ and the rate is r_2. We assume that $\overline{P}_1 = 0$ denotes the stoppage of the secondary transmission in the presence of an active primary user. Both transmission rates, r_1 and r_2, can be fixed or time-variant depending on whether the transmitter has channel side information or not. In general, we assume that $\overline{P}_1 < \overline{P}_2$. In the above model, the discrete-time channel input-output relation in the absence of primary users in the channel is given by

$$y(i) = h(i)x(i) + n(i), \quad i = 1, 2, \ldots \tag{5}$$

where $h(i)$ is the channel coefficient, i is the symbol duration. If primary users are present in the channel, the discrete-time channel input-output relation is given by

$$y(i) = h(i)x(i) + s_p(i) + n(i), \quad i = 1, 2, \ldots \tag{6}$$

where $s_p(i)$ represents the sum of the active primary users' faded signals arriving at the secondary receiver $n(i)$ is the additive thermal noise at the receiver and is zero-mean, circularly symmetric, complex Gaussian random variable with variance $E\{|n(i)|^2\} = \sigma_n^2$ for all i.

We assume that the receiver knows the instantaneous value $h(i)$, while the transmitter has no such knowledge. We have constructed a state-transition model for cognitive transmission by considering cases in which fixed transmission rates are greater or lesser than the instantaneous channel capacity values. In particular, the ON state is achieved if the fixed rate is smaller than the instantaneous channel capacity. Otherwise, the OFF state occurs.

We assume that maximum throughput can be obtained in the state transition model [1], given in Fig. 2. Four possible scenarios are associated with the model, namely:

1) channel is busy, detected as busy (correct detection),
2) channel is busy, detected as idle (miss-detection),
3) channel is idle, detected as busy (false alarm),
4) channel is idle, detected as idle (correct detection).

If the channel is detected as busy, the secondary transmitter sends with power $\overline{P}_1$. Otherwise, it transmits with a larger power, $\overline{P}_2$. In the above four scenarios, we have instantaneous channel capacity, namely

$$C_1 = B\log_2(1 + SINR_1 \cdot z(i)) \text{channel is busy, detected as busy} \qquad (7)$$

$$C_2 = B\log_2(1 + SINR_2 \cdot z(i)) \text{channel is busy, detected as idle} \qquad (8)$$

$$C_3 = B\log_2(1 + SINR_3 \cdot z(i)) \text{channal is idle, detected as busy} \qquad (9)$$

$$C_4 = B\log_2(1 + SINR_4 \cdot z(i)) \text{channel is idle, detected as idle} \qquad (10)$$

where B is the bandwidth available in the system, $z(i) = [h(i)]^2$, $SINR_i$ for $i = 1,\ldots,4$ denotes the average signal-to-noise ratio (SINR) values in each possible scenario.

Cognitive transmission is associated with the ON state in scenarios 1 and 3, when the fixed rates are below the instantaneous capacity values ($r_1 < C_1$ or $r_2 < C_2$). Otherwise, reliable communication is not obtained when the transmission is in the OFF state in scenarios 2 and 4. Thus, the fixed rates above are the instantaneous capacity values ($r_1 \geq C_1$ or $r_2 \geq C_2$). The above channel model has 8 states and is depicted in Fig. 2. In states 1, 3, 5 and 7, the transmission is in the ON state and is achieved successfully. In the states 2, 4, 6 and 8 the transmission is in the OFF state and fails.

3   Statistical QoS Guarantees

Real-time multimedia services such as video and audio require bounded delays, or guaranteed bandwidth. If a received real-time packet violates its delay, it will be discarded. The concept of effective capacity was developed to provide the statistical QoS guarantee in general real-time communication. Among others, in the paper by [5], it was shown that for a queuing system with a stationary ergodic arrival and service process, the queue length process $Q(t)$ converges to a random variable $Q(\infty)$ so that

$$-\lim_{x\to\infty} \frac{\log(Pr\{Q(\infty) > x\})}{x} = \theta \qquad (11)$$

Note that the probability of the queue length exceeding a certain value x decays exponentially as x increases. The parameter θ (*theta* > 0) gives the exponential decade rate of the probability of a QoS violation.

A framework of statistical QoS guarantees [6] was developed in the context of wireless communication [10]. In accordance with the effective bandwidth theory, effective capacity can be defined as

$$E_{cap}(\theta) \triangleq -\lim_{t\to\infty} \frac{1}{\theta t} \log\left(E\left[e^{-\theta S[t]}\right]\right) \qquad (12)$$

where $S[t] \triangleq \sum_{i=1}^{t} R[i]$ is the partial sum of the discrete-time stationary and ergodic service process $\{R[i], i = 1, 2, \ldots\}$.

The probability that the packet delay violates the delay requirement is given by

$$Pr\{Delay > d_{max}\} \approx e^{-\theta \delta d_{max}} \tag{13}$$

where d_{max} is the delay requirement, δ is a constant jointly determined by the arrival process and theirs service process, θ is a positive constant referred to QoS exponent.

4 Fuzzy Reinforcement Learning Methods for QoS Provisioning Transmission in Two-Tier Cognitive Femtocell Networks

We assume that each secondary user possesses three sensors: one to detect the required SINR, the second to detect the primary user transmission and the third to define channel quality. For a two-dimensional environment, all of the information obtained by the j-th secondary user is given by Fig. 3.

The current SINR is defined by the membership function plotted in the Fig. 3(a) and 3(b). Figs. 3(c) and 3(d) show the membership function associated with the required SINR. The levels of transmission acceptance realised by primary users are defined by the membership functions presented in Fig. 3(e) and Fig. 3(f). Figs. 3(g) and 3(h) show the membership function associated with the channel transmission rate defined by the current value of r.

A membership value defining the fuzzy state of the j-th SU with reference to the k-th transmission channel with respect to the current SINR for a two-dimensional environment is given by:

$$\mu_{state}^{(j)}(current\ SINR^{(k)}) = \mu_x^{(j)}(current\ SINR^{(k)}) \cdot \mu_y^{(j)}(current\ SINR^{(k)}) \tag{14}$$

A membership function defining the fuzzy state of thhe j-th SU with reference to the k-th transmission channel with respect to the required SINR for a two-dimensional environment is as follows:

$$\mu_{state}^{(j)}(required\ SINR^{(k)}) = \mu_x^{(j)}(required\ SINR^{(k)}) \cdot \mu_y^{(j)}(required\ SINR^{(k)}) \tag{15}$$

A membership function defining the fuzzy state of the j-th SU defining its acceptance level of transmission realized by the l-th PU for a two-dimensional environment is as follows:

$$\mu_{state}^{(j)}(PU\ acceptance^{(l)}) = \mu_x^{(j)}(PU\ acceptance^{(l)}) \cdot \mu_y^{(j)}(PU\ acceptance^{(l)}) \tag{16}$$

Similarly, the current transmission rate of j-th SU with reference to the k-th channel which also defines the fuzzy state for a two-dimensional environment is computed as:

$$\mu_{state}^{(j)}(trans.\ rate^{(m)} = \mu_x^{(j)}(trans.\ rate^{(k)} \cdot \mu_y^{(j)}(trans.\ rate^{(k)}) \tag{17}$$

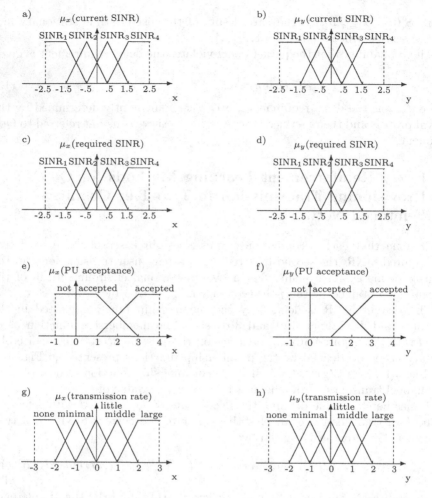

Fig. 3. Fuzzy sets for two-dimensional environment defining membership functions of *current SINR* (a, b), *required SINR* (c, d) with reference to the studied channel, *PU acceptance* (e, f) with respect to the nearest PU; *transmission rate* with reference to the studied channel

The system model is described by the multidimensional membership function, which can be treated as a multidimensional hypercube. The fuzzy state for the j-th SU can be defined by the fuzzy pair (s_n, a_n) for the n-th fuzzy variable, where s and a are respectively the state and action. Using the aggregation of the fuzzy state, we can achieve:

$$Q_{state}^{(j)}(s, a) \leftarrow Q_{state}^{(j)}(s, a) + \sum_{n=1}^{N} \alpha_n^{(j)} \cdot \mu_{state}^{(j)}(s_n, a_n) \qquad (18)$$

where N is the total number of fuzzy variables.

For the four exemplary fuzzy variables we have the Q-function for j-th SU, namely

$$Q_{state}^{(j)} \leftarrow Q_{state}^{(j)}(s, a)$$

$$+ \sum_{k=1}^{K} (\alpha_k^{(j)} \mu_{state}^{(j)}(current\ SINR^{(k)}) + \alpha_k^{(j)} \mu_{state}^{(j)}(required\ SINR^{(k)}))$$

$$+ \sum_{l=1}^{L} (\alpha_l^{(j)} \mu_{state}^{(l)}(PU\ acceptance^{(l)} + \sum_{k=1}^{K} \alpha_k^{(j)} \mu_{state}^{(k)}(trans.\ rate^{(k)}) \qquad (19)$$

where $\alpha_n^{(j)}$ is the learning rate for SU j with respect to n-th fuzzy variable, K is the total number of channels for j-th SU, L is the total number of PUs.

Let the radio transmitting range of the SU be equal to $\mathcal{R}$. Thus, we can again define the Q-function value as follows:

$$Q_{state}^{(j)}(s_{t+1}, a_{t+1}) \leftarrow \begin{cases} 0 & \text{if } j \notin \{J\} \\ Q_{state}^{(j)}(s_t, a_t) + \alpha_{state}^{(j)}(s_t, a_t) & \text{if } j \in \{J_{0<\tau \leq 0.5\mathcal{R}}\} \\ Q_{state}^{(j)}(s_t, a_t) + \beta^{(j)} Q_{state}^{(j)}(s_t, a_t) & \text{if } j \in \{J_{0.5\mathcal{R}<\tau \leq \mathcal{R}}\} \end{cases}$$

where $\{J\}$ is the set of SUs and PUs in the range of the PU observation with the radius equal to $\mathcal{R}$, $\{J_{0<\tau \leq 0.5 \cdot \mathcal{R}}\}$ and $\{J_{0.5 \cdot \mathcal{R} < \tau \leq \mathcal{R}}\}$ are the sets of SUs and PUs in the range of the SU observation with the radius equal to $0 < \tau \leq 0.5 \cdot \mathcal{R}$ and $0.5 \cdot \mathcal{R} < \tau \leq \mathcal{R}$, respectively. $\beta^{(j)}$ are learning rate factors.

The state space in reinforcement learning can be treated as a stochastic problem. In the standard approach, we can generalise the Q-value across states using the function approximation $Q(s, a, f)$ for approximating $Q(s, a)$, where f is the set of all learned fuzzy logic mechanisms [3], [4]. To handle all this information, we can use the data mining approach.

The function Q is computed by the Q-learning algorithm; this algorithm was presented by Watkins and Dayan [9]. We recall that the Q-function is given by:

$$Q_t(s_t, a_t) = (1 - \alpha)Q_t(s_t, a_t) + \alpha(r_t + \gamma \max_{a_t \in A} Q_t(s_t', a_t')) \qquad (20)$$

where A is the set of all the possible actions, α $(0 \leq \alpha < 1)$ and γ $(0 \leq \gamma \leq 1)$ denote the learning rate and the discount parameter, $Q_t(s_t', a_t')$ is the value of the Q function after the execution of action a_t'. Fig. 4 shows the raw form of the Q-learning algorithm. It can be seen that the Q-learning algorithm is an incremental reinforcement learning method. The choice of the action does not show how to obtain it. Therefore, the Q-learning algorithm can use other strategies that it learns, irrespective of the assumed strategy. This means that it does not need actions that would maximise the reward function.

5   Simulation Results

A simulation was implemented to benchmark the performance of the designed fuzzy Q-learning method. We simulated the cognitive framework as an extension

initialization $t = 0$, $r_T = (s_t, a_t) = 0$;
begin
for $\forall s_t \in S$ **and** $a_t \in A$ **do**
$\quad t := t + 1$; *access the current state* s_t;
$\quad a_t \leftarrow$ *choose_action*(s_t, Q_t);
$\quad$ *perform_action* a_t;
$\quad$ *compute:* $r_t(s_t, a_t)$, s_{t+1};
$\quad\quad \Delta Q_t \leftarrow (r_t + \gamma \max_{a_t}(Q_t(s_{t+1}, a_t)) - Q_t(s_t, a_t)$;
$\quad\quad Q_t(s_t, a_t) \leftarrow (1 - \alpha)Q_t(s_t, a_t) + \alpha \Delta Q_t$;
end;

Fig. 4. Q-learning algorithm estimates new state obtained by performing the chosen action at each time step

of 10 femtocells, 5 MUEs and 1 PU. We assumed that the radius of the macrocell is equal to 1 km. The maximum and minimum transmit power of femtocell are equal to 10 dBm and 5 dBm, respectively. The system bandwidth and subcarrier bandwidth are equal to 2 MHz and 15 KHz, respectively. In our OFDMA network, the modulation 64-QAM represents 6 bits per symbol. We assumed that only one femtocell user (FU) is associated with a femtocell and the service rate requirement of a femtocell user is equal to 1 Mbps. All macrocell users are always outdoor and femtousers are always indoor.

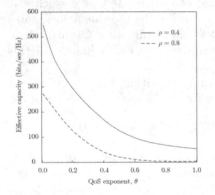

Fig. 5. Effective capacity of the femtocell versus QoS exponent θ

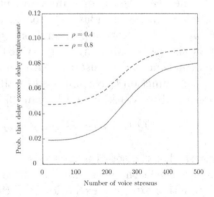

Fig. 6. Probability that the delay exceeds the delay requirement

In Fig. 5, we present the effective capacity of the femtocell depending on the values of the QoS exponent. As we see in Fig. 5, the effective capacity is decreasing with an increased system load. Fig. 6 shows the probability of delay bound violations for voice streams versus the number of voice streams. We can observe that the system load increases the probability for delay bound violations of the voice stream.

6   Conclusion

In this paper, we have explored the usage of a fuzzy Q-learning algorithm to assist the allocations of OFDMA resources for QoS provisioning in cognitive femtocell networks. Thus, two-tier cognitive femtocell network could be viewed as a promising option for the next generation wireless communication networks that achieves QoS performance benefits in terms of throughput and delay. We have proposed the usage of a fuzzy Q-learning algorithm to QoS provisioning in these networks. The effectiveness of the proposed method has been verified in the simulation study. Moreover, the obtained results demonstrate that the proposed approach can improve some parameters for video-streaming users, such as the maximum delay and shape parameters.

References

1. Akin, S., Gursoy, M.C.: Effective Capacity Analysis of Cognitive Radio Channels for Quality of Service Provisioning. IEEE Wireless Communications 9(11), 3354–3364 (2010)
2. Balakrishnan, R., Canberle, B.: Traffic-Aware Provisioning and Admission Control in OFDMA Hybrid Small Cells. IEEE Trans. on Vehicular Technology 63(2), 802–810 (2014)
3. Beon, H.R., Chen, H.S.: A Sensor-based Navigation for a Mobile Robot Using Fuzzy-Logic and Reinforcement Learning. IEEE Trans. SMC 25(3), 467–477 (1995)
4. Berenji, H.R., Vengerov, D.: Advantages of Cooperation Between Reinforcement Learning Agent in Difficult Stochastic Problems. In: Proc. The Ninth IEEE International Conference on Fuzzy Systems, San Antonio, TX, pp. 871–876 (2000)
5. Chang, C.-S.: Stability, Queue Length, and Delay of Deterministic and Stochastic Queueing Networks. IEEE Trans. on Automat. Control 39(5), 913–931 (1994)
6. Courcoubetis, C., Weber, R.: Effective Bandwidth for Stationary Souces. Probability in Engineering and Information Science 9(2), 285–294 (1995)
7. Gür, G., Bayhan, S., Alagoz, F.: Cognitive Femtocell Networks: an Overlay Architecture for Localized Dynamic Spectrum Access [Dynamic Spectrum Management]. IEEE Wireless Communications 17(4), 62–70 (2010)
8. Liang, Y.-S., Chung, W.-H., Ni, G.-K., Chen, I.-Y., Zhang, H., Kuo, S.-Y.: Resource Allocation with Interference Avoidance in OFDMA Femtocell Networks. IEEE Trans. on Vehicular Technology 61(5), 2243–2255 (2012)
9. Watkins, C.J.C.H., Dayan, P.: Technical Note: Q-learning. Machine Learning 8, 279–292 (1992)
10. Wu, D., Negi, R.: Effective Capacity: A Wireless Link Model for Support Quality of Service. IEEE Trans. on Wireless Comm. 2(4), 630–643 (2003)

A Neurologically Inspired Model of the Dynamics of Situation Awareness Under Biased Perception

Dilhan J. Thilakarathne$^{(\boxtimes)}$

Agent Systems Research Group, Department of Computer Science,
VU University Amsterdam, Amsterdam, The Netherlands
d.j.thilakarathne@vu.nl
http://www.few.vu.nl/~dte220/

Abstract. This paper presents a computational cognitive agent model of Situation Awareness (SA), which is inspired by neurocognitive evidences. The model integrates bottom-up and top-down cognitive processes, related to various cognitive states: perception, desires, attention, intention, awareness, ownership, feeling, and communication. The emphasis is on explaining the cognitive basis for biased perception in SA, which is considered to be the most frequent factor in poor SA (the reason for 76% of poor SA errors), through perceptual load. A model like this will be useful in applications which relay on complex simulations (e.g. aviation domain) that need computational agents to represent human action selection together with cognitive details. The validity of the model is illustrated based on simulations for the aviation domain, focusing on a particular situation where an agent has biased perception.

Keywords: Situation awareness · Perceptual load · Perception · Attention · Intention · Bottom-up · Top-down · Cognitive modeling · Simulation

1 Introduction

The relation between human awareness and action selection is a complex issue. Nevertheless; due to the developments in brain imaging and recording techniques, the insight in human brain processes is growing rapidly. Human cognitive processes are often grouped into conscious and unconscious processes. The understanding of the interplay between conscious and unconscious processes associated with action selection and related phenomena has improved a lot, especially thanks to the experimental framework proposed by Benjamin Libet and his colleagues [1] and later improvements made to it. In the literature, bottom-up cognitive processes have been mapped to unconscious action formation, whereas top-down processes have been related to conscious action formation (cf. [2]–[5]); it seems our action selection process initiates from unconscious phenomena, and that later we develop the conscious experience of this action selection. The unconscious neural activations in the brain seem to be a result of habitual tasks, through the effects of prior learning, which can be automatically activated when a relevant stimulus is perceived [6]. Nevertheless, conscious awareness of action selection also plays an important role (cf. [2]).

© Springer International Publishing Switzerland 2015
M. Ali et al. (Eds.): IEA/AIE 2015, LNAI 9101, pp. 84–94, 2015.
DOI: 10.1007/978-3-319-19066-2_9

Situation Awareness (SA) can be considered as a subjective quality or interpretation of the awareness of a situation a person is engaged in. When a person is engaged in a situation based on the information that he/she perceives, the attention that is allocated to that information based on his/her subjective desires will develop his/her subjective awareness of the situation. This is the reason why different individuals may have different interpretations of the same situation. The correctness of SA is always relative and its quality can be analyzed when a task is performed with an expert critiquing as a benchmark. Due to this complexity and subjective nature of SA, the concept has received many definitions in the literature (according to [7] there are more than fifteen definitions for SA); among those, the definition proposed by Endsley [8] became the most widely used. According to Endsley, SA is: "*the perception of the elements in the environment within a volume of time and space, the comprehension of their meaning, and the projection of their status in the near future*" [9, p. 36]

Based on this definition, Endsley highlighted three elements as the necessary conditions for SA; these are three levels of which one is followed by the other, in order to develop complete (subjective) awareness namely: Level 1: perception, Level 2: comprehension, and Level 3: projection. Furthermore, it has been found that, based on safety reports in the aviation domain, 76% of the errors related to SA were because of Level 1 (i.e., failure to correctly perceive information), 20.3% were Level 2 errors (i.e., failure to comprehend the situation), and 3.4% were Level 3 errors (i.e., failure to project situation into future) [10], [11]. Hence, this statistical information provides an indication of the relative importance of these three aspects of SA. Therefore, the biased or poor perception addressed by Level 1 has received most attention, due to its high frequency among all errors (cf. [10]). Furthermore, Endsley has indicated how attention, goals, expectations, mental models, long-term memory, working memory and automaticity contribute to situation assessment in terms of cognitive processes [9], [12]. The summary from Endsley in [9] provides some useful indications of how this definition (through her model) can be related to the neurocognitive literature.

Though there is a positive analogy between Endsley's model and its psychological basis, from a dynamic perspective the model leaves room for questions. In particular, she highlights that a person will first develop situation awareness, only then decision making will follow, and that finally the selected actions will be performed:

Environment → Situation Awareness < *Perception* → *Interpretation* → *Projection* > → Decision Making → Performance

Nevertheless, this particular linear transformation is not supported by the current viewpoints in neuroscience; instead, situation awareness, decision making, and performance of selected actions are viewed as one compound process in which all sub-processes dynamically interact, striving for actions with an optimal result. Endsley has explicitly separated situation awareness from the process itself, which she calls situation assessment [12] (in her terms, SA is a product and situation assessment is the process to make decisions which follow up on the developed SA). Having this fundamental concern, there is a necessity of better explaining SA in terms of current evidence from a cognitive science perspective, in such a way that the explanations can be used in relevant application domains (such as the aviation domain), for instance by simulating complex situations in a more realistic and detailed manner (cf. [13]).

2   Neurological Background

Cognitive processes of action formation are complicated and on the exact mechanisms involved have not been fully unraveled yet. Nevertheless, due to the developments in brain imaging and recording techniques, researchers are gaining more and more insight in human cognition. According to those insights, human action selection is for a main part determined by automatic unconscious processes such as habitual tasks (a task will become a habitual task through a learning process, depending on its frequency and recency (cf. [6])). Nevertheless, stimuli received from the environment contain far more information than a person can process in a given time. To cope with this, it seems that two main processes play a role, namely bottom-up and top-down control. Processes related to bottom-up effects are more automatic and are mainly data driven, triggered by factors external to the agent (humans are also agents (see [14])) such as salient features of a stimulus [15]. Top-down effects are internally guided based on prior knowledge, intentions, and long-term desires and they are internal to the observer and unrelated to the salient features of a stimulus [5], [15], [16], [17].

When investigating the brain circuits related to human cognition, it seems that these consist of complex loops, rather than linear chains [4]; therefore, higher order coupling among processes has been observed, rather than those processes being categorically independent. Similarly, the bottom-up and top-down processes are also not isolated; instead, there is overlap among these two, even at the neural level and many pieces of evidence have been found that demonstrate the interplay among these two in the context of attention and perception, together with other supportive cognitive states [5], [15], [17]–[19]. The bottom-up processes have many relations to perception and emotions, and more details of the cognitive basis of these have been separately presented in previous work (see [20]). The amygdala was noted as a key element in bottom-up processes, which include monitoring the salient features in stimuli and projecting them onto higher levels of cognitive processing (the amygdala has connectivity with eight of the cortical areas [21]). Furthermore it has been observed that the amygdala directly shapes perception when perceiving an emotionally salient stimulus [22]. Attention is another important cognitive state that is related to the interplay among bottom-up and top-down processes: in particular through bottom-up attention and top-down attention. In [15] it has been pointed out that the posterior parietal cortex (PPC) and prefrontal cortex (PFC) could be segregated for distinct roles in bottom-up and top-down attentional systems, and the close interaction of these regions with each other is highlighted to explain the constant influence of these two processes to orient the attention necessary for more sophisticated cognitive control processes (cf. [17], [23], [24]). In addition, the prefrontal cortex has long been assumed to play an important role in top-down driven cognitive control, as a temporal integrator. The higher order interconnectivity of the PFC with other cortical and subcortical areas has been interpreted as indicating a process that generates and maintains information when sensory inputs are weak, ambiguous, rapidly changing, novel and/or multiple options exist [18], [25]. Furthermore, neurocognitive evidence for some of the main factors of top-down processes (i.e., intention, attention, subjective desires and awareness) has been presented separately in previous work [26].

In the current paper, a specific interest is given to understand reasons for poor or incorrect perception. One suitable analogy for this is the literature on distraction, which explains why and how people get distracted from their current task (under a poor 'Level 1 SA', agents are unable to switch to proper perception due to the focus developed on selected items or data in the stimulus). Once a person is focusing on something, there are many reasons why he/she may sometimes be distracted and sometimes not. This is interesting to study the question what are the causes that prevent a person from processing other important cues in the environment (as an example from [27]: people attending to a ball game have failed to notice a woman walking across the pitch and holding up an umbrella). To explain this phenomenon, the load theory of attention and cognitive control [28] provides detailed information about early versus late selection schemes. Early selection is a perceptual selection mechanism associated with automatic (or passive) behavior [29]. The main reason behind this mechanism is the limited processing capacity of our perception under a high level of perceptual load. Because of this, an agent under high perceptual load is unable to shift his or her selection to other salient features in the environment. Instead, when the perceptual load is lower, the agent is capable of perceiving information in parallel (cf. [29]–[31]). Nilli Lavie's summary on this provides a complete overview for this (see [27, pp. 143–144]). Indeed, there is empirical evidence that shows a competition effect under low perceptual load, but not with high perceptual load [31]. Therefore, when receiving new stimuli with no perceptual load, all features may be processed completely, which will lead to the development of perception on the salient features, while at the same time the perceptual load will increase. Once the perceptual load is high, the agent seems to be unable to pay attention to additional features. Moreover, perception is related to early selection, whereas cognitive controlling through attention is related to late selection (late selection is not considered in this paper).

3 Description of the Computational Model

The model in this paper is an extension of the previous work presented in [13], but extended by incorporating the idea of perceptual load, to explain the cognitive reason for poor Level 1 SA. Fig. 1 presents the cognitive model and Table 1 summarizes the abbreviations used. The model takes inputs from two world states $WS(s_k)$ and $WS(b_i)$, where s is a stimulus (that can be either external or internal to the agent) that may lead to an action execution, and b_i represents the effects of the execution of an action a_i. The model accepts multiple inputs in parallel. Therefore, in this model, external input is a vector s_k, where k inputs are taken in parallel. The input $WS(s_k)$ leads to a $SS(s_k)$, and subsequently to a $SR(s_k)$. Moreover, the model includes both conscious and unconscious aspects. The states: $SR(b_i)$, $PD(b_i)$, $PA(a_i)$, $F(b_i)$, $Per(b_i, s_k)$, and $PO(a_i, b_i)$ are considered to be unconscious and contributing to bottom-up processes. In contrast, the states: $SD(b_i)$, $Att(b_i, s_k)$, $CInt(b_i, s_k)$, and $PAwr(a_i, b_i, s_k)$ represent more conscious influences, contributing to top-down processes. The dotted box represents the boundary between external and internal states: the states inside the dotted box represent internal cognitive states (those have a relatively higher firing rate).

The unconscious bottom-up process of action selection is modelled by combining Damasio's as if body loop (cf. [32]) (through $PA(a_i) \to SR(b_i) \to F(b_i)$) and James's body loop (cf. [33]) (through $PA(a_i) \to EA(a_i) \to WS(b_i) \to SS(b_i) \to SR(b_i) \to F(b_i)$). According to Damasio, the cognitive process of action selection is based on an internal simulation process prior to the execution of an action. Effects of each relevant action option $PA(a_i)$ (a stimulus s will have many options i=1..n) are evaluated (without actually executing them) by comparing the feeling-related valuations associated to their individual effects. Each preparation state $PA(a_i)$ for action option a_i suppresses preparation of

Table 1. Nomenclature for Fig. 1

WS(W)	world state W (W can be either stimulus s, or effect b)
SS(W)	sensor state for W
SR(W)	sensory representation of W
PD(b_i)	performative desires for b_i
SD(b_i)	subjective desires for b_i
PA(a_i)	preparation for action a_i
Per(b_i,s_k)	perception state for s_k on b_i
F(b_i)	feeling for action a_i and its effects b_i
PO(a_i, b_i)	prior ownership state for action a_i with b_i
Att(b_i,s_k)	attention state for s_k on b_i
CInt(b_i,s_k)	conscious intention state for s_k on b_i
PAwr(a_i,b_i,s_k)	prior-awareness state for action a_i with b_i and s_k
EA(a_i)	execution of action a_i
EO(a_i,b_i,s_k)	communication of ownership of a_i with b_i and s_k

all its complementary options $PA(a_j)$ with j≠i (see Fig. 1), and therefore by a kind of winner-takes-all principle, naturally the option that has the highest valuated effect felt by the agent will execute through the body loop. Furthermore, according to the literature, the predictive effect and sensed actual effect of the action are added to each

other through an integration process (cf. [2]); this is expected to be reflected in this model through the $SR(b_i)$. Through this, it is also possible to demonstrate the difference between when predictive and actual effects are the same and not. This process is further strengthened by embedding $PD(b_i)$ and Per(b_i, s_k). The $PD(b_i)$ facilitates short-term desire effects on action execution that has the ability to strengthen the current action selection based on its desires (as a bias injected to the process). In parallel to the action preparation

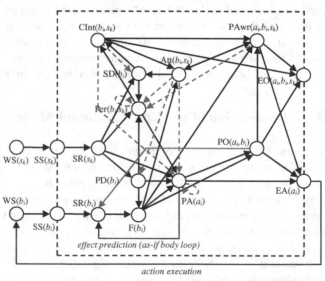

Fig. 1. Overview of the computational cognitive agent model. The arrow → represents a direct activation of state B by state A, arrow → represents a direct suppression of state B by state A, arrow ⋯→ represents a suppression of all the complements of the 'ith' state of B_i by state A_i (where 'i' represents an instance of a particular state), and ⋯→ represents a direct activation of state B_i by state A_i while suppressing all the complements of the 'ith' state of B_i.

process, the Per(b_i, s_k) also develops based on the salient features of the stimulus s; this will strengthen the bottom-up process which leads to even further strengthening of the action preparation process (due to the activation that spreads from Per(b_i, s_k) to PA(a_i)). Furthermore; by having a suppressive link from the Per(b_i, s_k) state to itself, the competition among perceptual entities as mentioned in Section 2 is represented (cf. [31]). Each suppressive link's negative effect (strength) is relatively proportional to the strength of that particular perceptual state and therefore the perception state for element i that has the highest activation suppresses its complements most strongly. As a result, the strongest candidate will dominate the competition and naturally will contribute for a stronger perceptual load. Due to this developed perceptual load, the agent will not automatically attend to other salient features (unless a particular attention is put on some salient feature intentionally). Furthermore, this selected perception will be strengthened by the agent's attention and subjective desires (see Fig. 1). While the agent is passively (unconsciously) performing action selection as explained in Section 2 (for more details see [26]), the agent starts to activate bottom-up attention (this is represented by the link from F(b_i) to Att(b_i, s_k)). The main functionality of the bottom-up attention is to pass current information into higher order cognitive states. Due to this bottom-up attention, the agent will activate its SD(b_i), which in turn leads to a CInt(b_i, s_k), and subsequently back to the attention state again. This cyclic process represents the transformation from bottom-up to top-down. Furthermore, this is in line with the idea of transforming Level 1 SA to Level 2 SA in Endsley's model (i.e., from perception to comprehension) in terms of a dynamic process [9], [11], [12]. Intention is considered to trigger goal directed preparation (see [26]) and therefore this model includes that effects via the Att(b_i, s_k). Once the attention (and its subjective aspects) has been developed, it injects conscious biases (through the top-down attention) into the action preparation and perception states. This is represented through the links from Att(b_i, s_k) to PA(a_i) and Per(b_i, s_k), and these links (purple dotted arrows) play a special role: while activating the matching option (i.e. i[th] option) they suppress all complements of the ith option. This emphasizes the conscious influence on action formation, and therefore attention will quickly enable the agent's concentration, which may shorten the time required for action selection. More importantly this will strengthen the current perception even further, and due to strong subjective feelings the agent may not be able to shift its attention easily (nevertheless, over longer time spans, attention will naturally get diluted; however, those effects are yet not included in this model).

Together with these processes, the agent will develop a state of ownership, which mainly determines to what extent an agent attributes an action to himself or to another agent. This particular aspect is important when it comes to situations where collaborative situation awareness plays a role, e.g. through collective decision making (although this is not in the scope of this paper). Also, as explained in previous works (see [13], [26]), the agent will develop an awareness state of action a_i that is related to effect b_i and stimulus s_k. According to Haggard [2], [4], there may be an influence from awareness states to action selection; therefore, this model includes a link from the PAwr(a_i, b_i, s_k) to the EA(a_i) (however, note that there are also claims that awareness of motor intentions does not have any influence on action execution, but emerges after action preparation and just before action execution [1], [34], [35]). Due to the empirical evidence that supports that awareness appears just before action execution, the current model includes that aspect by having awareness be affected mainly by higher order cognitive states; also, it does not affect many other states directly. The agent will execute the selected action a_i and then this action will have an effect in the environment (through WS(b_i)), and be sensed again, as explained earlier through the body loop. Finally, the agent has the ability to communicate the process (e.g., verbally) through state EO(a_i, b_i, s_k). In addition to the suppressive links mentioned here, a few more are included in the model; more details on those can be found in [13], [26].

Each connection between states has been given a weight value (where ω_{ji} represents the weight of the connection from state j to i) that varies between 1 and -1. To model the dynamics following the connections between the states as temporal–causal relations, a dynamical systems perspective is used, as explained in [36]. Furthermore, each state includes an additional parameter called speed factor γ_i, indicating the speed by which the activation level of the state 'i' is updated upon receiving input from other states. Two different speed factor values are used, namely fast and slow values: fast values are used for internal states and slow values for external states. The level of activation of a state depends on multiple other states that are directly attached to it. Therefore, incoming activation levels from other states are combined to some aggregated input and affect the current activation level according to differential equation (1). As the combination function for each state, a continuous logistic threshold function is used: see equation (2), where σ is the steepness, and τ the threshold value. When the aggregated input is negative, equation f(x) = 0 is used. To achieve the desired temporal behavior of each state as a dynamical system, the difference equation represented by equation (3) is used (where Δt is the time step size).

$$\frac{dy_i}{dt} = \gamma_i \left[f\left(\sum_{j \in s(i)} \omega_{ji} y_j \right) - y_i \right] \tag{1}$$

$$f(X) = th(\sigma, \tau, X) = \left(\frac{1}{1 + e^{-\sigma(X-\tau)}} - \frac{1}{1 + e^{\sigma\tau}} \right)(1 + e^{\sigma\tau}) \; when \; x > 0 \tag{2}$$

$$y_i(t + \Delta t) = y_i(t) + \gamma_i \left[th\left(\sigma, \tau, \sum_{j \in s(i)} \omega_{ji} y_j \right) - y_i(t) \right] \Delta t \tag{3}$$

4   Analysis of Level 1 SA by Simulation

According to the statistics provided by Endsley for the aviation domain (see [10], [11]), 76% of the errors related to poor SA were Level 1 errors, which are due to a failure to correctly perceive information (mainly due to poor design or limitations in user interfaces). The focus of this paper is to model the cognitive behavior as a process for Level 1 SA. The same generic example is used as was used to understand Level 1 SA in [13], which is taken from the Airbus Company. The example is summarized as "Focusing on recapturing the LOC and not monitoring the G/S". More specifically, it refers to a situation where a pilot is supposed to consider information from both devices (i.e., the LOC (Localiser) and the G/S (Glide Slope)), but due to biased perception (s)he only develops perception on LOC and not both on LOC and G/S. To simulate this example, two input stimuli (s_1 and s_2) have been used. One input triggers action selection based on reading the LOC data only, and the other one triggers action selection based on reading the data from both LOC and G/S combined. Details on all the input information and parameter values (Δt, γ, σ, τ, and weight values) for each state can be found in an external appendix[1]. In this simulation, the main cognitive state of interest is the perception state, and the process used to influence this is the perceptual load. Therefore, for all weight values in the model and for all options (two options have the potential to trigger: a_1 and b_1 & a_2 and b_2), identical values are used, except for the connection weights between {(SR(s_k), PD(b_i)), (PD(b_i), Per(b_i, s_k)), (SR(sk), Per(b_i, s_k)), (SD(b_i), Per(b_i, s_k)), (Att(b_i, s_k), Per(b_i, s_k))}. Also, in this particular simulation, the 'complements-suppressive' link from Per(b_i, s_k) to Per(b_j, s_k) (where $j \neq i$) has the same weight value (therefore no bias is introduced through this link, but the suppression is only proportional to the strength of the particular state). Upon receiving the two input stimuli, the agent will prepare for two action options PA(a_1) and PA(a_2), where action a_1 is based on information from the LOC device and action a_2 is based on information from both devices. From the simulation results in Fig. 2, it can be seen that the provided input stimuli have relatively large effects on SR(s_k) for both options, with the maximum of 0.53 per each. Nevertheless, the agent only generates a strong action preparation state for action option a_1: the level of PA(a_1) becomes very high (with a max of 0.85), just like that of perception state Per(b_1,s_1) (with a max of 0.85). Instead, for action option a_2 it has a very weak Per(b_2,s_2) (max of 0.03) that contributes to the development of a poor PA(a_2) (max

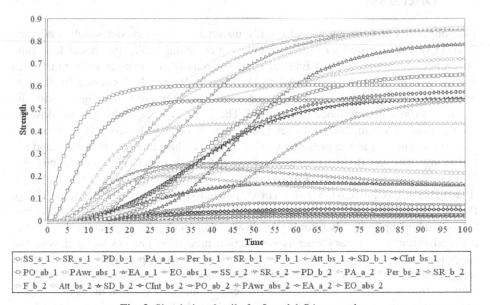

Fig. 2. Simulation details for Level 1 SA example

---

[1]   http://www.few.vu.nl/~dte220/IEAAIE15Appendix.pdf

of 0.12). Hence, merely through this effect of incomplete perception (as Endsley highlighted), the agent has not developed the right situation awareness (in this case the 'correct' awareness would have been PAwr(a_2,b_2,s_2)). Instead, the 'incorrect' awareness state PAwr(a_1,b_1,s_1) (max of 0.74) is generated, based on wrong perception.

More importantly, this simulation has illustrated that the model can reproduce the effect of perceptual load under early selection while having identical weights for each option, except the five modifications mentioned earlier. Therefore, a bias has been injected through the process from perception to action selection process, and mainly with the support of unconscious processes, the agent has moved towards action selection (note that Level 2 and 3 SA are assumed to be more conscious than Level 1 SA). Subsequently, the agent generates sufficient activation levels for all the other states related to option a_1, and finally executes the action EA(a_1) (max of 0.78) with a PAwr(a_1,b_1,s_1) of max 0.74. The maximum activation levels of the other states related to option 1 are: F(b_1) is 0.68, Att(b_1,s_1) is 0.84, SD(b_1) is 0.57, CInt(b_1,s_1) is 0.54, PO(a_1,b_1) is 0.65, and EO(a_1,b_1,s_1) is 0.53. This pattern is as expected based on previous works (see [26]). In addition, the agent has properly integrated its sensory representations and its feeling on predictive effects and sensed actual effects. This can be explained by the two-step sigmoid curve (for SR(b_1) there is a slight saturation at time point 40, and then with the execution of EA(a_1) this is increased again with a slightly higher steepness) [2]. If the agent's predicted effect and its sensed actual effect would not be the same, then there would not be such two-step sigmoid behavior.

5   Discussion

This paper has presented a neurologically inspired cognitive model which is adapted from [13] and extended with process behind perceptual load. This model has some differences compared to what Endsley suggests; mainly, it moves away from the idea that there is a causal chain from SA to Decision Making to Performance Evaluation. In the proposed model, these 3 aspects are still exist but are more aligned with the findings from a neuroscience perspective. The simulation example used to illustrate the behavior of this model (for Level 1 SA) is the same as was used in [13]. Nevertheless, the simulation in the previous work did not use the same weight values, and the following links had different values: {(PA(a_i), SR(b_i)), (SR(s_k), PD(b_i)), (Att*(b_i,s_k), PD(b_i)), (Att*(b_i,s_k), PA(a_i)), (PA*(a_i), PA(a_i)), (CInt*(b_i,s_k), PA(a_i)), (SR(b_i), F(b_i)), (PD(b_i), F(b_i)), (CInt(b_i,s_k), SD(b_i)), (F(b_i), Att(b_i,s_k)), (Per(b_i,s_k), CInt(b_i,s_k)), (SR(s_k), Per(b_i,s_k)), (SD(b_i), Per(b_i,s_k)), (PAwr(a_i,b_i,s_k), Per(b_i,s_k))}, where '*' represents 'complement' options. Therefore, in that approach (i.e. [13]) a perceptual bias was realized through a primary unconscious action prediction process together with the support of conscious states. Also, for the 'complement-suppressive' links, different weight values were used. Therefore, the biased behavior was represented through the combination of all these weight values. Instead, in this new version, the approach was further improved, since it models perception through the process of perceptual load. For each action option, exactly the same values were used, except for {(SR(s_k), PD(b_i)), (PD(b_i), Per(b_i,s_k)), (SR(s_k), Per(b_i,s_k)), (SD(b_i), Per(b_i,s_k)),

(Att(b_i,s_k), Per(b_i,s_k))} which are related to perception and performative desires. Hence, based on these simulation results, this model demonstrates the basic features of perceptual load and further explains the construct of poor Level 1 SA as a cognitive process in more realistic manner. A model like this will be useful mainly in complex simulations where cognitive details are essential (e.g. air traffic controlling situation under an emergency conditions). In multi-agent based simulations a main limitation is lacking of nature inspired realistic models to mimic the natural behavior and cognition behind it. Having a model like this will contribute to fill this gap and even be useful in training simulations to improve the human cognition though these processes.

Acknowledgment. I wish to thank Prof. Jan Treur and Dr. Tibor Bosse at VU University Amsterdam, for their great support and supervision in all the phases of this work. This work is part of the SESAR WP-E programme on long-term and innovative research in ATM. It is co-financed by Eurocontrol on behalf of the SESAR Joint Undertaking (SJU).

References

1. Libet, B., Gleason, C.A., Wright, E.W., Pearl, D.K.: Time of conscious intention to act in relation to onset of cerebral activity (readiness-potential). The unconscious initiation of a freely voluntary act. Brain **106**(3), 623–642 (1983)
2. Moore, J., Haggard, P.: Awareness of action: Inference and prediction. Conscious. Cogn. **17**(1), 136–144 (2008)
3. Engel, A.K., Fries, P., Singer, W.: Dynamic predictions: Oscillations and synchrony in top–down processing. Nat. Rev. Neurosci. **2**(10), 704–716 (2001)
4. Haggard, P.: Human volition: towards a neuroscience of will. Nat. Rev. Neurosci. **9**(12), 934–946 (2008)
5. Kiefer, M.: Top-down modulation of unconscious 'automatic' processes: A gating framework. Adv. Cogn. Psychol. **3**, 289–306 (2007)
6. Monsell, S.: Task switching. Trends Cogn. Sci. **7**(3), 134–140 (2003)
7. Dominguez, C.: Can SA be defined? In: Vidulich, M., Dominguez, C., Vogel, E., McMillan, G. (eds.) Situation Awareness: Papers and Annotated Bibliography (AL/CFTR-1994-0085), pp. 5–15. Armstrong Laboratory, Wright-Patterson AFB, OH (1994)
8. Endsley, M.R.: Design and evaluation for situation awareness enhancement. In: Proc. of the Human Factors Society 32nd Annual Meeting, Santa Monica, CA, pp. 97–101 (1988)
9. Endsley, M.R.: Toward a Theory of Situation Awareness in Dynamic Systems. Hum. Factors J. Hum. Factors Ergon. Soc. **37**(1), 32–64 (1995)
10. Endsley, M.R., Garland, D.G.: Pilot situation awareness training in general aviation. In: In Proc. of the 14th Triennial Congress of the Inter[nl]. Ergonomics Association and the 44th Annual Meeting of the Human Factors and Ergonomics Society, pp. 357–360 (2000)
11. Endsley, M.R.: Situation awareness and human error: designing to support human performance. In: Proc. of the High Consequence Systems Surety Conference (1999)
12. Endsley, M.R.: Theoretical underpinnings of situation awareness: a critical review. In: Endsley, M.R., Garland D.J. (eds.) Situation Awareness Analysis and Measurement. Lawrence Erlbaum Associates, Mahwah, NJ (2000)
13. Thilakarathne, D.J.: Neurologically inspired computational cognitive modelling of situation awareness. In: Ślęzak, D., Tan, A.-H., Peters, J.F., Schwabe, L. (eds.) BIH 2014. LNCS, vol. 8609, pp. 459–470. Springer, Heidelberg (2014)
14. Moore, J.W., Obhi, S.S.: Intentional binding and the sense of agency: A review. Conscious. Cogn. **21**(1), 546–561 (2012)

15. Katsuki, F., Constantinidis, C.: Bottom-Up and Top-Down Attention: Different Processes and Overlapping Neural Systems. The Neuroscientist, December 2013

16. Awh, E., Belopolsky, A.V., Theeuwes, J.: Top-down versus bottom-up attentional control: a failed theoretical dichotomy. Trends Cogn. Sci. 16(8), 437–443 (2012)

17. Baluch, F., Itti, L.: Mechanisms of top-down attention. Trends Neurosci. 34(4), 210–224 (2011)

18. Miller, E.K., Cohen, J.D.: An integrative theory of prefrontal cortex function. Annu. Rev. Neurosci. 24(1), 167–202 (2001)

19. Rigoni, D., Brass, M., Roger, C., Vidal, F., Sartori, G.: Top-down modulation of brain activity underlying intentional action and its relationship with awareness of intention: an ERP/Laplacian analysis. Exp. Brain Res. 229(3), 347–357 (2013)

20. Thilakarathne, D.J., Treur, J.: Modelling the dynamics of emotional awareness. In: Proceedings of the 21st European Conference on Artificial Intelligence: Front. Artif. Intell. Appl., vol. 263, pp. 885–890 (2014)

21. Pessoa, L.: Emergent processes in cognitive-emotional interactions. Dialogues Clin. Neurosci. 12, 433–448 (2010)

22. Pessoa, L.: Emotion and cognition and the amygdala: From 'what is it?' to 'what's to be done?'. Neuropsychologia 48(12), 3416–3429 (2010)

23. Poljac, E., Poljac, E., Yeung, N.: Cognitive Control of Intentions for Voluntary Actions in Individuals With a High Level of Autistic Traits. J. Autism Dev. Disord. 42(12), 2523–2533 (2012)

24. Bor, D., Seth, A.K.: Consciousness and the Prefrontal Parietal Network: Insights from Attention, Working Memory, and Chunking. Front. Psychol. 3 (2012)

25. Miller, E.K.: The prefontral cortex and cognitive control. Nat Rev Neurosci 1(1), 59–65 (2000)

26. Thilakarathne, D.J.: Modeling dynamics of cognitive control in action formation with intention, attention, and awareness. In: Proc. of the IEEE/WIC/ACM Inter[nl] Joint Conf. on Web Intelligence and Intelligent Agent Technologies, vol. 3, pp. 198–205 (2014)

27. Lavie, N.: Attention, Distraction, and Cognitive Control Under Load. Curr. Dir. Psychol. Sci. 19(3), 143–148 (2010)

28. Lavie, N., Tsal, Y.: Perceptual load as a major determinant of the locus of selection in visual attention. Percept. Psychophys. 56(2), 183–197 (1994)

29. Lavie, N., Hirst, A., de Fockert, J.W., Viding, E.: Load Theory of Selective Attention and Cognitive Control. J. Exp. Psychol. Gen. 133(3), 339–354 (2004)

30. Lavie, N.: The role of perceptual load in visual awareness. Brain Res. 1080(1), 91–100 (2006)

31. Lavie, N.: Distracted and confused?: Selective attention under load. Trends Cogn. Sci. 9(2), 75–82 (2005)

32. Damasio, A.R.: Self Comes to Mind: Constructing the Conscious Brain. Pantheon Books, NY (2010)

33. James, W.: What is an Emotion? Mind 9(34), 188–205 (1884)

34. D'Ostilio, K., Garraux, G.: Brain mechanisms underlying automatic and unconscious control of motor action. Front. Hum. Neurosci. 6 (2012)

35. Haynes, J.-D.: Decoding and predicting intentions. Ann. N. Y. Acad. Sci. 1224(1), 9–21 (2011)

36. Treur, J.: An integrative dynamical systems perspective on emotions. Biol. Inspired Cogn. Archit. 4, 27–40 (2013)

Knowledge Based Systems

A Real-Time Monitoring Framework
for Online Auctions Frauds

Samira Sadaoui[✉], Xuegang Wang, and Dongzhi Qi

Computer Science Department, University of Regina, Regina, SK, Canada
sadaouis@uregina.ca

Abstract. In spite of many advantages of online auctioning, serious frauds menace the auction users' interests. Today, monitoring auctions for frauds is becoming very crucial. We propose here a generic framework that covers real-time monitoring of multiple live auctions. The monitoring is performed at different auction times depending on fraud types and auction duration. We divide the real-time monitoring functionality into threefold: detecting frauds, reacting to frauds, and updating bidders' clusters. The first task examines in run-time bidding activities in ongoing auctions by applying fraud detection mechanisms. The second one determines how to react to suspicious activities by taking appropriate run-time actions against the fraudsters and infected auctions. Finally, every time an auction ends, successfully or unsuccessfully, participants' fraud scores and their clusters are updated dynamically. Through simulated auction data, we conduct an experiment to monitor live auctions for shill bidding. The latter is considered the most severe fraud in online auctions, and the most difficult to detect. More precisely, we monitor each live auction at three time points, and for each of them, we verify the shill patterns that most likely happen.

Keywords: Online auction frauds · Real-time monitoring of auctions · Fraud detection · Shill bidding patterns · Shill bidding metrics

1 Introduction

In auction websites, transactions are growing rapidly as a large number of users are bidding daily, e.g. 110 million auctioned items in e-Bay between December 2009 and July 2010 [7]. In spite of many advantages of online auctioning, serious frauds menace the users' interests. Auction frauds, representing the largest part of all Internet frauds, are on the rise because auctions give many opportunities for conducting misbehavior [12]. Often, the innocent bidders are not even aware about the committed frauds [12]. Both auctioneers and bidders utilize various strategies to commit frauds, such as shill bidding. Actually, fraudulent users can manipulate an auction to generate more profits; for example by creating fake identities, a malicious bidder can inflate the bidding price to let the winner pay more for the auctioned item [19]. Shill bidding has been recognized as one of the most dominant cheating activities in online auctions and also the hardest one to detect since there is no proof of its occurrence unlike other frauds [20]. Several empirical studies have demonstrated the presence of shills by

© Springer International Publishing Switzerland 2015
M. Ali et al. (Eds.): IEA/AIE 2015, LNAI 9101, pp. 97–108, 2015.
DOI: 10.1007/978-3-319-19066-2_10

analyzing a substantial amount of historical auction data from popular auction houses [21]. Consequently, monitoring auctions for frauds is becoming crucial. We categorize existing auction monitoring solutions into two types: offline and real-time. Offline or batch monitoring examines a huge volume of raw transactions of multiple past auctions [5], [19]. It is however too late because the infected auctions have already resulted in losses for honest bidders [20]. Moreover, processing the huge set of submitted bids of several auctions may be really time-consuming. Since the damage occurs during the auctions, it is essential to detect and stop frauds in run-time before it is too late. So far, all existing auction houses in the e-market do not offer monitoring services, and almost all the previous research works proposed services that detects frauds in offline auctions, and do not react against non-legitimate bidders.

We propose a generic framework that covers real-time monitoring of multiple live auctions. This practical framework observes the progressing auctions to be able to take actions on time and prevent fraudsters from succeeding. The monitoring is performed at different times (the real-time events) depending on the auction duration (in terms of hours or days) and fraud types. We divide the real-time monitoring functionality into threefold: detecting frauds, reacting to frauds, and updating users' clusters. The first task examines in run-time bidding activities in the running auctions by applying fraud detection mechanisms. The second one determines how to react to suspicious activities by taking proper run-time actions against the infected auctions and fraudsters. At last, every time an auction ends (successfully or unsuccessfully), participants' fraud scores and clusters are updated dynamically. Our monitoring system considers both bidding and auction properties. A fraud score represents the user's behavior in past auctions. More we include evidences of user's misbehavior the more we increase the accuracy of his fraud score. Another issue is how to merge those evidences to produce the current fraud score. Clusters (normal, suspicious and fraudster), which are shown to all to see, may be used as a deterrent for bidders to commit frauds as well as a basis for the reputation-based models. The fraud score is more reliable than those third-party ratings that can be easily falsified. Our goal is to prevent and minimize undesirable bidding behavior before it is too late in order to establish trust among users. Thanks to our real-time monitoring framework, we process a smaller number of bids to increase the system performance. The negotiation process should not be interrupted while the monitoring component is conducting the live analysis. Through a proper case study, we conduct an experiment to detect shill bidding. More precisely, we monitor each live auction at three time points, and for each of them, we verify the shill patterns that most likely happen.

2   Related Work

We studied many works on auction frauds as well as on shill bidding strategies and their detection algorithms. Also, some prevention approaches have been adopted to try to deter auction users from committing frauds, such as requiring credit card information when registering to the auction system [14], [23], or taking into account the users' past reputation [20]. The goal of shilling is to artificially increase the cost of goods or services in order to generate an interest for the auctioned items. For instance, the seller asks an accomplice to compete on his item, or would create an alternate auction account to commit the fraud [19]. Several shilling strategies have been identified, for example 6 in [17], 3 in [19], 9 in [20], 5 in [10], 3 in [9], 8 in [5], and 1 in [6]. Several patterns are similar across these papers because they appear more often in the auction data. We classify all these patterns into several classes: Security related (a Shill Bidder (SB) has alternate identities by using different accounts and IP addresses); Collusive behavior (a SB participates exclusively in auctions held by some particular sellers, colluding bidders work together to inflate the price, sellers place bids on each other's auctions, or users who live in a proximity area collude); Competitive shilling (a SB aggressively increases the price, or bids more often); Reputation manipulation (usually no feedback is provided for a SB from the seller, or users collude by helping each other building a good reputation by submitting positive ratings); Buy back shilling (the seller or his accomplice wins the auction to re-sell the item again in case when the current auction price is low); Discover-and-stop shilling (a SB increases the second highest bid until he figures out about the highest bid and then stops negotiating).

The shill detection mechanisms employ various models, including: a) Statistical methods, such as assigning metrics for shill patterns and then aggregate them to provide a final shill score [4], [17], [19], or using Dempster-Shafer theory to express patterns as evidences and then fuse them to produce the degree of belief of a bidder for shilling [5]; b) Formal methods to specify and verify shill patterns [20]; c) Machine learning methods based: on neural networks to predict the auction price according to the shill activities [4], or to search for the optimal weights for the shill patterns [18]; on decision trees to classify bidders into two groups: "Regular" and "Shill Bidder" [1], [23]; on hierarchical clustering to group users into several clusters [11], [16], [7]; on Bayesian graph models to calculate the probability to be a shill or not [9], [13].

Almost all the works on shill detection are done offline. Indeed, very few papers proposed real-time shill detection services. One of the limitations of offline approaches is the preparation and analysis of a huge amount of batch data, which are both very time consuming tasks. Besides, it is too late to react to shills. In the following, we just summarize the in-auction detection fraud techniques [8], [20, 21]. In [21], the authors developed a multi-agent trust management framework for real-time shill detection. Users' sub-roles (five groups) are dynamically assigned according to their behaviors. Based on his sub-role, a user can be granted or denied auction services and resources. Every time one shill pattern is detected, it is verified further. The reputation score (in terms of feedback ratings) is used to verify those detected shills. The bidder's shill score increases by one, and two actions can be taken (cancelling the auction or warning the bidder at fault). In our point of view, a bidder cannot be blamed only on few fraud patterns because they can be inexperienced, and the reputation is not the most suitable way to verify shills. In [20], Xu et al. introduced a formal approach to detect shills in live auctions. The approach employs three sources: the auction model which is updated dynamically as new bids arrive, Linear Temporal Logic (LTL) formulas representing the shill patterns, and a SPIN model checker that checks whether the

LTL formulas are violated or not. Nevertheless, monitoring an auction after every single submitted bid may make the detection process not efficient, especially for a large number of participants. [7] presented a real-time classifier that identifies efficiently suspicious bidders based on a neural network method. The latter is initialized with a training sample (produced by a hierarchical clustering mechanism), and then updated incrementally to adapt to new bidding data in real time. One of the difficult tasks is labeling manually the large training data sets of users. In [7, 20], three detection stages have been identified for shilling: Early stage i.e. at 25% of the elapsed auction time because SB places false bids very early in the auction to encourage others to participate, especially when the participation rate is low; Middle stage i.e. between 25% to 90% since most of the bidding activities happen at this stage (competitive shilling); Late stage i.e. in the last 10% since SB places very few bids. Bidding towards the end of an auction is risky as the fraudster could accidentally win.

3   A Real-Time Monitoring Framework for Online Auctions

To increase trust, every live auction should be systematically verified at different times as shown in Algorithm 1. These time points are defined by the auctioneer depending on the fraud types and auction duration. We consider both bidding and auction properties. In this way, the monitoring system will ensure better results and we will have a very good idea about each user's behavior (called liveFraudScore) in each live auction. When an auction ends (successfully or unsuccessfully), the current live-FraudScore of each participant is merged with his fraud score of past auctions, and his cluster is updated accordingly.

We design our auction system with three layers: a) The presentation layer interacts with users and displays information of running auctions; b) The application layer is composed of several monitoring components, each one assigned dynamically to an ongoing auction. Each component intercepts at different time points the submitted bids in the corresponding auction, examines the behaviors of users, then makes proper decisions; c) The data store layer consists of four main logs: User history log stores users' current fraud scores and clusters; Live auction log contains information such as auctionID, sellerID, productID, start time, duration, reserve price, users' submitted bids in different intervals and current live fraud scores; When an auction ends, it is then stored in the past auction log by adding new data, like final price, winner, total bids and status (unsuccessful or successful); Fraud pattern log stores the formulation of patterns, their corresponding time points and thresholds. To take into account new frauds, we just add their formulations into this log.

3.1   Real-Time Detection of Frauds

Some fraud patterns are most likely to occur in a certain auction interval [20]. Our detection technique, Algorithm 2, evaluates them against the bids submitted at that interval. A bidder cannot be blamed with only few fraud patterns, but a series of patterns should be used to improve the accuracy of the live fraud score. We quantify the fraud patterns with some metrics and aggregate them with a fusing method, such as the weighted mathematical average (we can associate weights to fraud patterns to denote their importance), Dempster-Shafer theory, or Bayesian average. The aggregation goal is to produce a comprehensive value, called liveFraudScore, which measures the level of fraud of each user at each running auction. We monitor each bidder as he can change his behavior in just one auction.

---

monitorLiveAuction

Input: a_i *//live auction*
Sources: *liveAuctionLog, userLog*
{ time = [T_1, .., T_n]; //time points defined w.r.t to auction duration and fraud type
   do *{*
      detectFraud(a_i, time);
      //compute liveFraudScore for each user u_j in auction a_i at time point
      reactToFraud(a_i, time);
      //analyze liveFraudScore to take actions at time point
      time = progressTime();   // progress to the next time point
   }
   while *($status_{ai}$ = "successful") //monitor as long as the auction was not terminated*
   updateUserCluster(a_i);
   //update participants' fraud scores and clusters by considering past and current
behaviours
}

Algorithm 1. Real-Time Monitoring of Auctions

---

detectFraud

Inputs: a_i *//live auction, time //time point*
Sources: *liveAuctionLog, fraudPatternLog*
*{ **forall** $u_j \in a_i$ **do***
  { setOfBids = extractBids(a_i, u_j, time);
  setOfPatterns = extractPatterns(time);
  forall $p \in setOfPatterns$ **do***
    { score = evaluatePattern(p, setOfBids);
    // evaluate a fraud pattern against a set of submitted bids
    liveFraudScore$_{Uj}$ = mergeScores1(liveFraudScore$_{Uj}$, score);
    // combine all the fraud metrics}
  }
}

Algorithm 2. Real-Time Fraud Detection

3.2   Real-Time Reaction to Frauds

The reaction component, Algorithm 3, decides on how to react to bidders' suspicious behaviors upon the value of the current live fraud score in each progressing auction. If the latter is in a certain range (decided by the auctioneer) or beyond it, we perform one of these two following actions:

- *warnUser*: send a warning to the suspicious bidder to bid more responsibly in the current auction. Additionally, notify all participants about the possibility of a suspicious activity.
- *cancelAuction*: cancel the infected auction because there is a high fraud activity, and change its status to "unsuccessful". In this case, the item is not sold.

---

reactToFraud

Inputs: a_i //live auction, time //time point
Source: *liveAuctionLog*
{ **forall** $u_j \in a_i$ **do**
    if *(livefraudScore$_{Uj}$ $\in$ range1)* **then**
        perform(warnUser, u_j, a_i, time);
    else if *(livefraudScore$_{Uj}$ $\in$ range2)* **then**
        {perform(cancelAuction, a_i, time); break;}
}

Algorithm 3. Real-Time Reaction to Frauds

---

updateUserCluster

Input: a_i //live auction
Sources: *liveAuctionLog, userLog*
{ **forall** $u_j \in a_i$ **do**
  { *fraudScore$_{Uj}$ = mergeScores2(liveFraudScore$_{Uj}$, fraudScore$_{Uj}$);*
  if *(fraudScore$_{Uj}$ $\in$ range1)* **then**
    cluster$_{Uj}$ = "suspicious";
  else if *(fraudScore$_{Uj}$ $\in$ range2)* **then**
    { *cluster$_{Uj}$ = "fraudster";*
      suspendAccount(u_j); }
  }
}

Algorithm 4. Real-Time Update of Users' Clusters

3.3 Real-Time Update of Users' Clusters

It is useful to assign to each user a cluster regarding his level of conduct: "normal", "suspicious", or "fraudulent". A new user will have a status of "normal". The cluster of each bidder is displayed to all to see. Every time an auction ends, the fraud score and cluster of all participants will be dynamically updated. In Algorithm 4, we merge the score of current analysis with the score of past analysis. We therefore obtain a value representing the fraud score in all the participated auctions. By merging the two values, we always give a user a chance to improve his fraud score. Only normal and suspicious bidders can negotiate in new auctions. The accounts of fraudsters are suspended.

4   Application to Shill Bidding

4.1   Shill Bidding Strategies

We compiled most of the following shill patterns from [17], [19], [20], [10], [9], [5], [6], [4]. We monitor shill bidding at three time points (at 25%, at 90% and at 100%), and we determine for each of them the shill patterns that most probably occur. For instance, at the auction end we verify some auction properties (#15, #16 and #17) to detect more fraud patterns and update more accurately the users' fraud scores.

T_{early}:
1. SB places bids very close to the auction starting time (**early bidding**).
2. SB participates exclusively in auctions conducted by some sellers. A normal bidder may negotiate in several concurrent auctions to find the best price, but SB deals with a limited range of sellers (**buyer tendency**).
3. SB submits a bid very close to the reserve price.
4. SB posts small bid increments with the minimum amount required by the auction.

T_{middle}:
5. SB outbids legitimate bids until he is satisfied, or he has reached the reserve price, or because of a very slow bidding (**bidding ratio**).
6. SB often bids successively to outbid oneself even when he is the current winner (**successive outbidding**).
7. Successive outbidding or bidding ratio is high when the current auction price is smaller than the reserve price; otherwise they are lower to reduce the risk of winning (**reserve price shilling**).
8. SB submits a bid within a short time interval (1 minute) of any new legitimate bids to give others more time to bid.
9. SB outbids any bid with a minimum of 10% to 20% of the current auction price.
10. SB participates in concurrent auctions with higher bidding prices rather with lower prices.

T_{end}:
11. SB stops negotiating early before the auction ends (avoids sniping) (**last bidding**).
12. Winner ratio of SB is very low even when his bids aggressively (**winning ratio**).
13. SB places no bids for high or medium value items, and very few bids for low ones.
14. SB submits low bid increment with the minimum amount required by the auction.
15. An auction with shills has more bids that the average number of bids in normal concurrent auctions (**auction bids**).
16. When the auction price is significantly higher than the expected price, there is a probability of 66.7% of the auction being infected [4].
17. Starting price of an infected auction is less than the average staring price of concurrent auctions.

Table 1. Fraud pattern metrics

Shilling	Expression	Description
earlyBidding (u_j, a_i, t_{early})	$1 - \dfrac{firstBidTime(u_j, a_i) - startTime_{ai}}{duration_{ai}}$	Source: liveAuctionLog.
buyerTendency (u_j, sk, t_{early})	$\dfrac{|auctionSeller(u_j, sk)|}{|auctionPart(u_j)|}$	Dividend is the number of auctions that u_j has participated in for seller sk in the past 6 months. Divisor is the total number of auctions that u_j has joined during the same period. Sources: pastAuctionLog and LiveAuctionLog for the past 6 months.
biddingRatio (u_j, a_i, t_{middle})	$\dfrac{totalBids(u_j, a_i)}{totalBids_{ai}}$	Number of bids should be collected only from the interval $[t_{early}, t_{middle}]$ of the auction. Source: liveAuctionLog.
succOutbid (u_j, a_i, t_{middle})	*succBid = 0;* *if (succOutbid($u_j,a_i,3$)) then* *{succBid=1; break;}* *else if (succOutbid(uj,ai,2)) then* *{succBid=0.5; break;}*	succOutbid($u_j,a_i,3$) and succOutbid($u_j,a_i,2$) checks for three or two successive bids respectively for the same user. Source: liveAuctionLog.
reservePriceShill (u_j, a_i, t_{middle})	*if auctionPrice$_{ai}$ < reservePrice$_{ai}$* *then* *if succOutbid(u_j, a_i) >= 0.5 or biddingRatio(u_j, a_i) > 0.5 then* *reservePriceShill = 1* *else reservePriceShile = 0*	Source: liveAuctionLog.
lastBidding (u_j, a_i, t_{end})	$\dfrac{endTime_{ai} - lastBidTime(u_j, a_i)}{duration_{ai}}$ [20]	Source: liveAuctionLog.
winningRatio (u_j, t_{end})	$1 - \dfrac{|auctionWon(u_j)|}{|auctionPartHigh(u_j)|}$ auctionPartHigh(u_j) = { a_i / biddingRatio(u_j, a_i, t_{middle}) > threshold}	Dividend is the number of auctions won by u_j. Divisor is the number of auctions participated by u_j with a high bidding ratio to remove non-active bidders. Source: pastAuctionLog for the past 6 months.
auctionBids (a_i, t_{end})	*if (averageBids < totalBids$_{ai}$) then* auctionBids(a_i,t_{end}) = $1 - \dfrac{averageBids}{totalBids_{ai}}$ *else* auctionBids(a_i,t_{end}) = 0	Average bids come from the concurrent auctions to a_i (selling identical items). Sources: pastAuctionLog and liveAuctionLog for the past 6 months.

4.2   Detection Mechanisms for Shill Bidding

We examine the bidders' activities against the metrics that we define in Table 1. Here we implement the metrics for only 8 shill frauds. We may note that we calculate the metrics from past and/or currently examined auctions. The higher the metric value, the more suspicious is the observed user.

We assign different weights to the eight fraud patterns: early bidding (EB) with a weight of 0.1; buyer tendency (BT) with 0.3; bidding ratio (BR) with 0.3; successive outbidding (SO) with 0.9; reserve price shilling (RPS) with 0.6; last bidding (LB) with 0.6; winning ratio (WR) with 0.6; auction bids (AB) with 0.6. For instance, BT has a lower weight due to the false tendency issue (a bidder has a tendency for a certain seller due to his good reputation or he is the only one selling the item [19]); WR has a medium weight because of the problem of one-time bidders; SO has a high weight because it is a good indicator of shilling. In Table 2, at each time point we show how to merge the produced pattern scores by using the weighted mathematical average.

Table 2. Fraud Score Combination

T_{early}	T_{middle}	T_{end}
$W_{early} = W_{EB} + W_{BT}$	$W_{middle} = W_{BR} + W_{SO} + W_{RPS}$	$W_{end} = W_{LB} + W_{WR} + W_{AB}$
$S_{early} = W_{EB}*EB + W_{BT}*BT$	$S_{middle} = W_{BR}*BR + W_{SO}*SO + W_{RPS}*RPS$	$S_{end} = W_{LB}*LB + W_{WR}*WR + W_{AB}*AB$
$LFS_{early} = \dfrac{S_{early}}{W_{early}}$	$LFS_{middle} = \dfrac{S_{early} + S_{middle}}{W_{early} + W_{middle}}$	$LFS_{end} = \dfrac{S_{early} + S_{middle} + S_{end}}{W_{early} + W_{middle} + W_{end}}$

5   An Experiment

We conduct an experiment based on simulated data of 2 live and 10 offline auctions, and 10 users. Both live auctions have a reserve price of \$1,200 and a duration of 48 hours. We set the thresholds range1 to $[0.5, 0.7)$ and range2 to $[0.7, 1.0]$.

In Table 3, there are five users participating in the first live auction. Its starting time is 9am, Nov. 1st 2014. T_{early}, T_{middle} and T_{end} correspond resp. to 21pm Nov. 1st, 4:12am Nov. 3rd, and 9am Nov 3rd. Users, U_4, U_7 and U_{10} are behaving normally. Even though U_2 has a high live fraud score in early time, after warning him, his live fraud score (LFS) declined in the following time points. U_1 always behaves aggressively during the whole auction. Despite this, U_1's behavior cannot result in cancelling the auction. At the end of this auction, the shill scores of these bidders are all updated (cf. Table 5). The starting time of the second live auction is 11am, Nov. 1st. T_{early}, T_{middle} and T_{end} correspond resp. to 23pm Nov. 1st, 6:12am Nov. 3rd, and 11am Nov 3rd. In Table 4, we can see that from early time, U_1 has already had relatively a high live shill score, and got a warning. Other users are quite honest as their fraud scores are always less than 0.5. In middle time, since U_1's LFS is high, more than 0.7, we cancel the auction, and update U_1's cluster from "suspicious" to "fraudulent" as shown in Table 5. As a result, U_1's account is suspended. In Fig. 1, we give an idea about the behavior of the 10 users in both live auctions.

Table 3. Monitoring Live Auction#1

User / Time		U_1	U_2	U_4	U_7	U_{10}
T_{early}	LFS	0.6738	0.6120	0.2914	0.3055	0.2483
	Action	WarnUser	WarnUser	No Action	No Action	No Action
T_{middle}	LFS	0.6475	0.1317	0.0734	0.0828	0.0656
	Action	WarnUser	No Action	No Action	No Action	No Action
T_{end}	LFS	0.54958	0.2531	0.2059	0.2088	0.2043
	Action	WarnUser	No Action	No Action	No Action	No Action

Table 4. Monitoring Live Auction#2

User / Time		U_1	U_3	U_5	U_6	U_9
T_{early}	LFS	0.6734	0.2506	0.2403	0.2511	0.3172
	Action	WarnUser	No Action	No Action	No Action	No Action
T_{middle}	LFS	0.8649	0.0607	0.0664	0.0684	0.0728
	Action	CancelAuction				

Fig. 1. Live Fraud Scores of Users

Table 5. Updating Users' Clusters

Past Auctions		Successful Auction#1		Past Auctions		Unsuccessful Auction#2	
	Cluster	Fraud Score	Cluster		Cluster	Fraud Score	Cluster
U_1	Suspicious	0.5696	Suspicious	U_1	Suspicious	0.7172	Fraudulent
U_2	Suspicious	0.2537	Normal	U_3	Normal	0.1321	Normal
U_4	Normal	0.2082	Normal	U_5	Normal	0.1339	Normal
U_7	Normal	0.2146	Normal	U_6	Normal	0.1409	Normal
U_{10}	Normal	0.1979	Normal	U_9	Normal	0.1294	Normal

6   Conclusion and Future Work

Online auctions are still not trustworthy due to the lack of real-time monitoring services. This lack allows participants to fake their identities and behave as they desire. Without a proper monitoring, auction systems will have a negative impact on innocent bidders. The proposed monitoring system examines bidders' activities at several auction time points. After detecting suspicious behaviors in live auctions, our system takes actions immediately by notifying the users at fault, or cancelling the infected auctions and suspending fraudsters' accounts. In this way, we increase the confidence of bidders for the auction services.

There are several research directions of this work. First of all, we would like to implement the entire real-time monitoring framework with the agent technology in order to be able to add or remove dynamically monitoring agents [2]. Also, we are interested in evaluating the research prototype on large datasets of auctions and users. Moreover, to observe a large set of running auctions, we will investigate how to apply in real-time well-known detection techniques, such as genetic algorithms and supervised machine learning, to the context of online auctions. The goal is to extract in run-time patterns of frauds. An example of these techniques is the powerful binary classifier SVM, which may also be used for imbalanced auction data classification [24]. The latter will significantly improve the monitoring performance.

References

1. Chau, D.H., Pandit, S., Faloutsos, C.: Detecting fraudulent personalities in networks of online auctioneers. In: Fürnkranz, J., Scheffer, T., Spiliopoulou, M. (eds.) PKDD 2006. LNCS (LNAI), vol. 4213, pp. 103–114. Springer, Heidelberg (2006)
2. Chen, B., Sadaoui, S.: A Generic Formal Framework for Constructing Agent Interaction Protocols. International Journal of Software Engineering & Knowledge Engineering 15(1), 61–85 (2005)
3. Dong, F., Shatz, S., Xu, H.: Combating online in-auction frauds: Clues, techniques and challenges. Computer Science Review 3(4), 245–258 (2009)
4. Dong, F., Shatz, S.M., Xu, H., Majumdar, D.: Price comparison: A reliable approach to identifying shill bidding in online auctions? Electronic Commerce Research and Applications 11(2), 171–179 (2012)
5. Dong, F., Shatz, S.M., Xu, H.: Reasoning under Uncertainty for Shill Detection in Online Auctions Using Dempster-Shafer Theory. International Journal of Software Engineering and Knowledge Engineering 20(7), 943–973 (2010)
6. Engelberg, J., Williams, J.J.: eBay's proxy bidding: A license to shill. Journal. Econom. Behav. Organ 72(1), 509–526 (2009)
7. Ford, B.J., Xu, H., Valova, I.: A Real-Time Self-Adaptive Classifier for Identifying Suspicious Bidders in Online Auctions. Computational Journal 56(5), 646–663 (2013)
8. Ford, B.J., Xu, H., Valova, I.: Identifying suspicious bidders utilizing hierarchical clustering and decision trees. In: IC-AI, pp. 195-201 (2010)
9. Goel, A., Xu, H., Shatz, S.M.: A multi-state bayesian network for shill verification in online auctions. In: Proc.22nd Int. Conf. Software Engineering and Knowledge Engineering, USA, pp. 279–285 (2010)

10. Kauffman, R., Wood, C.A.: Irregular bidding from opportunism: an explanation of shilling in online auctions. Information Systems Research **5**, 1–36 (2007)
11. Lie, B., Zhang, H., Chen, H., Liu, L., Wang, D.: A k-means clustering based algorithm for shill bidding recognition in online auction. In: 24th Chinese Control and Decision Conference, pp. 939–943. IEEE (2012)
12. Mamum, K., Sadaoui, S.: Combating shill bidding in online auctions. In: International Conference on Information Society, i-Society, pp. 174–180. IEEE (2013)
13. Manikanteswari, D.S.L., Swathi, M., Nagendranath, M.V.S.S.: Machine Learning Approach to Handle Fraud Bids. International journal for development of computer science & technology, **1**(5) (2013)
14. Mundra, A., Rakesh, N.: Online Hybrid model for online fraud prevention and detection. In: ICACNI, pp. 805–815 (2013)
15. Patel, R., Xu, H., Goel, A.: Real-time trust management in agent based online auction systems. In: SEKE, pp. 244–250 (2007)
16. Shah, H.S., Joshi, N.R., Sureka, A., Wurman, P.R.: Mining eBay: bidding strategies and shill detection. In: Zaïane, O.R., Srivastava, J., Spiliopoulou, M., Masand, B. (eds.) WebKDD 2003. LNCS (LNAI), vol. 2703, pp. 17–34. Springer, Heidelberg (2003)
17. Trevathan, J., Read, W.: Detecting shill bidding in online english auctions. In: Handbook of Research on Social and Organizational Liabilities in Information Security, pp. 446–470 (2006)
18. Tsang, S., Koh, Y., Dobbie, G., Alam, S.: Detecting online auction shilling frauds using supervised learning. Expert Syst. Appl. **41**(6), 3027–3040 (2014)
19. Trevathan, J., Read, W.: Investigating Shill Bidding Behaviors Involving Colluding Bidders. Journal of Computers **2**(10), 63–75 (2007). Academy Publisher
20. Xu, H., Bates, C., Shatz, S. M.: Real-time model checking for shill detection in live online auctions. In: SERP, pp. 134–140 (2009)
21. Xu, H., Shatz, S. M. Bates, C. K.: A framework for agent-based trust management in online auctions. In: ITNG, pp.149–155 (2008)
22. Yu, C., Lin, S.: Fuzzy rule optimization for online auction frauds detection based on genetic algorithm. Electronic Commerce Research **13**(2), 169–182 (2013)
23. Yoshida, T., Ohwada, H.: Shill bidder detection for online auctions. In: Zhang, B.-T., Orgun, M.A. (eds.) PRICAI 2010. LNCS, vol. 6230, pp. 351–358. Springer, Heidelberg (2010)
24. Zhang, S., Sadaoui, S., Mouhoub, M.: An Empirical Analysis of Imbalanced Data Classification. Computer and Information Science **8**(1), 151–162 (2015). doi:10.5539/cis.v8n1p151

Reasoning About the State Change of Authorization Policies

Yun Bai[✉], Edward Caprin, and Yan Zhang

Artificial Intelligence Research Group, School of Computing,
Engineering and Mathematics, University of Western Sydney, South
Penrith, Australia
{ybai,edward,yan}@scem.uws.edu.au

Abstract. Reasoning about authorization policies has been a prominent issue in information security research. In a complex information sharing and exchange environment, a user's request may initiate a sequence of executions of authorization commands in order to decide whether such request should be granted or denied. Becker and Nanz's logic of State-Modifying Policies (SMP) is a formal system addressing such problem in access control. In this paper, we provide a declarative semantics for SMP through a translation from SMP to Answer Set Programming (ASP). We show that our translation is sound and complete for bounded SMP reasoning. With this translation, we are able not only to directly compute users' authorization query answers, but also to specifically extract information of how users' authorization states change in relation to the underlying query answering. In this way, we eventually avoid SMP's tedious proof system and significantly simply the SMP reasoning process. Furthermore, we argue that the proposed ASP translation of SMP also provides a flexibility to enhance SMP's capacity for accommodating more complex authorization reasoning problems that the current SMP lacks.

Keywords: Access control · Authorization policies · Knowledge representation and reasoning · Logic programming · State change

1 Introduction

Reasoning about authorization policies has been a prominent issue in information security research, e.g., [5,6,10]. With the increasing complexity of large Internet based information sharing and exchange, such as various social media websites, developing sophisticated access control frameworks for information privacy protection is becoming a challenging task.

Traditionally, an access control framework mainly deals with a static reasoning problem about authorizations. For instance, a university bookstore offers discount for all its currently enrolled students. When a student is purchasing books, the student only needs to show his/her valid student card in order to get discount for his/her purchase. In this case, the student's request of getting discount is granted simply based on the evaluation on the existing policies. However,

© Springer International Publishing Switzerland 2015
M. Ali et al. (Eds.): IEA/AIE 2015, LNAI 9101, pp. 109–119, 2015.
DOI: 10.1007/978-3-319-19066-2_11

when we deal with a large and open information sharing and exchange platform, access control may be significantly complicated, where authorization states play an important role in evaluating an user's request. In this case, quite often, we not only need to handle static authorization reasoning as traditional situations, but also have to update the relevant authorization state when a request is successfully granted.

Let us consider a photography social media website where users post their images as well as view and comment on others' with different levels of privileges. Under a role based access control model, a user's request for viewing some images is not only related to the role that the user is assigned, but also related to certain actions that the user has to execute in order to be granted such request. For example, Alice is registered as a primary member and Bob is a senior member of the site. Alice requests to view the EXIF data of one of Bob's images, which the policy will deny Alice request as primary members are usually not allowed to view senior members' image EXIF data. However, what is important for Alice is to know *how* she is able to get her request granted. For instance, Alice could become a follower of Bob first, to comment 10 of Bob's images, in order to get her request granted. Once Alice executes these two actions, i.e., following Bob and then commenting Bob's 10 images, Alice will be able to view Bob's image EXIF data, and consequently, Alice's authorization state has been changed, i.e., Alice is a follower of Bob and has commented 10 of Bob's images. Then next time, when Alice wants to view the EXIF data of Bob's another image, her request will be granted straightaway.

The above scenario demonstrates an example of reasoning about the state change of authorization policies which only attracted researchers' attentions recently. Beck and Nanz proposed a logic of State-Modifying Policy (SMP) to deal with such problem [3]. SMP has a similar syntax of Datalog in which a *policy* is specified as a finite set of rules. These rules are classified into two different types: one is to represent implicit facts based on others, for instance, we may specify that if Alice is a member of curator team, then Alice is able to view every user's image EXIF data. Obviously, this rule is simply a normal Datalog rule representing implicit yet static information about Alice's privilege. The second type of rules in SMP, on the other hand, is so called *command rules*, which specify how commands (actions) can be executed by an agent. As an example, we may express in SMP that if Alice is a member of curator team, after Alice rates a Bob's image, then this image receives a rating score. Different from the previous rule, this rule specifies a dynamic execution of a command, and as such, an authorization state change will occur.

The semantics of SMP is provided via Transaction Logic [Bonner and Kifer 1994], in which reasoning about authorization state change is achieved based on the updates on authorization states when a sequence of commands is executed. Becker and Nanz further developed a sound and complete proof system that provides a goal-oriented algorithm for finding minimal sequences leading to a specified target authorization state [3].

In this paper, we propose a declarative translation of SMP into Answer Set Programming (ASP) [1]. The main motivation of this work is to develop a framework for efficient implementation of reasoning about authorization state change that is applicable in practical problem domains. While SMP has its own proof system, it has not been implemented. Furthermore, as indicated by the authors, SMP is designated to be a minimal language focusing solely on state change. It, however, does not deal with more complex issues in authorization reasoning, such as constraints, delegation, negation, aggregation etc. In this sense, there is a major limit for extending SMP to handle more general authorization reasoning problems (see [3], section 4).

On the other hand, ASP has been demonstrated its reasoning power in both expressiveness and effectiveness over the years. ASP solvers have also been used for many real world applications [9]. By translating SMP into ASP, we will be able to implement the SMP reasoning based on existing efficient ASP solvers [4,7,8]. Such translation also provides a flexibility for further extending SMP reasoning if we want to address more complex authorization specification and reasoning involving negation, nonmonotonicity, delegation, trust management, and aggregations.

2 A Logic for State-Modifying Policies (SMP): An Overview

In this section, we provide an overview of Becker and Nanz's logic for state-modifying policies [3]. We start with a first-order language without function symbols. A *vocabulary* consists of a finite set $\mathcal{C}$ of *constant symbols* and disjoint finite sets of *predicate symbols*, named *extensional* ($\mathcal{Q}_{ext}$) and *intensional* ($\mathcal{Q}_{int}$) predicate symbols, and *command* predicate symbols ($\mathcal{Q}_{cmd}$). Let $\mathcal{X}$ be a set of variables. Then the syntax of SMD is defined as follows:

Term	t	$::= X \mid a$	where $X \in \mathcal{X}, a \in \mathcal{C}$
Atom	P_τ	$::= p(t_1, \cdots, t_k)$	where $p \in \mathcal{Q}_\tau, \tau \in \{ext, int, cmd\}$
Literal	L	$::= P_{int} \mid P_{ext} \mid \neg P_{ext}$	
Effect	K	$::= +P_{ext} \mid -P_{ext}$	
Rule	Rl	$::= P_{int} \leftarrow L_1 \wedge \cdots \wedge L_m$	where $m \geq 0$
		$\mid \quad P_{cmd} \leftarrow L_1 \wedge \cdots \wedge L_m \otimes K_1 \otimes \cdots \otimes K_n$	where $m, n \geq 0$

As in Datalog [2], in SMP, the extensional predicates are defined by *extensional database* (EDB), which is a finite set of ground extensional atoms called *facts*. Then an *authorization state* **B** is a finite set of facts. There two types of rules in SMP: one is of the form

$$P_{int} \leftarrow L_1 \wedge \cdots \wedge L_m, \text{ where } m \geq 0, \tag{1}$$

where the *head* P_{int} is an intensional atom, and the *body* (possibly empty) is a conjunction of extensional or intensional literals. This type of rules is the same as rules in Datalog. The other is of the form

$$P_{cmd} \leftarrow L_1 \wedge \cdots \wedge L_m \otimes K_1 \otimes \cdots \otimes K_n, \text{ where } m, n \geq 0, \tag{2}$$

where the head P_{cmd} is a command atom, and the body (possibly empty) consists of two parts: $L_1 \wedge \cdots \wedge L_m$ called *conditions*, and $\otimes K_1 \otimes \cdots \otimes K_n$ called *effects*. Each L_i $(i = 1, \cdots, m)$ is an extensional or intensional literal, represents the conditions that make the command, i.e., P_{cmd}, be executed, while $\otimes K_1 \otimes \cdots \otimes K_n$ represents the sequence of *effects* on the authorization state by performing this command. In particular, each K_i $(i = 1, \cdots, n)$ is of the form $+P_{ext}$ or $-P_{ext}$, where P_{ext} is an extensional atom, to represent the addition of the fact P_{ext} to, or the removal of the fact P_{ext} from the authorization state, respectively. A rule is *well formed* if all variables of its effects also occur in the head[1]. Then a *policy* is a finite set of well formed rules.

Example 1. We consider an example from [3] to illustrate how SMP is used to represent the state change of policies. Let a policy consist of the following rules:

$r_1 : buy(X, M) \leftarrow +bought(X, M),$
$r_2 : play1(X, M) \leftarrow bought(X, M) \wedge \neg played1(X, M) \otimes +played1(X, M),$
$r_3 : play2(X, M) \leftarrow played1(X, M) \wedge \neg played2(X, M) \otimes +played2(X, M).$

Rule r_1 says that by doing command $buy(X, M)$, it causes the effect that X bought movie M. Then rule r_2 states that if X bought movie M and has not played it yet, then playing the movie will cause the effect $played1(X, M)$, while r_3 says that if X has played M once, and has not played it twice yet, then playing it again will cause the effect $played2(X, M)$. □

Now we introduce the semantics of SMP. Given a policy, i.e., a finite set of well formed rules, a sequence of authorization states $\overline{\mathbf{B}} = \mathbf{B}_1 \cdot \mathbf{B}_2 \cdots \mathbf{B}_n$, and a SMP formula ϕ, the *entailment* $\models_{\mathcal{P}}$ between $\overline{\mathbf{B}}$ and ϕ is defined as follows.

(pos) $\mathbf{B} \models_{\mathcal{P}} P_{ext}$ iff $P \in \mathbf{B}$,
(neg) $\mathbf{B} \models_{\mathcal{P}} \neg P_{ext}$ iff $P \notin \mathbf{B}$,
(and) $\overline{\mathbf{B}} \models_{\mathcal{P}} \phi \wedge \psi$ iff $\overline{\mathbf{B}} \models_{\mathcal{P}} \phi$ and $\overline{\mathbf{B}} \models_{\mathcal{P}} \psi$
(seq) $\mathbf{B}_1 \cdots \mathbf{B}_k \models_{\mathcal{P}} \phi \otimes \psi$ iff $\mathbf{B}_1 \cdots \mathbf{B}_i \models_{\mathcal{P}} \phi$ and
                $\mathbf{B}_i \cdots \mathbf{B}_k \models_{\mathcal{P}} \psi$ for some $i \in \{1, \cdots, k\}$
$(plus)$ $\mathbf{B}_1 \cdot \mathbf{B}_2 \models_{\mathcal{P}} +P$ iff $B_2 = B_1 \cup \{P\}$
(min) $\mathbf{B}_1 \cdot \mathbf{B}_2 \models_{\mathcal{P}} -P$ iff $B_2 = B_1 \setminus \{P\}$
$(impl)$ $\overline{\mathbf{B}} \models_{\mathcal{P}} Q$ iff $Q \leftarrow \phi$ is a ground instantiation of
                a rule in $\mathcal{P}$ and $\overline{\mathbf{B}} \models_{\mathcal{P}} \phi$
                where Q is not extensional

Then the reasoning problem with SMP is as follows: given an initial authorization $\mathbf{B}_0$, a policy $\mathcal{P}$, and a ground formula ϕ which we call *ground goal formula*, how can we find a sequence of ground commands $Q_{cmd_1} \otimes \cdots \otimes Q_{cmd_l}$ from which a sequence of authorization states $\mathbf{B}_0 \cdots \mathbf{B}_n$ is generated, such that $\mathbf{B}_0 \cdots \mathbf{B}_n \models_{\mathcal{P}} Q_{cmd_1} \otimes \cdots \otimes Q_{cmd_l}$ and $\mathbf{B}_n \models \phi$. In this case, we simply write $\mathbf{B}_0 \cdots \mathbf{B}_n \models_{\mathcal{P}} \phi$.

---

[1] This is somewhat like the *safe* notion in answer set programming [11].

Generally, SMP only considers the ground goal formula ϕ of the form $Q_1(\overline{a_1}) \wedge \cdots \wedge Q_k(\overline{a_k})$, where each $Q_i(\overline{a_i})$ $(1 \leq i \leq k)$ is a ground intensional or extensional atom. That is, such ϕ represents the user's authorization request. Then the command sequence $Q_{cmd_1} \otimes \cdots \otimes Q_{cmd_l}$ explains that in order to achieve such request, what commands (actions) should be undertaken (sequentially), while the sequence of authorization states represents the state change starting from $\mathbf{B}_0$ in terms of such command sequences.

Example 2. Example 1 continued. We consider the policy $\mathcal{P} = \{r_1, r_2, r_3\}$ as displayed in Example 1. Let $\mathbf{B}_0 = \emptyset$, and $\phi = played1(tom, star\_war)$. Then we would expect to derive a command sequence $buy(tom, star\_war) \otimes play1(tom, star\_war)$. Intuitively, we would expect to generate a sequence of authorization states $\mathbf{B}_0 \cdot \mathbf{B}_1 \cdot \mathbf{B}_2$, where $\mathbf{B}_1 = \mathbf{B}_0 \cup \{bought(tom, star\_war)\}$, and $\mathbf{B}_2 = \mathbf{B}_1 \cup \{played1(tom, star\_war)\}$. Then it is easy to see that $\mathbf{B}_0 \cdot \mathbf{B}_1 \cdot \mathbf{B}_2 \models_{\mathcal{P}}$ $buy(tom, star\_war) \otimes play1(tom, star\_war)$ and $\mathbf{B}_2 \models_{\mathcal{P}} played1(tom, star\_war)$. Therefore, we have

$$\mathbf{B}_0 \cdot \mathbf{B}_1 \cdot \mathbf{B}_2 \models_{\mathcal{P}} played1(tom, star\_war). \square$$

To deal with such reasoning problem involving authorization state change, Becker and Nanz provides a sound and complete proof system, from which a goal-oriented algorithm is proposed to implement such reasoning procedure [3].

3   Translating SMP to Answer Set Programming

In this section, we provide an alternative approach to implement SMP. In particular, we present a translation from SMP to answer set programming, in this way, a reasoning problem in SMP can be viewed as the problem of computing answer set of a given logic program. Comparing to Becker and Nanz's proof system, the advantage of this approach is that we can directly use existing ASP solver to compute the solutions to SMP reasoning problems.

3.1   Answer Set Programming

We first introduce general concepts and notions of answer set programming, readers may find other relevant information from [1]. Again, we consider a first-order language without function. A *rule* is of the form

$$A \leftarrow B_1, \cdots, B_m, not\ C_1, \cdots, not\ C_n, \tag{3}$$

where $A, B_1, \cdots, B_m, C_1, \cdots, C_n$ are atoms. For convenience, we also call each $not\ C_i$ $(i = 1, \cdots, n)$ a *negative atom*. A rule is called *ground* if no variable occurs in all atoms of this rule. Intuitively, rule (3) represents a type of reasoning with absent information: if $B_1, \cdots, B_m$ are true, and there is no evidence to support that $C_1, \cdots, C_n$ to be true, then we can conclude A to be true.

A *logic program* Π is a finite set of rules. A program is *ground* if all its rules are ground. Π is called *positive* if all its rule do not contain negative atoms, i.e., do not contain $not\ C_1, \cdots, not\ C_n$. By *grounding* of Π, denoted as $ground(\Pi)$, we mean a ground program in which each rule r of Π is replaced by the set of all ground rules obtaining from r in which each variable is substituted with a constant occurring in Π^2.

Now we define the answer set semantics of logic programs. Let Π be a positive ground program. A set S of ground atoms is called an *answer set* of Π if (1) for each rule of the form: $A \leftarrow B_1, \cdots, B_m$, $\{B_1, \cdots, B_m\} \subseteq S$ implies $A \in S$; and (2) S is a minimal set in terms of set inclusion that satisfies condition (1). Now we consider a ground program Π. For a given set S of ground atoms, the positive program Π^S is obtained via the following transformation: (1) deleting each rule from Π if a negative atom $not\ C$ occurs in the rule and $C \in S$; (2) for the rest of the rules in Π, deleting all negative atoms from the rules. Obviously, Π^S is a ground positive program. Then S is called an *answer set* of Π if it is an answer set of Π^S. For an arbitrary logic program Π that may contain variables in its rules, a set S of ground atoms is an *answer set* of Π if S is an answer set of $ground(\Pi)^S$.

3.2 The Translation

Now we present a translation from SMP to ASP. For doing this, we first need to formally define the reasoning problem under SMP framework.

Definition 1. *A SMP frame is a pair* $\mathbf{F} = (\mathcal{P}, \mathbf{B}_0)$*, where* $\mathcal{P}$ *is a policy, i.e., a finite set of well-formed rules, and* $\mathbf{B}_0$ *is an initial authorization state. Let* $\phi = Q_1(\overline{a_1}) \wedge \cdots \wedge Q_k(\overline{a_k})$ *be a ground goal formula. We say that* ϕ *is entailed by frame* $\mathbf{F}$*, denoted* $\mathbf{F} \models_{\mathrm{SMP}} \phi$*, iff there exists a sequence of authorization states* $\overline{\mathbf{B}} = \mathbf{B}_0 \cdots \mathbf{B}_n$ *such that* $\overline{\mathbf{B}} \models_{\mathcal{P}} \phi$*.*

Our basic idea of this translation is depicted as follows. For a given SMP frame $\mathbf{F}$ and a ground formula ϕ, we will construct a logic program Π, which encodes the policy $\mathcal{P}$, authorization state $\mathbf{B}_0$ and formula ϕ, such that behaviours of the underlying authorization state updates and reasoning in SMP are precisely simulated by program Π, and the resulting sequence of authorization states is generated by the corresponding answer sets of Π. It is important to note that here we only consider *bounded* SMP reasoning, in the sense that for any ground goal formula ϕ, $\mathbf{F} \models_{\mathrm{SMP}} \phi$ iff for the underlying sequence $\overline{\mathbf{B}}$ of authorization states, we have $|\overline{\mathbf{B}}| < t_{max}$ for some fixed integer $t_{max} \in \mathbb{N}$. This requirement is needed because if without such restriction, the translated ASP program $\Pi(\mathbf{F}, \phi)$ will not be finitely groundable, and consequently, we will not be able to apply existing ASP solvers to computing the answer sets of such program.

Definition 2. *Let* $\mathbf{F} = (\mathcal{P}, \mathbf{B}_0)$ *be a SMP frame and* ϕ *a ground SMP formula. For the SMP language of* $\mathbf{F}$ *and* ϕ*, we define a logic programming language as follows: for each n-ary extensional or intensional predicate symbol* P *in the*

---

^{2} If Π contains no constant, then we simply introduce a new constant in the language of Π.

SMP language of $\mathbf{F}$ and ϕ, we introduce two corresponding $(n+1)$-ary predicate symbols $\hat{P}$ and $\hat{P}_-$, where their first n arguments are the same as P's, and the $(n+1)^{th}$ arguments will be used to reference the authorization states. We also introduce a unary predicate symbol time to represent time. Then we specify the following groups of rules:

1. Firstly, there is the following rule:

$$time(0..t_{max}) \leftarrow, \qquad (4)$$

   where t_{max} is an integer representing the upper bound of the sequence of authorization states during the reasoning;

2. For each fact $P(\overline{a})$ in $\mathbf{B}_0$ (here $\overline{a}$ is a ground term consisting of a tuple of constants in the underlying SMP language), there is a rule of the form

$$\hat{P}(\overline{a}, 0) \leftarrow, \qquad (5)$$

3. For each ground extensional atom $P(\overline{a})$ not in $\mathbf{B}_0$, there is a rule of the form

$$\hat{P}_-(\overline{a}, 0) \leftarrow, \qquad (6)$$

4. For each rule in $\mathcal{P}$ of the form:

$$P_{int}(\overline{X}) \leftarrow L_1(\overline{X}_1) \wedge \cdots \wedge L_m(\overline{X}_m),$$

   there is a rule of the form

$$\hat{P_{int}}(\overline{X}, T) \leftarrow [\hat{L}_1(\overline{X}_1, T)|, \cdots, [\hat{L}_m(\overline{X}_m, T)], time(T), T < t_{max}, \qquad (7)$$

   where
$$[\hat{L}(\overline{X}, T)] = \begin{cases} \hat{P}(\overline{X}, T), & if\ L(\overline{X}) = P(\overline{X}), \\ \hat{P}_-(\overline{X}, T), & if\ L(\overline{X}) = \neg P(\overline{X}); \end{cases}$$

5. For each command rule in $\mathcal{P}$:

$$P_{cmd}(\overline{X}) \leftarrow L_1(\overline{X}_1) \wedge \cdots \wedge L_m(\overline{X}_m) \otimes K_1 \otimes \cdots \otimes K_n,$$

   there is a set of rules as follows:

$$\hat{P_{cmd}}(\overline{X}, T) \leftarrow [\hat{L}_1(\overline{X}_1, T)], \cdots, [\hat{L}_m(\overline{X}_m, T)], time(T), T < t_{max}, \qquad (8)$$

$$\{\hat{K}_i(\overline{X}, T+1)\} \leftarrow \hat{P_{cmd}}(\overline{X}, T), time(T), T < t_{max}, (1 \leq i \leq n) \qquad (9)$$

   where
$$\{\hat{K}(\overline{X}, T)\} = \begin{cases} \hat{P}(\overline{X}, T), & if\ K(\overline{X}) = +P(\overline{X}), \\ \hat{P}_-(\overline{X}, T), & if\ K(\overline{X}) = -P(\overline{X}); \end{cases}$$

6. *For each extensional or intensional predicate symbol P occurring in $\mathbf{F}$ where P is not a command predicate, there are two following rules:*

$$\hat{P}(\overline{X}, T+1) \leftarrow \hat{P}(\overline{X}, T), not\ \hat{P}_-(\overline{X}, T+1), time(T), T < t_{max}, \quad (10)$$

$$\hat{P}_-(\overline{X}, T+1) \leftarrow \hat{P}_-(\overline{X}, T), not\ \hat{P}(\overline{X}, T+1), time(T), T < t_{max}, \quad (11)$$

7. *Finally, for $\phi = Q_1(\overline{a}_1) \wedge \cdots \wedge Q_k(\overline{a}_k)$, there are following rules (constraints):*

$$\leftarrow not\ \hat{Q}_i(\overline{a}_i, t_{max}), \ where\ (1 \le i \le k). \quad (12)$$

We define a logic program $\Pi(\mathbf{F}, \phi)$ to be the set of all rules specified from (4) to (12), which we call the ASP translation *of $\mathbf{F}$ and ϕ.*

Now let us take a closer look at Definition 2. Firstly, rule (4) sets the boundary. In other words, the steps of authorization state changes will be bounded by this constant t_{max}. Rules (5) and (6) encoded the initial authorization state $\mathbf{B}_0$. Then rules (7), (8), and (9) are the translations of SMP rules including both general rules with intensional predicates in heads and command rules. Here for rule (9), the time argument in the rule head represents the *next state* after the command (action) is executed. Rules (10) and (11) are so-called *inertia rules* which represents those persistent facts not affected by the command execution. Finally, rule (12) encodes the SMP ground goal formula ϕ.

The following example shows how such an ASP translation handles the reasoning about the state change of authorization policies, as SMP does.

Example 3. We consider a frame $\mathbf{F} = (\mathcal{P}, \mathbf{B}_0)$, where the policy $\mathcal{P}$ is as shown in example 1 and let $\mathbf{B}_0 = \emptyset$. Furthermore, we specify $\phi = played1(tom, star\_war) \wedge played2(tom, star\_war)$. Now by setting the bound to be $time(0..4)$, according to Definition 2, we have the following ASP translation $\Pi(\mathbf{F}, \phi)$ as follows:

$r_1 : time(0..4) \leftarrow,$

$r_2 : bought\_(tom, star\_war, 0) \leftarrow,$

$r_3 : played1\_(tom, star\_war, 0) \leftarrow,$

$r_4 : played2\_(tom, star\_war, 0) \leftarrow,$

$r_5 : buy(X, M, T) \leftarrow time(T), T < 4,$

$r_6 : bought(X, M, T+1) \leftarrow buy(X, M, T), time(T), T < 4,$

$r_7 : play1(X, M, T) \leftarrow bought(X, M, T), played1\_(X, M, T),$
$\qquad\qquad\qquad time(T), T < 4,$

$r_8 : played1(X, M, T+1) \leftarrow play1(X, M, T), time(T), T < 4,$

$r_9 : play2(X, M, T) \leftarrow played1(X, M, T), played2\_(X, M, T),$
$\qquad\qquad\qquad time(T), T < 4,$

$r_{10} : played2(X, M, T+1) \leftarrow play2(X, M, T), time(T), T < 4,$

$r_{11} : bought(X, M, T+1) \leftarrow bought(X, M, T),$
$\qquad\qquad\qquad not\ bought\_(X, M, T+1), time(T), T < 4,$

$r_{12} : bought\_(X, M, T + 1) \leftarrow bought\_(X, M, T),$
$\qquad\qquad not\ bought(X, M, T + 1), time(T), T < 4,$
$r_{13} : played1(X, M, T + 1) \leftarrow played1(X, M, T),$
$\qquad\qquad not\ played1\_(X, M, T + 1), time(T), T < 4,$
$r_{14} : played1\_(X, M, T + 1) \leftarrow played1\_(X, M, T),$
$\qquad\qquad not\ played1(X, M, T + 1), time(T), T < 4,$
$r_{15} : played2(X, M, T + 1) \leftarrow played2(X, M, T),$
$\qquad\qquad not\ played2\_(X, M, T + 1), time(T), T < 4,$
$r_{16} : played2\_(X, M, T + 1) \leftarrow played2\_(X, M, T),$
$\qquad\qquad not\ played2(X, M, T+1), time(T), T < 4.$ □

3.3   Properties

The following theorem simply states that our ASP translation of SMP is sound and complete.

Theorem 1. *Let* $\mathbf{F} = (\mathcal{P}, \mathbf{B_0})$ *be a SMP frame and* ϕ *a ground goal formula. Suppose for any ground goal formula* ϕ, $\mathbf{F} \models_{\text{SMP}} \phi$ *implies that the length of the underlying sequence of authorization states is bounded by an integer* t_{max}, *i.e.,* $|\overline{\mathbf{B}}| < t_{max}$. *Then* $\mathbf{F} \models_{\text{SMP}} \phi$ *iff* $\Pi(\Pi, \phi)$ *has an answer set.*

Recall that under SMP framework, not only a sequence of authorization states will be generated, but also a sequence of commands is discovered which interprets how an authorization state is changed caused by a command, and eventually to achieve the user's request ϕ. We will show that in the translation $\Pi(\mathbf{F}, \phi)$, such important information of the underlying command sequence will also be possibly extracted from the answer sets of $\Pi(\mathbf{F}, \phi)$. The following theorem provides a result on this aspect.

Theorem 2. *Let* $\mathbf{F} = (\mathcal{P}, \mathbf{B_0})$ *be a SMP frame,* ϕ *a ground goal formula such that* $\mathbf{F} \models_{\text{SMP}} \phi$, $\overline{\mathbf{B}} = \mathbf{B_0} \cdots \mathbf{B_n}$ *(*$0 < n < t_{max}$*) a sequence of authorization states satisfying* $\overline{\mathbf{B}} \models_{\mathcal{P}} \phi$, *and* $\hat{Q}_{cmd} = Q_{cmd_1}(\overline{a_1}) \otimes \cdots \otimes Q_{cmd_l}(\overline{a_l})$ *a sequence of ground command atoms. Then we have*

$\overline{\mathbf{B}} \models_{\mathcal{P}} Q_{cmd_1}(\overline{a_1}) \otimes \cdots \otimes Q_{cmd_l}(\overline{a_l})$ *iff*
$\Pi(\mathbf{F}, \phi) \models \hat{Q}_{cmd_1}(\overline{a_1}, i) \wedge \cdots \wedge \hat{Q}_{cmd_l}(\overline{a_l}, i{+}l{-}1)$

for some $0 \leq i \leq (n - l + 1)$.

Example 4. (Example 3 continued). By computing $\Pi(\mathbf{F}, \phi)$'s answer set, we will show that $\Pi(\mathbf{F}, \phi)$ not only encodes the optimal reasoning, but also precisely captures the state changes with respect to $\mathbf{F} \models_{\text{SMP}} \phi$. Firstly, from SMP semantics, we are able to obtain a sequence of authorization states as follows: $\mathbf{B_0} \cdot \mathbf{B_1} \cdot \mathbf{B_2} \cdot \mathbf{B_3}$, where

$\mathbf{B}_0 = \emptyset$,
$\mathbf{B}_1 = \{bought(tom, star\_war)\}$,
$\mathbf{B}_2 = \{bought(tom, star\_war), played1(tom, star\_war)\}$, and
$\mathbf{B}_2 = \{bought(tom, star\_war), played1(tom, star\_war), played2(tom, star\_war)\}$.

Then can verify that

$\Pi(\mathbf{F}, \phi) \models bought(tom, star\_war, 3) \wedge played1(tom, star\_war, 3) \wedge played2(tom, star\_war, 3)$,
$\Pi(\mathbf{F}, \phi) \not\models bought(tom, star\_war, t) \wedge played1(tom, star\_war, t) \wedge played2(tom, star\_war, t)$
for all $t < 3$, although we do have
$\Pi(\mathbf{F}, \phi) \models bought(tom, star\_war, 1)$,
$\Pi(\mathbf{F}, \phi) \models played1(tom, star\_war, 2)$.

This means that $\Pi(\mathbf{F}, \phi)$ represents the optimal reasoning with respect to $\mathbf{F} \models_{SMP} \phi$. On the other hand, since we also have

$\mathbf{B}_0 \cdot \mathbf{B}_1 \cdot \mathbf{B}_2 \cdot \mathbf{B}_3 \models_{\mathcal{P}} buy(tom, star\_war) \otimes play1(tom, star\_war) \otimes play1(tom, star\_war)$,

then it can be checked that

$\Pi(\mathbf{F}, \phi) \models buy(tom, star\_war, 0) \wedge play1(tom, star\_war, 1) \wedge play2(tom, star\_war, 2)$,

which precisely reflects the state changes as represented via $\mathbf{B}_0 \cdot \mathbf{B}_1 \cdot \mathbf{B}_2 \cdot \mathbf{B}_3$. □

4   Concluding Remarks

In this paper, we have proposed a translation from SMP logic into ASP. We have showed that such ASP translation preserves the bounded SMP reasoning. We further implemented a system prototype that performs effective reasoning about state change of authorization policies.

Our work may be further extended in two major aspects. Firstly, for the sake of simplicity, our current ASP translation did not address the SMP *state constraints* that are used to a class of authorization states, i.e., the policy writer may like to know whether there exists a sequence of commands that leads to states satisfying (or not satisfying) particular state constraints. In fact, this ability may be easily represented in our translated ASP programs because state constraints may be naturally interpreted as a set of constraint rules. Secondly, since ASP with aggregations has been extensively studied in recent years, and ASP solvers augmented with aggregations have also been developed, it will be feasible to add aggregations into authorization specification and reasoning so that our framework is able to handle complex aggregations associating to various authorization rules and command rules. Similarly, other functionalities of ASP such as disjunctive information, negations and nonmonotonicity can also be added into our framework.

References

1. Baral, C.: Knowledge Representation, Reasoning and Declarative Problem Solving. MIT (2003)
2. Abiteboul, S., Hull, R., Vianu, V.: Foundations of Databases. Addison-Wesley Publishing (1995)
3. Becker, M.-Y., Nanz, S.: A logic for state-modifying authorization policies. ACM Transactions on Information System Security **13** (2010)
4. Calimer, F., Ianni, G., Ricca, F.: The third answer set programming system computation. Theory and Practice of Logic Programming (2012)
5. Dhia, I.B.: Access control in social networks: a reachability-based approach. In: Proceedings of the 2012 Joint EDBT/ICDT Workshops (EDBT-ICDT 2012), pp. 227–232 (2012)
6. Dimoulas, C., Moore, S., Askarov, A., Chong, S.: Declarative policies for capability control. In: Proceedings of CSF-2014 (2014)
7. Gebser, M., Kaminski, R., Kaufmann, B., Schaub, T., Schneider, M.T., Ziller, S.: A portfolio solver for answer set programming: preliminary report. In: Delgrande, J.P., Faber, W. (eds.) LPNMR 2011. LNCS, vol. 6645, pp. 352–357. Springer, Heidelberg (2011)
8. Gebser, M., Kaufmann, B., Neumann, A., Schaub, T.: Conflict-driven answer set solving: From theory to practice. Artificial Intelligence **187–188**, 52–89 (2012)
9. Grasso, G., Leone, N., Ricca, F.: Answer set programming: language, applications and development tools. In: Faber, W., Lembo, D. (eds.) RR 2013. LNCS, vol. 7994, pp. 19–34. Springer, Heidelberg (2013)
10. Hinrichs, T., Martinoia, D., Garrison, W.C., Lee, A., Panebianco, A., Zuck, L.: Application-sensitive access control evaluation using parameterized expressiveness. In: Proceedings of CSF-2013, pp. 145–160 (2013)
11. Lierler, Y., Lifschitz, V.: One more decidable class of finitely ground programs. In: Hill, P.M., Warren, D.S. (eds.) ICLP 2009. LNCS, vol. 5649, pp. 489–493. Springer, Heidelberg (2009)

Integration of Disease Entries Across OMIM, Orphanet, and a Proprietary Knowledge Base

Maori Ito[1], Shin'ichi Nakagawa[2], Kenji Mizuguchi[1], and Takashi Okumura[3(✉)]

[1] National Institute of Biomedical Innovation, 7-6-8 Saito-Asagi, Ibaraki-City,
Osaka 567-0085, Japan
[2] Research Institute of Info-Communication Medicine, 5-17-11-106 Honcho,
Koganei city, Tokyo 184-0004, Japan
[3] National Institute of Public Health, 2-3-6 Minami, Wako-shi,
Saitama 351-0197, Japan
taka@niph.go.jp

Abstract. Integration of disease databases benefits physicians searching for disease information. However, current algorithmic matching is not sufficiently powerful to automate the integration process. This paper reports our attempt to manually integrate disease entries spread across public disease databases, Online Mendelian Inheritance in Man and Orphanet, with a proprietary disease knowledge base. During the process, we identified that relations between synonyms require special handling, and a set of resolution rules are proposed. Situations encountered throughout the integration suggested that variations in the cross-references would facilitate future integration of distinct disease databases.

Keywords: Disease knowledge base · Semantic integration · Ontology alignment

1 Introduction

For the treatment of hard-to-diagnose cases, physicians need to collect various information of possible diagnosis from a variety of sources. However, this information, particularly for rare diseases, is stored across multiple databases. In order to mitigate the burden of performing the survey needed to find the information, a comprehensive disease knowledge source that covers the entire knowledge of known diseases is desirable. In reality, however, physicians are forced to search over different databases: for medical literature, PubMed [11]; for genetic disorders, Online Mendelian Inheritance in Man (OMIM) [8]; for rare diseases, Orphanet [7] and Google [16]. To eliminate this inefficiency, researchers are investigating technology for *cross-searching* databases that virtually integrates the independent databases [10].

The essential process of such a technology is the integration of entries across different databases. If a Disease A is included in both database α and database β, as Disease A and Disease X, the process needs to identify these entries as such

© Springer International Publishing Switzerland 2015
M. Ali et al. (Eds.): IEA/AIE 2015, LNAI 9101, pp. 120–130, 2015.
DOI: 10.1007/978-3-319-19066-2_12

to allow their integration. Although this process may appear simple, it involves various complexities, for example, database β may include only subtypes of X, and not X, or only Disease XYZ that consolidates Diseases X, Y and Z. The complexity of the process escalates as the number of similar diseases and target databases increases.

In the field of the knowledge processing, such techniques as *semantic integration* [5,13] and *ontology alignment* [6] have been studied. However, the interpretation of disease concepts requires highly-paid domain experts having a medical background, and thus, only a few research studies have addressed the alignment task of disease concepts [15]. Accordingly, this paper presents our case report to integrate disease entries located in different databases toward a complete disease knowledge base [14], in the aim of clarifying the issues that need to be considered to align entries in distinct disease databases.

The paper is organized as follows. First, in Section 2, related work is summarized. Section 3 introduces the materials, and the integration process is outlined. The actual integration processing is presented in Section 4 and Section 5 for our original database against public databases and for public databases with cross-reference information, respectively. Section 6 discusses the implication of the task, and Section 7 concludes the paper.

2   Related Work

The integration of databases has been investigated in more general contexts, using various methods, such as heuristic algorithms and machine learning techniques [13]. In life sciences, the integration of databases has been attempted with a structure, called the Resource Description Framework (RDF), more recently. This framework represents the relationship of data, using the Uniform Resource Identifier (URI), allowing integration of web resources. Examples include Bio2RDF [1] and Chem2Bio2RDF [4]. Integration through the RDF mechanism is straightforward, in that resources pointing to the same URI are virtually unified [5]. However, the excessive flexibility may also burden the system developers.

For the mapping of disease entries across distinct databases and vocabularies, several efforts have been made. In [9], an attempt was made to map the Orphanet thesaurus to MeSH [12]. The results showed that the string-match approach outperforms approaches that utilize only Unified Medical Language System (UMLS) [2] and ICD references. Likewise, in reference [3], the Orphanet terminology was mapped into standardized terminology, and 94.6% precision was achieved, with the help of human inspection. In [15], the authors attempted to map disease entries between databases, using links to UMLS terminology from these databases, to find unknown relationships through the mapped UMLS terms that are common to them. The advantage of their proposal is that the concept hierarchy of UMLS is used to detect *indirect* relationships.

Unfortunately, however, these methodologies are not sufficiently powerful to completely automate the integration, even with the cross-references. Accordingly, we were forced to manually integrate the databases, drawing lessons to align entries in distinct disease databases.

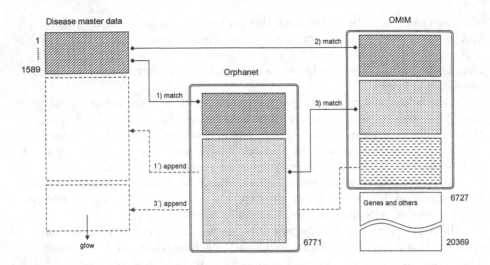

Fig. 1. Overview of the integration process

3   Materials and Methods

Simple unification to integrate disease databases would result in redundant disease entries, if identical diseases exist in multiple databases. Accordingly, methodical processing to identify overlapping of entries is the key here. From a microscopic point of view, this constitutes determining whether two entries are distinct or identical. However, the resulting entries will differ if multiple databases are integrated in different orders. Accordingly, from a macroscopic point of view, the order in which the integration processing is performed is also important.

Figure 1 outlines the process for the integration task. The disease master data constitutes the list of diseases included in a knowledge base we have been developing. These data include common diseases and other medically important diseases, amounting to 1,589 diseases. To improve the completeness of the knowledge base, we attempted to incorporate in the knowledge base an extensive collection of rare diseases in public databases.

Orphanet is a rare disease database, developed in Europe [7]. We obtained a snapshot of the database in March 2013, which contained descriptions of 6,771 diseases, in free-text format. OMIM is a database for genetic disorders, which includes records for diseases, as well as for genes and other related information [8]. In this study, we utilized 6,727 records for diseases out of 20,369 OMIM records, excluding records for genes and others information. These OMIM records also contain descriptions of diseases in free-text format. An Orphanet record may have a different number of cross-references to OMIM, which may vary from zero to a maximu of 73. Of the 6,771 Orphanet records, 3,913 included such cross-references to OMIM entries and 5,361 included the standardized disease codes of the International Classification of Diseases (ICD) [17].

The first step of the integration process is to perform matching between the original list and Orphanet (1 in Figure 1). The matched entries are overlaps with the original entries, and thus, are discarded, leaving the unmatched entries as candidates for future integration (1′ in Figure 1). Likewise, the matching between the original and OMIM is performed and the matched entries are discarded (2 in Figure 1). Finally, the unmatched entries in Orphanet and OMIM are matched against each other (3 in Figure 1). The matched entries are considered to be represented by Orphanet, and thus, the corresponding entries in OMIM were discarded, leaving the unmatched ones as candidates for future integration with the original list (3′ in Figure 1). In the following two sections, the processing is described in detail.

4   Matching of Original List Against OMIM/Orphanet

The matching of the original list against the public databases involved in this order: i) preprocessing; ii) algorithmic matching; iii) human inspection; and iv) synonym processing. The following subsections present each of the processes.

4.1   Preprocessing and Algorithmic Matching

Although the diseases in the original disease master data have quite simple names, OMIM and Orphanet use unique conventions for formulating their disease names and their synonyms. Accordingly, we normalized the names before the actual matching process to remove the differences in the naming conventions.

The normalization process comprised: i) removal of the definite article (*the*); ii) removal of hyphens and slashes; iii) removal of umlauts and other symbols; iv) modification of genitive cases (−'s); v) replacement of specific terms (*syndrome*, *disease*, and *infection*) with a special token; vi) replacement of Roman numerals with Arabic numerals; and vii) capitalization.

Utilizing the normalized disease names and the ICD codes of each entry, the process of matching entries across databases was performed. Then, according to the matching information found using two search keys, the relations between entries were classified into i) exact matching in the name strings (Exact-match), ii) similarity in the names and the ICD coding (Almost-exact-match), or iii) similarity in the names or in the ICD coding (Partial-match).

4.2   Human Inspection

The Almost-exact-match and Partial-match classes necessitated a significant amount of human intervention to avoid erroneous relations. Even after the normalization process, the following points had to be taken into consideration for name matching. First, inversion of terms is frequently found in the names, e.g., *acute intermittent porphyria* may be entered as *porphyria, acute intermittent* (OMIM 176000). Second, the names sometimes included unnatural modifiers, e.g., *sarcoidosis, susceptibility to, 1* (OMIM 181000). Third, the names include

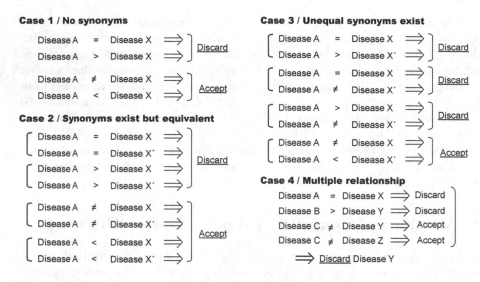

Fig. 2. Processing disease relationships

unexpected numbers, e.g. *amyotrophic lateral sclerosis 9* (OMIM 611895), which indicates genetic variants or other disease causes. Accordingly, relations in the Almost-exact-match and the Partial-match classes were manually inspected by physicians.

The purpose of the human inspection was to label the pairs in the Almost-exact-match class and Partial-match class as *match, no match,* or *don't know.* However, the trial session revealed many irregular cases. The *don't know* class was found to have two subclasses: "don't know the disease concepts", and "don't know the relationship". The former class may require a reference, whereas the latter requires that a literature survey be performed. Accordingly, we added another class, *give up,* to facilitate this labeling task by leaving the decision for the second round of inspection performed by another physician. Likewise, *the left hand side includes the right* labels and *the right hand side includes the left* labels were also added to facilitate the inspection process.

4.3   Processing of Synonyms

The results of the labeling task are used to classify all the candidate pairs into match ($=$), no match ($\neq$), *the left hand side includes the right hand side* ($>$), or *the right hand side includes the left hand side* ($<$). According to the classification, we attempted to decide which pair should be appended to the knowledge base. However, in this process, we found the processing of synonyms problematic. Suppose that Disease A has a synonym, Disease A'. In this case, the matching with a Disease X in another database causes two pairs, Disease A against Disease X and Disease A' against Disease X. The issue is simple if both pairs have the same label. However, in reality, synonyms can have different labels, and the issue is more complicated because a disease may have multiple synonyms.

Accordingly, we introduced rules to aggregate the synonym pairs and to decide whether to accept or discard Disease X (Figure 2). First, we classified the four labels into two major categories: Accept ($<$, $\neq$) and Discard ($>$, $=$). The former labels were accepted because the right hand side diseases were not included in the left hand side diseases. The latter labels were discarded because these diseases were included in the left hand side diseases. In the simplest case scenario (Case 1), where there were no synonyms, these rules were simply applied to the disease pairs. Even if synonyms existed, the decision still remained simple, provided that the synonyms were identical (Case 2).

Different decisions about synonyms caused conflicts between the rules for accepting or discarding. Accordingly, as shown in the examples (Case 3), we gave priority to *discard* over *accept*. If a disease pair with a synonym included a *discard* match, the disease on the right hand side was discarded, because it was included in a disease listed on the left hand side.

Finally, cases where a disease had a relationship with several diseases (Case 4) needed to be considered. In such cases, the disease on the right hand side was discarded, if a disease pair included at least one discard match. In the example shown, only Disease Z was accepted, because Disease A is identical to Disease X and Disease Y is included in Disease B.

4.4   Results and Analysis

These processes finally identified 887 diseases in Orphanet (1 in Figure 1) that were identical to diseases in our master data, leaving 5,884 diseases for future integration (1' in Figure 1). For OMIM diseases against the disease master data, the processing identified 716 matches (2 in Figure 1), leaving 6,011 diseases for matching with Orphanet (3 in Figure 1).

As discussed, diseases may take either one of the four major relations. "Equal ($=$)" is the simplest relation, which implies that two diseases are identical, or one is a synonym of the other. An example is *Kawasaki Disease* and *MucoCutaneous Lymph-node Syndrome (MCLS)*. "Unrelated ($\neq$)" is a relation between diseases that show no similarity from the medical point of view. In this case, the right-hand side disease is safely appended to the left hand side.

"Inclusive relation ($>$)" applies to a pair such as *lung cancer* and *small cell lung cancer*. If the more general term is listed in the left hand side list, the variants on the right hand side can be discarded. However, if the right-hand side term is more generic, for example, "hereditary breast and ovarian cancer syndrome" and "breast cancer", the decision may vary. Because we defined the processing rules to avoid overlapping between databases wherever possible, we discarded the right hand side term in this case. However, other researches may choose a different strategy, depending on the objective of the integration.

A debatable case is an example like $DiseaseA = DiseaseX$ and $DiseaseA < DiseaseX'$. This case implies that Disease X has a broader meaning than Disease A. In such a case, our rule excludes Disease X, because of the first pair. However, if the objective of the integration is to accept more diseases, Disease X might be accepted because the latter pair assures that Disease X has a broader meaning.

Table 1. Cross-references between Orphanet and OMIM

	With Cross-reference	Without Cross-reference	Total
Orphanet	3685	3086	6771
OMIM	4504	2223	6727

Table 2. Results of the integration with disease master data

	Discard	Accept	Total
Orphanet	887	5884	6771
OMIM	4599	2128	6727

During the classification task, we encountered various cases. Medical knowledge may be required to decide in the case, "Cushing's syndrome" against "Cushing disease, pituitary". The cases "Myasthenia Gravis" against "Myasthenia Gravis, limbgirdle" and "Neuralgic amyotrophy" against "amyotrophy, hereditary neuralgic" were labeled ">". The case "Neuroblastoma" against "Neuroblastoma with Hirschsprung disease" was labeled "≠". We also found a case, "Malaria" against "Malaria, susceptibility to", which was labeled "≠". These cases are mostly caused by the characteristic objectives of OMIM and Orphanet, as databases for specific purposes.

5   Matching of Orphanet Against OMIM

Because OMIM and Orphanet include a large number of diseases, the matching of OMIM and Orphanet involves many disease pairs, more than ten thousand. Even if we discard the diseases that are matched in the disease master data, the human inspection cost remains prohibitive. Accordingly, for the processing of OMIM and Orphanet matching, we attempted an automated matching of diseases using the cross-reference data between them, without applying the synonym processing proposed above.

5.1   Preprocessing of Cross-References

As described in Section 3, cross-reference data are available only for Orphanet entries. Accordingly, we obtained the Orphanet data in XML (Extensible Markup Language) at the website, and converted the data to infer the reverse relationships from the Orphanet data. In the XML file, the Orphanet ID is associated with IDs in an external database. In this preprocessing, non-disease entries are excluded, such as gene information.

The resulting number of cross references between OMIM and Orphanet is shown in Table 1. The difference in the cross-references between Orphanet and OMIM is caused by the nature of many-to-many mappings. The results show that more than half of Orphanet IDs are mapped to OMIM IDs, and almost two-thirds of OMIM IDs are mapped to Orphanet IDs.

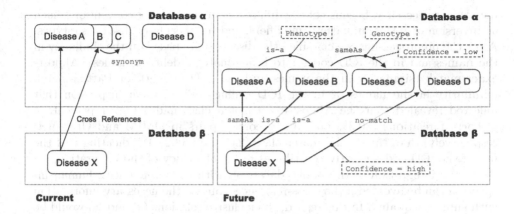

Fig. 3. Disease records and reference types

5.2 Results of Matching and Analysis

The results of the cross-reference matching are shown in Table 2. In Orphanet, we found 887 diseases matching entries in the disease master data (1 in Figure 1), leaving 5884 diseases as no match, which will be integrated into the disease master data (1′ in Figure 1). In OMIM, we found 4599 diseases that account for 2 and 3 in Figure 1, leaving 2128 diseases that will be integrated into the disease master data.

This result strongly depends on the cross-references in the Orphanet database against OMIM, and the result may change for several possible reasons. First, the cross-references may have incorrect relations that the human inspector may correct. It is noteworthy that the references can be incomplete, because it is costly to consistently maintain cross-references of large databases [15]. Second, there can be indirect relations between disease entries, and these relations can be used to detect hidden relations of equivalent diseases that will be removed in the integration. Finally, the number of accepts would change if the order of the integration was different. Suppose that Disease A, B, and C exist in Databases α, β, and γ, respectively, which have inclusive relations, $A > B > C$. Integration in the order $B > C$ and $A > B$ would result in a single disease entry for A. However, the order $A > B$ and $A \neq C$ would yield Disease A and C. The order in which the databases are integrated may also cause the order $B < A$ and $C < B$, which yields Disease A and B, even when applying the same integration rule as described in Section 4.

6 Discussion

The integration of disease databases in this study involved matching with and without cross-references. For the integration of the public databases against a proprietary knowledge base, we performed the matching with disease names and ICD codes, without cross-references. In this process, the normalization of the names was necessary, because of the differences in the naming conventions, such as inversion of terms, unnatural modifiers, and numbering of disease subtypes. After the preprocessing, we classified the disease pairs based on the similarity in the names and in the ICD codes into Exact-match (identical names), Almost-exact-match (similarity in the names and the ICD coding), or Partial-match (similarity in the names or in the ICD coding). The human inspection that followed revealed that the Almost-exact-match class had a higher percentage of correct relations (80.3% and 70.8% correct pairs in OMIM and Orphanet, respectively) than the Partial-match class (2.6% and 38.2%), indicating that the use of the ICD codes greatly contributes to the efficiency of the integration.

In the integration of the public databases with cross-references, human inspection can be avoided if the cross-references contain the necessary information with sufficient quality. In this regard, the inclusive relations ($<$ and $>$) would be the vital parts, particularly for relations across distinct databases. Nevertheless, such a relation is implicitly expressed with one-to-many mapping and synonym relations in the current settings (Figure 3–Current), and thus, there should be a formal method of expressing the relation, such as by an is-a and sameAs relationship across databases (Figure 3–Future). Likewise, because the nature of the human inspection is that it excludes the misleading no match ($\neq$) cases in the list of candidate pairs suggested by the algorithms, an external database for no match relations would further contribute to the efficiency of the integration.

Other possible relations include the variants of equivalency ($=$), such as similarity in the phenotype and genotype. In our identification task, we prioritized similarity in the phenotype over that in the genotype, because the resulting knowledge base targets clinical applications. However, to produce a database for a genetic research study a different approach would be taken. For example, a clinical knowledge base may not differentiate *familial Alzheimer's disease* and *Alzheimer's disease*, but a genetic database would certainly discriminate the two. In the case shown in Figure 3, Disease A and X can be *Alzheimer's disease* and Disease B a subtype of *Alzheimer's disease*, such as *familial Alzheimer's disease*. Disease C can be *early onset familial Alzheimer's disease*, which is also a variant of *Alzheimer's disease* and *may* have the same genetic background as Disease B. Disease D can be another type of dementia, such as vascular dementia, the etiology of which is completely different.

The integration of the databases necessitates that a combination of the suggested approaches be applied. This procedures involves processing various relations with different qualities. Accordingly, meta-knowledge of the relations, such as the confidence level, may facilitate the integration tasks. In the case of Figure 3–Future, the confidence level for the relation between Disease B and C is kept low, while the relation between Disease X and D is high. Using all the information available, the integration of Databases A and B is now automated. Toward the development of clinical applications, the unification processing would yield only Disease A/X and Disease D, and for the genetic database, the processing can choose Disease A/X, B, C, and D. It should be noted that Disease C can be omitted for a high-confidence database. By iteratively applying all the processing to each of the disease pairs, the suggested approach can be extended to integrate multiple databases.

7   Concluding Remarks

The ability to cross-search distinct disease databases would benefit physicians searching for abundant and detailed knowledge of diseases. However, because of the difference in objectives, distinct databases contain records of different granularity and contexts, the integration of which requires that various issues be considered.

Currently, algorithmic matching is not sufficiently powerful to automate the entire integration process, demanding human inspection by domain experts to decide whether terms are equivalent or not. Depending on the objective of the integration, the processing may prioritize matching by phenotypes, or by geno- type. In such a process, a variety of cross-reference types, other than simple similarity, would facilitate the integration of distinct databases and leverage the values of existing resources.

Acknowledgments. The authors sincerely thank Mr. Masanao Igarashi for perform- ing the data processing. Mr. Shinya Shimizu at the University of Tokyo contributed to the name matching algorithm, and Mr. Takao Kondo at Keio University helped us in the analysis of databases.

References

1. Belleau, F., Nolin, M.A., Tourigny, N., Rigault, P., Morissette, J.: Bio2RDF: towards a mashup to build bioinformatics knowledge systems. Journal of Biomed- ical Informatics **41**(5), 706–716 (2008)
2. Bodenreider, O.: The unified medical language system (UMLS): integrating biomedical terminology **32**, D267–D270 (2004)
3. Miličić Brandt, M., Rath, A., Devereau, A., Aymé, S.: Mapping orphanet termi- nology to UMLS. In: Peleg, M., Lavrač, N., Combi, C. (eds.) AIME 2011. LNCS, vol. 6747, pp. 194–203. Springer, Heidelberg (2011)
4. Chen, B., Dong, X., Jiao, D., Wang, H., Zhu, Q., Ding, Y., Wild, D.J.: Chem2Bio2RDF: a semantic framework for linking and data mining chemogenomic and systems chemical biology data. BMC bioinformatics **11**(1), 255 (2010)
5. Doan, A., Halevy, A.Y.: Semantic integration research in the database community: A brief survey. AI magazine **26**(1), 83 (2005)
6. Euzenat, J., Shvaiko, P., et al.: Ontology matching, vol. 18. Springer (2007)
7. INSERM SC11: Orphanet. http://www.orpha.net/
8. John Hopkins University: OMIM: Online Mendelian Inheritance in Man. http:// www.ncbi.nlm.nih.gov/omim
9. Merabti, T., Joubert, M., Lecroq, T., Rath, A., Darmoni, S.: Mapping biomedical terminologies using natural language processing tools and UMLS: mapping the Orphanet thesaurus to the MeSH. BioMedical Engineering and Research **31**(4), 221–225 (2010)

10. Morita, M., Igarashi, Y., Ito, M., Chen, Y.A., Nagao, C., Sakaguchi, Y., Sakate, R., Masui, T., Mizuguchi, K.: Sagace: A web-based search engine for biomedical databases in japan. BMC research notes **5**(1), 604 (2012)
11. National Center for Biotechnology Information, National Library of Medicine: PubMed. http://www.ncbi.nlm.nih.gov/pubmed/
12. National Library of Medicine: Medical Subject Headings. http://www.nlm.nih.gov/mesh/
13. Noy, N.F.: Semantic integration: a survey of ontology-based approaches. ACM SIGMOD Record **33**(4), 65–70 (2004)
14. Okumura, T., Tanaka, H., Omura, M., Ito, M., Nakagawa, S., Tateisi, Y.: Cost decisions in the development of disease knowledge base : A case study. In: 2014 International Workshop on Biomedical and Health Informatics, November 2014
15. Rance, B., Snyder, M., Lewis, J., Bodenreider, O.: Leveraging Terminological Resources for Mapping between Rare Disease Information Sources. Studies in health technology and informatics **192**, 529 (2013)
16. Tokuda, Y., Aoki, M., Kandpal, S.B., Tierney, L.M.: Caught in the web: E-diagnosis. Clinical Care Conundrums: Challenging Diagnoses in Hospital Medicine, pp. 169–175 (2013)
17. World Health Organization: ICD-10: International statistical classification of diseases and related health problems. World Health Organization (2011)

Sheet2RDF: A Flexible and Dynamic Spreadsheet Import&Lifting Framework for RDF

Manuel Fiorelli, Tiziano Lorenzetti, Maria Teresa Pazienza,
Armando Stellato[✉], and Andrea Turbati

ART Research Group, University of Rome,
Tor Vergata, Via del Politecnico, 1 00133 Rome, Italy
{fiorelli,pazienza,stellato,turbati}@info.uniroma2.it,
tiziano.lorenzetti@gmail.com

Abstract. In this paper, we introduce Sheet2RDF, a platform for the acquisition and transformation of spreadsheets into RDF datasets. Based on Apache UIMA and CODA, two wider-scoped frameworks respectively aimed at knowledge acquisition from unstructured information and RDF triplification, Sheet2RDF narrows down their capabilities in order to restrict the domain of acquisition to spreadsheets, thus taking into consideration their peculiarities and providing informed solutions facilitating the transformation process, while still exploiting their full potentialities. Sheet2RDF comes also bundled in the form of a plugin for two RDF management platforms: Semantic Turkey and VocBench. The integration with such platforms enhances the level of automatism in the process, thanks to a human-computer interface that can exploit suggestions by users and translate them into proper transformation rules. In addition, it strengthens this interaction by direct contact with the data/vocabularies edited in the platform.

Keywords: Human-computer interaction · Ontology engineering · Ontology population · Text analytics · UIMA

1 Introduction

As organizations and public institutions are exposing their data under permissive and open licenses, they often resolve to spreadsheets and other tabular formats to make data available. The motivations rely on the acquaintance of publishers with spreadsheet editors, and the easiness of producing and reading tabular data.

There is, however, a need for methodologies and systems that support publishers in lifting these tabular data to a semantically clear form, as it is enabled by representation languages and practices of the Semantic Web [1]. To satisfy this need, we propose Sheet2RDF, an integrated system that supports a streamlined process for the transformation and lifting of spreadsheets to RDF.

The paper is structured as follows. Section 2 surveys related systems and motivates our work. Section 3 describes the approach underlying Sheet2RDF combining a powerful transformation specification language with the automatic generation of skeletal

© Springer International Publishing Switzerland 2015
M. Ali et al. (Eds.): IEA/AIE 2015, LNAI 9101, pp. 131–140, 2015.
DOI: 10.1007/978-3-319-19066-2_13

specifications based on the recognition of known modeling patterns in the data. Section 4 presents the system architecture. Section 5 reports on some use cases of Sheet2RDF and the feedback of its users. In Section 6, we draw the conclusions.

2   Related Works

In this section, we briefly describe some of the systems and approaches for the triplication of spreadsheets, underlining their specificities. The systems may differ in their support to different data formats: e.g. Microsoft Office, Open Office, CSV, TSV. Obviously, spreadsheet editors provide a more convenient environment for manual editing, while plain-text serializations based on predefined delimiters often enable machine-to-machine data interchange. Some systems exploit additional metadata usually not available in plain-text serializations, thus requiring at least a further preprocessing step.

Any23 (https://any23.apache.org/dev-csv-extractor.html) implements a systematic transformation of CSV (and otherwise delimited value formats) into RDF. The transformation strategy assumes that the first row is a header, providing the properties, while subsequent rows describe individual resources, by providing in each cell the value for the property associated with the corresponding column. Any23 has limited choices for the representation of property values, either as URIs or as literals the type of which is inferred from the values themselves.

Systematic strategies for the conversion of tabular data produce a sort of raw RDF, which usually necessitate further processing to meet specific modeling patterns and improve their quality. The DataLift platform [2] uses this two-step approach, which initially turns various data formats (not only spreadsheets) into RDF, and then leverages SPARQL [3] Constructs to refactor the RDF data originally obtained.

Tarql (https://github.com/cygri/tarql) somehow conflates these two steps, by allowing the evaluation of SPARQL queries directly on CSV data. Tarql is a command line program, which provides no specific assistance for writing the SPARQL queries specifying the mapping.

The system csv2rdf4lod-automation [4] follows an iterative workflow, which begins with a raw transformation of CSV into RDF, and then proceeds with successive refinements, which are expressed in RDF through a dedicated conversion vocabulary. Available enhancements include: resource typing, adoption of user-supplied vocabularies, reification of property values, custom mapping of values to URIs, and the generation of more than one resource from each row. This system also lacks a dedicated user interface, especially to support the writing of mappings.

Spread2RDF (https://github.com/marcelotto/spread2rdf) uses an internal Ruby-based DSL to specify the transformation. This approach may be advantageous in terms of expressivity, while at same time forcing users to adopt a new syntax, sometimes really different from the ones they are acquainted with. Conversely, Tarql uses SPARQL, a standard and popular query language among Semantic Web practitioners, likely to be interested in this type of systems.

RDF123 [5] provides an application for writing mapping specifications, as well as a web service for executing them against spreadsheets. It moves away from the assumption that rows represent homogenous resources. Hence, it uses more complex mappings, which may possibly be conditioned to specific data contained in the rows.

TabLinker (https://github.com/Data2Semantics/TabLinker) does not commit on the existence of a predetermined structure in the spreadsheet, while leveraging annotations in the spreadsheet to properly interpret the data organization. It is specifically tailored to produce data conforming to the RDF Data Cube [6] vocabulary. TabLinker uses Open Annotation Core Data Model [7] and PROV [8] to add annotations and provenance information to the cells.

Spreadsheets and tabular data in general may be considered a somewhat limited form of relational data. As such, it is possible to leverage existing RDB to RDF converters. Sparqlify-CSV (http://aksw.org/Projects/Sparqlify.html), as an example, allows to load CSV files into Sparqlify, which is a middleware for viewing relational data as virtual RDF graphs.

The limitation of a completely automated conversion motivated the development of a system [9], which uses Semantic MediaWiki to crowdsource the generation of a mapping specification for Sparqlify-CSV.

LODRefine (https://github.com/sparkica/LODRefine) is a distribution of OpenRefine that includes several extensions related to Linked Open Data. Specifically, it complements the data cleaning features of OpenRefine with the ability to link values to entities from remote datasets. Moreover, from our viewpoint, it provides a graphical language to represent mappings from tabular data to RDF conforming to user provided vocabularies. The system seems not to have any mechanism for automatic generation of the mappings.

LODRefine is a standalone system to clean data and transform it to RDF. It is also possible to find import and transformation capabilities directly in some systems concerning with the actual management and development of RDF data. The thesaurus editor PoolParty is able to import Excel files (https://grips.semantic-web.at/display/POOLDOKU/Import+Excel+Taxonomies), which adhere to the one resource per row convention. In addition, PoolParty organizes spreadsheets, trying to visualize the conceptual taxonomy through the insertion of empty cells in the rows.

TopBraid Composer is an RDF development environment, which also offers a generic facility to import both conceptual and factual knowledge from a CSV file (http://www.topquadrant.com/composer/videos/tutorials/spreadsheets/import.html). A wizard-based user interface allows the customization of the import, by mapping to an existing ontology, or by providing additional information about the conceptual content, e.g. the range of the properties inferred by the column names.

The preceding review shows that the combination of expressiveness and convenience is quite uncommon among existing systems. If the system disburdens the user from heavy configuration, in many cases the reason is that the system implements a rigid transformation based on strict assumptions about the input data. Conversely, more capable systems rely on mapping specifications, which need to be completely specified by the user, often without the support of a dedicate user interface.

Sheet2RDF aims to fill this aspect, by combining a rich mapping specification language with an intelligent approach for the semi-automatic generation of skeletal mappings. Sheet2RDF follows the convention-over-configuration principle, when it tries to relieve the user from writing the mappings at hand, by automatically generating suitable transformations based on the recognition of known modeling patterns. Further user input supports the refinement of the generated mapping, possibly informed by an already existing   RDF dataset. However, the user may always resort to the underlying mapping specification language, if the spreadsheet requires a very specific transformation.

3   The Approach

The conversion of a spreadsheet to RDF may be considered an instance of the more general task of triplificating unstructured and semi-structured information. CODA [10,11] is one of the various systems supporting the achievement of this goal.

While such a system may be used as it is, we leverage the verticality of the domain (spreadsheets) to build a more focused layer on top of CODA. Fig. 1 represents the overall approach that Sheet2RDF implements to construct such an additional layer.

At the lower layer, CODA uses (in fact, it extends) UIMA [12] in order to analyze the input spreadsheet and, then, it executes a PEARL [13] document prescribing the transformation. The support of Sheet2RDF to deal with this setting is manifold. By first, Sheet2RDF defines a meta-type system that is then instantiated into a concrete

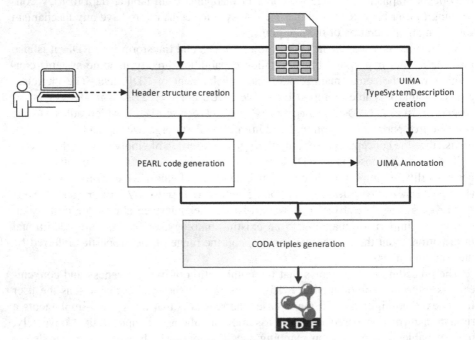

Fig. 1. Sheet2RDF approach

one for each spreadsheet. This type system models the data that will be read from the spreadsheet through a UIMA annotator provided by Sheet2RDF. At the same time, a mutable data structure mirroring the instantiated type system enables further customization from the user. In fact, user input requires a dedicated user interface, possibly providing visualizations based on already existing knowledge in the target semantic repository. Sheet2RDF uses this data structure, possibly adjusted by the user, to generate a skeleton of the PEARL transformation that lifts the spreadsheet to RDF. The user may refine this skeleton, until the PEARL document exactly matches the intended transformation. The generation of the skeleton is informed by a number of conventions that are looked for inside the data structure that reflects the spreadsheet header. A number of heuristics have been designed to match these conventions and refine the transformation rules. Ideally, the system should be able to generate a complete specification for a spreadsheet fully conforming to the foreseen conventions.

The use of an expressive transformation language guarantees the applicability to real-world scenarios, while at the same time facilities that automate the generation of the transformation specifications disburden the user from writing these specifications, as much as possible given the adherence of the input to specific conventions.

We have already developed a user interface for the system as extensions for the knowledge acquisition and management platform Semantic Turkey [14]. A user interface for the Collaborative Thesaurus Editor VocBench [15,16] is under development.

4   Sheet2RDF Architecture

Sheet2RDF consists of a library, a command-line utility and an extension for Semantic Turkey (and soon also for VocBench).

The library implements the core functionalities of Sheet2RDF, which include:

1. Generation of a skeletal PEARL transformation specification,
2. Refinement of the specification based on user input,
3. Use of CODA to execute the transformation that produces new RDF triples.

The Sheet2RDF library uses UIMA to analyze the spreadsheet (determining its header). It then produces a UIMA type system including a feature structure type that represents the rows and that has different features corresponding to each column name. Repeated columns are associated with a single feature, the value type of which is an array of elements. This type system is used by a further UIMA analyzer to read the actual content of the spreadsheet; consequently, it is also used as basis to write the corresponding PEARL transformation specifications.

As already said, the system also generates a skeleton of the transformation, by applying a set of heuristics to the aforementioned type system. The assumption is that roughly each row corresponds to an entity description, and that each triple in this description can be built from each column of the spreadsheet. The heuristics consider the entity corresponding to the row as the subject, the predicate as something inferable from (or explicitly associated to) the header of the column, and the object as (an RDF node obtained from) the value in the cell at the crossing between the row and the column. In some particular cases, the recognized property could fire the automatic generation of more complex triple patterns in the target transformation specification. As an example, SKOS-XL [17] labelling properties also require the reification of the label into a URI.

While the heuristics are meant to generate an almost complete specification, the system also assumes that the user may provide further input that helps to refine the generated transformation. Therefore, an additional data structure has been introduced to hold this additional information provided by the user. In the end, the transformation rules are produced by merging the information provided by the type system and the additional supporting data structure. The automatically generated transformation rules may be handcrafted to match the desired output, whenever the heuristics fail to produce a complete specification. However, a better approach is to allow the user filling missing information for the application of heuristics (e.g. associate a column with an RDF property), and customizing the details (e.g. whether to reify property values, as in case of skos:definitions) through a more convenient interface that avoids writing a transformation rule. Indeed, the command-line utility only supports the basic functionalities of Sheet2RDF, while this more advanced input requires a more complete user interface.

One such interface has been developed as an extension for Semantic Turkey, which enables the user to refine the proposed rules (without actually writing them), possibly by exploiting the already assessed information in an RDF dataset. The extension does not simply relay user input to the library (as it happens in the command-line utility), but it also integrates in the loop the user and the underlying triple store, which is managed by Semantic Turkey. This integration allows the user to refine the transformation skeleton by supplying additional hints, which are based on information already assessed in the triple store. As an example, the user may associate a column from the spreadsheet with an existing property. In this task, the user is supported with a hierarchical visualization that is populated with the information contained in the current ontology.

5   Use Cases and Community Feedback

The first and immediate use case we faced (actually the one that led to the development of the system) was the automatic generation of Thesauri and Knowledge Organization Systems (KOSs) [18] in general. The standard modeling language in the RDF family for representing KOSs is SKOS [19], together with its extension for reified labels SKOS-XL [17]. Thanks to SKOS and SKOS-XL, the communities of librarians, metadata developers and archivists opened up to the Semantic Web, though this shift was not painless. The processes that used to lead to the development of KOSs in the pre-RDF era were following the classical "domain experts handling something more or less understandable (mostly to them) to data geeks" workflow pattern. These craftworks caused a proliferation of in-house conventions and (mostly one-time) conversions to some formal (though no more standard than the original one) data representation.

No wonder that the preferred representation adopted by librarians was spreadsheets, as that solution offered a rare combination of high usability with a minimal underlying structure adopted to detail and organize the concept descriptions, their hierarchical organization and their relationships. Our experience as developers of Knowledge Management Systems (Semantic Turkey and VocBench), and especially the feedback we gathered from users, showed us that spreadsheet-import is a much desired feature, but its demand is an order of magnitude larger when considering KOS developers. The immediate problem with this feature is that it is perceived as an import, while it is actually, as already discussed in this paper, a transformation, in that most of the assumptions underlying the organization of data in the spreadsheet (which offer no more than a bidimensional matrix) are implicit in the mind of its creator.

We thus examined a few real use cases we received from organizations willing to bring their Excel tables into SKOS or SKOS-XL. These use cases emerged in different contexts, such as research projects, being directly involved in the consortium (SemaGrow [http://semagrow.eu/]) or through partnership with consortium members (as for agINFRA [http://aginfra.eu/]), or as direct feedback gathered through the community groups (mailing lists, forums..) of our aforementioned knowledge management platforms. In some cases we have been directly involved in the realization of the target RDF representation (Soil Data Linked Dataset, FAO Land and Water thesaurus, FAO Topics vocabulary), while in other cases we provided support through the mailing list to users directly using Sheet2RDF in importing their own data. A few characteristics emerged as recurring across the various examined cases:

1. *A concept-per-line approach*: in all cases, a concept description was expressed through a single line of the spreadsheet file. This is no wonder: while multiple lines could fit more closely (and completely) the structure of an RDF graph (e.g. each row representing an individual triple), the choice of spreadsheet is mostly dictated by a need for simplicity, creating a match between identity of concepts and lines in the spreadsheet was a natural path rather than a meditated choice.

2. *Concept references (especially hierarchy)*: thesauri often come as hierarchy of concepts, and in order to represent a hierarchy (or express any relationship between concepts) across linear descriptions, there is the need for identifiers. At the same time, usually these identifiers were not available or, in the better cases, not clearly expressed. Labels represented the most common case of implicit identifier. This cleared out any possibility for ambiguous names; however, in some of the analyzed cases there were an explicit terminological separation between "terms" (which uniquely define concepts in a domain) and alternative lexicalizations.

3. *Implicit bindings between subsets of values*: often, users represented as a linear array of independent characteristics, elements that were actually grouped into bound subsets of elements dependent from each other. Specific cases include:
   (a) Reified labels, lexical descriptions etc..: very often in thesauri, lexical descriptors of any kind bring metadata with them, such as editorial notes (related to the same lexical content rather than to concepts) date-of-creation/update, provenance (e.g. author of the description) and so on. As such, columns following a

reified label are usually not providing further characteristics of the concept but they describe its label. This binding may be guessed by a person by reading the spreadsheet content, while for sure it is not explicit to machines processing the spreadsheet.

(b) Information about atomic elements sparse across two or more cells: a common case are labels (even simple literals) and their associated language, most often when there are conceptual resources represented non-uniformly in a plethora of languages. While the common approach is to define the language of the labels in a column header (e.g. one or more columns for English labels, for French ones), users might avoid excessive proliferation of columns and data sparseness in the sheet, by having pairs of columns, holding the label content and language respectively. Usually new column pairs are just added upon need, when they are not enough to represent the concept with the larger number of labels.

Thanks to the expressiveness of PEARL, all of the above phenomena could be managed successfully. However, our intent was to cover as much as possible of the typical cases, minimizing the amount of human effort to be carried on the PEARL transformation. This can be dealt with either in a completely automatic fashion, by having sheet2rdf guessing what is needed in/how to handle a particular situation, or through human computer-interaction. In the latter case, the machine is able to detect potential patterns and suggest choices to be taken by the user, presenting them in an understandable way and still hiding the transformation technicalities

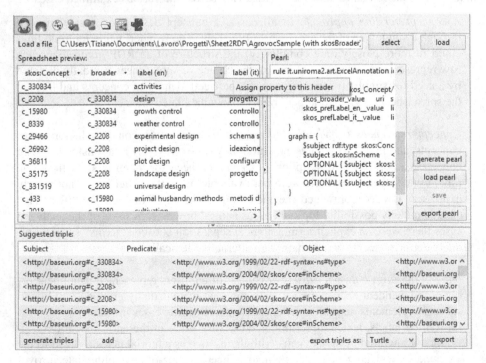

Fig. 2. Sheet2RDF user interaction

which the naïve user it not able to handle. In addressing these issues, we have followed a combination of both approaches. The first one has been pursued by defining a series of conventions that are recognized by the system and that automate the transformation generation (http://art.uniroma2.it/sheet2rdf/docu-mentation/heuristics.jsf). At the same time, by benefiting from the possibilities offered by Sheet2RDF integration into editing systems, we have plugged interactive wizards (see Fig. 2) raising issues/warnings to the user, or just signaling that human intervention is welcome (e.g. by highlighting the header of a column and providing menu boxes with choices) in order to disambiguate among different possible choices. We set default choices (made customizable per system and by the user), so to minimize even this interactive part by tailoring choices to systems/user needs.

6   Conclusion

In this paper, we presented Sheet2RDF, a system for importing and lifting spreadsheets to RDF. Sheet2RDF combines a flexible mapping specification language with the automatic generation of skeletal mappings, by recognizing patterns in the input spreadsheet. This approach filled a void in the landscape of currently available systems, which tend to be either minimally customizable or highly dependent on human intervention for the specification of the transformation.

Sheet2RDF supports human participation in the process by providing several levels of interaction, each one trading generality for ease of usage. Automatically generated mapping skeletons aim to cover commons cases, while manual refinement may be required for specific cases. A graphical user interface enables some refinements without the need to explicitly use the underling mapping specification language.

Acknowledgments. This research has been partially supported by the EU funded project SemaGrow (http://www.semagrow.eu/) under grant agreement no: 318497.

References

1. Berners-Lee, T., Hendler, J., Lassila, O.: The Semantic Web: A new form of Web content that is meaningful to computers will unleash a revolution of new possibilities. Scientific American **279**(5), 34–43 (2001)
2. Scharffe, F., Atemezing, G., Troncy, R., Gandon, F., Villata, S., Bucher, B., Hamdi, F., Bihanic, L., Képéklian, G., Cotton, F., Euzenat, J., Fan, Z., Vandenbussche, P.-Y., Vatant, B.: Enabling linkeddata publication with the datalift platform. In: AAAI Workshop on Semantic Cities (2012)
3. Prud'hommeaux, E., Seaborne, A.: SPARQL query language for RDF. In: World Wide Web Consortium - Web Standards, January 15 2008. http://www.w3.org/TR/rdf-sparql-query/

4. Lebo, T., Williams, G.: Converting governmental datasets into linked data. In: Proceedings of the 6th International Conference on Semantic Systems, New York, NY, USA, pp. 38:1–38:3 (2010)
5. Han, L., Finin, T.W., Parr, C.S., Sachs, J., Joshi, A.: RDF123: from spreadsheets to RDF. In: Sheth, A.P., Staab, S., Dean, M., Paolucci, M., Maynard, D., Finin, T., Thirunarayan, K. (eds.) ISWC 2008. LNCS, vol. 5318, pp. 451–466. Springer, Heidelberg (2008)
6. W3C: The RDF data cube vocabulary. In: World Wide Web Consortium (W3C), January 14 2014. http://www.w3.org/TR/vocab-data-cube/
7. W3C Open Annotation Community Group: Open annotation data model. In: World Wide Web Consortium (W3C), February 08 2013. http://www.openannotation.org/spec/core/
8. W3C: PROV-DM: the PROV data model. In: World Wide Web Consortium (W3C), April 30 2013. http://www.w3.org/TR/prov-dm/
9. Ermilov, I., Auer, S., Stadler, C.: CSV2RDF: user-driven CSV to RDF mass conversion framework. In: Proceedings of the ISEM 2013, Graz, Austria, September 04–06 2013
10. Fiorelli, M., Pazienza, M.T., Stellato, A., Turbati, A.: CODA: Computer-aided ontology development architecture. IBM Journal of Research and Development 58(2/3), 14:1–14:12 (2014)
11. Fiorelli, M., Gambella, R., Pazienza, M.T., Stellato, A., Turbati, A.: Semi-automatic knowledge acquisition through CODA. In: Ali, M., Pan, J.-S., Chen, S.-M., Horng, M.-F. (eds.) IEA/AIE 2014, Part II. LNCS, vol. 8482, pp. 78–87. Springer, Heidelberg (2014)
12. Ferrucci, D., Lally, A.: Uima: an architectural approach to unstructured information processing in the corporate research environment. Nat. Lang. Eng. 10(3–4), 327–348 (2004)
13. Pazienza, M.T., Stellato, A., Turbati, A.: PEARL: ProjEction of annotations rule language, a language for projecting (UIMA) annotations over RDF knowledge bases. In: LREC, Istanbul (2012)
14. Pazienza, M.T., Scarpato, N., Stellato, A., Turbati, A.: Semantic Turkey: A Browser-Integrated Environment for Knowledge Acquisition and Management. Semantic Web Journal 3(3), 279–292 (2012)
15. Caracciolo, C., Stellato, A., Rajbahndari, S., Morshed, A., Johannsen, G., Keizer, J., Jacques, Y.: Thesaurus maintenance, alignment and publication as linked data: the AGROVOC use case. International Journal of Metadata, Semantics and Ontologies (IJMSO) 7(1), 65–75 (2012)
16. Stellato, A., Rajbhandari, S., Turbati, A., Fiorelli, M., Caracciolo, C., Lorenzetti, T., Keizer, J., Pazienza, M. T.: VocBench: a Web Application for Collaborative Development of Multilingual Thesauri. In : The Semantic Web: Trends and Challenges. Springer International Publishing (2015) (accepted for publication)
17. World Wide Web Consortium (W3C): SKOS simple knowledge organization system eXtension for labels (SKOS-XL). In: World Wide Web Consortium (W3C), August 18 2009. http://www.w3.org/TR/skos-reference/skos-xl.html
18. Hodge, G.: Systems of Knowledge Organization for Digital Libraries: Beyond Traditional Authority Files. Council on Library and Information Resources, Washington, DC (2000)
19. World Wide Web Consortium (W3C): SKOS simple knowledge organization system reference. In: World Wide Web Consortium (W3C), August 18 2009. http://www.w3.org/TR/skos-reference/

An Approach to Dominant Resource Fairness in Distributed Environment

Qinyun Zhu[✉] and Jae C. Oh

Electrical Engineering and Computer Science Department,
Syracuse University, Syracuse, USA
qzhu02@syr.edu, joh@ecs.syr.edu

Abstract. We study the multi-type resource allocation problem in distributed computing environment. Current approaches that guarantee the conditions of Dominant Resource Fairness (DRF) are centralized algorithms. However, as P2P cloud systems gain more popularity, distributed algorithms that satisfy conditions of DRF are in demand. So we propose a distributed algorithm that mostly satisfies DRF conditions. According to our simulation results, our distributed dominant resource fairness algorithm outperforms a naive distributed extension of DRF.

Keywords: Dominant resource fairness · Distributed resource allocation · Distributed decision making

1 Introduction

Cloud computing platforms require fair resource allocation to improve user satisfaction and resource utilization. Popular systems, such as Hadoop fair scheduler, adopt centralized resource management and only consider resources of homogeneous type.

Later systems with centralized management improve fairness of resource allocation by considering heterogeneous settings. These systems like Mesos [3] and Hadoop Yarn [5] employ the Dominant Resource Fairness (DRF) [4] for resources of heterogeneous types. A user's dominant resource (the resource type that the user needs most) dictates the allocation procedure in DRF. Different users may have different dominant resource, which makes DRF take into account the heterogeneous resource demands of users. Although people developed different improvements of DRF like weighted DRF for group strategy-proofness [10] and dynamic dominant resource fairness for dynamic environment [11], these algorithms assume all resources are kept in a single huge computer. To address the resource allocation problem of multiple computers with heterogeneous resource capacities, a centralized Dominant Resource Fairness allocation in Heterogeneous environment (DRFH) [2] has been developed based on DRF.

In DRFH, a single resource manager has to trace resource availability for every resource provider and resource usages for every user. However, structures of future cloud systems, such as mobile cloud [7] and peer-to-peer cloud [8],

© Springer International Publishing Switzerland 2015
M. Ali et al. (Eds.): IEA/AIE 2015, LNAI 9101, pp. 141–150, 2015.
DOI: 10.1007/978-3-319-19066-2_14

introduce resource providers with different owners. These cloud systems are naturally decentralized and distributed. Moreover, as the scale of data center is growing to handle big data computations, it is difficult for a centralized resource manager to promptly track the status of computing nodes and users' tasks while making allocation decisions. Therefore, to address the needs of such systems, we develop a scheme with distributed resource manager. This scheme let allocation decisions be made locally with limited global information. A resource provider is a computing node providing resources in the distributed system. Unlike client-server structure in centralized management, the resource providers in our distributed setting communicate in a peer-to-peer manner. In addition, instead of tracking all users' resource usages like a centralized manager, local resource managers can only obtain status of their known users.

In this paper, we present the state-of-the-art in our research in fully distributed heterogeneous resource allocation scheme that is maximally optimal:

- We propose a distributed resource allocation model in a heterogeneous environment
- We design implementations of Distributed Dominant Resource Fairness (DDRF) based on a fitness heuristic and a task forwarding mechanism
- We simulate DDRF using sampled task demands and resource configurations from google-cluster-traces [6] and demonstrate performance of our algorithm in terms of fairness, efficiency and scalability

2 Dominant Resource Fairness with Centralized Management

In this section, we introduce previous resource allocation algorithms of centralized resource management.

First, we briefly introduce the basic algorithm– Dominant Resource Fairness (DRF) [4]– and fairness properties. Then we describe the Dominant Resource Fairness in Heterogeneous environments (DRFH) [2] formally. These resource allocation procedures consist the following major participants:

- Resource Provider: each resource provider is a computer that provides computational resources like CPU, memory. Let resource providers be $P = \{1, ..., k\}$ with m types of resources.
- User: each user is the person or computer that requests resource from the resource providers to schedule computing tasks. A set of users $U = \{1, ...n\}$ share the computing resources of the cluster. In addition, we assume that a user has infinite tasks and all tasks from the same user have same demand.

The Dominant Resource Fairness (DRF) views all resources as they are in one computer. For example, if we have 2 resource providers with [10 CPU, 5GB] and [5 CPU, 10 GB], DRF views them as a computer with total [15 CPU, 15 GB] of CPU and memory. DRF allocates resources to users in a non-wasteful way according to resource requirements of tasks. Suppose two users schedule tasks in the system. Each task of user A requires (2 CPU, 1 GB) and each task of user B requires (1 CPU, 2 GB). That is, each task of user A needs (2/15, 1/15) in terms of fraction (share) of resources in the system and each task of user B needs (1/15, 2/15). The dominant resource is the resource that requires maximum share. DRF tries to equalize users' shares of allocated dominant resources by maximizing the minimum share of dominant resource among the users. In our example, if both of the users have sufficient tasks to fill the system, each of them can get 10/15 share of their dominant resources under the constraints of total resource capacity. In other words, user A can schedule 5 tasks with a (10 CPU, 5 GB) allocation while user B can schedule 5 tasks with a (5 CPU, 10 GB) allocation. Thus, the allocation is done fairly in terms of four key properties [4]:

- *Strategy-proofness*: A user can achieve maximal benefits if it declares resource demands honestly.
- *Envy-freeness*: A user should not prefer any other users' allocation.
- *Pareto Efficiency*: It should not be possible to increase the total scheduled tasks of a user without decreasing the total scheduled tasks of at least one other user.
- *Sharing Incentive*: Each user should be better off getting resources according to their heterogeneous demands, rather than exclusively owning a equally split allocation (e.g. each of user A and B owns (7.5 CPU, 7.5 GB) out of [15 CPU, 15 GB]) in the system.

However, a system may have multiple resource providers with heterogeneous resource capacity. A **naive extension** is applying DRF separately on each resource provider, but it is inefficient because different resource capacities can cause fragmentation. For instance, resource provider P_1 has [10 CPU,5 GB] and resource provider P_2 has [5 CPU,10 GB]. If we run DRF separately on them, user A can have an allocation of (8 CPU, 4 GB) for 4 tasks and user B can have an allocation of (4 CPU, 8 GB) for 4 tasks in the system. So 3 CPU and 3 GB are not utilized in the system. But if we schedule all tasks of user A on P_1 and all tasks of user B on P_2, the system achieves the optimal utilization as 5 tasks of user A and 5 tasks of user B are scheduled. Therefore, [2] proposed DRFH.

For a resource provider $p \in P$, $c_p = (c_{p1}, ... c_{pm})^T$ is its normalized resource capacity, where $\sum_{p \in P} c_{pr} = 1, \forall r \in R$. For a user $i \in U$, $D_i = (D_{i1}, ... D_{im})^T$ is the demand vector, where D_{ir} is the share of resource r–the fraction of demanded resource r over total resource capacity of r in the system. So the global dominant resource of user i is $r_i^* \in argmax_{i \in R} D_{ir}$. User i's share of resource allocation on resource provider p is denoted by $A_{ip} = (A_{ip1}, ..., A_{ipm})^T$. So user i can schedule $min_{r \in R}\{A_{ipr}/D_{ir}\}$ tasks on resource provider p.

Let global dominant resource share that user i receives from resource provider p be $G_{ip}(A_{ip}) = min_{r \in R}\{A_{ipr}/D_{ir}\}D_{ir_i^*}$. Let $G_i(A_i) = \sum_{p \in P} G_{ip}(A_{ip})$, the problem of DRFH is defined as:

$$\max_A \min_{i \in U} G_i(A_i)$$

$$\text{s.t.} \sum_{i \in U} A_{ipr} \leq c_{pr}, \forall p \in P, r \in R \tag{1}$$

The DRFH aims to optimize fair dominant resource allocation globally. A solution of DRFH assures three key fairness properties: envy-freeness, strategy-proofness and Pareto efficiency. Yet, it assumes a centralized resource manager.

3   Model of Distributed Dominant Resource Fairness

In this section, we introduce the distributed environment of resource management and define the dominant resource fairness under this distributed setting.

3.1   Distributed Environment

Similar to the heterogeneous environment, different resource providers may have different resource capacities in distributed environment. However, in distributed environment, each resource provider has a local resource manager. These local resource managers can access local information of its affiliated resource provider. In our distributed setting, each resource provider is assigned a set of **Friends**–other resource providers that it can communicate with–when the system starts. Then a resource provider can communicate with its friends in a peer-to-peer manner. In addition, a local resource manager has a set of **Known Users** whose information can be readily accessible by that resource provider. Formally, we define the known users of a resource provider P as $U_p \subset U$ and $\bigcup_{p \in P} U_p = U$.

3.2   Naive Distributed DRF Extension

A naive distributed extension of DRF allocates resources on each resource provider independently according to its local resource capacity and local allocations of users. However, similar to the heterogeneous environment, a naive extension of DRF in distributed setting can result in leftover resources that might be less efficiency in resource utilization. What's more, the envy-freeness among all the users on a specific resource provider cannot be ensured because a resource provider only have access to information of its known users rather than all users. For example, three users–A,B and C–with the same task demand (1 CPU, 1 GB) schedule tasks on resource provider P_1 with [3 CPU, 3 GB] and resource provider P_2 with [6 CPU, 6 GB]. If $U_{P_1} = \{A, B, C\}$ while $U_{P_2} = \{A, B\}$, then both A and B can get (1 CPU, 1 GB) on P_1 and (3 CPU. 3 GB) on P_2. However, C can only get (1 CPU, 1 GB) on P_1 and nothing on P_2. The result is not fair to user C. It follows that the naive extension does not lead to global fairness among users under our distributed environment. We use this naive extension as the baseline for our distributed resource allocation algorithms.

3.3   Global Information

To build the model of DRF in the distributed environment, resource providers need some global information to improve the fairness of allocation. As shown in Figure 1, we assume the distributed environment provides limited global information, such as:

- Dictionary of all resource providers in the system
- Global resource capacity in the system
- Lookup of global resource allocations of specific users

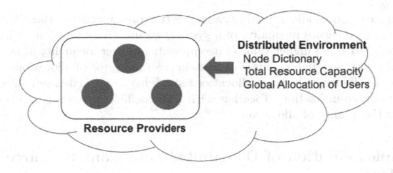

Fig. 1. Distributed Environment with Limited Global Information

To simplify our model, we assume global information pre-exists in the distributed environment. That is, some global name services and synchronization services are provided as the infrastructure of this system. Also, a user in the system is responsible for tracking the global allocation of that user and servicing the query of the information.

3.4 Distributed Dominant Resource Fairness

Distributed Dominant Resource Fairness considers the global information for fairness. For a resource provider p, the minimum dominant share on it is:

$$f(p) = \min_{i \in U_p} G_i(A_i) \tag{2}$$

Then the dominant resource fairness in distributed environment is achieved by maximizing the minimum output of (2):

$$\max_A \min_{p \in P} f(p)$$
$$\text{s.t. } \sum_{i \in U} A_{ipr} \le c_{pr}, \forall p \in P, r \in R \tag{3}$$

Since $\bigcup_{p \in P} U_p = U$, we have $\min_{p \in P} f(p) = \min_{p \in P} \min_{i \in U_p} G_i(A_i) = \min_{i \in U} G_i(A_i)$. It follows that $\max_A \min_{p \in P} f(p) = \max_A \min_{i \in U} G_i(A_i)$. Thus, like (1), the essential properties, such as envy-freeness, Pareto efficiency and strategy-proofness, are preserved for a solution of problem (3). Although a linear optimization solver may give a solution of equation 3, we need an approximation instead of exact solution due to the huge search space of possible allocations. We discuss our implementation in the next section.

3.5 Evaluation Metrics for Fairness and Efficiency

According to [4][2], envy-freeness and strategy-proofness are preserved by equalizing the dominant resource share among the users. Therefore, lower inequality of dominant resource allocation indicates a better preservation of these fairness properties. To measure inequality of a system, we use Gini-coefficient, which is range from 0 to 1 and higher value of it represents a higher inequality among the allocated resources. High resource utilization is an indicator of allocations satisfying Pareto efficiency, because allocations satisfying Pareto efficiency have the highest resource utilization. Together with Gini-coefficient, resource utilization measures the quality of allocation.

4   Implementation of Distributed Dominant Resource Fairness

Following [4][2], our implementation of approximated DDRF is based on progressive filling on dominant resource. However, the progressive filling algorithm can only guarantee the envy-freeness and strategy-proofness among the known users of a resource provider. To approximate global fairness, a resource provider allocates resources to its known users according to their global allocations and the lowest dominant resource share of friend resource providers as shown in Algorithm 1.

For optimal allocation of resources, our implementation should match different task requirements to a proper resource provider. Intuitively, the higher similarity between the available resources and the task demand indicates a better allocation because it leads to fewer fragmentation. So we use the cosine similarity as the heuristics to estimate the optimal solution of equation (3). Suppose the available resources on a resource provider p is $Available(p, r) = c_{pr} - \sum_{i \in U} A_{ipr}$, we define the fitness function of resource provider p and tasks from user u as:

$$Fitness(p, u) = \frac{\sum_{r \in R} D_{ur} Available(p, r)}{\sqrt{\sum_{r \in R} (Available(p, r))^2} \sqrt{\sum_{r \in R} (D_{ur})^2}} \tag{4}$$

If a resource provider fits the task of a user well, it is very likely that the resource provider favors this user's other tasks. Therefore, we define a **Primary User** as a user that shares resources on a resource provider. That is, a resource provider only allocates resources to its primary users. In addition, each user can only be the primary user of exactly one resource provider. Let PU_p be the set of primary users of resource provider p. We have $\bigcap_{p \in P} PU_p = \emptyset$.

With the newly designed primary users, we introduce task-forwarding mechanism. Algorithm 2 describes the procedures of DDRF with task forwarding. At the beginning, a resource provider gets a random set of users as its primary users and a random set of other resource providers as its friends. Then every resource provider keeps checking if any primary users' global dominant resource allocation are equal or below the lowest allocation of its known users and its friends' known users. Once a user with lowest allocation is found, the resource provider tries to allocate a task of that user according to its available resources and fitness threshold Thr. If a task fails being allocated locally, the resource provider will forward the task to a random friend and then remove the user from its user list. If the task

---

Algorithm 1. Local Resource Management Procedure without Task Forwarding

PROCEDURE Resource_Management:
loop
   for all $i \in U_p$ **do**
      if $D_{ir*} \leq \min_{j \in U_p} D_{jr*}$ and $D_{ir*} \leq \min_{q \in Friends(p)} \min_{j \in U_q} D_{jr*}$ And $\forall r \in$
      $R(D_{ir} \leq Ava(p, r))$ **then**
         $c_{ip} \leftarrow c_{ip} - D_i$
         $A_i \leftarrow A_i + D_i$
      end if
   end for
end loop

---

still fails in matching with any resource provider after several forwarding, the user may be assigned to a random resource provider again as primary user.

In the procedure on receiving of task forward message, if a task is forwarded fewer than *step_limit* times, the resource provider runs the first-fit algorithm. That is, it tries to allocate the task according to the fitness and the resource provider's available resources. If successful, the resource provider adds the user to its primary user list. Otherwise, it forwards the task to its friends again. On the other hand, if the number of task forwarding exceeds the step limit, the resource provider runs a best-fit algorithm, which tries to schedule the task on a resource provider with highest fitness value along the forwarding path. Also, whenever a resource provider receives a task, the resource provider will add the user information of that task to its list of known users.

5 Experiments

In this section, we present our experiments and the simulating results. Without loss of generality, we compare the performances of different resource allocation schemes by software simulation. Our simulation only focuses on the resource allocation procedures. In other words, we only simulate the resource allocation procedure on the resource providers and assume other services and information is available from the environment. So our simulation starts from the point that all resources are available and stop when no more tasks can be scheduled in the system.

We uniformly sample users' demands of tasks and configurations of resource providers from the Google cluster-usage traces [1]. The trace contains information of task demands from over 900 users and resource capacities of over 10000 servers. Since we assume that task demands from a user are identical and each user has infinite tasks, we sample a total of 20 tasks from different users. Also, we sample CPU and memory configurations of 100 machines. Therefore, we simulate a system with totally 20 users and 100 resource providers. Based on this setting, We run two simulations. First, we evaluate the stress of making decision/task forwarding for resource allocation. We compare the number of decision making under configurations of number of resource providers in range from 20 to 100 under different algorithms including centralized DRFH, distributed DDRF and

---

Algorithm 2. Local Resource Management Procedures with Task Forwarding

PROCEDURE Resource_Management:
loop
    for all $i \in PU_p$ **do**
        if $D_{ir*} \leq \min_{j \in U_p} D_{jr*}$ and $D_{ir*} \leq \min_{q \in Friends(p)} \min_{j \in U_q} D_{jr*}$ **then**
            $path \leftarrow \emptyset$
            if $Fitness(p, i) \geq Thr$ And $\forall r \in R(D_{ir} \leq Ava(p, r))$ **then**
                $c_{ip} \leftarrow c_{ip} - D_i$
                $A_i \leftarrow A_i + D_i$
            else
                $j \leftarrow Random(Friends(p))$
                $PU_p \leftarrow PU_p \setminus \{i\}$
                $Send(j, D_i, path \cup (i, Fitness(p, i)), 0)$
            end if
        end if
    end for
end loop

PROCEDURE On_Receive_Task_Forwarding:
$(D_i, path, step) \leftarrow$ Message
if $step < step\_limit$ **then**
    if $Fitness(p, i) \geq Thr$ And $\forall r \in R(D_{ir} \leq Ava(p, r))$ **then**
        $c_{ip} \leftarrow c_{ip} - D_i$
        $A_i \leftarrow A_i + D_i$
        $PU_p \leftarrow PU_p \cup \{i\}$
    else
        $j \leftarrow Random(Friends(p) \setminus path.keys)$
        $Send(j, D_i, path \cup (i, Fitness(p, i)), step + 1)$
    end if
else
    if $(Fitness(p, i) \geq Thr$ Or $Fitness(p, i) \geq max_{j \in path.keys}\{path[j]\})$ And $(\forall r \in R, D_{ir} \leq Ava(p, r))$ **then**
        $c_{ip} \leftarrow c_{ip} - D_i$
        $A_i \leftarrow A_i + D_i$
        $PU_p \leftarrow PU_p \cup i$
    else
        $path \leftarrow path \setminus (i, Fitness(p, i))$
        $j \leftarrow argmax_{j \in path.keys}\{path[j]\}$
        if $j \in \emptyset$ **then**
            $Send(j, D_i, path \setminus (i, Fitness(p, i)), step)$
        end if
    end if
end if
$U_p \leftarrow U_p \cup \{i\}$

---

the naive extension of DRF in distributed setting. Then we compare the fairness and efficiency among the naive extension of DRF under distributed environment, DDRF without task forwarding and DDRF with task forwarding.

Figure 2 shows that maximum stress of decision making in the centralized algorithm increases linearly along the scale of the cluster. On the other hand, distributed algorithms including the DDRFs and the naive extension of the DRF keeps the maximum stress among the resource providers stable, because a resource manager of a distributed algorithm only needs to deal with its local resources and keep connections to a limited number of friend resource providers. Therefore, DDRF is more scalable than the centralized DRFH.

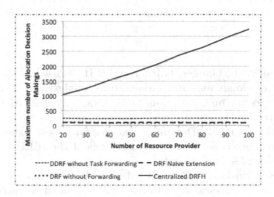

Fig. 2. Decision Making Stress VS. Number of Resource Providers

In the simulation of our two versions of DDRF algorithms, each resource provider is configured to have 10 friends. We set the step limit to 3 and fitness threshold to 0.8 in the simulation of DDRF with task forwarding. Figure 3 shows the two versions of DDRF have similar performance in fairness as the centralized DRFH and a better general resource utilization than other algorithms.

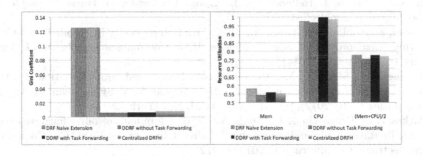

Fig. 3. Fairness and Efficiency of Different Resource Allocation Algorithms

6   Conclusion

In this work, we studied the multi-type resource allocation problem in distributed environment. We focused on the decentralized resource allocation procedures among the distributed resource providers. Our proposed formalization is based on DRFH, which has a centralized mechanism. Then we proposed several implementations of the distributed dominant resource fairness and compare them with a naive distributed extension of DRF and centralized DRFH. The simulation results show that our decentralized algorithms have a better scalability than the centralized DRFH. In addition, the DDRFs have comparable performance with the centralized DRFH and outperform the naive extension of DRF in terms of fairness and efficiency.

References

1. Reiss, C., Tumanov, A., Ganger, G.R., Katz, R.H., Kozuch, M.A.: Heterogeneity and dynamicity of clouds at scale: Google trace analysis. In: Proceedings of the Third ACM Symposium on Cloud Computing (SoCC 2012), Article 7, p. 13. ACM, New York (2012)
2. Wang, W., Li, B., Liang, B.: Dominant resource fairness in cloud computing systems with heterogeneous servers. In: Proceedings of the IEEE INFOCOM, 2014, pp. 583–591, April 27 – May 2 (2014)
3. Hindman, B., Konwinski, A., Zaharia, M., Ghodsi, A., Joseph, A.D., Katz, R., Shenker, S., Stoica, I.: Mesos: a platform for fine-grained resource sharing in the data center. In: Proceedings of the 8th USENIX Conference on Networked Systems Design and Implementation (NSDI 2011), pp. 295–308. USENIX Association, Berkeley (2011)
4. Ghodsi, A., Zaharia, M., Hindman, B., Konwinski, A., Shenker, S., Stoica, I.: Dominant resource fairness: fair allocation of multiple resource types. In: Proceedings of the 8th USENIX Conference on Networked Systems Design and Implementation (NSDI 2011), pp. 323–336. USENIX Association, Berkeley (2011)
5. Hadoop Capacity Scheduler. http://hadoop.apache.org/docs/current/hadoop-yarn/hadoop-yarn-site/CapacityScheduler.html
6. Reiss, C., Wilkes, J., Hellerstein, J.L.: Google Cluster-Usage Traces. https://code.google.com/p/googleclusterdata/
7. Sanaei, Z., Abolfazli, S., Gani, A., Buyya, R.: Heterogeneity in Mobile Cloud Computing: Taxonomy and Open Challenges. IEEE Communications Surveys & Tutorials 16(1), 369–392 (2014). First Quarter
8. Babaoglu, O., Marzolla, M., Tamburini, M.: Design and implementation of a P2P cloud system. In: Proceedings of the 27th Annual ACM Symposium on Applied Computing (SAC 2012), pp. 412–417. ACM, New York (2012)
9. Parkes, D.C., Procaccia, A.D., Shah, N.: Beyond dominant resource fairness: extensions, limitations, and indivisibilities. In: Proceedings of the 13th ACM Conference on Electronic Commerce (EC 2012) (2012)
10. Friedman, E.J., Ghodsi, A., Shenker, S., Stoica, I.: Strategyproofness, Leontief Economies and the Kalai-Smorodinsky Solution, Technical Report (2011)
11. Kash, I., Procaccia, A.D., Shah, N.: No agent left behind: dynamic fair division of multiple resources. In: Proceedings of the 2013 International Conference on Autonomous Agents and Multi-Agent Systems (AAMAS 2013), pp. 351–358. International Foundation for Autonomous Agents and Multiagent Systems, Richland (2013)

Mining SQL Queries to Detect Anomalous Database Access Using Random Forest and PCA

Charissa Ann Ronao[✉] and Sung-Bae Cho

Department of Computer Science, Yonsei University, 50 Yonsei-ro,
Seodaemun-gu, Seoul 120-749, South Korea
cvronao@sclab.yonsei.ac.kr, sbcho@cs.yonsei.ac.kr

Abstract. Data have become a very important asset to many organizations, companies, and individuals, and thus, the security of relational databases that encapsulate these data has become a major concern. Standard database security mechanisms, as well as network-based and host-based intrusion detection systems, have been rendered inept in detecting malicious attacks directed specifically to databases. Therefore, there is an imminent need in developing an intrusion detection system (IDS) specifically for the database. In this paper, we propose the use of the random forest (RF) algorithm as the anomaly detection core mechanism, in conjunction with principal components analysis (PCA) for the task of dimension reduction. Experiments show that PCA produces a very compact, meaningful set of features, while RF, a graphical method that is most likely to exploit the inherent tree-structure characteristic of SQL queries, exhibits a consistently good performance in terms of false positive rate, false negative rate, and time complexity, even with varying number of features.

Keywords: Database · Intrusion detection · Security · SQL query · Random forest · PCA

1 Introduction

The emergence of big data has led to the development of high performance data storage systems, called relational database management systems (RDBMS). Majority of companies and organizations rely on these databases to safeguard sensitive data [1]. Illegal tampering or misuse of these data will not only cost companies and organizations huge monetary losses, but also incur customer damages and legal sanctions. While RDBMS provides efficient and systematic storage of data, its traditional security mechanisms, however, are not enough to protect from malicious threats [2].

An important component of a strong security framework able to protect sensitive data in these databases is an intrusion detection system (IDS). These systems aim to detect intrusions as early as possible, to ensure that any compromise in integrity of data is reported and acted upon. Research on IDS has been going on for years, most of which are works focused on network-based and host-based IDS. However, neither network-based nor host-based IDS's can detect malicious behavior at the database level [3]. Malicious attacks specifically directed to the database are likely to be invisible at the network and operating systems level, and thus, invisible to the detectors on

© Springer International Publishing Switzerland 2015
M. Ali et al. (Eds.): IEA/AIE 2015, LNAI 9101, pp. 151–160, 2015.
DOI: 10.1007/978-3-319-19066-2_15

that level. Therefore, network-based and host-based IDS's are rendered useless in the face of database-specific attacks.

In line with this, anomaly-based IDS using data mining approaches are gaining more and more attention in the field of database anomaly detection because of their high intrusion detection accuracy, efficiency, and automation features [4]. Several research works in the field have modeled SQL queries to detect database anomalies; however, most of them have not taken into account the inherent tree-structure of the SQL language syntax. In this paper, we propose a combination of principal components analysis (PCA) and random forests (RF) for the task of query feature selection and database anomaly detection, respectively. We show that PCA produces a compact and meaningful set of features, while a graphical model like RF exploits the inherent tree-structure of SQL queries to achieve a consistently good performance and fast detection speed in comparison with other alternatives, even with varying number of features.

The paper is structured as follows: Section 2 discusses the related work, followed by Section 3 which focuses on the IDS's architecture, query parsing, feature extraction, and an overview of PCA and RF. Section 4 presents our experimental results, and lastly, we draw our conclusion in Section 5.

2   Related Work

There have been a number of research works that have proposed to use data mining techniques to implement an anomaly-based IDS. Hu et al. used classification rules, with the rationale that an item update does not happen alone and is accompanied by a set of other events that are also recorded in the database logs [5]. Srivastava et al. introduced an improved version of the former by considering attribute sensitivity [6]. Such methods, however impose the burden of identifying proper support and confidence values to the user [4].

Hidden Markov models (HMM) have also been used by Barbara et al. to capture the change in database's normal behavior over time [7]. Consecutively, Ramasubramanian and Kannan proposed two database intrusion detection frameworks based on artificial neural networks (ANN) [8][9]. Support vector machines and multilayer perceptrons were utilized by Pinzon and his colleagues to detect SQL injection attacks [10]. Although most of the above works have proposed comprehensive frameworks for a database IDS, they are either very impractical when applied to typical database sizes (which usually contains a large number of tables and attributes), or only focused on detecting outsider attacks. Outsider attacks (e.g. SQL injection) can usually be mitigated by defensive programming techniques; insider threats, however, are much more difficult to detect, and are potentially more dangerous [3][13].

Our work is most similar to [11], which took advantage of the presence of role-based access control (RBAC), a standard that is incorporated in most database implementations today, and made use of a naïve Bayes classifier, to detect anomalous insider query access. We believe that by making use of access control and SQL queries, we will be able to defend the system at the root of the problem, and therefore, block attacks when they are still trying to gain access into the system [10]. It is important to note, however, that the latter work did not take into consideration the inherent tree-structure of the SQL language syntax, and that graph-based methods will most likely exploit this attribute to produce a much better model [12].

The work by Bockermann et al. used tree kernels to enable them to exploit the structure of the SQL syntax [14]. They proved that this indeed has a good effect on performance; however, the method introduced a huge computational overhead, making anomaly detection drastically slow. In this paper, we show the advantage of using a suitable machine learning technique for the task at hand. We develop an anomaly detection system based on the random forest algorithm, an ensemble classifier composed of decision trees, which had been widely and successfully used in various IDS fields [15][16]. Partnering it with PCA, which we had found to perform very well with query feature vectors [12], we build a simple, practical, and efficient anomaly detection system specifically for the relational database.

3   Database Intrusion Detection System Using Random Forest and PCA

3.1   System Architecture

The main concept of our anomaly detection mechanism is to build profiles from normal RDBMS user access behavior and detect anomalous behavior using these profiles. Normal profiles are intrusion-free query sequences, and they can be obtained through parsing database audit logs. Given these normal profiles, we define an anomaly as an access pattern that deviates from these profiles [11].

Our method takes advantage of the presence of RBAC, which is already a standard that has been adopted in various commercial DBMS products [11]. This control mechanism assigns authorizations to users with respect to roles or profiles, instead of assigning them with respect to individual users. By making use of these roles in conjunction with an anomaly-based IDS, we can effectively reduce the number of profiles to maintain, which is a more efficient method when considering the problem of maintaining profiles for a large database user population.

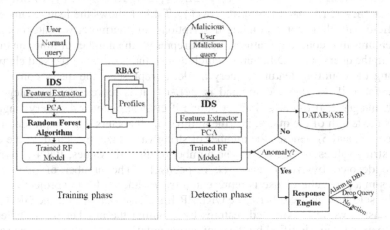

Fig. 1. Database IDS Architecture

Detecting anomalous queries, in conjunction with profiles defined in RBAC, can be considered as a supervised classification problem with two phases: the training phase and the detection phase, as shown in Fig. 1. During the training phase, SQL raw query logs representing normal user behavior are gathered from the database together with their respective profile annotations. Query parsing is then performed to divide the query into relevant SQL clauses. Query features are extracted from the parsed data so as to transform query logs into a form data mining algorithms can understand. Query data, now in the form of features, are then fed to the data mining classifier for supervised training, producing a statistical model which can predict and differentiate between profiles. On the other hand, during the detection phase, new queries from users go through the same query parsing and feature extraction process, and they are assessed by the trained model if they are anomalous or not. If the prediction of a particular query from a user does not match the user's role or profile, the query is said to be anomalous and an alarm is raised.

3.2 Query Parsing and Feature Extraction

Normal role behavior is represented by intrusion-free query logs for that particular role. Query logs, in turn, correspond to the SQL language syntax. We illustrate this through the SELECT command:

```
SELECT      <Projection attribute clause>
FROM        <Projection relation clause>
WHERE       <Selection attribute clause>
ORDER BY    <ORDER BY clause>
GROUP BY    <GROUP BY clause>
```

We parse queries in this manner, line-by-line, separating the above SQL clauses. We then use the parser output to extract query features.

We adopt the features extracted in [12]. A query log is represented in the query feature space by a feature vector Q with the following seven vector fields: Q (SQL-CMD[], PROJ-REL-DEC[], PROJ-ATTR-DEC[], SEL-ATTR-DEC[], ORDBY-ATTR-DEC[], GRPBY-ATTR-DEC[], VALUE-CTR[]). Table 1 shows the features under each vector field with their corresponding descriptions. Furthermore, we categorize these query features into counting features, which represent the number of times an element appears in the query, and ID features, which denote the position of the said element.

Among the counting features, query mode, c, represents the query commands in numeric form: if the query command is SELECT, it is represented by integer 1; if INSERT, integer 2; if UPDATE, integer 3; and if DELETE, integer 4. Query length, Q_L, signifies the length of the query, i.e., the number of characters that are present, including spaces. S_V and S_L denotes the number of string values and the length of the concatenated string values, respectively. The number of numeric values, JOINs, ANDs and ORs are denoted by N_V, J, and AO, respectively. The number of relations and attributes in a particular clause is represented by: P_R and P_A for the projection clause, S_A for the selection clause, O_A for the ORDER BY clause, and G_A for the GROUP BY clause. Lastly, we also extracted features by counting the number of attributes with respect to each table, signified by the convention notation $N_A[]$, which has the same

Table 1. Query vector fields, features, and their descriptions

Vector field	Description	Feature elements
SQL-CMD[]	Command features	query mode, c query length, Q_L
PROJ-REL-DEC[]	Projection relation features	Number of projected relations, P_R Position of projected relations, P_{RID}
PROJ-ATTR-DEC[]	Projection attribute features	$(P_A, P_A[], P_{AID}[])^a$
SEL-ATTR-DEC[]	Selection attribute features	$(S_A, S_A[], S_{AID}[])^a$
ORDBY-ATTR-DEC[]	ORDER BY clause features	$(O_A, O_A[], O_{AID}[])^a$
GRPBY-ATTR-DEC[]	GROUP BY clause features	$(G_A, G_A[], G_{AID}[])^a$
VALUE-CTR[]	Value counter features	Number of string values, S_V Length of string values, S_L Number of numeric values, N_V Number of JOINs, J Number of ANDs and ORs, AO

a. Convention $(N_A, N_A[], N_{AID}[])$:
N_A – number of attributes in a particular clause
$N_A[]$ – number of attributes in a particular clause counted per table
$N_{AID}[]$ – position of the attributes present in a particular clause, represented in decimal

number of elements as the number of tables in the schema. This last feature is applied to all four clauses, i.e., projection, selection, ORDER BY, and GROUP BY clauses.

ID features are also applied to all four clauses. The position of projected relations, $P_{RID}[]$, has a number of elements equal to the number of relations in the schema; i.e., if a certain relation/table is present in the query, it is denoted by binary number 1 in its corresponding position; otherwise, it is represented by 0 in that position. This produces a binary string, which is converted into its decimal form to get the final value of $P_{RID}[]$. The same logic is applied to the rest of the $N_{AID}[]$ features (the position of an attribute present in a particular clause), resulting in features which signify the position of the elements present in the query in decimal form.

3.3   Principal Components Analysis for Query Feature Selection

To alleviate the problem of scaling to the size of the database and to filter out meaningless features, we employ an efficient feature selection technique which effectively reduces feature dimensionality.

Principal components analysis is an unsupervised dimension reduction method which transforms a data set S into a new coordinate system to produce components $p \in P$. The top components, called principal components (PCs), are said to contain most of the variance of data set S. Clearly, trailing p's, which contain less of the variance of S, can be eliminated, effectively reducing dimensionality of S. Given a matrix of query features y and n query samples in data set S, S is centered on the means of each variable. The first principal component p_1 is given by the linear combination of the variables $Y_1, Y_2, \dots Y_y$,

$$p_1 = a_{11}Y_1 + a_{12}Y_2 + \dots + a_{1y}Y_y, \tag{1}$$

where the weights are governed by the constraint, $a_{11}^2 + a_{12}^2 + ... + a_{1y}^2 = 1$. The calculation of the second principal component, which accounts for the second highest variance of data set S, is pretty straightforward:

$$p_2 = a_{21}Y_1 + a_{22}Y_2 + ... + a_{2y}Y_y, \qquad (2)$$

with the condition that it is uncorrelated with p_1. This is performed until a number of y principal components has been calculated. The sum of the variances of all p's is always equal to the sum of the variances of all the variables of S, i.e., data are transformed but information is retained. The elements that fall in the diagonal of the variance-covariance matrix of the principal components P, are known as the eigenvalues, which denote the variance explained by each component p. We have shown that in our previous work [12], PCA is able to represent a large percentage of the total variance with only few components, i.e., it is lightweight yet effective. This is why it has also garnered a lot of attention in many fields, e.g., network intrusion detection, among others.

3.4   Random Forest for Anomaly Detection

Decision Trees. Decision trees (DT) are one of the most common and most simple techniques in data mining. A decision tree uses tree-like graph decisions based on information gain (IG). First, a query feature Y is selected for the root node, and the leaves of the trees are built according to a test function, which decides whether a query instance q of data set S should go to its left decision sub-tree A, or to its right decision sub-tree B. This is done recursively for each leaf until all query instances have the same class.

When building a DT, there are two basic goals: to get the smallest tree, and choose the attribute that produces the purest node. This attribute is determined by computing the IG of a given feature Y, which can be done by first calculating the entropy, or the measure of disorder of data, given by:

$$I(S) = -\sum_{k=1}^{K} \frac{|s_k|}{|s|} \log \frac{|s_k|}{|s|}, \qquad (3)$$

where s is the total number of query instances in data set S, s_k is the number of query instances in class $k \in K$. IG is then obtained by:

$$IG(Y) = I(S) - \sum_{m=1}^{M} \frac{|s_m|}{|s|} I(S_m), \qquad (4)$$

where the second term is the conditional entropy (the entropy with respect to feature Y), $I(S|Y)$, and s_m is the number of query instances in outcome $m \in M$. Understandably, the higher the IG of feature Y, the higher the chances for feature Y to be chosen at the root node, or the consecutive nodes after.

Decision trees are very easy to understand and can achieve high accuracy with little effort. However, they are prone to overfitting and feature bias. Ensemble methods, like random forest, were devised to alleviate these problems.

Random Forest. Random forest is an ensemble approach for classification which makes use of bagging and random feature selection to create numerous decision trees [17]. Given a random forest F composed of a collected of decision trees, $t=h(q, \Theta_y)$, $y \in Y$, where $\Theta_y \in Y$ are independent, identically-distributed random features, each tree t casts a unit vote for the most popular class/role r^* at input query instance $q \in S$, i.e.,

$$r^* = \arg\max_{r \in R} \sum_{t \in F} [h(q \mid t) = r]. \tag{5}$$

Combining the outputs of multiple decision trees into a single classifier supports the ability to generalize and diminishes the risk of overfitting. Moreover, about one-third of the query instances are left out of the bootstrap sample every time a tree is built. Using these left-out samples as test cases, an unbiased estimate of test set error, called out-of-bag (OOB) error, is obtained.

In addition to the OOB error, RF also gives good estimates of strength, correlation, and variable importance. Furthermore, it does not require tuning of many parameters and can handle very unbalanced data sets (such as database query access) [11][19].

4   Experiments

We have used the TPC-E database benchmark for all our experiments. TPC-E simulates the online transaction processing (OLTP) workload of a brokerage firm. The database schema is composed of 33 tables, with a total of 191 attributes [18]. The 11 TPC-E read-only and read/write transactions were modeled as user profile scenarios. For each transaction/profile, 1000 queries were generated based on its corresponding grammar file. This grammar file contains the rules and limits imposed for a given profile, according to its corresponding database footprint and pseudo-code found in [18]. We referenced the latter to set the tables, T, that a profile, r, is allowed to access, and the set of commands, C, that it is allowed to issue. We use the same probabilities in [12] that govern the rules and privileges for each role.

For our normal query log training set, 11,000 queries with its corresponding role labels were generated. Since we have built the system with insider threats in mind, we generated anomalous queries with the same distribution as the normal queries, only with the role annotation negated; i.e., if the role label of a query is role 1, we simply change it into any role other than role 1 [11]. 10-fold cross validation is performed for each run, transforming 30% of test instances for each fold to anomalous queries.

A total of 277 features have been extracted with respect to the TPC-E database schema. Applying PCA to our query log data set, a total of 144 PCs were obtained with 99.99% variance of the data. The run yielded a set of features Y_y in Eq. (1), namely: Q_L, AO, S_A, P_A, and P_R. Noticeably, only counting features were used to compute p_1. This denotes that counting features are the most basic variables and are essential for good enough classification performance [12].

Consequently, in [12], it was also found that although the PCA subsets produced by different threshold-determination methods greatly differ in the amount of features used, there was no significant change in the resulting false positive rates (FPR) and false negative rates (FNR). We further stress this point by plotting the true positive rate (TPR), true negative rate (TNR), FPR, FNR, and OOB error of generated random forest models with feature subsets of increasing number of PCs, as shown in Fig. 2.

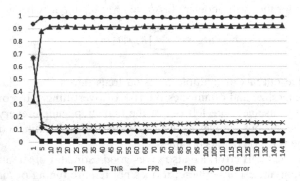

Fig. 2. TPR, TNR, FPR, and FNR with Increasing Number of Principal Components

Table 2. Confusion matrix of random forest using the PCA3 subset

	Role 1	Role 2	Role 3	Role 4	Role 5	Role 6	Role 7	Role 8	Role 9	Role 10	Role 11
Role 1	1000	0	0	0	0	0	0	0	0	0	0
Role 2	0	983	1	0	0	8	1	0	0	5	2
Role 3	0	2	987	1	4	0	0	5	0	1	0
Role 4	0	0	7	984	5	0	0	0	4	0	0
Role 5	1	10	1	7	893	12	0	71	1	3	1
Role 6	0	7	0	0	7	895	0	68	0	22	1
Role 7	0	1	2	0	7	0	828	3	5	16	138
Role 8	0	0	2	1	55	147	0	754	8	16	17
Role 9	0	3	11	4	2	0	7	10	950	0	13
Role 10	0	2	1	0	4	25	3	13	0	950	2
Role 11	0	3	0	0	13	3	173	32	35	40	701

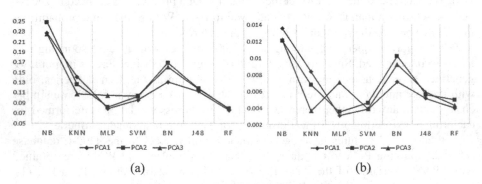

(a)   (b)

Fig. 3. Comparison of (a) FPR and (b) FNR of Different Classifiers

We increased the number of PCs by 5 and used 30 trees and 7 random features in each run (there was no change in FPR for tree configurations greater than 30, and varying the number of random features did not have a significant effect on the performance). All metrics showed the same dip in error rate (or peak, in terms of performance) around 10-15 PCs (in agreement to the number of features of the PCA3 subset in [12]). This shows that PCA was able to concentrate the variance of the whole data set in just a few leading PCs, while enabling us to safely eliminate other trailing ones.

Table 2 shows the confusion matrix of an RF 10-fold cross validation run using PCA3, which was obtained without injecting anomalies. Given that Roles 1-6 are read-only (RO) profiles and 7-11 are read/write (RW) profiles, it is evident that RO profiles are much easier to classify, since they only consist of SELECT commands. Those configured with INSERT, UPDATE, and DELETE commands are harder to distinguish from other RW roles. Some instances of Roles 9 and 10 were confused with other profiles because of their similarity in set of command C configurations. However, most of the other confused instances (of Roles 7, 8, and 11) were due to these profiles accessing almost the same set of tables.

The comparison of performance of RF with other data mining classifiers in terms of FPR and FNR is shown in Fig. 3. RF yielded the best performance in terms of FPR in all PCA subsets (PCA1 with 113 features, PCA2 with 63, and PCA1 with 13). On the other hand, in terms of FNR, MLP achieved the best results for PCA1 and PCA2, followed by SVM, with RF on third place. With PCA3, KNN produced the best FNR, followed by SVM and RF, while MLP's performance drastically dropped. It is interesting to note that both SVM's and RF's FNR performance remained consistent for all subsets.

It is clear that based on these results, the best contenders of RF are MLP and SVM. However, if we consider the build and detection time of these algorithms, RF wins by a mile. According to the results of our multiple cross validation runs, RF is almost 30 times faster than MLP in build time, and 5 times faster than SVM. Detection rate of RF is also 2 times faster than MLP, and almost 170 times faster than SVM. This goes to show that RF's computational overhead is very low compared to its contenders, and that it exhibits a good FPR and FNR trade-off performance on top of it.

5   Conclusion

In this paper, we have proposed the use of a graph-based algorithm, random forest, as the anomaly detection mechanism. To reduce the dimensionality of the query log data set, we applied principal components analysis, and showed that PCA effectively reduces the number of features while maintaining high classification performance. Moreover, RF, when compared to other data mining classifiers, exhibits competitive performance in terms of FPR, FNR, and time complexity.

Future works will include experimenting on more probability configuration for each profile to minimize the confusion between roles, and considering the sensitivity of tables and attributes in the database schema. Further work will also be geared to validate the usefulness of the proposed method with regards to SQL injection.

References

1. Lee, S.-Y., Low, W.L., Wong, P.Y.: Learning fingerprints for a database intrusion detection system. In: Gollmann, D., Karjoth, G., Waidner, M. (eds.) ESORICS 2002. LNCS, vol. 2502, pp. 264–279. Springer, Heidelberg (2002)
2. Huynh, V.H., Le, A.N.: Process mining and security: visualization in database intrusion detection. In: Chau, M., Wang, G., Yue, W.T., Chen, H. (eds.) PAISI 2012. LNCS, vol. 7299, pp. 81–95. Springer, Heidelberg (2012)
3. Jin, X., Osborn, S.L.: Architecture for data collection in database intrusion detection systems. In: Jonker, W., Petković, M. (eds.) SDM 2007. LNCS, vol. 4721, pp. 96–107. Springer, Heidelberg (2007)
4. Rajput, I.J., Shrivastava, D.: Data Mining based Database Intrusion Detection System: A Survey. Int'l Journal of Engineering Research and Applications (IJERA) 2(4), 1752–1755 (2012)
5. Hu, Y., Panda, B.: A Data Mining Approach for Database Intrusion Detection. ACM Symposium on Applied Computing, pp. 711-716 (2004)
6. Srivastava, A., Sural, S., Majumdar, A.K.: Database Intrusion Detection Using Weighted Sequence Mining. Journal of Computers 1(4), 8–17 (2006)
7. Barbara, D., Goel, R., Jajodia, S.: Mining Malicious Corruption of Data with Hidden Markov Models. In: Gudes, E., Shenoi, S. (eds.) Research Directions in Data and Applications Security, IFIP. IFIP, vol. 128, pp. 175–189. Springer, US (2003)
8. Ramasubramanian, P., Kannan, A.: Intelligent multi-agent based database hybrid intrusion prevention system. In: Benczúr, A.A., Demetrovics, J., Gottlob, G. (eds.) ADBIS 2004. LNCS, vol. 3255, pp. 393–408. Springer, Heidelberg (2004)
9. Ramasubramanian, P., Kannan, A.: A Genetic Algorithm Based Neural Network Short-term Forecasting Framework for Database Intrusion Prediction System. Soft Computing 10(8), 699–714 (2006)
10. Pinzón, C., Herrero, A., De Paz, J.F., Corchado, E., Bajo, J.: CBRid4SQL: a CBR intrusion detector for SQL injection attacks. In: Corchado, E., Graña Romay, M., Manhaes Savio, A. (eds.) HAIS 2010, Part II. LNCS, vol. 6077, pp. 510–519. Springer, Heidelberg (2010)
11. Kamra, A., Terzi, E., Bertino, E.: Detecting Anomalous Access Patterns in Relational Databases. The VLDB Journal 17(5), 1063–1077 (2008)
12. Ronao, C.A., Cho, S.-B.: A Comparison of Data Mining Techniques for Anomaly Detection in Relational Databases. Int'l. Conf. on Digital Society (ICDS), pp. 11-16 (2014)
13. Mathew, S., Petropoulos, M., Ngo, H.Q., Upadhyaya, S.: A data-centric approach to insider attack detection in database systems. In: Jha, S., Sommer, R., Kreibich, C. (eds.) RAID 2010. LNCS, vol. 6307, pp. 382–401. Springer, Heidelberg (2010)
14. Bockermann, C., Apel, M., Meier, M.: Learning SQL for database intrusion detection using context-sensitive modelling (Extended Abstract). In: Flegel, U., Bruschi, D. (eds.) DIMVA 2009. LNCS, vol. 5587, pp. 196–205. Springer, Heidelberg (2009)
15. Zhang, J., Zulkernine, M., Haque, A.: Random-Forests-Based Network Intrusion Detection Systems. Systems, Man, and Cybernetics 38(5), 649–659 (2008)
16. Elbasiony, R.M., Sallam, E.A., Eltobely, T.E., Fahmy, M.M.: A Hybrid Network Intrusion Detection Framework based on Random Forests and Weighted K-means. Ain Shams Eng'g. Journal 4(4), 753–762 (2013)
17. Breiman, L.: Random forests. Machine Learning 45(1), 5–32 (2001)
18. Transaction Processing Performance Council (TPC): TPC benchmark E, Standard specification, Version 1.13.0 (2014)
19. Cutler, A., Cutler, R., Stevens, J.R.: Tree-based methods. In: High-Dimensional Data Analysis in Cancer Research, pp. 1-19. Springer (2008)

Synthetic Evidential Study
for Deepening Inside Their Heart

Takashi Ookaki, Masakazu Abe, Masahiro Yoshino,
Yoshimasa Ohmoto, and Toyoaki Nishida(✉)

Graduate School of Informatics, Kyoto University, Sakyo-ku, Kyoto, Japan
{ookaki,abe,yoshino}@ii.ist.i.kyoto-u.ac.jp,
{ohmoto,nishida}@i.kyoto-u.ac.jp

Abstract. Synthetic evidential study (SES) is a novel technology-enhanced methodology that combines dramatic role play and group discussion to help people learn by spinning stories comprised of partial thoughts and evidence. The SES Support System combines a game engine and an augmented conversation technology to help with the facilitation of SES workshops. We performed a feasibility study with a partial implementation of SES and simplified SES sessions and obtained numerous supportive findings.

Keywords: Inside understanding · Group discussion and learning · Intelligent virtual agents · Theatrical role play · Narrative technology

1   Introduction

Inside understanding is deemed mandatory in establishing empathic relationships in a hybrid society that contains not only people but also social agents as first citizens. The long-term research goal underlying this paper is to establish a methodology that not only helps people achieve an inside understanding of each other but also endows artificial agents with the same ability, enabling both people and artificial agents to live and work in harmony with each other. Our approach relies on the sharing hypothesis [1], which implies that sharing the common ground is a key to empathy. Although group discussions and learning are powerful means of sharing knowledge about inside understanding, the success of such techniques depends largely on the cognitive ability of the participants, as the discussions are mostly based on indirect experiences brought about by secondary presentation materials. Moreover, the methodology is not directly applicable to help artificial agents feel empathy. We believe a solution to these problems is dramatic role playing, which not only helps people simulate subjective experience to delve deep inside a given role but also provides a useful way for artificial agents to obtain data about how people deepen their inside understanding. In fact, dramatic role playing has been used in various contexts of education [2]. Meanwhile, Cavazza et al [3] demonstrated that immersive interactive story telling is useful in gaining the first-person understanding of the story in the area of intelligent virtual agents. In order for such an approach to scale up, a significant work is necessary for effective content production.

© Springer International Publishing Switzerland 2015
M. Ali et al. (Eds.): IEA/AIE 2015, LNAI 9101, pp. 161–170, 2015.
DOI: 10.1007/978-3-319-19066-2_16

Synthetic evidential study (SES) combines dramatic role play and group discussion to help people spin stories by bringing together partial thoughts and evidence [4, 5]. The SES framework consists of SES sessions and interpretation archives. In each SES session, participants repeat a cycle consisting of dramatic role play, its projection into an annotated agent play, and a group discussion. In an SES workshop, one or more successive SES sessions are executed until participants come to an initial agreement.

In the dramatic role play phase, participants play respective roles to demonstrate their first-person interpretation in a virtual space. This allows them to interpret the given subject from the viewpoint of an assigned role. In the projection phase, an annotated agent play is produced on a game engine from the dramatic role play in the previous phase by applying oral edit commands (if any) to dramatic actions by the actors elicited from the behaviors of all the actors. We use this play for refinement and extension in the later SES sessions or even adapt it for reuse in a new context. In the group discussion phase, the participants or audience members share a third-person interpretation of the play performed by the actors for criticism. The actors then revise the virtual play until they are satisfied. The understanding of the given theme will be progressively deepened by repeatedly looking at embodied interpretation from the first- and third-person points of view.

The SES Support System provides computational support for the entire SES process. It combines a game engine and a conversation augmentation technology we developed for research on conversational informatics [6]. This system allows us to capture a group dramatic performance and create an agent play on Unity 3D so that participants can switch between objective and subjective views for each scene in an immersive environment consisting of large surrounding displays that provide a 360-degree view at any point in a virtual shared space.

We conducted a feasibility study with a partially implemented SES Support System. There were numerous encouraging results regarding the typical behavioral patterns of participants in role playing and how the inside understanding was deepened through the role play and discussion in the SES framework.

In this paper, we explain how SES is applied to deepen inside understanding in a group discussion. First we describe the conceptual framework of SES and discuss the technologies used to support SES with an emphasis on capturing group performance. We then present the insights obtained from the feasibility study. An interpretation archive is used to provide logistical support for the SES sessions. The annotated agent plays and stories resulting from SES workshops are decomposed into components for later use so that participants in subsequent SES workshops can adapt previous annotated agent plays and stories to use as a part of the present annotated agent play.

2   SES Support System

The aim of the SES Support System is to help facilitate the smooth management of an SES session cycle, as shown in Fig. 1. It supports the capture of human performances, which the system then converts into virtual agent behavior in the dramatic role play phase. Specifically, it provides agent role plays in an immersive virtual space, edits agent motions in the virtual space in the projection phase, and accumulates and structures potential knowledge from the discussion for evidential study.

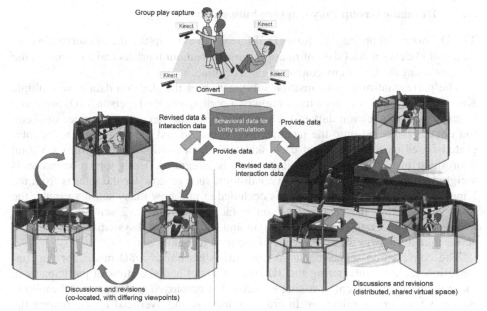

Fig. 1. SES Support System

2.1 The Core

The core of the SES Support System consists of an immersive collaborative interaction environment (ICIE) and a distributed elemental application linker (DEAL). ICIE is an immersive interaction environment made available by a 360-degree display and surround speakers and audio-visual sensors for measuring user behaviors. It consists of an immersive audio-visual display and plug-in sensors that capture user behaviors therein. The user receives an immersive audio-visual information display in a space surrounded by eight large displays about 2.5 m in diameter and 2.5 m in height. We have prototyped motion capture consisting of multiple depth sensors that can sense the user's body motion in this narrow space. This platform constitutes a "cell" that will allow users to interact with each other in a highly immersive interface. Cells can be connected with each other or with other (e.g., robot) interfaces so that multiple users can participate in interactions in a shared space.

DEAL is a software platform that permits individual software components to cooperate in providing various composite services for the interoperating ICIE cells. Each server has one or more clients that read/write information on a shared blackboard on the server. Servers can be interconnected on the Internet and the blackboard data can be shared. One can extend a client with plug-ins using DLL. Alternatively, a client can join the DEAL platform by using DLL as a normal application that bundles network protocols for connecting a server.

Leveraging the immersive display presentation system operated on the DEAL platform and the motion capture system for a narrow space allows the behaviors of human users to be projected onto those of an animated character who habits are shared in a virtual space. The ICIE+DEAL platform is coupled with the Unity (http://unity3d.com/) platform so that participants can work together in a distributed environment. This permits the participants to take the form of animated Unity characters in the virtual space.

2.2   Dramatic Group Play Capture Subsystem

The Dramatic Group Play Capture (DGPC) system can capture the 3D surface model data and skeleton model data of multiple persons without markers and can convert the data for using the Unity environment in a short time.

The participant-motion-estimation subsystem uses the skeleton data from multiple Kinect v2 sensors to estimate the motion of a participant. First, personal IDs are given to each piece of skeleton data of each Kinect v2 sensor. Second, each piece of skeleton data is projected onto the integrated coordination and the personal IDs are integrated based on the overlap of each skeleton's coordination data. Third, each joint coordinate of each piece of skeleton data is integrated. Each joint coordinate is weighted based on various heuristic conditions, such as how far the joint is from the Kinect v2 sensor, whether the joint is occluded or not, how many sensors capture the joint, and whether the captured person is facing the Kinect v2 sensor. In order to avoid any misunderstanding between right and left joints, a subsystem checks the time series data to confirm the consistency of the joint recognition.

The conversation-scene-reconstruction subsystem builds a 3D model for the conversation scene by integrating the 3D surface model and skeletons of participants in the conversation. Currently, a simple method is employed to integrate the multiple Kinect v2 sensor coordination. In order to decrease the overhead for calibrating the multiple Kinect v2 sensor coordination, the subsystem integrates them by using skeleton data and a depth map. First, a person is captured by multiple Kinect v2 sensors. Second, the skeleton data is used to calculate the relative positions of the Kinect v2 sensors. Third, measured 3D points near the matched skeleton points in the depth map are used for fine adjustment by using a least-square method. This method can recalculate the sensor coordination after the measuring by using the skeleton data from the measured scenes if the data has been correctly captured.

The system can capture multiple user interactions including up to four persons without markers at the same time. Each piece of skeleton data captured by the participant-motion-estimation subsystem can be output in real-time. In multiple user interactions, however, it is hard to identify adequate skeleton data because of occlusion and contact between participants. Therefore, in the conversation-scene-reconstruction subsystem, depth maps of each sensor are used for modifying the surface and skeleton model data. The processing takes 100 second to 300 second. Furthermore, the system cannot always correctly modify the data when the skeleton data is captured by less than two sensors. These problems should be improved in the future.

The projection subsystem can convert the captured skeleton data into character motion data in the Unity virtual environment and can play the motion data using the character. After the data has been converted, users can easily change the projection. For example, the same data can be shown on both a big flat display and a 360-degree immersive display. In addition, users can change the position of virtual cameras for showing the data on a screen. They can select an objective position for the overview an entire role play scene using a flat display in discussion and one of the role player's subjective position for experiencing the player's situation using an immersive display. Figure 2 shows multiple views of one shot of the role play data in an actual SES workshop. In this example, the blue character attacked the red character and the white character was a spectator. Figure 2 (a) shows a snapshot from a recorded video and

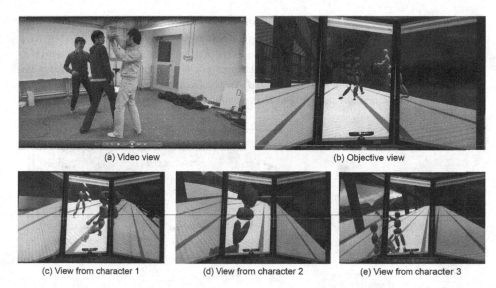

(a) Video view   (b) Objective view

(c) View from character 1   (d) View from character 2   (e) View from character 3

Fig. 2. Group play capture by SES Support System

Fig. 2 (b) shows the same view reconstructed in the Unity virtual environment. Figure 2 (c) (d) (e) shows the subjective views of each character. In the actual SES session, the scene was played as a movie on a life-size screen. As shown in Fig. 2 (d), the perception from the person who was attacked is very different from the objective view shown in Fig. 2 (b). The perception from the person who was a spectator is also different from the objective view. These indicate that different perspectives provide useful clues to investigate principles of human behavior in SES sessions.

3   A Feasibility Study

We conducted a feasibility study using one simplified SES workshop. The simplified SES session consisted of the group discussion phase based on document-based discussions (about 30 minutes) followed by the role play phase (about 20 minutes) and the projection phase (about 10 minutes) in which participants discussed with each other while switching between the objective and subjective views of one of the three roles in the scenario. A partial implementation of the SES Support System was used to support the feasibility study. Figure 3 shows an overview.

3.1   Experiment Setting

We used the "Matsu no Roka" ("Hallway of Pine Trees") passage of Chushingura as a sample scenario. This fictionalized account based on a historical incident roughly reads like this:

At the beginning of the 18th century, a feudal lord named Asano Takumi-no-kami Na-ganori was in charge of a reception for envoys from the Imperial Court in Kyoto. Another feudal lord, Kira Kozuke-no-suke Yoshinaka, was appointed to instruct Asano in the ceremonies. On the day of the reception, while Kira was talking with Yoriteru

Kajikawa, a lesser official, at "Matsu no Roka" ("Hallway of Pine Trees") in Edo Castle, Asano came up to them screaming "This is for revenge!!" and slashed Kira twice with a short sword. Soon after the incident, Kajikawa restrained Asano, who was then imprisoned. The reason for the attack was not known, though it was widely believed that Kira had somehow humiliated Asano. Ultimately Asano was sentenced to commit seppuku, a ritual suicide, but Kira went without punishment.[1]

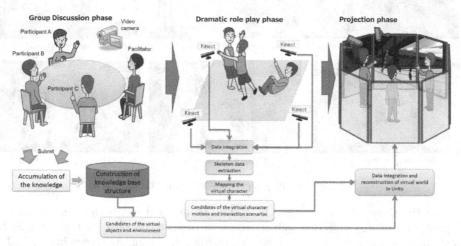

Fig. 3. Procedure for Feasibility Study

We hired three participants (males) who knew each other and asked them to serve as hypothetical inspectors of historical affairs and determine a sentence for Asano as a result of group discussions. The participants went through a simplified SES session consisting of three phases: group discussion, role play, and projection. The participants were asked over the course of these three phases to investigate the incident, propose a legal judgment for it, and explain their reason for reaching their decision. The conclusion had to be unanimous within the group. At the end of each phase, we asked the participants to write a brief explanation of how each of them interpreted the "Matsu no Roka" affair. We videotaped the discussions for later analysis. This simplified version of an SES session took about one hour.

1) Group Discussion Phase
We provided the participants with historical documents about Asano, Kira, and Kajikawa and had them conduct a group discussion about the affair for about 30 minutes. The documents were based on Wikipedia.

2) Role Play Phase
We allocated the roles of Asano, Kira, and Kajikawa to the three participants and had them reproduce the "Matsu no Roka" affair on the basis of their interpretation. We captured the dramatic role play by the DGPC subsystem to produce an agent play on Unity 3D. It took about 30 minutes.

---

[1] Based on multiple sources, e.g., http://en.wikipedia.org/wiki/Ch%C5%ABsh ingura.

3) Projection Phase
We used ICIE to show the participants the agent play reproduced from their role play. Objective and subjective views from each player were displayed repeatedly until the participants were satisfied. We did not allow the participants to revise any parts of this agent play. It took about 10 minutes.

3.2   Findings

We obtained five major findings from this feasibility study.

1) Participants' Behavior in Role Playing
The behavior of the participants in the role play phase can be classified into four patterns: (a) acting behavior, (b) commenting behavior, (c) oral editing behavior, and (d) idling behavior. Acting behavior is what the participants do when they are actually acting. Commenting behavior is a critique of the incidents and the acting, including reasoning, discussion and thinking aloud. The editing behavior is suggested revisions to the acting, e.g., "This action could have been better." It can be accompanied by gestures and examples of preferred acting behavior. Idling behavior includes all actions that are not classified above—in other words, irrelevant movement, such as stretching or making the motion of swinging a bat to loosen up the body.

Further analysis of the role play phase revealed that it could be roughly classified into two parts: rehearsal acting scenes and production acting scenes. A rehearsal acting scene is a scene in which acting and discussions are mixed. Acting a scene while checking the position and behavior of each character in the discussion and thinking aloud (essentially, a continuation of the discussion in the group discussion phase) was observed. In a rehearsal acting scene, special acting behavior such as repeatedly mimicking the acting of different roles was observed. It seems that mimicking the acting of others was an unconscious attempt to experience the viewpoint of others. A production acting scene is a scene in which the acting is performed from the beginning to the end, and discussion and thinking aloud are not observed. Switches between the rehearsal acting scene and production acting scene can be clearly identified by eye. For example, just before the production acting scene, explicit signaling behavior such as giving-a-cue was observed.

In a rehearsal acting scene, acting behavior, oral editing behavior, speaking his/her role, commenting behavior, thinking aloud, and idling behavior were observed. On the other hand, in a production acting scene, acting and speaking his/her role were observed while oral editing, commenting, thinking aloud, and idling were not.

Twelve detailed behaviors were observed in the role play phase: (1) acting, (2) commenting, (3) oral editing, (4) idling, (5) speaking his/her role, (6) acting + thinking aloud, (7) acting + commenting, (8) acting + oral editing, (9) acting + speaking his/her role, (10) idling + commenting, (11) idling + oral editing, and (12) idling + speaking his/her role.

2) Deepening by Playing
In the role play phase, progress of the discussion occurred. We confirmed details that are not usually described in historical materials by remarks from the participants such

(a) Acting is modified: "Kajikawa jumped over
the Kira that fall, and go straight toward Asano (rather than to take a roundabout path)."

(b) Discussions with gestures and body
movements

Fig. 4. Snapshots from the feasibility study

as, "He could probably help earlier", "Probably, he thrust with the sword rather than swinging the sword down", and "Perhaps, the distance between the two should have been closer", as shown in Fig. 4(a). However, the content of the discussion in the role play phase was biased toward content that had already been discussed in the group discussion phase and revealed through the acting of each person.

3) Deepening Based on Contrasting Objective and Subjective Views
In the projection phase, we observed contrasting objective and subjective views. After experiencing all perspectives on the action of Kira falling prone, one participant re-marked that "It seems strange in the objective viewpoint, but when I experience the subjective viewpoint, it looks like a natural movement", and everyone agreed. They considered it and remarked "I don't know. Which viewpoint is correct?" This event is evidence that the presentation of a subjective viewpoint provided the participants with a more intense experience than could be obtained from the objective viewpoint alone.

4) Deepening Based on Subjective View Transfer
In the projection phase, when the participants had experienced Kajikawa's subjective viewpoint, the participant who had played the role of Kira in the acting phase (the Kira player) remarked, "(Kajikawa was too slow to) try and hold (Asano) down after having been slashed" while gesturing (Fig. 4(b)). In contrast, the participants who played the roles of Kajikawa and Asano had no such feeling. This probably means that the Kira player thought that Kira had hoped to be helped earlier. We suspect that the memory of acting affects the subsequent thinking and discussion. In addition, soon after this, the Kira player remarked, "When (Asano) swung his sword up the first time, Kajikawa might have been farther away from Asano (and therefore, Kajikawa couldn't quickly hold down Asano)". This remark suggests that, through repeatedly experiencing Kajikawa's viewpoint, the Kira player could imagine the feeling of Ka-jikawa and thus change his opinion.

5) Implicit Opinion Expression by Acting
In the acting phase, we observed an interesting case. The participant playing Kira continued to repeat the action of falling prone in the production acting scene, in spite of the opinion that "It is unnatural for Kira to fall prone", which was shared with eve-ryone in the rehearsal acting scene. The Kira player himself remarked that falling prone is unnatural, and he actively tried to collapse on the spot instead. Furthermore, the other two participants demonstrated the action of staggering behind, but still he

never changed. We feel this phenomenon is an example of implicit opinion expression via acting. In the acting phase, the change of a participants' inner heart, which cannot be observed in the form of speech, may be expressed by acting.

According to the notes after the acting phase, the participants described Kira as having collapsed on the spot. However, in the memo after the projection phase, the Kira player wrote, "If the assumption that Kira fell prone is correct, I think that our acting is correct." This means that an opinion that was not observed previously had been exposed by reviewing their acting from the subjective/objective viewpoint. We feel that this implicit opinion expression has been overlooked in traditional workshops with theater. By modifying acting as editing behavior in the projection phase, further progress on the inner understanding of participants can be expected.

4   Discussions

The results of the feasibility study are very promising in that we have obtained plenty of evidence to support our initial expectation that SES is an effective method for deepening inside understanding. One concern during the initial discussions was the trade-off between physical role play and agent play. Physical role play appeared more realistic and depended too much on the skill of the actor. In this feasibility study, non-experienced participants were the actors, and their interpretation was not always properly expressed as far as the video analysis is concerned. For example, they did not wear historically accurate clothing, and their unintentional shyness appeared to hinder realistic acting from time to time. In contrast, although agent play was less expressive, it hid excessive details and allowed for soliciting space for interpretation in the discussion phase. Overall, the employment of agent play powered by Unity 3D was effective. In addition to our initial expectation, it seems that participants were also able to deepen their understanding of how other participants interpret the role he or she played.

There are several issues remaining for future research. We believe that the following are three of the most challenging. The first is intelligent role play support that can watch the role play like a stagehand and offer assistance if needed, e.g., setting-up/updating the background, virtually creating props or historical clothing, and virtually editing the film as the participants like, to name just a few. The second is story archiving that allows for community-wide sharing of agent plays so that interested members can refine or extend their interpretations. The third is discussion facilitation in which an intelligent agent mediator facilitates group discussions.

5   Conclusion

In this paper, we presented initial findings obtained in a feasibility study of SES. The results were quite encouraging. Major findings include participants' behavior in role playing, deepened understanding by playing, deepened understanding based on contrasting objective and subjective views, deepened understanding based on subjective view transfer, and implicit opinion expression by acting.

Acknowledgment. This study has been carried out with financial support from the Center of Innovation Program from JST, JSPS KAKENHI Grant Number 24240023, and AFOSR/AOARD Grant No. FA2386-14-1-0005

References

1. Nishida, T.: Towards Mutual Dependency between Empathy and Technology, 25th anniversary volume. AI & Society **28**(3), 277–287 (2013)
2. Hawkins, S.: Dramatic Problem Solving: Drama-Based Group Exercises for Conflict Transformation, Jessica Kingsley Publishers (2012)
3. Cavazza, M., Lugrin, J.L., Pizzi, D., Charles, F.: Madame Bovary on the holodeck: immersive interactive storytelling. In: MULTIMEDIA 2007: Proc. of the 15th International Conference on Multimedia, ACM, pp. 651-660 (2007)
4. Nishida, T., Abe, M., Ookaki, T., Lala, D., Thovuttikul, S., Song, H., Mohammad, Y., Nitschke, C., Ohmoto, Y., Nakazawa, A., Shochi, T., Rouas, J.-L., Bugeau, A., Lotte, F., Zuheng, M., Letournel, G., Guerry, M., Fourer, D.: Synthetic evidential study as augmented collective thought process – preliminary report. In: Nguyen, N.T., Trawiński, B., Kosala, R. (eds.) ACIIDS 2015. LNCS, vol. 9011, pp. 13–22. Springer, Heidelberg (2015)
5. Nishida, T., Nakazawa, A., Ohmoto, Y., Nitschke, C., Mohammad, Y., Thovuttikul, S., Lala, D., Abe, M., Ookaki, T.: Synthetic Evidential Study as Primordial Soup of Conversation. In: Chu, W., Kikuchi, S., Bhalla, S. (eds.) DNIS 2015. LNCS, vol. 8999, pp. 74–83. Springer, Heidelberg (2015)
6. Nishida, T., Nakazawa, A., Ohmoto, Y., Mohammad, Y.: Conversational Informatics–A Data-Intensive Approach with Emphasis on Nonverbal Communication, Springer (2014)

Multi-criteria Decision Aid and Artificial Intelligence for Competitive Intelligence

Dhekra Ben Sassi[1(✉)], Anissa Frini[2],
Wahiba Ben AbdessalemKaraa[1], and Naoufel Kraiem[1]

[1] RIADI Laboratory, National School of Computer Science, La Manouba, Tunisia
bensassi.dhekra@yahoo.fr, {wahiba.bak,naoufel.kraiem}@gmail.com
[2] Department of Management Sciences, University of Quebec at Rimouski,
Levis, Rimouski, Canada
Anissa_Frini@uqar.ca

Abstract. In this paper, we started by showing that artificial intelligence and multi-criteria decision aid are two closely concepts that could benefit from each other. Then, we propose to use this relation and complementarity to deal with competitive intelligence problem and issues. A discussion towards building the foundations of new domain knowledge named the multi-criteria intelligence aid (MCIA) is presented.

Keywords: Competitive intelligence · Multi-criteria decision aid · Artificial intelligence

1 Introduction

To define the research area, we start from the basic idea which supposes a relationship and a complementarity between multi-criteria analysis and artificial intelligence (AI) and from the fact that those two techniques could benefit from each other. Indeed, solving problems and making the best choice of action is a challenge for any decision maker. An appropriate decision is necessarily the result of a good situation awareness (related to the concept of intelligence) combined with a good decision-making pro-cess. So, it is obvious that there is a link to be established between artificial intelli-gence and decision making areas as specified [44].

After admitting this relationship, we start from the idea of integrating the ability of ar-tificial intelligence in managing large amount of technical data/information and in ex-tracting knowledge with the ability of multi-criteria decision aid (MCDA) methods in managing conflicting criteria for complex decision-making problems. To motivate the research, we propose a practical case in the context of competitive intelligence (CI) and more specifically for competitors' strategies prediction. First in this context, the competi-tor decision might be influenced by multiple conflicting criteria, not commensurable, measurable either on quantitative or qualitative scales. Second, to predict the strategies of competitors, we have to deal with large amount of open source data. In addition, several constraints and considerations related to the competi-tive and dynamic nature of the business environment must be considered simultaneously. Therefore, competitive intelli-gence appears an interesting application area for our research.

M. Ali et al. (Eds.): IEA/AIE 2015, LNAI 9101, pp. 171–178, 2015.
DOI: 10.1007/978-3-319-19066-2_17

This paper will be organized as follows: In the next section, we will argue our choice of studying the competitive intelligence. Section 3 will be devoted to the relevant literature review combining the two techniques i.e. multi-criteria analysis and artifi-cial intelligence. Section 4 will then present our proposed solution. Finally, some future research perspectives will be addressed in the last section.

2   Why the Competitive Intelligence

Competitive intelligence (CI) is a sub-domain or a sub-branch of business intelligence that deals with the competitive environment of the company. The term was defined by the Society of CI Professionals SCIP [49] as "a systematic and ethical program for gathering, analyzing and managing external information that can affect your company's plans, decisions and operations". After an exhaustive review of the literature, we found that no single definition of CI is likely to be precise and universally accepted [40, 10, 12, 46]. If we return to the origins of the concept, we found that it is related to the military domain) and its roots extends back over 5,000 years of Chinese history [51, 12]. Authors of CI consider that the earliest reference is "The Art of War" by Sun Tzu [26]. This problem differs from classical decision problems where the analyst intervenes to help a decision maker to clarify preferences, structuring the problem and offering a recommendation. It is a situation where the decision maker (the competitor) is not involved in the decision making process so its preferences and strategies will be predicted since they are unknown. So it is a context of reasoning under uncertainty. When doing the literature review of CI, we have found that all existing works define the concept of CI and propose a scheme of the CI process and its stages, but there is no works that touched the practical aspect of the field or developed a complete CI solution that can be delivered to the decision maker. This is due to the difficulty of information collection and action anticipation about its competitor. This difficulty is due to several challenges :

- Information is not available and the data collection could be difficult because the data to be collected is not internal but external to the company.
- The competitor, for whom the decision will be anticipated, is absent. There will be no negotiation or validations.
- The preferences of the competitor should be predicted and modeled. All stages of the anticipation of the decision will be made in a context of uncertainty.

3   State of Art

The literature review on the complementarily of AI and multi-criteria decision aid techniques, identified two lines of research in this context. On one hand, there are researches which involve the use of artificial intelligence techniques to enrich and improve the existing MCDA methods. On the other hand, other researches applied the MCDA methods to enrich the artificial intelligence techniques. These methods are indeed complementary and combining them adds value to each one. In [44], Pomerol argues that AI refers to cognitive processes and reasoning. Since any decision requires reasoning, one can easily understand why the relationship and complementarity of the two concepts is beneficial for both techniques.

3.1   The Use of MCDA Methods to Enrich artificial Intelligence Techniques

The combined use of artificial intelligence and multi-criteria decision aid is embodied in the "sensemaking" concept. For example, the method of case based reasoning has been enhanced by the use of multi-criteria analysis [5]. Case-Based Reasoning consists in searching among the stored cases in the knowledge base and see if there is a close confronted case. It is a kind of pattern matching. If a case is close enough to the current case, then the same actions made in the nearest case will be made very quickly. If no similarity is found, the decision problem is submitted to either new ad-hoc reasoning or a change according the highest recorded pocket case [44]. Authors [5] argue the complementarity between the MCDA and the "Case-Based Reasoning" to solve decision problems by proposing to use the multi-criteria methods ELECTRE I and II for finding a solution in the case base. The results show the strength of the proposed approach. In the same context, the AI techniques have been developed to deal with uncertainty in the context of prediction. For example, in [20], authors proposed a model that incorporates both the ability of Bayesian networks to resonate under uncertainty, and the ability of a multi-criteria approach to consider simultaneously multiple attributes. Other researchers have integrated MCDA and AI techniques such as the integration of multi-criteria in data mining and classification [36, 38].

3.2   Using Artificial Intelligence to Enhance the Multi-criteria Decision Aid

Several studies have been developed in this context such as the work of Greco et al. [25] which have extended the theory of rough sets to fit with multi-criteria decision aid. The theory of rough sets is a tool for extracting knowledge from uncertain and incomplete data. This extension in the theory of rough sets is due to the fact that the two concepts belong to two areas or two "worlds" which have no common language. More specifically, the terminology of rough sets theory does not include terms such as criteria, selection, classification, sorting, etc. On the other hand, the MCDA terminology does not use terms like approximation, reduction, decision rule, etc. Hence, there has been a need for a common language, which led to the birth of a new methodology for modeling and exploitation of preferences in terms of decision rules. These rules are induced from the preference information given by the decision maker as examples of decisions. The induction of rules from examples is a typical approach to artificial intelligence. Another more recent work [22] has been developed in 2007. Author used the techniques of artificial intelligence to solve a particular multi-criteria decision problem in which the expert expressed his knowledge in linguistic terms, so that only linguistic information was available. The authors used the techniques of natural language processing and data mining to manage linguistic information and the fuzzy set theory to quantify this information. Another technique of artificial intelligence was adopted by Chen [13] to enhance multi-criteria decision aid methods which is the "Screening". It is a multi-criteria decision process in which a large number of alternatives are reduced to a smaller set that contains the best choice. Chen has devel-oped a distance case based model to solve the problem of "screening".

3.3   The Combined Use of the Two Approaches

Several studies based on the combination of the two techniques have been conducted on the prediction of business failure. Business failure is a useful tool to ensure the financial health of companies. Statistical methods were used to predict business failures since the 1960s [7, 3]. With the development of artificial intelligence, intelligent methods have been introduced to predict business failures. Although these methods are costly in terms of time, they can produce more accurate predictions than the statistical methods. Decision trees [21], neural networks [60], case-based reasoning [32] support vector machines [47] were used for the prediction of business failure. Data mining has also been used to predict business failures [50] by finding hidden patterns that can distinguish failing of those healthy companies. Recently in [36], authors have attempted to integrate decision support techniques with intelligent and statistical techniques to implement Business failure prediction. The same team has developed [35] a system failure prediction by integrating over-classification approach to decision making, intelligent CBR technique and the statistical method of k-NN. For CI, at the best of our knowledge, the MCDA methods have never been applied for solving CI problem or for anticipating competitor future decisions.

After presenting the state of the art related to the combination of MCDA methods and AI techniques, we will return to our topic which is CI and the challenges related to the prediction of competitors decisions and actions. The next section will present our preliminary solution to address these challenges.

4   Proposed Solution

To deal with problems cited in the previous sections, we propose to develop a novel multi-criteria approach to anticipate the decisions of competitors. We aim to integrate the ability of MCDA methods to manage conflicting criteria in a complex environ-ment, with the ability of artificial intelligence to manage large amount of technical data/information in such context and their ability to extract knowledge which serve as inputs in the decision-making process. Our research objective is the development of a new multi-criteria approach to anticipate the decisions of competitors in the context of competitive intelligence. We believe that the techniques of artificial intelligence could be combined with multi-criteria methods to anticipate competitor decisions. The re-search could help building the foundations of new domain knowledge: multi-criteria intelligence aid (MCIA). This is an ongoing research project and so far the body of knowledge addressing this research topic is very limited. Let us note that Jakubchak [30] has conducted a research to estimate the value of using multi-criteria analysis for intelligence analysts. This research work tested empirically with controlled experiments the value of using MCDA to support analysts in their intelligence work and conclude that it is likely to be a valuable method to use when conducting intelligence analysis. Moreover, this interest towards combining the two areas is shared by the scientific community. Increasingly, multi-criteria community is interested in combin-ing multi-criteria decision methods with the artificial intelligence techniques. During the last year, the European Working Group on multi-criteria decision has devoted his biannual meeting specifically for this theme and the "International Journal of Multi-criteria Decision Making" has launched in September 2012 a call for proposals for articles on the theme "multi-criteria decision and artificial intelligence." There is therefore an increased commitment to this issue and an awareness of its relevance. To achieve the main goal and develop a multi-criteria CI method, we have several specif-ic objectives to address:

- Propose a method for competitor preferences prediction and modeling. For this component, the learning techniques of artificial intelligence and automated reasoning techniques will be explored and an approach will be proposed.
- Develop a multi-criteria aggregation method in the context of uncertainty that will anticipate potential competitor decisions. Fuzzy sets and linguistic variables will be explored.
- Develop an approach to competitor action generation. For this part, the artificial intelligence techniques, especially automated reasoning techniques will be explored and an approach will be proposed.

5   Conclusion

The basic idea of this paper was to integrate the capacity of multi-criteria methods to satisfy several conflicting criteria, resulting from the complex competitive environment, with the ability of artificial intelligence techniques to manage information, develop situation awareness, "sense making" and reasoning under uncertainty. In future research, the artificial intelligence techniques, especially automated reasoning techniques will be explored and an approach combining artificial intelligence and multi-criteria decision aid techniques will be proposed.

References

1. Kamal, A.: Competitive intelligence in business decision-An overview. IEEE Engineering Management Review **36**(3), 17–34 (2008)
2. Alden, B.: Competitive Intelligence: information, espionage and decision-making; a special report of businessman. Mass., C.I. Associates, Watertown (1959)
3. Altman, E.: Financial ratios, discriminant analysis and the prediction of corporate bankruptcy. Journal of Finance **23**, 589–609 (1968)
4. Andriole, S.: The collaborate/integrate business technology strategy. Communications of the ACM **49**(5), 85–90 (2006)
5. NegarArmaghan and Jean Renaud: An application of multi-criteria decision aids models for Case-Based Reasoning. Information Sciences **210**, 55–66 (2012)
6. Donald Ballou and GiriTayi: Enhancing data quality in data warehouse environment. Communications of the ACM **42**(1), 73–78 (1999)
7. Beaver, W.: Financial ratios as predictors of failure. Journal of Accounting Research **4**, 71–111 (1966)
8. Bernhardt, D.: I Want It Fast, Factual, Actionable' - Tailoring Competitive Intelligence to Executives' Needs. Long Range Planning **27**(1), 12–24 (1994)
9. Bilich, F., DaSilva, R., Ramos, P.: Flexibility in Multicriteria Competitive Intelligence for Operations Management. XI SIMPEP Bauru, SP, Brasil (2004)

10. Boncella, R.: Competitive intelligence and the web. Communications of the Association for Information Systems **12**, 327–340 (2003)
11. Bose, I.: Deciding the financial health of dotcoms using rough sets. Information Management **43**, 835–846 (2006)
12. Calof, J., Wright, S.: Competitive intelligence: A practitioner, academic and inter- disciplinary perspective. European Journal of Marketing **42**(7/8), 717–730 (2008)
13. Chen, Y., Hipel, M.K.K.: A casebased distance method for screening in multiplecriteria decision aid. Omega **36**, 373–383 (2008)
14. Clark, T., Jones, M., Armstrong, C.: The Dynamic Structure of Management Support Systems: Theory Development, Research Focus, and Direction. MIS Quarterly **31**(3), 579–615 (2007)
15. Cui, G.: Man Wong andHon-KwongLui. Machine learning for direct marketing response models: Bayesian network with evolutionary programming, Management Science **52**(4), 597–612 (2006)
16. DaSilva, R., Bilich, F.: Kwowledge management: multicriteria analysis of intellectual capital. FACEF Pesquisa **12**(3) (2009)
17. Dasilva, R., Gomes, L., Bilich, F.: Valuation and optimization of intellectual capital: a multicriteria analysis. REAd **12** (2) (2006)
18. Eckerson, W.: New ways to organize the BI team. Business Intelligence Journal **11**(1), 43–48 (2006)
19. Fan, W., Wallace, L.: Stephanie Rich andZhongju Zhang. Tapping the power of text mining. Communications of the ACM **49**(9), 77–82 (2006)
20. Fenton, N., Neil, M.: Making decision: using Bayesian nets and MCDA. Knowledge-Based Systems **14**, 307–325 (2001)
21. Frydman, H., Altman, E., Kao, D.-L.: Introducing recursive partitioning for financial classification: the case of financial distress. Journal of Finance **40**, 269–291 (1985)
22. Garcia-Cascales, S., Lamata, T.: Solving a decision problem with linguistic information. Pattern Recognition Letters **28**, 2284–2294 (2007)
23. Guy, H.: Gessnera and Linda Volonino. Quick response improves returns of business intelligence investments, Information Systems Management **22**(3), 66–74 (2005)
24. Goria, S.: Knowledge Management et Intelligence Economique: deux notions aux passés proches et aux futurs complémentaires, Informations, Savoirs. Décisions et Médiations (ISDM) **27**, 1–16 (2006)
25. Greco, S., Matarazzo, B., Slowinski, R.: Rough sets theory for multicriteria decision analysis. European Journal of Operational Research. **129**, 1–47 (2001)
26. Griffith, S.: Sun Tzu: The Art of War. Oxford University Press, New York (1971)
27. Guyton W.J.: A guide to gathering marketing intelligence, in Industrial Marketing. March, 84-88 (1962)
28. Clyde, W.: Holsapple and Mark Sena. ERP plans and decision-support benefits, Decision Support Systems. **38**, 575–590 (2005)
29. Claudia Imhoff. Keep your Friends Close, and your Enemies Closer, DM Review, 4 (13)4: 36-37 (2003)
30. Jakubchak, L.N.: The Effectiveness of MultiCriteria Intelligence Matrices in Intelligence Analysis, Mercyhurst College (2009)
31. Greenough, J.B: Business Wargames, Editions Abacus Press (1985)
32. Jo, H., Han, I., Lee, H.: Bankruptcy prediction using case-based reasoning, neural network and discriminant analysis for bankruptcy prediction. Expert Systems with Applications **13**, 97–108 (1997)

33. Jourdan, Z., Rainer, R.K., Marshall, T.E.: Business intelligence: an analysis of the litera-ture. Information Systems Management **25**(2), 121–131 (2008)
34. King, J.: 5 BI Potholes to Bypass, Computerworld, 41: 38-40, September (2007)
35. Li, H., Sun, J.: Business failure prediction using hybrid2 case-based reasoning (H2CBR). Computers & Operations Research **37**, 137–151 (2010)
36. Li, H., Sun, J.: Hybridizing principles of the Electre method with case-based reasoning for data mining: Electre-CBR-I and Electre-CBR-II. European Journal of Operational Re-search **197**, 214–224 (2009)
37. Lönnqvist, A., Pirttimäki, V.: The measurement of business intelligence. Information Sys-tems Management **23**(1), 32–40 (2006)
38. Ma, L.C.: A two-phase case-based distance approach for multiple-group classification problems. Computers & Industrial Engineering **63**, 89–97 (2012)
39. Massa, S., Testa, S.: Data warehousein-practice: exploring the function of expectations in organizational outcomes. Information Management **42**, 709–718 (2005)
40. Nasri, W.: Competitive intelligence in Tunisian companies. Journal of Enterprise Informa-tion Management **24**(1), 53–67 (2011)
41. Negash, S.: Business intelligence. Communications of the Association for Information Systems **13**, 177–195 (2004)
42. Novintel Inc. (2003). http://www.novintel.com/
43. Olszak, C.M., Ziemba, E.: Business intelligence systems in the holistic infrastructure de-velopment supporting decision-making in organizations. Interdisciplinary Journal of In-formation, Knowledge, and Management **1**, 47–58 (2006)
44. Pomerol, J.-C.: Artificial intelligence and human decision making. European Journal of Operational Research **99**, 3–25 (1997)
45. An empirical investigation: Thiagarajan Ramakrishnan, Mary C. Jones and Anna Sidorova. Factors influencing business intelligence (BI) data collection strategies. Decision Support Systems **52**, 486–496 (2012)
46. Shamel, C.: The Competitive Intelligence Process and How to Locate Information. San Diego Association of Law Librarians San Diego, California (2001)
47. Shin, K.S., Lee, T.S., Kim, H.J.: An application of support vector machines in bankruptcy prediction model. Expert Systems with Applications **28**, 127–135 (2005)
48. Simon, H.A.: The new science of management decision, Editions Harper et Row (1960)
49. Society of Competitive Intelligence Professionals (SCIP) (2008). www.scip.org/content. cfm?itemnumber¼2226&navI tem Number¼2227 (accessed 11 February)
50. Sun, J., Li, H.: Data mining method for listed companies' financial distress prediction. Knowledge-Based Systems **21**, 1–5 (2008)
51. Tao, Q., Prescott, J.: China: competitive intelligence practices in an emerging market envi-ronment. Competitive Intelligence Review **11**(4), 65–78 (2000)
52. Thomas, J. Group Inc.: (2003). http://www.mindspring.com/~jt-group/default.htm
53. Thomas, J.H. Jr.: business intelligence why, eAI Journal (2001)
54. Vitt, E., Luckevich, M., Misner, S.: Business Intelligence: Making Better Decisions Faster, p. 202. Microsoft Press, Washington (2002)
55. Viva Business Intelligence Inc.: Introduction to Business Intelligence. Helsinki: Pro-How Paper, vol. 1, p. 18 (1998)
56. Watson, H.J.: Three targets for data warehousing. Business Intelligence Journal **11**(4), 4–7 (2006)
57. Watson, H.J., Abraham, D., Chen, D., Preston, D., Thomas, D.: Data warehousing ROI: justifying and assessing a data warehouse, Business Intelligence Journal, 6–17 (2004)

58. Watson, H.J., Volonino, L.: Customer relationship management at Harrah's entertainment. In: Forgionne, G.A., Gupta, J.N.D., Mora, M. (eds.) DecisionMaking Support Systems: Achievements and Challenges for the Decade. Idea Group Publishing, Hershey, PA (2002)
59. Watson, H.J., Wixom, B.H., Hoffer, J.A., AndersonLehman, R., Reynolds, A.M.: Realtime business intelligence: Best practices in Continental Airlines. Business Intelligence **23**(1), 7–18 (2006)
60. Williams, S., Williams, N.: The Profit Impact of Business Intelligence, Morgan Kaufmann (2007). Wilson, R.L. and Sharda, R., Bankruptcy prediction using neural networks. Decision Support Systems, 11: 545–557, (1994)

Ontology-Based Fuzzy-Syllogistic Reasoning

Mikhail Zarechnev[(✉)] and Bora İ Kumova

Department of Computer Engineering, İzmir Institute of Technology,
35430, Urla/İzmir, Turkey
{mikhailzarechnev,borakumova}@iyte.edu.tr

Abstract. We discuss the Fuzzy-Syllogistic System (FSS) that consists of the well-known 256 categorical syllogisms, namely syllogistic moods, and Fuzzy-Syllogistic Reasoning (FSR), which is an implementation of the FSS as one complex approximate reasoning mechanism, in which the 256 moods are interpreted as fuzzy inferences. Here we introduce a sample application of FSR as ontology reasoner. The reasoner can associate up to 256 possible fuzzy-inferences with truth ratios in [0,1] for every triple concept relationship of the ontology. We further discuss a transformation technique, by which the truth ratio of a fuzzy-inference can increase, by adapting the fuzzy-quantifiers of a fuzzy-inference to the syllogistic logic of the sample propositions.

Keywords: Categorical syllogisms · Approximate reasoning · Ontologies · Fuzzy-logic

1 Introduction

Reasoning is the ability to make inferences and automated reasoning is an area of computer science and mathematical logic, dedicated to understanding different aspects of reasoning and concerned with building computing systems that automate this process. Although the term can be applied to various reasoning tasks, usually automated reasoning is considered with different forms of valid deductive reasoning, like in various applications of automated theorem proving or formal verification [11].

Reasoning with intermediate quantifiers is called fuzzy syllogistic reasoning [16], where a syllogism is an inference rule that consists of deducing a new quantified statement from one or several quantified statements.

Based on ideas of syllogistic reasoning we proposed the Fuzzy-Syllogistic System (FSS) [6], which attempts to integrate both approaches, approximate and exact reasoning, in one system.

The work is organized as follows: in chapters 2 and 3 we discuss syllogisms and their applications, in particular the concepts of the fuzzy-syllogistic system. In chapter 4 we provide a short description of ontology. Finally in chapters 5 we present the idea of ontology-based syllogistic reasoning.

2 Classical Syllogisms

As an inference scheme, a syllogism may generally be expressed in the form:

© Springer International Publishing Switzerland 2015
M. Ali et al. (Eds.): IEA/AIE 2015, LNAI 9101, pp. 179–188, 2015.
DOI: 10.1007/978-3-319-19066-2_18

$$\frac{\psi_1 A \; are \; B}{\psi_2 C \; are \; D}$$
$$\overline{\psi_3 E \; are \; F}$$

where ψ_1, ψ_2 and ψ_3 are numerical, or more general, fuzzy quantifiers (e.g. few, many, most), and A, B, C, D, E and F are crisp or fuzzy predicates. The predicates A, B, ... F are assumed to be related in a specific way, giving rise to different types of syllogisms [16].

A categorical syllogism can be defined as a logical argument that is composed of two logical propositions for deducing a logical conclusion, where the propositions and the conclusion each consist of a quantified relationship between two objects [1], [14].

2.1   Syllogistic Propositions

A syllogistic proposition or synonymously categorical proposition specifies a quantified relationship between two objects. We shall denote such relationships with the operator ψ. Four different types are distinguished $\psi=\{$A, E, I, O$\}$:

A	Universal Affirmative	All S are P		E	Universal Negative	All S are not P
I	Particular Affirmative	Some S are P		O	Particular Negative	Some S are not P

2.2   Syllogistic Figures

A syllogism consists of the three propositions: major premise, minor premise and conclusion. The first proposition consist of a quantified relationship between the objects M and P, the second proposition of S and M, the conclusion of S and P (Table 1).

Table 1. Syllogistic figures

Figure Name	I	II	III	IV
Major Premise	MψP	PψM	MψP	PψM
Minor Premise	SψM	SψM	MψS	MψS
Conclusion	SψP	SψP	SψP	SψP

Note the symmetrical combinations of the objects. Since the proposition operator may have four values for ψ, 64 syllogistic moods are possible for every figure and 256 moods for all four figures in total.

3   The Fuzzy Syllogistic System

The proposed fuzzy-syllogistic system (FSS) is a complex model for approximate reasoning. It may be used for constructing hybrid systems that can reason deductively over emergent data concepts and their relationships. Here we improve the mathematical model of syllogistic system, presented in [5], [6]   and [7].

3.1   Syllogistic System

For three sets, there are seven possible subsets in a Venn diagram (Fig. 1). These sub-sets or spaces constitute the basic data of modelling the syllogistic system (Table 2).

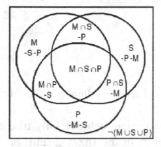

Fig. 1. Venn diagram for three sets

In an algorithmic implementation of the syllogistic system, we get 96 distinct space combinations, where everyone contains elements from all three sets simultaneously. We will refer to these combinations as the syllogistic cases.

Table 2. Identification of the seven possible subsets of three sets as distinct spaces

Space ID	δ_1	δ_2	δ_3	δ_4	δ_5	δ_6	δ_7
Subset	S-(P+M)	P-(M+S)	M-(S+P)	(M∩S)-P	(M∩P)-S	(S∩P)-M	M∩S∩P

Every proposition of everyone of the 256 moods matches some of these 96 syllo-gistic cases. By relating the number of true and false matching cases per mood we calculate the degree of validity (or truth ratio) of that mood. More precisely, for a given mood, all 96 cases are checked against the premises and the conclusion and the number of true and false matching cases are identified. The ratio between matched true and false cases becomes the truth ratio $\tau=[0,1]$. Based on τ, we can then judge about the accuracy of an inferred conclusion from given premises.

3.2   Fuzzification

Since the vast majority of syllogistic moods are invalid, we had to introduce the term of fuzzy-syllogistic reasoning (FSR) to generalize reasoning scheme and extend the possible number of valid syllogistic structures. According to the structure of syllog-ism we applied the fuzzification in two ways by using fuzzy quantifiers and defining fuzzy sets [8].

Let us consider mood AIA of figure 1:

A: All M are P   I: Some S are M   A: All S are P

According to the rules of classical logic this syllogism is not valid; there are only 6 valid moods for Figure 1 and AIA not in these list. For this mood we found 10 cases that satisfy premises, but only 4 of them satisfy conclusion and premises at the same time, so we can calculate true ratio τ=4/10 (Fig. 2 a). For elements in S that are not shared in M and P, the conclusion becomes wrong (Fig. 2 b)).

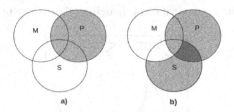

Fig. 2. True a) and false b) syllogistic cases of the mood AIA1

Using fuzzy-quantification we can achieve fully true conclusion. Indeed if we replace the universal quantifier A in the fuzzy-quantifier A': AlmostAll, the resulting syllogism became true:

A: All M are P   I: Some S are M   A': AlmostAll S are P

In certain situations, by taking some assumptions, we can replace one quantifier by another. More specifically, we can introduce another quantifier like A':AlmostAll or E':AlmostNone. Obviously, quantifier A' can be considered as a special case of quantifier I. Likewise, quantifier E' is a special case of quantifier O:NotSome.

Returning to the given example, "AlmostAll S are P" means that the proportion of elements of S being elements of P is "very very important". In other words, the proportion in S not being P is "very very weak". Where the proportion of elements of set S in the intersection with P to cardinality of S (refers to A') is close to 1   and proportion for   I' is close to 0.

Taking this into account we can replace quantifier A by I and E by O respectively, according to the cardinalities of the given sets.

Applying the fuzzy quantification to mood AIA1 and the given example, potentially we can obtain 3 modified moods such as AII1, III1, IIA1. AII1 is valid mood for figure 1. Thus potentially for the given mood we can increase τ from 0.4 to 1, which is the actual objective of our fuzzification approach.

3.3   Fuzzy-Syllogistic System (FSS)

Based on the ideas discussed above, we have designed the software system FSS. The working cycle of the system is specified as follows:

> Inductively accumulate sample instances of relationships between the objects M, P, S and classify them into the 96 distinct sub-sets.
> Calculate the truth values of all 256 moods for these M, P, S relationships.
> Apply fuzzy-quantification to all moods.
> Use approximate reasoning (fuzzy sets), if results from previous step do not meet given requirements.

Currently, the system can load experimental input data from specific XML files. Thus, FSR can reason with any ontology that is in XML format, like OWL or RDF(S).

4   Ontologies

The most popular definition of the concept ontology in information technology and AI community from a theoretical point of view is "A formal explicit specification of a shared conceptualization" or "An abstract view of the world we are modelling, describing the concepts and their relationships" [3].

Formally, an ontology may be defined as **O=(C, R, A, Top)**, in which **C** is the non-empty set of concepts (including relation concepts and the **Top**), **R** is the set of all assertions, in which two or more concepts are related to each other, **A** is the set of axioms and **Top** is the most top level concept in the hierarchy. **R** itself is partitioned to two subsets, **H** and **N**. **H** is the set of all assertions in which the relation is a taxonomic relation and **N** is the set of all assertions, in which the relation is a non-taxonomic relation[13].

Ontology construction is still an active topic of research. The major problems in building and using ontologies are the knowledge acquisition and the time-consuming construction [9].

5   Ontology-Based Fuzzy-Syllogistic Reasoning

Our objective is to implement syllogistic reasoning with ontologies and iteratively quantify ontological relationships with FSR. In this process, FSR does not directly produce an ontology. Concepts and relationships of a given ontology are evaluated and altered by the FSR.

Among the various possible ways to construct ontologies for a given domain, the most widely used approaches is the generation of an ontology from text-based sources [12]. There are several open-source tools for ontology generation from text corpora available for research purposes, such as Text2Onto [2], WebKB or DLLearner.

The most convenient tool for our purposes was Text2Onto, because it allows generating ontologies automatically and the generated ontology is sufficiently good.

5.1   Building a Source Ontology

For generating a source ontology it is necessary to prepare text corpus for the given domain. In case of Text2Onto, the text corpus may be a set of plain text documents, html pages and other unstructured or semi-structured text sources. The integration of this tool with the web search engine seems to be an optimal solution for collecting and preparing a text corpus for a given domain. Furthermore, as a result of stepwise synthesising concepts and properties, we obtain a domain ontology for the given corpus. The resulting ontology includes a set of nodes, which can represent terminal nodes or intermediate nodes, with linked relationships (Fig. 3 a).

5.2   Building Graph of Dependencies

For an existing ontology we can build a graph of syllogistic dependencies, which reflects quantitative relationships between concepts of the original ontology, such that FSR can be applied.

Such a graph of dependencies contains all concepts and concept attributes of the original ontology and additionally all quantities that had contributed to their conceptualisations during the ontology learning process. The result is a quantified ontology.

Every attribute of a concept is further decomposed, e.g. new sub-concepts (subclasses) are created for each attribute value. This helps revealing hidden dependencies. All subclasses constructed from attributes must be linked with their parent class via a direct link (from attribute-class to parent).

Now we can link all concepts (classes) in our graph according to the bellow procedure. Let the number of all (sub)classes (terminal and intermediate) be equal to N. For every subclass $SubCli$ we need to consider $N-1$ subclasses. Analysis should be performed on pairs. For each pair of subclasses there are following possible conditions:

$SubCli$ is subclass for $SubClj$ (or otherwise) (Fig. 3 a, Subclass2 and Subclass3): direct link from subclass to superclass on the graph of dependencies;
$SubCli$ and $SubClj$ have no shared subclasses (Fig. 3 a, Subclass1 and Subclass3): no link between the classes on the graph of dependencies;
$SubCli$ and $SubClj$ have shared subclasses (Fig. 3 a, Subclass3 and SubclassK): in this case we need to calculate fraction F of shared subclasses to number of subclasses for each of 2 subclasses (nodes), if $F=1$ for one of the nodes, then this node becomes subclass of the other node and we need to create a directed link from subclass to superclass; if $F<1$, we need to create a non-directed link (nodes have shared subclasses but no super classes of each other).

After performing these operations we will get the graph of dependencies, which is a reflection of the input ontology (Fig. 3 c).

There are four possible types of relationships between the classes (Fig. 3 b, 1-4 respectively):

directed link from CLASS_1 to CLASS_2: CLASS_2 includes all elements from CLASS_1, corresponds with A quantifier;
directed link from CLASS_2 to CLASS_1: CLASS_1 includes all elements from CLASS_2, corresponds with I quantifier (some elements of CLASS_1 in CLASS_2);

non-directed link from CLASS_1 to CLASS_2 (or otherwise): some elements from CLASS_1 in CLASS_2, at the same time, some elements from CLASS_2 in CLASS_1: corresponds with I, O quantifiers (appropriate quantifier can be selected according to cardinality of given sets);
no link between classes: corresponds with E quantifier.

FSR with such a dependency graph enables reasoning with 256 possible fuzzy inferences per triple concept relationships.

In some cases, the transitive concept of a triple can be removed, as that is not included in the conclusion. This helps reducing the complexity of the ontology and increases the level of abstraction over details that are no more required in reasoning.

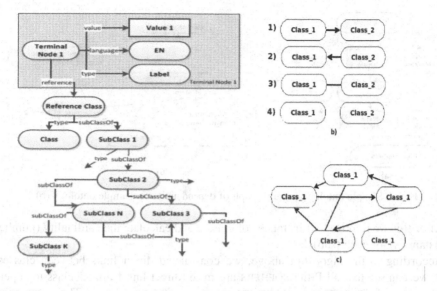

Fig. 3. Structural schema of a simple ontology (a); 1 – 4 types of possible class relationships (b) and a sample graph of dependencies (c)

5.3 Reasoning with Ontologies: Procedure

Our objective is to prove the feasibility of FSR for reasoning with ontologies. The procedure consists of a few steps, such as rebuilding a given ontology in a format, such that always triple concepts are selected for analyses by the FSR component:

1. Calculate truth ratios of all 256 moods.
2. For given ontology, build dependency graph as described bellow.
3. Select triple of sets for analysis and label them with M, P and S.
4. Construct four syllogistic figures and associate quantifiers approriate for the quantities of the premises from the graph of dependencies; apply all possible quantifiers on conclusions.
5. Calculate truth ratios for all possible moods.
6. Select the moods with the highest truth ratio τ; if $\tau<1.0$, try to apply fuzzification.

5.4 Sample Application

Let us perform the steps of the algorithm on a sample ontology (Fig. 4 a). One can see that there are four classes, Humans (with attribute #gender), Philosophers, Scientists and Artists. Also there are six instances of the class Humans.

First of all, we need to distinguish all attributes of each class as separate subclasses. As shown on (Fig. 4 b), we have created 2 subclasses of Humans, humans_gender_male and humans_gender_female. Subclasses have direct link to their superclass, because these operation can be considered as decomposition of superclass, so all subclasses are part of superclass and relation between subclass and superclass. In terms of syllogistic quantification this is a A: all relationship.

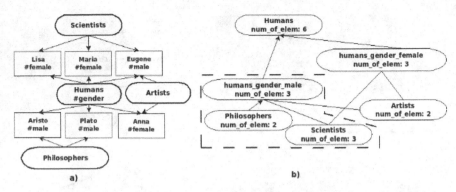

Fig. 4. Sample ontology (a) and graph of dependencies for sample ontology (b)

After this we removed all instances of classes and calculate the cardinality (number of instances) for each (sub)class.

According to the algorithm above, we constructed direct links between classes. Now we can see that all PHILOSOPHERS are male (direct link from subclass to superclass), some of SCIENTISTS and ARTISTS are women, some are man. The constructed graph is suitable for performing FSR.

Now we need to select 3 classes. For now it looks quite indefinably, but when we embed our system in a real application, like an intelligent agent, the selected classes will be determined by the logic of the agent.

Lets consider the relationships between SCIENTISTS and PHILOSOPHERS classes through people_gender_male class (Fig. 4 b, dashed area).

According to the structure of syllogisms, the middle term (M) is humans_gender_male class, predicate (P) is SCIENTISTS class and subject (S) is PHILOSOPHERS class.

So, for four syllogistic figures we have following combinations:

Figure 1	**Figure 2**	**Figure 3**	**Figure 4**
{I, O}: M P	{I, O}: P M	{I, O}: M P	{I, O}: P M
{A}: S M	{A}: S M	{I, O}: M S	{I, O}: M S
{?}: S P	{?}: S P	{?}: S P	{?}: S P

The problem is to find the most appropriate quantifier as conclusion. The quantifiers for the premises were selected according to the relationships between classes on the graph of dependencies. For example, direct link from Philosophers to humans_gender_male (S M) corresponds to the A quantifier, non-directed link from Scientists to humans_gender_male (P M) can be an I or O quantified relationship.

By calculating then truth ratios of all possible moods, we obtain the following results:

Figure 1		Figure 2		Figure 3		Figure 4
IAA=0.285	IAA=0.285	OIA=0.084	IIA=0.142	OIA=0.134	IIA=0.142
IAE=0.285	IAE=0.285	OIE=0.157	IIE=0.144	OIE=0.104	IIE=0.144
IAI=0.714	IAI=0.714	OII=0.845	III=0.885	OII=0.895	III=0.885
IAA=0.714	IAA=0.714	OIO=0.915	IIO=0.857	OIO=0.865	IIO=0.857
OAA=0.214	OAA=0.333	OOA=0.128	IOA=0.183	OOA=0.194	IOA=0.183
OAE=0.357	OAE=0.333	OOE=0.185	IOE=0.154	OOE=0.134	IOE=0.154
	OAI=0.666	OOI=0.814	IOI=0.845	OOI=0.865	IOI=0.845
OAI=0.642					
OAA=0.785	OAA=0.666	OOE=0.871	IOE=0.816	OOE=0.805	IOE=0.816

The highest truth ratio is OIO3=0.915. We cannot apply fuzzy-quantification to a given mood, because it does not contain A or E quantifiers. Based on the bellow results we can restore the most suitable syllogism for the given data:

O: Male Scientists   **I**: Male Philosophers   **O**: Philosophers Scientist

with truth ratio τ=0.915. Considering the OIO3 mood, we see that it has 71 cases, only 6 cases are false.

In analogy with previous example, consider following scenario: M= humans_gender_male class, S=Philosophers, P=Humans. So, we want to investigate the relationship between Philosophers and Humans.

Possible moods are listed below:

Figure 1	Figure 2	Figure 3	Figure 4
{A}: M P	{I, O}: P M	{A}: M P	{I, O}: P M
{A}: S M	{A}: S M	{I, O}: M S	{I, O}: M S
{?}: S P	{?}: S P	{?}: S P	{?}: S P

After calculating the truth ration for all moods, only three moods AAA1, AAI1 and AII3 have τ=1:

A: Male Humans A: Philosophers Male   {**A, I**}: Philosophers Humans
A: Male Humans I: Male Philosophers   **I**: Philosophers Humans

Actually, we can remove the link between PHILOSOPHERS and humans_gender_men and create directed (or non-directed) link between PHILOSOPHERS and HUMANS. Performing the same operation for SCIENTISTS and ARTISTS classes, it is possible to remove all links to humans_gender_men. Since there is no link, related with the class, we can simply delete this class from the graph.

6   Related Works

Approximate reasoning [15] with ontologies is becoming increasingly popular in semantic web applications [9]. However, rather modus ponens or tollens-based fuzzifications are employed than categorical syllogisms.

Work on intermediate quantifiers [8], [10] is mostly theoretical, applications with ontologies are not known.

The objective of probabilist ontologies is concerned with quality ontology generation [4], hence with improving reasoning as well, but these approaches too, do usually not involve categorical syllogisms.

7   Conclusions

We have discussed the fuzzy-syllogistic system (FSS) and introduced an application of fuzzy-syllogistic reasoning (FSR) with ontologies, which can be interpreted as a complex approximate reasoning approach that consists of 256 fuzzy inferences.

One future work is to expand the current restriction of two premises to n premises. FSR could then be applied transitively to all classes of the ontology, without decomposing the ontology into multiple triple relationships.

References

[1] Brennan, J.G.: A Handbook of Logic. Brennan Press (2007)
[2] Cimiano, P., Völker, J.: Text2Onto. In: Montoyo, A., Muñoz, R., Métais, E. (eds.) NLDB 2005. LNCS, vol. 3513, pp. 227–238. Springer, Heidelberg (2005)
[3] Gruber, T.R.: A translation approach to portable ontologies. Knowledge Acquisition 5(2) (1993)
[4] Ji, Q., Gao, Z., Huang, Z.: Reasoning with noisy semantic data. In: Antoniou, G., Grobelnik, M., Simperl, E., Parsia, B., Plexousakis, D., De Leenheer, P., Pan, J. (eds.) ESWC 2011, Part II. LNCS, vol. 6644, pp. 497–502. Springer, Heidelberg (2011)
[5] Kumova, B.İ., Çakır, H.: Algorithmic Decision of Syllogisms. Conference on Industrial, Engineering & Other Applications of Applied Intelligent Systems (IEA-AIE 2010). Springer LNAI (2010)
[6] Kumova, Bİ., Çakir, H.: The fuzzy syllogistic system. In: Sidorov, G., Hernández Aguirre, A., Reyes García, C.A. (eds.) MICAI 2010, Part II. LNCS, vol. 6438, pp. 418–427. Springer, Heidelberg (2010)
[7] Kumova, B.İ.: Symmetric Properties of the Syllogistic System Inherited from the Square of Opposition. Logica Universalis (2015)
[8] Murinová, P., Novák, V.: A formal theory of generalized intermediate syllogisms. Fuzzy Sets and Systems 186 (2012)
[9] Pan, J.Z., Thomas, E.: Approximating OWL-DL Ontologies. American Association for Artificial Intelligence (2009)
[10] Peterson, P.: On the logic of 'few', 'many', and 'most'. Notre Dame Journal of Formal Logic 20 (1979)
[11] Robinson, A., Voronkov, A.: Handbook of Automated Reasoning. Elsevier (2001)
[12] Shamsfard, M., Barforoush, A.A.: The State of the Art in Ontology Learning: A Framework for Comparison. The Knowledge Engineering Review 18(4) (2003)
[13] Sowa, J.F.: Knowledge Representation: Logical, Philosophical and Computational Foundations. Brooks/Cole (2000)
[14] Wille, R.: Contextual Logic and Aristotle's Syllogistic. Springer (2005)
[15] Zadeh, L.A.: Fuzzy Logic and Approximate Reasoning. Syntheses 30 (1975)
[16] Zadeh, L.A.: Syllogistic reasoning in fuzzy logic and its application to usuality and reasoning with dispositions. IEEE Transactions on Systems, Man and Cybernetics 15(6) (1985)

Optimization

Stability in Biomarker Discovery: Does Ensemble Feature Selection Really Help?

Nicoletta Dessì and Barbara Pes[(✉)]

Università degli Studi di Cagliari, Dipartimento di Matematica e Informatica,
Via Ospedale 72 09124, Cagliari, Italy
{dessi,pes}@unica.it

Abstract. Ensemble feature selection has been recently explored as a promising paradigm to improve the stability, i.e. the robustness with respect to sample variation, of subsets of informative features extracted from high-dimensional domains including genetics and medicine. Though recent literature discusses a number of cases where ensemble approaches seem to be capable of providing more stable results, especially in the context of biomarker discovery, there is a lack of systematic studies aiming at providing insight on when, and to which extent, the use of an ensemble method is to be preferred to a simple one. Using a well-known benchmark from the genomics domain, this paper presents an empirical study which evaluates ten selection methods, representatives of different selection approaches, investigating if they get significantly more stable when used in an ensemble fashion. Results of our study provide interesting indications on benefits and limitations of the ensemble paradigm in terms of stability.

Keywords: Ensemble feature selection · Feature selection stability · Biomarker discovery · High-dimensional data

1   Introduction

An increasing number of real-world applications produce datasets with a huge number of attributes (or features). In this context, feature selection [1] is a necessary step for making the analysis more manageable and for extracting useful knowledge on the domain of interest. A wide literature is currently available on the strengths and weaknesses of different feature selection methods [2, 3], the choice of the "best" method being dependent on the specific problem at hand.

Research on feature selection has only recently addressed the issue of *stability*, i.e. the capacity of a selection method of producing the same (or almost the same) results regardless of changes in the dataset composition [4]. A task where the stability of the selection process plays a crucial role is the discovery of biomarkers from high-dimensional genomics datasets, e.g. the identification of a subset of genes that are related to the diagnosis of a disease or its treatment [5].

In this domain, the high dimensionality (thousands of genes) is often coupled with a small sample size (a few dozens of biological samples), causing the stability of

© Springer International Publishing Switzerland 2015
M. Ali et al. (Eds.): IEA/AIE 2015, LNAI 9101, pp. 191–200, 2015.
DOI: 10.1007/978-3-319-19066-2_19

feature selection to decrease [6]. Unstable gene selection often involves the non-reproducibility of the reported markers, which is one of the major obstacles to clinical applications [7]. Hence, increasing attention has been paid in the last few years to the stability of protocols used in biomarker discovery applications.

One of promising frameworks for achieving more stable feature sets, especially in the context of high-dimensional/small sample size domains, is *ensemble feature selection* [8]. Originally devised to enhance the performance of classification models [9], the ensemble paradigm can be applied in the context of feature selection by combining the outcome of different selectors into a single, hopefully more stable, feature list. Basically, there are two ways of creating different selectors [6, 7], i.e. by applying a single feature selection method to different perturbed versions of a dataset (*data-diversity* approach) or by applying different selection methods to the same dataset (*functional-diversity* approach). There are also several ways of aggregating the outcome of the different selectors [10].

Though recent literature discusses a number of cases where ensemble approaches have been successfully applied in the context of biomarker discovery [11, 12], there is not a consensus on the superiority of ensemble methods over simple ones [13] nor clear indications on when the adoption of an ensemble method, despite its inherently higher computational cost, may be convenient in terms of stability.

This paper aims to give a contribution in this direction by presenting an empirical analysis that evaluates, with and without an ensemble setting, ten selection methods widely used in the context of biomarker discovery. Specifically, we explored the effects of a data-diversity ensemble strategy, that has been suggested as a primary avenue for improving stability [8]. Our results on a public genomics benchmark show that the ensemble approach induces a gain in stability that, in some way, is "inversely proportional" to the stability of the original methods, suggesting that the use of an ensemble strategy is not always and necessarily beneficial in itself. Nevertheless, it can be very useful (also in dependence on the size of the selected gene subsets) when using selection algorithms that are intrinsically less stable.

The rest of the paper is organized as follows. Section 2 describes the methodology used in the empirical study, explaining it in the context of the underlying background on stability evaluation and ensemble feature selection. Experimental settings and results are presented in section 3, while section 4 contains concluding remarks.

2   Background and Methodology

As recognized by recent literature [6–8], stability of feature selection is a very important issue, especially if subsequent analysis or validation of selected feature subsets are costly, as in the context of biomarker discovery. In the last years, research activity has focused both on defining suitable protocols and metrics for stability evaluation as well as on developing feature selection approaches, such as ensemble techniques, that can achieve more stable results. In this work, we investigate the potential of ensemble feature selection using a methodological approach that leverages on best practices from literature.

Specifically, we focus on feature ranking techniques, or *rankers*, that in practice are often preferred to more expensive selection methods [2], especially in the context of high dimensional problems. Ranking techniques give as output a *ranked list* where features appear in descending order of relevance: this list is usually cut at a proper threshold point in order to obtain a subset of highly predictive features. Since there are not specific criteria for setting the "optimal" cut-off, it is worth studying the performance of rankers within a suitable range of thresholds. The methodology we adopt for evaluating rankers stability (sub-section 2.1) takes into account this dependence on the threshold and enables to derive a "baseline" to investigate if, and for which threshold values, rankers benefit from the adoption of an ensemble approach (sub-section 2.2).

2.1   Stability Evaluation

To evaluate the degree of stability of a ranking method R, i.e. its sensitivity to perturbations of the training data, we adopt a sub-sampling based strategy. Specifically, we extract from the original dataset D, with M instances and N features, a number K of reduced datasets D_i ($i = 1, 2, \ldots, K$), each containing $f \cdot M$ (with $f \in (0, 1)$) instances randomly drawn from D.

R is then applied to each D_i ($i = 1, 2, \ldots, K$) in order to obtain a ranked list L_i ($i = 1, 2, \ldots, K$) and, after cutting the list at a suitable threshold t, a subset S_i^t ($i = 1, 2, \ldots, K$) of t features. The resulting K subsets are compared, in a pair-wise fashion, using a proper similarity index [14] which basically expresses the degree of overlapping among the subsets (with a correction term that takes into account the probability that a feature is included in both subsets simply by chance):

$$Sim_{ij}(t) = \frac{\left| S_i^t \cap S_j^t \right| - t^2/N}{t - t^2/N} \tag{1}$$

To obtain an overall evaluation of the degree of similarity among the K subsets, we average over all the involved pair-wise comparisons:

$$S_{avg}(t) = \frac{2 \sum_{i=1}^{K-1} \sum_{j=i+1}^{K} Sim_{ij}(t)}{K(K-1)} \tag{2}$$

S_{avg} can be assumed as a measure of stability: the more similar (in average) the subsets are, the more stable the ranking method R. Since S_{avg} refers to a specific threshold t, i.e. to a specific subset size, a stability pattern can be easily derived for feature subsets of increasing size.

2.2   Ensemble Feature Ranking

The ensemble paradigm involves two fundamentals steps, i.e. (i) creating a set of different feature selectors (ensemble components) and (ii) aggregating the outputs of the different selectors into a single final decision. At the first step, we use a *data perturbation* approach to inject diversity into the ensemble components, i.e. we apply a given ranking algorithm to different perturbed versions of a dataset so as to obtain a number of ranked lists that are potentially different from each other, though produced by the same algorithm (their diversity comes from the diversity of the training data). At the second step, these lists are aggregated into a single *ensemble list* where the rank (ranking position) of each feature is calculated as a function of the feature's rank in the lists being combined. As aggregation function we use the *mean*, i.e. we simply average the rank of a feature across all the lists. The use of more complicated (and costly) aggregation strategies seems to not significantly improve the ensemble performance [15].

Since our objective is to evaluate the ensemble stability, the above strategy is applied to each of the reduced datasets D_i ($i = 1, 2, \ldots, K$) previously obtained by sampling (without replacement) the original dataset D. In more detail, each D_i is further sampled, but with replacement, to get a set of *bootstrap* datasets D_{is} ($s = 1, \ldots, P$). A ranking method R is then applied to each D_{is} to obtain a set of ranked lists L_{is} ($s = 1, \ldots, P$) to be aggregated into an ensemble list $L_{i\ ensemble}$. By cutting this list at a threshold point t, a feature subset $S_{i\ ensemble}^t$ of t features can be derived. Finally, the K subsets $S_{i\ ensemble}^t$ ($i = 1, 2, ..., K$) are compared, in a pair-wise fashion, according to the approach described in section 2.1: their average similarity gives a measure of the ensemble stability. Spanning a suitable range of thresholds, we can then evaluate if the stability pattern of a ranking method R changes significantly when the method is used in an ensemble fashion.

3   Experimental Study

According to the methodology presented in section 2, we conducted an experimental study involving ten selection methods widely used in the context of biomarker discovery. A brief description of these methods is provided in sub-section 3.1, along with a brief description of the genomic dataset used as benchmark, while the experimental results are presented and discussed in sub-section 3.2.

3.1   Dataset and Ranking Techniques

We performed extensive experiments on a genomic dataset deriving from DNA micro-array experiments, the *Colon Tumor* dataset [16], which contains 62 biological samples distinguished between tumor colon tissues and normal colon tissues (40 and 22 samples, respectively). Each sample is described by the expression level of 2000 genes: the task, in terms of feature selection, is to identify the genes most useful in discriminating among cancerous tissues and normal ones.

In our experiments, we studied the stability pattern of ten rankers that are representative of different classes of selection methods. In particular, we considered both univariate approaches, where each single feature is weighted independently from the others, and multivariate approaches, where the inter-dependencies among features are taken into account [17]. For all of them we leveraged the implementation provided by the WEKA machine learning package [18].

As representatives of the univariate approaches, we chose: *Chi Squared* (χ^2) [19], that evaluates features individually by measuring their chi-squared statistic with respect to the class; *Information Gain* (IG) [20], *Symmetrical Uncertainty* (SU) [21] and *Gain Ratio* (GR) [22], that are grounded on the concept of entropy; and finally *OneR* (OR) [23], that uses a simple rule-based classifier to derive the level of significance of each feature.

As representatives of the multivariate approaches, we considered *ReliefF* (RF) and *ReliefF-W* (RFW) [24] that evaluate the relevance of features based on their ability to distinguish between instances that are near to each other (with a weighting mechanism, in RFW, to account for the distance between instances). Moreover, we employed *SVM-embedded feature selection* [25] that basically relies on a linear SVM classifier to derive a weight for each feature. Based on their weights, the features can be simply ordered from the most important to the less important (SVM_ONE approach). Otherwise, a backward elimination strategy can be adopted that iteratively removes the features with the lowest weights and repeats the overall weighting process on the remaining features (SVM_RFE approach). The computational complexity of the method is greatly influenced by the percentage of features removed at each iteration: in our experiments we set this parameter as 10% (SVM_RFE10 approach) and 50% (SVM_RFE50 approach).

3.2   Results and Discussion

Based on the sub-sampling approach described in sub-section 2.1, we first evaluated the stability of each ranking method across $K = 50$ reduced datasets, each containing a fraction $f = 0.9$ of instances of the original Colon dataset. The more similar the gene subsets selected from the different training sets, the higher the stability of the method. In our experiments, we considered subset sizes (i.e. threshold values) ranging from 10 to 150 since biologists are typically interested in selecting a small fraction of potentially predictive genes.

The stability of each ranker was then compared with the stability of the corresponding ensemble, implemented according to a *data diversity* strategy (sub-section 2.2): for each of the $K = 50$ reduced training sets, a gene subset is selected by aggregating ranking results over a number $P = 50$ of bootstraps, and the similarity of the resulting subsets is used as a measure of the ensemble stability. The outcome of this analysis is shown in Fig. 1 and Fig. 2, respectively for univariate and multivariate ranking methods.

Specifically, for each of the univariate rankers considered in this study (χ^2, IG, SU, GR, OR), Fig. 1 shows the comparison between the stability pattern of the ranker itself (simple ranking) and the stability pattern of the ensemble version of the same ranker (ensemble ranking). As we can see, the most stable methods, i.e. χ^2, IG and SU, slightly benefit from the ensemble approach only for larger threshold values (> 30 for χ^2 and SU, > 100 for IG). For GR, which turns out less stable than the other entropic methods, ensemble ranking is better than simple ranking for all threshold values. The positive impact of the ensemble approach is still more evident for OR, which is the least stable among the univariate methods.

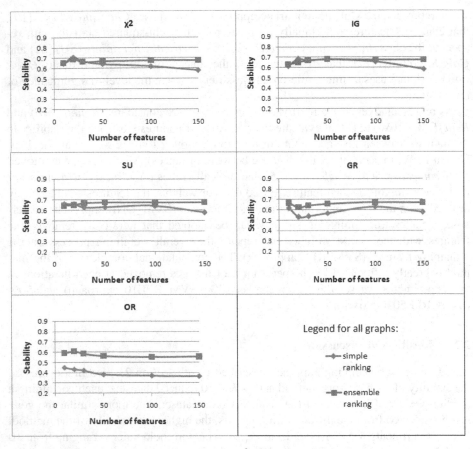

Fig. 1. Stability of univariate ranking methods (χ^2, IG, SU, GR, OR): simple ranking vs ensemble ranking

As regards multivariate ranking methods (RF, RFW, SVM_ONE, SVM_RFE10, SVM_RFE50), the comparison between simple ranking and ensemble ranking (Fig. 2) reveals that the most stable methods (i.e. RF and RFW) don't benefit at all from the adoption of an ensemble strategy. Conversely, the methods that are less stable in the simple version take great advantage of the ensemble approach.

Hence, both Fig. 1 and Fig. 2 seem to indicate that the ensemble approach induces a gain in stability that, in some way, is "inversely proportional" to the stability of method itself. This may explain the (apparently) discordant findings in recent literature, where different studies seem to achieve different conclusions about the beneficial impact of ensemble feature selection [11, 13, 26]. Indeed, the use of an ensemble strategy is not always and necessarily beneficial in itself, but only in dependence on the "intrinsic" stability of the considered method (as well as in dependence of the size of the selected subset).

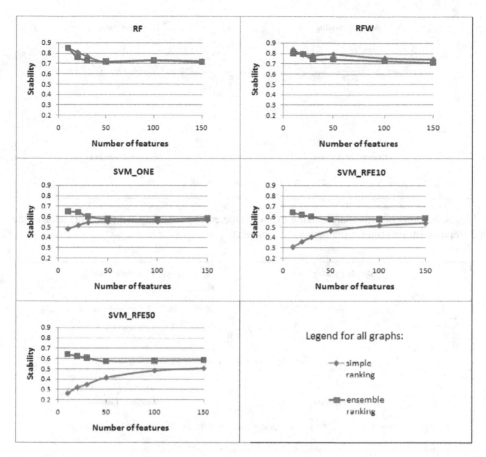

Fig. 2. Stability of multivariate ranking methods (RF, RFW, SVM_ONE, SVM_RFE10, SVM_RFE50): simple ranking vs ensemble ranking

Additional insight can be obtained from the graphs in Fig. 3 and Fig. 4 which give a summary view of our experimental results. Specifically, Fig. 3 shows the stability pattern of all univariate methods in their simple form (on the left), as well as the stability pattern of the corresponding ensembles (on the right). Interestingly, in the ensemble setting, the differences among the original methods are significantly reduced: the GR behavior is almost the same as that of χ^2, IG and SU, and even OR (the least stable method) gets more close to the other univariate rankers. A similar trend can be observed for multivariate methods (Fig. 4): when used in an ensemble fashion, they get more similar to each other. In particular, the stability patterns of all SVM-based selection methods (SVM_ONE, SVM_RFE10 and SVM_RFE50) become almost coincident, despite the significant differences observed in the simple ranking setting.

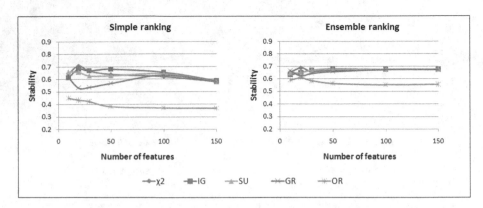

Fig. 3. Stability of univariate ranking methods: the ensemble setting (on the right) reduces the differences among the simple rankers (on the left)

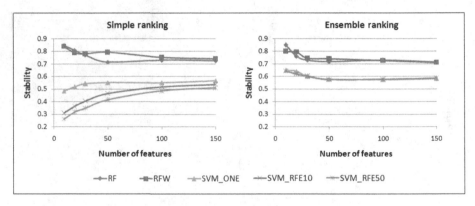

Fig. 4. Stability of multivariate ranking methods: the ensemble setting (on the right) reduces the differences among the simple rankers (on the left)

We can then achieve the conclusion that the main effect of an ensemble strategy as the one here considered (based on injecting diversity into the training data) is to make "weak" methods significantly more stable, hence narrowing their gap with the strongest methods, which in turn seem not to benefit from the adoption of a computationally expensive ensemble setting.

4   Concluding Remarks

This paper investigated the potential of the ensemble paradigm in the context of the extraction of stable feature sets from high-dimensional domains. Specifically, we focused on biomarker discovery from high-dimensional genomics data: this is a task, indeed, where the adoption of stable feature selection protocols is of paramount importance as the degree of stability of selected gene sets may greatly influence subsequent biological validations.

Aiming at understanding if gene selection really benefits from the adoption of an ensemble strategy, we conducted an empirical analysis involving different classes of selection methods, both univariate (that evaluate the relevance of each single gene) and multivariate (that measure the joined relevance of set of genes). Our results give a number of interesting indications that we plan to further validate in future, more extensive, experiments.

As a first point, it emerges that only methods that are intrinsically weak (in term of stability) significantly improve their performance when used in an ensemble fashion. Hence, the ensemble approach seems not to have a generalized positive effect on the stability of a selection method. But, interestingly, we observed that the adoption of an ensemble strategy narrows the differences among the stability patterns of the original methods: even rankers that exhibit a quite different behavior become similar to each other when trained in a sufficiently diversified data space (as implied by the ensemble paradigm here considered). These findings, in our opinion, can help to provide a more comprehensive understanding of the potential and the implications of the ensemble selection paradigm. Future studies will be devoted to investigate if what observed for stability has a counterpart in terms of predictive accuracy (i.e. which methods improve their predictive performance when used in an ensemble fashion?).

Acknowledgments. This research was supported by RAS, Regione Autonoma della Sardegna (Legge regionale 7 agosto 2007, n. 7), in the project *"DENIS: Dataspaces Enhancing the Next Internet in Sardinia"*.

References

1. Guyon, I., Elisseeff, A.: An introduction to variable and feature selection. Journal of Machine Learning Research **3**, 1157–1182 (2003)
2. Saeys, Y., Inza, I., Larranaga, P.: A review of feature selection techniques in bioinformatics. Bioinformatics **23**(19), 2507–2517 (2007)
3. Bolón-Canedo, V., Sánchez-Maroño, N., Alonso-Betanzos, A.: A review of feature selection methods on synthetic data. Knowledge and Information Systems **34**(3), 483–519 (2013)
4. Kalousis, A., Prados, J., Hilario, M.: Stability of feature selection algorithms: a study on high-dimensional spaces. Knowledge and Information Systems **12**(1), 95–116 (2007)
5. Dessì, N., Pascariello, E., Pes, B.: A Comparative Analysis of Biomarker Selection Techniques, BioMed Research International 2013, Article ID 387673, p. 10 (2013)
6. Awada, W., Khoshgoftaar, T.M., Dittman, D., Wald, R., Napolitano, A.: A review of the stability of feature selection techniques for bioinformatics data. In: IEEE 13th International Conference on Information Reuse and Integration, pp. 356–363. IEEE (2012)
7. Zengyou, H., Weichuan, Y.: Stable feature selection for biomarker discovery. Computational Biology and Chemistry **34**, 215–225 (2010)
8. Saeys, Y., Abeel, T., Van de Peer, Y.: Robust feature selection using ensemble feature selection techniques. In: Daelemans, W., Goethals, B., Morik, K. (eds.) ECML PKDD 2008, Part II. LNCS (LNAI), vol. 5212, pp. 313–325. Springer, Heidelberg (2008)
9. Dietterich, T.G.: Ensemble methods in machine learning. In: Kittler, J., Roli, F. (eds.) MCS 2000. LNCS, vol. 1857, pp. 1–15. Springer, Heidelberg (2000)

10. Wald, R., Khoshgoftaar, T.M., Dittman, D., Awada, W., Napolitano, A.: An extensive comparison of feature ranking aggregation techniques in bioinformatics. In: IEEE 13th International Conference on Information Reuse and Integration, pp. 377–384. IEEE (2012)

11. Abeel, T., Helleputte, T., Van de Peer, Y., Dupont, P., Saeys, Y.: Robust biomarker identification for cancer diagnosis with ensemble feature selection methods. Bioinformatics **26**(3), 392–398 (2010)

12. Yang, F., Mao, K.Z.: Robust Feature Selection for Microarray Data Based on Multicriterion Fusion. IEEE/ACM Transactions on Computational Biology and Bioinformatics **8**(4), 1080–1092 (2011)

13. Haury, A.C., Gestraud, P., Vert, J.P.: The Influence of Feature Selection Methods on Accuracy, Stability and Interpretability of Molecular Signatures. PLOS ONE **6**(12), e28210 (2011)

14. Kuncheva, L.I.: A stability index for feature selection. In: 25th IASTED International Multi-Conference: Artificial Intelligence and Applications, pp. 390–395. ACTA Press, Anaheim (2007)

15. Wald, R., Khoshgoftaar, T.M., Dittman, D.: Mean aggregation versus robust rank aggregation for ensemble gene selection. In: 11th International Conference on Machine Learning and Applications, pp. 63–69 (2012)

16. Alon, U., Barkai, N., Notterman, D.A., Gish, K., et al.: Broad Patterns of Gene Expression Revealed by Clustering Analysis of Tumor and Normal Colon Tissues Probed by Oligonucleotide Arrays. PNAS **96**, 6745–6750 (1999)

17. Dessì, N., Pes, B.: Similarity of feature selection methods: An empirical study across data intensive classification tasks. Expert Systems with Applications **42**(10), 4632–4642 (2015)

18. Bouckaert, R.R., Frank, E., Hall, M.A., Holmes, G., et al.: WEKA - Experiences with a Java Open-Source Project. Journal of Machine Learning Research **11**, 2533–2541 (2010)

19. Liu, H. Setiono, R.: Chi2: Feature selection and discretization of numeric attributes. In: IEEE 7th International Conference on Tools with Artificial Intelligence, pp. 338–391 (1995)

20. Quinlan, J.R.: Induction of decision trees. Machine Learning **1**(1), 81–106 (1986)

21. Witten, I.H., Frank, E., Hall, M.A.: Data Mining: Practical Machine Learning Tools and Techniques. Third Edition. Morgan Kaufmann Publishers (2011)

22. Quinlan, J.R.: C4.5: Programs for Machine Learning. Morgan Kaufmann Publishers (1993)

23. Holte, R.C.: Very simple classification rules perform well on most commonly used datasets. Machine Learning **11**, 63–91 (1993)

24. Robnik-Sikonja, M., Kononenko, I.: Theoretical and empirical analysis of relieff and rrelieff. Machine Learning **53**(1–2), 23–69 (2003)

25. Rakotomamonjy, A.: Variable selection using SVM based criteria. Journal of Machine Learning Research **3**, 1357–1370 (2003)

26. Yang, P., Zhou, B.B., Yang, J.Y., Zomaya, A.Y.: Stability of feature selection algorithms and ensemble feature selection methods in bioinformatics. In: Biological Knowledge Discovery Handbook: Preprocessing, Mining, and Postprocessing of Biological Data. John Wiley & Sons (2014)

Integrating TCP-Nets and CSPs:
The Constrained TCP-Net (CTCP-Net) Model

Shu Zhang, Malek Mouhoub$^{(\boxtimes)}$, and Samira Sadaoui

Department of Computer Science, University of Regina, Regina, Canada
{mouhoubm,sadaouis}@uregina.ca

Abstract. In this paper, a new framework for constraint and preference representation and reasoning is proposed, including the related definitions, algorithms and implementations. A Conditional Preference Network (CP-Net) is a widely used graphical model for expressing the preferences among various outcomes. While it allows users to describe their preferences over variables values, the CP-Net does not express the preferences over the variables themselves, thus making the orders of outcomes incomplete. Due to this limitation, an extension of CP-Nets called Tradeoffs-enhanced Conditional Preference Networks (TCP-Nets) has been proposed to represent the relative importance between variables. Nonetheless, there is no research work reporting on the implementation of TCP-Nets as a solver. Moreover, the TCP-Net only deals with preferences (soft constraints). Hard constraints are not explicitly considered. This is a real limitation when dealing with a wide variety of real life problems including both constraints and preferences. This has motivated us to propose a new model integrating TCP-Nets with the well known Constraint Satisfaction Problem (CSP) framework for constraint processing. The new model, called Constrained TCP-Net (CTCP-Net), has been implemented as a three-layer architecture system using Java and provides a GUI for users to freely describe their problem as a set of constraints and preferences. The system will then solve the problem and returns the solutions in a reasonable time. Finally, this work provides precious information for other researchers who are interested in CSPs and graphical models for preferences from the theoretical and practical aspects.

Keywords: Constraint Satisfaction · Qualitative preferences · CP-Nets · TCP-Nets

1 Introduction

Preference elicitation, representation and reasoning plays an important role in many real life applications, such as collaborative filtering, product configuration, automated decision making system and recommender systems. In most cases, we cannot require the users to have enough patience and knowledge to make a decision or a choice. Thus, helping the users to make a decision efficiently and correctly is important as discussed by the researchers in the field.

© Springer International Publishing Switzerland 2015
M. Ali et al. (Eds.): IEA/AIE 2015, LNAI 9101, pp. 201–211, 2015.
DOI: 10.1007/978-3-319-19066-2_20

While past research work has focused on the quantitative representation of preferences through a utility function based on the well-known Multi-Attribute Utility Theory(MAUT) for example [1] or the C-semiring framework for preferences in temporal problems [2], it is more natural to describe preferences in a qualitative way. Conditional Preference Networks(CP-Nets)[3] is a qualitative graphical model for representing qualitative preference information to reflect the conditional dependency of preferences under **ceteris paribus** (all else being equal) interpretation. It provides a convenient and intuitive tool for specifying the problem, and in particular, the decision maker's preferences. Tradeoffs-enhanced Conditional Preference Network(TCP-nets)[4] is introduced by extending CP-nets, allowing users to describe their relative importance on variables, thus improving the limitations of CP-Nets. However, research on a TCP-Net is very scarce as this latter is a very new model initially proposed in 2006[4]. Furthermore, to our best knowledge there is no implementation of TCP-Nets. In other words, we do not know the performance of TCP-Nets for real wold applications. Moreover, hard constraints are not considered. In this paper, we extended the TCP-Net with constraints, producing a more powerful and novel model called Constrained TCP-Net (CTCP-Net) to address problems under constraints and preferences. The CTCP-Net allows users to add their unconditional and conditional constraints into a Condition-Constraint Table (CCT). The Constraint Satisfaction Problem (CSP) framework [5] has been used to filter the variables values which do not satisfy the CCT. This filter greatly reduces the search space, thus enhancing the efficiency of the solving algorithm. Our proposed model comes with a friendly GUI and can constitute the core of an online shopping system as long as it connects to the corresponding online shopping databases, for electronic selections, air ticket selections and more.

The rest of the paper is organized as follows. Section 2 reviews the related research work on constraints and preferences representation and reasoning. Basic concepts, CSPs, CP-Nets and TCP-Nets are discussed in details in this section. In Section 3, our proposed CTCP-Net model is defined. The solving algorithm is then presented in section 4. Finally, section 5 concludes this research works and prospects possible future works.

2   Background

2.1   CSPs

A Constraint Network (CN) includes a finite set of variables with finite domains, and a finite set of constraints restricting the possible combinations of variable values [5]. Given a CN, a Constraint Satisfaction Problem (CSP) consists of finding a set of assigned values to variables that satisfy all the constraints. A CSP is known to be an NP-hard problem in general[1], and is solved with a backtrack search algorithm of exponential time cost. In order to reduce this

---

[1] There are special cases where CSPs are solved in polynomial time, for instance, the case where a CSP network is a tree [6, 7].

cost in practice, constraint propagation techniques have been proposed [5,6,8]. The idea here is to reduce the size of the search space before and during the backtrack search. In the past four decades the CSP framework, with its solving techniques, has demonstrated its ability to efficiently model and solve a large size real-life applications, such as scheduling and planning problems, configuration, bioinformatics, vehicle routing and scene analysis [9].

2.2   CP-Nets and TCP-Nets

A Conditional Preference Network (CP-Net) is a graph model for representing and reasoning on conditional *ceteris paribus* preferences in a compact, intuitive and structural manner. This model allows users to express their preferences in a qualitative way, which is more natural and comfortable for users compared to quantitative descriptions. Non-conditional preferential independence is used to represent the fact that customers' preference relation over values of a given feature is the same regardless of the values given to other features. The definition [4] is shown as follows.

Definition 1. *[4] Let $x_1, x_2 \in D(X)$ for some $X \subseteq V$ and $y_1, y_2 \in D(Y)$, where $Y = V - X$. We say that X is preferentially independent of Y iff, for all x_1, x_2, y_1, y_2 we have that $x_1 y_1 \succ x_2 y_1 \leftrightarrow x_1 y_2 \succ x_2 y_2$*

In reality, the domains of features and customers' preference are much more complex. In most cases, the preferential independence relies on a certain value of other features, hence we call it conditionally preferentially independent and the corresponding definition [4] is as follows.

Definition 2. *[4] Let X, Y and Z be a partition of V and let $z \in D(Z)$. X is conditionally preferentially independent of Y given z iff, for all x_1, x_2, y_1, y_2 we have that $x_1 y_1 z \succ x_2 y_1 z \iff x_1 y_2 z \succ x_2 y_2 z$*

Moreover, we can say that X is conditionally preferentially independent of Y under Z if the above formula is satisfied for every value of Z.

In order to address the limitations of CP-Nets, tradeoffs-enhanced CP-nets (TCP-Nets) have been proposed by extending the relative preferences to the variables themselves through the non-conditional and conditional relative importance properties [4] defined below.

Definition 3. *[4] Let a pair of variables X and Y be mutually preferentially independent given $W = V - X, Y$. We say that X is more important than Y, denoted by $X \triangleright Y$, if for every assignment $w \in D(W)$ and for every $x_i, x_j \in D(X), y_a, y_b \in D(Y)$, such that $x_i \succ x_j$ given w, we have that: $x_i y_a w \succ x_j y_b w$.*

The above definition also works if $y_b \succ y_a$ given w.

In general, customers describe their preference under certain conditions, hence the conditional relative importance is more commonly used.

Definition 4. *[4] Let X and Y be a pair of variables from V, and let $Z \subseteq W = V - X, Y$. We say that X is more important than Y given $z \in D(Z)$ iff, for every assignment w' on $W' = V - (\{X, Y\} \bigcup Z)$ we have: $x_i y_a z w' \succ x_j y_b z w'$ whenever $x_i \succ x_j$ given zw'. We denote this relation by $X \rhd_Z Y$. Finally, if for some $z \in D(Z)$ we have either $X \rhd_Z Y$, or $Y \rhd_Z X$, then we say that the relative importance of X and Y is conditioned on Z, and write $RI(X, Y|Z)$.*

Accordingly, if for some $z \in Z$, we can also find $Y \rhd_Z X$, then we can conclude that the relative importance of X and Y relies on Z, denoted as $RI(X, Y \mid Z)$.

3   Constrained TCP-Nets (CTCP-Nets)

3.1   Definition of the CTCP-Net

Following the definition of the TCP-Nets in [4] we define the CTCP-Net T as a tuple $< G, cc, cp, i, ci, cct, cpt, cit >$, where:

1 **G** is the set of nodes corresponding to a set of problem variables $\{X_1, X_2, \ldots, X_n\}$.
2 **cc** is the set of directed $cc - arcs\{\alpha_1, \alpha_2, \ldots, \alpha_n\}$ corresponding to conditional constraints. A $cc - arc[\overrightarrow{X_i, X_j}]$ means that X_j is restricted by a given condition on X_i's values.
3 **cp** is the set of directed $cp-arcs\{\beta_1, \beta_2, \ldots, \beta_n\}$ corresponding to conditional preferences. A $cp-arc < \overrightarrow{X_i, X_j} >$ means that the preferences over the values of X_j depend on the actual value of X_i.
4 **i** is the set of directed i-arcs $\{\gamma_1, \gamma_2, \ldots, \gamma_n\}$ corresponding to non-conditional relative importance relations. An i-arc $(\overrightarrow{X_i, X_j})$ corresponds to the following relative importance: $X_i \rhd X_j$.
5 **ci** is the set of undirected $ci - arc\{\lambda_1, \lambda_2, \ldots, \lambda_3\}$, where ci stands for conditional relative importance. A $ci - arc(\widehat{X_i, X_j})$ is in $Tiff$ there is $\mathcal{RI}(X_i, X_j|Z)$ for some $Z \subseteq G - \{X_i, X_j\}$.
6 **cct** associates with conditional constraint table with every node $X \in G$. A cct is from $D(Cd(X))$ (ie..assignments to X's conditional nodes) to constraint value over $D(X)$. Where $Cd(X)$ is $X's$ corresponding conditional variable.
7 **cpt** associates a Conditional Preference Table (CPT) with every node $X \in G$. $CPT(X)$ is a mapping from $D(Pa(X))$ (ie., assignments to X's parents nodes) to strict partial order over $D(X)$. $Pa(X)$ is $X's$ conditional (dependent) variable.
8 **cit** associates with every $ci - arc$ $\gamma = (\widehat{X_i, X_j})$, a (possibly partial) mapping $CIT(\gamma)$ from $D(S(X_i, X_j))$ to orders over the set $\{X_i, X_j\}$.

Note that the sub tuple $< cp, i, ci, cpt, cit >$ corresponds to a TCP-Net. Moreover, if the sets **i** and **ci** are empty then we have a CP-Net.

3.2   Constraint Propagation for CTCP-Nets

We define two types of constraints: non-conditional and conditional and we propagate them respectively using our node and arc consistency algorithms in the preprocessing stage as well as the backtrack search of our solving algorithm.

A non-conditional constraint is a constraint where the condition clause is null. We define it as follows: $NULL \implies rel$. rel denotes a relationship between an attribute value and a constant and is defined as $X \; rel \; a$, where X is an attribute and a is a constant. There are 5 relationships in our model, that is $rel \in \{=, \neq, \prec, \succ, \preceq, \succeq\}$. For this type of constraints, we remove the unsatisfied values according to each constraint. After enforcing node consistency, all the unsatisfied values will be removed from the corresponding domains. In other words, the remaining values satisfy all the non-conditional constraints.

A conditional constraint cc_i is defined as follows.

$$cc_i = (rel_1) \; \text{and}|\text{or} \; (rel_2) \; \text{and}|\text{or} \; \ldots (rel_m) \implies rel \tag{1}$$

We denote by $varsCd(cc_i)$ the set of variables in the condition of cc_i and by $varCc(cc_i)$ the variable present in the conclusion of cc_i.

Here rel is dependent on the condition (rel_1) and $\backslash$ or (rel_2) and $\backslash$ or ,..., and $\backslash$ or (rel_m). If this latter condition is true then rel takes effect. eg. "I do not want to buy a black laptop if the brand is Sony". The "brand is Sony" is the condition here while no black laptop is the constraint, which can be noted as $(Brand = Sony) \implies Color \neq black$. Note that $m = 0$ corresponds to a non conditional constraint.

The following is the general procedure we use to manage non conditional and conditional constraints.

- **m=0**. This is a non-conditional constraint. We process it as a unary constraint in the pre-processing step to restrict the values that each attribute can take. This is done by removing any attribute value that is inconsistent with the unary constraint.
- **m=1**. This is conditional constraint involving one relation in the premise of the condition. We process it as a form of binary constraint using Algorithm 1 in the pre-processing step to remove the inconsistent values. Note that Algorithm 1 enforces directional arc consistency between the variable in the premise and the one in the conclusion of the conditional constraint. For instance, if the conditional constraint is $X \neq a \implies Y \neq b$ and value a has been removed from the domain of X then b has to be removed from the domain of Y.

   As well, the conditional constraint is also used to propagate the effect of an assignment during the backtrack search following the look ahead strategy. For instance, if the conditional constraint is $X = a \implies Y \geq b$ and the current assignment is $X = a$ then all values from the domain of Y that are less than b should be removed.
- **m > 1**. This is a conditional constraint involving more than one relation in the condition of the conditional constraint. We process it similarly to the

case where $m = 1$ but using generalized directional arc consistency as we are dealing with a form of n-ary constraint in this particular case. For instance, let us assume we have the following conditional constraint: $A \neq a$ or $B \neq b \Longrightarrow C < c$ then if a is removed from the domain of A or b is removed from the domain of B then all the values from C's domain that are greater or equal to c should be removed. This propagation can happen in the pre processing stage as well as the search phase. For example, if during the backtrack search A (or B) is assigned a value other than a (or b) then C cannot be assigned a value that is greater or equal to c.

Algorithm CTCP-GAC
1. Given a CTCP-Net $T = < G, cc, cp, i, ci, cct, cpt, cit >$
   (X: set of variables, C: set of constraints between variables)
2. $Q \leftarrow \{(i,j) \mid i \in cc \wedge j \in varCc(i)\}$
3. **While** $Q \neq Nil$ **Do**
4. $Q \leftarrow Q - \{(i,j)\}$
5.   **If** $REVISE(i,j)$ **Then**
6.   **If** $Domain(j) = Nil$ **Then** return false
7. $Q \leftarrow Q \sqcup \{(k,l) \mid k \in cc \wedge j \in varsCd(k)$
8. $\wedge\, l \in varCc(k) \wedge k \neq i \wedge j \neq l\}$
9.   **End-If**
10. **End-While**
11. Return true

Function $REVISE(i,j)$
(REVISE for bound consistency)
1. $domainSize \leftarrow |Domain(j)|$
2. **While** $|Domain(j)| > 0$
   $\wedge \neg seekSupportArc(i,j,min(j))$ **Do**
3.   remove $min(j)$ from $Domain(j)$
4. **End-While**
5. **While** $|Domain(j)| > 1$
   $\wedge \neg seekSupportArc(i,j,max(j))$ **Do**
6.   remove $max(j)$ from $Domain(j)$
7. **End-While**
8. Return $domainSize \neq |Domain(j)|$

Function $REVISE(i,j)$
(REVISE for arc consistency)
1. $REVISE \leftarrow false$
2. $nbElts \leftarrow |Domain(j)|$
3. **For** each value $a \in Domain(j)$ **Do**
4.   **If** $\neg seekSupport(i,j,a)$ **Then**
5.   remove a from $Domain(j)$
6. $REVISE \leftarrow true$
7.   **End-If**
8. **End-For**
9. Return Revise

Fig. 1. CTCP-GAC algorithm for CTCP-Nets, updated from [10]

Arc consistency is enforced with an arc consistency algorithm [5,8]. Since we are dealing with n-ary constraints, we use an adapted version of the Generalized Arc Consistency (GAC) algorithm presented in [10]. This latter is a revised version of the original GAC algorithm proposed in [11] as well as a modified version of the bound consistency algorithm for discrete CSPs in the case of inequality relations [12]. More precisely, bounds consistency is first used through inequality relations to reduce the bounds of the different domains of variables. The adapted GAC is then used to further reduce the domains of the variables. Let us describe now the details of our method.

The modified GAC algorithm that we call $CTCP - GAC$ is described in figure 1. This algorithm enforces arc consistency on all variables domains. $CTCP - GAC$ starts with all possible pairs (i,j) where j is a variable involved by the constraint i. Each pair is then processed, through the function $REVISE$ as follows. Each value v of the domain of j should have a value supporting it (such that the constraint j is satisfied) on the domain on every variable involved by i otherwise v will be removed. If there is a change in the domain of j (after removing values without support) after calling the function $REVISE$ then this change should be propagated to all the other variables sharing a constraint with j. When used as a bound consistency algorithm, cc involves inequality relations and the $REVISE$ function (the function that does the actual revision of the domains) is defined as shown in figure 1 [12]. In the other case, the $REVISE$ function is defined as shown in the bottom right of figure 1 [11]. In the function $REVISE$ (for bound consistency) of figure 1, the function $seekSupportArc$ (respectively the function $seekSupport$ of $REVISE$ for semantic constraints in figure 1) is called to find a support for a given variable with a particular value. For instance when called in line 2 of the function $REVISE$ for bound consistency, the function $seekSupportArc$ looks, starting from the lower bound of j's domain, for the first value that has a support in i's domain. When doing so, any value not supported will be removed.

We also propose a Condition-Constraint Table (CCT) in our model to clearly present all the unconditional and non-conditional constraints. It consists of two columns, conditions and constraints, and each consists of the following parts: variables, relations, value and connectives (if applicable). The fist column can be empty (Null). Each row is a full description of constraints, which must be completely satisfied. For a certain attribute X_i, its CCT is noted as $CCT(X_i)$. The CCT including all the constraints is called $CCT(U)$, where U is set of all variables.

An example is presented in Table 1. After enforcing node and arc consistency in the preprocessing stage of our proposed solving method, we run a backrack search algorithm with a look ahead strategy [5] to find the Pareto optimal solutions of a given CTCP-Net. In order to improve the time performance of the backtrack search, variables are first ordered following the most constrained variables first heuristic [13]. Some of these variables will then be reordered according to the dependencies imposed by the CTCP-Net. In this regard, variables need to be sorted after their respective parents in the corresponding conditional constraint, conditional preference or unconditional relative importance relation. In addition to this static variable ordering that occurs before the backtrack search, some variables are rearranged dynamically during the backtrack search according to the conditional relative importance relations (anytime the variable, the relative importance relies on, is assigned a particular value). Variables values are ordered according to the CPTs. Note that, like for variable ordering, some of these orders depend on values assigned to some other variables and this is done dynamically during the backtrack search.

We adopt the Forward Check strategy [6] as the constraint propagation technique during the backtrack search. Anytime a variable (that we call current variable) is assigned a value during the search, we propagate this decision to the non assigned variables using our CTCP-GAC algorithm as described above in our general procedure. In addition to reducing the size of the search space, this propagation will also detect earlier later failure. For instance, if one of the domains of the non assigned variables becomes empty then we assign another value to the current variable or backtrack to the previously assigned variable if there are no more values to assign to the current one. This backtrack search method will continue until all the variables are assigned in which case we obtain an optimal solution or the search space is exhausted. In this latter case the CTCP-Net is inconsistent.

Table 1. Condition-Constraint table

Conditions	Constraints
$Null$	$Color \neq black$
$Brand = sony$	$Price \leq 1000$
$(Color = red) \wedge (Price > 1000)$	$Brand = sony$
$(Brand = dell) \vee (Brand = sony)$	$Color = silver$

4   Experimentation

In order to evaluate the time performance of our solving method, we conducted several experiments on real data selected from Kjiji.ca[2]. The experiments are conducted on a PC with the following specifications: Inter(R) Core(TM)i7-4500U CPU @1.8GHz and 16GB RAM; and running Windows 8 64-bit operating system. The test platform is MyEclipse 8.5. The application corresponds here to purchasing a vehicle online. 50 products are used for the experiments and for each product the following attributes are considered: brand, model, year, engine size, color, milage, price, transmission, body type and seller name. The constraints and preferences are represented in our model as shown below.

- **Non-conditional constraints (NCC):**
(ncc1) $Null \rightarrow Saleby \neq capital$
(ncc2) $Null \rightarrow Saleby \neq roadway$
(ncc3) $Null \rightarrow Kilometers \leq 150000$
(ncc4) $Null \rightarrow Brand \neq Kia$
(ncc5) $Null \rightarrow Year \geq 2002$
- **Conditional constraints (CC):**
(cc1) $Saleby = nelson \rightarrow Bodytype \neq hatchback$
(cc2) $Bodytype = hatchback \rightarrow Saleby \neq owner$
(cc3) $(Brand = Honda)or(Bodytype = suv) \rightarrow Color \neq red$
(cc4) $(Kilometer \geq 13)and(Transmission = manual) \rightarrow Brand \neq Ford$
(cc4) $(Price \geq 8000) \rightarrow Kilometers \leq 120000$
- **Non-Conditional preferences (NCP):**
(ncp1) $Null \rightarrow Year(descendingorder)$
(ncp2) $Null \rightarrow Price(ascendingorder)$
(ncp3) $Null \rightarrow Kilometers(ascendingorder)$
(ncp4) $Null \rightarrow Transmission(auto \succ manumatic \succ manual)$
- **Conditional preferences (CP):**
(cp1) $(Year \geq 2005)and(Kilometers \leq 15) \rightarrow Brand(Toyota \succ Pontiac \succ Nissan$
$\succ Mazda \succ Kia\ succHyundai \succ Honda \succ Ford \succ Dogde \succ Chrysler \succ$
$Chevrolet \succ Buick \succ Benz \succ BMW \succ Audi); otherwise : Brand(Audi$
$\succ BMW \succ Benz \succ Buick \succ Chevrolet \succ Chrysler \succ Dodge \succ Ford \succ$
$Honda \succ Hyundai \succ Kia \succ Mazda \succ Nissan \succ Pontiac \succ Toyota)$

---

[2] http://www.kijiji.ca/b-cars-vehicles/regina-area/c271l700194

(cp2) $(Saleby \neq owner) or (Transmission = manual) \rightarrow Color(yellow \succ white \succ silver \succ red \succ grey \succ green \succ gold \succ brown \succ blue \succ black); otherwise : Color(black \succ blue \succ brown \succ gold \succ green \succ grey \succ red \succ silver \succ white \succ yellow)$

(cp3) $Brand = Honda \rightarrow Bodytye(convertible \succ wagon \succ suv \succ coupe \succ sedan \succ truck \succ van succhatchback); otherwise : Bodytye(hatchback \succ coupe \succ sedan \succ suv \succ truck \succ van \succ wagon \succ convertible)$

(cp4) $Price \leq 10000 \rightarrow Saleby(owner \succ nelson \succ capital \succ roadway); otherwise : Saleby(nelson \succ capital \succ owner \succ roadway)$

– **Non-conditional relative importance(NCIR):**

(ncri1) $Null \rightarrow Price \rhd Year$

(ncri2) $Null \rightarrow Price \rhd Saleby$

(ncri3) $Null \rightarrow Year \rhd Brand$

(ncri4) $Null \rightarrow Brand \rhd Bodytype$

– **Conditional Relative importance(CIR):**

(cri1) $Saleby \neq owner \rightarrow Year \rhd Kilometers; otherwise : Kilometers \rhd Year$

(cri2) $Transmission = auto \rightarrow Color \rhd Bodytype; otherwise : Bodytype \rhd Color$

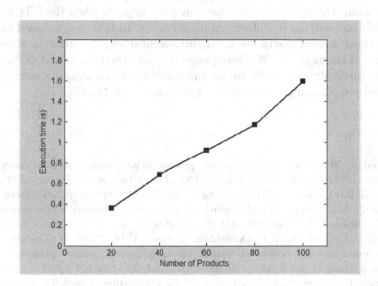

Fig. 2. Execution time(s) vs number of products

Figure 2 reports the running time required in seconds to return the optimal solution when varying the number of products from 20 to 100. For each experiment, 30 run are conducted and the average running time is taken. As we can see from the figure our proposed method is capable of provide an answer in less than 2 seconds even when the number of products is 100.

5   Conclusion and Future Work

Constraints and preferences handling is a complex but interesting problem that is related to a wide variety of real world applications. Our proposed CTCP-Nets has the ability to represent and solve these constraint problems under preferences by returning one or more solutions satisfying all the constraints and maximizing the preferences. This process can be done in a very efficient running time, thanks to the constraint propagation techniques that we propose.

The proposed CTCP-Net has been implemented with a generic design that offers the flexibility for future maintenance and extensibility. It will be therefore possible in the future to add other modules dealing with new features and properties such as the case of cyclic CTCP-Nets.

In the near future we intend to deal with dynamic CTCP-Nets in the case of the addition or retraction of constraints and preferences. The case of constraint of preference addition can be relevant when looking for all k *Pareto* solutions. k can however be very large and not practical. In this particular situation, instead of returning all these optimal solutions we can ask the user to input more information in order to distinguish between them. These information can be in the form of new constraints or preferences. The system should then use these new information to search in an incremental way for new Pareto optimal solutions. On the other hand, the retraction of constraints can happen when the CTCP-Net is inconsistent and solving it will return no solutions. In this case we need to think about relaxing some constraints in an incremental way in order to obtain the consistency of the network. We have previously proposed constraint propagation as well as search algorithms for managing constraints in a dynamic environment [14,15] and are planning to adapt these techniques for the CTCP-Net.

References

1. Sadaoui, S., Shil, S.K.: Constraint and qualitative preference specification in multi-attribute reverse auctions. In: Ali, M., Pan, J.-S., Chen, S.-M., Horng, M.-F. (eds.) IEA/AIE 2014, Part II. LNCS, vol. 8482, pp. 497–506. Springer, Heidelberg (2014)
2. Mouhoub, M., Sukpan, A.: Managing temporal constraints with preferences. Spatial Cognition & Computation 8(1–2), 131–149 (2008)
3. Boutilier, C., Brafman, R.I., Domshlak, C., Hoos, H.H., Poole, D.: Cp-nets: A tool for representing and reasoning with conditional ceteris paribus preference statements. Journal of Artificail Intelligence Research 21, 135–191 (2004)
4. Brafman, R.I., Domshlak, C., Shimony, S.E.: On graphical modeling of preference and importance. Journal of Artificial Intelligence Research 25, 389–424 (2006)
5. Dechter, R.: Constraint Processing. Morgan Kaufmann (2003)
6. Haralick, R., Elliott, G.: Increasing tree search efficiency for Constraint Satisfaction Problems. Artificial Intelligence 14, 263–313 (1980)
7. Mackworth, A.K., Freuder, E.: The complexity of some polynomial network-consistency algorithms for constraint satisfaction problems. Artificial Intelligence 25, 65–74 (1985)
8. Mackworth, A.K.: Consistency in networks of relations. Artificial Intelligence 8, 99–118 (1977)
9. Meseguer, P., Rossi, F., Schiex, T.: Soft constraints. In: Handbook of Constraint Programming (2006)
10. Mouhoub, M., Feng, C.: CSP techniques for solving combinatorial queries within relational databases. In: Chiong, R., Dhakal, S. (eds.) Natural Intelligence for

Scheduling, Planning and Packing Problems. Studies in Computational Intelligence, vol. 250, pp. 131–151. Springer (2009)

11. Lecoutre, C., Szymanek, R.: Generalized arc consistency for positive table constraints. In: Benhamou, F. (ed.) CP 2006. LNCS, vol. 4204, pp. 284–298. Springer, Heidelberg (2006)

12. Lecoutre, C., Vion, J.: Bound consistencies for the csp. In: Proceeding of the Second International Workshop "Constraint Propagation and Implementation (CPAI 2005)" Held With the 10th International Conference on Principles and Practice of Constraint Programming (CP 2005), Sitges, Spain, September 2005

13. Mouhoub, M., Jashmi, B.J.: Heuristic techniques for variable and value ordering in csps. In: Krasnogor, N., Lanzi, P.L. (eds.) GECCO, pp. 457–464. ACM (2011)

14. Mouhoub, M.: Dynamic path consistency for interval-based temporal reasoning. In: 21st International Conference on Artificial Intelligence and Applications (AIA 2003), pp. 10–13 (2003)

15. Mouhoub, M., Sukpan, A.: Conditional and composite temporal csps. Applied Intelligence (2010)

Multiobjective Optimization for the Stochastic Physical Search Problem

Jeffrey Hudack[1,2]([⊠]), Nathaniel Gemelli[1], Daniel Brown[1],
Steven Loscalzo[1], and Jae C. Oh[2]

[1] Air Force Research Laboratory, Rome, NY, USA
{jeffrey.hudack,nathaniel.gemelli,daniel.brown.81,
steven.loscalzo}@us.af.mil
[2] Syracuse University, Syracuse, NY, USA
jcoh@ecs.syr.edu

Abstract. We model an intelligence collection activity as multiobjective optimization on a binary stochastic physical search problem, providing formal definitions of the problem space and nondominated solution sets. We present the Iterative Domination Solver as an approximate method for generating solution sets that can be used by a human decision maker to meet the goals of a mission. We show that our approximate algorithm performs well across a range of uncertainty parameters, with orders of magnitude less execution time than existing solutions on randomly generated instances.

Keywords: Path planning · Planning under uncertainty · Multiobjective optimization · Stochastic search

1 Introduction

We address the challenge of utilizing Intelligence, Surveillance, and Reconnaissance (ISR) assets to locate and identify Anti-satellite weapons (ASAT), which we refer to as AS-ISR. Sites that support ASAT require fixed infrastructure to deploy, but can house mobile equipment that may be located at any site. Construction of the proper infrastructure only indicates that the site is capable of housing the weapon, but provides no guarantee that it will be located there. Due to the high cost of the equipment, and a motivation to not have a small easily identified set of target, there are often more sites than platforms. Therefore, there is uncertainty surrounding the existence of an ASAT that can only be dispelled by sending a sensor-laden platform to the site. This problem can be represented as a planar graph, with vertices representing the site of interest and edges denoting the travel cost between sites.

The AS-ISR problem seeks solutions that minimize cost and minimize the probability of failure, objectives that are in conflict. Additionally, conducting these missions requires human oversight to assure objectives are met. The decision maker

© Springer International Publishing Switzerland 2015 (outside the US)
M. Ali et al. (Eds.): IEA/AIE 2015, LNAI 9101, pp. 212–221, 2015.
DOI: 10.1007/978-3-319-19066-2_21

may have a preference prior to search, allowing the search to focus on a single objective. But commonly, the decision maker may wish to evaluate a set of alternatives after analyzing a set of alternatives against their own preference [9]. In the latter case, a *solution set* needs to be generated that spans the objective space.

If the decision maker is able to provide additional budget or probability of failure constraints, the goal is to find a path that meets one of two objective functions, both of which have been shown to be NP-complete [8] on general graphs:

- **Minimize budget**: given a required probability of failure p^*_{fail}, minimize the budget necessary to ensure an active site is located with certainty of at least $1 - p^*_{fail}$. This answers the question, "How much budget will I need to ensure the risk of failure is below acceptable levels?"
- **Minimize probability of failure**: given a fixed starting budget B^*, minimize the probability of failing to find an active site, while ensuring the sum of travel and purchase costs do not exceed B^*. This addresses, "What risk of failure can I expect given this limited budget?"

A user may want to be presented with a range of options representing the trade-off between budget and risk. For example, the mission may be a part of a larger operation and is one of many competing objectives, which requires evaluating the overall efficiency of the search process to determine if it should be carried out at all, or if additional resources are needed. If a decision maker has a requirement for a maximum travel cost, or minimum risk, we can optimize the solution for the other objective. Otherwise, we must present the user with a set of alternatives that they can choose from. While similar formulations exist [8] [5], to our knowledge this is the first work to provide solutions for this multiobjective optimization problem.

2   Related Work

This problem is similar to a Traveling Salesman Problem and its variations (see [4] for a comprehensive review), but varies in signficant ways. In this work, each site has a distinct probability of failure that differentiates it from other sites, imposing a preference order. More recent work extended the TSP to include features that assign value to visited locations [1] [11], but these models assume that costs are fixed and known. The Orienteering Problem with Stochastic Profits (OPSP) introduces a distribution over the profits at each site and seeks to meet a profit threshold within a given time limit [12].

Another key difference of this domain from prior work is the minimization of multiple competing objectives. Multi-objective optimization for path planning has been approached using evolutionary algorithms [6], but has often been limited to simple cycles on graphs. The Traveling Purchaser Problem with Stochastic Prices [8] introduces the model used in this work, but generates solutions by generalizing the cost distribution at each site as the expected cost, which we compare against in our evaluation in Section 6. More recent work has provided solutions for maximizing probability with a fixed budget and minimizing budget with a fixed probability [5] [2], but provides solutions for a path.

3   Problem Formulation

We formulate the AS-ISR problem as a stochastic physical search problem (SPSP), where we are given an undirected network $G(S^+, E)$ with a set of sites $S^+ = S \cup \{o, d\}$ where $S = \{s_1, ..., s_m\}$ is the set of m sites that may be active, o and d are the origin and destination locations. We are also given a set of edges $E = \{(i, j)] : i, j \in S^+\}$. Each (i, j) has the cost of travel t_{ij} that is deterministic and known. An agent must start at origin node o, visit a subset of sites in S to find an active site, and then end at the destination point d.

The cost of identifying an ASAT at each site $s_i \in S$ is an independent random variable C_i with an associated probability mass function $P_i(c)$, which gives the probability that determining if a site is active will cost c at site s_i. This cost may be infinity, which indicates the site is not active and no information can be gained, or 0 to indicate there is no cost. We refer to this specialization as a *Binary SPSP*. A site s_i is active with some probability p_i, or inactive with probability $1 - p_i$, and the objective is to minimize the cost and probability of failure for visiting an active site.

A solution is a path π, a list of length n where $\pi(i)$ and $\pi(i+1)$ are consecutive locations along a path, with $\pi(1) = o$ and $\pi(n) = d$. We define $\pi(i, j)$ as the the sub-path of π containing consecutive path locations $\pi(i)$ through $\pi(j)$. In Figure 1 we show a small instance of an AS-ISR problem with 5 sites and the associated pareto set of path solutions. The longest solution path, the shortest path that visits all sites, is also the solution to the Traveling Salesman Path Problem [4].

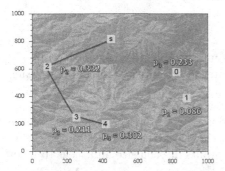
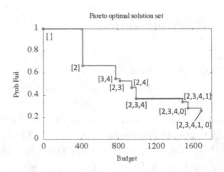

Fig. 1. An AS-ISR instance (left) with starting location s and 5 sites. Below each site is the probability that the site is active. The set of solutions (right) contains a collection of nondominated paths and their associated cost and probability. Each point is labeled with the path, which is the order of site visits starting from s. A single path (blue) is shown on both graphs, for reference.

4   Multiobjective Optimization of SPSP

A multi-objective optimization problems (MOP) represents a trade-off between competing objectives. Efficient exploration of the solution space is difficult in the case where MOP criteria share dependencies on a common set of variables, such as budget and prob of failure, in this case. The following definitions are derived from multiobjective optimization literature [13], with adaptations for the specifics of the B-SPSP.

In general, a multi-objective problem is characterized by a vector of r decision variables, $x = (x_1, x_2, ..., x_r) \in X$, and k criteria. There is an evaluation function $F : X \to A$, where $A = (a_1, a_2, ...a_k)$ represents the k attributes of interest. We represent the fitness of vector x as $F(x) = (f_1(x), f_2(x), ...f_k(x))$, where $f_i(x)$ is the mapping from the decision variable vector to a single attribute a_i.

Definition 1.1 (Multiobjective Optimization for SPSP). We define a set of decision variables x denoting a path, a set of 2 objective functions for cost, $f_1 : X \to \mathbb{R}$, and probability of failure, $f_2 : X \to [0, 1]$, and a set of m constraints that require a decision vector be a valid path. We use probability of failure for clarity of definition, allowing us to minimize both objectives. The optimization goal is to

$$\begin{aligned} \text{minimize} \quad & F(x) = (f_1(x), f_2(x)) \\ \text{subject to} \quad & e(x) = (e_1(x), e_2(x), ..., e_m(x)) \leq 0 \\ \text{where} \quad & x = (x_1, x_2, ..., x_n) \in X \end{aligned} \quad (1)$$

The *feasible set* X_f is defined as the set of decision vectors in X that form a valid path, with each decision vector x satisfying the path constraints $e(x)$, and such that $X_f = \{x \in X | e(x) \leq 0\}$. While we would prefer solutions that provide the minimum probability of failure at the minimum cost without violating the constraints, the objectives do not have optima that correspond to the same solution.

Definition 1.2. For a pair of objective vectors u and v,

$$\begin{aligned} u = v \text{ iff } \forall i \in \{1, 2, ..., k\} : u_i = v_i \\ u \leq v \text{ iff } \forall i \in \{1, 2, ..., k\} : u_i \leq v_i \\ u < v \text{ iff } u \leq v \wedge u \neq v \end{aligned} \quad (2)$$

The relations $\geq$ and $>$ are defined similarly.

These definitions impose a partial order on solutions. When none of these relations hold between u and v, we can only state that $u \not\geq v$ and $u \not\leq v$, meaning neither is superior. A solution that provides higher probability and higher cost than another solution is just as strong and may prove more effective if the additional budget is available to spend.

Definition 1.3 (Pareto Dominance). For any decision vectors b and c,

$$b \succ c \text{ (b dominates c)} \qquad \text{iff } F(b) < F(c)$$
$$b \succeq c \text{ (b weakly dominates c) iff } F(b) \leq F(c) \qquad (3)$$
$$b \frown c \text{ (b is indifferent to c)} \quad \text{iff } F(b) \nleq F(c) \wedge F(b) \ngeq F(c)$$

The definitions for "dominated by" $(\prec, \preceq, \frown)$ are analogous.

An optimal solution is one that can not be improved by any other solution in the feasible set. This does not include solutions that are indifferent, as they are not comparable within the partial order imposed by pareto dominance. A decision vector $x \in X_f$ is nondominated with respect to a set $D \subseteq X_f$ iff $\nexists a \in D : a \succ x$. We refer to a solution x as Pareto optimal iff x is nondominated by X_f. Because the set of all solutions may be a partial order, there can be two or more Pareto optimal solutions for a given problem instance. The collection of all nondominated solutions with respect to the set of all solutions is referred to as the **nondominated solution set**.

Let $D \subseteq X_f$. The function $n(D) = \{d \in D : d \text{ is nondominated regarding } D\}$ outputs the set of nondominated decision vectors in D. The objective vectors in $f(n(D))$ is the *nondominated front* regarding D. The set $X_n = n(X_f)$ is referred to as the Pareto-optimal set (Pareto set). We use the term **solution set** to denote both nondominated decision set and their associated objective values. Solution sets represent the output of a solver that can be used to evaluate performance on problem instances.

4.1   Comparing Solution Sets

We desire a measure to compare solution sets, so that we can determine the trade-offs between using different solution techniques. Because the budget values are not normalized, and will vary greatly with respect to the number of sites, it is important that any measure be scale-independent. We use a measure similar to the $\mathcal{S}$ metric [13] on a two-dimensional space.

Definition 1.4 (Size of solution set). Let $D = (d_1, d_2, ..., d_m) \subseteq X$ be a set of decision vectors, ordered by f_1. The function $\mathcal{S}(D)$ returns the union of the area enclosed by the objective values for each vector d_i. The area enclosed by a single decision vector d_i is a rectangle defined by the points $(f_1(d_{i+1}), 0)$ and $(f_1(d_i), f_2(d_i))$. If $i = m$, then let $f_1(d_{i+1}) = f_1(d_i)$.

Using the $\mathcal{S}$ metric, we can derive a measure of error with respect to the optimal solution.

Definition 1.5 (Solution set error). Let $B^* = (x_1^*, x_2^*, ..., x_m^*) \subseteq X_f$ be the optimal set of decision vectors. Given a decision vector $D = (x_1, x_2, ..., x_m) \subseteq X$ we define the error of C as

$$\mathcal{E}(D) = \left(\frac{\mathcal{S}(D)}{\mathcal{S}(B^*)} \right) - 1 \qquad (4)$$

A solution set should seek to minimize the $\mathcal{E}$ metric, with $\mathcal{E}(D) = 0$ indicating solution set D is optimal.

5   Approach

The Iterative Domination Solver (IDS) starts by generating all paths containing one market and adding them to the active set. Each iteration of of the algorithm consists of two phases: the *search phase* and the *domination phase*. In the search phase, remaining sites are added to all paths in the active set. This generates a new set of candidate paths that are added to the active set. In the domination phase, the solver evaluates new candidate paths and determines if they are dominated or dominate any other solutions. Depending on the filter depth parameter, dominated solution are removed from the active set, preventing any further search using that path. This procedure allows us to focus search on high value sub-paths without expending significant effort on paths that are low value and unlikely to be found on a non-dominated path.

5.1   Search Phase

A site is added to the previous path immediately after the start site or previous to the destination site, which ever adds lower total travel cost. This operation prevents insertions that break up edges from the last iteration, preserving inter-site edges that exist in sub-paths. An example of this operation is shown in Figure 2. This process continues, iteratively adding new markets to the candidate paths, until we have the path that visits all sites in the problem, at which point the process terminates and the non-dominated set of intervals is returned.

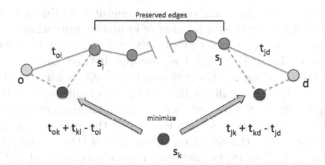

Fig. 2. Illustration of the insertion step for the IDS algorithm. A new node s_k is added to the search path after the origin site or before the destination site, whichever has the lower added path cost. This preserves the edges from the previous path in the search process, exploiting nondominated substructure.

We can reduce the search space via a *filter depth* (fd) parameter, which is path length at which we start pruning out dominated solutions. For example, if we set $fd = 1$, only a single path consisting of (o, s_i, d) will remain after initial path generation, and all paths of length 2 must contain s_i. Then, if the only nondominated size 2 path contains s_i and s_j, all resulting paths from the next iteration must contain that ordered pair, and so on. Using filter depth is an effective pruning strategy because it prefers edges with shorter path lengths and there is a diminishing return on probability as the problem size increases.

5.2 Domination Phase

A quad tree [3] is a rooted data structure where a node may have up to 4 children. For our purposes, each child represents a quadrant in the 2D plane of budget and probability of failure. This is defined recursively, making it an extension of a binary tree to 2 dimensions. The quad tree data structure allows us to store the value pairs (cost, probability) and their associated paths, while also detecting when a new solution is dominated or dominates another solution.

Because the number of possible paths is exponential with respect to the number of sites [10], storing all solutions is unreasonable. When a domination relation is detected, we use the structure to search for entries that should be removed or preserved and restructure the tree as necessary. This adds additional overhead during storage of individual solutions, but allows us to limit the space complexity of storing candidate solutions. The efficiency of searching this structure is sensitive to proper selection of the initial root node, an optimization still being investigated. In this work we choose to seed the tree with a random path using half of the markets, but better methods are certain to exist.

6   Experimental Results

We evaluate our approach using randomly generated problem instances and comparing performance against a number of intuitive approaches to solving the B-SPSP. Site networks are formed by placing sites randomly in a 1000×1000, with travel cost t_{ij} being the Euclidean distance between sites s_i and s_j. The probability of failure at each sites is drawn from a Gaussian distribution with varying mean and variance. All results are based on 100 instances, with error bars indicating a 95% confidence interval. We use a naming convention based on the chosen filter depth. If $fd = k$ then we refer to the algorithm as IDS-k.

In order to determine the relative performance of IDA, we compare against a solver for these problem types that minimizes Expected Cost (EC), adapted from the approach outlined in [8]. Changes to the algorithm are the acquisition of only a single 'item', and the removal of item cost, which is 0 in this domain.

At each step, EC will provide a path the minimizes the overall cost of the complete path, without consideration for budget or some fixed probability of success. In order to generate a solution set from this single path, we include all sub-paths of the solution beginning at the starting site. This gives us a step-wise set of intervals that still meet the criterion of minimizing expected cost. We also include a greedy solution that generates a set of solutions by selecting the lowest cost edge available.

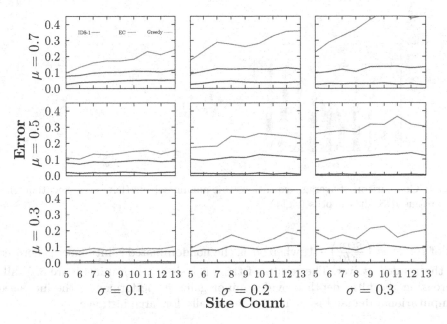

Fig. 3. Comparison of error on random instances with respect to mean and variance of probability of failure on sites

Using the $\mathcal{E}$ metric defined in Section 4.1, we can compute the error of each approach with respect to optimal. In Figure 3, we show the effectiveness of IDS with respect to both changes in the mean and variance of the probability of failure. As the mean probability of failure increases, all 3 algorithms perform closer to optimal. This is due mostly to each site individually moving closer to 0% failure, leaving less room for error, in general.

As the variance on the probability is increased, expected cost seems relatively unaffected, the greedy solver performs significantly worse, and IDS sees significant improvement. Increased variance distinguishes the site probabilities, making higher cost edges feasible options for paths that may be non-dominated. This makes the single path with low average cost less likely to cover the space of non-dominated solutions.

In Figure 4 we show the execution time of IDS with expected cost, optimal and a random solver that generates 30,000 randomly selected paths. IDS completes search orders of magnitude faster than optimal and EC, mainly due to being able to terminate search before expanding the search tree on paths that are not of good quality.

Because the optimal solution requires exponential time to solve as the problem size increases, it is difficult to compare performance against optimal for large instances. To show that the performance of IDS is maintained for large problem instances, we compare against a random solver, which generates 30,000 random paths and generates a solution set from all enumerated sub-paths. We define the

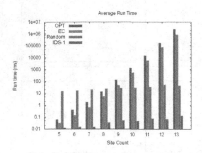 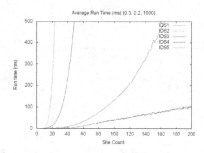

Fig. 4. Comparison of average execution time among solvers (left) and execution time for various IDS filter depths (right)

lift of set D as $\left(\frac{S(D)}{S(R)}\right) - 1$, where R is the nondominated solution set generated by the random solver. The results of this analysis are shown in Figure 5. While increasing the filter depth provides minor gains in performance, the increased computational demand is significant, especially for large instances.

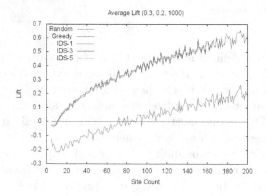

Fig. 5. Lift of IDS and greedy versus random sampling

7   Conclusions and Future Work

We have presented B-SPSP as a method for solving AS-ISR problems, providing analysis of the problem space and a characterization of nondominated solution sets. We have also presented the Iterative Domination Solver as an approximate method for generating solution sets. We have shown that IDS performs better than existing approaches across a wide range of problem parameters with orders of magnitude less execution time on randomly generated instances. While the filter depth can be increased for minor performance gains on instances of the AS-ISR problem, in most cases IDS-1 will prove sufficient.

Future work will focus on solution sets for general SPSP instances with multiple costs per site, which will significantly increase the complexity of finding optimal solutions due to the branching of search on cost realizations at each site. We plan to explore how IDS can be adapted to solve these instances using the same iterative approach with pruning, and we expect to see larger gains from increased filter depth. Additionally, a number of solution concepts have been developed for TSP that could, in some cases, be adapted for this problem formulation.

References

1. Arora, S., Karakostas, G.: A 2+ ε approximation algorithm for the k-mst problem. In: Proceedings of the Eleventh Annual ACM-SIAM Symposium on Discrete Algorithms, pp. 754–759. Society for Industrial and Applied Mathematics (2000)
2. Brown, D.S., Hudack, J., Banerjee, B.: Algorithms for stochastic physical search on general graphs. In: Workshops at the Twenty-Ninth AAAI Conference on Artificial Intelligence (2015)
3. Finkel, R.A., Bentley, J.L.: Quad trees a data structure for retrieval on composite keys. Acta informatica 4(1), 1–9 (1974)
4. Gutin, G., Punnen, A.P.: The traveling salesman problem and its variations, vol. 12. Springer (2002)
5. Hazon, N., Aumann, Y., Kraus, S., Sarne, D.: Physical search problems with probabilistic knowledge. Artificial Intelligence 196, 26–52 (2013)
6. Jozefowiez, N., Glover, F., Laguna, M.: Multi-objective Metaheuristics for the Traveling Salesman Problem with Profits. Journal of Mathematical Modelling and Algorithms 7(2), 177–195 (2008). http://link.springer.com/10.1007/s10852-008-9080-2
7. Kambhampati, S.: Model-lite planning for the web age masses: the challenges of planning with incomplete and evolving domain models. In: Proceedings of the National Conference on Artificial Intelligence, pp. 1601–1604 (2007)
8. Kang, S., Ouyang, Y.: The traveling purchaser problem with stochastic prices: Exact and approximate algorithms. European Journal of Operational Research 209(3), 265–272 (2011)
9. Nguyen, T.A., Do, M., Gerevini, A.E., Serina, I., Srivastava, B., Kambhampati, S.: Generating diverse plans to handle unknown and partially known user preferences. Artificial Intelligence 190, 1–31 (2012)
10. Roberts, B., Kroese, D.P.: Estimating the Number of s - t Paths in a Graph. Journal of Graph Algorithms and Applications 11(1), 195–214 (2007)
11. Snyder, L.V., Daskin, M.S.: A random-key genetic algorithm for the generalized traveling salesman problem. European Journal of Operational Research 174(1), 38–53 (2006)
12. Tang, H., Miller-Hooks, E.: A tabu search heuristic for the team orienteering problem. Computers & Operations Research 32(6), 1379–1407 (2005)
13. Zitzler, E.: Evolutionary Algorithms for Multiobjective Optimization: Methods and Applications (30)

Towards Multilevel Ant Colony Optimisation for the Euclidean Symmetric Traveling Salesman Problem

Thomas Andre Lian[(✉)], Marilex Rea Llave, Morten Goodwin,
and Noureddine Bouhmala

Department of ICT, University of Agder, Grimstad, Norway
angeland_89@hotmail.com

Abstract. Ant Colony Optimization (ACO) metaheuristic is one of the best known examples of swarm intelligence systems in which researchers study the foraging behavior of bees, ants and other social insects in order to solve combinatorial optimization problems.

In this paper, a multilevel Ant Colony Optimization (MLV-ACO) for solving the traveling salesman problem is proposed, by using a multilevel process operating in a coarse-to-fine strategy. This strategy involves recursive coarsening to create a hierarchy of increasingly smaller and coarser versions of the original problem. The heart of the approach is grouping the variables that are part of the problem into clusters, which is repeated until the size of the smallest cluster falls below a specified reduction threshold. Subsequently, a solution for the problem at the coarsest level is generated, and then successively projected back onto each of the intermediate levels in reverse order. The solution at each level is improved using the ACO metaheuristic before moving to the parent level. The proposed solution has been tested both in circular and randomized environments, and outperform single level counterparts.

1 Introduction

Complex optimization problems arise in several areas of artificial intelligence and computer science. In their full generality, these problems are NP-complete and consequently algorithmically intractable. With the growing popularity of natured inspired algorithms, several researchers have applied ACO [1] metaheuristic in various extensive fields including scheduling [2], vehicle routing [3], to timetabling [4], and continuous optimization [5]. Designing efficient optimization search techniques requires a tactical interplay between diversification and intensification. The former refers to the ability to explore many different regions of the search space, whereas the latter refers to the ability to obtain high quality solutions within those regions.

ACO algorithms, like other metaheuristics, offer the advantage of being flexible as they can be applied to any problem whether discrete or continuous. Nevertheless, even ACOs may still suffer from either slow or premature convergence [6]. The performance of ACOs as well as other available optimization techniques

© Springer International Publishing Switzerland 2015
M. Ali et al. (Eds.): IEA/AIE 2015, LNAI 9101, pp. 222–231, 2015.
DOI: 10.1007/978-3-319-19066-2_22

deteriorates very rapidly mostly for two reasons; Firstly, the complexity of the problem usually increases with its size, and secondly, the solution space of the problem increases exponentially with the problem size. Consequently, optimization search techniques tend to spend most of the time exploring a restricted area of the search space preventing the search to visit more promising areas, and thus leading to solutions of poor quality.

This paper introduces MLV-ACO, an ACO algorithm operating in the multilevel context for the Euclidean symmetric traveling salesman problem.

The paper is organized as follows. Section 2 defines the traveling salesman problem. Section 3 explains the single level ACO. Section 4 continues with discussing the generic multilevel approach. Section 5 introduces the novel MLV-ACO metaheuristic. Section 6 presents the experimental results while section discusses the obtained results. Section 7 concludes the paper with guidelines for future work.

2 Traveling Salesman Problem (TSP)

The traveling salesman problem (TSP) is one of the classical problems of combinatorial optimization. It is concerned with a salesman who must visit a number of cities (vertices) once and return to the city of origin. In this paper we shall concentrate on the Euclidean symmetric TSP in which the distance from vertex v_i to city v_j is the same as the distance from vertex v_j to city v_i. The problem is defined with N vertices having coordinates in the 2D plane and a symmetric distance matrix $D = [d(v_i, v_j)]$ which gives the Euclidean distance between any two cities v_i and v_j. The goal in the TSP is to find an ordering π of the cities that minimizes the distance $L(\pi)$ expressed in equation 1.

$$L(\pi) = \sum_{i=1}^{N-1} d(v_{\pi(i)}, v_{\pi(i+1)}) + d(v_{\pi(N)}, v_{\pi(1)}).$$ (1)

The TSP belongs in the class of combinatorial optimization problems known as NP-hard [7] and has been extensively studied due to its simplicity and applicability. Problems having the TSP structure occur, for example, in circuit board drilling applications, instances arising in VLSI fabrication, and sometimes as a subproblem for the vehicle routing problem. The simplicity of stating the problem coupled with its intractability makes it an ideal platform for exploring new algorithmic techniques. Due to the combinatorial explosion nature of the TSP, the researchers have proposed various approaches to solve this problem with the aim usually being just to find near optimal solution.

3 Ant Colony Optimisation (ACO)

Swarm Intelligence algorithms, such as ACO was first used to find shortest path from a source to a sink in a bidirectional graph. It has later increased in popularity due to its low complexity and its ability to work in dynamic environments, including its ability to solve the TSP [3,8,9].

Finding the path in a graph $G(V, E)$ using ACO in its simplest form works as follows. Artificial ants move from vertex to vertex. When an ant finds a route s from the source $v_s \in V$ to the sink $v_t \in V$, the ant release pheromones $\tau_{i,j}$ corresponding all edges $e_{i,j} \in s$. The pheromones for all ants m is defined as

$$\tau_{i,j} \leftarrow (1-p)\tau_{i,j} + \sum_{k=1}^{m} \Delta\tau_{i,j}^{k} \tag{2}$$

where is $\Delta\tau_{i,j}^{k}$ represent the quality of the route, e.g. the shortest path [10]. The amount of pheromone released represent quality of the solution, which is in turn used to guide the ants as they walk randomly with a preference towards pheromones.

4   Multilevel Techniques

Multilevel techniques aim at producing smaller and smaller problems that are easier to solve than the original one. These techniques were first introduced when dealing with the graph partitioning problem (GCP) [11] and have proved to be effective in producing high quality solutions at a lower cost than single level techniques. Recently, a memetic algorithm integrating a new hierarchical crossover operator and a perturbation-based tabu algorithm has been introduced in [12] for GCP. Experimental studies showed that the proposed approach performs far better than any of the existing graph partitioning algorithms in terms of solution quality. The traveling salesman problem (TSP) was the second combinatorial optimization problem to which the multilevel paradigm was applied [13] and has shown a clear improvement in the asymptotic convergence of the solution quality. A tabu search operating in a multilevel context has been developed for the feature selection problem in biomedial data [14]. The empirical results showed that the approach obtained more accurate and stable classification than those obtained by using the other feature selection techniques. The multilevel paradigm has been combined with the memetic algorithm [15] for the satisfiability problem and the broad conclusions drawn from this work was that the multilevel context can either speed up the problem to be solved or improve its asymptotic convergence. A recent survey over existing hierarchical techniques can be found in [16].

5   MLV-ACO: Combining ACO with the Multilevel Paradigm

The implementation of the multilevel strategies requires four components: (1) an initial solution, (2) a coarsening phase, (3) a projection phase and (4) an optimization algorithm to be used across the different levels. Figure 1 shows a simple example of how this is carried out with MLV-ACO with a nearest neighbour coarsening scheme (MLV-ACO NN).

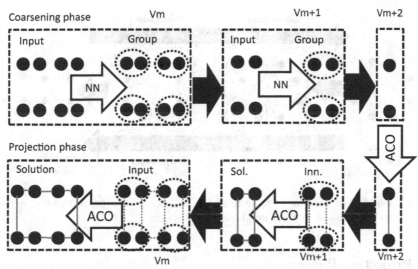

Fig. 1. Example of MLV-ACO with nearest neighbour coarsening phase, and ACO optimisation algorithm in the projection phase

5.1 Initial Solution

The algorithm is applied to a connected graph $G(V, E)$. The initial solution, s is a random ordering of vertices.

5.2 Coarsening Phase

The coarsening phase groups vertices together so that the problem to be solved contains fewer vertices and is in turn simpler.

Let V_m (the subscript represents the level of problem scale) be a constructed grouping of the vertices so that the graph at level m is represented by $G(V_m, E_m)$. The next coarser level, V_{m+1}, is constructed from V_m by merging variables. Two variants of the coarsening phase are implemented:

- **MLV-ACO R:** Multilevel ACO with Random Pairing. For each vertex $v_i \in V_m$ randomly select another vertex in $v_i \in V_m$ to pair with. Move the pair up to V_{m+1}.
- **MLV-ACO NN:** Multilevel ACO with Nearest Neighbour Pairing. For each vertex in $v_i \in V_m$ select a vertex $v_j \in V_m$ where v_j is the vertex closest to v_i.

Figure 2 presents an example of the coarsening phase moving from 8 vertices to 4.

If there is an odd number of vertices, any unmerged vertex is simply copied to the next level. Any such vertex is than itself a cluster of size 1.

The new formed clusters are used to define a new and smaller problem and recursively iterate the reduction process until the size of the problem reaches some desired threshold.

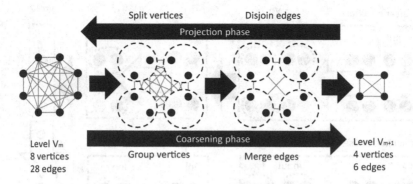

Fig. 2. Example of coarsening — from level V_m to V_{m+1} — and project phase — from level V_{m+1} to V_m —using nearest neighbour pairing

5.3 Projection Phase

The projection phase refers to the inverse process as the coarsening phase. In the projection, the problem is moved from V_{m+1} to V_m. Figure 2 includes the project phased moving from 4 vertices to 8.

When performing projection, moving from a level V_{m+1} to V_m, the process is as following:

1. Apply ACO until a stop criteria (see section 5.5).
2. Split up vertices and disjoin edges.
3. Project the solution at level $m + 1$ using the solution from level m so that any optimisation at level m has a good starting point. Upon disjoining edges, the pheromones are spread out equally.
4. If not at lowest level, goto 1.

5.4 Optimisation Algorithm

The optimisation algorithm is ACO applied at for finding an ordering of the vertices. In contrast to normal ACO, the MLV-ACO has a distinct advantage due to the fact that there exists a solution in coarser phases — with an exception for the very highest level.

Thus, upon applying MLV-ACO at level $G(m + 1, e + 1)$, the results from $G(m, e)$ are used as a starting point. This means that MLV-ACO always starts out with a coarse solution which it makes more fine grained for every iteration of the projection phase.

5.5 Level Timing

The ACO metaheuristic has no built in stop criteria. It is therefore interesting to know how varying the number of ants affects the performance. Figure 3a shows the MLV-ACO NN with different number of ants released before moving to the next level in the projection phase. The only exception are single level ACO which does not move to any coarser level, and therefore has no intention of stop criteria. Interestingly, the figure shows that the algorithm performs very similar independent of the when the algorithm is stopped.

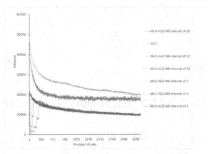

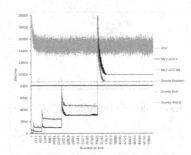

(a) MLV-ACO NN with where coarsening takes place at different intervals

(b) MLV-ACO NN (nearest neighbour) and MLV-ACO R (random pairing) in a randomized environment with 100 vertices 55 000 ants compared with greedy and random

Fig. 3. Results from running single and MLV-ACO in in a randomized environment

6   Experiments

To test the proposed both MLV-ACO variants, we apply the them in two distinct environments; a circular and a randomized.

It should be noted that in these experiments the distance is calculated based on vertices for the current level of the solution. This could be wrongly interpreted as a jump in the performance of the multilevel algorithm, but is caused moving by $level_m$ to $level_{m+1}$ which produces additional vertices and edges, and in turn yields a longer distance. The true comparison of distance between the solutions is at the last level after ~ 600 ants

6.1   Circular Environments

A circular environment is constructed so that all vertices lie in a perfect circle and in this way form a Hamiltonian cycle. This is chosen because it is the go-to test for TSP [17], and the optimal solution will be a route in a perfect circle. Hence, in this environment it is easy to verify both whether an algorithm finds the optimal solution and how far away from the optimal solution the algorithm is.

Figure 4 shows an illustration of how single level and MLV-ACO NN is able to find a solution for TSP. Figure 5a shows the performance, as total distance of the cycle of 500 vertices.

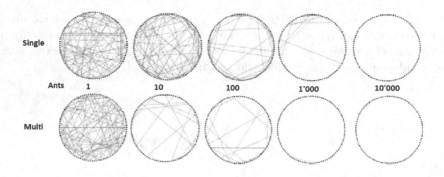

Fig. 4. Example of single level ACO and MLV-ACO NN in a circular environment

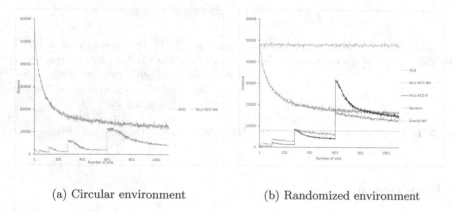

(a) Circular environment   (b) Randomized environment

Fig. 5. Experiment results in environments with 500 vertices and 1200 ants

6.2 Randomized Environments

The second set of experiments are carried out in randomized environments. Fig 6a and 6b show examples of pheromone distribution after multileveled and single level ACO have been applied. The thickness of the line represents the pheromone weight. The figure indicates that by clustering the problem the MLV-ACO is able to solve the grouped problems. This is opposed to the single level solution which uses much longer time to converge.

Figure 5b shows the two variants of MLV-ACO in a randomized environment and compares it to single level ACO, a random solution, and a greedy nearest neighbour.

An interesting aspect seen here is that the MLV-ACO R starts out by performing similar to ACO. However, the distance drops fairly quickly. The MLV-ACO NN on the other hand has a higher distance than both single level ACO and MLV-ACO R, but in turn converges on a solution somewhat quicker.

In the randomized environment, the MLV-ACO variants are not able to find as good solutions as in the circular environment. An explanation for this might

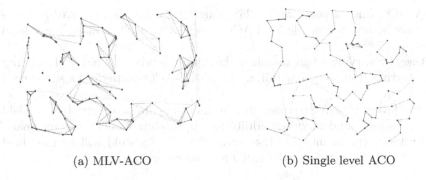

| (a) MLV-ACO | (b) Single level ACO |

Fig. 6. Example of pheromones after single and MLV-ACO has been applied in a randomized environment

be because the naïve nearest neighbour picks good pairs in the beginning, but when fewer vertices are available it might end up with pairing vertices which are not very well suited together. This is less of a problem in the circular environment because pairs are always close by in the circle.

Despite this, MLV-ACO R starts out better than MLV-ACO NN, but MLV-ACO NN is able to find a much shorter distance. The MLV-ACO NN once again displays superiority over single level ACO, and MLV-ACO R also performs equally good or better than single level ACO.

Figure 3b show some interesting aspects when running the algorithms over time. The single level ACO struggles to fit to a path, it finds only local optimas. However, MLV-ACO R and MLV-ACO NN find a more global optima. This is despite the fact that the ACO-part of the algorithms are exactly the same code for all three variants. This indicates that clustering of the problem gives the MLV-ACO variants an advantage so that they are able to improve beyond local optimas.

7   Conclusion

In this paper an applications displaying the results for both single and multilevel approaches of the Ant Colony Optimization (ACO) over Traveling salesman problems. The paper introduces two novel variants of multilevel ACO; MLV-ACO R, where pairs are randomly selected, and MLV-ACO NN, where pairs are selected in line with nearest neighbour.

The algorithms are tested on both a circular and randomized traveling salesman environments.

Both MLV-ACO variants are applied to the circular environment, yielding much better solutions than the single level counterparts. In fact the distance using multilevel is about half the length compared to single level ACO. Further, in randomized environments the same trends are visible, but MLV-ACO does not outperform the single level ACO variants to the same degree.

When running MLV-ACO with 55 000 simulated ants it reaches the same performance as a much more complex greedy algorithm, which indicates that the MLV-ACO finds a path that is 25% longer than the shortest path. This is significantly better than single level ACO which locks itself to much worse local optimas.

Further, by varying when the algorithms jump levels, this paper shows that already starting from very few initial ants, MLV-ACO outperforms single level ACO.

For our future research we plan to fine-tune the parameter variables to yield even better results, and carry out additional approaches of clustering and coarsening. Further experiments on other areas than TSP should will be examined for an example shortest path within 2D and 3D environments.

References

1. Dorigo, M., Stützle, T.: Ant colony optimization: overview and recent advances. In: Handbook of Metaheuristics, pp. 227–263 (2010)
2. Merkle, D., Middendorf, M., Schmeck, H.: Ant colony optimization for resource-constrained project scheduling. IEEE Transactions on Evolutionary Computation 6(4), 333–346 (2002)
3. Bell, J.E., McMullen, P.R.: Ant colony optimization techniques for the vehicle routing problem. Advanced Engineering Informatics 18(1), 41–48 (2004)
4. Socha, K., Sampels, M., Manfrin, M.: Ant algorithms for the university course timetabling problem with regard to the state-of-the-art. In: Raidl, G.R., Cagnoni, S., Cardalda, J.J.R., Corne, D.W., Gottlieb, J., Guillot, A., Hart, E., Johnson, C.G., Marchiori, E., Meyer, J.-A., Middendorf, M. (eds.) EvoIASP 2003, EvoWorkshops 2003, EvoSTIM 2003, EvoROB/EvoRobot 2003, EvoCOP 2003, EvoBIO 2003, and EvoMUSART 2003. LNCS, vol. 2611, pp. 334–345. Springer, Heidelberg (2003)
5. Socha, K., Dorigo, M.: Ant colony optimization for continuous domains. European Journal of Operational Research 185(3), 1155–1173 (2008)
6. Stützle, T., Hoos, H.H.: Max-min ant system. Future Generation Computer Systems 16(8), 889–914 (2000)
7. Garey, M.R., Johnson, D.S.: A guide to the theory of np-completeness. San Francisco (1979)
8. Manfrin, M., Birattari, M., Stützle, T., Dorigo, M.: Parallel ant colony optimization for the traveling salesman problem. In: Dorigo, M., Gambardella, L.M., Birattari, M., Martinoli, A., Poli, R., Stützle, T. (eds.) ANTS 2006. LNCS, vol. 4150, pp. 224–234. Springer, Heidelberg (2006)
9. Stützle, T., Dorigo, M.: Aco algorithms for the traveling salesman problem. In: Evolutionary Algorithms in Engineering and Computer Science, pp. 163–183 (1999)
10. Goodwin, M., Granmo, O.C., Radianti, J.: Escape planning in realistic fire scenarios with ant colony optimisation. Applied Intelligence, 1–12 (2014)
11. Hendrickson, B., Leland, R.W.: A multi-level algorithm for partitioning graphs. SC 95, 28 (1995)
12. Benlic, U., Hao, J.K.: A multilevel memetic approach for improving graph k-partitions. IEEE Transactions on Evolutionary Computation 15(5), 624–642 (2011)
13. Walshaw, C.: A multilevel approach to the travelling salesman problem. Operations Research 50(5), 862–877 (2002)

14. Oduntan, I.O., Toulouse, M., Baumgartner, R., Bowman, C., Somorjai, R., Crainic, T.G.: A multilevel tabu search algorithm for the feature selection problem in biomedical data. Computers & Mathematics with Applications **55**(5), 1019–1033 (2008)
15. Bouhmala, N.: A multilevel memetic algorithm for large sat-encoded problems. Evolutionary Computation **20**(4), 641–664 (2012)
16. Blum, C., Puchinger, J., Raidl, G.R., Roli, A.: Hybrid metaheuristics in combinatorial optimization: A survey. Applied Soft Computing **11**(6), 4135–4151 (2011)
17. Whitley, D., Starkweather, T., Shaner, D.: The traveling salesman and sequence scheduling: Quality solutions using genetic edge recombination. Colorado State University, Department of Computer Science (1991)

On the Relationships Between Sub Problems in the Hierarchical Optimization Framework

Marouene Chaieb$^{(\boxtimes)}$, Jaber Jemai, and Khaled Mellouli

LARODEC Institut Supérieur de Gestion de Tunis, 41 rue de la liberté, Le Bardo 2000, Tunisie
Chaieb.marouene@live.fr, jaber.jemai@amaiu.edu.bh,
khaled.mellouli@ihec.rnu.tn

Abstract. In many optimization problems there may exist multiple ways in which a particular hierarchical optimization problem can be modeled. In addition, the diversity of hierarchical optimization problems requires different types of multilevel relations between sub-problems. Thus, the approximate and accurate representations and solutions can be integrated. That is, to address the how partial solutions of sub-problems can be reintegrated to build a solution for the main problem. The nature of relations between components differs from one decomposition strategy to another. In this paper, we will investigate the possible links and relationships that may appear between sub-problems.

Keywords: Hierarchical optimization · Relationships between sub problems · Parallel processing · Sequential processing · Gradually mixed optimization · Totally mixed optimization · Stackelberg strategy

1 Introduction

Many hierarchical optimization problems can be viewed as a particular combination and ordering of other optimization problems. There, the solution of the initial problem can be rebuilt by combining solutions of its sub-problems. Generally, such sub-decisions have to be taken in a particular sequence (order) due to the fact that the solution of the upper level will define the level of optimality of its following levels. Such kind of problems was defined as hierarchical optimization problems and initially presented by Bracken and McGill [13]. Other names may be found like multi-level optimization problems [22], dynamic optimization [4]. In the hierarchical optimization problems the decision making process is divided into different dependent levels. The decisions have to be taken in a particular precedence order. That is, a decision or a solution of the first sub-problem will affect the quality of the solution found in the subsequent level. Moreover, trying to optimize the overall problem solution needs a review of all taken decision and not a solution to particular sub-problem on a particular level. The diversity of hierarchical optimization problems requires different types of multilevel relations between sub-problems. Some of them were presented in the literature. We set up the sub-problems relations framework to resume all possible relations between sub problems. We considered that relations between components of

© Springer International Publishing Switzerland 2015
M. Ali et al. (Eds.): IEA/AIE 2015, LNAI 9101, pp. 232–241, 2015.
DOI: 10.1007/978-3-319-19066-2_23

the master problem are divided into two basic categories: dependent sub-problems and independent sub-problems. Each category is divided into two classes. The dependent category is divided into sequential approach and parallel approach. The second category is divided into gradually mixed approach and totally mixed approaches. This paper will be organized as follows: In the next section, we will present the motivations and benefits of hierarchical optimization modeling approach. Section 3 will be devoted to detailing the relation-ships between sub-problems in the hierarchical optimization framework by presenting the Stackelberg strategy and different possible relations between components of the global problem. The proposed framework detailed in section 4 will be supported by a set of examples from the relevant literature. The paper will then be concluded and some future research perspectives will be presented in the last section.

2   Motivations and Benefits

In this section, we present a new modeling technique for complex optimization problems; it is based on the application of the Divide and Conquer strategy. The modeling process aims to identify a set of sub-problems interconnected in such a way to represent all the requirements of the main problem. The proposed optimization problems modeling alternative permits the following benefits detailed in the following subsections.

2.1   Time Minimization

The multilevel optimization consists of solving a set of sub problems and then combining the obtained partial solutions to find global solution. Sub problems are supposed to be easier to solve than the initial problem; thus the required time to solve each sub problem separately and then integrate partial solutions will be significantly less than the time required for solving the initial problem as a unit. Many works showed that in the best case using an abstraction hierarchy in problem-solving can yield an exponential speedup in search efficiency. Such a speedup is predicted by various analytical models developed in the literature and efficiency gains of this order have been confirmed empirically. This was illustrated in a number of works like the work of Bacchus and Yang [6]. Moreover, Kretinin et al. [7] modeled the problem of Fan design as a multilevel optimization problem and they showed by their experimentations that hierarchical optimization gain a considerable reduction in CPU time.

2.2   Multidisciplinary

Large-scale problems require multidisciplinary decision making at multiple levels of a decision hierarchy. The multilevel optimization facilitates the modeling of problems in which different disciplines interact. Hierarchical optimization allows designers to incorporate all relevant disciplines simultaneously. These techniques have been used in a number of fields, including automobile design, naval architecture, electronics, architecture, computers, and electricity distribution, etc. Also, such multi-discipline applications were presented in some works in literature like the work of Mesarovic et al. [16] and the work of Sobieski and Hafka [11].

2.3   Parallel Processing

Parallel processing is the ability to carry out multiple operations or tasks simultaneously. The multilevel optimization allows parallel processing in which sub problems can be solved in the same time in a parallel computing environment to guarantee a high performance computing. We can refer here to the work of Azarm and Li [19] which proves that hierarchical optimization allows parallel processing which reduces the implementation time.

2.4   Reduction of Search Space

By decomposing the initial problem into a set of sub problems we will intuitively transform the initial, generally very large search spaces into a reduced search spaces. In the literature, many works prove this motivation. As stated by Newell et al. [2] and Marvin [14], the identification of intermediate sub problems which decompose a problem can significantly reduce search and empirical evidence of the net benefit.

2.5   Reusability

After decomposing the principle problem, the resulting sub problems can be resolved iteratively or recursively by applying the same process at different levels (on different data sets). The reusability of the toolbox of programs to resolve the sub problems guarantee the consistency, extensibility and modularity.

2.6   Organization

In some cases, because of the organization of people involved in modeling and optimization, or simply for convenience, it may be easier to organize the problem as a collection of subsystems with well-defined interfaces rather than attempt to pose a single monolithic problem statement. In addition, to model complex systems, it is not possible or desirable to have a single decision-maker in charge of all decisions.

3   Relations Between Sub-problems

In this section, we will investigate the possible links and relationships that may appear between sub-problems. That is, to address the how partial solutions of sub-problems can be reintegrated to build a solution for the main problem. There are many ways in which the approximate and accurate representations and solutions can be integrated. The nature of relations between components differs from one decomposition strategy to another. In the following we present some possible relations to coordinate between sub-problems to form consistent and optimal model for the overall problem. Thus, we first introduce the Stackelberg strategy which starts from an economic point of view to demonstrate the nature of influence of sub-problem of a high level on a sub-problem a low level. Then we introduce different possible links between components of the global problem.

3.1   The Stackelberg Strategy

The Stackelberg Strategy is named by the German the economist Heinrich Freiherr von Stackelberg in 1934. In economics, the Stackelberg model is a strategic game in which the leading firm moves first and then the follower firms move sequentially. Computer science and a wide range of fields benefited from this strategy. In hierarchical optimization, the Stackelberg strategy is present in defining the type of relations between different components of the original problem. The decisions made by each sub-problem in the basic problem affect the decisions made by the others and their objectives. One set of sub-problems has the authority to strongly influence the preferences of the other sub-problems. Here we can refer to some related works in the literature; Reyniers et al. [4] examines supplier-customer interactions in quality control using the Stackelberg equilibrium approach and derives optimal strategies. According to Leitmann [8], the concept of Stackelberg strategy for a nonzero-sum two-person game is extended to allow for a non-unique rational response of the follower. They defined a generalized Stackelberg strategy, then they gave a simple example. The idea of a generalized Stackelberg strategy and strategy pair is then applied to the situation of one leader and many rational followers. Korzhyk et al. [5], attempt a study of how competition affects network efficiency by examining routing games in a flow over time model. They gave an efficiently computable Stackelberg strategy for this model (routing games in allow over time) and showed that the competitive equilibrium under this strategy is no worse than a small constant times the optimal, for two natural measures of optimality. Bhaskar et al. [23], attempt a study of how competition affects network efficiency by examining routing games in allow over time model. They present an efficiently computable Stackelberg strategy for this model and show that the competitive equilibrium under this strategy is no worse than a small constant times the optimal, for two natural measures of optimality. Also, Stackelberg strategies have been used in computer science literature to manage the efficiency loss at equilibrium like the works of Korilis et al. [24], Roughgarden [21] and Swamy [3].

3.2   Sub-problems Relationships Framework

As mentioned in the introduction of this paper, multiple relations can exist between sub-problems. The following framework present a resume of all possible relations between sub-problems which will be exhaustively detailed and argued by a set of examples in next sections.

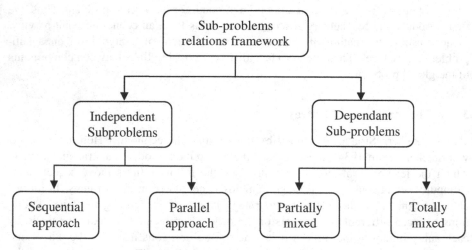

Fig. 1. Sub-problems relations framework

Independent Sub-problems Category
The independent sub-problems category supposes that each sub-problem is executed separately without any mixing with other sub-problems. Reviewing the literature, we remarked that this category was divided into two possible approaches: the sequential approach and the parallel approach. In the following two subsections we detail each approach.

Sequential Approach
In this approach, the optimization process is started using the least accurate level of representation, then after a certain set number of function evaluations, the optimization on this level is stopped and the results used as starting points for the next more accurate level. This is carried on sequentially and the number of function evaluations is decreased from one level to the next until the most accurate level is reached where fewest function evaluations are carried out. We can refer here to some works which used this approach such as the work of approach like the work of El-Beltagy and Kean [15] in which they presented empirical results. Kim et al., [9] named the same approach decomposition method and used it to cooperate between sub-problems in the Analytical Target Cascading (ATC).

Parralel Approach
In this approach, sub-problems of the principle problem interact in a collaborative form. The goal of collaboration is to allow an easy interaction amongst sub-problems from different levels. Thus, the complex problem is hierarchically decomposed into a number of sub-problems which interacts by a system-level coordination process. This form of interaction between components is well used specifically in case of mul-tidisciplinary environment. Braun and Kroo [18] enumerate some advantageous of collaborative optimization like reducing the amount of information transferred between disciplines and removing of large iteration-loops. Reviewing the literature, we

remarked that this approach has multiple terminology like "all at once approach", "the recursive approach", "the iterative approach" in which designers talk about a process in which top problem targets are cascaded down to the lowest level in one loop. More details are presented in the work of Kim et al. [9] in which authors presented the following figure to demonstrate the looping aspect of the parallel approach.

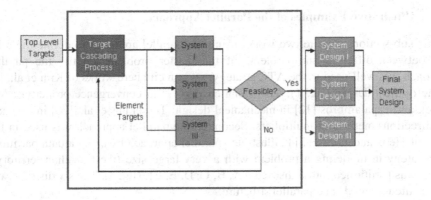

Fig. 2. Analytical Target Cascading (ATC)

Dependent Sub-problems Category

As mentioned above, this approach supposes that two or more sub-problems are executed simultaneously. Two possible ways can exist in the dependent categories which are the partially mixed multilevel approach or the totally mixed optimization approach. Each approach will be detailed in the next two subsections. Two types of mixed multilevel optimization approaches have been proposed depending on the scheduling algorithm used to organize the order and the level to reach in the solving of each sub-problem which are the gradually mixed multilevel optimization and the totally mixed optimization.

Gradually Mixed Optimization

The optimization procedure is carried out with multiple levels which are heterogeneously mixed through-out the optimization process. For example, we suppose that our initial problem is composed by three sub-problems. The master problem is gradually mixed optimized if two components are executed simultaneously in a totally mixed form and the third is executed independently. El-Baltegy and Kean [15], gives a practice example of the use of this method of relation between sub-problems, in which they presented different number of evaluations in each level.

Totally Mixed Multilevel Optimization

In totally mixed optimization, the probability of using a particular level is constant throughout most of the optimization process, i.e. all components are executed simultaneously in a totally mixed form. For example, if we suppose that our initial problem is composed by three sub-problems, if the three sub-problems are executed simultaneously in a totally mixed form then the initial optimization problem is totally mixed.

4   Illustrative Examples

In the following we try to validate the sub-problems relations framework by presenting a brief review of the current literature in models and methodologies used to model hierarchical optimization problems.

4.1   Illustrative Examples of the Parallel Approach

In this sub-section, we cite two works using the parallel approach to define the relation between different sub-problems of the master problem. Firstly, the parallel approach was well used in the ATC strategy, we can cite here works of Kim et al., [9] in the optimal design (fig. 1); Michelena et al., [17] in convergence properties; Michalek and Papalambros [12] in mechanical design; Tosserams et al., [20] in alternating directions method of multipliers. Secondly, the parallel approach was used in the work of Hertz and Lahrichi [1] illustrates parallel approach by using a data partitioning strategy in modeling a problem with a very large size (the Canadian territory), which was partitioned into 6 districts {A, B, C, D, E, F}. (fig. 3). The six districts will be executed (treated) in a parallel structure.

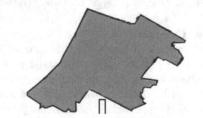

After data partitioning strategy process

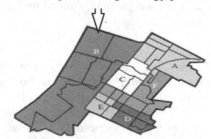

Fig. 3. The six districts of the Canadian territory

4.2   Illustrative Examples of the Sequential Approach

In this section, we basically cite the Home health care scheduling problem which a well-known combinatorial optimization problem. The home health care is an emergent kind of health care service given at home. It consists of visiting patients at their homes and performs the required treatments without a need for moving patients.

This type of service is generally given to elderly, handicapped and with special needs peoples. Patients suffering from long term maladies like Alzheimer. Moreover, some particular post-operational (after surgery) treatments can be completed at home without a need to carry them at the hospital. The hospital has a set of skilled caregivers able to perform the required task at patient homes. Basically, patients ask caregivers with particular skills to do the required treatments. Once assigned to patients, caregivers will move following specified routes to patients. The home care service dervices its importance from the considerable reduction in cost that may be incurred for patients and also for caregivers companies. The patient will have just to pay the service at home without extra charges related to hospitalization, transportation and overload of hospital facilities. The hospital or the company providing at home health care will benefit in term of patients satisfaction, low shortage rate of hospital facilities, etc. The home health care problem asks, then, for finding the set of nurses assigned to each patient and also it needs to know the routes to be followed by each team of caregivers to reach its destination. Clearly, each decision comes as answer of a particular optimization problem. The first problem is an assignment problem where the question is on which caregiver will help which patient. The second problem is on the routes to be followed by vehicles transporting nurses to reach their already assigned patients. Then, the solution of the assignment problem is an input to the routing problem. Consequently, the HCSP can be viewed and modeled as a sequential hierarchical optimization problem. The literature on the HCSP shows three types of studies of the problem: in first class the focus is on the assignment problem, in the second class of papers the main studied part is the routing problem and in some recent papers the problem is handled without omitting or hiding one of its two components. Jaber et al. [10] presented in their paper a near exhaustive literature review of the HCSP and the possible ways to model it.

4.3   Illustrative Examples of Gradually Mixed Approach

El- Baltegy and Kean [15], gives a practice example of the use of this method of relation between sub-problems, in which they presented different number of evaluations in each level. The following figure presents the rate of mixing different levels during the execution process.

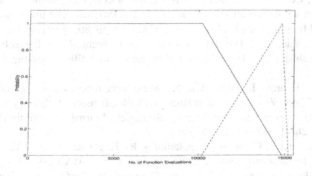

Fig. 4. Gradually mixed approach

4.4   Illustrative Examples of the Totally Mixed Approach

In the same work, El-Beltagy and Kean [15] gives an example of using the totally mixed approach in which the first level has a probability of 82.22 % of the total number of evaluations, the second has a probability of 16.44 % and the third has a probability of 1.315%.

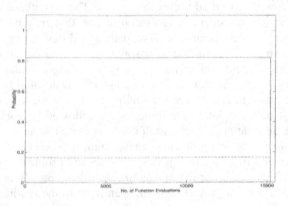

Fig. 5. Totally mixed approach

References

1. Hertz, A., Lahrichi, N.: A patient assignment algorithm for home care services. Journal of the Operational Research Society **60**, 481–495 (2009)
2. Newell, A., Shaw, C., Simon, H.: The process of creative thinking. In: Gruber, H.E., Terrell, G., Wertheimer, M. (eds.) Contemporary Approaches to Creative Thinking, pp. 63–119. Atherton, New York (1993)
3. Swamy, C.: The effectiveness of Stackelberg strategies and tolls for network congestion games. In: SODA, pp. 1133–1142 (2007)
4. Reyniers, D., Tapiero, C.: The Delivery and Control of Quality in Supplier-Producer Contracts. Management Science **41**(10), 1581–1590 (1995)
5. Korzhyk, D, Conitzer, V., Parr, R.: Complexity of computing optimal stackelberg strategies in security resource allocation games. In: The Procedings of the National Conference on Artificial Intelligence (AAAI), Atlanta, GA, USA pp. 805–810 (2002)
6. Bacchus, F., Yang, Q.: The expected value of hierarchical problem-solving. In: AAAI 1992 Proceedings of the Tenth National Conference on Artificial Intelligence, pp. 369–374 (1992)
7. Kretinin, K., Egorov, I., Fedechkin, K.: Multi-level robust design optimization fan. In: Workshop CEAS, VrijeUniversiteit Brussels (VUB), Brussels, Belgium (2010)
8. Leitmann, G.: On general Stackelberg Strategies. Journal of optimization theory and applications **26**(4), 637–643 (1978)
9. Kim, H., Kumar, D., Chen, W., Papalambros, P.: Target feasibility achievement in enterprisedriven hierarchical multidisciplinary design. In: AIAA-2004-4546, 10th AIAA/ ISSMO Multidisciplinary Analysis and Optimization Conference, Albany, New York (2004)

10. Jemai, J., Chaieb, M., Mellouli, K.: The home care scheduling problem: A modeling and solving issue. In: Proceedings of the 5th International Conference on Modeling, Simulation and Applied Optimization (ICMSAO) (2013)
11. Sobieski, J., Hafka, R.: Interdisciplinary and multilevel optimum design. In: Mota Soares, C.A. (ed.) Computer Aided Optimal Design: Structural and Mechanical Systems NATO ASI Series. Springer, Berlin, Heidelberg Berlin (1987)
12. Michalek, J., Papalambros, P.: Weights, norms, and notation in analytical target cascading. Journal of Mechanical Design **127**(3), 499–501 (2005)
13. Bracken, J., McGill, J.: Mathematical programs with optimization problems in the constraints. Operations Research **21**, 37–44 (1973)
14. El-Beltagy, M., John Keane, J.: A comparison of various optimization algorithms on a multilevel problem. Engineering Applications of Artificial Intelligence **12**(8), 639–654 (1999)
15. Mesarovic, M., Takahara, Y., Macko, D.: Theory of Hierarchical Multilevel Systems. Academic Press, New York, USA (1970)
16. Michelena, N., Park, H., Papalambros, P.: Convergence properties of analytical target cascading. AIAA Journal **41**(5), 897–905 (2003)
17. Braun, R., Kroo, I.: Development and application of the collaborative optimization architecture in a multidisciplinary design environment multidisciplinary design optimization: State of the art. SIAM, 98–116 (1995)
18. Azarm, S., Li, W.-C.: Multi-level design optimization using global monotonicity analysis. Journal of Mechanical Design **111**(2), 259–263 (1989)
19. Tosserams, S., Etman, L., Rooda, J.: An augmented Lagrangian relaxation for analytical target cascading using the alternating directions method of multipliers. Structural and Multidisciplinary Optimization **31**(3), 176–189 (2006)
20. Roughgarden, T.: Stackelberg scheduling strategies. SIAM J. Comput. **33**(2), 332–350 (2004)
21. Paul Ramasubramanian, P., Kannan, A.: Intelligent Multi-Agent Based Multivariate Statistical Framework for Database Intrusion Prevention System. International Arab Journal of Information Technology **2**(3), 239–247 (2005)
22. Bhaskar, U., Fleischer, L., Anshelevich, E.: A stackelberg strategy for routing flow over time. In: Proceedings of the Twenty-Second Annual ACM-SIAM Symposium on Discrete Algorithms, pp. 192–201 (2010)
23. Korilis, Y., Lazar, A., Orda, A.: Achieving network optima using stackelberg routing strategies. IEEE/ACM Trans. Netw. **5**(1), 161–173 (1997)

Decomposability Conditions of Combinatorial Optimization Problems

Marouene Chaieb[✉], Jaber Jemai, and Khaled Mellouli

LARODEC Institut Supérieur de Gestion de Tunis, 41 rue de la liberté, Le Bardo 2000, Tunisie
Chaieb.marouene@live.fr, jaber.jemai@amaiu.edu.bh,
khaled.mellouli@ihec.rnu.tn

Abstract. Combinatorial Optimization Problems (COP) are generally complex and difficult to solve as a single monolithic problem. Thus, the process to solve the main initial COP may pass through solving intermediate problems and then combining the obtained partial solutions to find initial problem's global solutions. Such intermediate problems are supposed to be easier to handle than the initial problem. To be modeled using the hierarchical optimization framework, the master problem should satisfy a set of desirable conditions. These conditions are related to some characteristics of problems which are: multi-objectives problem, over constrained problems, conditions on data and problems with partial nested decisions. For each condition, we present supporting examples from the literature where it was applied. This paper aims to propose a new approach dealing with hard COPs particularly when the decomposition process leads to some well-known and canonical optimization sub-problems.

Keywords: Hierarchical optimization · Decomposability conditions · Complex problems · Problem with nested decisions · Large scale data sets

1   Introduction

Hierarchical optimization consists of dividing an optimization problem into two or more sub-problems; each sub-problem has its own objectives and constraints. These sub-problems are usually interconnected in a hierarchical structure where a sub-problem in level i coordinates with a sub-problem of level i-1. Hierarchical optimization can be viewed as an application of the divide and conquer strategy for handling complex and hard optimization problems. The final solution of the main problem is produced by combining in some way the solutions of different sub-problems through a set of links and relationships between sub-problems. The integration schema and the nature of links between sub-problems show simply the hierarchical structure proposed to represent the initial problem as a set of sub-problems. Modeling and solving above complex and large size problems as hierarchical optimization problem provide multiple benefits such as reducing the time, reducing the sub-problems search spaces, increasing the performance and reducing the implementation cost. Thus, two main questions should be answered; first when can a complex problem be decomposed into 'smaller' sub-problems? This question was addressed in the literature by Halford et al.

© Springer International Publishing Switzerland 2015
M. Ali et al. (Eds.): IEA/AIE 2015, LNAI 9101, pp. 242–251, 2015.
DOI: 10.1007/978-3-319-19066-2_24

[4]. Secondly, how to identify the required sub-problems? This question was addressed in the literature by Phillips [9]. Many characters aid us to precise if a particular problem can be considered as a hierarchical problem or not. We identified four characteristics of a particular complex problem to be modeled using the hierarchical optimization framework, which are:

Multi-objective problems.
Over-constrained problems.
Very large instances.
Problems with partial nested decisions.

We will detail the four decomposability conditions and argue them by a set of examples. First, multi-objectives optimization problems involve satisfying a set of objectives which facilitate its splitting into a set of interconnected sub-problems and modeled it hierarchically. The second condition to model a particular problem hierarchically is about the over constrained level characteristic of certain problems. The nature of data presents also a third decomposability condition based on two data characteristics; first the size of the date sets which, in special case, requires a data partitioning process to facilitate the solving of the whole data sets by iteratively applying the same solving process to the small data sets. Second, based on the categorical data characteristic, the large scale data sets are divided into a set of categories based on classes (patients/nurses, teacher/student, etc). Finally, Problem with nested decisions, require a certain number of decision makers, which needs a hierarchical structure to solve the complex model. All these conditions to model a particular problem hierarchically will be discussed in next sections. This paper will be organized as follows: the next section will be devoted to detail the necessary conditions to model an optimization problem hierarchically. The conditions presented in section 2 will be supported by a set of examples in section 3. The paper will then be concluded and some future research perspectives will be presented in the last section.

2   Decomposability Conditions of Optimization Problems

Problems' modeling is the main step in optimization problems handling process; it will help to prove the correct understanding and represent in a different form that facilitates its solving. In this work, we stipulate that a hierarchical decomposition of complex problems can yield to more effective solutions. However, some conditions shall be verified to model the problem using the hierarchical structure. Such conditions are problems' characteristics that will help to identify if a COP can be modeled hierarchically and they are detailed in the following subsections (see fig. 1)

2.1   Condition 1: Multi-objective Problem

Multi-objective Optimization Problems (MOOP) has been existing in many fields of science, including healthcare, economics, finance and logistics. Generally, multi-objective problems aim to realize multiple and often conflicting objectives to be

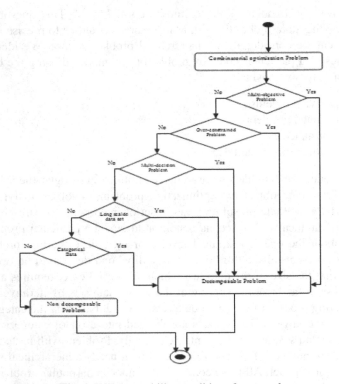

Fig. 1. Decomposability conditions framework

optimized. Consequently, a solving approach for a MOOP should provide a set of solutions with the best compromise between all required objectives. Optimizing with multiple different and conflicting objectives is an additional level of complexity of optimization problems. Then, it is possible to decompose the master problem into two or more sub-problems based on objectives; each sub-problem has its own objectives and constraints. To illustrate this idea, let's cite the work of Begur et al. [2] in which authors studied the Home Care Scheduling Problem (HCSP) and considered it like a multi-objectives problems that aims to satisfy three objectives to know:

The first objective is to assign patient visits to specific weekly time during a 16-or-so-week horizon,
The second objective is to allocate the visits planned for a given patient to a specific day of the week,
The third objective is to assign the patient visits scheduled for a given day to a particular nurse.

2.2   Condition 2: Over Constrained Problems

In many real-life applications (logistics, transportation, finance, etc) most optimization problems are highly constrained where different types of constraints have to be satisfied in the final solution. Over constrained problem are complex and difficult to solve like a single monolithic problem. To handle this problem, a constraint relaxation mechanism becomes necessary which consist to approximate of a difficult problem by nearby problems that are easier to solve by relaxing complicating constraints and recalling them after. Thus, the initial complex problem is modeled and solved in multiple levels, in each level a set of constraints will be satisfied until gratifying all required (hard) constraints and as much as possible satisfying preferential (soft) constraints. In this context, constraint hierarchies is a new concept proposed to describe high constrained problems by specifying constraints with hierarchical strengths or preferences, i.e. required and preferential constraints, most important and less important (preferences) constraints... Moreover, constraint hierarchies allow "relaxing" of constraints with the same strength by applying weighted-sum, least- squares or similar comparators [6] [1].

2.3   Condition 3: Conditions on Data

Two data sets characteristics can help to define the main problem to solve can be modeled using the proposed hierarchical framework. First, if the size of the data set of the instance is very large then it will be possible to divide it into two or more subsets based on particular criterion (geographic for example). Second, in some problems the data is classified into different types (level of proficiency of nurses, level of patients' illness, etc). For such problems the data sets can be partitioned following the defined data types. In the following we detail each condition and give some illustrative examples.

Large Scale Data Sets
Large scale data sets are collections of so large and complex data inputs that it becomes difficult to process traditionally as one batch. With very large data sets, experiments may face ambiguous situations and may not end. The data decomposition is the other primary form of breaking up monolithic processing into chunks that can be farmed out to multiple cores for parallel processing. The size of the problem space is one of the most obvious candidate measures of complexity which involve modeling a particular problem hierarchically. Problem difficulty was thought to vary with the size of the problem space. In COOP context, Hertz and Lahrichi [5] presented an illustration of data-partitioning decomposition strategy to model and solve the huge size of the problem space of the HCSP. Considering the very large size of the Canadian territory and to balance the work load of nurses while avoiding long travels to visit the clients, authors partitioned the Canadian territory into 6 districts.

Categorical Data
Another attribute of data sets of complex problem that can be modeled within the hierarchical framework is the existence of types and categories. Categorical data divide implicitly the input data into classes like types of customers in banking (VIP, Important, Ordinary) or type of employees following their skills (Expert, Skilled, Basic), etc. Such characteristics help to organize the main data set into smaller subsets following the proposed categories which define a set of sub-problems. The main

problem will be consequently solved via solving each component and then merging the obtained partial solutions to form the main solution. Hertz and Lahrichi [5] propose to decompose the instance of the HCSP, following categories of patients and nurses. Similarly, Mullinax and Lawley [7] decompose the data set into three sub-groups based on the patient level of illness, into patients that require minimal care; patients that require close attention and critically ill patients.

2.4   Condition 4: Problems with Partial Nested Decisions

Generally, complex problems are multi-decision problems where some intermediate decisions must be taken to reach a final solution of the main problem. Multi-decisions problem solving process embeds the solving of sub-problems at different times by different decision makers at different levels. The final solution will be built by combining in some way the partial solutions of intermediate sub-problems. Consequently, the structure of the initial problem can be modeled as a particular combination of sub-problems. Such sub-problems are intuitively easier to handle and to solve than the main problem for different reasons: reduced search space and data sets, uncomplicated combinatorial structure, adapted solving approaches may be already known and solving tools (software) are available. For the above advantages, it is possible to represent multi-decisions problems by its components organized in such a way to fulfill the requirements of the initial problems. For instance, consider the HCSP where the question is about scheduling to serve patients at their home subject to different types of constraints and optimizing some objectives. Finding such a solution for the HCSP, passes through determining which nurse will serve which patient, then how their medical teams will be formed to move together and finally which routes will be followed to reach patients' homes in the transportation network [6] (see fig. 2).

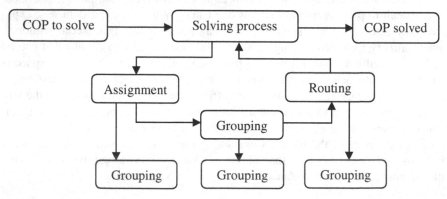

Fig. 2. Illustrative example of a solving process based on the multi-decision problem condition

3   Illustrative Examples

As mentioned in the introduction of this paper, it is necessary to validate our proposed approach through a set of research works from the literature. In the following section, a set of papers which modeled optimization problems using the hierarchically structure will be presented.

3.1   Multi-objectives Optimization Problems

The decomposition by objectives consists of dividing the basic problem into a set of sub-problems based on targets of the initial problem. For each problem at a given level, an optimization sub-problem is formulated to satisfy a set of constraints and achieve a certain objective function. In this context, Mutingi and Mbohwa [8] considered the initial problem as a multi-objective optimization problem with three conflicting management goals. The first sub-problem aims to minimize the schedule cost associated with the trips which is influenced by the nature of the routes assigned to healthcare workers to fulfill the demand requirements. The second component aims to maximize worker satisfaction which entails meeting the worker preferences to the highest degree possible and especially by ensuring the fairness in workload. The last sub-problem concerns the maximization of client satisfaction which can be expressed as a function of the violations of time windows preferred by the clients. The multi-objective formulation is achieved by optimizing the three objective functions jointly and the three components were solved in a parallel processing form.

3.2   Over-Constraints Optimization Problems

This approach consists of relaxing a set of complicating constraints in order to obtain a more tractable model. In this context, Bartk [1], thinks that constraint hierarchies allows to specify declaratively not only the constraints that are required to hold (hard constraints), but also weaker, and a finite hierarchy and order of satisfying. According to authors, weakening the strength of constraints helps to find a solution of previously over- constrained system of constraints. To illustrate the constraint hierarchy concept, we cite the work of Clark and Walker [3], in which authors considered the problem as an over constrained problem. They developed two models for the problem for comparison purposes. They called the first model "the individual days model" because the model assign individual shifts to each of the nurses and give more exibility over the schedules produced. The second one called the patterns model because it uses predefined weekly patterns which form a set covering problem. In the first model, authors try to satisfy simple (important ,hard) constraints in a first level like ensuring that every shift has the correct number of nurses assigned to it, calculating any shortage or surplus of nurses, obliging each nurse to work no more than shift per day and to have at least 2 days off every week. In the next levels of the hierarchy, to improve the quality of the solution, authors decided to add some new constraints preferred by nurses. This addition was made in two stages and in each stage the system became more complex. In the same context, Clark and Walker [3] added two constraints to the model to improve the perception of nurses of the shifts which in turn improves morale and has a positive effect on nursing staff. The first constraint insure that each nurse gets at least one complete week end off in every four weeks, the second one ensure that a nurse works no more than five days in a row. After adding these two constraints, the authors remark that the computational time remains reasonable and model complexity is not an issue for the user, so they decided to add more constraints to improve the model and its perceived fairness. The first added constraint ensures that night shifts and worked weekends are evenly distributed among nurses. The second one ensures that the day a night shift ends should be considered as a working day when calculating weekends off. The third constraint ensures working five consecutive days and isolated days off. In the second model, which named patterns model, authors used the same strategy to model the nurse scheduling problem using the constraints relaxing hierarchy. The patterns model selects from an input set of predefined shift patterns to identify shifts for each nurse. In the first stage they developed a simple model with two essential constraints, the first one ensure that every shift has the correct number of nurses assigned to it, calculating any shortage or surplus of nurses, and the second one makes sure that only one shift pattern per week is assigned to each nurse. Clark and Walker [3] remarked that they still need to additional constraints for week-linking because the patterns were one week long. Thus, they added four additional constraints to the model to checks whether 2 patterns can be adjacent and ensure that no nurse works more than 5 consecutive days. The second constraint ensures that each nurse gets at least one weekend off in 4. The two last constraints make sure that prohibited patterns are not assigned to nurses on non-full-time contracts.

3.3   Conditions on Data

In this subsection, we propose a set of illustrative examples of research works from the literature for illustrate both data decomposability conditions; the large scale data sets condition and the categorical data condition.

3.4   Large Scale Data Sets

The large scale data is a term used to describe a data collection of so large and complex that it becomes difficult to process traditionally (in a monolithic form). However, with very large data sets, experiments faced ambiguous situations and still not sure what to do. Thus, dividing the large scale data sets into smaller data sets can provide a good solution to reduce the complexity of the problem. In this context, the work of Hertz and Lahrichi [5] presented a perfect illustration of using the data-partitioning decomposition strategy to facilitate the modeling and the resolution of the huge size of the problem space of the HCSP. Given the large size of the Canadian territory, authors partitioned it into 6 districts (each one being constituted by several basic units) to balance the work load of nurses while avoiding long travels to visit the clients. According to authors [5], this data partitioning of the territory increased the efficiency in terms of client assignment, reduced transportation time, and therefore allowed for more time for direct patient care.

Categorical Data

Another data decomposability measure of hierarchy is to check if the data sets of the complex problem can be divided into small data sets based on its types (category, characteristics). Divided into small data sets, the complex problem will be recursively solved by iterating the solving approach into small sub-problem search space. Hertz and Lahrichi [5] used a data partitioning process based on categories of patients and nurses to simplify the main problem. The set of nurses was decomposed into three sets based on the level of patients' illness. The first sub-set was named case manager nurses (who typically hold a Bachelor's degree in nursing) which give care to patients requiring more complex experience like coordination of visits and ensuring links with doctors and specialists, organizing the activities of daily living. The second sub-set of nurses was named nurse technicians (who typically hold a community college degree in nursing) which give cares to the short-term clients or long-term clients needing punctual nursing care. The third sub-set of nurses present a surplus team that is not assigned neither any patient nor any district. Their role is to deliver specific nursing care treatments for the client. Authors cited a set of particular intervention of this group of nurses (handle nursing visits that the team nurses are unable to absorb visits that are needed outside regular working hours). It is important to mention here that these three types of groups are a part of the six multidisciplinary teams. The last data partitioning decomposition was based on patients' types which affect the work load of patients because the nurse volume of depends on the time needed to treat his/her specific patients. Thus, authors identified five category of clients based on the type of patients (see Fig. 3).

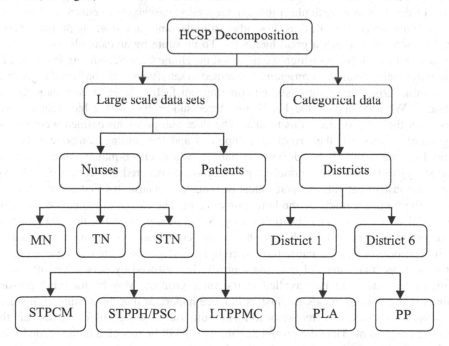

Fig. 3. Illustrative example of a solving process based on the multi-decision problem condition

The first category contains short-term clients that do not require case management (STPCM), the second category contain short-term clients that need post-hospitalization or post-surgery care (STPPH/ STPPPSC), the third category contain long-term clients needing punctual nursing care (LTPPMC), the fourth category contain clients with loss of autonomy (PLA) and the last category contain palliative patients (PP). A second example to cite here is the work of Mullinax and Lawley [7] in which authors decompose the data set (infants) into three sub-groups based on the patient level which means the type (state) of the patient. They grouped patients into three levels: level I patients that require minimal care; level II patients that require close attention, Level III care for critically ill patients. Then, a neonatal acuity system consists of 14 models contained scores were developed to precise to which level a newborn belong. The data partitioning strategy was also used in the second sub-problem (the assignment sub-problem) in which the newborn homes were divided into a number of physical zones. The following descriptive schema summarizes different data partitioning based on data conditions applied to the HCSP in the work of Hertz and Lahrichi [5].

3.5   Problems with Partial Nested Decisions

Generally, complex problems are multi-decision problems; in another words to reach the final solution of the problem, a set of decisions must be taken in a particular order to solve the principle problem. Multi-decision problems present a discipline of operations research (OR) that explicitly considers multiple decision-making in different level of decision in a particular problem. Generally, the targets to ensure in a particular problem are conflicted, thus, it is advised to decompose the whole problem based on decisions to establish a good hierarchy. To illustrate by an example, we cite here our previous work [6], in which we modeled the Home Care Scheduling Problem like a hierarchical optimization problem. We aimed to satisfy a set of conflicted objectives such minimizing travelling costs, maximizing satisfaction level of both patients and nurses... We decomposed the HCSP into three sub-problems in a hierarchical form based on the order of decisions to take. The three sub-problems defined were the assignment component, the grouping component and the routing component. Before modeling this problem we defined semantically a decision-making map (decision hierarchy) to facilitate the modeling process. We considered that first of all, the assignment task of patients to nurses must be modeled because this first component will affect the decisions made by the latter components. The choice of the next component to model depends on our decision making. Semantically, we think that the grouping component must be defined before the routing component to minimize fuel and by result minimizing costs because the last component depends of the routing component (it is so logic, that after defining the routes to be followed by nurses, sets of nurses with same routes will be travelled in the same groups). May be for other persons thinks that grouping nurses in clusters is more important that the routing component because for authors, it is important to form multidisciplinary group of nurses than the routing component. Thus the grouping component will be modeled in the second level and the routing component in the last level.

4 Conclusions and Perspectives

The hierarchical decomposition frameworks to model complex optimization problems are based on their decomposition into a set of interconnected sub-problems easier to handle. It's an application of the Divide and Conquer strategy to facilitate the handling of difficult problems. We detailed through this paper the different decomposability conditions to precise if a particular problem can be modeled hierarchically or not. The fourth measures were illustrated by a set of research papers from the literature. In the forthcoming project, we will attempt to develop a hierarchical decomposition strategy framework based on the four decomposability conditions detailed in this paper. It is important to mention here that the set of derived sub-problem should be linked and their partial solution should participate to build the final solution of the main initial problem. Thus, we aims in a future work to present the possible relationships that may combine the set of sub-problems to obtain a final solution for the global problem.

References

1. Bartk, R.: Constraint hierarchy networks. In: Proceedings of the 3rd ERCIM/Compulog Workshop on Constraints, Lecture Notes in Computer Science. Springer (1998)
2. Begur, S., Miller, D., Weaver, J.: An integrated spatial DSS for scheduling and routing home-health-care nurses. Interfaces 27(4), 35–48 (1997)
3. Clark, A., Walker, H.: Nurse rescheduling with shift preferences and minimal disruption. Journal of Applied Operational Research 3(3), 148–162 (2011)
4. Halford, G., Wilson, W., Phillips, S.: Processing capacity defined by relational complexity: Implications for comparative, developmental, and cognitive psychology. Behavioral & Brain Sciences 21(6), 803–864 (1998)
5. Hertz, A., Lahrichi, N.: A patient assignment algorithm for home care services. Journal of the Operational Research Society 60, 481–495 (2009)
6. Jemai, J., Chaieb, M., Mellouli, K.: The home care scheduling problem: a modeling and solving issue. In: Proceedings of the 5th International Conference on Modeling, Simulation and Applied Optimization (2013)
7. Mullinax, C., Lawley, M.: Assigning patients to nurses in neonatal intensive care. Journal of the Operational Research Society 53, 25–35 (2002)
8. Mutingi, M., Mbohwa, C.: A satisficing approach to home healthcare worker scheduling. In: International Conference on Law, Entrepreneurship and Industrial Engineering (2013)
9. Phillips, S.: Measuring relational complexity in oddity discrimination tasks. Noetica 3(1), 1–14 (1997)

Learning Low Cost Multi-target Models by Enforcing Sparsity

Pekka Naula[✉], Antti Airola, Tapio Salakoski, and Tapio Pahikkala

University of Turku, Turku, Finland
{pekka.naula,antti.airola,tapio.salakoski,tapio.pahikkala}@utu.fi

Abstract. We consider how one can lower the costs of making predictions for multi-target learning problems by enforcing sparsity on the matrix containing the coefficients of the linear models. Four types of sparsity patterns are formalized, as well as a greedy forward selection framework for enforcing these patterns in the coefficients of learned models. We discuss how these patterns relate to costs in different types of application scenarios, introducing the concepts of extractor and extraction costs of features. We experimentally demonstrate on two real-world data sets that in order to achieve as low prediction costs as possible while also maintaining acceptable predictive accuracy for the models, it is crucial to correctly match the type of sparsity constraints enforced to the use scenario where the model is to be applied.

Keywords: Multi-target learning problems · Sparsity patterns · Feature selection · Extractor cost · Extraction cost

1 Introduction

Linear models are often preferred in real world prediction tasks due to their simplicity, understandability and ease of evaluation. A linear model

$$Y = w_0 + w_1 X_1 + w_2 X_2 + \ldots + w_d X_d + \epsilon,$$

describes a relationship between the output variable Y and input variables $\{X_i\}_{i=1}^d$. The unknown regression coefficients $\{w_i\}_{i=0}^d$ have to be estimated before the model can be used in prediction tasks such as classification or regression. Typically, the regression coefficients are determined by learning from data that are acquired, depending on the application in question, for example by making measurements using physical sensors, sending out questionnaires, performing medical experiments, following user behaviour online, or buying from a data provider. Usually the data does not come for free [9], rather there are costs associated with the features needed during model construction and when making predictions. Measuring the data takes time, hardware components needed for measurements are expensive, data providers may charge money for each variable recorded etc. Thus often it is beneficial if one can produce accurate predictions using as few input variables as possible, that is, sparsity of the model is preferred.

M. Ali et al. (Eds.): IEA/AIE 2015, LNAI 9101, pp. 252–261, 2015.
DOI: 10.1007/978-3-319-19066-2_25

In single-target regression or classification tasks, where the model is represented as a vector of coefficients, the degree of sparsity for the model can be defined as the number of non-zero components in the coefficient vector. Formally, $\|w\|_0 = |\{w_i \neq 0, i = 0, \ldots, d\}|$ where $\| \cdot \|_0$ denotes l_0-norm. Sparse models contain less variables than dense ones and are therefore easier to interpret. Sometimes too complex models are prone to overfitting the data, enforcing sparsity may prevent this as a form of model regularization. However, the advantage considered in this work is the cost reduction [7] gained through removal of redundant or irrelevant features.

Many sparse modeling methods have been developed in order to remove unnecessary features from a set of candidate features. One of the most popular method is the lasso [8] which is often used to encourage sparsity by regularizing with l_1-norm. Sometimes prior knowledge about expected sparsity patterns can be used in feature selection, which motivates for example the use of the group lasso [10] that encourages sparsity in group level. The other popular family of techniques for inducing sparseness in coefficients of the model, is to implement a search over the power set of available features [2]. Considering all the possible feature subsets is computationally infeasible, rather suboptimal search heuristics are applied. In practice, the most widely used search approach is the greedy forward selection method (see e.g. [11]). We note that many standard dimensionality reduction techniques, such as principal component analysis, are not suitable for learning sparse models. While they project the data into a low-dimensional space, they still require all the original features to perform the projection.

Typically, rather than requiring only a single prediction, many real-world problems constitute rather multi-target prediction tasks. Here one may based on same variables make predictions on multiple related outcomes, using several linear models. Examples of such settings include multi-class classification with one-versus-all encoding [6], multi-label classification problems [3] and multivariate regression problems. More generally we may consider multi-task prediction problems where the tasks still share a feature representation, but unlike in the previous cases the training data for different tasks may be gathered from different sources, and predictions are not necessary needed for all possible targets during prediction time. For multi-target prediction problems, the coefficients of the models can be represented in a matrix form, where rows correspond to different targets and columns correspond to features. A matrix is considered sparse, if it consists primarily of zero elements. However, compared to the single-target case the concept of matrix sparsity allows for different sparsity patterns, depending whether the non-zero elements are distributed freely, or along rows or columns in the matrix. Depending on the application, two coefficient matrices with exactly the same number of non-zero elements may lead to drastically different evaluation costs.

In this paper, we study how one can lower the costs of making predictions for multi-target learning problems by enforcing sparsity on the matrix containing the coefficients of the linear models. We consider different types of costs that lead to favoring different types of sparsity patterns in the predictive models, and show how such sparsity can be obtained. The goal of this work is not to propose completely new algorithms, since the optimization problems needed for learning sparse predictors can already be solved sufficiently well using existing greedy or lasso type of methods. Rather, the goal of this work is to explore what are the types of sparsity patterns best suited for different types of application settings. For our experiments we implement greedy regularized least-squares methods, as they have been recently shown in a comprehensive experimental comparison to have state-of-the art performance when learning linear multi-target models with sparsity constraints [3].

2   Sparsity Patterns

In this section, we consider a setting where several targets have to be predicted under a strict budget. We assume that features or groups of features have unit costs (a natural extension is to allow varying costs, but for the simplicity of presentation and lack of suitable data this case is not considered in this work). By *feature extractor* we refer to a source that generates features. Such an extractor can for example be a sensor embedded into a device, a medical test, a question in a questionnaire etc. We consider two settings, one where there is a one-to-one mapping between features and extractors, and one with one-to-many mapping, where each extractor produces a group of features. Correspondingly, by *feature extraction* we refer to the procedure, where the extractors are used to obtain the features. We assume that the available extractor or extraction budgets are essentially unlimited when gathering the training data and building the model, but the final models should be as cheap to use as possible. This is a fairly reasonable assumption for example in product development in industry, where one may accept large initial R&D costs when building model prototypes, but the per unit cost of the final products should be as low as possible.

Thus, we may distinguish between two different costs, the extractor or the extraction costs. The extractor cost refers to the price of the feature extractors needed for computing the predictions. The cost could be for example the amount of money needed to manufacture the sensors embedded in a device or the number of medical devices needed in a hospital ward. The extractor cost can be considered as an initial investment made before the predictor can be used. In contrast, extraction cost is paid every time a prediction is made. This can be, for example, the monetary costs of performing blood tests for a patient, the time spent filling a questionnaire or the time spent for computing the predictions. In the single-target case there is no real need to distinguish between these two settings, as minimizing the number of non-zero coefficients in the model (possibly subject to variable costs and group structure) will lead to good solutions for both goals. However, for multi-output prediction problems the settings clearly differ.

The coefficients of the linear models of the predictors are denoted in matrix form, with rows and columns corresponding to targets and features, respectively. By *Sparsity patterns* we refer to the different ways non-zero coefficients can be distributed in the matrix. In this study we consider four types of sparsity patterns (see Fig. 1, where blue cells denote non-zero coefficients). Type I

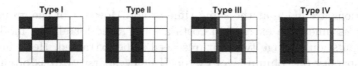

Fig. 1. Sparsity patterns. Rows correspond to targets and columns to features. The columns are ordered into three groups: $\{1, 2\}$, $\{3, 4\}$ and $\{5\}$.

presents a pattern where non-zero coefficients are distributed freely over the rows and columns. Type II denotes the pattern where non-zero coefficients are set columnwise, reflecting a setup where all the targets require the same features for prediction. Type III and IV extend these two basic cases with the additional assumption that the features are ordered in groups and they are selected a whole group at a time. Next, we discuss how these sparsity patterns relate to different types of extractor or extraction costs that may be encountered in practical settings.

2.1 Predicting Multiple Outputs Simultaneously

In the standard setting for multi-class or multi-label classification, with binary encoding of the classes or labels, or in multi-output regression, one needs to provide predictions for all possible outputs simultaneously. For example, one may analyze a picture aiming to detect multiple properties of it simultaneously (outside or inside, contains a face, nature or city), or assign a news document to one or several possible categories (sports, weather, business...). Since the models are evaluated at the same time, clearly any feature that has a non-zero coefficient on any of the rows needs to be extracted.

First, let us consider the feature extractor costs for the case where each feature is generated by its own extractor (see Fig. 1). Here, the cost of the Type I is 5, since we need to be able to generate every single feature for computing the predictions. In contrast, the cost for Type II and IV is only 2, since only two features is needed for prediction. If we further assume the previously defined group structure, then Type I needs all 3 extractors, Types II and III two of them and Type IV only one of them. In the latter setting the feature extraction costs coincide with the extractor costs, since every feature that is present in the model for even one target will be used during prediction.

2.2 Predicting Outputs Independently

Prediction may also be done independently for the outputs. For example, mobile phone applications may use the same set of available feature extractors (sensors and basic software on phone), but the predictions are made independently at different time points (one app for health monitoring while running, other for managing diabetes while at home).

Here (see Fig. 1 again), the feature extractor costs are the same as in the previous case, since the number of required extractors does not depend on whether the predictions are made at the same time or not. However, if the group structure is not assumed the extraction costs are the same for all the models. For all the outputs, on average two features need to be extracted. Here, instead of enforcing sparsity on column-level (Type II or IV sparsity), it is often advantageous to select features independently for each linear model (Type I or III sparsity), since allowing this freedom results in no extra costs. As a summary, if the extractor costs are the main concern, then this setting corresponds to the one considered in the previous subsection. However, in case one needs to consider only the extraction costs, then the models corresponding to different targets may be independently learned using standard sparsity-enforcing methods such as (group) lasso or greedy search methods.

3   Framework for Sparse Multi-target Problems

In the following considerations we reserve bold lowercase and uppercase letters for vectors and matrices, respectively. Moreover, we denote by $A_{\mathcal{I}}$ the submatrix of A containing only the columns indexed by $\mathcal{I}$.

3.1   Optimization Framework

In this section we present a framework for learning linear multi-target predictors under budget constraints. Let T be the number of targets and k and be the budget. Let $\mathbf{X}^t \in \mathbb{R}^{n_t \times d}$ be a design matrix for tth problem, where the rows correspond to training instances and the columns to feature values, and n_t denotes the number of samples for the target t and d is the number of features, and let $\mathbf{y}^t \in \mathbb{R}^{n_t}$ be the output vector for the tth target. We also divide the features into disjoint groups $G_i \subset \{1, \ldots, d\}$, $i = 1, \ldots, J$ and $G_i \cap G_j = \emptyset, i \neq j$.

The aim is to create a linear model for each of the T targets:

$$\mathbf{y}^t = \mathbf{X}^t \overline{\mathbf{w}}^t + \epsilon^t, t = 1, \ldots, T,$$

where $\overline{\mathbf{w}}^t \in \mathbb{R}^d$ are true regression coefficients that are going to be estimated and ϵ^t are error terms. Let $\mathbf{w}^t$ be estimated regression coefficients for target t and let $\mathbf{W}$ be a matrix that contains these vectors of coefficients in rows. Applying the square loss for least squares regression, we can write the multi-target optimization problem in the following form:

$$\underset{\mathbf{W} \in \mathbb{R}^{T \times d}}{\operatorname{argmin}} \sum_{t=1}^{T} \|\mathbf{y}^t - \mathbf{X}^t \mathbf{w}^t\|_2^2 + \lambda_1 \Omega_1(\mathbf{W}) + \lambda_2 \Omega_2(\mathbf{W}), \qquad (1)$$

where $\Omega_1, \Omega_2 : \mathbb{R}^d \to \mathbb{R}$ are penalties, and λ_1, λ_2 are regularization parameters.

Constraint function $\Omega_1(\mathbf{W})$ is not mandatory in (1) but to shrink the regression coefficients to achieve some regularization effect we use quadratic constraint

$$\Omega_1(\mathbf{W}) = \sum_{t=1}^{T} \|\mathbf{w}^t\|_2^2$$

Table 1. Sparsity inducing constraints

$\Omega_2(\mathbf{W})$	separate	joint
no group	$\sum_{t=1}^{T} \|\mathbf{w}^t\|_0$	$|\{j \mid \exists i, \mathbf{W}_{i,j} \neq 0\}|$
group	$\sum_{t=1}^{T} |\{\mathbf{w}_{G_i}^t \neq \mathbf{0}, i = 1, \ldots, J\}|$	$|\{G_i \mid \exists t, \mathbf{w}_{G_i}^t \neq \mathbf{0}\}|$

over all the targets. The choice of constraint function $\Omega_2(\mathbf{W})$ depends on the type of sparsity we wish to enforce, the different alternatives are presented in Table 1. If features are not generated in groups Type I sparsity (1st row, 1st column) or Type II sparsity may be enforced (1st row, 2nd column). If the group structure is known in advance, we may either enforce Type III sparsity (2nd row, 1st column), or Type IV sparsity (2nd row, 2nd column).

3.2 Algorithms

The optimization problem (1) cannot be optimally solved in polynomial time due to the presence of the sparsity constraint Ω_2. However, it has been previously shown that in such settings approximative greedy forward selection procedure can still produce quite good solutions (see e.g. the recent results in [3] on enforcing Type II sparsity). We consider a greedy search heuristic that starts from an empty set of features and on each iteration chooses one additional feature such that provides the lowest mean squared error via cross-validation, when added to the model. The new chosen feature is then added into the current selected feature set and same procedure is continued until budget limit k is reached.

Algorithm 1. greedy separate feature selection

1: **for** $t \in \{1, \ldots, T\}$ **do** ▷ go through all the targets
2:     $S \leftarrow \emptyset$ ▷ The set of selected features.
3:     **while** $|S| < k$ **do**
4:         $b := \operatorname{argmin}_{r \in \{1,\ldots,d\} \setminus S} \left\{ \mathcal{L}(\mathbf{X}_{S \cup \{r\}}^t, \mathbf{y}^t) \right\}$ ▷ Find the best new feature.
5:         $S \leftarrow S \cup \{b\}$
6:     $\mathbf{w}^t \leftarrow \mathcal{A}(\mathbf{X}_S^t, \mathbf{y}^t)$ ▷ Update coefficients.

Algorithm 1 presents a pseudocode for enforcing Type I sparsity. Inside the **while**-loop, we go through all the remaining features $r \in \{1, \ldots, d\} \setminus S$ and calculate the mean squared cross-validation error in the procedure denoted by $\mathcal{L}(.)$ that trains predictor for target t using features included in the set $S \cup \{r\}$. The feature b that returns the lowest error will be selected and added into feature set S. After budget limit k is reached, we train the final model using only the features in the set S in the procedure denoted by $\mathcal{A}(.)$.

Algorithm 2 selects jointly a set of common features for all the targets, enforcing Type II sparsity. Otherwise it behaves similarly as the Algorithm 1 except

---

Algorithm 2. greedy joint feature selection

1: $S \leftarrow \emptyset$ ▷ The set of selected features.
2: **while** $|S| < k$ **do**
3:     $b := \text{argmin}_{r \in \{1,\dots,d\} \setminus S} \left\{ \sum_t \mathcal{L}(\mathbf{X}^t_{S \cup \{r\}}, \mathbf{y}^t) \right\}$ ▷ Find the best new feature.
4:     $S \leftarrow S \cup \{b\}$
5: $\mathbf{w}^t \leftarrow \mathcal{A}(\mathbf{X}^t_S, \mathbf{y}^t), t = 1, \dots, T$ ▷ Update coefficients.

Algorithm 3. greedy group separate feature selection

1: $G_i \subset \{1, \dots, d\}, i = 1, \dots, J$ ▷ Init the disjoint feature groups
2: **for** $t \in \{1, \dots, T\}$ **do**   ▷ go through all the targets
3:     $F \leftarrow \emptyset$ ▷ The set of selected feature groups.
4:     $S \leftarrow \emptyset$ ▷ The set of selected features.
5:     **while** $|F| < k$ **do**
6:       $b := \text{argmin}_{r \in \{1, \dots, J\} \setminus F} \left\{ \mathcal{L}(\mathbf{X}^t_{S \cup G_r}, \mathbf{y}^t) \right\}$ ▷ Find the best feature group.
7:       $F \leftarrow F \cup \{b\}$
8:       $S \leftarrow S \cup G_b$
9:       $\mathbf{w}^t \leftarrow \mathcal{A}(\mathbf{X}^t_S, \mathbf{y}^t)$ ▷ Update coefficients.

Algorithm 4. greedy group joint feature selection

1: $G_i \subset \{1, \dots, d\}, i = 1, \dots, J$ ▷ Init the disjoint feature groups
2: $F \leftarrow \emptyset$ ▷ The set of selected feature groups.
3: $S \leftarrow \emptyset$ ▷ The set of selected features.
4: **while** $|F| < k$ **do**
5:     $b := \text{argmin}_{r \in \{1, \dots, J\} \setminus F} \left\{ \sum_t \mathcal{L}(\mathbf{X}^t_{S \cup G_r}, \mathbf{y}^t) \right\}$ ▷ Best new feature group.
6:     $F \leftarrow F \cup \{b\}$
7:     $S \leftarrow S \cup G_b$
8: $\mathbf{w}^t \leftarrow \mathcal{A}(\mathbf{X}^t_S, \mathbf{y}^t), t = 1, \dots, T$ ▷ Update coefficients.

---

that the mean squared error is calculated over all the targets. Algorithm 3 and Algorithm 4 works similarly than Algorithm 1 and Algorithm 2, respectively, but instead of selecting features they select feature groups.

A straightforward implementation of the Algorithms 1-4 based on black-box solver that would compute the regularized-least squares cross-validation estimate for each tested feature at a time is computationally demanding, as the training needs to be done for each tested feature, target and round of cross-validation. It has recently been shown that both Algorithm 1 [4,5] and Algorithm 2 [3] can be performed in linear time, using linear algebra shortcuts. An interesting direction of future research, that falls outside the scope of this work, would be to extend these speed-ups also for Algorithms 3 and 4.

4 Experiments

In the experiments we study how enforcing different sparsity patterns reduces the prediction costs on two real-world problems.

4.1   Datasets and Setup

We perform our experiments on two real world datasets, the Flags (43 features, 19 feature groups, 7 labels) and the Kddcup10 (118 features, 41 feature groups, 9 labels), available at the Mulan library[1] and at the UCI KDD archive[2], respectively. We modify both data sets to make them suitable for our test settings. First, the Kddcup10 and the Flags data sets are transformed into multi-target data sets by considering each label as its own binary classification task. In order to create feature groups for the setting where we compare group separate feature learning and group joint feature learning schemes, we code each categorical variable as a group of binary features, one for each possible value. Because the Kddcup10 is a quite unbalanced dataset that includes only a few samples for some rare labels, we consider only the 9 most common labels in our experiments.

The experiments are based on nested 5-fold cross-validation [1]. Parameter and feature selection is performed in an inner cross-validation loop, and the final model trained on four folds is always tested on the independent test fold that was not used for feature or parameter selection. Both data sets are tested over regularization parameter values in grid $[2^{-4}, 2^{-2}, 2^0, 2^2, 2^4]$ and the best value for each budget is chosen based on the lowest 4-fold internal cross-validation error. We report the average AUC (area under the ROC curve) over all the targets.

4.2   Results

Fig. 2 contains the average AUC performance curves with respect to costs over all the targets on the Flags and the Kddcup10 datasets. The figures on the first and the second row are based on the assumption that there are no feature groups, whereas the last two rows show the case where the features are divided into groups. Moreover, on the left column we plot the average costs for each target, while on the right column are the joint costs over all the targets.

First, we consider the case where the predictions are done independently for the targets (left column) and extraction costs dominate. This corresponds to the application setting where the models may be jointly trained but the predictions are not done at the same time. In this case better classification performance can be gained with smaller extraction costs by favoring Type I sparsity (or in group case Type III) on the Kddcup10 dataset whereas benefits on the Flags dataset are dependent on the budget size, maybe due to the small sample size of the dataset. Thus, the results support the notion that if the costs are not shared between the targets, then sharing common features might not be beneficial, though apparently it can sometimes still be helpful.

Second, we consider the case where the costs are shared between the targets (right column), either because the extractor costs dominate, or because the predictions are always done jointly, so that the extraction costs are shared. Here, the methods that perform feature selection jointly, enforcing Type II or IV sparsity,

---

[1] http://mulan.sourceforge.net

[2] http://kdd.ics.uci.edu

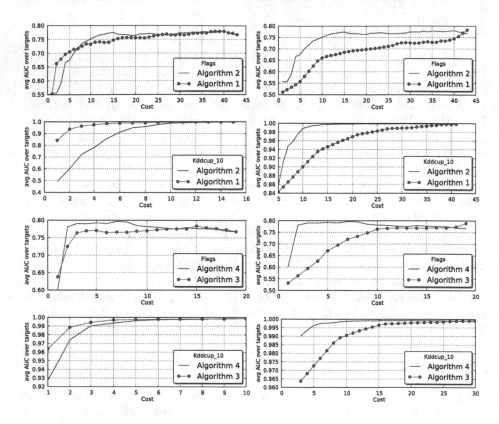

Fig. 2. Performance curves

are clearly beneficial as they allow much higher predictive accuracies especially with small budget sizes. Thus, the results match the intuition behind what sparsity patterns are best suited for which type of setting, though for the Flags data, enforcing joint sparsity proves to be beneficial even if from budget perspective there is not a specific reason why the sharing should be needed.

5   Conclusions and Future Work

In this study, we considered the costs of making predictions for multi-target problems defining four types of sparsity patterns. The results also demonstrate how these patterns relate to extractor and extraction costs resulting in lower prediction costs in different application domains. We presented also greedy forward selection algorithms that enforce such sparsity in the coefficients of the models resulting in lowered prediction costs.

References

1. Ambroise, C., McLachlan, G.J.: Selection bias in gene extraction on the basis of microarray gene-expression data. Proceedings of the National Academy of Sciences of the United States of America **99**(10), 6562–6566 (2002)
2. Kohavi, R., John, G.H.: Wrappers for feature subset selection. Artificial Intelligence **97**, 273–324 (1997)
3. Naula, P., Airola, A., Salakoski, T., Pahikkala, T.: Multi-label learning under feature extraction budgets. Pattern Recognition Letters **40**, 56–65 (2014)
4. Pahikkala, T., Airola, A., Salakoski, T.: Speeding up greedy forward selection for regularized least-squares. In: Draghici, S., Khoshgoftaar, T.M., Palade, V., Pedrycz, W., Wani, M.A., Zhu, X. (eds.) Proceedings of the Ninth International Conference on Machine Learning and Applications (ICMLA 2010), pp. 325–330. IEEE (2010)
5. Pahikkala, T., Okser, S., Airola, A., Salakoski, T., Aittokallio, T.: Wrapper-based selection of genetic features in genome-wide association studies through fast matrix operations. Algorithms for Molecular Biology **7**(1), 11 (2012)
6. Rifkin, R., Klautau, A.: In defense of one-vs-all classification. Journal of Machine Learning Research **5**, 101–141 (2004)
7. Shalev-Shwartz, S., Srebro, N., Zhang, T.: Trading accuracy for sparsity in optimization problems with sparsity constraints. SIAM Journal on Optimization **20**(6), 2807–2832 (2010)
8. Tibshirani, R.: Regression shrinkage and selection via the lasso. Journal of the Royal Statistical Society, Series B **58**, 267–288 (1994)
9. Turney, P.D.: Types of cost in inductive concept learning. In: Dietterich, T., Margineantu, D., Provost, F., Turney, P.D. (eds.) Proceedings of the ICML 2000 Workshop on Cost-Sensitive Learning (2000)
10. Yuan, M., Lin, Y.: Model selection and estimation in regression with grouped variables. Journal of the Royal Statistical Society, Series B **68**, 49–67 (2006)
11. Zhang, T.: On the consistency of feature selection using greedy least squares regression. Journal of Machine Learning Research **10**, 555–568 (2009)

Interactive Interface to Optimize Sound Source Localization with HARK

Osamu Sugiyama[1]([✉]), Ryosuke Kojima[1], and Kazuhiro Nakadai[1,2]

[1] Tokyo Institute of Technology, 2-12-1 Ookayama, Meguro-ku, Tokyo, Japan
{sugiyama.o,kojima}@cyb.mei.titech.ac.jp, nakadai@mei.titech.ac.jp
[2] Honda Research Institute Japan Co., Ltd., Saitama, Japan

Abstract. In this study, we designed and developed an interactive interface to optimize sound source localization with the multi-channel robot audition software, HARK. With the developed interface, the system can lighten the loads of optimizing parameters and supports users easily to handle the parameter optimization in sound source localization. In order to properly handle the multi-channel sounds, it is better dynamically to indicate the parameter from both temporal and spatial perspectives, though almost all of the software can only indicate a static threshold. We developed an interactive interface, with which the user can create or delete the sound source on the MUSIC spectrum and can set up an appropriate parameter settings for the environment. We also conducted an evaluation of the software and revealed that our proposed interface was superior than that of the current HARK interface from the view points of intuitiveness and visibility.

1 Introduction

These days, the studies of sound source separation and sound source localization based on microphone arrays are getting to be one of the most active fields for improving the performance of the speech recognition softwares. Especially in the robotics, these technologies are pursued to be studied in robot audition fields for better sound environmental understanding using robot equipped with microphones [1][2][3].

So far, various techniques were proposed. And also the open-sourced robot audition software HARK (Honda Research Institute Japan Audition for Robots with Kyoto University) was released in 2008 as "OpenCV in robot audition." [4] HARK is a compilation of the achievements, one of which includes a function to listen to simultaneous utterances [5].

The software consists of a lot of modules—including modules for multichannel audio input, sound-source localization [6][7], sound-source tracking, sound-source separation, and recognition of separated speech [8]—that are based on the data-flow-oriented software programming environment FlowDesigner [9]. By combining these modules, a user can easily acquire from a microphone array the separate sources of multi-channel sound signals and can use this information for understanding the environment as well as recognizing the utterances in the environment. However, utilizing such software with a good performance is actually

© Springer International Publishing Switzerland 2015
M. Ali et al. (Eds.): IEA/AIE 2015, LNAI 9101, pp. 262–271, 2015.
DOI: 10.1007/978-3-319-19066-2_26

not so easy. Particularly problematic is the tuning in localization and separation processes. Each process has several parameters and optimal value depends on the circumstances and microphone array geometry.

In common, sound source localization and separation are processed in each time frame. In other words, it only utilizes local information. After all it is dead-peak detection in spatial orientation. There are also source separation algorithms utilizing the relationship between frequencies. However, it also focuses on the local information of time, space and frequency on the audio signal though it is known that there is a certain continuity in audio signals in the temporal and spatial perspectives. We need to integrate these spatial and temporal perspectives for the better understanding of the auditory environment. To develop such algorithm that handle the whole auditory analysis automatically is an ultimate goal for auditory environments understandings. It is quite difficult to develop such a system at a time. The practical way is that we manually tune the parameters with trials and errors. However, it is also required the specialized knowledge and experiences to handle these optimization. Hybrid methodology with which users can tune the optimal value with system support is required.

In this study, we designed and developed an interactive interface to optimize the parameters of the multi-channel robot audition software, HARK. With the developed interface, the system can lighten the loads of optimizing parameters as well as support users easily to handle optimization in sound source localization with multi-perspective operations. We also conducted an experiment to evaluate the effectiveness of the developed interface and revealed that the developed system is superior than the current HARK interface in intuitiveness and visualization ability.

2   Current Limitation of Optimization in HARK and Its Underlying Requirements

The robot audition software, HARK is featured by the following points:

1. GUI-based flexible customization and integration based on dataflow-oriented programming environment, Flowdesigner.
2. Various functional modules for robot audition.
3. Online and real-time processing.

Currently, it supports Linux OS (e.g. Ubuntu 14.04), and Windows. As for input, it supports new sound devices like Kinect. Required functions for robot audition depend on the target applications. This means that robot audition software should enable users to be able to flexibly select required functions and easily combine them. Flowdesigner supports both GUI-based integration and batch-based execution so that it is utilized as middleware to fulfill this requirement, and robot audition functions are implemented as modules for Flowdesigner [4].

In order to build a sound source localization application with HARK, all we have to do is to select necessary modules from the list of developed modules and connect them with each other. Figure 1 shows the current operation flow of executing the sound source localization with HARK. Currently we activate the sound source localization application by executing the network file constructed by Flowdesigner (Fig. 1, a) and confirm the results of the localization rendered in the interface (Fig. 1, b). If the results are different from what we expected, we can optimize them by setting the parameter, such as the power threshold to isolate the sound source from environmental noise (Fig. 1, c).

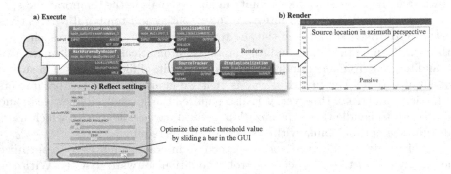

Fig. 1. Current sound source localization network

However, with the current interface, we can only set up the parameter statically and have to repeat the parameter tuning until we can find an optimal value depending on the environment. The iterations to find the optimal value is quite frustrating since we cannot acquire any clues to set up the parameters for localization. MUSIC spectrum [10], which is a power representation of the multi-channel audio signal in the azimuth domain, is useful to decide the optimal value. However, it is not enough to decide the exact value of the parameter since the spectrum only illustrates the power transition in the azimuth domain with color distribution. We need efficient system support to decide the optimal value of the parameter.

3   Development of an Interactive Interface for Optimizing Sound Source Localization with HARK

Our developed interactive interface lightens the load of these iterations in the optimization. Figure 2 shows the operation flow with our developed software. We first activate the interactive interface (Fig. 2, a) and execute the network of Flowdesigner by clicking the execution button (Fig. 2, b). Under execution, the interface receives the results of the localization from HARK and renders them on its viewer (Fig. 2, c). We can see the integration results of the MUSIC spectrum and the sound source localization on its viewer (See Fig. 3 viewer and editor region).

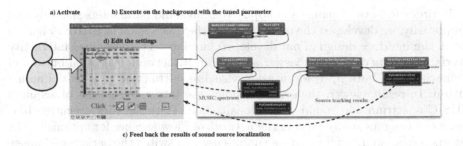

Fig. 2. Operations with the developed interactive interface

All the users have to do to optimize the parameter is to create or to delete the sound sources on the MUSIC spectrum (Fig. 2, d). In every operation of creation and deletion of the sound sources, the system reflects its results toward the internal filter matrix and utilizes it with the next iteration of the sound source localization. By repeating these processes (Fig. 2, b-d), the users can effectively optimize the parameter of sound source localization.

3.1   Design of the Developed Interactive Interface

How many steps are required to optimize the parameter for sound source localization? Usually we first visualize the MUSIC spectrum, which is the time-azimuth power representation of multi-channel audio, to find the peak of the recorded audio. Then, next, we set up the power thresholds to isolate the sound sources from an environmental noise for tracking the sources by observing the MUSIC spectrum.

In most of the cases, we utilized a single threshold to isolate the target sound signal from the noise. Actually, it is an easy way to set up the parameter and does not require a complex interface for it. It might be enough in the experimental condition, where there are less sound signals and the SNR of them is high enough to isolate the sources from the environmental noise. However, in real world applications, there are many target sound signals to be traced and the SNR of these signals are not so high due to the noisy environment. The power differences of the target sounds are also critical. If one of the sources is quite louder than others, the smaller sound signals have possibilities to be undetected. The flexible setup of the power threshold per target sound signals in spatial perspective is required.

From the other perspective, the time differences in the sound signal are also essential to trace the target sounds. Suppose if the target is moving around and the positions of the target will be changed. The power of the target sound will be weaker if the distance from the microphone array gets far away. The single setup of the threshold cannot handle this kind of situation. The weaker the power of the sound is, the higher the possibility to lose sight of the targets would be. In order to handle this kind of situation, the dynamic setup of the power threshold from the temporal perspective is also required.

In order to both visualize and adjust the threshold parameters spatially and temporarily, we developed the interactive interface as a node in HARK. Figure 3 shows the interface design of our developed interface. The developed interactive interface consists of a) result viewer and editor, b) network execution button, c) a binarize filter button, d) localization visualize button and e) a spatial adjust button. In result viewer, the system renders the integrated results of acquired MUSIC spectrum and sound source localization. With network execution button, users can repeatedly execute the network of Flowdesigner for optimizing the parameter. A binarize filter button enables users to switch the background image of the interface from the MUSIC spectrum to the binarized image of the spectrum, which is the subtraction of the MUSIC spectrum and optimized power thresholds. A localization visualize button is for switching whether to visualize the sound source localization results. Lastly the spatial adjust button is to activate a secondary spatial adjust window.

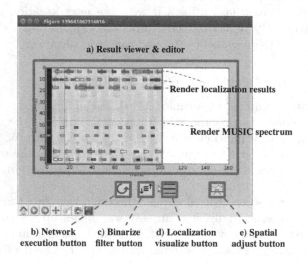

Fig. 3. Design of developed interactive interface

The features with *c), d), e)* are mainly utilized for optimizing the parameter in spatial domain. And the direct editing features with *a)* are utilizing for optimizing the parameter in temporal domain. Detailed explanations of the features follow.

3.2   Spatial Optimization of the Parameter

The developed spatial setup features are shown in Fig. 4. By clicking the binarize filter button, we can see the binarize image of the target MUSIC spectrum, which is the subtraction of the matrix of the MUSIC spectrum and that of the dynamic power thresholds. By typing *up* and *down* arrow keys on the keyboard, we can increase and decrease the threshold value of the entire matrix plane. We can see the reflection of our operations through the transitions of the binarized image (See Fig. 4, left).

In case that it is not enough to adjust the threshold in the entire matrix plane, we can press the spatial adjust button in the interface. Then the window for adjusting dynamic thresholds pops up (See Fig. 4, right). In the window, we can see the power transitions of the azimuth perspective. And we can dynamically optimize the power threshold to isolate the sound source of the environmental noise. The setting way of the dynamic thresholds is quite simple. First, like an auto-shape line feature in the Microsoft Office, there is a horizontal line representing the initial power threshold on the window. Next, we can increase the nodes on the line by double-clicking it. Lastly, the values between the nodes are spline-interpolated.

The system can also provide the dynamic threshold candidates with the current sound source localization results (See section 3.4 in details.) By re-shaping the line of the threshold, we can flexibly set up the power threshold in the azimuth perspectives, as well as confirm the results with the binarized image on the window.

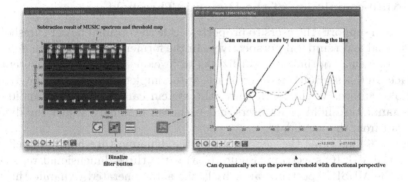

Fig. 4. Developed features to optimize parameters spatially

3.3 Temporal Optimization of the Parameter

The developed temporal setup features of the parameters are shown in Fig. 5. The optimization of the power threshold in the temporal perspective would be done by creating or deleting the rectangles representing the sound source localization results.

We can create a rectangle representing the location of the sound source by dragging and releasing the mouse on the MUSIC spectrum viewer (See Fig. 5, a). With this operation, system automatically decreases the thresholds in the peripheral region around the sound source. On the contrary, we can remove a rectangle by right-clicking the rectangle on the MUSIC spectrum viewer (See Fig. 5, b). At this time, the system automatically increases the thresholds in the peripheral region around the indicated sound source. The sum of the threshold transition is interpolated as b-spline surface and reflected in the sound source localization.

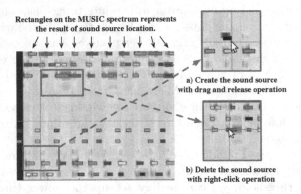

Fig. 5. Developed features to optimize parameters temporally

3.4   Auto-completion of the Dynamic Threshold

The system supports the users to properly decide the optimal power thresholds in both spatial and temporal perspectives. In the former subsections, we described the way of manual optimization utilizing the developed features of the interactive interface. In this section, we describe the auto completion feature of the dynamic threshold surface. With this feature, the system can provide the candidate of the dynamic threshold to its users and users can manually adjust the dynamic threshold from the candidate.

Figure 6 illustrates the generation of the dynamic threshold and reflection toward the MUSIC spectrum. Figure 6, a) is the three-dimensional representation of the MUSIC spectrum and, b) is the auto-generated dynamic threshold surface to isolate the sound sources of the environmental noise. Figure 6, c) is the subtraction result of the MUSIC spectrum and the dynamic threshold surface. The upper part of the figure is the three dimensional representation of each data and the lower part of the figure is a bird's eye view of the data respectively.

The auto-generated dynamic threshold surface is the mixture of spatial and temporal threshold surfaces. The spatial threshold surface is generated based on the maximum power vector in azimuth domain, b_i, and the minimum power value, τ.

The maximum power factor in azimuth domain, b_i, is given by,

$$b_i = \max_j(a_{i,j}),\ i \in A,\ j \in T \tag{1}$$

Where $a_{i,j}$ is a matrix value of the MUSIC spectrum. A is a set of columns in the azimuth domain and T is a set of rows in the temporal domain. On the other hand, the minimum power value, τ, is given by,

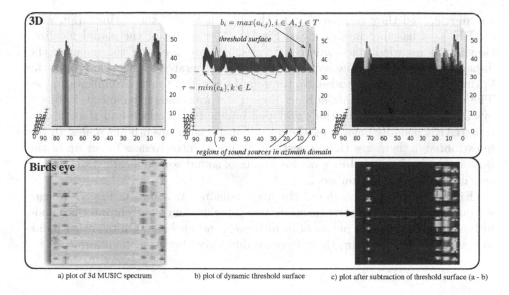

a) plot of 3d MUSIC spectrum   b) plot of dynamic threshold surface   c) plot after subtraction of threshold surface (a - b)

Fig. 6. Auto generation of threshold surface

$$\tau = \min_k(c_k), \ k \in L \tag{2}$$

where c_k is a matrix value in the peripheral regions of the sound localization results and L is a point set of the peripheral regions. The spatial threshold surface is interpolated as a spline curve [11] of d_i, given by,

$$
\begin{aligned}
d_i &= clip(b_i) \\
clip(x) &= \begin{cases} x & \text{if } x > \tau \text{ and } x \notin P \\ \tau & otherwise \end{cases}
\end{aligned} \tag{3}
$$

where P is a set of the matrix value in the peripheral regions of the existing sound source localization targets. The transitions of the power except the peripheral regions of the sound sources will be the spatial threshold surface.

The temporal threshold surface is interpolated as the b-spline surface of the peripheral regions of the sound sources, which are created or deleted by the users' operations [12]. The mixture of those two different threshold surfaces is proposed as the candidate of threshold surface for the users. With the combination of user manual operations and system proposal, users can effectively and intuitively tune the optimal value of the dynamic power thresholds for the sound source localization in HARK.

4   Experiment

In order to confirm the effectiveness of the developed interactive interface in HARK, we conducted an experiment. In the experiment, seven university students were instructed the way of using the interactive interface by watching the movie. And they compared it to the current way of the optimization with HARK as a baseline method (See Fig. 1). The source of the sound localization is the recorded voice where four members are speaking the menu of the Japanese restaurant at the same time. And the operator in the video tunes the optimal threshold value to isolate the sound sources of the environmental noise. The experiment was within-subject design and was counter-balanced. Subjects evaluated the usability of the developed interactive interface and that of current HARK node by answering the questionnaire with seven-point scales. In the questionnaire, there are three questions asking a) intuitiveness to set up of the power threshold, b) visibility of the results of sound source localization, and c) continuity of the iteration work.

Figure 7 shows the results of the questionnaire. As shown in Fig. 7, We can see our developed interactive interface was superior than a current interface node in HARK from the view points of intuitiveness and visibility. On the other hand, in the continuous domain, there were no differences between the interfaces.

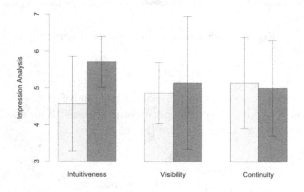

Fig. 7. Experimental results

From the results of intuitiveness and visibility, it was revealed that our proposed interactive interface can lighten users' loads in optimizing the parameter. As for continuity of the iteration work, the part of the subjects supported the current interface in HARK since it is easier to adjust static parameters. In the free-form answer, the subjects also answered that the operation to tune the optimal value is far more accurate with our proposed interactive interface. Our proposed method sometimes had to handle the two different windows together and thus the subjects found some difficulties in the operation with the proposed interface. Currently we are planning to evaluate the system with the trained users who got accustomed to execute sound source localization with HARK. And, we believed, the effectiveness of the developed system would be evaluated higher with those people.

5   Conclusion

In this study, we designed and developed an intuitive interface to optimize the parameters of the multi-channel robot audition software, HARK. With the developed interface, the system can lighten the loads of tuning parameters and supports users easily to handle parameter optimization in sound source localization with HARK. Through the experiment, we revealed that the developed interactive interface could be superior than those of the current HARK interface with the viewpoints of intuitiveness and visibility.

Acknowledgments. This work was supported by KAKENHI-No. 24220006.

References

1. Asono, F., et al.: Sound source localization and signal separation for office robot "Jijo-2". In: Proceedings of 1999 IEEE/SICE/RSJ International Conference on Multisensor Fusion and Integration for Intelligent Systems. MFI 1999, pp. 243–248. IEEE (1999)
2. Nakadai, Kazuhiro, et al.: Active audition for humanoid. AAAI/IAAI (2000)
3. Hara, I., ct al.: Robust speech interface based on audio and video information fusion for humanoid HRP-2. In: Proceedings of 2004 IEEE/RSJ International Conference on Intelligent Robots and Systems, (IROS 2004), vol. 3. IEEE (2004)
4. Nakadai, K., et al.: An open source software system for robot audition HARK and its evaluation. In: 2008 8th IEEE-RAS International Conference on Humanoid Robots. Humanoids 2008. IEEE (2008)
5. Takahashi, T., et al.: Improvement in listening capability for humanoid robot HRP-2. In: 2010 IEEE International Conference on Robotics and Automation (ICRA). IEEE (2010)
6. Nakamura, K., et al.: Intelligent sound source localization for dynamic environments. In: 2009 IEEE/RSJ International Conference on Intelligent Robots and Systems, IROS 2009. IEEE (2009)
7. Nakamura, K., et al.: Real-time super-resolution sound source localization for robots. In: 2012 IEEE/RSJ International Conference on Intelligent Robots and Systems (IROS). IEEE (2012)
8. Mizumoto, T., et al.: Design and implementation of selectable sound separation on the Texai telepresence system using hark. In: 2011 IEEE International Conference on Robotics and Automation (ICRA). IEEE (2011)
9. Ct, Carle, et al.: Code reusability tools for programming mobile robots. In: 2004 Proceedings of IEEE/RSJ International Conference on Intelligent Robots and Systems, (IROS 2004), vol. 2. IEEE (2004)
10. Quinn, B.G., et al.: The estimation and tracking of frequency (Vol. 9). Cambridge University Press
11. Salomon, D.: Curves and surfaces for computer graphics. Springer, New york (2007)
12. Catmull, E., Clark, J.: Recursively generated B-spline surfaces on arbitrary topological meshes. Computer-aided design 10(6), 350–355 (1978)

Web and Social Networks

Retrieval of Highly Related Biomedical References by Key Passages of Citations

Rey-Long Liu[✉]

Department of Medical Informatics, Tzu Chi University, Hualien, Taiwan
rlliutcu@mail.tcu.edu.tw

Abstract. Biomedical researchers often need to carefully identify and read multiple articles to exclude unproven or controversial biomedical evidence about specific issues. These articles thus need to be *highly related* to each other. They should share similar *core* contents, including research goals, methods, and findings. However, given an article *r*, existing search engines and information retrieval techniques are difficult to retrieve highly related articles for *r*. We thus present a technique KPC (key passage of citations) that extracts key passages of the citations (*out-link* references) in each article, and based on the key passages, estimates the similarity between articles. Empirical evaluation on over ten thousand biomedical articles shows that KPC can significantly improve the retrieval of those articles that biomedical experts believe to be highly related to specific articles. The contribution is of practical significance to the writing, reviewing, reading, and analysis of biomedical articles.

1   Introduction

Biomedical literature has been accumulating a huge and ever-increasing amount of evidence found by researchers. The evidence is an essential resource for biomedical research. As information needs of biomedical researchers are often about specific topics or issues, retrieval of the references (articles) that are *highly related* to the information needs is a fundamental step for the researchers. For example, a huge amount of evidence about specific gene-disease associations have been published in the literature. For each gene-disease pair, researchers need to carefully identify and read multiple articles to exclude unproven or controversial evidence.[1] These articles thus need to be highly related to each other so that the researchers can focus their limited effort on reading and judging relevant evidence.

In this paper, we present a technique to retrieve candidate articles that are highly related to a given target biomedical article *r*. A candidate article *d* is said to be highly related to a target article *r* if *d* and *r* share similar *core* contents, including research goals, methods, and findings. Retrieval of these highly related articles is quite

---

[1] For example, Genetic Home Reference (GHR) maintains an online database of gene-disease associations published in literature. Experts in GHR need to routinely find and read many articles to curate the evidence for each gene-disease pair, see http://ghr.nlm.nih.gov/about#leading-edge.

© Springer International Publishing Switzerland 2015
M. Ali et al. (Eds.): IEA/AIE 2015, LNAI 9101, pp. 275–284, 2015.
DOI: 10.1007/978-3-319-19066-2_27

challenging. Although complicated natural language processing techniques were tested in extracting health data (e.g., opinions of patients [4]), no content-based techniques were developed to properly and efficiently extract the core contents of a biomedical article and estimate the similarity between two articles based on their core contents. Moreover, in practice, biomedical search engines do not retrieve highly related articles either. For example, PubMed is a popular biomedical search engines.[2] It measures the relatedness between two articles r and d by considering the terms in the titles and abstracts of r and d, and d can get a high relatedness score if many in-frequent terms occur in both r and d many times.[3] Obviously, the relatedness score can only indicates how terms co-occur in r and d. It cannot properly indicate the match between the core contents of r and d.

Therefore, we employ the citations (*out-link* references) of two articles as the basis to estimate the similarity between the two articles. Out-link citations in an article x are selected by the author(s) of x to indicate what other studies are related to the core content of x. Therefore, two articles that cite similar sets of references should share some core contents. Note that, we do not consider "in-link" citations of an article (i.e., those that cite the article), since a huge amount of biomedical articles do not have many in-link citations (especially those articles that are newly published, being written, or less visible). Moreover, authors of two articles may cite different references for the same core contents, since there may be many articles about the same research issues. Therefore, for each out-link citation c in an article x, we extract a *key passage* from x to indicate how author(s) of x describes c. Similarity between the key passages is used to improve the retrieval of highly related articles. Our technique is thus named KPC (key passages of citations). We will show that, given a target article r, KPC can be used to significantly rank high those articles that biomedical experts believe to be highly related to r. The contribution is of practical significance to curators, authors, reviewers, and readers of biomedical evidence.

2   Related Work

Previous studies employed citation relationships among research papers for several purposed, such as visualization of citation links for biomedical researchers [2] and prediction of research topic tends [11]. However, they did not employ the citation relationships to retrieve highly related biomedical articles. Main challenges of KPC include (1) *extraction* of the key passage for each citation in an article, and (2) *estimation* of the similarity between two articles based on key passages of their citations. To our knowledge, no previous studies tackled the challenges.

In traditional information retrieval studies, many techniques have been developed to employ citation relationships to estimate the similarity between articles. Typical examples of such techniques included *bibliographic coupling* [13] and *co-citation* [14]. However, they did not consider key passages of citations. Bibliographic

---

[2] PubMed is available at http://www.ncbi.nlm.nih.gov/pubmed.

[3] The pre-computation of the relatedness score by PubMed is summarized in http://www.ncbi.nlm.nih.gov/books/NBK3827/#pubmedhelp.Computation_of_Related_Citati.

coupling is based on the expectation that two articles $d1$ and $d2$ may be related to each other if they cite a similar set of articles (i.e., $d1$ and $d2$ share similar *out-link* citations). Bibliographic coupling was shown to be useful in classifying scientific papers [5] and retrieving similar legal judgments [12]. Equation 1 is a typical way to estimate the bibliographic coupling similarity (BC) between $d1$ and $d2$ [3][5], where O_{d1} and O_{d2} are the sets of articles that cited by $d1$ and $d2$ respectively (i.e., out-link citations of $d1$ and $d2$ respectively). When both O_{d1} and O_{d2} are empty, BC is set to 0.

$$BC(d1,d2) = \frac{|O_{d1} \cap O_{d2}|}{|O_{d1} \cup O_{d2}|} \tag{1}$$

The idea of bibliographic coupling was extended by considering the common articles (entities) that were cited (linked) by $d1$ and $d2$ *indirectly* (i.e., starting from $d1$ and $d2$, one can reach the common entities through a certain amount of citation links, [17]). Bibliographic coupling could also be used to detect plagiarism, especially for paraphrased or translated plagiarism that could be detected by considering the common sequence of out-link citations in two articles [6]. It was also used to measure the professional similarity between authors, based on the expectation that two authors a_1 and a_2 may have similar research experiences if papers of a_1 and a_2 often cite the same articles [8].

On the other hand, co-citation is based on the expectation that two articles $d1$ and $d2$ may be related to each other if they are co-cited by many other articles (i.e., $d1$ and $d2$ share similar *in-link* citations). Co-citation could be useful in classifying webpages [3][5]. Equation 2 is a typical way to estimate the co-citation similarity (CC) between $d1$ and $d2$ [2][3], where I_{d1} and I_{d2} are the sets of articles that cite $d1$ and $d2$ respectively (i.e., in-link citations of $d1$ and $d2$ respectively). When both I_{d1} and I_{d2} are empty, CC is set to 0.

$$CC(d1,d2) = \frac{|I_{d1} \cap I_{d2}|}{|I_{d1} \cup I_{d2}|} \tag{2}$$

Co-citation could also be used to measure the professional similarity between authors, based on the expectation that two authors a_1 and a_2 may have similar research fields and interest if papers of a_1 and a_2 are often cited together in articles [15]. To improve co-citation, proximity of $d1$ and $d2$ in the citing articles was found to be helpful in estimating the similarity between $d1$ and $d2$ [7]. Another approach to improving co-citation was to iteratively propagate the similarity among the citing articles, in the hope to deal with the case where an article is cited by very few articles [9].

Therefore, bibliographic coupling and co-citation are based on out-link and in-link citations respectively. Previous studies found that bibliographic coupling was more useful in finding similar scientific papers [5] and legal judgments [12], while co-citation was more helpful in finding similar webpages [3][5]. As the out-link and in-link citations can be used to indicate different kinds of similarity, their integration can produce a new similarity measure. The integration was achieved by several approaches, including counting both the common out-links and in-links [1], treating out-links and in-links as undirected links [16], and iteratively propagating the similarity among the citing and the cited articles [18].

When compared with the above studies, KPC has two main features: (1) KPC considers *out-link* (rather than in-link) citations so it can properly respond to the fact that a huge amount of biomedical articles do not have many in-link citations; (2) KPC considers *key passages* of out-link citations but all the previous techniques did not. Another contribution of KPC lies on the retrieval of *highly related* biomedical articles. Previous techniques did not aim at retrieving highly related articles. They often retrieved those articles that cited some common papers [9], shared some terms in their titles [9][17], or belonged to the same category [5] or the same chapter of a book [16]. Obviously, these articles were not necessarily highly related to each other, and hence the previous studies did not evaluate the performance of their techniques in retrieving highly related biomedical articles.

3   Similarity Estimation by Key Passages of Citations

KPC consists of two components that respectively work for (1) *extraction* of the key passage for each citation in an article, and (2) *estimation* of the similarity between two articles based on key passages of their citations. The key passage of a citation c in an article x consists of two parts: (1) the title of x and (2) the text immediately before each position p where c is mentioned in the main body of x. Both parts can provide strong information about how the author(s) of x discussed c from the perspective of the core content of x. The former can indicate the *context* of the discussion, since the title indicates the general goal of the article. The latter can indicate the *specific* issue of the discussion, since the text immediately before p should aim at providing specific comments on c (from the perspective of the core content of x). More specifically, the key passage (KP) of a citation c in an article x is defined in Equation 3. KP of a citation c in an article x is actually the union of distinct words in these two parts.[4]

$$KP(c,x) = \{distinct\ words\ in\ title\ of\ x\} \cup \bigcup_{\substack{p \in \{\ positions\ where \\ c\ appears\ in\ x\}}} \{distinct\ words\ in\ \alpha\ words\ before\ p\} \quad (3)$$

It is interesting to note that, KP of c will have more words if c is discussed in more different ways. In such case, c is discussed more diversely. Conversely, KP of c will have fewer words if c is discussed using simlar words in the title. In such case, c should be more related to the core content of x. We set the parameter α to 20, as authors tend to discuss a citation with a sentence whose length should be about 20.

The second component of KPC aims at estimating the similarity between two articles by the KPs of the citations in the articles. Given two articles $d1$ and $d2$, KPC produces a larger similarity value for them if more KPs of the citations in $d1$ match the KPs of the citations in $d2$, *and* vice versa. In such case, $d1$ is said to be more similar to $d2$ even though they have no citations in common. More specifically, KPC estimates the similarity between $d1$ and $d2$ by Equation 4, where O_{d1} and O_{d2} are the sets of out-link citations in $d1$ and $d2$ respectively, and *PM* (passage match) is the degree of match between two KPs (defined in Equation 5).

---

[4] Stopwords are removed from the article before the KP is extracted.

$KPCsimilarity(d1,d2) =$

$$\frac{\sum_{o1 \in O_{d1}} Max_{o2 \in O_{d2}} PM(KP(o1,d1), KP(o2,d2)) + \sum_{o2 \in O_{d2}} Max_{o1 \in O_{d1}} PM(KP(o2,d2), KP(o1,d1))}{|O_{d1}| + |O_{d2}|} \quad (4)$$

$$PM(kp1, kp2) = \frac{|kp1 \cap kp2|}{|kp1| + |kp2|} \quad (5)$$

This way of similarity estimation does not rely on any in-link citations of the articles, and hence is suitable for those articles that happen to obtain very few (or even no) in-link citations (e.g., those articles that are being written, newly published, or less visible). The similarity estimation strategy also has several interesting features:

(1) Even though two citations in $d1$ and $d2$ are totally different from each other, they can contribute to the similarity between $d1$ and $d2$ if their KPs have some words in common.

(2) Those citations whose KPs in $d1$ ($d2$) share more words with the *title* of $d2$ ($d1$) tends to have a higher degree of passage match with citations in $d2$ ($d1$), and such citations will contribute more to the similarity between $d1$ and $d2$.

(3) Those citations that are discussed more *diversely* in $d1$ ($d2$) tends to have a lower degree of passage match with the citations in $d2$ ($d1$), and such citations will contribute less to the similarity between $d1$ and $d2$.

4   Empirical Evaluation

A case study on over ten thousand biomedical articles had been conducted to empirically evaluate the contribution of KPC.

4.1   Experimental Data

To evaluate KPC, the experimental data needs to consist of a large number of articles that biomedical experts have identified as *highly related*. Therefore, we extracted gene-disease pairs from DisGeNET, which is an online platform that collects and integrates information on hundreds of thousands of gene-disease associations from several data sources and the biomedical literature.[5] For each gene-disease pair, DisGeNET lists the information about the PubMed articles that are closely related to the pair. DisGeNET also lists the source of the information. We selected 53 gene-disease pairs that fulfill two requirements: (1) they have the largest number of related PubMed articles among which at least two are included PubMed Central (PMC)[6], which provides the full texts of the articles for our systems to estimate similarity; and

---

[5] DisGeNET is available at www.disgenet.org/web/DisGeNET/v2.1/home.

[6] PubMed Central (PMC) is a database of full texts of biomedical papers. It is available at http://www.ncbi.nlm.nih.gov/pmc/.

(2) their sources of information are GAD (Genetic Association Database)[7] or CTD_human (Comparative Toxicogenomics Database for human[8]), which employ biomedical experts to select highly related articles to annotate the gene-disease pairs. The articles selected for a gene-disease pair can thus be expected to be highly related to each other.

For each gene-disease pair, we selected the most recently published article as the *target* article, and the other articles as the *candidate* articles that are highly related to the target. Moreover, in practice a search engine often needs to rank a large number of candidate articles that are *not* highly related to the target but share some contents with the target. Therefore, for each gene-disease pair $<g, d>$, we sent two queries to PubMed Central: "g NOT d" and "d NOT g", which retrieved at most 200 "near-miss" candidate articles (as they mentioned g or d but not both). A better retrieval system should rank higher (lower) those candidate articles that are highly related (near-miss) to the target article. For the 53 gene-disease pairs, we obtained 10,119 articles (including 53 targets and 10,066 candidates). When considering their out-link citations, there were totally 313,571 articles involved in the experiment.

4.2   The Article Ranking Systems

As noted in Section 2, we are concerned with the contribution of KPC to those systems that rely on out-link citations to retrieve highly related articles. Therefore, in addition to KPC, we implemented BC (ref. Equation 1), which was routinely used to detect plagiarism [6] and find similar scientific papers [5] and legal judgments [12]. We also implemented several systems that integrated BC and KPC in three ways:

(1) *Linear* fusion: The similarity between two articles was the weighted sum of the similarity values produced by BC and KPC (the weight of BC is b; while the weight for KPC was 1-b). The system was named BC+KPC;

(2) *Sequence* fusion: BC was used to rank articles first, and KPC was used only when two articles had the same BC similarity. The system was named BC→KPC.

(3) *Machine-learning* based fusion: We employed RankingSVM [10], which is one of the best techniques routinely used to integrate multiple factors with SVM (Support Vector Machine) to achieve better ranking. We employ SVM^{rank} to implement RankingSVM.[9] The 53 pairs were evenly split into 5 parts so that 5-fold experiments were conducted: one part was used for testing and the other for training; and the process repeated five times. The system was named BC+KPC_SVM.

---

[7] GAD is a curated database of genetic association data from complex diseases and disorders. It is available at http://geneticassociationdb.nih.gov/.

[8] CTD is a curated database about the environmental chemicals on human health, including the associations between diseases and genes/proteins. It is available at http://ctdbase.org/.

[9] SVM^{rank} is available at http://www.cs.cornell.edu/People/tj/svm_light/svm_rank.html.

By observing the performance of the integrative systems, we can comprehensively investigate whether, how, and to what extend KPC can be used to improve BC.[10]

4.3   Evaluation Criterion

Mean average precision (MAP) was employed as the criterion to evaluate the systems. MAP is commonly employed to evaluate text rankers. It is defined in Equation 6, where $|Q|$ is the number of gene-disease pairs, k_i is number of articles that are believed (by the experts) to be highly related to the target article r for the ith gene-disease pair, and $Arc_i(j)$ is the number of articles whose ranks are higher than or equal to that of the jth highly related article for the ith gene-disease pair.

$$MAP = \frac{\sum_{i=1}^{|Q|} AP(i)}{|Q|}, \quad AP(i) = \frac{\sum_{j=1}^{k_i} \frac{j}{Arc_i(j)}}{k_i} \tag{6}$$

Therefore, $AP(i)$ is actually the average precision (AP) for the ith gene-disease pair. Given a target article r for a gene-disease pair, if a ranker can rank higher those articles that are highly related to r, AP for the gene-disease pair will be higher. MAP is simply the average of the AP values for all gene-disease pairs.

4.4   Results

Figure 1 shows the performance of the systems produced by linearly integrating BC and KPC (i.e., BC+KPC). Best MAP was achieved when the integration weight (b) was in the range of [0.5, 0.8]. To make sure whether the performance difference between two systems was *statistically significant*, we conducted significance test: for the AP values of the two systems on the 53 gene-disease pairs, we conducted two-sided and paired t-test with 95% confidence level. The results showed that the performance difference between BC and KPC was *not* statistically significant ($p=0.5374$), however BC+KPC performed significantly better than BC ($p<0.005$) when the integration weight (b) was 0.5 or 0.8. The results thus suggested that KPC should be coupled with other techniques such as BC. As KPC is novel in assessing similarity by key passages of out-link citations, it can be used to complement other similarity measurement techniques. The results also indicated that proper integration weights are not quite difficult to set, as the performance of the system did not significantly change when the weight is in the range of [0.5, 0.8].

Figure 2 compares different fusion strategies for BC and KPC. All the fusion strategies successfully produced the systems that performed significantly better than BC ($p<0.05$), reconfirming that KPC can be used to significantly improve BC. Moreover, BC+KPC with integration weight (b) set to 0.5 performed best, however it did *not* perform statistically better than the other two integrated systems ($p>0.09$). Therefore, integration by the machine learning approach SVM was not necessarily needed.

---

[10] In case where a system (including PKC, BC, and the integrative ones) assigned the same similarity value to two candidate articles, the more recently published one was rank higher.

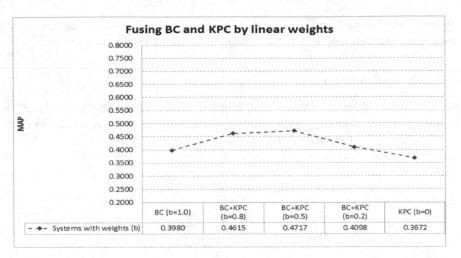

Fig. 1. Linear combination of BC and KPC: Performance difference between BC and KPC was *not* statistically significant ($p=0.5374$), however when the integration weight (b) was 0.5 or 0.8, BC+KPC performed significantly better than BC ($p<0.005$)

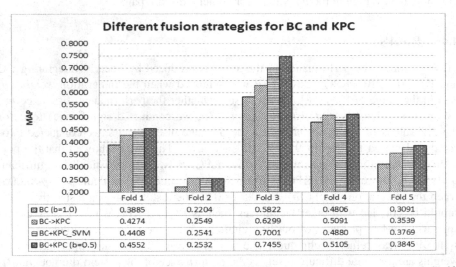

Fig. 2. Different fusion strategies for BC and KPC: All the integrated systems performed significantly better than BC ($p<0.05$), and BC+KPC with integration weight (b) set to 0.5 performed best, however it did *not* perform statistically better than the other two integrated systems ($p>0.09$)

5   Conclusion and Future Extensions

We have presented a novel technique KPC that can significantly improve the retrieval of highly related biomedical articles, which are quite difficult to retrieve by simply counting the common words in the articles. Contribution of KPC indicates that similarity on the *key passages* of *out-link* citations in biomedical articles can be a good indicator to identify highly related articles. The contribution is of practice significance to the biomedical researchers, who often need to carefully identify and read multiple articles to exclude unproven or controversial evidence. The highly related articles recommended by KPC can support more timely and complete analysis and curation of the biomedical evidence already published in the literature. The highly related articles are helpful for authors as well. When an author is writing an article r, KPC can support the author in checking those articles that are highly related to r. KPC can also be used as a tool to support the review of new articles. When an article r is being reviewed, the articles that are highly related to r are helpful for the reviewers to verify the contribution of r and check the possible problem of plagiarism.

KPC can be extended by considering the *importance* of each citation in an article. Authors of an article tend to discuss many citations in the article, however the citations are not necessarily closely related (important) to the article. It is thus reasonable to expect that two articles that share more important citations should be more related to each other. Therefore, by estimating the degree of importance of each citation, KPC can be further improved. To estimate the importance, several types of information may be helpful, including the position and the frequency of a citation in an article. The extension is of technical significance to information retrieval studies.

Acknowledgment. This research was supported by the Ministry of Science and Technology of the Republic of China under the grant MOST 103-2221-E-320-006.

References

1. Amsler, R.A.: Application of citation-based automatic classification. Technical report, Linguistics Research Center, University of Texas at Austin (1972)
2. Belew, R.K., Chang, M.: Purposeful retrieval: Applying domain insight for topically-focused groups of biologists. In: Proceedings of the SIGIR 2004 Bio Workshop: Search and Discovery in Bioinformatics (2004)
3. Calado, P., Cristo, M., Moura, E., Ziviani, N., Ribeiro-Neto, B., Goncalves, M.A.: Combining link-based and content-based methods for web document classification. In: Proc. of the 2003 ACM CIKM International Conference on Information and Knowledge Management (CIKM 2003), New Orleans, Louisiana, USA (2003)
4. Cambria, E., Hussain, A., Havasi, C., Eckl, C., Munro, J.: Towards crowd validation of the UK national health service. In: Proc. of Web Science Conference, Raleigh, NC, USA (2010)
5. Couto, T., Cristo, M., Goncalves, M.A., Calado, P., Nivio Ziviani, N., Moura, E., Ribeiro-Neto, B.: A Comparative study of citations and links in document classification. In: Proc. of the 6th ACM/IEEE-CS joint conference on Digital libraries, pp. 75–84 (2006)
6. Gipp, B., Meuschke, N.: Citation pattern matching algorithms for citation-based plagiarism detection: greedy citation tiling, citation chunking and longest common citation sequence. In: Proc. of 11th ACM Symposium on Document Engineering, Mountain View, CA, USA (2011)

7. Gipp, B., Beel, J.: Citation proximity analysis (CPA) – a new approach for identifying related work based on Co-citation analysis. In: Proc. of the 12th International Conference on Scientometrics and Informetrics, vol. 2, pp. 571–575 (2009)

8. Heck, T.: Combining social information for academic networking. In: Proc. of the 16th ACM Conference on Computer Supported Cooperative Work and Social Computing (CSCW 2013), San Antonio, Texas, USA

9. Jeh, G., Widom, J.: SimRank: a measure of structural-context similarity. In: Proc. of the Eighth ACM SIGKDD International Conference on Knowledge Discovery and Data mining, pp. 538–543 (2002)

10. Joachims, T.: Optimizing search engines using clickthrough data. In: Proceedings of ACM SIGKDD, Edmonton, Alberta, Canada, pp. 133–142 (2002)

11. Jung, S., Segev, A.: Analyzing future communities in growing citation networks. Knowledge-Based Systems **69**, 34–44 (2014)

12. Kumar, S., P. Reddy, K., Reddy, V.B., Singh, A.: Similarity analysis of legal judgments. In: Proc. of the Fourth Annual ACM Bangalore Conference (COMPUTE 2011), Bangalore, Karnataka, India

13. Kessler, M.M.: Bibliographic coupling between scientific papers. American Documentation **14**(1), 10–25 (1963)

14. Small, H.G.: Co-citation in the scientific literature: A new measure of relationship between two documents. Journal of the American Society for Information Science **24**(4), 265–269 (1973)

15. White, H.D., Griffith, B.C.: Author cocitation: A literature measure of intellectual structure. Journal of the American Society for Information Science **32**(3), 163–171 (1981)

16. Yoon, S.-H., Kim, S.-W., Park, S.: A link-based similarity measure for scientific literature. In: Proc. of The 19th International World Wide Web Conference (WWW 2010), North Carolina, USA

17. Zhang, M., He, Z., Hu, H., Wang, W.: E-Rank: a structural-based similarity measure in social networks. In: Proc. of IEEE/WIC/ACM International Conferences on Web Intelligence and Intelligent Agent Technology (2012)

18. Zhao, P., Han, J., Sun, Y.: P-Rank: a comprehensive structural similarity measure over information networks. In: Proc. of the International Conference on Information and Knowledge Management, pp. 553–562 (2009)

Summarization of Documents by Finding Key Sentences Based on Social Network Analysis

Su Gon Cho and Seoung Bum Kim[(⊠)]

Department of Industrial Management Engineering, Korea University, Seoul, Korea
{sugoncho,sbkim1}@korea.ac.kr

Abstract. Finding key sentences or paragraphs from a document is an important and challenging problem. In recent years, the amount of text data has grown astronomically and this growth has produced a great demand for text summarization. In the present study, we propose a new text summarization process by text mining and social network methods. To demonstrate the applicability of the proposed summarization procedure, we used Martin Luther King, Jr's public speech

Keywords: Text summarization · Social network · Centrality degree · Text mining

1   Introduction

With the rapid growth of the World Wide Web and on-line information services, huge amounts of information are generated and accessible. This explosion of information has resulted in a well-recognized information overload problem [1, 2]. There is no time to read everything, and we have to make critical decisions based on available information. In recognition of this need, research on summarizing text has become the focus of considerable interest in the field of text mining. As part of the effort to better organize this information for users, researchers have investigated the problem of document summarization.

Document summarization is the process of automatically creating a shorten version of a given text that provides useful information with readers. Generally, two types of document summarizations exist: abstractive and extractive summarization. The goal of the abstractive summarization is to represent the main concepts and ideas of a document by paraphrasing of the source document in natural language. The goal of extractive summarization is to extract the most meaningful parts of the documents (e.g., sentences and paragraphs) to represent the main concepts of the document. Various algorithms have been proposed for abstractive text summarization. One idea is to summarize text using machine learning techniques such as a naive-Bayes classifiers [3], decision trees [4], and hidden Markov models [5, 6]. Further, natural language analysis models [7] and topic-driven summarizations based on the maximal marginal relevance measure [8] have been used for abstractive summarization.

Several studies have been conducted using social network models [9, 10]. These studies constructed social networks by selecting words of nodes and introduced links

© Springer International Publishing Switzerland 2015
M. Ali et al. (Eds.): IEA/AIE 2015, LNAI 9101, pp. 285–292, 2015.
DOI: 10.1007/978-3-319-19066-2_28

between the two words if and only if these two words appeared in the same sentence: these are termed co-occurrence social networks. However, they only focused on centroid or classifier algorithms for finding key sentences. These existing approaches cannot meet the needs of practical applications. Therefore, it is important that new processes be developed for document summarization. The main goal of this study is to construct a new summarization process with the weighted closeness centrality degree of social network theory.

The rest of the study is organized as follows. Section 2 presents the proposed process of document summarization including the closeness centrality degree of network theory. Section 3 presents the main steps of our approach for one particular example, the 'I Have a Dream' public speech by Martin Luther King, Jr, and American activist. Section 4 provides the limitations and future directions of this research.

2   Text Summarization Process

In this section, we describe the proposed document summarization. Fig. 1 illustrates an overview of the proposed process. Given a document to summarize, the textual contents are segmented into sentences and these are stored in a database. A social network based on the sentences is then created from the predefined keywords. From the network created, we calculate the score of each sentence. Finally, we use these scores to identify the key sentences in a document.

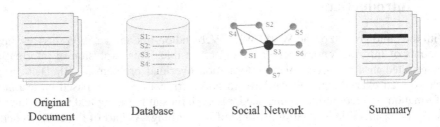

| Original Document | Database | Social Network | Summary |

Fig. 1. The overall process of the proposed document summarization

2.1   Social Network of Sentences

A social network is a collection of objects called 'nodes' and 'links' that connect pairs of nodes. A social network provides a clear way of analyzing the structure of entire social entities [11]. From the perspective of text summarization, each node represents a sentence of a document and the links represent pairwise keyword relations between the sentences. To create a social network of sentences, one needs to determine keywords. In this study, we used a term frequency–inverse document frequency (TF-IDF) algorithm that computes the co-occurrence networks of sentences.

TF-IDF is a statistic that reflects how important a word is to a document in a collection or corpus. TF-IDF is often used as a weighting factor in information retrieval and text mining [12]. The TF-IDF value increases proportionally to the number of times a word appears in the document, but is offset by the frequency of the word in the corpus. This helps mitigate the fact that some words are generally more common than others.

TF-IDF is the product of two statistics, term frequency and inverse document frequency. In the case of the term frequency $TF(t,d)$, the simplest choice is to use the raw frequency of a term in a document (i.e. the number of times that term t occurs in the document d). The inverse document frequency (IDF) is a measure of whether the term is common or rare across all documents. IDF can be calculated by the following equation:

$$IDF_i = \log (N / n_i) ,$$ (1)

where N is the total number of the documents in a collection, and n_i is the number of documents in which the word i occurs. For example, the words that are most likely to occur tend to have an IDF value close to zero while infrequent words (e.g., medical terms and proper nouns) typically have higher IDF values.

Finally, TF-IDF can be calculated by

$$TF \text{-} IDF = TF(t, d) \times IDF(t, D) .$$ (2)

It can be seen that a high value of TF-IDF can be obtained by a term that occurs with a high frequency in the given document, but is found in a few of the other documents in the corpus.

A co-occurrence network is the collective interconnection of terms based on their paired presence within a specified unit of text. In this study, we define a unit of text to be a sentence. A social network of these sentences is generated by connecting pairs of keywords identified by the TF-IDF algorithm. For example, if keyword 'A' appears one time in each of the sentences 'S1' and 'S2', the two sentences may be said to 'co-occur' and linked by keyword 'A' as shown in Fig. 2. To create a social network of the sentences, this co-occurrence network theory can be applied to all pairs of the sentences using the keywords.

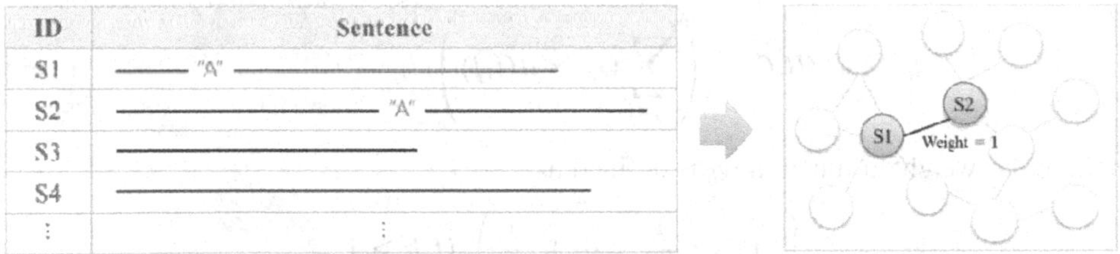

ID	Sentence
S1	———— "A" —————————
S2	———————————— "A" —————————
S3	—————————————
S4	—————————————
⋮	⋮

Fig. 2. An example of the co-occurrence keywords network

2.2   Summarization Score Function

A summarization score is computed for each sentence in the document to quantify its importance. In this study, we used the centrality degree to calculate the score of each sentence.

Centrality is an important concept in social network models [13, 14, 15]. It measures how central an individual's position is in a network. In graph theory and network analysis, various centrality measures have been proposed to quantify the relative importance of a node within a network [16]. Among these, four of them are widely used. These are degree centrality, betweenness centrality, closeness centrality, and eigenvector centrality. In this paper, we focus on the degree of closeness centrality.

The closeness centrality of a node in a network is the inverse of the average shortest-path distance from the node to any other nodes in the network. It can be viewed as the efficiency of each node in spreading information to all other nodes [11]. The larger the closeness centrality of a node, the shorter the average distance from the node to any other nodes, and thus the better positioned the node is to spread information to other nodes. The closeness centrality of all nodes can be calculated by solving the all pairs shortest-paths problem. For example, the closeness centrality of the ith node (CC_i) can be calculated as follows:

$$CC_i = \left(\sum_{\forall j \neq i} d(i,j) \right)^{-1}, \tag{3}$$

where $d(i, j)$ is the number of links in the shortest path from node i to node j.

Conventional closeness centrality does not consider all the weight values of links in a social network; it considers only whether a link connects or not. However, many social networks in the real world have various weight values that represent the power of links between nodes. Therefore, we propose here weighted closeness centrality.

Similar to conventional closeness centrality, weighted closeness centrality finds the shortest path from the node to any other nodes. However, it also takes the inverse values that can be used as the weights of each link, and adds up all these values. For example, the weighted closeness centrality of the ith node (WCC_i) can be calculated as follows :

$$WCC_i = \left(\sum_{\forall j \neq i} w_{ij} \times d(i,j) \right)^{-1}, \tag{4}$$

where the weighted function w_{ij} is defined as:

$$w_{ij} = \begin{cases} \min_{\forall h} \left(\dfrac{1}{C_{ih}} + \cdots + \dfrac{1}{C_{hj}} \right) & if\ h \geq 1, \\ \dfrac{1}{C_{ij}} & \text{otherwise,} \end{cases} \tag{5}$$

where h are any intermediary nodes on the paths between node i and j and C_{ij} is the number of co-occurrence keywords between two nodes.

The weighted closeness centrality degree can be viewed as the efficiency of each node in spreading information to all other nodes. The larger the weighted closeness centrality of a node, the shorter the weighted average distance from the node to any other node, and thus the better positioned the node is to spread information to other nodes, an aspect not considered in the conventional algorithm.

3   Experiment

3.1   Data

'I Have a Dream' is a 17-minute public speech by Martin Luther King, Jr. delivered on August 28, 1963. In this speech, he called for a racial equality and an end to discrimination in the United States. King's speech invoked many historical events such as the Emancipation Proclamation, the Declaration of Independence, and the United States Constitution. The speech was ranked the top American speech of the 20th century by a 1999 poll of scholars of public address. This document was composed of 85 sentences and 1,579 words as shown in Table 1.

Table 1. Document overview of King's speech

Sentences	85
Words	1,579
Different words	523
Words per sentence	18.57

This document was split into 85 sentences and saved to a database. After this process, we applied a TF-IDF algorithm to identify the 16 keywords; America, brother, dream, faith, free, god, hope, justice, land, men, mountain, nation, negro, people, swelter, together. These keywords was used to build an 85 by 85 co-occurrence matrix for social network analysis.

3.2   Results

Fig. 3 shows the result of social network analysis performed using the co-occurrence matrix of 85 sentences. We used NetMiner 4 [17] to generate the social network of sentences and used the proposed weighted closeness centrality degree for calculating the summarization scores. It can be observed from the network that the role of each sentence and relationship among the sentences could be identified by the location. The size of the circles represents the frequency of keywords in the sentences, and the thickness of the lines indicates the strength of the co-occurrence keywords. Of these, 21 sentences have no relationship with other sentences and the others have at least one more connections via the co-occurrence keywords.

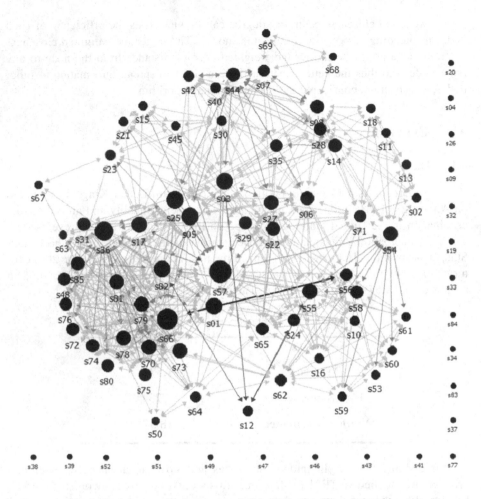

Fig. 3. Social network of sentences in Martin Luther King, Jr.'s address

Table 2 shows the top five extracted sentences using the summarization score function. The first-ranked sentence in our result is the 12th sentence in the entire document and has a summarization score of 0.0650. This list of highly scored sentences indicates the key sentences in King's speech from the point of view of the weighted diffusion of information. However, this result could have a different interpretation because the evaluation of document summarization involves human judgment. Unfortunately, there is no universally accepted method for evaluating summaries because even humans produce summaries with a wide variance.

Table 2. Summarization result of King's speech

Rank	ID	Score	Sentence
1	S12	0.0650	This note was a promise that all men, yes, black men as well as white men, would be guaranteed the "unalienable Rights" of "Life, Liberty and the pursuit of Happiness."
2	S44	0.0450	We cannot be satisfied as long as a Negro in Mississippi cannot vote and a Negro in New York believes he has nothing for which to vote.
3	S36	0.0425	And they have come to realize that their freedom is inextricably bound to our freedom.
4	S69	0.0405	Land where my fathers died, land of the Pilgrim's pride
4	S68	0.0405	My country 'tis of thee, sweet land of liberty, of thee I sing.
⋮	⋮	⋮	⋮

4   Conclusions

In this paper, we introduce the new approach to identify the key sentences of a text document. We utilized a text mining approach based on a TF-IDF algorithm to find keywords and used the proposed weighted closeness centrality to effectively compute the importance of sentences in a network. We demonstrated the applicability of our approach by using Martin Luther King's speech.

In our future studies, we will explore additional useful factors to improve the accuracy of document summarization. We hope this paper will stimulate further investigation on the application of appropriate document summarization tools in both academia and industry.

References

1. Mani, I., Maybury, M.T.: Advances in automatic text summarization, vol. 293. MIT press, Cambridge (1999)
2. Hemp, P.: Death by information overload. Harvard business review **87**(9), 82–89 (2009)
3. Pera, M.S., Ng, Y.K.: A Naive Bayes classifier for web document summaries created by using word similarity and significant factors. International Journal on Artificial Intelligence Tools **19**(04), 465–486 (2010)

4. Bhargavi, P., Jyothi, B., Jyothi, S., Sekar, K.: Knowledge Extraction Using Rule Based Decision Tree Approach. IJCSNS International Journal of Computer Science and Network 296 Security, pp. 296–301 (2008)

5. Mu, X., Hao, W., Chen, G., Zhao, S., Jin, D.: Research based on concept maps and hidden Markov model for multi-document summary. In: 2011 4th IEEE International Conference Broadband Network and Multimedia Technology (IC-BNMT), pp. 611–614 (2011)

6. Conroy, J.M., O'leary, D.P.: Text summarization via hidden Markov models. In: Proceedings of the 24th annual international ACM SIGIR conference on Research and development in information retrieval, pp. 406–407. ACM (2001)

7. Pourvali, M., Abadeh, M.S.: Automated text summarization base on lexicales chain and graph using of wordnet and wikipedia knowledge base. arXiv preprint arXiv:1203.3586 (2012)

8. Kurmi, R., Jain, P.: Text summarization using enhanced MMR technique. In: 2014 International Conference Computer Communication and Informatics (ICCCI), pp. 1–5 (2014)

9. Ganesan, K., Zhai, C., Han, J.: Opinosis: a graph-based approach to abstractive summarization of highly redundant opinions. In: Proceedings of the 23rd International Conference on Computational Linguistics, pp. 340–348 (2010)

10. Zheng, H.T., Bai, S.Z.: Graph-Based Summarization without Redundancy. In: Web Technologies and Applications, pp. 449–460 (2014)

11. Wasserman, S.: Social network analysis: Methods and applications, vol. 8. Cambridge university press (1994)

12. Chowdhury, G.: Introduction to modern information retrieval. Facet publishing (2010)

13. Freeman, L.C.: Centrality in social networks conceptual clarification. Social networks 1(3), 215–239 (1979)

14. Newman, M.E., Park, J.: Why social networks are different from other types of networks. Physical Review E 68(3) (2003)

15. Opsahl, T., Agneessens, F., Skvoretz, J.: Node centrality in weighted networks: Generalizing degree and shortest paths. Social Networks 32(3), 245–251 (2010)

16. Koschützki, D., Lehmann, K.A., Peeters, L., Richter, S., Tenfelde-Podehl, D., Zlotowski, O.: Centrality indices. In: Brandes, U., Erlebach, T. (eds.) Network analysis. LNCS, vol. 3418, pp. 16–61. Springer, Heidelberg (2005)

17. Netminer 4, http://www.netminer.com

Dynamic Facet Hierarchy Constructing
for Browsing Web Search Results Efficiently

Wei Chang and Jia-Ling Koh[✉]

Department of Information Science and Computer Engineering,
National Taiwan Normal University, Taipei, Taiwan, Republic of China
jlkoh@csie.ntnu.edu.tw

Abstract. In this paper, a method is proposed to dynamically construct a
faceted interface to help users navigate web search results for finding required
data efficiently. The proposed method consists of two processing steps: 1)
candidate facets extraction, and 2) facet hierarchy construction. At first, the
category information of entities in Wikipedia and a learning model are used to
select the query-dependent facet terms for constructing the facet hierarchy.
Then an objective function is designed to estimate the average browsing cost of
users when accessing the search results by a given facet hierarchy. Accordingly,
two greedy based algorithms, one is a bottom-up approach and another one is a
top-down approach, are proposed to construct a facet hierarchy for optimizing
the objective function. A systematic performance study is performed to verify
the effectiveness and the efficiency of the proposed algorithms.

Keywords: Faceted search · Wikipedia · Search result organization

1 Introduction

Keyword based search is a popular way for discovering required data of interest from a
huge collection of resources. The effectiveness of data retrieval mainly depends on
whether the given queries properly describe the information needs of users. However,
it is not easy to give a precise query because most queries are short (less than two
words on average) and many query words are ambiguous. Using a general keyword
with broad semantics as a query usually causes a huge amount of data returned. Most
of the search services return results as a ranked list, but it is difficult for users to
explore and find objects satisfying their search needs from a long list of results.
Accordingly, how to automatically group search results into meaningful "topics" has
become a significant issue to improve the usability of search results.

Faceted interface is a common feature of e-commerce sites, where the objects are
structured data with attributes. Users can select an attribute and specify a constraint
on the attribute value, such as the price or category of products, to filter the search
results. Recently, several search engines also provide a set of facets to the search

This work was partially supported by the R.O.C. M.O.S.T. under Contract No. 103-2221-E-
003-033.

M. Ali et al. (Eds.): IEA/AIE 2015, LNAI 9101, pp. 293–304, 2015.
DOI: 10.1007/978-3-319-19066-2_29

interface, which are usually the structural attribute of data such as location, time, size of document, etc. Faceted interfaces provide a convenient and efficient way to navigate the search results. However, most of the systems have to define the facets of their search interfaces in advance or assume that a prior taxonomy exists. Inspired by the recent works on automatic facets generation [7], in this paper, we would like to dynamically find keyword sets from the search results and construct a facet hierarchy of the search results in order to help users find the required data more efficiently.

Imagine that a user is exploring the news about "iPhone" and give a keyword query. The search engine returns a ranked list of search results with snippets. The snippet of a search result is a short text segment, which summarizes the important content of the search result. Our goal is to dynamically create a faceted interface for covering the top k search results, where the interface consists of a hierarchy of categories for the topic words in the search results. For example, a search result of the news of iPhone talks about "BlackBerry Messenger coming to iPhone and Android: ... ." Accordingly, the facets include "BlackBerry(company)_mobile_phones", "information applications", and so on. The user can navigate the category path "information applications/Instant_ messaging/BlackBerry Messenger" to find the related results.

In order to eliminate the network transmission time of downloading the entire webpages of the search results, our proposed method extracts the topic terms from the snippets of search results as the candidates of facet terms on the lowest level of the facet hierarchy. Besides, the categories of the topic terms are looked up from Wikipedia to be the candidate facets on the higher levels. However, a facet term may belong to multiple categories in Wikipedia but some categories are not semantically related to the search results. To deal with this problem, we perform a filtering step which applies a learning-to-rank strategy to select the categories which are highly semantics related to the search results. Furthermore, we define an objective function of a facet hierarchy to estimate the browsing cost of users for finding a search result from the facet hierarchy. Then we propose two greedy-based algorithms, a top-down approach and a bottom-up approach, to construct a facet hierarchy from the candidate facet terms for optimizing the objective function. A systematic performance study is performed to verify the effectiveness and the efficiency of the proposed algorithms.

The rest of the paper is organized as follows. In the next section, a brief overview of the related works is introduced. The formal problem definition of constructing a facet hierarchy of query results is given in Section 3. Section 4 introduces the proposed method of facet terms extraction and two algorithms of constructing a facet hierarchy. The performance evaluation on the proposed algorithms is reported in Section 5. Finally, we conclude this paper in Section 6.

2   Related Works

Many existing works [2][3][4][9][13] studied how to separate the search results into semantics related groups, which are called search results clustering. Topic modeling is a technique for extracting latent topics from a set of unlabeled documents. LDA [2][4] is a popular approach to construct a topic model, whose goal is to assign the documents representing the same topic concept into a cluster. After performing the unsupervised training phase, each obtained cluster is labeled manually. Then the constructed LDA model is used to assign the search results to the corresponding topic clusters. Many researches [1][8][11][12] applied the LDA approach to perform topic discovery from text data for various applications. However, the disadvantage of LDA is that it is required to give the number of clusters manually. It is not easy to determine a proper number of topic clusters in advance.

On the other hand, some works [9][13] applied knowledge bases, such as Wikipedia, to find topic clusters of search results. These works extracted the topic terms from the search results and find the corresponding entities in the knowledge bases. Then the categories of the entities in the knowledge bases are used to perform topic clustering. However, in [9], it provided only single level of clusters for the search results.

In e-commerce systems and digital libraries, faceted search is a popular way for searching objects with different attributes. The faceted search system provides a simple interface for users to filter and browse objects easily according to the selected faceted-value pair. However, most of the faceted search interface is constructed for the structured data where the faceted-value pairs are predefined. In recent years, some works [6][7][11] studied how to extract facets and facet terms from data dynamically from unstructured or semi-structured data. In [6], a supervised approach was proposed to recognize query facets from noisy candidates. This approach applied a directed graphical model, which learns how to recognize the facet terms in search results and how to group the terms into a query facet together. However, this approach didn't provide the semantic concept of the corresponding query facet. [11] extended LDA topic model to find facets from the messages posted in Twitter. However, this approach only extracts five specific types of facet terms, i.e. locations, organizations, persons, time distributions and general terms.

The Facetedpedia system proposed in [7] was a faceted retrieval system designed for information discovery and exploration in Wikipedia. The system automatically and dynamically constructs facet hierarchy for navigating the set of Wikipedia articles resulting from a keyword query. This work used the hierarchical categories of the Wikipedia articles as the candidate facets and designed a ranking algorithm for these facets. The proposed algorithm aims to select diverse facets for minimizing the navigation cost of search results. Although our goal is similar to the one of Facetedpedia, it has more challenges to generate a facet hierarchy for organizing the web search results than generating a facet hierarchy of the Wikipedia search results. The Facetedpedia system downloaded all the articles in the Wikipedia search results and built a category index for each article by performing an offline process. However, for generating a facet hierarchy of the web search results, it is not feasible to download and process the whole web pages of the search results. Accordingly, our approach processes the snippets instead of the web pages of the search results. The snippet of a search result is usually short. How to extract facet terms from the snippets and the knowledge base for effectively organize search results is the main issue.

3   Problem Definition

Given a query q, let $R_q = \langle s_1, s_2, \ldots, s_k \rangle$ denote the ranked top-k list of the search results with snippets which are returned by a search engine. Given a search result snippet s_i, some terms which correspond to the named entities in Wikipedia are extracted as the candidate facet terms (the extraction method is introduced in Sec 4.1). Given R_q, let F^0 denote the set of candidate facet terms $\{f_1^0, f_2^0, \ldots, f_{|F^0|}^0\}$, where each f_j^0 appears in at least one snippet in R_q. For example, given a query "bull", $R_{bull} = \langle s_1, s_2, \ldots, s_7 \rangle$ denotes the top 7 snippets in the search result. The set of terms extracted from the snippets is F^0 ={"Bull(Company)", "Chicago Bull", "Red Bull(Drink)", "Derrick Rose", "Spanish Fighting Bull"}. If a candidate facet term f_j^0 appears in a snippets s_i in R_q, we call that f_j^0 *covers* s_i. Besides, E^0 is used to denote the cover relationship between F^0 and R_q.

As defined in [7], the Wikipedia *category hierarchy* is a connected and directed acyclic graph $\mathcal{H}(r_{\mathcal{H}}, C_{\mathcal{H}}, \mathcal{E}_{\mathcal{H}})$, where $r_{\mathcal{H}}$ denotes the root node, the node set $C_{\mathcal{H}}$consists of the categories in Wikipedia, and the edge set $\mathcal{E}_{\mathcal{H}}$ denotes the set of category-subcategory relationships in Wikipedia. Given R_q, the category hierarchy of the facet terms $FC(C_F, \mathcal{E}_F)$ is a connected subgraph of the category hierarchy $\mathcal{H}(r_{\mathcal{H}}, C_{\mathcal{H}}, \mathcal{E}_{\mathcal{H}})$, where $C_F \subseteq C_{\mathcal{H}}$ and $\mathcal{E}_F \subseteq \mathcal{E}_{\mathcal{H}}$, and $c \in C_F$ if c is the category of a candidate facet term in F^0. For example, suppose R_q contains the top-7 query result snippets, Fig. 1(a) shows the covering relationships between the 5 candidate facet terms and R_q, and the category hierarchy of the facet terms up to two levels.

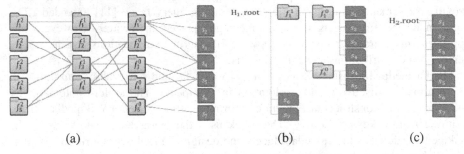

(a)   (b)   (c)

Fig. 1. Example of facet hierarchies

Definition 1 (A facet hierarchy of query results). A facet hierarchy $H(H.root, N, E)$ of the query result R_q is an ordered tree structure, where $N \subseteq R_q \cup F^0 \cup C_F$, $E \subseteq E^0 \cup \mathcal{E}_F \cup (\{H.root\} \times N)$. Besides, $R_q \subseteq N$.

[Example 1]. Fig. 1(b) and (c) show two examples of facet hierarchy. Fig. 1(c) shows the case that no facet term is selected to construct the facet hierarchy H_2, which corresponds to a ranked list of the search results. On the other hand, there are two facet terms and a category facet term selected to construct the facet hierarchy H_1, where s_6 and s_7 are the remained search results not covered by the selected facet terms.

A facet hierarchy shows an organized hierarchy of showing the search results, similar to the directory of a file system. A browsing path simulates a browsing behavior of users to retrieve the snippet of a search result. We assume that a user will browse the children nodes of each node one-by-one according to the sorted order. Moreover, after selecting a node, the user will recursively browse the children nodes until finding the required result.

Definition 2 (A browsing path of a snippet). A browsing path of a snippet s_i on the Hierarchy H is a node visiting sequence from H.root to the leaf node representing s_i. The set of all browsing paths of a snippet s_i on the hierarchy H is denoted as $path^H(s_i)$.

Definition 3 (The browsing cost of a snippet). The browsing cost of a snippet s_i in a facet hierarchy H is the number of nodes in the shortest browsing path of s_i in H except H.root.

$$cost_H(s_i) = \min\{len(path)|\ path \in path^H(s_i)\},$$

where $len(path)$ denote the number of nodes in $path$ except H.root.

[Example 2]. As shown in Fig. 1(b), there is only one browsing path to retrieve the snippet s_2, denoted as $path^{H_1}(s_2) = \{f_1^1 \to f_1^0 \to s_1 \to s_2\}$. The browsing path simulates the user behavior of selecting facet term f_1^1, selecting facet term f_1^0, denying s_1, and then confirming s_2. The browsing cost of s_1 is 4. For the snippet s_4, $path^{H_1}(s_4)$ contains two browsing paths: $f_1^1 \to f_1^0 \to s_1 \to s_2 \to s_3 \to s_4$ and $f_1^1 \to f_1^0 \to f_4^0 \to s_4$. Therefore, $cost_{H_1}(s_4) = 4$. In Fig. 1(c), the browsing cost of s_k equals to k for k=1, ..., 7, respectively.

According to the definition of the browsing cost of a snippet on a facet hierarchy, our goal is to find a facet hierarchy which provides the least expected browsing cost of the query result.

Definition 4 (Facet hierarchy construction with minimum browsing cost problem). Given the query result R_q and $P(s_k)$, which denote the access probability of query result s_k, where $P(s_n) \geq P(s_m)$ for $n < m$ and $\sum_{i=1}^{k} P(s_i) = 1$. The facet hierarchy construction with minimum browsing cost problem is to find a facet hierarchy H of R_q such that H=arg min $(\sum_{s_i \in R_q} P(s_i) \cdot cost_{H_j}(s_i))$.

4   Topic Term Extraction

4.1   Entity Mention Annotation

Based on our observation, most important topic words belong to nouns or noun phrases. Accordingly, for each snippet s_i, we use Stanford POS tagging tool and the NER library to parse the title and description of the snippet. Besides, we use the TAGME API provided on the official website [14] for finding the entity mentions.

Let $s_i.title$ and $s_i.description$ denote the set of Wikipedia entities extracted from the title and the description of s_i, respectively. From the terms in $s_i.title$ and $s_i.description$, a term t_j in $s_i.title$ is selected to be a *topic term* of s_i and inserted into $s_i.facets$ if t_j is a noun or t_j is recognized as a named entity by the Stanford NER library. According to the same rule, a term t_j in $s_i.description$ is selected to be a *context term* of s_i and inserted into $s_i.context$, where the context terms are used to provide more semantic information for selecting the categories of facet terms.

4.2   Candidate Facet Terms and Categories Generation

After extracting the topic terms of each snippet, those topic terms are the candidate facet terms on the lowest level of the hierarchy. Accordingly, F^0 is set to be the union of $s_i.topic$ for each s_i in R_q. Besides, the categories of the topic terms in Wikipedia are looked up to find the candidate category facet terms.

For each facet term f_j^0, let $C^1(f_i^0)$ denote the set of categories of f_i^0 in Wikipedia. By recursively looking up the directory structure of Wikipedia, $C^l(f_i^0)$ denote the union of the parent categories of $C^{l-1}(f_i^0)$ for $l > 1$. For example, for a candidate facet term f_1^0 ="BlackBerry Messenger", $C^1(f_1^0)$ consists of "BlackBerry(company)", "BlackBerry_software", "Instant_messaging" and "Instant_messaging_clients". Besides, $C^2(f_1^0)$ consists of "BlackBerry(company)_mobilephone", "Canadian brands", etc." Among the categories of the topic terms, only some categories are highly semantics related to the search result. For example, "BlackBerry", "Instant messaging" and "Cloud clients" are highly semantics related to the search result, but "2007 introduction" and "Digital audio players" are not. In order to address this problem, we perform a learning-to-rank approach to select the categories with highly semantic relatedness to the search result.

Let C^l(s) denote the union of $C^l(f_i^0)$ for each f_i^0 extracted from snippet s. We extract two groups of features: the *context aware features* and *popularity features* for each category in C^l(s). The context aware features of a category aim to evaluate the semantics related degree between the category and the query result s. Moreover, the popularity features of a category represent the popularity of the category.

We apply the method proposed in [5] to select two sets of children articles and split articles to represent the semantics of a category c. Besides, we apply the formula provided by [5] to compute the semantic relatedness between two Wikipedia articles. For each category c in C^l(s), the 9 features shown in Table 1 are extracted and used to rank the categories at level l. A parameter *max_l* can be given to control the number of levels in the constructed facet hierarchy.

Table 1. The features for ranking the categories of facet terms

Features	Description	Notation
Context Aware Features	Average semantic relatedness between children articles of the category and the topic terms of the search result	CT
	Average semantic relatedness between children articles of the category and the context terms of the search result	CC
	Average semantic relatedness between the split articles of the category and the topic terms of the search result	ST
	Average semantic relatedness between the split article of the category and the context terms of the search result	SC
Popularity Features	Number of children articles of the category	NC
	Word length of the category title	WL
	Character length of the category title	CL
	Number of split articles / word length of the category title	SP
	Have the corresponding main article	MA

In the training phase, for a snippet s, a manually ordered list of the categories in $C^l(s)$ for $l \geq 1$ is given according to their relevance with s. Then we use rankSVM to learn a model denoted $model^l$ for ranking the categories at level l. For each snippet s_i in R_q, the $model^1$ is used to select the top-K categories from $C^1(s_i)$ for each candidate facet term of s_i. Recursively, the $model^l$ is applied to select the top-K categories of a category facet term in F^{l-1} as F^l for $l > 1$, which are used to be the candidate category facet terms for constructing the facet hierarchy.

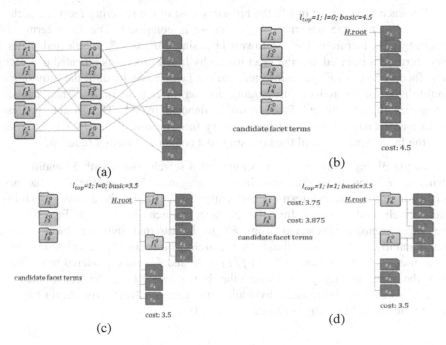

(a)

(b)

(c)

(d)

Fig. 2. An example of the FH-BU algorithm

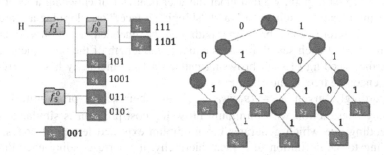

Fig. 3. The correspondence between a facet hierarchy and a binary encoding tree

5   Facet Hierarchy Construction

5.1   The Bottom-Up Approach

The bottom-up approach, which is named the FH-BU algorithm, applies a greedy approach to incrementally select the facet terms and categories into the facet hierarchy. Initially, the facet hierarchy $H(H.root, N, E)$ has $N = R_q$ and E $=\{H.root\} \times R_q$. Let $f.cover$ denote the search results in R_q covered by a facet term f in F^0. For each facet term f in F^0, the browsing cost of the resulting facet hierarchy H' by inserting f as the parent node of $f.cover$ is computed. The facet term whose resulting facet hierarchy has the lowest browsing cost, say f^*, is selected. After the facet term f^* is inserted into the facet hierarchy H, the process is repeated to select the next facet term from F^0 to be inserted into the facet hierarchy until the browsing cost couldn't be further reduced by adding another facet term in F^0. Then the set of categories of the selected facet terms is denoted by F^1. The similar process is performed recursively to select the category facet term in F^l for $l \geq 1$ to be inserted into the facet hierarchy until the browsing cost couldn't be further reduced.

[**Example 3**]. Fig. 2(a) shows an example of 8 search results with 5 candidate facet terms in F^0 and their corresponding 5 categories. We assume that the access probability of each search result is uniformly distributed, i.e. the access probability of each search result is 0.125. Initially, the facet hierarchy is shown as Fig. 2(b). Then the facet terms are selected from F^0 to be inserted into the facet hierarchy incrementally, that leads to the facet hierarchy shown in Fig. 2(c). After that, the categories of the facet terms f_1^0 and f_4^0, i.e. f_1^1 and f_4^1, are considered to be inserted into the facet hierarchy. However, the browsing cost of the constructed facet hierarchy could not be reduced by adding any category facet term. Accordingly, the constructed facet hierarchy is shown as Fig 2(d).

5.2   The Top-Down Approach

For each browsing path, we can label the user behavior of choosing a facet term and confirming a search result as "1", and "skipping a facet" or "skipping a search result" as "0". Therefore, each browsing path can be encoded as a bit sequence. The browsing cost of each search result is equal to the length of the bit sequence. In other words, the encoding of each browsing path of a facet hierarchy has a corresponding binary encoding tree as shown in Fig. 3.

Accordingly, given $R_q = \langle s_1, s_2, ..., s_k \rangle$ and their access probabilities, the facet hierarchy construction with minimum browsing cost problem is similar to construct an encoding tree which generates the minimum expected length of codes. Besides, according to the definition of a facet hierarchy, the corresponding encoding tree T_q must satisfy the following two requirements:

(1)   The encode bit sequence of each leaf node in T_q is ended with "1".
(2)   Each internal node in T_q has a corresponding facet term in $F^0 \cup C_F$.

We apply the concept of Shannon–Fano coding method [10] to measure the information entropy of the access probabilities for the snippets covered and not-covered by a facet term f as following:

$$Split(f, S) = -(f.p_r(S) \times log_2(f.p_r(S)) + f.p_u(S) \times log_2(f.p_u(S))),$$
$$f.p_r(S) = \frac{f.p\_cover(S)}{f.p\_sum(S)}, \quad f.p_u(S) = \frac{f.p\_uncover(S)}{f.p\_sum(S)},$$

where $f.p\_cover(S) = \sum_{s_i \in S \wedge s_i \in f.cover} p(s_i)$, $f.p\_uncover(S) = \sum_{s_i \in S \wedge s_i \notin f.cover} p(s_i)$, and $f.p\_sum(S) = f.p\_cover(S) + f.p\_uncover(S)$.

A higher *Split* value of f implies that the access probabilities of the covered and not-covered search results are more balanced. In other words, it is more possible to get a lower expected browsing cost by using f as a facet term for S.

Accordingly, the top-down approach for constructing the facet hierarchy, named the FH-TD algorithm, selects the facet term which has the highest *Split* function value to split the search results in R_q into two subsets. The same procedure is performed recursively to split the two subsets of the search results. However, the *Split* function only estimates the distribution of the access probabilities of the covered and not-covered search results of one level. In order to improve the effectiveness of estimation, a positive integer parameter d can be given to further compute the *Split* function values on the following d levels of facet terms under the current facet term.

Initially, S is equal to R_q. For each facet term f in $F^{max\,-1}$, f and the following d levels of facet terms under f are evaluated by the Split function and compute the average value. The facet term, say f', and its decedent facet terms with the highest average *Split* value is selected into the facet hierarchy. Then S is separated into $f'.cover$ and $(S- f'.cover)$. Accordingly, the same procedure is performed on $f'.cover$ and $(S- f'.cover)$ recursively to find the other facet terms to be inserted into the facet hierarchy until finishing the insertion of the facet terms in F^0.

6   Performance Evaluation

In this section, we compare the two proposed methods with the method used in Facetedpedia [7]. The algorithms were implemented using JAVA in the Eclipse platform and performed on a personal computer under the Microsoft Windows 7 environment with 8 GB RAM. The test query set is provided by "Web Track 2012 and 2013" in the TREC competition, where there are 50 queries in both "Web Track 2012" and in "Web Track 2013", respectively. Given a query q, the snippets of the search results are collected by the Google search engine for each test query. Moreover, the access probability p_i of each snippet s_i is simulated according to a uniform distribution and an exponential distribution, respectively.

The default setting of the parameters in the experiments is as follows. The level of facet categories *max_l* is 2. The maximum number of facet terms at the top-level, i.e. level 2, is set to be 8. The look-ahead parameter *d* of the top-down method is set to 2.

[Exp. 1]. The effect of changing the number of search results

In this experiment, the expected browsing costs of the constructed facet hierarchies are compared by changing the number of search results. The baseline is the browsing cost of the ranked list of search results without using a facet hierarchy. The results performed on the query results with uniformly distributed probabilities and exponentially distributed probabilities are shown in Fig. 4(a) and 4(b), respectively.

From the results, it shows that the facet hierarchies constructed by the proposed methods have lower expected browsing costs than the ones constructed by the Facetedpedia. When the access probabilities of the search results are uniform distribution, the facet hierarchies constructed by FH-BU and FH-TD have similar browsing costs. However, when performed on the access probabilities with exponential distribution, the performance of FH-TD is better than FH-BU. When the number of search results is more than 50, the constructed facet hierarchies can save more than half of the browsing costs of the baseline method. The gain of browsing cost saving is more significant when the number of search results increases.

[Exp. 2]. The effects of changing the maximum number of categories selected at the highest-level of the facet hierarchy.

In this experiment, the expected browsing costs of the constructed facet hierarchies are compared by changing the maximum number of categories at the highest level. The number of search results is set to be 100.

Fig. 4(c) and 4(d) show the results performed on the query results with uniformly distributed probabilities and exponentially distributed probabilities, respectively. From the results, it demonstrates that the browsing costs of the constructed facet hierarchies will decrease as the maximum number of categories at the highest level increase. For the FU-BU, the browsing costs keep stable when the maximum number of categories at the highest level is 8. It implies that the FU-BU couldn't add another category facet term to further reduce the expected browsing cost of the constructed facet hierarchy. On the other hand, the expected browsing costs of the facet hierarchies constructed by FU-TD continue to decrease until the maximum number of categories at the highest level is larger than 12.

[Exp. 3]. Evaluation on the execution time

In this experiment, the execution time of the proposed algorithms is observed. The number of search result is fixed to 100 and the maximum number of categories at the highest level is set to be 8.

The FH-TD estimate the expected browsing cost by using the entropy function without tracing the hierarchy and calculating the expected browsing cost of the resultant facet hierarchy. Accordingly, as shown in Fig. 4(e) and 4(f), the execution speed of FH-TD is much faster than the other twos.

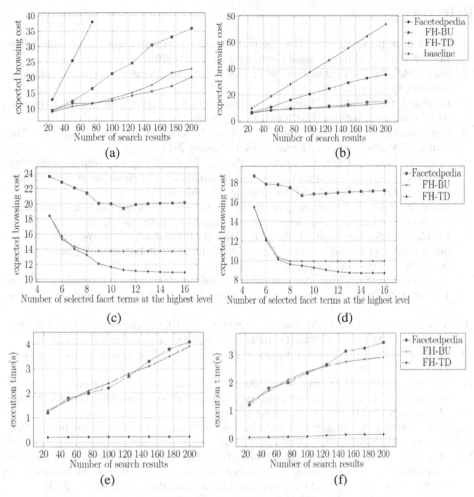

Fig. 4. The experimental results

7   Conclusion

In this paper, we propose a method to dynamically construct a faceted interface for browsing the search results efficiently. A learning-to-rank approach is used to select the query-dependent facet terms for constructing the facet hierarchy. Moreover, two greedy based algorithms, one is a bottom-up approach and another one is a top-down approach, are proposed to construct a facet hierarchy for minimizing the expected browsing cost as possible. The experimental results show that the facet hierarchies constructed by the two proposed methods both save more than half of the browsing costs of the baseline method. Moreover, the top-down approach can achieve better performance on the constructed facet hierarchy and spend less computing time than the bottom-up approach.

References

1. Agarwal, D. B., Chen, C.: fLDA: matrix factorization through latent dirichlet allocation. In: The Third ACM International Conference on WSDM (2010)
2. Blei, D.M., Ng, A.Y., Jordanm, M.I.: Latent dirichlet allocation. Journal of Machine Learning Research (2003)
3. Carpineto, C., Osiński, S., Romano, G., Weiss, D.: A survey of web clustering engines. ACM Computing Surveys (CSUR) 41(3), July 2009
4. Hoffman, M.D., Blei, D.M., Wang, C., Paisley, J.: Stochastic variational inference. The Journal of Machine Learning Research 14(1), January 2013
5. Jiang, P., Hou, H., Chen, L., Chen, S., Yao, C., Li, C., Wang, M.: Wiki3C: exploiting wikipedia for context-aware concept categorization. In: The Sixth ACM International Conference on Web Search and Data Mining (WSDM) (2013)
6. Kong, W., Allan, J.: Extracting query facets from search results. In: The 36th International Conference on Research and Development in Information Retrieval (2013)
7. Li, C., Yan, N., Roy S. B., Lisham, L., Das, G.: Facetedpedia: dynamic generation of query-dependent faceted interfaces for wikipedia. In: The 19th International Conference on World Wide Web (WWW), pp. 651–660 (2010)
8. Mei, Q., Shen, X.C., Zhai, X.: Automatic labeling of multinomial topic models. In: The 13th ACM SIGKDD International Conference on Knowledge Discovery and Data Mining (2007)
9. Scaiella, U., Ferragina, P., Marino, A., Ciaramita, M.: Topical clustering of search results. In: The Fifth ACM International Conference on Web Search and Data Mining (2012)
10. Shannon, C.E.: A mathematical theory of communication. Bell System Technical Journal 27, 379–423 (1948)
11. Vosecky, J., Jiang, D., Leung, K. W., Ng, W.: Dynamic multi-faceted topic discovery in twitter. In: The 22nd ACM International Conference on Information and Knowledge Management (2013)
12. Zhu, J., Ahmed, A., Xing, E.P.: MedLDA: maximum margin supervised topic models. The Journal of Machine Learning Research (2012)
13. Zhu, X., Ming, Z.Y., Zhu, X., Chua, T.S.: Topic hierarchy construction for the organization of multi-source user generated content. In: The 36th ACM International Conference on Research and Development in Information Retrieval (SIGIR) (2013)
14. http://tagme.di.unipi.it/tagme_help.html

Filtering Reviews by Random Individual Error

Michaela Geierhos[(✉)], Frederik S. Bäumer, Sabine Schulze,
and Valentina Stuß

Heinz Nixdorf Institute, University of Paderborn,
Fürstenallee 11, D-33102 Paderborn, Germany
{michaela.geierhos,fbaeumer,sabine.schulze,valentina.stuss}@hni.upb.de

Abstract. Opinion mining from physician rating websites depends on the quality of the extracted information. Sometimes reviews are user-error prone and the assigned stars or grades contradict the associated content. We therefore aim at detecting random individual error within reviews. Such errors comprise the disagreement in polarity of review texts and the respective ratings. The challenges that thereby arise are (1) the content and sentiment analysis of the review texts and (2) the removal of the random individual errors contained therein. To solve these tasks, we assign polarities to automatically recognized opinion phrases in reviews and then check for divergence in rating and text polarity. The novelty of our approach is that we improve user-generated data quality by excluding error-prone reviews on German physician websites from average ratings.

Keywords: Data quality improvement · Error-prone review detection · Text-rating-inconsistency

1 Introduction

People are encouraged to do their best when writing reviews. Even when they make mistakes, it is probably due an oversight, and not with intent to defraud. Since fraud prevention and detection of fake product reviews are big issues in online retail, the detection of fake reviews has attracted increasing attention in recent years both from business and research community [12]. In this paper, we analyze a different form of non-conforming (inconsistent) reviews caused by random individual error [19]. Such errors produce disagreement in polarity of the numerical ratings and the corresponding narrative comments within a review. An example therefore is 'The Dr. clearly failed in diagnosing my lung cancer' in combination with an excellent grade in the rating category 'treatment'.

Thus, it is not only (1) the natural language processing of the review texts itself that is addressed here, but also (2) the filtering of reviews by random individual error to provide consistent high-quality data. Therefore, we introduce a method to detect these inconsistencies on physician rating websites (PRWs) because a divergence between numerical ratings and review texts can be annoying for consumers and at the same time reduce the value of the rating portal [11]. Thus, the basic idea is first to identify relevant opinion phrases that describe

© Springer International Publishing Switzerland 2015
M. Ali et al. (Eds.): IEA/AIE 2015, LNAI 9101, pp. 305–315, 2015.
DOI: 10.1007/978-3-319-19066-2_30

service categories and to determine their polarity. Subsequently, the particular phrase has to be assigned to its corresponding numerical rating category before checking the (dis-)agreement of polarity values. For this purpose, several local grammars for the syntactic analysis as well as domain-specific dictionaries for the recognition of entities, aspects and polarities were used. Our approach is special in that we apply a rule-based method for the detection of reviews on German PRWs containing individual random error. Our goal thereby is to automatically filter reviews that are not consistent with their ratings in order to provide less noisy data for further analysis.

The paper is structured as follows: The next section provides an overview on the related work. It is followed by the research design in Section 3 which introduces the data set (3.1), describes the data preprocessing (3.2) and presents our approach for the detection of individual random error (3.3). In Section 4, we evaluate our approach and discuss our results before we conclude in Section 5.

2   Background and Related Work

In this section, we present background information relevant to our work. We also survey related work and point out their relationship to our approach.

Due to the fact that Web 2.0 basically decreased the cost of communication, patients can compare medical services without effort, evaluate the performance of physicians and their staffs and share their personal experiences with the health care system [1]. Hence, the number of PRWs is rising rapidly [16,17] and patients are increasingly turning "to online physician ratings, just as they have sought ratings for other products and services" [7]. These "online physician reviews are a massive and potentially rich source of information [for] capturing patient sentiment regarding healthcare" [18].

However, some reviews may be biased and then the detection of deceptive and fake opinions becomes a hot topic [12]. In order to provide high-quality research data, it is essential to discover fake reviews "to reflect genuine user experiences and opinions" [12]. There are two main approaches for the filtering of fake reviews: supervised and unsupervised learning [12]. Since "it is very hard if not impossible to reliably label fake reviews manually" [12], it is difficult to obtain labeled training data, especially in the health care domain. Moreover, research on opinion spam detection is mainly conducted for product [10] or restaurant reviews [2]. Other domains such as physician reviews haven't been yet investigated due to missing or very little training data for machine learning. Consequently, we apply rule-based methods which have the advantage that they (1) do not need training data and (2) can be easily elaborated by (manual) domain-specific rule adjustments. We concentrate on the detection of random individual error in reviews rather than on the identification of reviews faked on purpose. Possible explanations for such random individual errors in review texts and ratings are that in "most of the practical cases [...] users do not carefully rate items, they either forget to rate the given items (i.e., missing value problem) or they make a mistake on the precise evaluation (i.e. noisy rating problem)" [14]. Furthermore, "ratings are influenced by subjective factors (from user) and objective factors (from system) together. While user factors include various psychological effects, e.g., attitude, mentality, and satisfaction, the system factors are trust, interests, user interfaces and so on" [14].

Islam (2014) therefore proposes a "unified rating system" [8] in order to resolve possible inconsistencies. For this purpose, he infers sentiment from texts to generate numeric ratings based on the polarity of the full-text review. The overall rating is then defined as the "average of the rating done by sentiment analysis and the star rating given by the users" [8]. However, we do not distill numerical ratings from the texts in order to provide consistent reviews. We aim at identifying random individual error by comparing the polarity of the text to the corresponding polarity of the users' grades or stars. "Ratings that do not match the actual text of the comments" [3] are, for example, discovered by WisCom. Although Fu et al. (2013) also analyze inconsistencies in reviews, they concentrate on mobile App marketplaces such as Google Play Store [3]. These reviews are somehow different from physician reviews because they "are generally shorter in length since a large portion of them are submitted from mobile devices on which typing is not easy" [3]. Unlike Fu et al. (2013), we do not apply a regression model either for the detection of inconsistencies between ratings and review texts. And Mudambi et al. (2014) try to detect misalignment between review texts and star ratings in order to understand why and where such inconsistencies are most likely to occur [11]. In particular, they use a machine learning approach to determine if certain texts are specific for product reviews of n stars on amazon.com. Thus, the "classification algorithm can identify a 'typical' 5-star review, 3-star review, and so on" [11]. Jang et al. (2014) also infer sentiment scores from texts in order to face the issue of "the inconsistency between textual evaluation (review content) and scoring evaluation (review rating)" [9]. However, they conduct their survey on hotel reviews. Since we focus on a different domain (German PRWs), we cannot apply machine learning techniques because only few physician reviews of the entire data set are affected by random individual error (cf. Table 2).

3 Research Design

In this section, we describe the data collection and preprocessing as well as our approach towards filtering physician reviews by random individual error. For this purpose, we created an extensive data set of physician reviews (cf. Section 3.1), developed domain-specific local grammars for information extraction and sentiment analysis (cf. Section 3.2) and describe how to filter physician reviews by random individual error in Section 3.3.

3.1 Data Set

A data set is usually designed for a particular purpose. Because of no available ready-made corpora of physician reviews, we have to build our own specialized corpus. For corpus creation, we collected texts from two different German PRWs:

jameda.de and docinsider.de. Both sites provide enough data for a balanced corpus because of their user popularity and traffic volume and their great amount of physician reviews. Furthermore, many medical subjects are covered (representativeness). By gathering data from jameda.de and docinsider.de between October 2013 and December 2014, we built a specialized corpus containing 702,323 individual physician reviews in total, where each review consists of a qualitative and a quantitative part (cf. Figure 2). While the textual information includes the title, review text and metadata (e.g. patient's personal data), there are also up to 16 numeric rating criteria (e.g. quality of treatment, equipment, organization). All reviews were not edited, i.e. no spell checking or the like was applied. This data set covers the time period from January 2009 to December 2013, where the average length of a review text is 48 words and the longest one consists of 348 words. The descriptive metadata (e.g. age, type of health insurance) attached to each review provides classificatory information for a better understanding of the reviewer's background. In total, 61 % of all physician reviews contain details about the statutory health insurance (SHI) or the private medical insurance (PMI). Besides, 63 % provide information about the reviewer's age distributed in three main categories: 'younger than 30', 'between 30 and 50' and 'older than 50'. Because only five rating categories on jameda.de and six categories on docinsider.de are mandatory fields, the awarded number of rating criteria differs per review. Moreover, jameda.de uses a grading system (best: 1.0 – worst: 6.0) and docinsider.de applies star rating (best: 5 stars – worst: 1 star).

3.2 Data Preprocessing

Figure 1 illustrates the general workflow of our approach.

First of all, information extraction is performed [5]. We therefore apply a method that automatically identifies and extracts relevant information from user-generated physician reviews and transforms it into a structured representation (i.e. predefined templates). What should be recognized is defined by domain-specific rules (i.e. local grammars [6]). So our information extraction

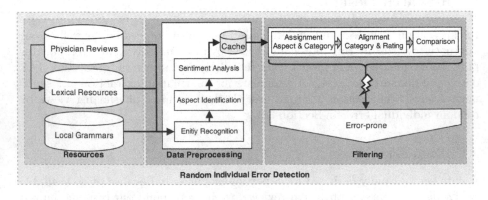

Fig. 1. Pipeline for random individual error detection in reviews

system works with predefined patterns for entity recognition and aspect identification. In our case, relevant expressions for the enrichment of the dictionaries and relevant phrases for the creation of local grammars are determined by frequency analysis on the physician reviews. We therefore generated n-grams (up to 5-grams) and build frequency lists. The most frequent n-grams of the length 5 build then the seed list for the construction of local grammars.

For instance, when the opinion phrase 'Dr. Foo is incompetent' occurs in the reviews, then a local grammar is able to detect the entity ('Dr. Foo'), the aspect ('incompetent') and substitute the adjective 'incompetent' by a variety of other negative adjectives such as 'unfriendly' or 'disinterested' belonging to the same category (here: 'treatment') of the given service category ontology (cf. Table 2). These words are taken from our lexical resources, especially the polarity dictionary SentiWS [15] and our own dictionaries (cf. Section 3.2) containing expressions that have already been identified in previous preprocessing iterations by n-grams on our data set. In general, information extraction is an iterative process that is able to identify relevant opinion phrases in our data set by means of dictionary and local grammar usage (cf. Section 3.2 and 3.2).

Finally, the polarity of each opinion phrase is compared to the polarity of the respective numerical rating in order to detect error-prone reviews.

Lexical Resources are crucial in most NLP tasks. Imagine, for instance, the following review text:

> *'The doctor is quite time-efficient. Without beating around the bush and with a mischievous smile, he told me that I have incurable leukemia. Then he said goodbye and went on to the next patient. All in all, a quite efficient patient handling!'*

Its polarity is unclear. Is this really a positive review? When placing an order on amazon.com for example, an 'efficient handling' is desirable. Also 'not beating around the bush' is preferable during a business meeting. But during consultation between physician and patient, less straightforwardness and time-efficiency are appreciated, especially for the receiver of bad news. Hence, almost nothing is entirely positive or negative [13].

Since existing polarity lexical resources for German, e.g. SentiWS [15], do not reflect domain-specific lexical usage, we have to enrich these dictionaries and use context information for polarity classification. SentiWS "contains several positive and negative polarity bearing words weighted within the interval of [-1; 1] plus their part of speech tag, and if applicable, their inflections" [15]. For our purpose, especially adjectives from SentiWS play a significant role. Moreover, we added the most frequently used adjectives in our data set to the corresponding lists for positive (P) and negative (N) uni- and bigrams. Furthermore, we use pattern dictionaries for the recognition of evaluative expressions (see bi- and trigrams in Table 1) and gazetteers work as specialized dictionaries (48,046 doctor's names) to support initial tagging. On top of this, 1,377 domain-specific (medical) terms for diagnoses, syndromes and treatment were collected and organized in their own dictionary. This is necessary for the sentiment analysis in order to assign aspects to their corresponding entities (from medical terminology). Table 1 gives an overview on the created lexical resources which are grouped by pattern (n-gram) and polarity.

Table 1. Polarity lexical resources for sentiment analysis

n-gram	description	example	polarity	amount
n=1	single adjective	'qualified'	P	33,327
n=1	single adjective	'busy'	N	25,081
n=2/n=3	adjective phrase	'clean and tidy'	P	1,373
n=2/n=3	adjective phrase	'very caring'	N	297

Neither these dictionaries nor their extensions can cover all variants of relevant opinion phrases. We therefore have to abstract from literal expressions and create semantically enriched syntactic patterns represented by local grammars. An example can be '<A> doctor's office', where <A> refers to an adjective phrase in the dictionary such as 'very caring' or other adjectives in the left context of 'doctor's office'.

Local Grammars describe semantic-syntactic structures that cannot be formalized in electronic dictionaries. They are implemented as finite state transducers (FST) [4]. These transducers produce output in terms of semantic annotations (i.e. labels) for recognized rating categories and evaluative expressions in the review texts. The grammar rules were instantiated with high-frequent n-grams and then generalized. For each of the five categories in Table 2 (assurance, reliability, responsiveness, tangibility and time), a local grammar was developed based on pattern dictionaries.

3.3 Detecting Random Individual Error in Reviews

We define random individual error in the course of this work as divergent polarities for each of the five categories (assurance, reliability, responsiveness, tangibility and time; cf. Table 2) in the qualitative and quantitative part of the same review. For this reason, we consider the disagreement of good ratings per category (grades 1 to 3 and 4 to 5 stars respectively) and negative patient's opinion in the review text as well as the mismatch of bad ratings (grades 4 to 6 and 1 to 3 stars respectively) and positive patient's statements as individual errors.

Figure 2 shows a sample review from the data set introduced in Section 3.1. It is divided in a qualitative and a quantitative part. Furthermore, the overall grade (2.6) is calculated as the average grade of the five mandatory rating categories. The voluntary ratings are not considered for the arithmetic averaging.

This example contains a lot of individual inconsistency occurrences. First, the review is entitled 'Unsatisfied' which indicates a bad rating. While the title

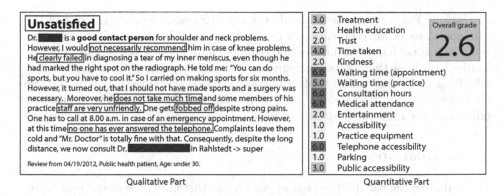

Unsatisfied	3.0 Treatment
Dr. ▓▓▓ is a **good contact person** for shoulder and neck problems. However, I would not necessarily recommend him in case of knee problems. He clearly failed in diagnosing a tear of my inner meniscus, even though he had marked the right spot on the radiograph. He told me: "You can do sports, but you have to cool it." So I carried on making sports for six months. However, it turned out, that I should not have made sports and a surgery was necessary. Moreover, he does not take much time and some members of his practice staff are very unfriendly. One gets fobbed off despite strong pains. One has to call at 8.00 a.m. in case of an emergency appointment. However, at this time no one has ever answered the telephone. Complaints leave them cold and "Mr. Doctor" is totally fine with that. Consequently, despite the long distance, we now consult Dr. ▓▓▓▓▓▓ in Rahlstedt -> super	2.0 Health education

Qualitative Part   Quantitative Part

Fig. 2. Sample review with labeled random individual errors

expresses negative sentiment, the overall grade of 2.6 on a scale that ranges from 1.0 (best) to 6.0 (worst) still implies satisfaction. Moreover, the first sentence of the review text refers to a positive past experience with this physician. It is therefore likely that the scores represent a mixture of several visits to the physician, which is a possible explanation for the following inconsistencies: Opinion phrases such as 'not necessarily recommend him' and 'clearly failed' indicate the disturbance of trust and poor treatment. But the grade 2.0 for the trust in the physician-patient relationship expresses satisfaction and the treatment is graded quite positive (3.0), too.

Referring to our definition of random individual error in reviews, we observed divergent polarities. In the category 'reliability', there is a disagreement of the strongly negative statements 'not necessarily recommend him' combined with the grade 2.0 for 'trust' and 'clearly failed' combined with the grade 3.0 for 'treatment' (cf. Table 2). Even more clearly is the contradiction between the given grade 2.0 for 'kindness' as subcategory of 'assurance' and the obviously negative remark 'very unfriendly'.

Table 2. Empirical probability of individual inconsistencies occurring in the data set

category	subcategories	occurrence of individual inconsistency
assurance	kindness, trust, ...	4.50 %
reliability	health education, ...	11.20 %
responsiveness	waiting time, ...	12.03 %
tangibility	entertainment, ...	8.41 %
time	time taken	2.95 %

In order to provide more detailed information about how often individual inconsistencies appear in the whole data set, we assigned all sentiment scores per review text to their corresponding rating categories (e.g. time taken, health education, kindness) in order to detect divergences. Then, we calculated the arithmetic mean for each category (assurance, reliability, responsiveness, tangibility and time) based on the identified random individual errors per subcategory. Table 2 outlines their relative frequency per rating category in order to show which categories are more (individual) error-prone (i.e. noisy). It is noticeable that the category 'responsiveness' is worst affected (12.03 %) while 'assurance' only contains 4.5 % inconsistent reviews.

4   Evaluation

We evaluated the reliability of our approach and therefore present our evaluation methodology in the next section. In Section 4.2, we provide our evaluation results of the random individual error detection which are then discussed in Section 4.3.

4.1   Evaluation Methodology

The recognition rate of random individual error in reviews highly depends on the local grammar-annotated phrases. For evaluation purposes, the data set introduced in Section 3.1 was split into a training set (66.6 % of the total size) and a test set (33.4 %). We therefor use traditional evaluation measures, i.e. precision, recall and the balanced F-score. While true positives are hits (i.e. correctly classified error-prone reviews), false positives count the number of falsely assumed erroneous reviews. If our pipeline in Figure 1 failed to detect the occurrence of random individual error, we call this a false negative.

$$\text{precision} = \frac{\text{true positive}}{\text{true positive} + \text{false positive}} \tag{1}$$

$$\text{recall} = \frac{\text{true positive}}{\text{true positive} + \text{false negative}} \tag{2}$$

$$F_1 = 2 \cdot \frac{\text{precision} \cdot \text{recall}}{\text{precision} + \text{recall}} \tag{3}$$

4.2   Evaluation Results

We assumed that it is a hard challenge to consider all random individual errors and therefore expected bad results for our first test run. Nevertheless, we could show that it is possible to automatically uncover inconsistencies within a patient review although there is evidence for improvement.

In Table 3, the average precision per category is shown. Here, the precision indicates the recognition rate of the (correctly identified) polarity disagreement of grades and opinion phrases within a single review. While the precision value for 'responsiveness' is 50 %, only 3 % of all assumed inconsistency occurrences in the category 'reliability' were correctly identified. Moreover, all existing occurrences of random individual errors in the test set were found over all categories. Thus, the recall is 100 % per each category.

Table 3. Evaluation results of individual inconsistency analysis

category	precision	recall	F₁-score
assurance	8 %	100 %	15 %
reliability	3 %	100 %	6 %
responsiveness	50 %	100 %	67 %
tangibility	28 %	100 %	44 %
time	14 %	100 %	25 %

4.3   Discussion

The strengths and weaknesses of the proposed approach are analyzed as follows. On the one hand, it was shown that it is possible to find random individual errors by means of local grammar-based analysis of opinion phrases. Local grammars perfectly fit our needs because of their domain-specific character and problem-modularity. On the other hand, the evaluation results show that our preprocessing step in Section 3.2 has the same shortcomings as other pattern-based approaches: The reliability of our inconsistency analysis depends on the predefined patterns for opinion phrases to be recognized. Once a pattern becomes too general within a local grammar, too many phrases will match this pattern (i.e. overgeneration) – even some refer to past visits or other treating physicians named within the same review (see (d) in Figure 2) – and will be assigned to a false category. However, too specific patterns reduce the recall, because we cannot conduct any inconsistency analysis on opinion phrases we did not identify during the preprocessing (i.e. overfitting). Additionally, in user-generated reviews, we have to deal with lots of misspellings. Then, no approximate matching is possible with local grammars and therefore opinion phrases with spelling errors are ignored. Furthermore, some opinion phrase patterns can belong to more than one category of the five we defined as assurance, reliability, responsiveness, tangibility and time. This means that, for example, 'strange practice' can be assigned to 'tangibility' because it is related to the facilities of a practice or to 'assurance' when it describes the patient's overall feeling about the practice and the treatment.

5   Conclusion and Future Work

The increasing amount of patients, that share their personal experiences with the health care system on PRWs, generates a large amount of valuable information that is also error-prone. In this paper, we have tackled two issues that come along with filtering physician reviews by random individual error: (1) Natural language processing of user-generated reviews as well as (2) the identification of disagreement in the polarity of review text and its corresponding numerical ratings.

Since such individual errors are annoying, frustrating and confusing for patients seeking for information on PRWs, we developed an approach to detect these inconsistencies and remove error-prone reviews. The basic idea was first to identify relevant opinion phrases and to determine their polarity. Subsequently, the particular phrase was assigned to its corresponding category (assurance, reliability, responsiveness, tangibility and time) and then aligned to its grade before checking the (dis-)agreement of both (textual and numerical) polarity values. For this purpose, several local grammars for the pattern-based analysis as well as domain-specific dictionaries for the recognition of entities, aspects and polarities were applied on 702,323 physician reviews from the German PRWs jameda.de and docinsider.de. Thus, our approach is special in that we automatically filter review texts that disagree with their ratings in order to reduce noise in user-generated data. Our results show that individual errors exist not only in product but also in physician reviews. These results are useful to solve the rating inference problem. That is, when we know that for the majority of patients, for example, a waiting time of thirty minutes corresponds to the grade 3, then this is a valuable information for the interpretation of user-generated reviews or rating predictions respectively. Furthermore, our research contributes to content quality improvement of PRWs because we provide a technique to detect inconsistent reviews that could be ignored for the computation of average ratings.

However, our approach has to be further improved. First of all, our local grammars have to be extended to cover a greater variety of opinion phrases in future work. Another challenge is the pattern enhancement – while manual refinement is promising but time-consuming, we have to experiment with several machine learning approaches for domain-specific pattern acquisition to avoid overfitting or overgeneration. Moreover, our approach is still monolingual. We intend to adapt it to further languages and will therefor create additional domain-specific dictionaries. A further fascinating task for future research is to resolve wrong assignments of patterns to specific categories. For instance, how to decide that the expression 'strange practice' belongs to the behavior of the staff and not to the practice facilities. Furthermore, we have to develop a strategy how to deal with off-topic opinion phrases referring to other physicians than the one that is on-topic in the review.

References

1. Cambria, E., Hussain, A.: Sentic Computing: Techniques, Tools, and Applications, vol. 2. Springer, Dordrecht (2012)
2. Chen, R.Y., Guo, J.Y., Deng, X.L.: Detecting fake reviews of hype about restaurants by sentiment analysis. In: Chen, Y., Balke, W.-T., Xu, J., Xu, W., Jin, P., Lin, X., Tang, T., Hwang, E. (eds.) WAIM 2014. LNCS, vol. 8597, pp. 22–30. Springer, Heidelberg (2014)
3. Fu, B., Lin, J., Li, L., Faloutsos, C., Hong, J., Sadeh, N.: Why people hate your app: making sense of user feedback in a mobile app store. In: Proceedings of the 19th ACM SIGKDD International Conference on Knowledge Discovery and Data Mining, KDD 2013, pp. 1276–1284. ACM (2013)
4. Geierhos, M.: BiographIE - Klassifikation und Extraktion karrierespezifischer Informationen, Linguistic Resources for Natural Language Processing, vol. 5. Lincom, Munich (2010)

5. Grishman, R.: Information extraction: techniques and challenges. In: Pazienza, M.T. (ed.) SCIE 1997. LNCS, vol. 1299, pp. 10–27. Springer, Heidelberg (1997)
6. Gross, M.: Local grammars. In: Roche, E., Schabes, Y. (eds.) Finite-State Language Processing, pp. 330–354. MIT Press, Cambridge (1997)
7. Hanauer, D.A., Zheng, K., Singer, D.C., Gebremariam, A., Davis, M.M.: Public Awareness, Perception, and Use of Online Physician Rating Sites. JAMA **311**(7), 734–735 (2014)
8. Islam, M.R.: Numeric rating of apps on Google play store by sentiment analysis on user reviews. In: Proceedings of the International Conference on Electrical Engineering and Information & Communication Technology (ICEEICT), pp. 1–4. IEEE (2014)
9. Jang, W., Kim, J., Park, Y.: Why the online customer reviews are inconsistent? textual review vs. scoring review. In: Digital Enterprise Design & Management, p. 151. Springer (2014)
10. Lu, Y., Zhang, L., Xiao, Y., Li, Y.: Simultaneously detecting fake reviews and review spammers using factor graph model. In: Proceedings of the 5th Annual ACM Web Science Conference, pp. 225–233. ACM (2013)
11. Mudambi, S.M., Schuff, D., Zhang, Z.: Why aren't the stars aligned? an analysis of online review content and star ratings. In: Proceedings of the 47th Hawaii International Conference on System Sciences (HICSS), pp. 3139–3147. IEEE (2014)
12. Mukherjee, A., Venkataraman, V., Liu, B., Glance, N.: Fake review detection: Classification and analysis of real and pseudo reviews. Tech. rep., UIC-CS-2013-03, University of Illinois at Chicago (2013)
13. Olsher, D.J.: Full spectrum opinion mining: integrating domain, syntactic and lexical knowledge. In: 12th International Conference on Data Mining Workshops (ICDMW), pp. 693–700. IEEE (2012)
14. Pham, H.X., Jung, J.J.: Preference-based user rating correction process for interactive recommendation systems. Multimedia tools and applications **65**(1), 119–132 (2013)
15. Remus, U., Quasthoff, R., Heyer, G.: SentiWS - a publicly available German-language resource for sentiment analysis. In: Proceedings of the Seventh Conference on International Language Resources and Evaluation (LREC 2010). European Language Resources Association (ELRA), pp. 1168–1171 (2010)
16. Sabin, J.E.: Physician-rating websites. Virtual Mentor **15**(11), 932–936 (2013)
17. Terlutter, R., Bidmon, S., Röttl, J.: Who Uses Physician-Rating Websites? Differences in Sociodemographic Variables, Psychographic Variables, and Health Status of Users and Nonusers of Physician-Rating Websites. Journal of medical Internet research **16**(3) (2014)
18. Wallace, B.C., Paul, M.J., Sarkar, U., Trikalinos, T.A., Dredze, M.: A large-scale quantitative analysis of latent factors and sentiment in online doctor reviews. Journal of the American Medical Informatics Association **21**(6), 1098–1103 (2014)
19. Whitely, S.E.: Individual inconsistency: Implications for test reliability and behavioral predictability. Applied Psychological Measurement **2**(4), 571–579 (1978)

Building Browser Extension to Develop Website Personalization Based on Adaptive Hypermedia System

Hendry[1], Harestu Pramadharma[2], and Rung-Ching Chen[1(✉)]

[1] Chaoyang University of Technology, Taichung, Taiwan
hendry.honk@gmail.com, crching@cyut.edut.tw
[2] Satya Wacana Christian University, Salatiga, Indonesia
harestupramadharma@gmail.com

Abstract. Website is a media that can be used to find information. Every website visitors have different characteristics that makes them interested in the different parts of information. The information is often displayed in the form of a picture or a link, by allowing user to manipulate images or links based on their needs, it will make the website more personalized. This technology is then applied to a browser extension called W-Changer extension. W-Changer extension works by injecting its functions into a website page and allows the user to manipulate images and links in the website page according to their needs. Then, the results of the manipulation can be saved in web storage by applying HTML5 web storage technology. It can be implemented for website which is not use *DOMContentLoaded* event.

Keywords: Adaptive hypermedia system · Web personalized · Browser extensions

1 Introduction

Website is a media that can be used to find information. There are variety of website visitors (user) with different characteristics who accessing the website every time. Those different characteristics make every user interested in the different parts of the information are displayed by the website. The information is often displayed in the form of a picture or a link, by allowing user to manipulate images or links based on their needs, it will make the website more personalized [1,2].

Adaptive Hypermedia System is a system developed to increase the functionality of hypermedia by making it personalized [3]. This system will be very useful when accessed by user who may be interested in different pieces of information presented by the system.

The implementation of Adaptive Hypermedia System in developing website personalization is one way to overcome the problem mentioned before. Website personalization is one way to adjust the information provided in the web to the needs and interests of the user [4]. Website personalization is divided into four models, which are memorization, customization, guidance or recommender system, and task performance support.

© Springer International Publishing Switzerland 2015
M. Ali et al. (Eds.): IEA/AIE 2015, LNAI 9101, pp. 316–325, 2015.
DOI: 10.1007/978-3-319-19066-2_31

At last, the system and the technology that have been explained previously will be implemented in a browser extension. Browser extension is small size software which can modify and enhance the functionality of a browser [5]. W-Changer is a browser extension that applies Adaptive Hypermedia System to develop website personalization, which allows the user to manipulate both, images and links in the website according to their needs. This research focus is implementation of the technology.

2   Background

In the development of W-Changer extension, some technologies, such as Adaptive Hypermedia System, Website Personalization, Browser Extension, HTML 5 Web Storage, and Javascript Regular Expression technology, are used. Those five technologies are then integrated in such a way that it can be used to make W-Changer extension usable for user to manipulate images and links on websites and save every manipulation process in a database.

2.1   Adaptive Hypermedia System

Adaptive Hypermedia System is a system which has been developed for a long time. This system is created to increase the functionality of hypermedia by making it personalized. In Adaptive Hypermedia System, there are some types of adaptation technology.

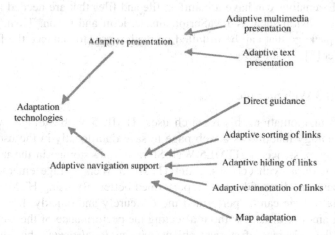

Fig. 1. Types of Adaptation Technologies [3]

In Figure 1, types of adaptation technology are explained in a diagram. There are two main types of Adaptation technology, which are Adaptive Presentation and Adaptive Navigation Support – each of them are divided into sub-categories. Adaptive Presentation is a kind of Adaptation Technology which is used to adjust the website's content accessed by the user [3]. The sub-category from Adaptive Presentation which is used in this research is Adaptive Multimedia Presentation that can display adapted multimedia information to the user. The second main type of Adaptation Technology is Adaptive Navigation Support. This type is used to help user found their needs in hyperspace by providing links which are suitable for their needs. The sub-category of Adaptive Navigation Support which is applied in this research is Adaptive Hiding of Links, which can be used to hide links that are irrelevant to the user's needs.

2.2   Website Personalization

Website personalization is a kind of technology which provides information that is adjusted to the web user's interest, role, and needs. Website personalization has four kinds of models, which are memorization, customization, guidance or recommender system, and task performance support model. Personalization model that will be applied in this research is the memorization one.

Memorization is one personalization model used to memorize or keep information from the website user [6]. Information is saved by cookies or session in the web server; in this case it is saved by HTML5 web storage.

2.3   Browser Extension

Browser extension is software that enables developer to add functionality of browser. User interface of browser extension is arranged such not directly that relate to the look of content a website. The addition of browser extension can be done in an explorer bars and add entries into context menu or tools also on the menu.

Browser Extensions can have a manifest file and files that are needed for the functionality of extension such as JavaScript, image, icon and so on. Then, all files are made into a package that can be installed on the browser to enhance the functionality of the browser [3].

2.4   HTML5 Web Storage

To save data manipulation, this research uses HTML 5 Web Storage, which is one kind of technology that allows a web page to save data locally in the user's browser [7]. Before the existence of HTML5 Web Storage, data storage in the user' browser (client side) is done with cookies. After HTML 5 Web Storage emerged, the data storage process on client side can be performed better. By using HTML5 Web Storage, the data storage can be performed more securely and quickly. It also works for storing a big amount of data without affecting the performance of the website. In addition, with the existence of storage object, *window.localstorage* that allows storing data without expire date (data is not lost when the tab is closed), HTML5 web storage can really outperform cookies.

2.5   Javascript Regular Expression

This research applies Javascript Regular Expression technology to insert its functions into the program codes of a website page. Javascript Regular Expression is an object that can perform a text searching by using Javascript pattern. Javascript Regular

Expression has several matching pattern categories, which are position matching, special literal character matching, character classes matching, repetition matching, alternation, and grouping matching and back reference matching [8].

3   Implementation

This part will discuss about the implementation of the technologies that will be used in the making of W-Changer extension. Those technologies are Javascript Regular Expression, HTML5 web storage, and website personalization.

3.1   Implementation of Regular Expression

Basically, W-Changer extension works by injecting its functions into the programming codes of a website page. Those functions are made "on the fly" through the execution of the programming codes in W-Changer content script. One of the processes of injecting function is the injection of the *onMouseOver* function into the tag *"img"* in the programming codes of a website. The general picture of the injection process of *onMouseOver* function can be seen in Figure 2.

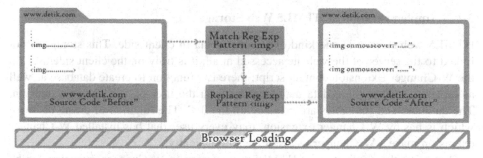

Fig. 2. Injection Process of *Onmouseover* Function

As an example – function injection to the programming codes of a website. As it can be seen in Figure 3, the first step to be done is by reading the programming codes from www.detik.com site and pattern matching tag *img* using Javascript Regular Expression. After finding the text that matches the pattern in the programming codes, for every text that is matched, it is replaced by using Javascript Regular Expression. The replacement of the text is done to exchange the previous tag img with the new tag img that has been injected with onmouseover function from W-Changer extension. By performing match and replace with Javascript Regular Expression, the programming codes in www.detik.com site for every tag img have got the onmouseover function from W-Changer. The whole process is done while the browser process is loading. The algorithm to inject our script in to the web page is:

```
Program Injection Page (new Tag)
Begin
  var dok = document.body.innerHTML;
  var reg1 = /\<img .+?\>/ig;
  var getimg = dok.match(reg1);
  if(getimg){
    for (every m in getimg){
      repp = getimg[m];
    repp = repp.replace("<img","<img onmouseov-
er=\"mover(\'"+getId+"\');\"   ..... ");
    document.body.innerHTML = docu-
ment.body.innerHTML.replace(getimg[m],repp);
End.
```

For every tag in the webpage that contains tag *img* would be collected by the regular expression and be saved in the array getting. The next process is to replace are the tag *img* with the new tag *img* with *onMouseOver* event, which is our function to the web page. We also get the ID for every tag that we modify and give default ID for some of the tag that does not have it.

3.2   Implementation of HTML5 Web Storage

HTML5 web storage is one kind of storage media in client side. This storage is not linked to the server of the website accessed at all; it is truly on the client side/user. In the W-Changer Extension content script; there is a function to create database as well as tables needed to store data manipulation from the user of W-Changer extension. The database and the tables are created "on the fly". The function of create database which is has by W-Changer extension, may every user that has installed W-Changer extension make a W-Changer web storage automatically.

Generally, the application of HTML5 web storage in W-Changer extension can be seen in Figure 3. It shows a web browser that has been installed W-Changer extension. As it was mentioned before, when the installation process of W-Changer is done then the making of W-Changers web storage will also be performed automatically because the execution of the function in W-Changer content script. W-Changer web storage is then created in the client side storage. This web storage cannot be seen in form of database like cookies for the sake of the web storage's security.

The initialize web storage in client web browser as shown in the program below:

```
Program Initialize Web Storage (DB)
Begin
  var initDatabaseImg = document.createElement('script');
  initDatabaseImg.setAttribute('id', 'initDatabaseImg');
  initDatabaseImg.innerHTML = 'function initDatabaseImg()
    if (!window.openDatabase)
```

```
      alert(Local Databases are not supported by your
browser. Please use a Webkit browser for this demo);
   else
      shortName = WCDB;
      version = 1.0;
      displayName = WCDB Image;
      maxSize = 100000;
      WCDB = openDatabase(shortName, version, display-
Name, maxSize);
   End if.
   document.head.appendChild(initDatabaseImg);
End.
```

This will assign HTML5 web storage in our client side browser, so when the application or w-changer need to save the personalized web setting of the user, it doesn't need to contact the server side. All of the personalized and note that user did in their browser will be save permanently here. Every user is unique according to their activity.

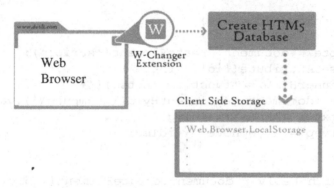

Fig. 3. The Application of HTML5 Web Storage Technology

3.3   Implementation of Web Personalization

In this part, one of the examples of personalization process that can be done by W-Changer extension will be discussed. The personalization process in this case is the process of adding note to the image in a website.

The general picture of adding note to an image can be seen in Figure 4, which depicts that personalization happens when a website page with image contents is given an action by W-Changer extension. When onmouseover is performed, then W-Changer extension icons will appear in the bottom of the image. After that, choose the icon to add note to the image.

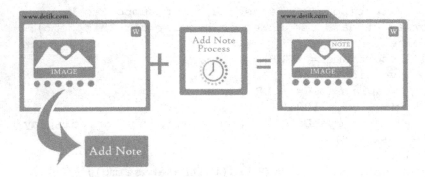

Fig. 4. Personalization of Adding Note to Image

After the process of adding note is carried out, an image with the additional note on the top right side of the image will appear as the result of the personalization. That note can be hidden, saved, and deleted, as previously designed. The program to create a note in the w-changer as shown below:

```
Program addNote (divnote)
Begin
  var fnote = document.createElement('script');
  fnote.setAttribute('id', 'fnote');
  fnote.innerHTML = 'function fnote(){'+
  'idImg = document.getElementById(\'tampung\').value;'
  var nmtxt = \'note\'\+idImg;'+
  var txtyop = \'yopnote\'\+idImg;'+
  .
  .
      'var noteDiv = document.createElement(\'div\');'+
    </div>
    <div class="notes">
      <textarea id=\"\'\+txtyop\+\'\" class=\"txtArea\"
cols=\"20\" rows=\"3\">
      </textarea>
    </div>\';'+
  'document.body.appendChild(noteDiv);'
  }
  document.head.appendChild(fnote);
End.
```

This will add a note to be displayed in the image, and then we could note some data in the text area. The note will be save in the HTML5 web storage.

4   Results

This part will discuss on the implementation results of W-Changer extension towards 30 samples of websites. The all thirty samples of the websites were chosen randomly. The implementation process was done by trying out the functions of W-Changer extension on images and links in a website. The results of the trials can be seen in Table 1.

Table 1. Testing Results From 30 Website Samples

FUNCTIONS	FUNCTION DESCRIPTION	SUCCESFUL WEBSITE	FAILED WEBSITE
Show W-Changer Image Icon	To show W-Changer icon for context image	28	2
Rotate and Flip Image	To rotate and flip *image*	28	2
Image Filter	To apply ten filters that W-Changer extension has on images	28	2
Hide Image	To hide images	28	2
Add Note	To add notes on an image	28	2
Manipulate Note	To hide, save, and remove notes	28	2
Show Hidden List	To show image lists and links hidden	28	2
Close W-Changer Image Icon	To hide W-Changer extension icon	28	2
Show W-Changer Link Icon	To show W-Changer icon on context link	28	2
Manipulate Link	To disable, enable, hide, and block links	28	2
Show Hidden List	To show image lists and links hidden	28	2
Close W-Changer Link icon	To hide W-Changer extension icon	28	2
Unhide Image dan Link	To unhide either images or links hidden	28	2

Based on Table 1, there are 28 websites[1] succeeded in getting the injections and implemented functions from W-Changer extension, and two websites which were failed. After the trial was done, it can be seen that the websites that can be injected by W-Changer extension are the websites using Load event, while the websites that cannot be injected are the websites using *DOMContentLoaded* event, such as

---

[1]   http://www.detik.com, http://www.cyut.edu.tw, etc

Google[2] and Facebookx[3]. The basic difference from these two events are that *DOM-ContentLoaded* event is triggered when the parsing of Document Object Model from a website page has been done, while Load event is triggered when all the files in the page, such as resource, image, etc. has finished loading. The injection process done by W-Changer extension occurs during the loading process of a website page. In Load event, the injection process succeeds because when the website page is loading, all the resource and image are also loading, so that it can be injected functions from W-Changer extension. It is different from *DOMContentLoaded* event which only does the parsing of Document Object Model.

Fig. 5. Dropdown Icons of Edit Icon

Fig. 6. Implementation effect blur of w-changer

in this section will also discuss the result of W-changer extension's design that has been done. W-changer extension has four different user interfaces, the first user interface for context image. User interface consists of six main icon, namely edit icon, filter icon, hide icon to conceal image, add note icon to add note on image, hidden list icon to display a list of image which has been concealed, and also close icon. Edit icon has four dropdown icons, rotate clockwise 90o icon, rotate counter clockwise 90o icon, flip horizontally icon and flip vertically icon as seen in Figure 5.

Those fourth icons are used to change image's orientation. Filter icon has ten dropdown icons used to give a filter / effect for an image that is highlighted. The tenth

---

[2]　http://www.google.com
[3]　http://www.facebook.com

filter were grayscale, sepia, blur, invert, brightness, hue-rotate, contrast, opacity, saturate and default. The snapshot of the w-changer implementation in the website could be seen in Figure 6.

5   Conclusions

Based on the research, it can be concluded that: 1) The browser extension built can only be applied to websites using load event; 2) W-Changer extension allows the user to manipulate images and links in a website and make it personalized; 3) The application of HTML5 web storage keep the manipulation that has been done by using W-Changer in the storage.

Suggestions to develop this application are: 1) The application of the implementation not only can be done for websites using load event, but also DOMContentLoaded event; 2) The functions to manipulate images and links can be further developed according to the user's needs.

Acknowledgement. This research was supported by the Ministry of Science and Technology, Taiwan, R.O.C., under contract number MOST103-2632-E-324-001-MY3 and MOST 103-2221-E-324 -028.

References

1. Tarmo, R., Ahto, K.: Learning from users for a better and personalized web experience. In: Proceedings of PICMET 2012. IEEE, Vancover (2012)
2. Zhongyun, Y., Zhurong, Z., Fengjiao, H., Guofeng, Z.: Research on personalized web page recommendation algorithm based on user context and collaborative filtering. In: 4th IEEE International Conference on Software Engineering and Service Science (ICSESS), Beijing (2013)
3. Brusilovsky, P.: Methods and techniques of adaptive hypermedia. In: Brusilovsky, P., Kobsa, A., Vassileva, J. (eds.), pp. 1–44. Springer Science+Business Media B.V., Springer Netherlands (1998)
4. Rakesh, K., Aditi, S.: Personalized web search using browsing history and domain knowledge. In: IEEE International Conference on Issues and Challenges in Intelligent Computing Techniques (ICICT), Ghaziabad (2014)
5. Google Documentation. http://developer.chrome.com/extensions/overview
6. Nasraoui, O.: World Wide Web Personalization. United States of America: Department of Computer Engineering and Computer Science. University of Louisville (2005)
7. W3School HTML5 Web Storage. http://www.w3schools.com/html/html5_webstorage.asp
8. Google, Introductory Guide to Regular Expression. http://www.javascriptkit.com/javatutors/re.shtml

Discovering Potential Victims Within Enterprise Network via Link Analysis Method

Yu-Ting Chiu[✉], Shun-Te Liu, Hsiu-Chuan Huang, and Kai-Fung Hong

Information and Communication Security Laboratory,
Chunghwa Telecom Laboratories, Taoyuan City 32661, Taiwan
{gloriac,rogerliu,pattyh,kfhong}@cht.com.tw

Abstract. Potential cyber victim detection is an important research issue in the domain of network security. During an adjacent period of time, cyber victims or even potential cyber victims within an enterprise have several common patterns to the currently seized victims. Hence, this paper applies the link analysis method and proposes a hybrid method to automatically discover potential victims through their behavioral patterns hidden in the network log data. In the experiment, the proposed method has been applied to reveal potential victims from a big data (6,846,097 records of proxy logs in 1.7G and 84,693,445 records of firewall logs in 9.3G). Afterward, a ranking list of potential victims can consequently be generated for stakeholders to understand the safety condition within an enterprise. Moreover, the hierarchical connection graph of hosts can further assist managers or stakeholders to find out the potential victims more easily. As a result, the safety and prevention practice of the information security group in an enterprise would be upgraded to an active mode rather than passive mode.

Keywords: Link analysis · Social network analysis · Network threats · Potential victim detection

1 Introduction

Potential cyber victim detection is an important research issue in the domain of network security. This paper proposes a new method to discover and detect potential cyber victims compromised by attackers, especially in a closed network within an enterprise. A potential victim indicates a machine has been hacked but has not yet been found, or a machine has not been hacked in the present but will possibly be infected or intruded in the near future.

Link analysis method has been applied in a wide range of domains. Via an analogy between the digital forensics domain with the criminal justice domain, link analysis in digital forensics is regards as the identification, analysis, and visualization of connections and relations between every pair of cyber victims. One of the major objectives in digital forensics is to maintain the network security by seizing all the cyber criminals.

In order to seize criminals in a criminal case like homicide, crime investigators need to read all the relevant documents to get clues and associations among the crime entities (people, locations, vehicles) [1]. Homicide cases, for example, usually consist

© Springer International Publishing Switzerland 2015
M. Ali et al. (Eds.): IEA/AIE 2015, LNAI 9101, pp. 326–335, 2015.
DOI: 10.1007/978-3-319-19066-2_32

of a relationship between victims and perpetrators. Perpetrators can mostly be found through the clues in victim's social network. Similarly, forensic experts need to investigate all the files and logs in a victim computer to seize the attacking source, virus, or malware. In other words, a cyber-criminal or an attacker can be found through the clues he/she left in the victim's computer and its corresponding network. Furthermore, it can also infer that the other potential victims harmed by the same attacking group or the similar attacking approaches can be discovered through the proved victim.

According to our observation, cyber victims within an enterprise (during an adjacent period of time) sometimes have several common patterns. Even the potential cyber victims, which are the hosts might be compromised in the near future, have common patterns to the current victims. Hence, investigating the characteristics and behaviors of hosts would be a great and critical assistant to sustain the discovery of potential victims. However, recognizing behavioral patterns of computers is a time consuming task with the crucial requirement of domain knowledge in information security and the expertise in digital forensics. Besides, it is almost impossible to consistently collect users' networking behaviors directly from their local machines. Network log data, as a substitution, is another way to remotely understand and reveal network activities and behaviors of users in a quantitative and rapid manner. Thus, this paper intends to propose a hybrid method to automatically discover potential victims through their behavioral patterns hidden in the log data.

2   Related Work

This section will be divided into four parts to review relevant researches: network threats, network log data, data standardization, and link analysis.

2.1   Network Threats

Modern business nowadays contains a lot of networking tasks, such as checking your emails, booking an e-order, and googling information. Accordingly, the network traffic is usually in a huge volume and would imply valuable and confidential information of a company. Advanced Persistent Threat (APT) is a type of network threat hiding itself under modern business model. Usually, individuals within a company would not notice the threat since it stealthily and silently performs some harmful activities on an individual computer and even in an enterprise network [2]. Command and control (C&C) site is a control center for infected computers to report its status. Botnets, which contain infected computers (bots) and C&C, are one of the most major threats, and have evolved from a centralized model towards a decentralized, highly scalable architecture [3, 4, 5].

2.2   Network Log Data

Since valuable assets nowadays are mostly stored as digital files, the network becomes the main channel for attackers to access these assets in an organization illegally [6]. In order to protect and secure the valuable assets, it is necessary to monitor network traffic and detect the attacks from the network. Hence, the network log data is collected and analyzed to detect the threats or even attacks within a close network so as to recognize network threats. An enterprise network is a kind of close network.

2.3 Data Standardization

While analyzing data, the analysis result would usually encounter the problem of data bias. Take the order prediction of a company canteen as an example, if the features for analysis are: total number of employees, average height of employees, and average weight of employees. Since total number of employees may be more than 1000, height values are mostly in the range of 150 to 200 cm and weight values are mostly in the range of 40 to 100 kg, using these values directly without preprocessing might raise a problem of data bias. It is important to take into consideration of data measurement scale [7]. In order to avoid the problem of data bias, there is a statistical method to represent scores extracted from different sources or features by a unified basis. The statistical method, data standardization, calculates a z-score for each original value through its variance extent in the standard normal distribution of its belonging data set. The distribution of z-score is the standard normal distribution with a mean of zero and a variance of one, as in the following equation [8].

$$f(x) = \left(\frac{1}{\sigma\sqrt{2\pi}} \right) e^{-(x-\mu)^2/(2\sigma^2)} \tag{1}$$

The z-score will be used in the preprocessing procedure of this research for the data cleaning and the elimination of bias effect.

2.4 Link Analysis

Link analysis methods provide a tool and a visualized presentation to discover patterns of interest from a set of data, and the matched as well as the violated objects according to the pattern [9, 10]. Conceptually, discovering new patterns of interest from data using link analysis methods is similar to that using social network analysis method in data mining. Link analysis is a set of techniques to operate data and represent them as nodes and links [11, 12]. The function of linkage degree between two nodes in social network analysis is stated as: $a(P_i, P_k)$, where 1 denotes an existed link between P_i and P_k (0 otherwise) [13].

Link analysis has been widely applied in the field of social network analysis, search engines, viral marketing, law enforcement, and fraud detection [10, 14]. Chen et al. [15] adopted the concept of link analysis to design a new method to detect relationships between two individual in social network, and to additionally discover and differentiate different relation types among people. Link analysis method is also used in the field of crime recognition to facilitate crime investigations by searching associations between crime entities and large datasets [1, 16].

3 Research Design

This research is intended to design a hybrid method to discover potential cyber victims within an enterprise network. The proposed method is composed of feature standardization as well as potential victim discovery and will be applied to monitor the data of proxy logs and firewall logs. The following subsections describe the features of proxy log and firewall log, the operation of standardization, and the steps of discovering potential victims.

3.1 Features of Proxy Log and Firewall Log

In the proposed method, there are two types of features to be extracted from log data: one is for a single node (denoted as "s") and the other is for a pair of nodes or dual nodes (denoted as "d"). The features in type "s" are indicated as $F_s=\{f_1, f_2, \ldots, f_8\}$, and the features in type "d" are indicated as $F_d=\{f_9, f_{10}\}$. The following table states the ten features with their corresponding descriptions.

Table 1. Types and descriptions of features.

no.	Type	Log	Name	Description
1	s	proxy	f_1	Avg. # of ports for a IP connecting to URLs
2	s	proxy	f_2	Avg. connections for a IP connecting to URLs
3	s	proxy	f_3	Avg. # of paths for a IP connecting to URLs
4	s	proxy	f_4	Avg. denied connections for a IP connecting to URLs
5	s	proxy	f_5	Avg. # of denied URLs for a IP
6	s	firewall	f_6	Avg. # of ports for a IP connecting to destinations
7	s	firewall	f_7	Avg. connections for a IP connecting to destinations
8	s	firewall	f_8	Avg. # of destinations via a specific destination port for a IP
9	d	n/a	f_9	Score for a pair of IPs whether it belongs to the same IP range
10	d	proxy	f_{10}	Co-occurrence in browsing activities (connecting to the same URL) for a pair of IPs

3.2 Operation of Standardization

In order to ensure the research design is appropriate for the type of data in this research, a data observation process has been executed before the design of our method. The data set we planned to use has a huge value span and is necessary to be solved or at least mitigated. Thus, a natural-logarithm transformation is conducted as the first stage in the operation of standardization. Additionally to avoid the problem of data bias, a z-score transformation (as mentioned in the Subsection 2.3) is conducted for each of the ten features as the second stage in the operation of standardization. The calculating steps are as the following equations, which the jth value in feature i is denoted as f_{ij}, and the average value for the value set of feature i is denoted as $\overline{f_i}$. S_i denotes the standard deviation of feature f_i.

$$\overline{f_i} = (1/m) \cdot \left(\sum\nolimits_{j=1}^{m} f_{ij} \right) \tag{2}$$

$$S_i = (1/(m-1)) \cdot \left(\sqrt{\sum\nolimits_{j=1}^{m} \left(f_{ij} - \overline{f_i} \right)^2} \right) \tag{3}$$

$$z_{ij} = \left(f_{ij} - \overline{f_i} \right) / S_i \tag{4}$$

Since there are two types of features in our hybrid model, each type of features will firstly be computed separately to obtain their original values. Then, all the features are presented as z-scores in the range of -4 to 4 after the two stages of standardization. In order to mitigate the bias occur from small numerical variance, the exact z-scores z_{ij} are transferred into discrete values and store in the set $Dis_{s,i}$. The original z-scores lower than -1 standard deviation are transformed to "Low", the original z-scores in the range of -1 and 1 are transformed to "Neutral", and the original z-scores higher than 1 are transformed to "High".

3.3   Potential Victim Discovery

For features in type "s", a host can be presented as h_a with its own discrete value $Dis_{s,i}(h_a)$, which i is the ith feature in type "s". Since the main objective of this research is to capture and recognize the relationships among host computers. It is essential to have a computation process to obtain correlation scores for paired computers from the single score of each computer. The correlation score $Cor_{s,i}(h_a, h_b)$ denotes the likeliness or linkage degree of the ith single feature in the set F_s between host h_a and host h_b as the following:

$$Cor_{s,i}(h_a, h_b) = \begin{cases} 1 \text{ if } Dis_{i,a} \text{ of } h_a \text{ is the same as } Dis_{i,b} \text{ of } h_b \\ 0 \text{ otherwise} \end{cases} \tag{5}$$

The summarized correlation score $Cor_s(h_a, h_b)$ is the accumulated score for host h_a and host h_b of the ten features F_s.

$$Cor_s(h_a, h_b) = \frac{\sum_{i=1}^{|F_s|} Cor_{s,i}(h_a, h_b)}{|F_s|} \tag{6}$$

For features in type "d", a pair of hosts h_a and h_b would have their own linkage scores $SC_{d,i}(h_a, h_b)$ for the ith dual features in F_d. A summarized score $Cor_d(h_a, h_b)$ for a pair of hosts linking via all dual features is computed as:

$$Cor_d(h_a, h_b) = \frac{\sum_{i=1}^{|F_d|} SC_{d,i}(h_a, h_b)}{|F_d|} \tag{7}$$

Eventually, the final score $FS(h_a, h_b)$ of these two types of scores for a pair of computers h_a and h_b would be merged together via the product computation to estimate similarity and can be used to predict the probability of being a victim.

$$FS(h_a, h_b) = Cor_s(h_a, h_b) \times Cor_d(h_a, h_b) \tag{8}$$

The final score FS would then be used to draw a connection graph. In order to discover the potential victims, an accumulated score for each host is needed. The total score $TS(h_a)$ of a host h_a can be calculated as follows.

$$TS(h_a) = \sum_{\forall i; not\ a} FS(h_a, h_i) \tag{9}$$

Using the accumulated score $TS(h_a)$, a ranking list of potential victims with corresponding risk degree can be derived.

4   Experiment and Explanations

In this section, an experiment has been carried out to reveal the idea of the proposed method using enterprise log data in the real world. The following subsections will describe the details of data used in the experiment, and the results of the experiment.

4.1   Data for the Experiment

In order to reveal the research design and to check its performance, we collected proxy logs as well as firewall logs during office hours. We randomly pinpointed to a certain date to gather those log data during an hour. After data collection, we have gathered 1.7G of proxy logs with 6,846,097 records and 9.3G of firewall logs with 84,693,445 records. There are totally 21,844 unique hosts in the collected proxy log. A host mentioned in this research is a machine, either a computer or a server, within the enterprise network. In other words, it would be demonstrated by an inner IP, like 10.1.1.1. Each one record in the proxy log for a specific host (i.e., IP) is calculated as a connection count for that IP. Using the factual data from an enterprise can be viewed as a special characteristic and one of the contributions of this research.

In order to present the visualization graph more friendly and easy-to-read, only part of data is retained and demonstrated in the graph. The selected data included in the graph are the hosts with the largest amount of connections comparing with other hosts. Each record in the collected proxy log is calculated as one connection for the specific host recorded in that record. The top one host of the collected data contains 130,254 records (i.e., nearly 2% of the whole gathered proxy log), and links to 40 different machines in an hour.

4.2   Experimental Results

The visualization graph created by the proposed hybrid method intends to demonstrate the relationships among hosts. The hosts can be further divided into two types, which are: compromised computers and the most active computers. In the experiment, there are three randomly-picked known victims regarded as compromised computers, and top 20 hosts with most records within the gathered dataset regarded as most active computers. The relationships in this visualized graph would be used to estimate and predict the potentiality and probability of a host which may become the next victim. The reason for demonstrating only top k hosts in the visualization graph can be explained as that the more data points retained in a diagram may generate the more noises in the diagram. Otherwise the core part or the main concern of recognizing and predicting potential victims may be defocused. Thus, only top k active computers and a few compromised computers are displayed in the graph so as to keep the network clear, succinct, and easy-to-read.

The following diagram is the visualization graph produced via the proposed method in the experiment.

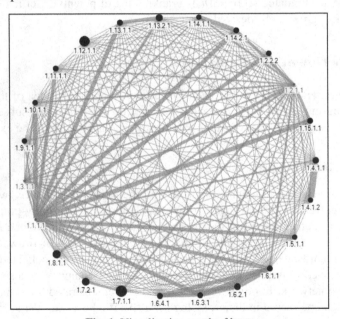

Fig. 1. Visualization graph of hosts

A node in the diagram denotes a host, and its size represents the amount of records belonging to that host. The bigger the node, the more active it is. A link in the diagram denotes a relation between a pair of hosts, and its thickness represents the degree of connectivity between the two specific nodes. The thickness of each link can also be viewed as the similarity degree of the two nodes on each side of that specific link. The thicker the link, the more similar the pair is. Furthermore, the compromised computers are denoted as red points and the most active computers are denoted as black points respectively. All the computer IP addresses in this diagram are scrubbed for the confidential reason.

However, the visualization graph (Fig. 1) is not clear enough to find out a potential victim with higher possibility. It is assumed that a link with higher *FS* score implies stronger and more robust relevance for the two nodes on each side of the link. On the contrary, a link with lower *FS* score implies a weaker relevance and possibly more noises. There are two stages to discover the most potential victims: the first one is to prune the weaker links of a host, and the second is to calculate the risk score of a host (i.e., $TS(h_a)$).

Pruning the weaker linkage can not only eliminate the noise within the network but also emphasize the core part and the main concern of the network. Hence, we refined the visualization graph and produce a hierarchical connection graph to improve the result of visualization. Meanwhile, a ranking list of potential victims can be generated via accumulating the score of linkages (i.e., $FS(h_a, h_b)$) to obtain the total score (i.e., $TS(h_a)$). In the ranking list, the *TS* score of top one host (1.10.1.1) is 1.043, of second one (1.6.1.1) is 0.962, of third one (1.14.2.1) is 0.933, and so on.

In the experiment, the original visualization graph is then reconstructed with a proper threshold value set to 0.0375 (the maximum of *FS* score is 0.375 and the minimum is 0), and the reconstructed diagram is named as a hierarchical connection graph shown in Fig. 2.

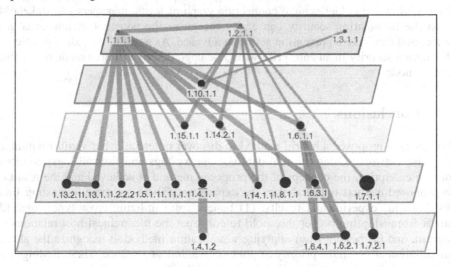

Fig. 2. Hierarchical connection graph of hosts

According to the above Fig. 2, the hierarchy consists of five levels. The top level in red color in the hierarchy is the known-victim layer. Since there are three compromised computers in the gathered dataset for the experiment, the top level contains the three red nodes of known victims.

The second level in orange color in the hierarchy is the most-potential-victim layer. The node in this layer is the most closely related to the known victims. A node in the most-potential-victim layer directly and completely links to the victim nodes. The nodes in this layer indicate that they are quite similar to known victims in some ways and would be the next host to be compromised. Take the host 1.10.1.1 as an example, it holds three linkages to the known-victim layer, and is possibly to be the next victim.

The third level in yellow color in the hierarchy is the high-potential-victim layer. The node in this layer is closely related to the known victim. A node contains over than 60% of linkages that directly links to the nodes in known-victim layer. It is assumed that nodes in this layer are secondly prioritized to be dealt with.

The fourth level in blue color in the hierarchy is the middle-potential-victim layer. The node in this layer is a bit relevant to the victim node. In this layer, a node contains at least 30% of linkages that directly links to the nodes in known-victim layer. The linkage to the nodes from a layer other than the known-victim layer doesn't count since it is not strong enough to determine the danger potentiality of a node while the node is indirectly link to the known victims. The host 1.14.1.1, for example, is located in the middle-potential-victim layer because only one of the two linkages for this host goes to the known-victim layer.

The bottom level in green color in the hierarchy is the low-potential-victim layer. The hosts in this layer have no direct connection to the victim machines in the top layer. Comparing to other hosts, the hosts in this layer are relatively in a safer condition.

Therefore, the most potential victim within the enterprise network in the near future can be discovered and decided through this proposed hierarchical connection graph. Utilizing this hierarchical connection graph of hosts, managers and stakeholders to the information security can easily recognize the potential victim hosts at a glance, and then take a prevention action in advance. As a result, the safety practice of information security in an enterprise would be upgraded to an active mode rather than passive mode.

5   Conclusions

This research proposed a hybrid method to discover potential cyber victims within an enterprise network. Using the real data (i.e., proxy logs and firewall logs) collected from an enterprise, the concept of the proposed method is achieved and the result of the proposed method is revealed in the experiment. There are three interesting findings from the experimental results: (1) benefit from utilizing cross types logs, (2) benefit from adopting proper threshold to construct the hierarchical host relationship diagram, and (3) benefit from applying visualization method to recognize the abnormal situation or grasp the priority of potential victims at a glance. These findings can also be viewed as the contributions of this research.

Since we utilized more than one type of network logs, the equation of calculating victim probability becomes a more comprehensive measure to predict the next potential victim. A pruning procedure with a proper threshold improves the overall result of discovering potential cyber victims. Based on the proposed method, every host will have a probability value for considering its potentiality of being a victim host. However, applying visualization method to produce the hierarchical connection graph of hosts provides a more convenient way for overseeing risky parts in the enterprise network. The research limitation is the lack of real data set with victim or non-victim class labels for evaluation. In the future, a possible direction is to design a new accumulated function or a linkage score function to improve the performance of prediction.

References

1. Schroeder, J., Xu, J., Chen, H., Chau, M.: Automated Criminal Link Analysis Based on Domain Knowledge. Journal of the American Society for Information Science and Technology **58**(6), 842–855 (2007)
2. Daly, M.K.: The advanced persistent threat. In: USENIX (ed.) 23rd Large Installation System Administration Conference. USENIX, Baltimore (2009)
3. Gu, G., Porras, P., Yegneswaran, V., Fong, M., Lee, W.: BotHunter: detecting malware infection through ids-driven dialog correlation. In: Proc. 16th USENIX Security Symposium (2007)
4. Gu, G., Perdisci, R., Zhang, J., Lee, W.: Botminer: clustering analysis of network traffic for protocol- and structure independent botnet detection. In: Proc. 17th USENIX Security Symposium (2008)
5. Gu, G., Zhang, J., Lee, W.: BotSniffer: detecting botnet command and control channels in network traffic. In: Proc. 15th Annual Network and Distributed System Security Symposium (2008)
6. Security laboratory. The 6 categories of critical log information (2013), version 3.0.1
7. Tan, P.N., Steinbach, M., Kumar, V.: Introduction to data mining. Addison-Wesley (2006)
8. Mendenhall, W., Beaver, R.J., Beaver, B.M.: Introduction to probability and statistics. Cengage Learning (2013)
9. Link analysis. http://cn.wikipedia.org/wiki/Link_analysis
10. Donoho, S.: Link analysis. In: Maimon, O., Rokach, L. (eds.) Data Mining and Knowledge Discovery Handbook, 2nd edn., pp. 355–368. Springer, Heidelberg (2010)
11. Wu, I.C., Wu, C.Y.: Using internal link and social network analysis to support searches in Wikipedia–A model and its evaluation. Journal of Information Science **37**(2), 189–207 (2011)
12. Chiu, T.F.: A proposed IPC-based clustering method for exploiting expert knowledge and its application to strategic planning. Journal of Information Science **40**(1), 50–66 (2014)
13. Freeman, L.C.: Centrality in social networks: Conceptual clarification. Social Networks **1**, 215–239 (1979)
14. Kim, G., Faloutsos, C., Hebert, M.: Unsupervised modeling of object categories using link analysis techniques. In: CVPR (2008)
15. Chen, Y.L., Chuang, C.H., Chiu, Y.T.: Community detection based on social interactions in a social network. Journal of the Association for Information Science and Technology **65**(3), 539–550 (2014)
16. Si, Y.W., Cheong, S.H., Fong, S., Biuk-Aghai, R.P., Cheong, T.M.: A layered approach to link analysis and visualization of event data. In: Seventh International Conference on Digital Information Management (ICDIM2012), pp. 181–185. IEEE Press (2012)

A Simple and Efficient Algorithm for Lexicon Generation Inspired by Structural Balance Theory

Anis Yazidi[✉], Aleksander Bai, Hugo Hammer, and Paal Engelstad

Institute of Information Technology, Oslo and Akershus University
College of Applied Sciences, Oslo, Norway
anis.yazidi@hioa.no

Abstract. Sentiment lexicon generation is a major task in the field of Sentiment Analysis. In contrast to the bulk of research that has focused almost exclusively on Label Propagation as primary tool for lexicon generation, we introduce a simple, yet efficient algorithm for lexicon generation that is inspired by Structural Balance Theory. Our algorithm is shown to outperform the classical Label Propagation algorithm.

A major drawback of Label Propagation resides in the fact that words which are situated many hops away from the seed words tend to get low sentiment values since the inaccuracy in the synonym-relationship is not taken properly into account. In fact, a label of a word is simply the average of it is neighbours. To circumvent this problem, we propose a novel algorithm that supports better transitive sentiment polarity transferring from seed word to target words using the theory of Structural Balance theory.

The premise of the algorithm is exemplified using the enemy of my enemy is my friend that preserves the transitivity structure captured by antonyms and synonyms. Thus, a low sentiment score is an indication of sentimental neutrality rather than due to the fact that the word in question is located at a far distance from the seeds.

The lexicons based on thesauruses were built using different variants of our proposed algorithm. The lexicons were evaluated by classifying product and movie reviews and the results show satisfying classification performances that outperform Label Propagation. We consider Norwegian as a case study, but the algorithm be can easily applied to other languages.

Keywords: Sentiment analysis · Structural Balance Theory

1 Introduction

With the increasing amount of unstructured textual information available on the Internet, sentiment analysis and opinion mining have recently gained a groundswell of interest from the research community as well as among practitioners. In general terms, sentiment analysis attempts to automate the classification

© Springer International Publishing Switzerland 2015
M. Ali et al. (Eds.): IEA/AIE 2015, LNAI 9101, pp. 336–347, 2015.
DOI: 10.1007/978-3-319-19066-2_33

of text materials as either expressing positive sentiment or negative sentiment. Such classification is particularity interesting for making sense of huge amount of text information and extracting the "word of mouth" from product reviews, and political discussions etc.

Possessing beforehand a sentiment lexicon is a key element in the task of applying sentimental analysis on a phrase or document level. A sentiment lexicon is merely composed of sentiment words and sentiment phrases (idioms) characterized by sentiment polarity, positive or negative, and by sentimental strength. For example, the word 'excellent' has positive polarity and high strength whereas the word 'good' is a positive having lower strength. Once a lexicon is built and in place, a range of different approaches can be deployed to classify the sentiment in a text as positive or negative. These approaches range from simply computing the difference between the sum of the scores of the positive lexicon and sum of the scores of the negative lexicon, and subsequently classifying the sentiment in the text according to the sign of the difference.

In order to generate a sentiment lexicon, the most obvious and naive approach involves manual generation. Nevertheless, the manual generation is tedious and time consuming rendering it an impractical task.

Due to the difficulty of manual generation, a significant amount of research has been dedicated to presenting approaches for automatically building sentiment lexicon. To alleviate the task of lexicon generation, the research community has suggested a myriad of semi-automatic schemes that falls mainly under two families: dictionary-based family and corpus-based family. Both families are semi-automatic because the underlying idea is to bootstrap the generation from a short list of words with polarity manually chosen (seed words), however they differ in the methodology for iteratively building the lexicon.

The majority of dictionary-based approaches form a graph of the words in a dictionary, where the words correspond to nodes and where relations between the words (e.g. in terms of synonyms, antonyms and/or hyponyms) may form the edges. A limited number of seed words are manually assigned a positive or a negative sentiment value (or label), and an algorithm, such the Label Propagation mechanism proposed in [1], is used to automatically assign sentiment scores to the other non-seed words in the graph.

The main contribution presented in this paper is the introduction of a novel algorithm that provides enhancements over existing work.

Another latent motivation in this article is to investigate the potential of generating lexicon in an automatic manner without any human intervention or refinement. We try to achieve this by increasing the sources of information, namely three different thesauruses, instead of solely relying on a single thesaurus as commonly done in the literature. In fact, we suggest that by increasing the number of thesauruses we can increase the quality of the generated lexicon.

Finally, while most of research in the field of sentiment analysis has been centered on the English language, little work has been reported for smaller languages where there is a shortage of good sentiment lists. In this paper, we tackle the problem of building sentiment lexicon for a smaller language, and use the Norwegian language as an example.

1.1   Background Work

Dictionary-based approaches were introduced by Hu and Liu in their seminal work [2]. They stop the generation when no more words can be added to the list. Mohammad, Dunne and Dorr [3] used a rather subtle and elegant enhancement of Hu and Liu's work [2, 4] by exploiting the antonym-generating prefixes and suffixes in order to include more words in the lexicon. In [5], the authors constructed an undirected graph based on adjectives in WordNet [6] and define distance between two words as the shortest path in WordNet. The polarity of an adjective is then defined as the sign of the difference between its distance from the word "bad" and its distance from the word "good". While the strength of the sentiment depends on the later quantity as well as the distance between words "bad" and "good".

Blair and his colleagues [7] employs a novel bootstrapping idea in order to counter the effect of neutral words in lexicon generation and thus improve the quality of the lexicon. The idea is to bootstrap the generation with neutral seed in addition to a positive and a negative seed. The neutral seeds are used to avoid positive and negative sentiment propagation through neutral words.

Rao and Ravichandran [8] used semi-supervised techniques based on two variants of the Mincut algorithm [9] in order to separate the positive words and negative words in the graph generated by means of bootstrapping. In simple words, Mincut algorithms [9] are used in graph theory in order to partition a graph into two partitions minimizing the number of nodes possessing strong similarity being placed in different partitions. Rao and Ravichandran [8] employed only the synonym relationship as a similarity metric between two nodes in the graph. The results are encouraging and show some advantages of using Mincut over label propagation.

In [10], Hassan and Radev use elements from the theory of random walks of lexicon generation. The distance between two words in the graph is defined based on the notion of hitting time. The hitting time $h(i|S)$ is the average number hops it takes for a node i to reach a node in the set S. A word w is classified as positive if $h(w|S_+) > h(w|S_-)$ and vice versa, where S_+ denotes the set of positive seed words, and S_- refers to the set of negative seed words.

Kim and Hovy [11] resorts to the Bayesian theory for assigning the most probable label (here polarity) of a bootstrapped word.

In [12], a quite sophisticated approach was devised. The main idea is to iteratively bootstrap using different sub-sets of the seed words. Then to count how many times a word was found using a positive sub-set seed and how many times the same word was found using a negative sub-set seed. The sentiment score is then normalized within the interval $[0, 1]$ using fuzzy set theory.

It is worth mentioning that another research direction for building lexicon for foreign language is based on exploiting the well-developed English sentiment lexicon. A representative work of such approaches is reported in [13]. In [13], Hassan and his co-authors employ the hitting time based bootstrapping ideas introduced in [10] in order to devise a general approach for generating a lexicon for a foreign language based on the English lexicon.

1.2   Transitive Relation and Structural Balance Theory

Structural Balance Theory involves modeling a social network using a graph where edges are annotated edges with positive or negative signs. A positive edge joining two nodes denotes a friendship relationship, while a negative edge denotes antagonism.

Structural Balance Theory goes beyond the classical concept of the friend of my friend is my friends, and introduces the notion of the enemy of my enemy is my friend.

The theory Structural Balance has recently found direction applications in social networks, resulting in the emergence of signed social networks that are able to describe friendship and antagonism relationships, [14,15], instead of solely relying on positive relationships (friendship).

Tidal Trust Algorithm In order to put our work in the right perspective, we shall review a representative work in the field of trust networks. The Tidal Trust algorithm is due to Golbeck [16]. It should be mentioned that it was applied in Film Trust, an online social network for rating movies. The Tidal algorithm belongs to the class of webtrust algorithms and its philosophy is based on the existence of a chain of trust between the user and the movie. In Tidal Trust, each user assesses the confidence that he has in each of his acquaintances using a score between 1 and 10. The participants of the reputation system and their relationships are modeled in terms of a graph, with each node being a participant or an object being rated. An edge describes a relationship between two nodes, in turn signifying the level of trust or rating, depending on the kind of nodes being involved. The algorithm works as follows:

- If a node s has rated the movie m directly, it returns its rating for m.
- Else, the node asks its neighbors in the graph for their recommendations.

Later, Golbeck [16] introduced two improvements to the original Tidal Trust algorithm as follows:

- The first enhancement was to consider only short paths from the source to the sink by specifying a maximum length of the trust chain. This improvement was motivated by the observation that trust values inferred through shorter paths tended to be more accurate.
- The second enhancement was to discard the paths that involved untrustworthy agents, i.e, whose trust falls below a user-specified threshold. In this manner, the system maintains only the paths with strong trust values.

2   The Novel Approach

A major shortcoming of the Label Propagation algorithm [1] is that while the seed values are propagated throughout the word graph of synonyms, the algorithm does not take into account that for each edge between two synonyms, there is an introduction of inaccuracy. The source of inaccuracy is primarily due to the fact that the meaning of two synonyms is typically not overlapping. A synonym of a word might carry a different and skewed meaning and sentiment compared to the meaning and sentiment of the word. Thus, the sentiment value of a word is not a 100% trustworthy indicator of the value that should be propagated to its synonyms. Problems are typically observed in the middle area between the positive and the negative seed words, i.e. typically several hops away from any seed word. While other works have introduced fixes to this problem (e.g. by introducing neutral seed words as described above), we take a different approach to the solution.

The implication of the inaccuracy introduced between two neighbouring synonyms in the word graph is that the value propagated from a seed word should have high trustworthiness and thus high weight in its close proximity in the word graph. Similarly, the value should be weighted comparably lower at longer synonym-hops away from the seed word. To account for this effect, we propose a novel approach that combines the ideas from Label Propagation and the Tidal Trust algorithm, and focusing on the shortest path between a word and the seed words. This is described in the following.

Assume that a word w is related to a set of seed words S in the graph. Then the sentiment score $score(w)$ for the word w is defined as:

$$score(w) = \frac{\sum_{s \in S} W_{s \to w} \cdot (-1)^{N(s \to w)} score(s)}{\sum_{s \in S} W_{s \to w} \cdot (-1)^{N(s \to w)}}, \tag{1}$$

- $N(s \to w)$ denotes the number of negative edges along the shortest path from seed s to node w.
- $W_{s \to w}$ is a weighting function. The weight function should capture the idea that: *the closer a seed to the word, the higher its associated weight*. In this sense, the weight of a seed should decay as the distance to the word increases.
- $score(s)$ is the score of the seed that we manually annotated.

The above rules describe a multiplication operation exemplified by the phrase the enemy of my enemy is my friend. For example, a chain of three positive edges exemplify the principle that the friend of my friend is my friend, whereas those with one positive and two negative edges capture the notions that the friend of my enemy is my enemy, the enemy of my friend is my enemy, and the enemy of my enemy is my friend.

An even number of antonyms will preserve the polarity of the seed word when transferred to the target word to be annotated since $(-1)^{N(s \to w)}$ will be equal to 1 in this case. While an odd number of antonyms will invert the polarity of the seed word when transferred to the target word to be annotated since $(-1)^{N(s \to w)}$ will be equal to -1 in this case.

Furthermore, we borrow ideas from Trust theory so that to account for decrease in "score certainty" as the distance between the seed and the word to be annotated increases.

Different Variants of Weighting Functions Let $l_{s \to w}$ denotes the path length from s to w in terms of number of hopes. We shall propose different weighting function that captures the idea that the weight decreases as the distance from seed word increases.

Multiplicative Decay $W_{s \to w} = \alpha^{l_{s \to w}}$ where α denotes a factor that controls the decaying such as $\alpha < 1$. In the experimental results, we tried different values of α, namely, 0.9, 0.7, $\exp(-0.5) = 0.61$, 0.5 and $\exp(-1) = 0.37$.

Inversely Proportional to Distance A possible variant: the weight is inversely proportional to the distance from the seed. There $W_{s \to w} = \frac{1}{l_{s \to w}}$

Blocking Contribution based on a Distance Threshold We propose a variant of our algorithm where only shortest paths that are shorter than a certain distance l_{min} are considered. The motivation behind this variant is to block the contribution from far away seeds. In the experimental results Section, we shall report result for $l_{min=4}$ and $l_{min} = 3$. The reader should note that this simple idea is also present in Tidal trust algorithm.

3 Linguistic Resources and Benchmark Lexicons for Evaluation

3.1 Building a Word Graph from Thesauruses

We built a large undirected graph of synonym and antonym relations between words from the three thesauruses. The words were nodes in the graph and synonym and antonym relations were edges. The graph where buildt from three norwegian thesauruses [17]. The full graph consisted of a total of 6036 nodes (words) and 16475 edges.

For all lexicons generated from the word graph, we started with a set of 109 seed words (51 positive and 57 negative). The words were manually selected where all used frequantly in the norwegian language and spanned different dimensions of positve ('happy', 'clever', 'intelligent', 'love' etc.) and negative sentiment ('lazy', 'aggressive', 'hopeless', 'chaotic' etc.).

3.2 Benchmark Lexicons

We generated sentiment lexicons from the word graph using the Label Propagation algorithm [1] which is the most popular method to generate sentiment lexicons. The Label Propagation algorithm initial phase consists of giving each positive and negative seed a word score 1 and -1, respectively. All other nodes in the graph are given score 0. The algorithm propagates through each non-seed words updating the score using a weighted average of the scores of all neighboring nodes (connected with an edge). When computing the weighted average, synonym and antonym edges are given weights 1 and -1, respectively. The algorithm is iterated to changes in score is below some threshold for all nodes. The resulting score for each node is our sentiment lexicon.

As a complement to the graph-generated lexicons, we generated a benchmark sentiment lexicon by translating the well-known English sentiment lexicon

AFINN [18] to Norwegian using machine translation (Google translate). We also generated a second lexicon by manually checking and correcting several different errors from the machine translation.

3.3 Text Resources for Evaluation

We tested the quality of the created sentiment lexicons using 15118 product reviews from the Norwegian online shopping sites www.komplett.no, mpx.no and 4149 movie reviews from www.filmweb.no. Each product review contained a rating from one to five, five being the best and the movie reviews a rating from one to six, six being the best.

4 Evaluation of Novel Approach Against Benchmark Lexicons

We generated a sentiment lexicon by applying our novel approach in Section 2 to the same word graph used for the Label Propagation benchmark lexicon, and by using the same seed words.

The quality of this lexicon was evaluated by comparing it against the quality of the different benchmark sentiment lexicons, including the lexicon generated from Label Propagation. The quality of a sentiment lexicon is measured as the classification performance of the www.komplett.no, mpx.no and www.filmweb.no reviews.

For each lexicon, we computed the sentiment score of a review by simply adding the score of each sentiment word in a sentiment lexicon together, which is the most common way to do it [19]. If the sentiment shifter 'not' ('ikke') was one or two words in front of a sentiment word, sentiment score was switched. E.g. 'happy' ('glad') is given sentient score 0.8, while 'not happy' ('ikke glad') is given score −0.8. Finally the sum is divided by the number of words in the review, giving us the final sentiment score for the review. We also considered other sentiment shifter, e.g. 'never' ('aldri'), and other distance between sentiment word and shifter, but our approach seems to be the best for such lexicon approaches in Norwegian [20].

Classification Method We divided the reviews in two parts, one part being training data and the other part for testing. We used the training data to estimate the average sentiment score of all reviews related to the different ratings. The computed scores could look like Table 1. We classified a review from the

Table 1. Average computed sentiment score for reviews with different ratings

Rating	1	2	3	4	5
Average sentiment score	−0.23	−0.06	0.04	0.13	0.24

test set using the sentiment lexicon to compute a sentiment score for the test review and classify to the closest average sentiment score from the training set. E.g. if the computed sentiment score for the test review was −0.05 and estimated averages were as given in Table 1, the review was classified to rating 2. In some rare cases the estimated average sentiment score was not monotonically increasing with the rating. Table 2 shows an example where the average for rating 3, is higher than for the rating 4. For such cases, the average of the two

Table 2. Example were sentiment score were not monotonically increasing with rating

Rating	1	2	3	4	5
Average sentiment score	−0.23	−0.06	0.18	0.10	0.24

sentiment scores were computed, $(0.10 + 0.18)/2 = 0.14$, and classified to 3 or 4 if the computed sentiment score of the test review was below or above 0.14, respectively.

Classification Performance We evaluated the classification performance using average difference in absolute value between the true and predicted rating for each review in the test set

$$\text{Average abs. error} = \frac{1}{n} \sum_{i=1}^{n} |p_i - r_i| \tag{2}$$

where n is the number off reviews in the test set and p_i and r_i is the predicted and true rating of review i in the test set. Naturally a small average absolute error showing that the sentiment lexicon performed well.

Note that the focus in this article is not to do a best possible classification performance based on the training material. If that was our goal, other more advanced and sophisticated techniques would be used, such as machine learning based techniques. Our goal is rather to evaluate and compare the performance of sentiment lexicons, and the framework described above is chosen with respect to that.

Classification Performance We evaluated the classification performance using average difference in absolute value between the true and predicted rating based on sentiment lexicons. For details, see [17].

5   Results

This section presents the results of classification performance on product and movie reviews for different sentiment lexicons. The results are shown in Tables 3 and 4. AFINN and AFINN_M refer to the translated and manually adjusted

AFINN sentiment lists. LABEL refers to the Label propagation algorithm. The α values refer to the base in the multiplicative decay approach and finally TRESH refers to the blocking approach with distance threshold. And finally INVERSE refers to the inverse proportional method. Training and test sets were created by

Table 3. Classification performance for sentiment lexicons on `komplett.no` and `mpx.no` product reviews. The columns from left to right show the sentiment lexicon names, the number of words in the sentiment lexicons, mean absolute error with standard deviaton and 95% confidence intervals for mean absolute error.

	N	Mean (Stdev)	95% conf.int.
AFINN	2161	1.4 (1.39)	(1.37, 1.43)
AFINN_M	2260	1.45 (1.33)	(1.42, 1.48)
$\alpha = 0.37$	6036	1.59 (1.47)	(1.55, 1.62)
$\alpha = 0.7$	6036	1.59 (1.44)	(1.56, 1.63)
INVERSE	6036	1.6 (1.45)	(1.56, 1.63)
$\alpha = 0.61$	6036	1.6 (1.46)	(1.57, 1.64)
$\alpha = 0.5$	6036	1.61 (1.47)	(1.57, 1.64)
$\alpha = 0.9$	6036	1.61 (1.47)	(1.58, 1.64)
TRESH_3	6036	1.64 (1.5)	(1.6, 1.67)
TRESH_4	6036	1.64 (1.49)	(1.6, 1.67)
LABEL	6036	1.67 (1.49)	(1.63, 1.7)

randomly adding an equal amount of reviews to both sets. All sentiment lexicons were trained and tested on the same training and test sets, making comparisons easier. This procedure were also repeated several times and every time the results were in practice identical to the results in Tables 3 and 4, documenting that the results are independent of which reviews that were added to the training and test sets.

For both review types we observe some variations in classification performance ranging from 1.39 to 1.67 for product reviews and from 1.98 to 2.25 for movie reviews. Comparing Tables 3 and 4, we see that the classification performance is poorer for movie reviews than for product reviews. It is known from the literature that sentiment analysis of movie reviews is normally harder than product reviews [19]. E.g. movie reviews typically contain a summary of the plot of the movie which could contain many negative sentiment words (sad movie), but still the movie can get an excellent rating.

For both review types we see that the sentiment lexicon with a rapid multiplicative decay ($\alpha = 0.37$) performs best among the automatically generated lists which indicates that the quality of the information in the graph decays rapidly. It performs significantly better than the Label propagation algorithm for both review types. Paired T-tests result in p-values $= 1.32 \cdot 10^{-4}$ and 0.018.

Overall the sentiment lexicons generated from Norwegian thesauruses using state-of-the-art graph methods do not outperform the automatically translated AFINN-lists. This shows that machine translation of linguistic resources in English can be used successfully in other smaller languages like Norwegian language.

Table 4. Classification performance for sentiment lexicons on `filmweb.no` movie reviews. The columns from left to right show the sentiment lexicon names, the number of words in the sentiment lexicons, mean absolute error with standard deviaton and 95% confidence intervals for mean absolute error.

	N	Mean (Stdev)	95% conf.int.
$\alpha = 0.37$	6036	1.98 (1.15)	(1.93, 2.03)
AFINN_M	2260	2.00 (1.12)	(1.96, 2.05)
$\alpha = 0.5$	6036	2.03 (1.15)	(1.98, 2.08)
AFINN	2161	2.06 (1.14)	(2.01, 2.11)
LABEL	6036	2.06 (1.1)	(2.02, 2.11)
INVERSE	6036	2.1 (1.12)	(2.05, 2.15)
TRESH_4	6036	2.14 (1.18)	(2.09, 2.19)
$\alpha = 0.7$	6036	2.16 (1.16)	(2.11, 2.21)
$\alpha = 0.61$	6036	2.2 (1.15)	(2.15, 2.25)
TRESH_3	6036	2.21 (1.14)	(2.16, 2.26)
$\alpha = 0.9$	6036	2.25 (1.13)	(2.2, 2.3)

6   Conclusion

In this paper, we have presented a simple but efficient algorithm for sentiment analysis that is based on elements from the Structural Balance Theory and Trust theory. In the same vein as Structural Balance Theory, the rational of our algorithm is to enable transitive transfer of polarity from seed words through positive and negative relationships, which are synonyms and antonyms in this case. Furthermore, we borrow ideas from Trust theory so that to account for decrease in "score certainty" as the distance between the seed and the word to be annotated increases. Thus, our algorithm possesses plausible properties that circumvent a major drawback of Label Propagation, namely the fact that nodes far away from the seed words (typically in the close-to-zero-value area in the word graph) get values that are incorrect since the inaccuracy in the synonym-relationship is not taken properly into account. In addition, our algorithm requires only a single run while Label Propagation should run iteratively until convergence of the scores. We conducted extensive experimental results on real life data that demonstrate the feasibility of our approach and its superiority to Label Propagation. A promising research direction that we are currently investigation is to adopt our algorithm as an alternative to Label Propagation in other Semi-Supervised Machine Learning problems.

References

1. Zhu, X., Ghahramani, Z.: Learning from labeled and unlabeled data with label propagation. Technical report, Technical Report CMU-CALD-02-107, Carnegie Mellon University (2002)
2. Hu, M., Liu, B.: Mining opinion features in customer reviews. In: Proceedings of AAAI, pp. 755–760 (2004)

3. Mohammad, S., Dunne, C., Dorr, B.: Generating high-coverage semantic orientation lexicons from overtly marked words and a thesaurus. In: Proceedings of the 2009 Conference on Empirical Methods in Natural Language Processing, vol. 2, pp. 599–608. Association for Computational Linguistics (2009)
4. Hu, M., Liu, B.: Mining and summarizing customer reviews. In: Proceedings of the ACM SIGKDD Conference on Knowledge Discovery and Data Mining (KDD), pp. 168–177 (2004)
5. Kamps, J., Marx, M., Mokken, R.J., De Rijke, M.: Using wordnet to measure semantic orientations of adjectives (2004)
6. Miller, G.A.: Wordnet: a lexical database for english. Communications of the ACM **38**(11), 39–41 (1995)
7. Blair-Goldensohn, S., Hannan, K., McDonald, R., Neylon, T., Reis, G.A., Reynar, J.: Building a sentiment summarizer for local service reviews. In: WWW Workshop on NLP in the Information Explosion Era, p. 14 (2008)
8. Rao, D., Ravichandran, D.: Semi-supervised polarity lexicon induction. In: Proceedings of the 12th Conference of the European Chapter of the Association for Computational Linguistics, pp. 675–682. Association for Computational Linguistics (2009)
9. Blum, A., Lafferty, J., Rwebangira, M.R., Reddy, R.: Semi-supervised learning using randomized mincuts. In: Proceedings of the Twenty-First International Conference on Machine Learning, p. 13. ACM (2004)
10. Hassan, A., Radev, D.: Identifying text polarity using random walks. In: Proceedings of the 48th Annual Meeting of the Association for Computational Linguistics, pp. 395–403. Association for Computational Linguistics (2010)
11. Kim, S.M., Hovy, E.: Automatic identification of pro and con reasons in online reviews. In: Proceedings of the COLING/ACL on Main Conference Poster Sessions, pp. 483–490. Association for Computational Linguistics (2006)
12. Andreevskaia, A., Bergler, S.: Mining wordnet for a fuzzy sentiment: sentiment tag extraction from wordnet glosses. In: EACL, vol. 6, pp. 209–215 (2006)
13. Hassan, A., Abu-Jbara, A., Jha, R., Radev, D.: Identifying the semantic orientation of foreign words. In: Proceedings of the 49th Annual Meeting of the Association for Computational Linguistics: Human Language Technologies: Short Papers, vol. 2, pp. 592–597. Association for Computational Linguistics (2011)
14. Leskovec, J., Huttenlocher, D., Kleinberg, J.: Predicting positive and negative links in online social networks. In: Proceedings of the 19th International Conference on World Wide Web, pp. 641–650. ACM (2010)
15. Kunegis, J., Lommatzsch, A., Bauckhage, C.: The slashdot zoo: mining a social network with negative edges. In: Proceedings of the 18th International Conference on World Wide Web, pp. 741–750. ACM (2009)
16. Golbeck, J.A.: Computing and Applying Trust in Web-based Social Networks. PhD thesis, College Park, MD, USA (2005) AAI3178583
17. Hammer, H., Bai, A., Yazidi, A., Engelstad, P.: Building sentiment lexicons applying graph theory on information from three norwegian thesauruses. In: Norweian Informatics Conference (2014)
18. Nielsen, F.Å.: A new ANEW: Evaluation of a word list for sentiment analysis in microblogs. CoRR abs/1103.2903 (2011)

19. Bing, L.: Web Data Mining. Exploring Hyperlinks, Contents, and Usage Data. Springer (2011)
20. Hammer, H.L., Solberg, P.E., Øvrelid, L.: Sentiment classification of online political discussions: a comparison of a word-based and dependency-based method. In: Proceedings of the 5th Workshop on Computational Approaches to Subjectivity, Sentiment and Social Media Analysis, pp. 90–96. Association for Computational Linguistics (2014)

Machine Learning

Heuristic Pretraining for Topic Models

Tomonari Masada[1](✉) and Atsuhiro Takasu[2]

[1] Nagasaki University, 1-14 Bunkyo-machi, Nagasaki 8528521, Japan
masada@nagasaki-u.ac.jp
[2] National Institute of Informatics, 2-1-2 Hitotsubashi,
Chiyoda-ku, Tokyo 1018430, Japan
takasu@nii.ac.jp

Abstract. This paper provides a heuristic pretraining for topic models. While we consider latent Dirichlet allocation (LDA) here, our pretraining can be applied to other topic models. Basically, we use collapsed Gibbs sampling (CGS) to update the latent variables. However, after every iteration of CGS, we regard the latent variables as observable and construct another LDA over them, which we call *LDA over LDA (LoL)*. We then perform the following two types of updates: the update of the latent variables in LoL by CGS and the update of the latent variables in LDA based on the result of the preceding update of the latent variables in LoL. We perform one iteration of CGS for LDA and the above two types of updates alternately only for a small, earlier part of the inference. That is, the proposed method is used as a *pretraining*. The pretraining stage is followed by the usual iterations of CGS for LDA. The evaluation experiment shows that our pretraining can improve test set perplexity.

1 Introduction

Since the inaugural work by Blei et al. [2] proposed latent Dirichlet allocation (LDA), topic models have been widely used in many applications [11][4][10]. Leaving practical aspects aside, progress in inference methods has been also an important contribution. The inaugural work [2] provides a variational Bayesian inference, and the work by Griffiths et al. [5] collapsed Gibbs sampling (CGS). While these are often used, collapsed variational Bayesian inference (CVB) [9], zero-order approximation of CVB [1], and expectation propagation [6] are also remarkable inference methods for topic models. Some of them have the advantage of better predictive power, which can be measured by test set perplexity. However, to achieve a better perplexity, some methods appeal to sophisticated mathematical machineries. In contrast, we propose a heuristic *pretraining* to achieve a better perplexity. While we consider LDA in this paper, our method may work for other models as long as they have a sequence of latent variables.

When modeling documents, we use a sequence of observable variables $x_d \equiv (x_{d1}, x_{d2}, \ldots)$ to represent a word token sequence in each document. For example, $x_{di} =$ "model" means that a token of the word "model" appears at the ith position of the dth document. LDA attaches to each x_{di} a latent variable z_{di}, whose value is the topic to which x_{di} is assigned. In this paper, we use CGS [5] to

© Springer International Publishing Switzerland 2015
M. Ali et al. (Eds.): IEA/AIE 2015, LNAI 9101, pp. 351–360, 2015.
DOI: 10.1007/978-3-319-19066-2_34

sample the values of the z_{di}s from the posterior distribution. There is a substantial difference between the x_{di}s and the z_{di}s, because the z_{di}s are not observable. However, both are a sequence of variables. Therefore, we can build another LDA over the z_{di}s by regarding them as observable. This is our main idea. We call the LDA built over the original LDA *LDA over LDA (LoL)*. LoL has its own sequence of latent variables $\bar{z}_d \equiv (\bar{z}_{d1}, \bar{z}_{d2}, \dots)$ for each document. Therefore, we can obtain the values of the $\bar{z}_{di}$s by CGS. However, simply by running CGS for the $\bar{z}_{di}$s, we have no effects on the z_{di}s. Therefore, we draw the values of the z_{di}s from the conditional distributions where we regard the values of the $\bar{z}_{di}$s as given. We can use the resulting values of the z_{di}s as the initial state of the succeeding CGS for the z_{di}s in LDA. In short, we conduct the following two steps alternately:

1. One iteration of CGS for the z_{di}s in LDA.
2. One iteration of CGS for the $\bar{z}_{di}$s in LoL, followed by an update of the z_{di}s based on the conditional distributions where the $\bar{z}_{di}$s are regarded as given.

In CGS for LDA, we only perform Step 1 repeatedly. In our approach, we perform Step 1 and Step 2 alternately for a small, earlier part of the inference, say for the first 100 iterations, and then perform only Step 1 repeatedly for the remaining part. Since LoL is used only for a small, earlier part of the inference, we call our approach *pretraining*. The experiment using four corpora revealed that our pretraining could improve test set perplexity for large corpora.

This paper is organized as follows. Section 2 describes the details of our pretraining. Section 3 contains the procedures and the results of the experiment for evaluating LoL. Section 4 presents the related work that gave an important intuition to us. Section 5 concludes the paper with future work.

2   Method

2.1   Collapsed Gibbs Sampling for LDA

For self-containedness, we first describe collapsed Gibbs sampling (CGS) for LDA[5]. Let D and W denote the numbers of documents and different words, respectively, in a given corpus. We denote the word set as $\mathcal{V} = \{v_1, \dots, v_W\}$. Let $\boldsymbol{x}_d = (x_{d1}, \dots, x_{dn_d})$ be the observable variable sequence representing the word token sequence of the dth document. $x_{di} = v_w$ means that a token of the word v_w appears at the ith position of the dth document. To each x_{di}, LDA attaches a latent variable z_{di} whose domain is the topic set $\mathcal{T} = \{t_1, \dots, t_K\}$. z_{di} represents the topic to which x_{di} is assigned. LDA generates a corpus as follows:

1. For each topic $t_k \in \mathcal{T}$, draw a word multinomial distribution parameter $\boldsymbol{\phi}_k = (\phi_{k1}, \dots, \phi_{kW})$ from the corpus-wide Dirichlet prior distribution $\text{Dir}(\boldsymbol{\beta})$.
2. For $d = 1, \dots, D$, draw a topic multinomial distribution parameter $\boldsymbol{\theta}_d = (\theta_{d1}, \dots, \theta_{dK})$ from the corpus-wide Dirichlet prior $\text{Dir}(\boldsymbol{\alpha})$.

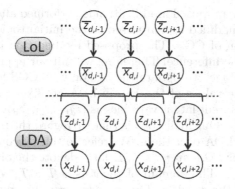

Fig. 1. Relationships among variables

(a) For $i = 1, \ldots, n_d$, draw a topic from the topic multinomial distribution Multi($\boldsymbol{\theta}_d$) and set the value of z_{di} to the drawn topic.
(b) For $i = 1, \ldots, n_d$, draw a word from the word multinomial distribution Multi($\boldsymbol{\phi}_k$) s.t. $z_{di} = k$ and set the value of x_{di} to the drawn word.

We denote the set of word sequences $\{\boldsymbol{x}_1, \ldots, \boldsymbol{x}_D\}$ as $\boldsymbol{X}$ and the set of topic sequences $\{\boldsymbol{z}_1, \ldots, \boldsymbol{z}_D\}$ as $\boldsymbol{Z}$. After integrating the multinomial parameters out, we can obtain the joint probability distribution of $\boldsymbol{X}$ and $\boldsymbol{Z}$ as follows:

$$p(\boldsymbol{X}, \boldsymbol{Z}; \boldsymbol{\alpha}, \boldsymbol{\beta}) = \prod_{d=1}^{D} \frac{\Gamma(\sum_{k=1}^{K} \alpha_k) \prod_{k=1}^{K} \Gamma(n_{dk} + \alpha_k)}{\prod_{k=1}^{K} \Gamma(\alpha_k) \; \Gamma(n_d + \sum_{k=1}^{K} \alpha_k)}$$
$$\times \prod_{k=1}^{K} \frac{\Gamma(\sum_{w=1}^{W} \beta_w) \prod_{w=1}^{W} \Gamma(n_{kw} + \beta_w)}{\prod_{w=1}^{W} \Gamma(\beta_w) \; \Gamma(n_k + \sum_{w=1}^{W} \beta_w)} \;, \tag{1}$$

where n_{dk} is the number of the word tokens of the dth document that are assigned to the topic t_k, and n_{kw} is the number of the tokens of the word v_w that are assigned to t_k. n_k is defined by $n_k \equiv \sum_w n_{kw}$. Since the posterior $p(\boldsymbol{Z}|\boldsymbol{X}; \boldsymbol{\alpha}, \boldsymbol{\beta})$ obtained from Eq. (1) is intractable, we use CGS in this paper and update the z_{di}s based on the following distribution:

$$p(z_{di} = t_k | \boldsymbol{Z}^{\neg di}, \boldsymbol{X}; \boldsymbol{\alpha}, \boldsymbol{\beta}) \propto (n_{dk}^{\neg di} + \alpha_k) \times \frac{n_{kw}^{\neg di} + \beta_{kw}}{n_k^{\neg di} + \sum_w \beta_{kw}} \;, \tag{2}$$

where we assume $x_{di} = v_w$. The notation $\neg di$ means that the statistics are calculated after removing the ith token of the dth document. Each iteration of CGS gives a sample from the posterior $p(\boldsymbol{Z}|\boldsymbol{X}; \boldsymbol{\alpha}, \boldsymbol{\beta})$. In the experiment, we also estimate the hyperparameters $\boldsymbol{\alpha}$ and $\boldsymbol{\beta}$ by Minka's method [7].

2.2   Heuristic Modification of Topic Assignments

We propose a heuristic modification of the z_{di}s as *pretraining* to improve test set perplexity. We perform CGS for LDA and the proposed modification of the z_{di}s *alternately*. That is, the proposed modification is performed after every iteration of CGS. Further, the modified z_{di}s are used as the initial topic assignments for the succeeding iteration of CGS. The proposed modification is used only for a small, earlier part of the inference. Therefore, we call our approach *pretraining*. After the pretraining stage, we update the z_{di}s only by CGS for LDA.

We assume that the values of the z_{di}s in LDA are fixed and then compose ordered pairs of two adjacent latent variables as $(z_{d1}, z_{d2}), (z_{d2}, z_{d3}), \dots, (z_{dn_d-1}, z_{dn_d})$ for each document. Our main idea is that we regard the pairs as observable and construct another LDA over them. We refer to this "second-layer" LDA by *LoL*, an abbreviation of *LDA over LDA*. Let $\bar{x}_{di}$ denote the observable variable composed as (z_{di}, z_{di+1}). The domain of the $\bar{x}_{di}$s is $\mathcal{T} \times \mathcal{T}$, which is the set of words in LoL. We call the words in LoL *pseudowords*. The number of different pseudowords is K^2. The reason we compose pairs of latent variables is that we would like to make the number of pseudowords close to that of words. As in LDA, we attach a latent variable $\bar{z}_{di}$ to each $\bar{x}_{di}$ and refer to the domain of the $\bar{z}_{di}$s by $\mathcal{S} = \{s_1, \dots, s_L\}$, whose elements are the topics in LoL, which we call *pseudotopics*. Relationships among variables are presented in Fig. 1. In the evaluation experiment, we set L, i.e., the number of pseudotopics, to 10, because no larger numbers were required to achieve good results.

We perform CGS also for LoL to update its latent variables, i.e., the $\bar{z}_{di}$s. However, if we only perform CGS for LoL, we have no effect on the latent variables in LDA, i.e., the z_{di}s. Therefore, we modify the z_{di}s based on a result of CGS for LoL as follows. We assume that the values of the $\bar{z}_{di}$s are given as a result of CGS for LoL and consider the corresponding pair (z_{di}, z_{di+1}) of latent variables in LDA. We decompose the joint probability $p(x_{di}, x_{di+1}, z_{di}, z_{di+1}, \bar{z}_{di})$ as follows:

$$p(x_{di}, x_{di+1}, z_{di}, z_{di+1}, \bar{z}_{di}) = p(z_{di}, z_{di+1}|\bar{z}_{di})p(x_{di}|z_{di})p(x_{di+1}|z_{di+1}) \quad (3)$$

We estimate the first term $p(z_{di}, z_{di+1}|\bar{z}_{di})$ in Eq. (3) by regarding z_{di} and z_{di+1} as an ordered pair. Let $\bar{n}_{ljk}$ represent how many tokens of the pseudoword (t_j, t_k) are assigned to the pseudotopic s_l. We then obtain an estimation of $p(z_{di}, z_{di+1}|\bar{z}_{di})$ as follows:

$$p(z_{di} = t_j, z_{di+1} = t_k|\bar{z}_{di} = s_l) \propto \bar{n}_{ljk} + \bar{\beta}_{jk}, \quad (4)$$

where the $\bar{\beta}_{jk}$s are the hyperparameters in LoL. The second term $p(x_{di}|z_{di})$ in Eq. (3) can be interpreted as a per-topic word probability in LDA. Therefore, we can estimate it as follows:

$$p(x_{di} = v_w|z_{di} = t_j) = \frac{n_{jw} + \beta_{jw}}{n_j + \sum_{w=1}^{W} \beta_{jw}}. \quad (5)$$

By combining Eq. (4) and Eq. (5), we obtain a conditional distribution of z_{di} as follows:

$$p(z_{di} = t_j|x_{di} = v_w, z_{di+1} = t_k, \bar{z}_{di} = s_l) \propto (\bar{n}_{ljk} + \bar{\beta}_{jk}) \times \frac{n_{jw} + \beta_{jw}}{n_j + \sum_w \beta_{jw}}, \quad (6)$$

Table 1. Specifications of corpora

	# docs	# words	# training/test tokens
DBLP	2,504,988	496,393	20,703,210/2,302,048
MOV	27,859	222,666	16,705,099/1,860,813
DIET	1,391,298	116,743	156,461,549/17,361,032
MED	1,087,572	2,314,733	192,265,331/21,355,378

where we regard z_{di+1} as fixed. We update z_{di} based on Eq. (6). Let t_j and $t_{j'}$ denote the old and new values of z_{di}. We update the following relevant statistics: $\bar{n}_{ljk}$, $\bar{n}_{lj'k}$, n_{jw}, $n_{j'w}$, n_j, and $n_{j'}$. Also for the other variable z_{di+1} in the pair (z_{di}, z_{di+1}) we consider, we sample its value based on a similar conditional distribution by regarding z_{di} as fixed and update the relevant statistics. In this manner, we update both of the paired latent variables (z_{di}, z_{di+1}). We perform this update for all pairs of the latent variables in LDA.

However, a preliminary experiment showed that we modified the z_{di}s too often to obtain a good perplexity. Therefore, by flipping a coin, we perform the modification only with probability 1/2. It should be noted that our method regards the z_{di}s in each document as forming an *ordered sequence*. Therefore, our method requires an implementation of CGS that outputs the z_{di}s as an ordered sequence, though document ordering is irrelevant.

3   Evaluation

3.1   Data Sets

The four corpora in Table 1 were used for the evaluation. DBLP is a set of the paper titles extracted from the DBLP XML records `dblp.xml`[1] downloaded on February 7, 2014. Since we regard each title as a document, all documents are short, though the number is large. MOV is a data set containing movie reviews[2], where the number of documents is not that large, but their lengths are fairly long. DIET is a subset of the minutes of the National Diet of Japan, where we collect the transcriptions whose dates range from January 30, 2001 to November 14, 2012. We regard a sequence of consecutive utterances by the same person as a document. Since all documents in DIET are written in Japanese, we use MeCab[3] for morphological analysis. MED is a set of the abstracts, whose dates range from 2012 to now, extracted from the 2014 data files of the MEDLINE®/PUBMED®, a database of the U.S. National Library of Medicine.

We split each data set into ten subsets of almost equal size. When we used one among the ten as a test set for computing perplexity, we merged the other nine subsets into one training set, on which the inference was performed. Consequently, we obtained ten perplexities for each data set. The rightmost column

---

[1] http://dblp.uni-trier.de/xml/

[2] http://www.cs.cornell.edu/people/pabo/movie-review-data/

[3] http://mecab.sourceforge.jp/

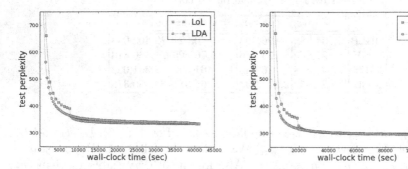

Fig. 2. Test perplexities for DBLP data set. The left and the right panels give the results for $K = 500$ and $K = 800$, respectively.

of Table 1 presents the numbers of the training and testing word tokens. While we only present a number pair corresponding to one among the ten train/test splits for each corpus, the number pairs obtained from the other splits are not widely different from the presented one. Note that DIET and MED are fairly large corpora.

We give an example of the wall-clock time measured in the experiment. When the number of topics is 800 and that of pseudotopics is 10, i.e., $K = 800$ and $L = 10$, DIET required around 13,400 seconds for each iteration in the pretraining stage and around 4,600 seconds for each iteration of CGS for LDA on the Intel Core i7 CPU X 990 @ 3.47GHz processor.

3.2 Results

We evaluated the proposed pretraining in terms of test set perplexity, a measure of the predictive power of topic models. We assume that the n_{kw}s, α, and β are given as a result of a CGS performed on the training set. For evaluation, we perform a folding-in CGS on the test set for 20 iterations and calculate test set perplexity. Let $\tilde{D}$ be the number of test documents, $n_{\tilde{d}}$ the length of the $\tilde{d}$th test document, $x_{\tilde{d}i}$ the ith word token in the $\tilde{d}$th test document, and $z_{\tilde{d}i}$ the latent variable attached to $x_{\tilde{d}i}$. We calculate test set perplexity as:

$$\frac{1}{\sum_{\tilde{d}} n_{\tilde{d}}} \sum_{\tilde{d}=1}^{\tilde{D}} \sum_{i=1}^{n_{\tilde{d}}} \log \Big(\sum_{k=1}^{K} p(z_{\tilde{d}i} = t_k) p(x_{\tilde{d}i} | z_{\tilde{d}i} = t_k) \Big) , \tag{7}$$

where the $z_{\tilde{d}i}$s are sampled by the folding-in CGS. Smaller perplexity means better predictive power. The relevant probabilities are defined as:

$$p(z_{\tilde{d}i} = k) \equiv \frac{n_{\tilde{d}k} + \alpha_k}{n_{\tilde{d}} + \sum_k \alpha_k} \text{ and } p(x_{\tilde{d}i} = v_w | z_{\tilde{d}i} = k) \equiv \frac{m_{kw} + n_{kw} + \beta_w}{m_k + n_k + \sum_w \beta_w} , \tag{8}$$

where $n_{\tilde{d}k}$ is the number of the tokens in the $\tilde{d}$th test document that are assigned to the topic t_k, m_{kw} is the number of the test tokens of the word v_w that are

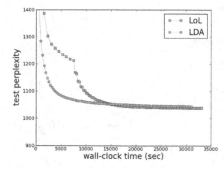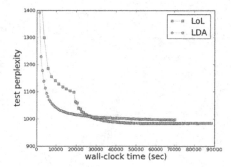

Fig. 3. Test perplexities for MOV data set. The left and the right panels give the results for $K = 500$ and $K = 800$, respectively.

assigned to the topic t_k, and m_k is defined as $\sum_w m_{kw}$. The m_{kw}s and the m_ks are determined by the folding-in CGS. The n_{kw}s and the n_ks are the statistics coming from the CGS performed on the training set and thus are fixed.

We calculated test set perplexity after every ten iterations. Figures 2, 3, 4, and 5 illustrate how test set perplexity changed in the course of inference for the four corpora. The horizontal axis represents wall-clock time in seconds, and the vertical axis represents test set perplexity. Every plotted perplexity is the mean of the ten perplexities obtained from the ten train/test splits. For each corpus, we present two charts, each corresponding to the cases $K = 500$ (left) and $K = 800$ (right). In each chart, the green line with square markers gives the perplexities for the inference using the proposed pretraining, and the gray line with circle markers gives the perplexities for the inference only using CGS for LDA. For convenience, we denote the inference using the proposed pretraining as LoL and the inference only using CGS for LDA as LDA. For DBLP and MOV, relatively small corpora, the pretraining was performed for the first 200 iterations from the 1,000 iterations in total. For DIET and MED, the pretraining was performed for the first 100 iterations from the 500 iterations in total. In all cases, we used the pretraining only for the first fifth of the total number of iterations. For the green line in each chart, the time intervals between two adjacent markers of the pretraining iterations were longer than those of the remaining iterations.

While LoL required longer execution time than LDA for the same number of iterations, LoL could give a better perplexity within the same wall-clock time for larger corpora as follows. Figures 2 and 3 show that our pretraining could not achieve any significant improvements for DBLP and MOV corpora. While the charts for MOV show a little improvement, the difference is not significant, because the standard deviation is around 5.0. In contrast, for DIET and MED, we could achieve significant improvements. For DIET, 500 iterations of CGS for LDA required around 161,000 seconds and gave the test set perplexity of 552.2 ± 0.9 when $K = 500$. Within the same length of time, LoL achieved the perplexity of 497.6 ± 1.9. A similar improvement was obtained for $K = 800$. For MED, 500 iterations of CGS for LDA required around 170,000 seconds and gave

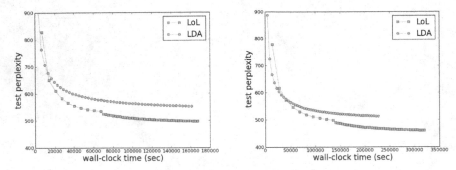

Fig. 4. Test perplexities for DIET data set. The left and the right panels give the results for $K = 500$ and $K = 800$, respectively.

the perplexity of 907.0 ± 1.3. In contrast, LoL gave the perplexity of 864.9 ± 3.0 within the same length of time. While the charts show the results ending at the 500th iteration, LDA couldn't surpass LoL even for a larger number of iterations. Note that, after the pretraining stage, both LoL and LDA performed completely the same inference on the same training set. It may be concluded that our pretraining guided the inference to a better local optimum.

3.3   Discussion

First, we discuss the time complexity of inference. The complexity of the proposed pretraining is proportional to the length of the corpus, i.e., the total number of word tokens, multiplied by the number of topics. This complexity is the same as that of the collapsed Gibbs sampling (CGS) for LDA in its order. However, the constant factor depends on the settings. The figures give the wall-clock time required for inference. LoL is applied only for the first fifth of the iterations. Therefore, when the running time of LDA is T, the running time of the inference accompanied with our pretraining is $0.2cT + 0.8T$. The constant factor c can be estimated based on the charts in the figures as follows. For example, in Fig. 4, the running time of LDA is around 230,000 sec for the DIET data set when $K = 800$, and that of LoL is around 325,000 sec. Therefore, the constant c satisfies $0.2c + 0.8 = 325000/230000$. Consequently, c is around 3.1. This means that the running time of each iteration of the proposed pretraining is larger than that of CGS for LDA by the factor of 3.1. However, we only perform the pretraining for the first 20% of the iterations. Therefore, the total running time shows a not that large increase.

Second, we discuss why our method works. By closely inspecting the experimental result, we can observe a "rich-get-richer" effect for the cases where our pretraining works. To be precise, we count the number of word tokens assigned to each topic and plot the distribution of those numbers. Then LoL gives a highly skewed distribution when compared with LDA. We guess that this feature is the reason for the perplexities being smaller than LDA. For example, when we set

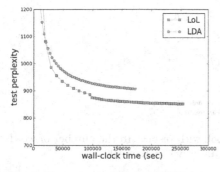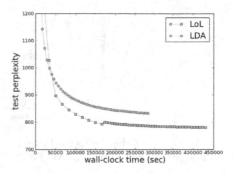

Fig. 5. Test perplexities for MED data set. The left and the right panels give the results for $K = 500$ and $K = 800$, respectively.

$K = 800$ for the MEDLINE data set, the number of word tokens assigned to the most dominant topic increases by a factor of 11 when we use LoL. That is, 11 times more word tokens are assigned to the most dominant topic (cf. Fig. 6). This kind of rich-get-richer effect could be observed when we used LoL.

Third, we characterize our work in relation to the topic modeling approach. We propose not a new probabilistic model, but a pretraining method. Therefore, the proposed approach can also be applied to e.g. HDP-LDA [8]. That is, we can also propose HDP-LDA over HDP-LDA based on the idea given in this paper. When a topic model has a sequence of latent variables that corresponds to a sequence of observable variables, we can propose a pretraining method for the model based on the idea given in this paper by regarding the sequence of latent variables as a sequence of observable variables.

4 Related Work

An important part of our intuition comes from the work by Bo et al. [3]. This work proposes a multi-layered sparse factor analysis for image data. Their probabilistic model regards the original image to be analyzed as the first layer data and obtains the second layer data by max pooling and collecting of latent coefficients obtained from the first layer. The layers are stacked by regarding the result of max pooling and collecting of coefficients from the $(l-1)$-th layer data as the observable variables for the l-th layer feature extraction. Since the observable variables for the l-th layer have the same tensor structure as the latent variables in the l-th layer, the authors can obtain a multi-layer stacked construction for their generative model. We follow the same line and regard the sequence of latent variables of an LDA as the sequence of observable variables of another LDA. We do not repeat the stacking, because we here aim to obtain a pretraining method and would like to avoid unnecessary complications.

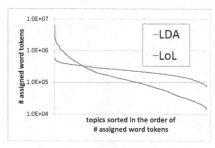

Fig. 6. A "rich-get-richer" effect achieved by LoL with respect to the number of word tokens assigned to each topic for MEDLINE data set when $K = 800$.

5   Conclusions

In this paper, we propose a heuristic pretraining for LDA, where we build another LDA over the sequence of latent variables of LDA by regarding the latent variables as observable. The experimental results show that the proposed pretraining worked for large document sets. It is an important future work to evaluate the performance of our technique when applied to other topic models. To explain why our method works in a more formal manner is also an important future work.

References

1. Asuncion, A., Welling, M., Smyth, P., Teh, Y.W.: On smoothing and inference for topic models. In: UAI 2009, pp. 27–34 (2009)
2. Blei, D.M., Ng, A.Y., Jordan, M.I.: Latent Dirichlet allocation. JMLR **3**, 993–1022 (2003)
3. Chen, B., Polatkan, G., Sapiro, G., Blei, D., Dunson, D., Carin, L.: Deep learning with hierarchical convolutional factor analysis. IEEE Trans. Pattern Anal. Mach. Intell. **35**(8), 1887–1901 (2013)
4. Gerrish, S., Blei, D.M.: A language-based approach to measuring scholarly impact. In: ICML 2010, pp. 375–382 (2010)
5. Griffiths, T.L., Steyvers, M.: Finding scientific topics. PNAS **101**(suppl. 1), 5228–5235 (2004)
6. Minka, T.P., Lafferty, J.: Expectation-propagation for the generative aspect model. In: UAI 2002, pp. 352–359 (2002)
7. Minka, T.P.: Estimating a Dirichlet distribution (2000). http://research.microsoft.com/en-us/um/people/minka/papers/dirichlet/
8. Teh, Y.W., Jordan, M.I., Beal, M.J., Blei, D.M.: Hierarchical Dirichlet processes. J. Amer. Statist. Assoc. **101**, 1566–1581 (2006)
9. Teh, Y.W., Newman, D., Welling., M.: A collapsed variational Bayesian inference algorithm for latent Dirichlet allocation. In: Advances in Neural Information Processing Systems, vol. 19 (2007)
10. Zhao, W.X., Jiang, J., He, J., Song, Y., Achananuparp, P., Lim, E.P., Li., X.: Topical keyphrase extraction from Twitter. In: HLT 2011, pp. 379–388 (2011)
11. Zhao, W.X., Jiang, J., Weng, J., He, J., Lim, E.-P., Yan, H., Li, X.: Comparing Twitter and traditional media using topic models. In: Clough, P., Foley, C., Gurrin, C., Jones, G.J.F., Kraaij, W., Lee, H., Mudoch, V. (eds.) ECIR 2011. LNCS, vol. 6611, pp. 338–349. Springer, Heidelberg (2011)

Fast Online Learning to Recommend a Diverse Set from Big Data

Mahmuda Rahman[✉] and Jae C. Oh

Department of Electrical Engineering and Computer Science,
Syracuse University, Syracuse, NY 13210, USA
mrahma01@syr.edu, jcoh@ecs.syr.edu

Abstract. Building a recommendation system to withstand the rapid change in items' relevance to users is a challenge requiring continual optimization. In a Big Data scenario, it becomes a harder problem, in which users get substantially diverse in their tastes. We propose an algorithm that is based on the UBC1 bandit algorithm to cover a large variety of users. To enhance UCB1, we designed a new rewarding scheme to encourage the bandits to choose items that satisfy a large number of users. Our approach takes account of the correlation among the items preferred by different types of users, in effect, increasing the coverage of the recommendation set efficiently. Our method performs better than existing techniques such as Ranked Bandits [8] and Independent Bandits [6] in terms of satisfying diverse types of users.

Keywords: Recommendation system · Online learning · Diversity · Multi armed bandit · Upper confidence bound

1 Introduction

Recommendation systems in Big-data era need to address not only the huge amount of users to satisfy but also the rapidly changing data patterns of preferences. With such a rapid change and continuous shift of users' preferences, recommendation system needs to learn quickly from the patterns of choices in previous users to suggest items to the new user. As diverse tastes of the incoming users induces a fast turn over time for items, online recommendation systems get little prior knowledge about the distribution of the preference on items among the user population. Moreover, most recommendation systems need to pick a limited number of items to suggest to a user, still require that at least one of these suggested items can satisfy her taste. An online learning algorithm tries to produce a substantially small but a diverse recommendation set from a large number of items, at the same satisfying different user types.

In this paper we defined users by their choice of items, therefore, users of similar preference can be considered of the same user type. We aim to develop a methodology that can discover correlations among the items so that the related items are associated with the same user type. Our approach has two main effects

© Springer International Publishing Switzerland 2015
M. Ali et al. (Eds.): IEA/AIE 2015, LNAI 9101, pp. 361–370, 2015.
DOI: 10.1007/978-3-319-19066-2_35

on the recommendation system: (1) It reduces the complexity within the data, making it capable of analyzing Big data; and (2) It allows the recommendation system to satisfy more diverse user types.

If a recommendation system (with a fixed small number of items to recommend) could pick only one from a group of related items, it would be able to satisfy all users of that user type. This approach improves the coverage of the recommendation system. Our work focuses on developing an online learning mechanism for recommendation systems under the constraint of limited time and number of items to suggest in Big data environments.

In this section, we also list some of the challenges which made our problem interesting. Then we state our contribution to overcome those challenges. The challenges are

- The problem of choosing the optimal set of recommended items for a given user population is an NP hard problem even if all the user vectors are given offline. Because the problem is equivalent to the maximum coverage problem.
- There is a greedy optimal solution to this problem that can be known given the entire user vector offline. But to identify that optimal set, each k-subset of all n items needs to be tried as a potential candidate, so there are exponentially many options for trial [8].
- It is hard for the recommendation system to determine user-type of an incoming user in an online learning paradigm. Hence it is difficult for a recommendation system to diversify the set to satisfy different user-types.

We achieved our goal of minimizing the abandonment by maximizing the payoff of the recommendation system in the following manner:

- We came up with a mechanism for our recommendation system where it learns the dependency structure among items though a simple rewarding scheme.
- We tested our method by recommending a small set of item from real big data sets where there are hundreds of users, each having thousands of choices.
- On an average, our proposed mechanism outperforms existing recommendation techniques in terms of coverage of different user types efficiently.

2   Problem Description

Consider a recommendation System where n items $\{i_1, i_2, i_3, ... i_n\}$ are given. When a user arrives, the system needs to show her a set of k items where $n >> k$. If she finds any one of them relevant, the system gets a payoff of 1. If none of them is relevant to her interest, then it gets a payoff of 0. A user vector is defined as the following:

Definition 1. *Each user j can be represented by a $\{0,1\}^n$ vector X^j where $X_i^j = 1$ indicates that user j found item i relevant.*

Distribution of these vectors X^j is unknown to the system but these vectors are assumed to express the "types" of a users. We assume that there is a large degree of correlation between these vectors. Therefore, thousands of users might be of only a few types. For that reason, we want to eliminate the redundancy of picking multiple items for a specific user-type in our recommendation set and diversely select items to satisfy as many user-types as possible. That would be the optimal set. In the main iteration of our algorithm, at time t, a recommendation set of

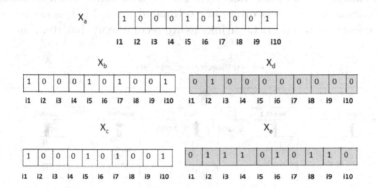

Fig. 1. Users X_b and X_c are of same user type as user X_a, whereas user X_d and X_c forms different user type based on their preference of items

k items is presented to a random user, then that user vector is disclosed to the system to evaluate whether it could satisfy that user. The target is to maximizes its pay off as defined while constructing the recommendation set for a random unknown user. As a result, it minimizes the probability that a user finds none of the recommended items interesting – a situation which is coined as %no by the IR community and 'abandonment' by machine learning researchers.

3   Preliminaries

We used Multi armed bandit algorithm [11] as a solver for our problem. MAB solves the problem a gambler faces given a slot machines with multiple levers (arm), deciding which arm to play, how many times to play a specific arm and the order to play them with an objective that the decision will maximize the sum of rewards earned through a sequence of arms played. Playing each arm provides a random reward from a distribution specific to that arm, unknown to the gambler.

For a recommendation System, at each time t, from n available items (arms), k items needs to be picked by the algorithm. So k instances of multi armed bandits are created, each having n arms to select from. Thus a set of k items is constructed. A random user's choice is compared with this recommended set and accordingly reward is fed back to the associated bandits.

For each slot i of k-item set we run a stochastic MAB algorithm to pick the one from the n-options which maximizes $r_i + \sqrt{\frac{2\log(f)}{f_i}}$ where r_i denotes the current average reward of the option i and f_i denotes the number of times option i has been played so far whereas f stands for $\Sigma_{i=1}^{n}(f_i)$ (total number of times all the arms are selected)

This is called Upper confidence bound (UCB1) [2] because this value can be interpreted as the upper bound of a confidence interval, so that the true average reward of each item i is below this upper confidence bound with high probability. If we have tried an item less often, our estimated reward is less accurate so the confidence interval is larger. It shrinks as we recommend that item more often.

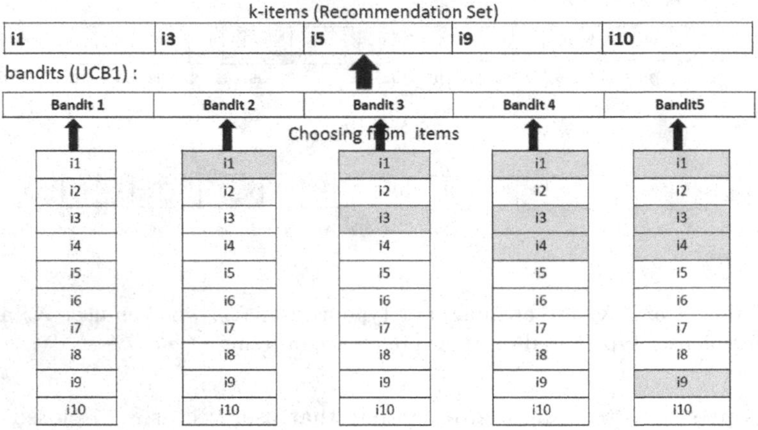

Fig. 2. 5 items are picked out of 10 items by 5 different bandits (k = 5, n = 10): once an item get picked by a bandit, it becomes unavailable to the rest of the bandits (shaded) by the IBA approach of using UCB1, we adopted similar strategy

Suppose, there is only one item to select from two: $i, j \in A$ and both of them have achieved same average reward (r_i and r_j are same) after some random number of trials. Now, if arm i has been tried more often than arm j, then $f_i > f_i$ with same f as a numerator. Then $\sqrt{\frac{2\log(f)}{f_i}} < \sqrt{\frac{2\log(f)}{f_j}}$. So confidence bound shrinks for i more than j. Again, this bound grows if f gets higher.

Provided that UCB1 algorithm have tried enough of each items to be reasonably confident, it rules out the chance that a selected item would be suboptimal or inferior in terms of achieving reward. While we would like to include this apparently superior arm (item), we have to make sure that the other arms (items) are sampled enough to be reasonably confident that they are indeed inferior. UCB1 does that for us, but unfortunately UCB1 assumes all n items are independent. We came up with a method to leverage it where items are dependent.

4 Leveraging UCB1

In the existing work, two different approaches has been found to build a recommendation system by using UCB1 bandits. We discuss their advantages and disadvantages before introducing our proposed method in this section.

4.1 Previous Approaches

The Ranked Bandit Algorithm (RBA) [8] used each item in the recommendation set to satisfy a different type of user and hence came up with a consensual set based on diverse users. It used strong similarity measures (dependency) between items and takes into account only the first item selected by an user from the recommendation set to represent the group of similar type of user. But as their algorithm strives to produce an ordered set, the learning is slow. It specifically holds i^{th} bandit responsible for i^{th} item in the recommendation set. But performance of ith bandit is actually dependent on picking the appropriate item (in proper order) by on all other bandits preceding i. As a consequence of this cascading effect, the learning for i^{th} item cannot really start before $\Omega(n^{i-1})$ time steps [6]. According to their setting, the probability of user $x \in X$ selecting the ith item from the recommendation set is denoted as p_i which is conditional on the fact that the user did not select any of the items in that set presented in any earlier positions. Formally, $p_i = Pr(x^i = 1 | x^{i-1} = 0)$ for all $i \in k$ where the binary value $\{0, 1\}$ of x^i denotes the probability of selecting the item i by the user x

On its attempt to maximize the marginal gain of ith bandit, where each bandit is a random binary variable, it forms a Markov chain where the later bandits have to wait for an earlier one to converge. To speed up the process, Independent Bandit Algorithm (IBA) [6] assumed independence between items and used Probability Ranking Principle (PRP) [9] as a greedy method to select items to recommend. PRP allows them to rank items in decreasing order of relevance probability without considering the correlations between them. So they count all the item a user select within the recommendation set and each bandit responsible for selecting an item of that user's choice gets a reward of 1. The overall payoff for the recommendation set is 1 even more than one items are selected by that user. This solution is sub-optimal in minimizing abandonment because diverse users are likely to be a part of the minority, which might not be covered by the top-k items PRP selects. So it often fails to capture diversity.

5 Our Approach

We present our proposed method in Algorithm 1. We initialize k number of bandits $UCB1_1(n), ... UCB1_k(n)$ to construct a set of k items where bandit i gets priority on selecting an item over bandit $i + 1$. Similar to [6], once an item gets selected by a preceding bandit, it becomes unavailable to any later bandits (ref line 6 and 7 of the algorithm). After the recommendation set S^t is created this way, it is compared with a random user vector X^t picked in time t (in line 9-11). The novelty of our approach is in the rewarding scheme for the bandits (shown in line 12,13). If the recommendation set contains more than one items preferred by the user, the first bandit responsible for picking the preferred item get a much higher reward as Fit for that item than any other bandits who picked other items of that user's preference. We set that higher reward C to be equal to the accumulated reward of all bandits who picked a preferred item for that user in that recommendation set. In this way, we uplift the average reward for the bandit who picked the first item the user preferred. This creates a bias towards the first item a user prefer and helps recommending users of same user type with one preferred item which is highly likely to satisfy all of them.

For example, if there is a total of 100 users in the system, each of them is represented by a user vector of size 10 (expressing their items of choice out of 10 available items). Lets assume we can only recommend 2 items out of 10. Say, some of these users prefer *item1, item3, item5, item7 and item9* together (**user type-1**) whereas some prefer *item2, item4, item6, item8 and item10* together (**user type-2**) and remaining prefer *item4 and item6* together(**user type-3**). UCB1 ensures that an item is recommended enough number of times to be reasonably confident about their chance of getting rewarded. Now if we reward a bandit with 5 (accumulated from *items 1,3,5,7 and 9*) for selecting *item1*, the probability of selecting *item1* will be higher for our recommendation set which can eventually cover any user of user type-1. Next item in our recommendation set would be *item2* for adopting the same rewarding strategy with a hope to cover the user type-2. But unfortunately item2 cannot cover user type-3 who prefer *item4 and item6*, not *item2*. If we could pick *item4* instead of *item2* we could cover both user types which is the optimal recommendation. But as recommending item2 fails to satisfy the users only preferring *item4* and *item6* together, according to UCB1 policy, average reward for a bandit picking *item2* will be decreased if more of the incoming users are of user type-3. Eventually average reward of a bandit selecting item4 will beat that of selecting *item2*. Although the basic idea behind this design follows PRP like IBA, but unlike IBA, unequal rewarding for bandits on selecting an item makes the highly rewarded bandit choose a representative item covering all other items that are correlated to it. Ultimate goal of this mechanism is to reduces the chance of selecting more than one item in our recommendation set preferred by the same user type.

6   Experiment Results

In this section we show the performance boost occurred over the existing methods by adopting our proposed technique. We used real data sets our experiments.

6.1   Data Set

Publicly available dataset of Jester Project [5] and MovieLens Dataset [7] has been used for this purpose. Jester Project has a small dataset is a collection of

Algorithm 1.. Proposed Method

1: **Input: n items**
2: **Output: k items**
3: **Initialize** $UCB1_1(n), ... UCB1_k(n)$
4: **for all** $t \in T$ **do**
5:     $S_0^t \leftarrow \emptyset$
6:     **for all** $i = 1$ **to** k **do**
7:         $select(UCB1_i, N \setminus S_{i-1}^t)$
8:     **end for**
9:     **Pick a random user vector from the dataset**
10:     **Display** S^t **to user for and receive feedback vector** X^t
11:     $C =$ **total number of items clicked by the user from** S^t
12:     **Feedback:**
13:     $F_{it} = \begin{cases} C, & \text{if } X_i^t = 1 \text{ for the } i \in S^t \text{ which is the first click} \\ 1, & \text{if } X_i^t = 1 \text{ for any } i \in S^t \text{ which is not the first click} \\ 0, & \text{otherwise.} \end{cases}$
14:     $update(UCB1_i, F_{it})$
15: **end for**

---

user ratings on jokes where the range of rating runs from -10.0 (not funny at all) to 10.0 (very funny). It consist of more than 17000 users rating on 10 jokes (which most users have rated). The real valued ratings has been converted to binary (relevant or not) by using a threshold rule of exceeding θ which is set to 3.5 for our experiment. MovieLens Dataset consist of more than 900 users rating almost 1700 movies which we picked based on most rated (rating ranges as discrete numbers between 1 to 5) and for that θ value has been conservatively set to 2.

6.2   Measurement Metrics

Arrival of a random user at time t is simulated by choosing a vector X_j iid from the unknown distribution D of user vector X. Then the recommendation system present a set of k items S_t without even observing X_t. To measure the fitness of the recommended set, we used the definition of set relevance function F from [6]:

Definition 2. $F(X_t, S_t)$ *is the payoff for showing* S_t *to user with relevance vector* X_t. *Characterization of user click event is done by the following conditions: if* $X_j^i = 1$ *for some i in* S_t *then* $F(X_t, S_t) = 1$ *else* $F(X_t, S_t) = 0$

So the value of displaying a set S_t to all the users, is the expected value $E[F(S_t, X)]$ where the expectation is taken over realization of the relevance vector X from the distribution D. Once realized, for a fixed S, it is denoted as $E[F(S)]$ (the fraction of users satisfied by at least one item in S). The main goal is to minimize abandonment, so we need to maximize click through rate by:

$$\text{maximize} \quad E[F(S)]$$
$$\text{subject to} \quad |S| \leq k$$

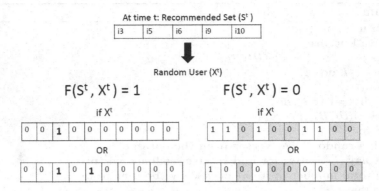

Fig. 3. Payoff for a recommendation set according to users click response

In online setting we update S when it user vector is disclosed and try to maximize $E[F(S)]$ for a the new user arrived.

6.3   Empirical Evaluation

According to the results shown in Fig 4 and 5 our method clearly outperforms RBA in terms of efficiency and accuracy for both the data sets. It is competitive with IBA for Jester dataset [Ref. Fig 4(left)] when we tried recommending 5 out of available 10 high rated jokes. But our method achieves much better result than IBA when we shrink the recommendation set size to 3 [Ref. Fig 4(right)]. Naturally the overall performance got scaled down as we decrease k. The Y-axis of both the plots (drawn in same scale) projects the % of users satisfied by the recommendation set at the time stamp denoted by X-axis. Each point in both the graph shows an average of 5 runs and went upto 100,000 time steps.

For the other experiment [Ref. Fig 5] we ran with Movielens data set, we decided to keep the recommendation set size (k) fixed as 10, but we had to deal with a much higher volume of available choices S (almost 1700 movies to pick from) for a user. This makes our method run to solve for more than a thousand armed bandit problem - which, to the best of our knowledge has never been tried before. It also shows similar trend of outperforming RBA and IBA on an average, as we simulated till 550,000 time steps. The point in the graph shows the results reached at 10K, 50K, 100K, 200K, 500K, 510K and 550K steps.

7   Conclusion

Unlike personalized recommendation systems [10] which often use collaborative filtering [4], content based filtering citecontb or a hybrid of these two [3], we came up with a recommendation system that fairly satisfies various user types. Previous works [6] argued that dependency among items can be ruled out by

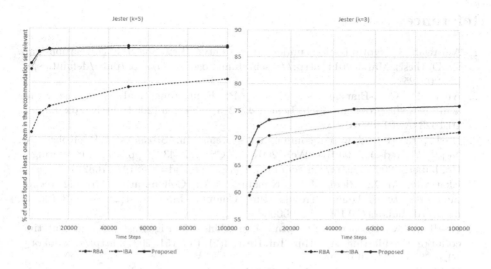

Fig. 4. Performance of recommending 5 jokes out of 10 high rated jokes form Jester dataset of 17000 users and performance of recommending 3 jokes for the same dataset

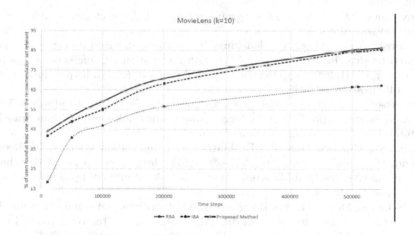

Fig. 5. Performance of recommending 10 movies out of 1700 from MovieLens dataset of around 1000 users

ignoring the correlation gap [1] . But we show that, the correlation between items is an important criteria to identify diversity in terms of user types. We propose a simple rewarding mechanism which aims to incorporate this diversity and our empirical evaluation shows that it outperforms existing techniques for big real data sets in terms of covering different user types without the loss of accuracy, introducing no additional complexity. In future, we want to extend this solution to a distributed recommendation system to facilitate a scalable and decentralized decision making for Big Data.

References

1. Agrawal, S.: Optimization under uncertainty: Bounding the correlation gap. Ph.D. thesis, March 2011. http://research.microsoft.com/apps/pubs/default.aspx?id=200425

2. Auer, P., Cesa-Bianchi, N., Fischer, P.: Finite-time analysis of the multi-armed bandit problem. Mach. Learn. **47**(2–3), 235–256 (2002). http://dx.doi.org/10.1023/A:1013689704352

3. Burke, R.: Hybrid web recommender systems. In: Brusilovsky, P., Kobsa, A., Nejdl, W. (eds.) Adaptive Web 2007. LNCS, vol. 4321, pp. 377–408. Springer, Heidelberg (2007). http://dl.acm.org/citation. cfm?id=1768197.1768211

4. Ekstrand, M.D., Riedl, J.T., Konstan, J.A.: Collaborative filtering recommender systems. Found. Trends Hum.-Comput. Interact. **4**(2), 81–173 (2011). http://dx.doi.org/10.1561/1100000009

5. Goldberg, K., Roeder, T., Gupta, D., Perkins, C.: Eigentaste: A constant time collaborative filtering algorithm. Inf. Retr. **4**(2), 133–151 (2001). http://dx.doi.org/10.1023/A:1011419012209

6. Kohli, P., Salek, M., Stoddard, G.: A fast bandit algorithm for recommendation to users with heterogenous tastes. In: desJardins, M., Littman, M.L. (eds.) AAAI. AAAI Press (2013). http://dblp.uni-trier.de/db/conf/aaai/aaai2013.html#KohliSS13

7. MovieLens dataset. http://www.grouplens.org/data/ (as of 2003). http://www.grouplens.org/data/

8. Radlinski, F., Kleinberg, R., Joachims, T.: Learning diverse rankings with multi-armed bandits. In: Proceedings of the 25th International Conference on Machine Learning, ICML 2008, pp. 784–791. ACM, New York (2008). http://doi.acm.org/10.1145/1390156.1390255

9. Robertson, S.E.: The probability ranking principle in IR. In: Readings in Information Retrieval, pp. 281–286. Morgan Kaufmann Publishers Inc., San Francisco (1997). http://dl.acm.org/citation.cfm?id=275537.275701

10. Shani, G., Gunawardana, A.: Evaluating recommendation systems. In: Recommender Systems Handbook, pp. 257–297 (2011). http://scholar.google.de/scholar.bib?q=info:AW2lmZl44hMJ:scholar.google.com/&output=citation&hl=de&as_sdt=0,5&ct=citation&cd=0

11. Vermorel, J., Mohri, M.: Multi-armed bandit algorithms and empirical evaluation. In: Gama, J., Camacho, R., Brazdil, P.B., Jorge, A.M., Torgo, L. (eds.) ECML 2005. LNCS (LNAI), vol. 3720, pp. 437–448. Springer, Heidelberg (2005)

Learning from Demonstration Using Variational Bayesian Inference

Mostafa Hussein$^{(\boxtimes)}$, Yasser Mohammed, and Samia A. Ali

Assiut University, Asyut, Egypt
{eng.mostafa_2005,yasserm}@aun.edu.eg, Samya.hassan@eng.au.edu.eg

Abstract. Learning from demonstration (LFD) is an active area of research in robotics. There are many approaches to LFD. One of the most widely used approaches is the combination of Gaussian Mixture Model learning for modeling and Gaussian Mixture Regression for behavior generation (GMM/GMR) due to its advantages including easy learning using Expectation Maximization and the simplicity of serializing learned behaviors as well as the ability to model internal correlations and constraints within the task.

A critical parameter that affects the accuracy of learned behavior in GMM/GMR is the number of components in the mixture. A handful of approaches for selecting this number can be found in the literature including classical model selection methods like Bayesian Information Criteria and Akaik Information Criteria and more advanced methods including Dirichlet Process modeling. These approaches are either wasteful of computational resources or hard to implement. This paper introduces a LfD approach which uses GMM with a variational Bayesian Inference (VB) approach to select the number of Gaussians that best fit the data. The proposed method is compared to classical model selection approaches and a recently proposed symbolization based method and is shown to provide an appropriate balance between execution speed and model accuracy.

Keywords: Learning from demonstration · Model selection · Variational bayesian inference

1 Introduction

LFD is a promising methodology that offers a natural mechanism that allows a robot to learn new skills through human guidance, and it is gaining more popularity in robotics due to its promise of providing a human-friendly technique for teaching robots new skills by robotics-naive users [12].

LFD systems can be divided into two categories: Low level trajectory and complex behavior (high level behavior) learning. A high level behavior learning system learns the generation of complex behaviors based on demonstrations given a predefined set of basic actions. An example of this approach can be found in [11].

© Springer International Publishing Switzerland 2015
M. Ali et al. (Eds.): IEA/AIE 2015, LNAI 9101, pp. 371–381, 2015.
DOI: 10.1007/978-3-319-19066-2_36

This paper is interested in low level trajectory learning of motion primitives. There are four main approaches to LFD in this context. The earliest of these methods is Inverse Optimal Control. The main idea is to have the robot learn a model of the task from its own attempts to perform it then using the single demonstration provided to the robot to learn a performance criterion to be optimized. Optimal control can then be used to execute the demonstrated behavior [1]. Another related approach is Inverse Reinforcement Learning (IRL). In IRL, the demonstrations are used to infer a reward function that is then used to learn a policy for the robot [2][6].

Recently, two main approaches to LFD gained popularity within the robotics research community. The first is Dynamic Movement Primitives [7] (DMP) and the second is based on statistical modeling (e.g. Gaussian Mixture Modeling/ Regression). DMP provides a framework in which motion is modeled by a dynamical system with a point attractor or a limit cycle that is modulated by a nonlinear function to move on a trajectory similar to the demonstrated trajectory.

The main advantage of this method is the flexibility in changing the motion's starting position, goal or speed, and main disadvantage of this method is that it is difficult to build complex tasks from segmented demonstrations by serializing them.

The final approach is based on probabilistic modeling of the demonstration in a way that facilitates regeneration of the learned behavior using several possible representation schemes including Hidden Markov Models (HMM) [14] , Gaussian processes (GP) and (GMM/GMR).

A technique of special importance in this category that found a widespread utilization is the Gaussian Mixture Modeling/Gaussian Mixture Regression (GMM/GMR) [8]. The first step in GMM modeling is to use Dynamic Time Wrapping (DTW) to unify the length of the time series. GMM modeling is then achieved using Expectation Maximization (EM) after initialization of the model using for example K-means [8].

One challenge to this approach is finding an optimal number of Gaussians appropriate to the data being modeled without over or under fitting.

In this paper, we propose using Variational Bayesian Inference (VB)[9] to solve this problem. The paper compares this approach with classical methods of model complexity selection like Bayesian Information criteria (BIC) as well as a recently proposed approach that utilize Symbolic Aggregate approXimation (SAX) [4].

The rest of this paper is organized as follows: Section 2 gives the problem statement and related work. Section 3 details the VB algorithm which is evaluated in section 4. The paper is then concluded.

2   Problem Statement and Related Work

Given a set of two or more multidimensional time series x_n of length T_n with $D+1$ dimensions, 1 temporal and D spatial, representing the different occurrences of the demonstration to be learned, we need to find an appropriate number of

Gaussian components K that fits this data efficiently and build a GMM that represents this data set for future reproduction using GMR.

A mixture model is defined by a probability density function given in equation 1.

$$p(x_n) = \sum_{k=1}^{K} p(K)p(x_n|K) \tag{1}$$

where x_n is a data point, $p(k)$ is the prior, and $p(x_n|k)$ is the conditional probability density function. For a Gaussian Mixture Model we have:

$$p(K) = \pi_k \tag{2}$$

$$p(x_n|k) = N(x_n; \mu_k; \Sigma_k) \tag{3}$$

$$p(x_n|k) = \frac{1}{\sqrt{(2\pi)^D|\Sigma_k|}} \exp -\frac{1}{2}((x_n - \mu_k)^T \Sigma_k^{-1}(x_n - \mu_k) \tag{4}$$

The parameters of this GMM are: π_k, μ_k, Σ_k.

Several approaches have been used to find the optimal number of component (k) that fit the data. The first and most used approach is Bayesian Information Criterion (BIC). BIC tries to balance the complexity of the learned model and the log-likelihood of predicting training data. It is composed of two components (see equation 5). The first one is the log-likelihood function which indicates how much the data fits the model and the second is the cost function of model complexity which tries to penalize overly complex models. The GMM modeling is done repeatedly in a range of K values then the one that minimizes BIC is selected.

$$BIC = -l + \frac{n_p}{2} log(N) \tag{5}$$

where N is the number of data points of D dimensions. The likelihood l and model parameter number n_p are calculated according to the following equations:

$$l = \sum_{j=1}^{N} log(p(X_n)) \tag{6}$$

$$n_p = (K - 1) + K \left(D + \frac{D(D+1)}{2} \right) \tag{7}$$

The main disadvantage of this approach is that it is time consuming to find K especially if the range searched for K is wide.

The second approach is to use Dirichlet process GMR models [10]. This is an efficient algorithm to set the number of components. It is effectively a infinite mixture of components that utilizes the discrete property of Dirichlet processes to lead to an effective small number of components in the model. The approach though is difficult to implement and in this paper we propose another variational approach that is much easier to implement while being able to achieve an accuracy comparable to BIC based model selection. This simpler approach would be more appropriate for low dimensional data like the case when learning is conducted in the operation space rather than the configuration space.

The third approach is to use Symbolic Aggregate approXimation Gaussian Mixture Model (SAX GMM) [4] which first uses DTW to unify time-series lengths then uses an extension of Z-score normalization to normalize the multidimensional data points. A multidimensional extension of SAX is utilized to produce a string representation for the model and finally a piecewise linear fit of the symbols of the generated string is used to estimate a number of components (K). GMM is finally called on the original data with the discovered K.

The advantage of this approach is that it can handle distortions and confusions in the demonstrations and work faster than the previous approaches, but it is less accurate than BIC in estimating an appropriate number of Gaussians.

3 Proposed Algorithm

At first Like all algorithms we start with DTW to unify the length of all demonstrations [13] Then we start VB. The method used here was first introduced in [9].

Algorithm 1 Proposed algorithm

1: Use DTW to equalize the length of all demonstrations.
2: Initialize VB using k-means .
3: *loop* :
4:   Evaluate $\mathbb{E}[z_{nk}] = r_{nk}$ ▷ Variational E step
5:   Recalculate the Distribution parameters $\alpha_k, \beta_k, W_k, \nu_k$ ▷ Variational M step
6:   Calculate the lower bound $L(Q)$ ▷ Stop criteria
7: **if** $L(Q)$ increases **then**
8:      **goto** *loop*.
9: Calculate exact value of π, μ and Λ using EM .
10: Generate Model using GMR .

Equation 1 defines the GMM model. Let us introduce a binary random variable z_n for each data point x_n comprising a $1-of-K$ that leads to one particular element z_k equaling 1 while all other x_k values equaling zero. The value of z_k satisfies $z_k \in \{0, 1\}$ and $\sum_k z_k = 1$ so we can write the conditional distribution of z ,given the mixing coefficient π ,in the form

$$p(Z|\pi) = \prod_{n=1}^{N} \prod_{k=1}^{K} \pi_k^{z_{nk}} \tag{8}$$

and the conditional distribution of the observed data ,given the latent variable, will be:

$$p(X|Z, \mu, \Lambda) = \prod_{n=1}^{N} \prod_{k=1}^{K} N(x_n|\mu_k, \Lambda_k^{-1})^{z_{nk}} \tag{9}$$

where Λ is the precision matrix and μ is the mean.

To evaluate $p(x|\pi)$ we have to marginalize the following equation with respect to z, μ, Λ which is analytically intractable . so we will use a variational approximation to evaluate an approximate value of it [15] [16].

Bishop et al.[9] have introduced a novel algorithm to solve this problem that we will use in this section. The VB algorithm has two steps similar to the Expectation and Maximization steps of the EM algorithm (and with only slightly higher computational overhead). The first is the E step where we use the current distribution to calculate $\mathbb{E}[z_{nk}] = r_{nk}$

$$\mathbb{E}_{\mu_k, \Lambda_k}[(X_n - \mu_k)^T \Lambda_k (X_n - \mu_k)] = D\beta_k^{-1} + v_k(X_n - m_k)^T W(X_n - m_k) \quad (10)$$

$$Ln\tilde{\Lambda}_k = \mathbb{E}[ln|\Lambda_k|] = \sum_{i=1}^{D} \phi(\frac{v_k + 1 - i}{2}) + DLn2 + Ln|W_k| \quad (11)$$

$$Ln\tilde{\pi}_k = \mathbb{E}[ln\pi_k] = \phi(\alpha_k) - \phi(\widehat{\alpha}) \quad (12)$$

$$Ln\rho_{nk} = \mathbb{E}[Ln\pi_k] + \frac{1}{2}\mathbb{E}[Ln|\Lambda_k|] - \frac{D}{2}Ln(2\pi) - \frac{1}{2}\mathbb{E}_{\mu_k, \Lambda_k}[(X_n - \mu_k)^T \Lambda_k (X_n - \mu_k)] \quad (13)$$

$$r_{nk} = \frac{\rho_{nk}}{\sum_{j=1}^{K} \rho_{nj}} \quad (14)$$

Where β_0 and W_0 are the parameters of a Gaussian-Wishart distribution which represents the mean and precision of the Gaussian component and α_0 is the parameter of Dirichlet process which represent π, The second step is to recalculate the Distribution parameters form the new r_{nk} as follows:

$$N_k = \sum_{n=1}^{N} r_{nk} \quad (15)$$

$$\alpha_k = \alpha_0 + N_k \quad (16)$$

$$\beta_k = \beta_0 + N_k \quad (17)$$

$$m_k = \frac{1}{\beta_k}(\beta_0 m_0 + N_k \bar{X}_k) \quad (18)$$

$$v_k = v_0 + N_k \quad (19)$$

$$\bar{X}_k = \frac{1}{N_k} \sum_{n=1}^{N} r_{nk} X_n \quad (20)$$

$$S_k = \frac{1}{N_k} \sum_{n=1}^{N} r_{nk}(X_n - \bar{X}_k)(X_n - \bar{X}_k)^T \quad (21)$$

$$W_k^{-1} = W_0^{-1} + N_k S_k + \frac{\beta_0 N_k}{\beta_0 + N_k}(\bar{X}_k - m_0)(\bar{X}_k - m_0)^T \quad (22)$$

At each iteration we have to calculate the lower bound $L(Q)$ (you can find equation at Bishop book[3]). A finite difference is used to check that each update gives a maximum bound otherwise we stop the algorithm and move to the next step.

At the end of the algorithm we will have an approximate value for the priors π, μ and Λ that represent the Gaussian components. To get an exact value of these parameters we use the standard EM algorithm to calculate them. finally we use GMR to reproduce the demonstrated action.

4   Evaluation

We have tested the three algorithms BIC-GMM,SAX-GMM and the proposed VB-GMM in two sets of data: Arabic letters and English letters. These two set

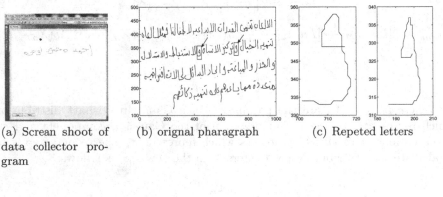

(a) Screan shoot of data collector program

(b) orignal pharagraph

(c) Repeted letters

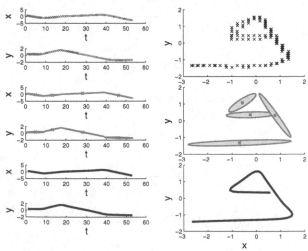

(d) Final result

Fig. 1. Steps of proposed algorithm

of data have been collected using a small program executed in a tablet laptop with a pen. The trajectory of the data point written by the pen was stored in a text file to be used for learning.

Fig. 1(a) shows a screen-shot of the program that we used to collect data. Three subjects were recruited. Each one was asked to write a paragraph and then all this data was saved as one time series and entered into a modified motif discovery algorithm that segmented these words into small time series and found the occurrences of the letters [5]. Fig. 1 shows the steps of collecting data ,segmenting it,finding repeated letters and finally learning it by each of the three compared algorithms.

Two criteria were used for comparing the three algorithms: Firstly, the Kullback Leibler Divergence (KLD) between the learned model and that of BIC was calculated. To do so, we used the fact that GMR returns for each timestep both the mean and covariance matrix of a Gaussian. By averaging the KLD at each point we get an estimate of the total difference between the model learned by each algorithm and the – assumed optimal – BIC. As shown in Fig. 2 the VBGMM algorithm leads to models that are more similar to the BIC model than SAX GMM.

$$DKL = \frac{1}{2}(tr(\Sigma_1^{-1}\Sigma_0) + (\mu_1 - \mu_0)^T \Sigma_1^{-1}(\mu_1 - \mu_0) - K - Ln(\frac{det(\Sigma_0)}{det(\Sigma_1)})) \quad (23)$$

The second criterion for the evaluation is the speed of the three algorithms. As shown in Fig. 3, VB is much faster than the BIC and slower than SAX. This difference increases with increased motion complexity (translated to an increase in the maximum allowable value for k in all algorithms). This can be confirmed by fitting these three curves by a line and calculating the slope of the three resulting lines. we find the slope of BIC-GMM = .5746 and VB-GMM = .1603 and SAX-GMM was almost equal zero. This means that SAX GMM has the best scalability in the three algorithms while BIC has the worst scalability The proposed algorithm fits in the middle of these two extremes.

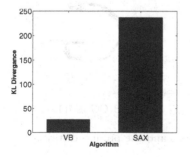

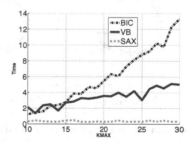

Fig. 2. DKL   Fig. 3. TIME

Taken together these results show that the proposed algorithm provides an appropriate balance between accuracy and speed outperforming SAX GMM in both aspects and providing similar results to BIC with much faster execution speed. Fig. 4 shows four examples of letters discovered and modeled in this experiment.

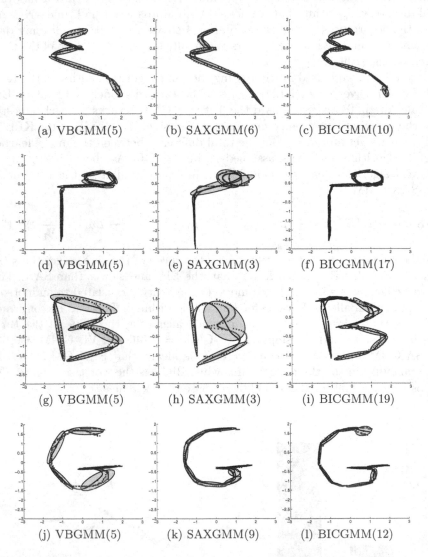

Fig. 4. Arabic and English letters demonstrations and the produced trajectory

The second experiment aimed at comparing the performance of the three algorithms with *intended* motion of the demonstrator rather than normalizing with BIC output. To focus on the LfD aspect of the system, subjects in this experiment were instructed to draw separate letters. The subject selects a letter and draws one example of it on a laptop with a touch screen which serves as the ground truth. The subject is then asked to repeat the drawing of the same character four times on top of a grayed version of this *ground-truth* figure as fast yet as accurately as possible. This allows us to have some thing like ground truth for each letter (i.e. the first drawing). Each subject was asked to write seven sets of letters each containing ground truth and 4 demonstrations for both Arabic and English letters. Finally we got 28 demonstrations for each language then we randomly choose 5 Arabic and 5 English letters and ran the three algorithms on them.

Two criteria were used for evaluating the results. Firstly, we calculate the Euclidean Distance between the ground truth points (a, b) and the mean of the GMR output (c, d) .as shown in Equation 24.

$$ED = \sum_{n=1}^{N} \sqrt{(a_n - c_n)^2 + (b_n - d_n)^2} \qquad (24)$$

where N is number of points. As shown in Fig. 5(a) BIC produced the most faithful reproduction of the demonstration's mean ($ED_{bic} = 16$), followed by the proposed algorithm ($ED_{vb} = 22.9$) and both were much better than SAX GMM which had a distance $ED_{sax} = 54.1$.

The second criterion used the fact that GMR returns the mean and covariance matrix of a Gaussian at each time-step. A likelihood of the ground truth under the output of each algorithm can be calculated by calculating the cumulative probability of the GMR output in a small area around each point of the ground truth (x_{gt}) at each time-step t. This gives us the probability of producing the given ground truth point under the distribution produced by the GMR output. Averaging again over all points of the ground truth drawing, we get a single similarity measure between the GMR output and the ground truth. The results of this evaluation are shown in Fig. 5(b). Again, and as expected, BIC

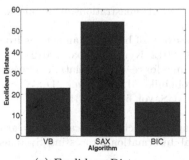

(a) Euclidean Distance

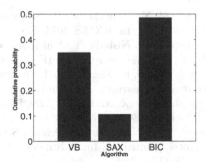

(b) Cumulative probability

Fig. 5. Results of the second experiment comparing the learned model with ground-truth data.

gave the best (e.g. maximum) result with a similarity measure of 0.48 followed closely by the proposed VB algorithm with a similarity score of 0.34 while SAX-GMM gave a much smaller similarity score of 0.11. The results of this second experiment confirms the findings in the first experiment. The proposed VB approach achieves comparable performance to BIC for model selection while having much better scalability and speed. On the other hand, the proposed approach, while slower, provides much improvement over SAX-GMM in terms of similarity to BIC output (first experiment) as well as modeling of the ground-truth demonstration.

5   Conclusions

This paper presented an evaluation of using Variational Bayesian Inference (VB) for selecting an appropriate number of Gaussians to be used in the GMM/GMR framework for learning from demonstrations. The proposed approach provides an appropriate balance between execution speed and accuracy leading to faster execution speed compared with the most commonly used method for model selection in this context (i.e. BIC) and more accurate models compared with the recently proposed faster SAX GMM method. This approach have been tested on 2D Data representing Arabic and English letters after extracting it form continuous streams of writing motions and could successfully learn basic letters existing in the time-series.

References

1. Ratliff, N., Ziebart, B.D., Peterson, K., Bagnell, J.A., Hebert, M., Dey, A., Srinivasa, S.: Inverse optimal heuristic control for imitation learning. In: Intl. Conf. on Artificial Intelligence and Statistics (AIStats) (2009)
2. Abbeel, P., Ng, A.Y.: Apprenticeship learning via inverse reinforcement learning. In: Proc. Intl. Conf. on Machine Learning (ICML) (2004)
3. Bishop, C.M.: Pattern Recognition and Machine Learning. Springer, New York (2006)
4. Mohammad, Y., Nishida, T.: Robust learning from demonstrations using multidimensional SAX. In: ICCAS 2014, Korea (2014)
5. Mohammad, Y., Nishida, T.: Unsupervised discovery of basic human actions from activity recording datasets. In: IEEE/SICE SII 2012, Kyushu, Japan (2012)
6. Lopes, M., Melo, F., Montesano, L.: Active learning for reward estimation in inverse reinforcement learning. In: Buntine, W., Grobelnik, M., Mladenić, D., Shawe-Taylor, J. (eds.) ECML PKDD 2009, Part II. LNCS, vol. 5782, pp. 31–46. Springer, Heidelberg (2009)
7. Ijspeert, A.J., Nakanishi, J., Schaal, S.: Learning attractor landscapes for learning motor primitives. In: Advances in Neural Information Processing Systems, pp. 1523–1530 (2002)
8. Calinon, S., Guenter, F., Billard, A.: On Learning, Representing and Generalizing a Task in a Humanoid Robot. IEEE Transactions on Systems, Man and Cybernetics, Part B **37**(2), 286–298 (2007). Special issue on robot learning by observation, demonstration and imitation

9. Bishop, C.M., Corduneanu, A.: Variational bayesian model selection for mixture distributions. In: Jaakkola, T., Richardson, T. (eds.) Artificial Intelligence and Statistics 2001, pp. 27–34 (2001)
10. Chatzis, S.P., Korkinof, D., Demiris, Y.: A nonparametric bayesian approach toward robot learning by demonstration. Robotics and Autonomous Systems **60**(6), 789–802 (2012)
11. Niekum, S., Osentoski, S., Konidaris, G., Barto, A.G.: Learning and generalization of complex tasks from unstructured demonstrations. In: IEEE/RSJ International Conference on Intelligent Robots and Systems, pp. 5239–5246 (2012)
12. Argall, B., Chernova, S., Veloso, M., Browning, B.: A survey of robot learning from demonstration. Robotics and Autonomous Systems **57**(5), 469–483 (2009)
13. Sakoe, H., Chiba, S.: Dynamic programming algorithm optimization for spoken word recognition. IEEE Transactions on Acoustics, Speech and Signal Processing **26**(1), 43–49 (1978)
14. Calinon, S., D'halluin, F., Sauser, E.L., Caldwell, D.G., Billard, A.G.: Learning and reproduction of gestures by imitation: An approach based on Hidden Markov Model and Gaussian Mixture Regression. IEEE Robotics and Automation Magazine **17**(2), 44–54 (2010)
15. Bishop, C.M.: Variational principal components. In: Proceedings Ninth International Conference on Artificial Neural Networks, ICANN 1999, vol. 1, pp. 509–514. IEEE (1999)
16. Jordan, M.I., Ghahramani, Z., Jaakkola, T.S., Saul, L.K.: An introduction to variational methods for graphical models. In: Jordan, M.I. (ed.) Learning in Graphical Models, pp. 105–162. Kluwer (1998)

Modeling Snow Dynamics Using a Bayesian Network

Bernt Viggo Matheussen[1][(✉)] and Ole-Christoffer Granmo[2]

[1] Agder Energi AS, University of Agder, Kristiansand, Norway
bermat@ae.no
http://www.ae.no/
[2] University of Agder, Grimstad, Norway
ole.granmo@uia.no
http://www.uia.no/

Abstract. In this paper we propose a novel snow accumulation and melt model, formulated as a Dynamic Bayesian Network (DBN). We encode uncertainty explicitly and train the DBN using Monte Carlo analysis, carried out with a deterministic hydrology model under a wide range of plausible parameter configurations. The trained DBN was tested against field observations of snow water equivalents (SWE). The results indicate that our DBN can be used to reason about uncertainty, without doing resampling from the deterministic model. In all brevity, the DBN's ability to reproduce the mean of the observations was similar to what could be obtained with the deterministic hydrology model, but with a more realistic representation of uncertainty. In addition, even using the DBN *uncalibrated* gave fairly good results with a correlation of 0.93 between the mean of the simulated data and observations. These results indicate that hybrids of classical deterministic hydrology models and DBNs may provide new solutions to estimation of uncertainty in hydrological predictions.

Keywords: Hydrology · Hydropower · Forecasting · Runoff · Snowmelt

1 Introduction

Understanding of uncertainty in hydrological predictions is crucial for hydropower operations and water resources management. In addition to errors related to future climate and initial conditions, uncertainty also stems from the computational models (parameters and model structure) and the observed data itself (climate forcings and streamflow data) [1]. To cope with uncertainty in hydrological simulations, various techniques and analysis frameworks have been presented. In [1–4], several examples and references to work on hydrologic uncertainty are provided. They include Bayesian uncertainty analysis methods, model error analysis methods, neural network techniques, and several more.

A review of uncertainty modeling techniques developed and applied within the scientific field of hydrology reveals interesting similarities and tendencies.

© Springer International Publishing Switzerland 2015
M. Ali et al. (Eds.): IEA/AIE 2015, LNAI 9101, pp. 382–393, 2015.
DOI: 10.1007/978-3-319-19066-2_37

First of all, most of the techniques require multiple runs of (or sampling from) deterministic models. To facilitate multiple runs, the deterministic models used are often simplistic, with low computational demands. With the exception of [5], integration of uncertainty in the hydrologic model-core itself is almost non-existing. Furthermore, the majority of the uncertainty studies presented cover warm or temperate climates. The more challenging case of snow dominated river systems, on the other hand, is sparsely studied.

As opposed to the approaches found in the literature, this paper introduces a hybrid hydrological model that extend current deterministic physical models with the advantages offered by Dynamic Bayesian Networks (DBNs) [6]. DBNs are widely used for knowledge representation and reasoning under uncertainty. The advantages offered by DBNs are manifold, including the ability to propagate uncertainty throughout a complex model structure. Such a method thus possesses several advantages compared to classical hydrologic approaches with Monte Carlo sampling from deterministic hydrological models. Furthermore, the DBN's ability to perform diagnostic reasoning (explanation of system behavior), may potentially be a tool for decomposition of global uncertainty in physically based hydrologic modeling. According to [1], this is very difficult, or even impossible, with the majority of methods found in the literature.

The rest of this paper is organized as follows. We present our methodology in Sect. 2, exposing its advantages over sampling based approaches. Then, in Sect. 3, we provide information about the field observations used to test our methodology in realistic settings. Furthermore, results from the simulations with the DBN and a deterministic hydrology model is presented in Sect. 4. We finally discuss the results, draw some conclusions, and provide pointers for further work.

2   A Novel Dynamic Bayesian Network for Snow Accumulation and Melt Modeling

Since the emergence of digital computers a large number of snow accumulation and melt models have been developed. They range in complexity from simple linear approaches [7], to the more complex physically based models ([8], and many others).

In its simplest form a snow accumulation and melt model is an expression of the mass balance of a snow pack. This is shown in Eq. 1. SWE means the total amount of water stored in the snow (vertical depth). The change in SWE is a function of snow accumulation (SP_t) and melt (M_t). The model is typically applied as a time series model with a daily time step length. Rainfall, sublimation, horizontal mass transport and other physical processes are ignored.

$$SWE_t = SWE_{t-1} + SP_t - M_t \tag{1}$$

$$SP_t = \begin{cases} Prcp_t \cdot PCorr & \text{if } T_{airt,t} < 0.0 \\ (Prcp_t - (0.5 \cdot T_{airt,t} \cdot Prcp_t)) \cdot PCorr & \text{if } 0.0 \leq T_{airt,t} < 2.0 \\ 0.0 & \text{if } T_{airt,t} \geq 2.0 \end{cases} \tag{2}$$

The snow accumulation (SP_t) In Eq. 1 is the amount of total precipitation that can be assumed to be snowfall. Usually this is estimated from observed records of precipitation and air temperature. According to the work of [9], who tested several precipitation phase determination schemes, snow and rain fractions can be calculated using a simple linear formula (Eq. 2).

In Eq. 2, $Prcp_t$ means daily total precipitation and $T_{airt,t}$ is the daily average air temperature in a time step. Units are in millimeters and Celsius, respectively. We can see that when air temperature is below zero, all precipitation is counted as snow. And vice versa, when the air temperature is above 2.0 C, all precipitation is assumed to be rain. Between 0.0 C and 2.0 C, precipitation is split into both rain and snow fractions. Note that the catch of precipitation in the gauge is highly dependent on windspeed. This results in the gauge measuring less precipitation than it should. A simple correction of the catch efficiency is therefore used. This is represented by $PCorr$ in Eq. 2.

$$M_t = \begin{cases} C_x \cdot T_{airt,t} & \text{if } T_{airt,t} > T_{thres} \text{ and } SWE_{t-1} > C_x \cdot T_{airt,t} \\ SWE_{t-1} & \text{if } T_{airt,t} > T_{thres} \text{ and } C_x \cdot T_{airt,t} \geq SWE_{t-1} \\ 0.0 & \text{if } T_{airt,t} \leq T_{thres} \end{cases} \quad (3)$$

A well used approach for calculation of snow melt is the degree-day method [7], shown in Eq. 3. It can be seen that snow melt (M_t) is calculated as a linear function of air temperature $(T_{airt,t})$ and the degree-day factor $(C_x$, see also Table 1). The idea behind this method is that melt can only occur when the air temperature is above a certain threshold (T_{thres}). We can also se from Eq.3, that the amount of melt is constrained by the snow availability.

Eq. 1 to 3, with the assumptions described above, represent a point snow model in its simplest form. The model is typically initialized in the warm season when there is no snow on the ground. Uncertainty is not accounted for in any way, so the model can be classified as being purely deterministic. Altogether there are three parameters in the model. They are shown in Table 1, together with a description, units, and plausible ranges of the parameter values. The range of each parameter has been investigated by several researchers ([7], and others), and is comparable to the parameter ranges shown in Table 1.

Table 1. Parameters and their range used in the snow model

Parameter	Description	Units	Range [Min,Max]
C_x	Degree day factor	$\frac{mm}{day \cdot C^\circ}$	[1.0, 6.0]
T_{thres}	Threshold air temperature (start melt)	C°	[0.0, 2.5]
$PCorr$	Gauge correction of precip.	-	[1.0, 1.6]

In the present work we implement two versions of the snow accumulation and melt model described in Eqs. 1 to 3. At first we coded the equations as a deterministic hydrological model (HM), that is, as a time series model without uncertainty considerations included. The majority of the models presented in the field of hydrology is coded like this. Throughout the rest of this paper we refer to this deterministic implementation as HM.

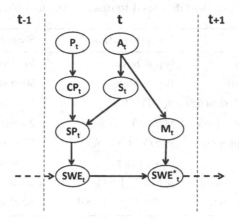

Fig. 1. Schematic of the DBN used to model snow accumulation and melt

In the second version, we introduce our novel snow model, taking advantage of Dynamic Bayesian Networks (DBNs) [6]. A DBN integrates concepts from graph theory and probability theory, capturing conditional independencies between a set of random variables by means of a directed acyclic graph (DAG). Each directed edge in the DAG typically represents a cause-effect relationship. This allows the joint probability distribution of the variables to be decomposed based on the DAG. Thus, the full complexity of the joint probability distribution can be represented by a limited number of simpler conditional probability tables. Leveraging the power of the DBN framework, we compactly encode hydrological knowledge, explicitly representing uncertainty. Such an approach has similarities to the work [5], however, they focus on flood alarming and manual DBN construction, while we here focus on automatic construction of snow accumulation and melt models.

In Fig. 1 the DBN structure that we propose is summarized. For each node shown in Fig. 1, the state space of the associated hydrological variable is defined in the discrete domain, with causal relationships among the variables being expressed as conditional probability tables (CPT).

Table 2 provides further details on each node presented in Fig. 1. With the exception of the two input nodes (P_t and A_t), each node defines a variable or process conditioned on the other nodes. For example the SWE in one time step is conditioned on the SWE in the previous time step, amounts of new snow and melt in the current step. So in essence it is the CPT in each node that represent the dynamics of the physical processes. For each of the CPTs every variable/process must be split into a set of distinct states. The right column in Table 2 illustrates how many states, and the assumed range [Min,Max] that each variable can handle. For most variables the units are in millimeter, except for air temperature that uses Celsius.

Table 2. Elements of the DBN used to describe the hydrological processes

Node/Process	Variables	Causal relation	States/[Min,Max]
P_t	Precip.	Input node	51 states, [0,250]
A_t	Airtemp.	Input node	50 states, [-0.5, 24.0]
CP_t	Corrected precip.	$p(CP_t \mid P_t)$	51 states, [0,250]
S_t	Snow fraction	$p(S_t \mid A_t)$	2 states, snow, no snow
SP_t	Snow precip.	$p(SP_t \mid CP_t, S_t)$	51 states, [0,250]
M_t	Melt	$p(M_t \mid A_t)$	41 states, [0,200]
SWE_t	SWE + snow	$p(SWE_t \mid SP_t, SWE_{t-1}^*)$	1000 states, [0,5000]
SWE_t^*	SWE - melt	$p(SWE_t^* \mid M_t, SWE_t)$	1000 states, [0,5000]

The number of states , and the width of each state, was selected based upon several criteria. Ideally it is desirable to have as few states as possible to limit computational expense. On the other hand, a certain level of detail is necessary to capture the main dynamics of the physical processes involved. Further on the range of the states have to cover all plausible outcomes of the processes studied. For precipitation, we assumed a maximum daily value of 250 millimeters, of which none of the historical data exceeded. Similar thinking was applied when setting the number of states and ranges for the other nodes.

Before the DBN can be used in simulations and inference, each of the CPTs presented in Table 2 need to encode appropriate hydrological knowledge. This can be done with the help of expert opinion, observed field data, or by use of other models. In this work, to leverage existing literature and data sets, we chose to train some of the CPTs by using a Monte Carlo analysis with the deterministic hydrology model (HM), while other tables were populated using expert opinion.

The CPT for the node S_t in Fig. 1 was populated using expert opinion. In [9], a simple model (Eq. 2) was recommended for separating precipitation into snow and rain. So the node CPT S_t was set manually as a direct translation of Eq. 2. An interval of 0.5 degrees Celsius was assumed to provide a compromise between a need for physical detail, and limit the computational expense. In the CPT an air temperature of 1.0 C, corresponds to a probability of 50 % for snow and no-snow fractions.

The training of the nodes SWE_t, and SWE_t^* (Fig. 1) was simple. The reason is that this is just the mass balance shown in Eq. 1, expressed as a CPT. In the node SWE_t the snow precipitation (SP_t) is added to the snow pack. In the node SWE_t^* the melt is subtracted from the snow pack. Based on the mass balance, the probabilities were set manually. Also in the node SP_t, the CPT could be set manually since the node S_t only uses two states.

2.1 Training of the Conditional Probability Tables

The CPTs in the two nodes CP_t and M_t (Fig. 1) were trained using a Monte Carlo method where 5000 samples were taken from the deterministic hydrology model (HM). For each of the runs the value of the three parameters C_x, T_{thres}, and $PCorr$ were assumed to be distributed uniformly with the minimum and maximum values shown in Table 1. Further on, for each run, each parameter set was given a weight depending on the HM's performance. In total three different weighting methods were tested. So for each simulation in the Monte Carlo analysis the weight of the parameter set was set in the corresponding place in the CPT. After all of the 5000 runs were carried out, each column in the CPT was normalized to unity.

$$p(M \mid z) = \frac{p(z \mid M) \cdot p(M)}{p(z)} \tag{4}$$

$$p(z \mid M) = f(z_t \mid M_{\mu,t}, M_{\sigma,t}) \tag{5}$$

$$W_p = \sum_{t=1}^{t=stps} \log \left[f(z_t \mid M_{\mu,t}, M_{\sigma,t}) \right] \tag{6}$$

We applied three weighting methods. We refer to the first one as BAYES, after Bayes' theorem [2], and was adopted from the work of [2] (pages 5-6). In Eq. 4, the left hand side can be thought of as the probability of the model M being correct given the data. In this work we include model structure, parameters, state variables and outputs, as being part of the "model", but only the effects of parameter uncertainty is included in the analysis.

On the right hand side of Eq. 4, $p(z \mid M)$ is the probability (likelihood) that the observations are true, given the model M. Further on, we follow [2], stating that $p(M)$, is the prior probability that the model M is correct. This is essentially a constant and can be ignored in the calculations. According to [2], $p(z)$ in Eq. 4, is a constant that normalize the posterior probability mass to unity. So the challenge in solving Eq. 4, is essentially the problem of finding a solution for the term $p(z \mid M)$.

Eq. 5 means in practice that for each time step in the simulations we assume a Gaussian distribution for the model (HM), and we find the value of the probability density function (PDF), at the location of z_t. So, $M_{\mu,t}$ and z_t, represents the daily change in SWE for the HM and the observations, respectively. Further on, we assumed a standard deviation of 100 for the simulated data ($M_{\sigma,t}$, eq. 5 and 6). The number was chosen based on the assumption that the model would always have more uncertainty than the observed data. Initial simulations also indicated that a standard deviation of 100 seemed reasonable. These considerations will be addressed in more detail in the following sections.

To avoid taking the product of many small numbers, we use the log transformation of the Gaussian PDF when calculating the weight for each parameter set. This is shown in Eq. 6. W_p is the weight of each parameter set.

We refer to the second weighting method as NASH, named after the Nash-Sutcliffe criterion [10]. This measure has similarities to correlation and calculates a score that is never higher than 1. It is used widely within the hydrological literature. In NASH, each parameter set was given a weight equal to the square of the Nash-Sutcliffe criterion, calculated between simulated and observed daily change in SWE. The reason for using the square was that we wanted to emphasize parameter sets with the highest scores.

Finally, the UNIFORM weighting scheme assigns weights uniformly across the parameter sets".

3   Study Site and Observed Data

In this work it was chosen to use data from the MtHood field site located in northern Oregon, USA (latitude: 45.32, longitude -121.72, elevation 1637 meters above sea level, site number 651). The site is operated by National Resources Conservation Service and collects daily observations of precipitation, air temperature and snow water equivalent (SWE). The data are open and distributed on the web. The site was chosen since it has a relatively steady winter snow pack, and is located at a fairly high elevation.

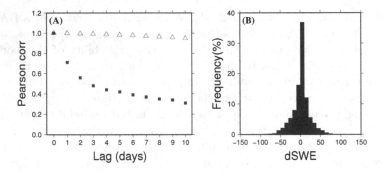

Fig. 2. (A) presents a lagged correlation plot of SWE (triangles) and the first derivative (dSWE, black squares). In (B) we can see a histogram of the dSWE data for all days in November to June.

Figure 2 (a) presents the autocorrelation in SWE (triangles) and in the daily change in SWE (black squares). We can see that the lagged correlation falls rapidly after a day. In Fig. 2(b), we can see that the frequency plot of the change in SWE is somewhat symmetric. In this work it was therefore assumed that the daily change in SWE was Gaussian, without any further investigations. The standard deviation of the observed data shown in Fig. 2(b), was calculated to be 22.

4   Results

Fig. 3 presents the results from the training of the CPT in node M_t (left side) and CP_t (right side), in Fig. 1. The node M_t represent the dependency of air temperature on melt. CP_t is the equivalent to $PCorr$ in Eq. 2, which in practice means the correction of precipitation due to gauge efficiency. In Fig. 3 (a, c) and (b, e) the results of the CPT when training method UNIFORM and BAYES were applied. And in Fig. 3 (c) and (f), the results from NASH training scheme is shown. On the horizontal axis of Fig. 3 (a), (b) and (c), air temperature in Celsius is shown. The horizontal axes in Fig. (d), (e) and (f) represent the levels of observed precipitation, which is input to the model.

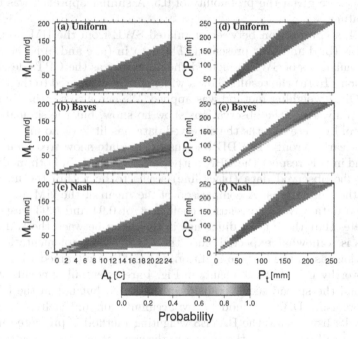

Fig. 3. CPT for node M_t and CP_t in the DBN illustrated in Fig 1. In (a, d) the CPT when trained with method UNIFORM, in (b, e) and (c, f) with method BAYES and NASH, respectively.

In Fig. 3 the colored areas indicate probabilities larger than zero, while the white area has probability of zero. One difference between the results is that BAYES seem to have a sharper or more distinct area where the probabilities are higher, compared to UNIFORM and NASH. From Fig. 3 (e) it can be seen that the values of CP_t are in the range of 1.0 to 1.1. For UNIFORM and NASH (d,f), the correction of precipitation (CP_t) had a more wide range (1.0-1.6), with no distinct regions with high probabilities. This pattern can be seen for both nodes and is caused by the method BAYES giving more weights to the "best" parameter sets. Whilst UNIFORM and NASH seem to weigh the parameters more equal.

After the CPTs had been trained, the DBN shown in Fig. 1 was used to simulate a time series of SWE for the MtHood field site. The DBN was implemented in C++ using the SMILE library [11]. The simulation period was set to 1 October 2000, until 30 September 2010, and the time step length was 24 hours. Due to this research being a "proof-of-concept", no independent validation period was chosen. The runtime for the final DBN was a few seconds and several magnitudes lower compared with the HM using a standard labtop.

The input to the DBN was a daily time series of air temperature and precipitation from the MtHood field site presented in Sect. 3. In every time step the probabilities in two states were set in the input nodes P_t and A_t. For example if the input of precipitation in a time step was 7.5 mm, then two states (5 mm and 10mm), were given the probability of 0.5. A similar approach was used for air temperature.

In Fig. 4, a comparison between simulated SWE from the HM (a, c and e) and the DBN (b, d and f) is presented. The data in (a, c and e) is the result of the 5000 simulations of SWE, weighted the same way as the CPTs described in earlier sections. In (a) the results show a wider spread compared to the DBN (b). The UNIFORM weighing method was applied. We can also see that in some of the months July, Aug, the observations show no snow, but the simulated SWE is above zero. In these months the observed data has little or no uncertainty, so the HM is clearly wrong. The DBN seems to simulate snow free ground fairly correct. And in this respect the DBN outperforms the HM. In both models (HM and DBN) the observed data (black-line) seems to be within the uncertainty bounds of the simulations. A comparison of the mean of the HM and DBN to the observed data gave a correlation coefficient of 0.94 and 0.93, respectively. This indicates that there is no difference in the mean between the HM and the DBN. This is somewhat expected since the DBN is an implementation of the same equations as in the HM, in addition, the CPTs are trained with the HM. More noteworthy is it that the results in Fig. 4 are essentially uncalibrated, but the mean and the spread seems realistic in the DBN, but not in the HM. This clearly favors using DBNs to model snow accumulation and melt.

The results from using the BAYES weighting method is presented in Fig. 4 (c and d). It can be seen that the spread in the simulations are not as wide as in (a) and (b). The correlation between the mean in the models and the observed SWE was calculated to be 0.94 for both the HM and the DBN models. This is similar compared to the UNIFORM weighting methods. In general the results in both the HM and DBN show that the mean is simulated fairly well, but there are larger differences in the spread. In the DBN there seems to be a more realistic simulation of the snow free conditions during the warm season. This favors using DBN compared to HM for simulation of snow accumulation and melt. Noteworthy in (c) and (d), is that for a few seasons the observed SWE, at the time of high accumulation, is outside the uncertainty bounds of the DBN. This happens not as often in the HM as in the DBN. The reason for this is not fully understood and more research is needed. On the other hand, the HM

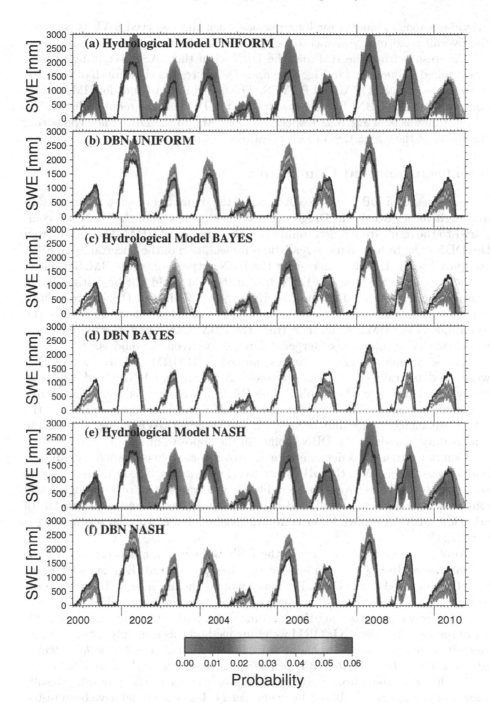

Fig. 4. Simulated uncertainty bounds of SWE. In (a, c, e) the results from the HM is plotted, while the the results from the DBN is shown in (b, d, f). The black line is the observed SWE data from MtHood field site.

provides wrong estimates for long periods when the observed SWE is zero. So, the overall performance seems to be poorer with HM than with the DBN.

The results from the HM and the DBN when the NASH weighting method was applied is illustrated in Fig. 4 (e) and (f). The results show much of the same patterns as in UNIFORM and BAYES. The uncertainty bounds for HM is wider compared to the DBN, but the results for the DBN seems more realistic. The correlation was 0.94 and 0.93 respectively for HM and the DBN when comparing the mean of the simulations to observations.

5   Discussion and Conclusions

In this research a DBN was used to model the dynamics of snow accumulation and melt for a field site in Oregon (USA). The results from the DBN was compared to the result from a deterministic hydrology model (HM). The results from the DBN seem to have a more realistic representation of the uncertainty in SWE compared to the HM. In this respect the DBN outperformed the HM. The mean of the simulated SWE in the DBN and in the HM could not be distinguished. One of the reasons for this may be that the CPTs in the DBN were trained with the results from a Monte-Carlo analysis with the HM. So in essence the dynamics of the HM was used to train the DBN. Other performance measures than correlation, may reveal larger differences between the models.

Three different weighing schemes, named UNIFORM, NASH and BAYES, were used to train the CPTs. The method NASH and UNIFORM had relatively equal effects on the results while BAYES gave sharper and more narrow uncertainty bounds. This may favor the BAYES weighing method. Despite the latter advantage, some difficulties with the observed SWE falling outside the uncertainty bounds of the DBN, points in the opposite direction.

Compared to using a deterministic hydrology model in operational prediction and forecasting of SWE, the DBN has several advantages. The DBN uses discrete states in the CPTs. This is considered to be an advantage since hydrological and climatological data can be argued being weakly Gaussian and multi-modal. In addition DBNs also have lower computational demands since no re-sampling is needed.

Another interesting property of the DBN is the way the uncertainty in SWE is constrained. In Fig. 4 it can be seen that the spread in SWE is much larger for the HM compared to the DBN. For many months the uncertainty bounds are not realistic due to the observations showing snow free conditions. The DBN seem to have a way of constraining the outcome which makes the uncertainty in SWE more realistic. For the UNIFORM weighing method this is an interesting finding, since it is running a model without any calibration. *This illustrates the potential of using DBNs to estimate SWE in regions without any field measurements.*

In this work data from only one field site was used. The promising results shown in this paper should not be generalized before the model have been tested against a broader range of field sites and climatic conditions.

In our further work, we intend to investigate how our current DBN can be extended to the domain of distributed models by introducing a spatial DBN structure. Furthermore, we will investigate how our DBN can be used to drive decisions in hydropower and water resource management, including schemes for guiding information collection to minimize state uncertainty.

Acknowledgments. This work was supported by the Norwegian Research Council and Agder Energi AS. We thank two anonymous reviewers for constructive feedback on this research paper.

References

1. Montanari, A.: Uncertainty of hydrological predictions. In: Wilderer, P. (ed.) Treatise on Water Science, vol. 2, pp. 459–478. Academic Press, Oxford (2011)
2. Liu, Y., Gupta, H.V.: Uncertainty in hydrologic modeling: Toward an integrated data assimilation framework. Water Resour. Res. **43**, W07401 (2007). doi:10.1029/2006WR005756
3. Vrugt, J.A., ter Braak, C.J.F., Diksd, C.G.H., Schoupse, G.: Hydrologic data assimilation using particle markov chain Monte Carlo simulation: Theory, concepts and applications. Advances in Water Resources **51**, 457–478 (2013). doi:10.1016/j.advwatres.2012.04.002 (35th Year Anniversary Issue)
4. Shrestha, D.L., Kayastha, N., Solomatine, D.P.: A novel approach to parameter uncertainty analysis of hydrological models using neural networks. Hydrology and Earth System Sciences **13**, 1235–1248 (2009)
5. Garrote, L., Molina, M., Mediero, L.: Probabilistic forecasts using bayesian networks calibrated with deterministic rainfall-runoff models. In: Extreme Hydrological Events: New Concepts for Security. NATO Science Series, vol. 78, pp. 173–183 (2007)
6. Murphy, K.: Dynamic Bayesian Networks: Representation, Inference and Learning. PhD Thesis, UC Berkeley, Computer Science Division, July 2002
7. Rango, A., Martinec, J.: Revisiting the degree-day method for snowmelt computations. Water Resources Bulletin **31**(4), August 1995
8. Wigmosta, M.S., Vail, L., Lettenmaier, D.P.: A distributed hydrology-vegetation model for complex terrain. Wat. Resour. Res. **30**, 1665–1679 (1994)
9. Feiccabrino, J., Gustafsson, D., Lundberg, A.: Surface-based precipitation phase determination methods in hydrological models. Hydrology Research **44**(1), 44–57 (2013). doi:10.2166/nh.2012.158
10. Nash, J.E., Sutcliffe, J.V.: River flow forecasting through conceptual models part I A discussion of principles. J. of Hydrology **10**(3), 282–290 (1970)
11. SMILE (Structural Modeling, Inference, and Learning Engine), Decision Systems Laboratory, University of Pittsburgh. https://dslpitt.org/genie/

Evaluation of the Rate Constants of Reactions of Phenyl Radicals with Hydrocarbons with the Use of Artificial Neural Network

V.E. Tumanov[✉] and B.N. Gaifullin

Laboratory of Information Support for Research, Institute of Problems of Chemical Physics RAS, 142432, Chernogolovka, Semenov ave, 1, RUSSIAN FEDERATION
tve@icp.ac.ru,http://www.icp.ac.ru

Abstract. This paper discusses the use of feed-forward artificial neural network to predict the rate constants of the bimolecular radical reactions $Ph + R_1H$ in the liquid phase on the experimental data. The hybrid lalgorithm of calculation of rate constants of bimolecular radical reactions on the experimental thermochemical data and an empirical index of the reactionary center is offered. This algorithm uses an artificial neural network for prediction of classical barrier of bimolecular radical reactions at a temperature of 333 K, a database of experimental characteristics of reaction and Arrhenius's formula for calculation of rate constant. Results of training and prediction of the network are discussed. Results of comparison of logarithms of the calculated and experimental rate constants are given.

Keywords: Artificial neural network · Algorithm · Empirical index of the reactionary center · Rate constants · Free radical reaction · Reactivity of organic molecules · Phenyl radical

1 Introduction

Currently artificial neural network (ANN) is widely used in solving applied problems of automated processing of scientific data. The main fields of ANN application in chemical, biochemical and informatics studies are given in [1-5]. Most works in this area are devoted to the correlation between the structure of chemical compounds and the physicochemical properties or biological activity they showed. In physical chemistry, the main directions of ANN application are the simulation of chemical processes and the simulation of dynamic properties of molecules and systems.

One of the urgent tasks is prediction of reactivity of molecules in chemical reactions (activation energy and rate constant). A number of research teams is engaged in the study of prediction of reactivity of organic molecules in radical reactions.

The prediction of reactivity was done using various approaches: using general regression neural network, quantum chemical descriptors, functional density theory (DFT) taking as a basis a ratio of «structure-property» (quantitative structure-activity relationship) [6]; using multi-layer perception, chemical descriptors, experimental

© Springer International Publishing Switzerland 2015
M. Ali et al. (Eds.): IEA/AIE 2015, LNAI 9101, pp. 394–403, 2015.
DOI: 10.1007/978-3-319-19066-2_38

data taking as a basis a ratio of «structure-property» [7]; using regular feed-forward neural network with a backpropagation training algorithm, experimental data described by kinetic differential equations [8,9]; using feed-forward neural network, kinetic curve, without a kinetic model [10]; using feed-forward neural network with a backpropagation training algorithm, experimental thermochemical and kinetic data [11].

In physical chemistry of radical reactions a large amount of experimental data on reactivity (specific reaction rate or activation energies) of molecules in radical reactions in the liquid phase was accumulated [12]. Knowledge of reactivity of organic molecules in radical reactions is necessary for the development of new organic materials, the design of new drugs, the design of technological processes, the planning and the conducting of scientific experiment, the training of students and graduate students. Therefore, the development of ANN based on existing experimental data to predict reactivity of organic molecules in radical reactions is the vital task.

This paper discusses the use of feed-forward artificial neural network to predict reactivity of organic molecules in the bimolecular radical reactions Ph + R_1H in the liquid phase on the experimental data. The real work is continuation of researches [11,13].

2   Problem Formulation

Experimentally, the activation energy (E) or the classical potential barrier (Ee) determines the reactivity of organic molecules in radical reactions:

$$E_e = E + 0.5(hLv_i - RT)$$ (1)

v_i is the frequency of the stretching vibrations for the bond being broken, R is the gas constant, h is the Planck constant, L is the Avogadro number, and T is the reaction temperature (K).

Specific rate constant (k) of chemical reaction is calculated by the formula:

$$k = nA_0 \exp(-E/RT)$$ (2)

where: A_0 is the collision frequency per one equireactive bond, n is the number of equireactive bonds in a molecule.

The works [14, 15] proposed empirical models of elementary radical reactions, which allowed constructing non-linear correlation dependences between the classical potential barrier of the radical reaction and its thermochemical properties:

- approximation of the above mentioned dependence in the work [14] by the parabola:

$$br_e = \alpha\sqrt{E_e - \Delta H_e} - \sqrt{E_e}$$ (3)

- approximation of the above mentioned dependence in the work [15] in the form of the tacitly set curve:

$$br_e = D_{ei}^{1/2} \ln(\frac{D_{ei}^{1/2}}{D_{ei}^{1/2} - E_e^{1/2}}) + \alpha D_{ef}^{1/2} \ln(\frac{D_{ef}^{1/2}}{D_{ef}^{1/2} - (E_e - \Delta H_t)^{1/2}}) \tag{4}$$

Under the proposed empirical models assuming the harmonic stretching vibrations, the reaction of the radical abstraction Ph + R_1H → PhH + $R°_1$ (where Ph is phenyl radical·and $R°_1$ are alkyl radicals, and PhH is benzene and R_1H are hydrocarbon molecules) has the following parameters [14,15]:

1. The enthalpy $\Delta H_e = D_i - D_f + 0.5 (hLv_i - hLv_f)$ including the energy difference of zero-point vibrations of broken and formed bonds (it represents a change in the potential energy of the system). Here v_i is the frequency of vibration of the molecule along the broken bond, v_f is the frequency of vibration of the molecule along the formed bond, D_i is the bond dissociation energy of the broken bond, $D_{ei} = D_i + 0.5hLv_i$, D_f is the bond dissociation energy of the formed bond, $D_{ef} = D_f + 0.5hLv_f$.

2. The classical potential barrier of the activation E_e (1), which includes the zero-point energy of the broken bond.

3. The parameters b = $\pi(2\mu_i)^{1/2} v_i$ and $b_f = \pi(2\mu_f)^{1/2} v_f$, that describe the potential energy dependence of the atoms vibration amplitude along the breaking (i) and the forming (f) valence linkage. $2b^2$ is the force constant of the linkage, μ_i is the reduced mass of the atoms due to bond breaking, μ_f is the reduced mass of the atoms due to bond forming.

4. The parameter r_e, which is the integrated stretching of breaking and forming bonds in the transition state.

5. The pre-exponential factor A_0 per equireactive bond in the molecule.

Thus, we can assume that the dependence of the classical potential barrier E_e of thermochemical characteristics of reagents and kinetic characteristics of radical reactions can be represented as the functional relation:

$$E_e = f(D_{ei}, D_{ef}, br_e, \alpha) \tag{5}$$

Then the task of ANN work for predicting the values of the classical potential barrier E_e as a functional relation of thermochemical and kinetic characteristics of the reagents and reaction center of radical reaction with subsequent calculation of the activation energies and specific reaction rate by the formulas (1) and (2) reduce to the approximation of unknown functional relation (5).

3   Problem Solution

3.1   Approximation of the Classical Potential Barrier of Radical Bimolecular Reactions in an Artificial Neural Network

The experimental sample includes 97 radical reactions of phenyl radicals with various hydrocarbons, of which 20 radical reactions are the control sample. The rate constants were obtained from a database [12], the dissociation energies of C-H bonds - from [16].

For this sample, $D_f = 474$ kJ/mole [16] and $\alpha = 0.945$ are constants. Therefore dependence (5) takes the form:

$$E_e = \varphi(D_{ei}, br_e) \qquad (6)$$

In this paper the analysis of the experimental data suggests the presence of weak parabolic trend with the pair correlation coefficient ($R_c = 0.7291$) and the dispersion relation ($F = 2.1122$) Fisher, according to the enthalpy of radical bimolecular reactions ($D_{ei} - 474.0$) from the square root of the value of the classical potential barrier (Fig.1):

$$D_{ei} = 1.5 \times 10^{-4} \left(\sqrt{E_e}\right)^2 - 0.102\sqrt{E_e} + 23.249 \qquad (7)$$

This fact allows to use the kinetic parameter br_e as an experimental index of the reaction center of radical bimolecular reactions and to use this parameter in the ANN learning process.

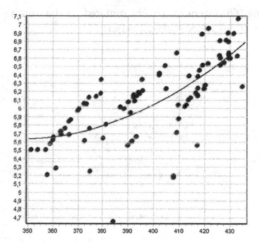

Fig. 1. Parabolic trend according to the enthalpy of considered radical bimolecular reactions from the square root of the value of their classical potential barrier

To approximate the dependence (6) we used a feed-forward artificial neural network [17] with a typical architecture shown in Fig. 1. We used the ANN having 3 inputs, 1 inner layer (7 neurons) and 1 output.

The ANN work is set by the formulae:

$$NET_{jl} = \sum_i w_{ijl} x_{ijl},$$

$$OUT_{il} = \Phi(NET_{jl} - \theta_{jl}),$$

$$x_{ij}(l+1) = OUT_{il},$$

$$\delta = 0.5 \sum_j \sum_k \left(y_j^k - d_j^k\right)^2. \qquad (8)$$

where the index i will always denote the input number, j — is the number of neurons in the layer, l — is the number of the layer; x_{ijl} — is the i- th input of the j-th neuron in the layer l; w_{ijl} — is the weighting factor of the i-th input of the neuron number j in the layer l; NET_{jl} — is the signal NET of the j-th neuron in the layer l; OUT_{jl} — is the output signal of the neuron; θ_{jl} — is the threshold of the neuron j in the layer l; x_{jl} — is the input column vector of the layer l.

The ANN input vector is set as the vector $x_0=\{D_i, br_e\}$, the output data is equal to E_e.

The method of back propagation of the error [17] was used as a training procedure. The activation function is the sigmoid function and is set by the following formula:

$$f(x) = \frac{1}{1 + e^{-x}} \tag{9}$$

For the ANN training 9700 iterations were required on the training set of 97 samples. The error of training of the ANN makes 1.63×10^{-19}.

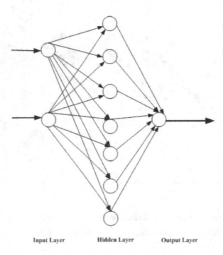

Input Layer   Hidden Layer   Output Layer

Fig. 2. Typical architecture of feed-forward artificial neural networks

Then the analysis of the kinetic parameter values br_e was performed for various reaction centers and the experimental index of the reaction center $br_{e\text{-class}}$ was calculated, as shown in Table 1. We used $br_{e\text{-class}}$ calculated for the certain reaction centers for prediction of the classical potential barrier.

Table 1. An empirical index of the reaction centers

Compounds	Reaction centres	br_{e_class}
alkanes	$-CH_2C°H_2$	16.96
	$-CH_2C°HC(CH_3)$	16.88
	$(CH_3)_3C°$	17.48
	cyclo-$[C°H(CH_2)_k]$	17.10
	cyclo-$[C°(CH_3)(CH_2)_k]$	16.85
alkenes	$CH_2=CHC°HCH_2-$	18.39
	$CH_2=C(CH_3)C°H-$	18.45
	$-C=C(CH_3)C°H_2$	18.53
	cyclo-	
	$[CH=CHC°H(CH_2)_k]$	18.38
alkynes	$-C≡CC°H_2$	18.51
	$-C≡CC°H-$	18.51
arenas	$PhC°H_2$	17.83
	$PhC(CH_3)_2C°H_2$	16.85
	$X-C_6H_4C°H_2$	17.8
	$(CH_3)_5C_6C°H_2$	17.87
	$(PH)_2C°H$	18.36
	$(Ph)_3C°$	17.27
alcohols	$C°H_2OH$	17.29
	$CH_3C°HOH$	15.82
	$C°H_2(CH_3)_2COH$	16.05
	$(CH_3)_2C°OH$	15.89
	$PhC°HOH$	16.01
aldehydes	$CH_3C°(O)$	17.60
	$PhC(CH_3)_2C°(O)$	17.61
	$X-C_6H_4C°(O)$	17.87
ketones	$-C(O)C°H_2$	17.38
	$PhC(O)C°H2$	16.80
	$-CH_2OC°H-$	17.27
	$=CHOC°(CH_3)_2$	17.44
ethers	$-OC°H2$	17.21
	cyclo-$[OC°H_2(CH_2)_k]$	16.44
	$PhOC°H_2$	17.81
esters	$-C(O)OC°H_2$	17.84
	$C°H[C(O)O-]_2$	17.92
	$PhC(O)OC°H_2$	17.01
nitriles	$C°H2CN$	17.86
	$-C°HCN$	17.86
	$PhC(CH_3)(C°H_2)CN$	17.05
nitro	$C°H_2NO_2$	17.76
	$-C°HNO2$	16.76
	$=C°NO2$	17.76

The ANN was retrained taking into account the obtained values of the experimental index of the reaction center. Learning outcomes for both ANN control samples are shown in Table. 2.

Table 2 shows the comparison of predictions of the values of the classical potential barrier of reactions using the ANN (E_{ANN}) when br_e was calculated by the formula (3), predictions of the values of the classical potential barrier of reactions using the ANN with an empirical index of the reaction center (E_{ANN2}) when br_e was taken from Table 1, and the values of the classical potential barrier (E_e) calculated by the formula (1).

Table 2. Training results of ANN

Reaction	E_e	E_{ANN}	E_{ANN2}
$CH_3(CH_2)_4CH_3$	44.2	44.5	44.1
$(CH_3)_3CCH_2CH_3$	42.9	43.6	42.8
$cyclo$-$[(CH_2)_7]$	42.8	43.6	42.2
$(CH_3)_2CH(CH_2)_2CH_3$	38.2	39.1	38.2
$CH_2=CH(CH_2)_2CH_3$	34.2	34.3	34.1
$CH_2=C(CH_3)CH_2CH_3$	33.5	33.7	34.2
$cyclo$-$[CH=CH(CH_2)_4]$	31.1	31.3	32.2
$CH\equiv CCH_2CH_3$	33.3	33.4	34.6
$PhCH_3$	38.1	38.5	37.3
$PhCH_2CH_3$	33.9	33.9	33.6
CH_2Ph_2	32.3	32.6	33.1
4-Cl-$C_6H_4CH_3$	37.9	39.1	37.5
CH_3CH_2OH	31.0	31.1	29.2
$CH_3CH(O)$	35.5	36.3	34.7
$CH_3C(O)CH_3$	47.6	46.8	46.5
CH_3OCH_3	46.3	46.0	45.6
$CH_3C(O)OCH_3$	48.4	47.3	46.8
$CH_3C(O)OH$	47.6	47.8	46.5
CH_3CH_2CN	42.4	42.6	42.0
$(CH_3)_2CHNO_2$	36.3	36.9	35.1

As shown in Table 2, the ANN predicts with good accuracy the value of the classical barrier E_{ANN2} for reactions of phenyl radicals with hydrocarbons in the liquid phase. The mean square error on all selection is 0.79±0.67 kJ/mol. It is in a good agreement with the error of experimental methods of determination of activation energy for such reactions (±4 kJ/mol).

3.2   Prediction of Rate Constants of Reactions Ph + R₁H

The estimation of reaction rate constants Ph + R_1H in the liquid phase is based on the formulas (1), (2) where the classical potential barrier is calculated using the ANN based on the dissociation energy of the broken bond and the experimental index of the reaction center. The general scheme of the algorithm is shown in Fig. 3. The algorithm uses the database that contains the pre-exponential factor for one equivalent reaction bond and the experimental index of reaction center for various groups of hydrocarbons.

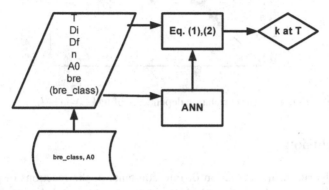

Fig. 3. Basic scheme of the hybrid algorithm of application of artificial neural networks for evaluation of rate constants of radical bimolecular reactions

In Figure 4 for the experimental rate constants k_{exp} and the calculated rate constants k_{cal}, there is the diagram of $\lg(k_{exp})$ on $\lg(k_{cal})$, which is described by the linear correlation equation:

$$\lg(k_{exp}) = 0.044 + 1.002 \times \lg(k_{cal}) \tag{10}$$

The pair correlation coefficient for the correlation ratio (12) is 0.952, and it indicates a good agreement between calculated and experimental values of logarithms of the rate constants. The mean square error for the entire sample is 0.14±0.11. The mean square error of relative error in calculation of rate constants is 38%.

It is necessary to pay attention to the problem of computational accuracy in estimating the reactivity of organic molecules from the experimental data. The relative error in determination of the rate constants of bimolecular reactions of radicals is anywhere from 10 to 35%, and it affects directly the quality of the experimental sample. The absolute error in determination of activation energy is from 2 to 4 kJ / mol. So, if the error is 3 kJ / mol in activation energy, this leads to the relative error 120% in calculation of the rate constant, that is more than an order of magnitude.

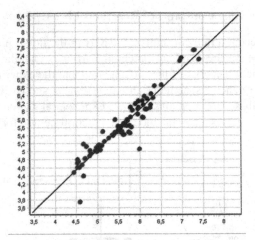

Fig. 4. Linear correlation dependence of $\lg(k_{exp})$ on $\lg(k_{cal})$

4   Conclusion

The hybrid algorithm of calculation of rate constants of the reactions of phenyl radicals with hydrocarbons in the liquid phase on the experimental thermochemical data and the empirical index of the reaction center with the use of the artificial neural network is offered.

The error of the values prediction of the classical potential barrier of radical reactions using the ANN in the control samples (of 20 samples) was $\approx \pm 1$ kJ / mol. which is within the experimental error (± 4 kJ / mol).

The rate constants of Ph + R_1H reactions in the liquid phase are estimated by proposed algorithm with the relative error of approximately 38%, and it can serve as a good estimate of the reactivity of reagents in such reactions.

Thus, this developed method for estimating the reactivity of reactions of phenyl radicals with various hydrocarbons in the liquid phase using the ANN is a good theoretical implement in researching of such reactions.

References

1. Gasteiger, J., Zupan, J.: Neural networks in chemistry. Angew. Chem. Int. Ed. Engl. **32**, 503–527 (1993)
2. Himmelblau, D.M.: Applications of Artificial Neural Networks in Chemical Engineering. Korean J. Chem. Eng. **17**(4), 373–392 (2000)
3. El-Bakry, H.M., Mastorakis, N.: A new fast neural network model, recent researches in applied computers and computational science. In: ACACOS 2012 Proceedings of the 11th WSEAS International Conference on Applied Computer and Applied Computational Science). Rovaniemi, Finland, pp. 224-231 (2012)

4. Reinaldo, F., Siqueira, M., Camacho, R., Reis, L.P.: Multi-strategy learning made easy. In: Proceedings of the 10th WSEAS International Conference on COMPUTERS, Vouliagmeni, Athens, Greece. pp. 278-284 (2006)
5. Slavova, A.: Reaction-diffusion cellular neural network models, advanced topics on neural networks. In: Proceedings of the 9th WSEAS International Conference on NEURAL NETWORKS (NN 2008), Sofia, Bulgaria, pp. 63-66 (2008)
6. Xu, Y., Yu, X., Zhang, S.: QSAR Models of Reaction Rate Constants of Alkenes with Ozone and Hydroxyl Radical. J. Braz. Chem. Soc. **24**(11), 1781–1788 (2013)
7. Dutot, A.-L., Rude, J., Aumont, B.: Neural network method to estimate the aqueous rate constants for the OH reactions with organic compounds. Atmospheric Environment. **37**, 269–276 (2003)
8. Kovacs, B., Toth, J.: Estimating Reaction Rate Constants with Neural Networks. World Academy of Science, Engineering and Technology. **26**, 13–17 (2007)
9. Ahmadi, M., Nekoomanesh, M., Arabi, H.: New Approach in Modeling of Metallocene-Catalyzed Olefin Polymerization Using Artificial Neural Networks. Macromol. Theory Simul. **18**, 195–200 (2009)
10. Bas, D., Dudak, F.C., Boyacı, I.H.: Modeling and optimization III: Reaction rate estimation using artificial neural network (ANN) without a kinetic model. Journal of Food Engineering. **79**, 622–628 (2007)
11. Tumanov, V., Gaifullin, G.: Application of the artificial neural networks for the prediction of reactivity of molecules in radical reactions, mathematical modelling and simulation in applied sciences. In: Proc. of the 3rd International Conference on Energy, Environment, Devices, Systems, Communications, Computers (INEEE 2012). Rovaniemi, Finland, pp. 62-65 (2012)
12. Tumanov, V., Gaifullin, B.: Subject-Oriented Science Intelligent System on Physical Chemistry of Radical Reactions. In: Ding, Wei, Jiang, He, Ali, Moonis, Li, Mingchu (eds.) Modern Advances in Intelligent Systems and Tools. SCI, vol. 431, pp. 121–126. Springer, Heidelberg (2012)
13. Tumanov, V.E.: Hybrid algorithm of application of artificial neuronets for an evaluation of rate constants of radical bimolecular reactions, advances in neural networks, fuzzy systems and artificial intelligence. Recent Advances in Computer Engineering Series. Jerzy Balicski (Eds) WSEAS Press. Gdansk, Poland, 21, 58-61 (2014)
14. Denisov, E.T.: New empirical models of free radical abstraction reactions. Uspekhi Khimii. **66**, 953–971 (1997)
15. Denisov, E.T., Tumanov, V.E.: Transition-State Model as the Result of 2 Morse Terms Crossing Applied to Atomic-Hydrogen Reactions. Zhurnal Fizicheskoi Khimii. **68**, 719–725 (1994)
16. Luo, Y.-R.: Comprehensive Handbook of Chemical Bond Energies. CRC Press, Boca Raton, London - New York. p. 1655 (2007)
17. Larose, D.T.: Data Mining Methods and Models. John Wiley & Sons, Inc Publication. p. 340 (2006)

Predicting SET50 Index Trend Using Artificial Neural Network and Support Vector Machine

Montri Inthachot[1(✉)], Veera Boonjing[2], and Sarun Intakosum[1]

[1] Department of Computer Science, King Mongkut's Institute
of Technology Ladkrabang, Bangkok 10520, Thailand
montri@rsu.ac.th, kisarun@kmitl.ac.th
[2] International College, King Mongkut's Institute of Technology Ladkrabang,
Bangkok 10520, Thailand
kbveera@kmitl.ac.th

Abstract. Finding ways to making better prediction of stock market trend has attracted a lot of attention from researchers because an accurate prediction can substantially reduce investment risk and increase profit gain for investors. This study investigated the use of two machine learning methods, Artificial Neural Network (ANN) and Support Vector Machine (SVM), for predicting the trend of Thailand's emerging stock market, SET50 index. Raw SET50 index records from 2009 to 2013 were converted into 10 widely-accepted technical indicators that were then used as input for model construction and testing. Our test results showed that the accuracy of the ANN model outperforms that of the SVM model.

Keywords: Artificial neural network · Support vector machine · Stock market prediction · SET50 index

1 Introduction

Predicting stock market trend has gained much attention from researchers because an accurate prediction method can help investors, especially short-term traders, to make good trading decision, reduce trading risk, and earn substantial profit [1]. The simplest prediction method is by looking at the movement of the index in a graph and making a prediction. More sophisticated methods use statistics and machine learning to analyze data and give a better prediction.

Artificial Neural Network (ANN) and Support Vector Machine (SVM) are two of machine learning methods that are reasonably successful at predicting stock index [2, 3]. In 1990, Kimoto et al. [4] started to apply Modular Neural Network to predict the movement of stock index of the Tokyo Stock Exchange to find the right time to buy and sell stocks. Thereafter, machine learning techniques became widely used for predicting stock index in established stock markets of well-developed, strong-economy countries in Europe and the United States of America. For example, Huang et al. [5] used SVM to predict NIKKEI 225 index. Manish et al. [1] used two techniques, SVM and Random Forest, to predict S&P CNF NIFTY Market index and found that SVM gave better predictions. Zhang and Wu [6] used IBCO together with BPNN to predict

© Springer International Publishing Switzerland 2015
M. Ali et al. (Eds.): IEA/AIE 2015, LNAI 9101, pp. 404–414, 2015.
DOI: 10.1007/978-3-319-19066-2_39

S&P 500 index. Guresen et al. [7] used ANN to predict NASDAQ index. Lastly, Bollen et al. [8] used data posted on Twitter to predict Dow Jones index. Machine learning techniques have been applied to predict the index of emerging stock markets in developing countries as well. For example, Birgul et al. [9] used ANN to predict ISE index. Chaigusin et al. [10] used BPNN to predict SET index. Sutheebanjard et al. [11] also used BPNN to predict SET index. Lastly, Kara et al. [3] used ANN and SVM to predict ISE 100 index.

The Stock Exchange of Thailand (SET) is an emerging stock market in the TIP group (Thailand, Indonesia, and the Philippines). SET started to operate on 30th April 1975 with 16 registered public companies. By 2013, the number of registered companies grew into over 500. SET50 index is the index calculated from the stock prices of the top 50 companies in SET in terms of large market capitalization and high liquidity. Accurate prediction of SET50 index not only helps general investors in making sound trading decision on these 50 stocks but also benefits short-term traders in their investment in SET50 Futures and SET50 Index Options of the TFEX Futures Market.

The objectives of this study were to construct accurate ANN and SVM models for predicting SET50 index movement by training them with a partial set of collected real data and then to compare their prediction performances based on the whole set of the collected data.

2   Literature Review

There have been various kinds of studies on stock market trend prediction. This section focuses only on the studies related to machine learning.

Atsalakis and Valavanis [12] used Adaptive Neuro Fuzzy Inference System (ANFIS) to predict the movement of the stock index of Athens Stock Exchange (ASE) and New York Stock Exchange (NYSE). Their predictions of ASE and NYSE indices were 63.21% and 62.32% accurate, respectively.

Lee [13] used SVM with a feature selection method called F-score and Supported Sequential Forward Search (F_SSFS) to predict NASDAQ stock index. Lee used 30 input variables: 9 variables of futures contracts, 11 variables of currency exchange rates, 9 variables of major stock markets' indices, and 1 variable of NASDAQ's the-day-before index. The accuracy of Lee's 5-fold cross-validated predictions from his SVM + F_SSFS model was 87.3% while that from a Back-propagation Neural Network (BPNN) + F_SSFS model was only 72.5%.

Kyoung-jae Kim [14] used SVM to predict the trend of Korean Composite Stock Price Index (KOSPI). He used 10 technical indicators as his model's input: Stochastic K%, stochastic D%, slow D%, momentum, ROC, Williams' R%, A/D oscillator, disparity5, disparity10, price oscillator, commodity channel index (CCI), and relative strength index (RSI). The hit rates for his and other models were the following: 64.7526% for his model with the training data set; 57.8313% for his model with the test data set; 58.5217% for a BPNN model with the training data set; 54.7332% for the BPNN model with the test data set; and 51.9793% for a Case-based Reasoning (CBR) model with the test data set.

Kara et al. [3] used ANN and SVM to predict the yearly index trend of Istanbul Stock Exchange (ISE) from 1997 to 2007. They used the following technical indicators as their model's input: simple moving average, weighted moving average, momentum, stochastic K%, stochastic D%, RSI, moving average convergence divergence (MACD), Williams' R%, A/D oscillator, and CCI. Their ANN model was 99.27% accurate when tested with their training data set and 76.74% accurate when tested with their test data set, while their SVM model was 100% and 71.52% accurate when tested with the training and test data sets, respectively.

Patel et al. [15] used ANN, SVM, Random Forest, and Naive-Bayes models to predict the movement of Indian stock market index. They also proposed to use deterministic input variable for inputting data into these models. A new layer of calculation was added to convert the 10 continuous input variables used by Kara et al. (Kara et al., 2014) into deterministic input variables. Using these deterministic variables as input, the degrees of prediction accuracy of all 4 models were consistently higher than those using the conventional continuous input variables; the degrees of accuracy were 86.69% for ANN, 89.33% for SVM, 89.33% for Random Forest, and 90.19% for Naive-Bayes, whereas the highest degree of accuracy obtained by using continuous input variables was only 83.56% from the Random Forest model.

As for studies pertaining to Thailand's stock market, most of them were about stock index prediction and individual stock price prediction. In 2005, Rimcharoen et al. [16] proposed using an adaptation of evolution strategies with evolving functional form and coefficients to predict the movement of SET index. In 2009, Sutheebanjard and Premchaiswadi [17] added factor analysis to Rimcharoen's model, and in 2010, they used BPNN to predict SET index during July 2, 2004–December 30, 2004 (124 days) and obtained predictions with a mean square error (MSE) of 234.68 and a mean absolute percentage error (MAPE) of 1.96% [11].

3   Methodology

3.1   Data and Technical Indicators

The data set used in this study was the daily SET50 index from January 5, 2009 to December 27, 2013 (a time period of 1,219 days). During this period, the index moved up 660 times (54.14%) and moved down 559 times (45.86%), as shown in Table 1.

For finding the 3 best parameter settings for ANN models and another 3 for SVM models, 20% (244 records) out of the total number of 1,219 daily index records were sampled and used; the sampled records were the first record and every fifth increment, i.e., 1st, 6th, 11th, 16th, 21st, … . For 5-fold cross validation of training results, these 244 records were sequentially divided into 5 groups, 49 records in the first four groups and 48 records in the last group. Each group was used 1 time as a test data set and 4 times as a training data set in 5 runs of cross-validation.

After the 3 ANN models, each with a different setting of the 3 best parameter settings found, and the similar 3 SVM models were obtained, they were run on all of the collected daily SET50 index records in order to set the models' weights for prediction of the next day's stock index movement. Again, for 5-fold cross validation runs, the whole set of records was divided into 5 groups and used in the same way as the sampled set of records was divided and used as described above, except the number of records in the first 4 groups was 244 while that of the last group was 243.

Table 1. The number of up-and-down movements of SET50 index during 2009-2013

Year	Up (times)	Up (%)	Down (times)	Down (%)	Total
2009	137	56.38	106	43.62	243
2010	138	57.02	104	42.98	242
2011	119	48.77	125	51.23	244
2012	140	57.14	105	42.86	245
2013	126	51.43	119	48.57	245
Total	660	54.14	559	45.86	1219

Table 2. Technical indicators used in this study and their equations [3]

Indicator	Equation
Simple n-day moving average	$\dfrac{C_t + C_{t-1} + \cdots + C_{t-n-1}}{n}$
Weighted n-day moving average	$\dfrac{(n)C_t + (n-1)C_{t-1} + \cdots + C_{t-(n-1)}}{n + (n-1) + \cdots + 1}$
Momentum	$C_t - C_{t-n}$
Stochastic K%	$\dfrac{C_t - LL_{t-(n-1)}}{HH_{t-(n-1)} - LL_{t-(n-1)}} \times 100$
Stochastic D%	$\dfrac{\sum_{i=0}^{n-1} K_{t-i}\%}{n}$
Relative Strength Index (RSI)	$100 - \dfrac{100}{1 + (\sum_{i=0}^{n-1}\frac{UP_{t-i}}{n})/(\sum_{i=0}^{n-1}\frac{DW_{t-i}}{n})}$
Moving Average Convergence Divergence (MACD)	$MACD(n)_{t-1} + \dfrac{2}{n+1} \times (DIFF_t - MACD(n)_{t-1})$
Larry William's R%	$\dfrac{H_n - C_t}{H_n - L_n} \times -100$
A/D (Accumulation / Distribution) Oscillator	$\dfrac{H_t - C_{t-1}}{H_t - L_t}$
Commodity Channel Index (CCI)	$\dfrac{M_t - SM_t}{0.015 D_t}$

where $n = 10$ in n-day moving average; C_t is closing price; L_t is low price at time t; H_t is high price at time t; $DIFF = EMA(12)_t - EMA(26)_t$; EMA is exponential moving average; $EMA(k)_t = EMA(k)_{t-1} + \propto (C_t - EMA(k)_{t-1})$; $\propto$ is smoothing factor $= \frac{2}{1+k}$; $k = 10$ in k −day exponential moving average; LL_t and HH_t are the lowest low and highest high in the last t days, respectively; $M_t = \frac{H_t + L_t + C_t}{3}$; $SM_t = \frac{\sum_{i=1}^{n} M_{t-i+1}}{n}$; $D_t = \frac{\sum_{i=1}^{n} |M_{t-i+1} - SM_t|}{n}$; UP_t is upward index change at time t, DW_t is downward index change at time t.

Input variables for the models were calculated from the collected SET50 index using Microsoft Excel. These input variables were 10 technical indicators widely-accepted by investors for predicting the trends of stock index and stock price [3], [14].

The equations for calculating each indicator are shown in Table 2. All input variables were given equal weight by normalization to the range [-1, 1]. The output variable could take only 2 values, 0 or 1. The output value of 0 signified that the movement of the predicted next-day SET index was level or downward, while the output value of 1 signified that the movement was upward.

3.2   Prediction Models

Artificial Neural Network. Artificial neural network (ANN) mimics human learning from past experience to make better prediction of the future. ANN is one of the most widely-used machine learning methods for predicting stock market trend, stock price, and stock index [2]. This study used three-layered feedforward ANN model for making prediction of the next day's SET50 index. The model consisted of 3 layers: an input layer, a hidden layer, and an output layer. The 10 technical indicators used were input into the input layer. The hidden layer consisted of n neurons where n is a parameter tested in the finding-suitable-parameter phase of model construction. The transfer function of its neurons was tan sigmoid. The output layer consisted of 1 neuron with log sigmoid transfer function. Its output was between 0 and 1; any value equaled to or less than 0.5 signified downward index movement, while any value more than 0.5 signified upward movement.

The architecture of the ANN is three-layered feedforward. The connections between the neurons of adjacent layers were weighted. The initial weights were selected randomly but then the weights got adjusted by the method of gradient descent with momentum in the training phase.

The parameters of our ANN model consisted of the number of hidden layer neurons (n), learning rate (lr), momentum constant (mc), and the number of iterations (ep). In the finding-suitable-parameter phase, 10 values of n, 9 values of mc, and 10 values of ep were tested, while lr was kept fixed at 0.1. All of these parameter values are shown in Table 3. The number of 5-fold cross validation runs needed to test all combinations of these parameter values was 10 x 9 x 10 x 5 = 4,500. All runs were performed using the neural network toolbox of MATLAB software.

Table 3. Parameter values tested in the finding-suitable-parameter phase of ANN model construction

Parameter	Value
Number of neurons (n)	10, 20, ..., 100
Epoch (ep)	1000, 2000, ..., 10000
Momentum constant (mc)	0.1, 0.2, ..., 0.9
Learning rate (lr)	0.1

Table 4. Parameter values tested in the finding-suitable-parameter phase of SVM model construction

Parameter	Value (polynomial)	Value (radial basis)
Degree of kernel function (d)	1, 2, 3, 4	-
Gamma in kernel function (γ)	0.5, 1.0, 1.5, ..., 5.0	0.5, 1.0, 1.5, ..., 5.0
Regularization parameter (c)	1, 10, 100	1, 10, 100

Support Vector Machine. Support vector machine was developed by Vapnik [18] for solving classification problems. Its operating principle is transformation of the original input vector space into a higher dimensional feature space by using a kernel function. The kernel function can be a polynomial function, a radial function, etc. Support vectors are the vectors closest to the optimal hyperplane in the higher dimensional feature space that maximizes the margin between them.

An example of 2-class classification problem solved with SVM is as follows. Given a set of input vectors $x_i \in \mathbb{R}^d, i = 1, 2, ..., N$ and class labels $y_i \in \{+1, -1\}, i = 1, 2, ..., N$, SVM maps the input vectors $x_i \in \mathbb{R}^d$ into a higher dimensional feature space $\Phi(x_i) \in \mathbb{H}$. The mapping $\Phi(\cdot)$ is performed by a kernel function $K(x_i, x_j)$ which defines an inner product in the space $\mathbb{H}$. The decision function is defined by Eq. (1) [19] below.

$$f(x) = sgn(\textstyle\sum_{i=1}^{N} y_i \alpha_i \cdot K(x, x_i) + b) \tag{1}$$

The coefficients α_i are obtained by solving the quadratic programming problem defined by equations (2)-(4) below.

$$Maximize \ \textstyle\sum_{i=1}^{N} \alpha_i - \frac{1}{2}\sum_{i=1}^{N}\sum_{j=1}^{N} \propto_i \propto_j \cdot y_i y_j \cdot K(x_i, x_j) \tag{2}$$

$$Subject \ to \ 0 \leq \propto_i \leq c \tag{3}$$

$$\textstyle\sum_{i=1}^{N} \propto_i y_i = 0 \quad i = 1, 2, ..., N \tag{4}$$

In Eq. (3), c is a regularization parameter that controls the degree of tradeoff between margin and misclassification error.

The kernel functions used in this study were the polynomial function and radial basis function defined by equations (5)–(6) below.

$$Polynomial \ function: K(x_i, x_j) = (\gamma x_i \cdot x_j + 1)^d \tag{5}$$

$$Radial \ basis \ Function: K(x_i, x_j) = \exp\left(-\gamma\|x_i - x_j\|^2\right) \tag{6}$$

Where d is the degree of polynomial function and γ is a constant of polynomial function and an exponent of radial basis function.

Different values of 3 parameters were tested in the finding-suitable-parameter phase: 4 values of d, the degree of polynomial function; 10 values of γ, the constant of the polynomial function and the exponent of the radial basis function; and 3 values of c, the regularization constant of the quadratic programming constraint equation. All of these values are tabulated in Table 4. Six hundred 5-fold cross validation runs of the model with different combinations of the polynomial function parameters were done and the accuracies of their prediction results were compared. In the same way, 150 runs with different combinations of the radial basis function parameters were performed and their accuracies were compared. All runs were performed using MATLAB software together with an SVM library developed by Chang and Lin [20].

3.3   Performance Evaluation

The models' performances were determined by using 2 performance measures: accuracy and f-measure. Accuracy was calculated from prediction results of true positive (TP), false positive (FP), true negative (TN), and false negative (FN), as shown in equation 9. F-measure was also calculated from these 4 types of prediction results, but they needed to be first converted into 2 intermediate calculation terms–precision and recall–shown in equations (7)-(8), then these terms were used to calculate f-measure from equation (10). All performance calculations were done using a program written by the authors.

$$Precision = \frac{TP}{TP+FP} \tag{7}$$

$$Recall = \frac{TP}{TP+FN} \tag{8}$$

$$Accuracy = \frac{TP+TN}{TP+TN+FP+FN} \tag{9}$$

$$F - measure = \frac{2 \times Precision \times Recall}{Precision + Recall} \tag{10}$$

4   Results and Discussion

ANN and SVM models were constructed for prediction of SET50 index and their prediction performances were evaluated. In the construction phase or finding-suitable-parameter phase, models with different combinations of parameter values were run with a partial set of the collected SET50 index records. In the performance evaluation phase, 3 models of ANN with the best 3 combinations of parameter values found in the finding-suitable-parameter phase were run with the whole set of the collected index records. In the same way, 3 models of SVM with the best 3 combinations of parameter values were also run.

All model runs were 5-fold cross validation runs. The performance measure of each of the 3 best ANN models are shown in Table 5 as average values of 5 runs. The column 'average of both data sets' shows the averages of the performance measures in the first 2 columns; these averages had been used for selecting these 3 models from 4,500 models in the first place. It can be seen in the table that the model with the combination of parameters $n = 100$, $ep = 8000$, $mc = 0.1$, and $lr = 0.1$ was the best ANN model in terms of accuracy, at 98.16% and 59.50% for the training data set and the test data set respectively.

In the next phase, the performance evaluation phase, the best 3 ANN models were run on the whole SET50 index records collected. The runs were performed in the same way as those in the first phase, but only the performance measures on the test data set are reported in Table 6 on a year-by-year basis. The lowest degree of prediction accuracy was 52.54% for the year 2010 and the highest was 59.86% for the year 2011. For prediction of the whole set of records of 5 years, the ANN model with the parameters $n = 100$; $ep = 8000$; $mc = 0.1$ was the most accurate, yielding an average degree of accuracy of 56.30%.

In the same way that the 3 best SVM models were obtained, the best 3 polynomial SVM models, each with one of the 3 best combinations of the parameter values, were obtained from 600 5-fold cross validation runs in the finding-suitable-parameter phase, and the 3 best radial basis SVM models were similarly obtained from 150 runs. All of their performance measures are shown in Table 7.

Table 5. Performance measures of the 3 best ANN models from the finding-suitable-parameter runs

No	Parameters				Training data set		Test data set		Average of 2 data sets	
	n	ep	mc	lr	Accuracy	F-measure	Accuracy	F-measure	Accuracy	F-measure
1	100	8000	0.1	0.1	0.9816	0.9834	0.5950	0.6181	0.7882	0.8008
2	70	6000	0.3	0.1	0.9508	0.9568	0.6081	0.6545	0.7794	0.8056
3	60	9000	0.4	0.1	0.9662	0.9705	0.5911	0.6245	0.7787	0.7975

Table 6. Prediction performances of 3 ANN models from the performance evaluation runs

Parameter combination (n; ep; mc)						
Year	(100; 8000; 0.1)		(70; 6000; 0.3)		(60; 9000; 0.4)	
	Accuracy	F-measure	Accuracy	F-measure	Accuracy	F-measure
2009	0.5602	0.6067	0.5644	0.6112	0.5600	0.6169
2010	0.5257	0.5919	0.5254	0.5812	0.5297	0.5876
2011	0.5986	0.5794	0.5777	0.5565	0.5780	0.5694
2012	0.5592	0.6093	0.5510	0.6167	0.5510	0.6212
2013	0.5714	0.6003	0.5796	0.5893	0.5347	0.5510
Average	0.5630	0.5975	0.5596	0.5916	0.5507	0.5892

Table 7. Performance measures of the 6 best SVM models from the finding-suitable-parameter runs

No	Parameters				Training data set		Testing data set		Average of both sets	
	Kernel	d	γ	c	Accuracy	F-measure	Accuracy	F-measure	Accuracy	F-measure
1	Polynomial	4	3	1	0.9959	0.9963	0.5330	0.5884	0.7644	0.7924
2	Polynomial	4	1	100	0.9959	0.9963	0.5248	0.5788	0.7604	0.7924
3	Polynomial	4	2	10	0.9990	0.9991	0.5283	0.5675	0.7599	0.7833
4	RBF	-	2.5	10	1.0000	1.0000	0.4961	0.5568	0.7480	0.7784
5	RBF	-	2.5	100	1.0000	1.0000	0.4920	0.5521	0.7460	0.7761
6	RBF	-	3.5	10	1.0000	1.0000	0.4878	0.5690	0.7439	0.7845

As can be seen in the table, the best polynomial SVM model that showed the highest average accuracy was with the parameters $d=4$, $\gamma=3$, and $c=1$. Its accuracies of prediction on the training data set and test data set were 99.59% and 53.30%, respectively. The best radial basis SVM model was with the parameters $\gamma=2.5$ and $c=10$, and its accuracy of prediction on the training data set and test data set were 100% and 49.61%, respectively.

Similar to the way ANN model's performance measures were obtained, the performance measures of the 3 best polynomial SVM models were obtained from runs in the performance evaluation phase. These measures are shown in Table 8, while those of the 3 best radial basis SVM models are shown in Table 9.

It can be seen that the polynomial SVM models gave the most accurate prediction for the whole index records of 5 years, at 53.24%, was with the parameters $d=4$, $\gamma=3$, and $c=1$. Also, the radial basis SVM models gave the most accurate prediction for the whole index records of 5 years, at 51.94%, was with the parameters $\gamma=2.5$, and $c=10$.

Table 8. Prediction performances of the 3 best polynomial SVM models from the performance evaluation runs

Parameter combination (d; γ; c)					
Year	(4; 3; 1)		(4; 1; 100)		(4; 2; 10)	
	Accuracy	*F-measure*	*Accuracy*	*F-measure*	*Accuracy*	*F-measure*
2009	0.5310	0.5879	0.5126	0.5755	0.5230	0.5696
2010	0.5043	0.5365	0.5126	0.5434	0.5209	0.5480
2011	0.5409	0.5566	0.5492	0.5564	0.5534	0.5587
2012	0.5429	0.5956	0.5429	0.5963	0.5265	0.5800
2013	0.5429	0.5392	0.5265	0.5271	0.5306	0.5239
Average	0.5324	0.5632	0.5288	0.5597	0.5309	0.5560

Table 9. Prediction performances of the 3 best radial basis SVM models from the performance evaluation runs

Parameter combination (γ; c)					
Year	(2.5; 10)		(2.5; 100)		(3.5; 10)	
	Accuracy	*F-measure*	*Accuracy*	*F-measure*	*Accuracy*	*F-measure*
2009	0.4901	0.5550	0.4800	0.5504	0.4860	0.5692
2010	0.4966	0.5677	0.4800	0.5545	0.4926	0.5725
2011	0.5777	0.5642	0.5736	0.5668	0.5491	0.5321
2012	0.5429	0.6061	0.5469	0.6121	0.5265	0.6009
2013	0.4898	0.4683	0.4980	0.4745	0.4857	0.4621
Average	0.5194	0.5523	0.5157	0.5517	0.5080	0.5474

Table 10. The best prediction accuracies from ANN, SVM and Naive models

	ANN	SVM	Naive
Testing data set	56.30%	53.24%	50.22%

Based on these accuracy figures, the best SVM model was the polynomial model with the parameter settings of $d = 4$, $\gamma = 3$, and $c = 1$; it was 53.24% accurate.

An issue to note is the difficulty of comparing prediction results of models investigated in different studies. However, a widely accepted way to compare prediction results does exist. It is done by comparing a study's result to the corresponding result from a Naive model [21]. A Naive model predicts the next-day's index using only today's index. Using a Naive model on our collected index records, its prediction accuracy was 50.22%.

Comparison of the prediction results of these 3 types of models showed that both the best ANN and SVM models yielded higher accuracies than that of the Naive model, and the highest accuracy was obtained from the best ANN model, at 56.30%, as shown in Table 10. However, to make sure that ANN outperforms SVM significantly, we applied T-test at 0.05 level of significance to two pairs of models: ANN and polynomial SVM, and ANN and radial basis SVM. The P-values of right-tailed T-test were 0.0290 and 0.0367, respectively. This means that ANN predicted significantly more accurately than SVM did (at $P < 0.05$).

5   Conclusion

Accurate index trend prediction is very important for investors especially short-term traders because they can use it to make sound investment on the stocks of that market and its futures market. Machine learning methods are popular as a technical analysis tool for making stock index trend prediction because their parameters are highly adjustable. This study used 2 machine learning methods–ANN and SVM–to predict Thailand's SET50 stock index trend in a five-year period of 2009–2013. Suitable parameter settings for ANN and SVM models were sought and found.

The chosen input variables for these models were 10 widely-accepted technical indicators converted from the raw index records. The output was the up or down movement of the next day's index. Results from performance evaluation runs showed that the best ANN model was 56.30% accurate while the best SVM model was 53.24% accurate. Both models performed better than a Naive model. Even though they were better than the Naive model, their prediction accuracies were still somewhat low probably due to highly fluctuated index movements. In future studies, improvements will be made by changing some input variables and adding more basic factor variables.

Acknowledgements. This study was supported by Rangsit University.

References

1. Manish, K., Thenmozhi, M.: Forecasting Stock Index Movement: A Comparison of Support Vector Machines and Random Forest. Social Science Research Network, Rochester, NY (2005)
2. Atsalakis, G.S., Valavanis, K.P.: Surveying stock market forecasting techniques – Part II: Soft computing methods. Expert Syst. Appl. **36**, 5932–5941 (2009)
3. Kara, Y., Acar Boyacioglu, M., Baykan, Ö.K.: Predicting direction of stock price index movement using artificial neural networks and support vector machines: The sample of the Istanbul Stock Exchange. Expert Syst. Appl. **38**, 5311–5319 (2011)
4. Kimoto, T., Asakawa, K., Yoda, M., Takeoka, M.: Stock market prediction system with modular neural networks. In: 1990 IJCNN International Joint Conference on Neural Networks, 1990, vol. 1, pp. 1–6 (1990)
5. Huang, W., Nakamori, Y., Wang, S.-Y.: Forecasting stock market movement direction with support vector machine. Comput. Oper. Res. **32**, 2513–2522 (2005)
6. Zhang, Y., Wu, L.: Stock market prediction of S&P 500 via combination of improved BCO approach and BP neural network. Expert Syst. Appl. **36**, 8849–8854 (2009)
7. Guresen, E., Kayakutlu, G., Daim, T.U.: Using artificial neural network models in stock market index prediction. Expert Syst. Appl. **38**, 10389–10397 (2011)
8. Bollen, J., Mao, H., Zeng, X.: Twitter mood predicts the stock market. J. Comput. Sci. **2**, 1–8 (2011)
9. Birgul, E., Ozturan, M., Badur, B.: Stock market prediction using artificial neural networks. In: Proceedings of the 3rd Hawaii International Conference on Business (2003)
10. Chaigusin, S., Chirathamjaree, C., Clayden, J.: The use of neural networks in the prediction of the stock exchange of thailand (SET) index. In: 2008 International Conference on Computational Intelligence for Modelling Control Automation, pp. 670–673 (2008)
11. Sutheebanjard, P., Premchaiswadi, W.: Stock exchange of thailand index prediction using back propagation neural networks. In: 2010 Second International Conference on Computer and Network Technology (ICCNT), pp. 377–380 (2010)
12. Atsalakis, G.S., Valavanis, K.P.: Forecasting stock market short-term trends using a neuro-fuzzy based methodology. Expert Syst. Appl. **36**, 10696–10707 (2009)
13. Lee, M.-C.: Using support vector machine with a hybrid feature selection method to the stock trend prediction. Expert Syst. Appl. **36**, 10896–10904 (2009)
14. Kim, K.: Financial time series forecasting using support vector machines. Neurocomputing. **55**, 307–319 (2003)
15. Patel, J., Shah, S., Thakkar, P., Kotecha, K.: Predicting stock and stock price index movement using Trend Deterministic Data Preparation and machine learning techniques. Expert Syst. Appl. (2014)
16. Rimcharoen, S., Sutivong, D., Chongstitvatana, P.: Prediction of the Stock Exchange of Thailand using adaptive evolution strategies. In: 17th IEEE International Conference on Tools with Artificial Intelligence, 2005. ICTAI 05. p. 5 pp.–236 (2005)
17. Sutheebanjard, P., Premchaiswadi, W.: Factors analysis on Stock Exchange of Thailand (SET) index movement. In: 2009 7th International Conference on ICT and Knowledge Engineering, pp. 69–74 (2009)
18. Vapnik, V.N.: The Nature of Statistical Learning Theory. Springer-Verlag New York Inc, New York, NY, USA (1995)
19. Hua, S., Sun, Z.: Support vector machine approach for protein subcellular localization prediction. Bioinformatics. **17**, 721–728 (2001)
20. Chang, C.-C., Lin, C.-J.: LIBSVM: A Library for Support Vector Machines. ACM Trans Intell Syst Technol. **2**, 27:1–27:27 (2011)
21. Hellström, T., Holmström, K.: Predicting the Stock Market (1998)

Mobile Application Recommendations Based on Complex Information

Shuotao Yang, Hong Yu$^{(\boxtimes)}$, Weiwei Deng, and Xiaochen Lai

School of Software, Dalian University of Technology,
Tuqiang street. 321, Dalian 116620, China
ssdut_yst@163.com,
hongyu@dlut.edu.cn

Abstract. Due to the huge and still rapidly growing number of mobile applications, it becomes necessary to provide users an application recommending service. In this work we present a recommendation system that recommends new applications for users according to their outdated applications. The recommender assumes that each owned application has complex information containing both descriptions and API information. The proposed approach mines application descriptions from publicly available online specifications and identifies APIs from the downloaded APK(Android PacKage) files. Text mining and incremental diffusive clustering(IDC) algorithm are utilized to generate common features. And APIs are extracted by disassembly technology. Then the complex information of applications can be represented by the features and APIs. In the processing of recommending, the k-Nearest-Neighbor algorithm based on the self-adaptive similarity(SS-KNN) is adopted to generate candidate sets of applications, and then the coverage-weighted similarity is utilized to select the final recommendations from the candidates. Extensive experiments are conducted on different application categories and the experimental results illustrate the effectiveness and efficiency of the approach.

Keywords: Mobile applications · Recommendation system · Clustering

1 Introduction

Recent years have witnessed the research trends in mining software repositories [1]. With the develop of the mobile devices, mobile applications play an important role for users. With so many applications providing different services, it is difficult for users to find suitable applications just by typing several keywords online. In consequence, it is necessary to provide some effective application recommendations to mobile users.

Modern recommenders provide services across many different item categories[2,3] Most conventional items(e.g books) are seen as one-shot consumption[4]. When users finish consuming outdated items (e.g reading books), they are willing to find some similar and new items. Mobile users also pay attention to the

© Springer International Publishing Switzerland 2015
M. Ali et al. (Eds.): IEA/AIE 2015, LNAI 9101, pp. 415–424, 2015.
DOI: 10.1007/978-3-319-19066-2_40

similarity and novelty of a new application when replacing an outdated application with a new one. However, different from item recommendation, when a user choose an alternative application, her downloaded application will affect her choice. Because once an application is downloaded, it provides user continuous service and may affect the users decision. Thus, an application recommendation service should provide users some new recommendations which are similar to outdated one but more innovative than outdated one.

To address the problem, researchers[4][5] recommend similar applications by utilizing the features of applications. However, the features are in free text format and described informally. It is difficult to make similar recommendations preciously only by utilizing text information. Besides the text information, the packages of applications also contain effective information[6][7][8]. In the process of recommending, the complex information of the outdated one is utilized to find some new and similar applications. The approaches above fail to explore the full range of features and do not help users a lot to find relatively similar and new applications.

In this paper, we present a novel approach utilizing the complex information of both features and APIs. The SS-KNN algorithm is employed to identify the similar application candidates. In contrast to previous methods which only utilize features to compute similarity, our approach also uses APIs which greatly enhance the accuracy of recommending similar applications. And the coverage-weighted similarity is used to make sure that the recommendations selected from candidates are novel enough.

The remainder of this paper is structured as follows. Section 2 provides a general overview of our approach, while Section 3 describes feature modeling and API extracting. Sections 4 describes the two main phases of our approach which to generate recommendations, namely the SS-KNN algorithm and the coverage-weighted similarity. Section 5 describes evaluations and empirical experiments. Finally we summarize up our work.

2   Overview

Our recommendation system includes several steps as illustrated in Figure 1. We briefly introduce the formation of recommendation system and how the system works. The complex information which consists of the dominant information and the latent information is extracted from applications in Googles application store. The dominant information is the features extracted from product specifications which contain the purposes, functions and characteristics of applications, while the latent information means the APIs extracted from APK(Android PacKage) files which are invisible for users. The complex information will serve the application recommendation system after preprocessing.

The preprocessing of raw information contains two parts. One part is the preprocessing of dominant information. The specifications are preprocessed by breaking up the paragraphs into sentences and phrases, removing stop words and representing each descriptor as a distribution of vocabulary. In this work,

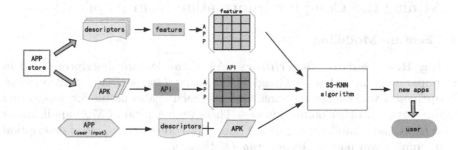

Fig. 1. Preprocessing and Recommendations

we focus on measuring the similarity of two descriptors by the number of common words. The more of the common words two descriptors contain the better two descriptors match. Indeed, semantically related words can also measure the similarity, we leave finding semantically related words to our future work. Then an incremental diffusive clustering (IDC) algorithm[9][10] is utilized to cluster similar and variant descriptors into features. After naming the features, an application-by-feature matrix is generated. So the features of each application can be represented as a distribution in the matrix. The other part is the preprocessing of latent information. The APKs are mined by decompiling technique and the APIs are extracted from APK files. The APIs of each application also can be represented as a distribution of common and variant APIs. Thus, an application-by-API matrix is generated. Therefore, the fully information of applications is included in the two matrices above.

Our recommendation system is initialized by the targeted application. For the targeted application which is input into our system by user, we assume that the targeted application have a certain amount of features and APIs. Our system will create a targeted application profile by extracting information from the target automatically. If the descriptions of the targeted application is not complete enough, the distribution of features may be sparse. On the contrary, if the descriptions contain several key features, the profile of the dominant information will be more clearly. And the situation is the same for the latent information. Finally, the recommendation system utilizes the SS-KNN algorithm and the coverage-weighted similarity to make recommendations for users.

The similarity between each two applications is computed according to the complex information. If we only use the dominant information to measure the similarity of each two applications, there will be a considerable deviation, because not all the descriptions are complete enough, some of them are incomplete and sparse. The latent information not only enriches the completeness of the dominant information, but also enhances the accuracy and efficiency of recommendation system. The complex information which contains both APIs and features makes the system do a better work.

3   Mining the Complex Information from Applications

3.1   Feature Modeling

Mining Raw Feature Descriptors. All of the feature descriptors used in this paper were mined from Google Play Store. Google Play Store provides descriptions of various applications. Our approach is also usable for application descriptions from other online sources. There were a total of 3799 applications mined from more than 20 categories. Figure 2 provides an example of description for the application named *Accounting Dictionary*.

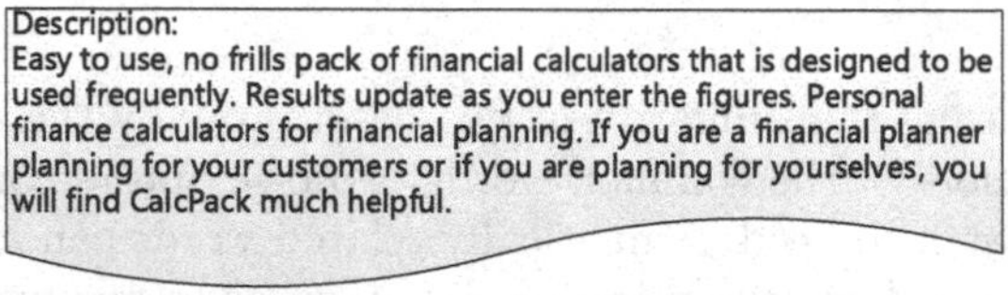

Description:
Easy to use, no frills pack of financial calculators that is designed to be used frequently. Results update as you enter the figures. Personal finance calculators for financial planning. If you are a financial planner planning for your customers or if you are planning for yourselves, you will find CalcPack much helpful.

Fig. 2. An example of description

Preprocessing. In preparation for feature extraction, all of the feature descriptors were preprocessed by breaking specifications down into sentences, removing stop words and stemming the remaining words to their root form (i.e., terms). And the remaining descriptor was modeled as a vector of terms. All of the terms constitute the vocabulary. And each dimension of vector corresponds to one of the terms in the vocabulary. In this paper, our feature mining process is based on[11], and we make the corresponding adjustment according to the property of mobile application.

We use the $tf - idf$ (term frequency - inverse document frequency) scheme to assign a weighting to term t in descriptor d as follows:

$$w_{t,d} = tf_{t,d} * log(\frac{N}{df_t}) \, . \tag{1}$$

where, $tf_{t,d}$ represents the number of times that term t occurs in d, df_t is the number of descriptors that contain term t, and N is the total number of descriptors. Thus, each descriptor can be represented by a weight vector.

Incremental Diffusive Clustering. In this paper, an incremental diffusive clustering (IDC) algorithm[9] is employed to identify common and variant features from descriptors. The algorithm repeatedly executes the following steps to incrementally identify features.(i) generating the candidate features by clustering the descriptors, (ii) identifying and reserving the best feature, (iii) selecting high-frequency terms and removing them from all descriptors to generate new reduced descriptors, (iv) repeating steps i-iii using the increasingly reduced descriptors until the produced features reach the targeted number. The incremental diffusive clustering (IDC) method uses a consensus clustering approach to generate a predefined number of candidate clusters (in this case 25). And then the final clusters are integrated by a voting scheme[9].

Clustering and Naming Features. In the process of feature extraction, there are a certain number of similar descriptors representing the same feature. These similar descriptors contain a common set of keywords. In order to represent the descriptors more clearly, similar descriptors are clustered into a group and represented by a feature chosen from the group. The similarity of each pair of descriptors is computed by the inner product of their corresponding vectors.

Our experiment combines consensus clustering methods with SPK-means algorithm. Dumitru et al.[11] find that consensus clustering methods used in conjunction with SPK outperform other methods for feature clustering. Therefore, we adopt a consensus based approach for our application recommendation system. First of all, 75 % of the raw descriptors are selected and clustered into candidate clusters by using the standard SPK algorithm. And then the remaining descriptors are classified into their most similar clusters. The final set of clusters are generated by utilizing IDC algorithm to reconcile these candidate clusters.

Finally, the dominant information of each application is identified by a set of features. All of the dominant information constitutes an application-by-feature matrix $A \times F$ in which the rows of the matrix correspond to applications and the columns correspond to features. The $(i, j)^{th}$ entry of this matrix can take a value of 0 or 1 to represent whether the jth feature is included in the ith application or not.

Meaningful names are selected for the mined features. The feature is a descriptor that is most representative of the features theme. Features are named by identifying the medoid. The medoid is computed by the cosine similarity between each descriptor and the centroid of the cluster. For all values above a certain threshold(0.1) in the term vector of descriptor, we sum up the different weighted values. Both scores are normalized and then added together for each descriptor. Finally, we select the descriptor with the highest score as the feature name. This approach produces quite meaningful names.

3.2   Preprocessing of APIs

The APIs are latent information for users, because of its technical and non-visual property. Some observations from the recent studies[6][12] indicate that inheritance is essentially unused in applications, that applications heavily rely on 3rd-party APIs. Ruiz et al. [13] discovered most of applications were entirely reused by other applications in the same category and they used the same APIs. Very few classes are unique to a mobile application. McMillan et al.[6] used API calls as semantic anchors to compute similarities among applications. Therefore, these studies indicate that the similarity of two applications can be measured by their common APIs partly.

We focused our study on the applications available in Google Play android market. And most packages reused are part of Google APIs. App stores usually do not provide source code. However, they do provide APK files which can be downloaded using a third party library. And the APIs can be extracted by the method of decompilation in[14]. In this paper, we focus only on the official Android APIs used by several applications. Therefore, the same decompiled APIs have their common name. The total number of API classes is 1762.

4 Application Recommendation System

Once features and APIs have been extracted, an application-by-API matrix and an application-by-API matrix are generated. The similarities between the targeted application and others are computed according to their cosine similarities of features and APIs. There are several steps executed to generate a set of recommendations. These include creating an initial profile of targeted application provided by user, computing the self-adaptive similarity of the target, using the KNN algorithm to make top N most similar initial recommendations and finally utilizing the coverage-weighted similarity to recommend similar and novel applications.

4.1 Creating a Targeted Application Profile

The initial features and APIs are provided by the targeted application from user. In preparation for application recommending, the targeted application profile should be created by preprocessing. The feature descriptors are generated by IDC algorithm, while the APIs are extracted by decompiling the APK files. Each individual descriptor is matched to previously mined features using the cosine similarity metric. And we use the same cosine similarity metric as to each API. If the targeted application is outside the knowledge of the recommendation system or contains insufficient information to launch a recommendation, the system will find insufficient matches and generate no recommendations. Thus, we established a threshold score of 0.6 for the self-adaptive similarity, and then required at least 4 features and 4 APIs to be matched to known features and APIs.

4.2 Generating Candidate Sets by SS-KNN Algorithm

The recommender generates candidate sets of the targeted applications utilizing the self-adaptive similarity based on a standard KNN algorithm. The most similar applications are considered as candidate neighbors of the target. The process of detecting most similar recommendations depends on dominant features and latent APIs. We adopt a self-adaptive similarity method to measure the similarity between the targeted application and an existing application, which consists of a feature-based similarity and an API-based similarity.

The feature-based similarity $appFeaSim(a, n)$ is computed using the binary equivalent of the cosine similarity as follows[15]:

$$AppFeaSim\,(a, n) = \frac{|(F_a \cap F_n)|}{\sqrt{|F_a| \cdot |F_n|}}. \tag{2}$$

where, F_a denotes the set of features of targeted application a. n denotes an existing application in our repository. $AppFeaSim(a, n)$ is computed between the targeted application and all previously mined applications. The result of feature-based similarity is between 0 and 1. The application with score 1 contains identical features, while the application with score 0 contains no common features.

Analogously, the computing method of API-based similarity is the same as the feature-based similarity. The API-based similarity between a targeted application a and each existing application n is computed as follows:

$$AppAPISim\,(a, n) = \frac{|(A_a \cap A_n)|}{\sqrt{|A_a| \cdot |A_n|}}. \tag{3}$$

where, A_a denotes the set of APIs of application a.

Finally, the self-adaptive similarity is computed as follows:

$$appSim\,(a, n) = \alpha appFeaSim\,(a, n) + \beta appAPISim\,(a, n)$$
$$\text{s.t.} \quad \alpha + \beta - 1, 0 \le \alpha \le 1, 0 \le \beta \le 1; \tag{4}$$

where, $appSim$ is the self-adaptive similarity between the targeted application a and a previously mined application n. $\alpha = p_F/(p_F + p_A)$ and $\beta = p_A/(p_F + p_A)$. In the constraints, $p_F = n_{F_a}/N_{Favg}$ is the normalized value of the feature set F_a based on the average feature number N_{Favg}, where $N_{Favg} = N_{Fa}/N_{app}$, N_{app} is the total number of applications. Obviously, $p_A = n_{A_a}/N_{Aavg}$ is the normalized value of the API set A_a based on the average API number of each application N_{Aavg}. α and β are the normalized proportions of the features and APIs in the targeted application a. Hence, $\alpha + \beta = 1$.

Then, we generate the candidate neighbor set of the target by the standard KNN algorithm based on the self-adaptive similarity.

4.3   Ranking of the Candidate Recommendations

However, users won't be interested in similar applications simply. The recommendations which contain more functions and novel characters will seize users heart. Thus, the coverage-weighted similarity is utilized to rank candidate neighbors. According to the targeted application a, the coverage of application i is computed as follows:

$$AppCover\,(i, a) = \alpha \frac{n_{F_i}}{n_{F_a}} + \beta \frac{n_{A_i}}{n_{A_a}}. \tag{5}$$

where, n_{F_i} is the number of the ith candidate neighbor's features and $0 < i \le k$. While n_{A_i} is the number of the ith candidate neighbor's APIs. n_{F_a}, n_{A_a}, α

and β are computed the same way as in the self-adaptive similarity. Thus, the candidate neighbors are ranked according to the $AppNewSim$ as follows:

$$AppNewSim\,(i,a) = AppCover\,(i,a) * AppSim\,(i,a)\;. \tag{6}$$

The candidate applications with high $AppNewSim$ scores are not only more similar to the target, but also have more new functions than the target. The application with an $AppNewSim$ score bigger than a threshold (1.0) is selected as a final recommendation.

5   Evaluation

The performance of the mobile application recommendation system is generally measured by some specific evaluating criteria, such as Precision and Recall. However, there are quite few data that is available in the field of mobile at present. So the situation makes it difficult to the experiment of mobile recommendation system. In order to the goal of assessment, researchers in this area generally organize volunteers to use the system in practice, and then we analyze the feedback through questionnaires from volunteers.

In this experiment we also adopt the assessment of volunteers voting for each recommendation of our system. Each volunteer judged whether each automatically generated recommendation represented (i) A boring recommendation with no similar functions or novel points, (ii) A ordinary recommendation with more similar functions and less novel points, (iii) A novel recommendation with more different functions and less similar points, or (iv) A prefect recommendation with similar and novel functions. The situations (i) and (ii) cannot attract volunteers, while (iii) and (iv) can arouse interest of volunteers. Then volunteers are also asked to rate the recommendations according to the corresponding situations above as(i) Completely inadequate, (ii) Poor, (iii) Good, or (iv) Excellent. The scores of i-iv is corresponded to 1, 2, 3 and 4. This experiment evaluates the accuracy of recommendation system by calculating an average value of the evaluators ratings and assesses the stability by analyzing the distribution of the ratings.

5.1   Experiment

The five-fold, cross-validations was applied in this experiment. The applications from each category were randomly divided into five parts,and for each test, one part was taken as test set while the remaining four parts were combined into a source set. For each category, we implemented 5 tests so that each part of applications was served as the test set once. In our experiment, there were a total of 3799 applications in our test data set mined from more than 20 categories in Google Play store randomly. We chose the application data set from 4 categories to evaluate our experiment, namely $music$, $finance$, $chat$ and $shopping$. As an exploratory research, our method is not compared with a random test, because of the random test has a strong uncertainty.

Table 1. Ratio of every score

Category	4	3	2	1
Music	0.2807	0.3333	0.2983	0.0877
Finance	0.4706	0.2353	0.2353	0.0588
Chat	0.4203	0.2029	0.2754	0.1014
Shopping	0.3333	0.2292	0.2917	0.1458

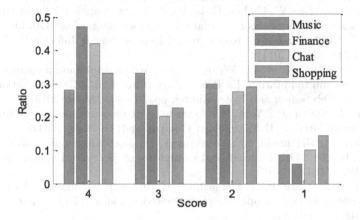

Fig. 3. The distributions of evaluators' voting scores

20 volunteers are selected to vote our recommendations, the ratios of voting scores and the distributions are exhibited in Table 1 and Figure 3. The evaluators voting of score 3 and 4 are accounted for about 60% of all voting. And for different test sets, the accuracy rates all make high scores relatively. Results from the experiment have shown that our recommendation system has the ability to recommend similar and novel applications for users. And the system perform well with high stability.

6   Conclusion and Future Work

In this work, we proposed the mobile application recommendation system utilizing both features and APIs to recommend similar and novel applications. The SS-KNN algorithm is proposed to generate the candidate sets of recommendations. The latent information of APIs enhances the accuracy of detecting similar applications.

The popularity of a mobile application is attributed to not only its function points but also its online time, practicability, stability etc. It is our future work to analyze applications' online time, users' review and more applications' information to make accurate recommending.

References

1. Xuan, J., Jiang, H., Ren, Z., Zou, W.: Developer prioritization in bug repositories. In: 2012 34th International Conference on Software Engineering (ICSE), pp. 25–35. IEEE (2012)
2. Chen, H.-C., Chen, A.L.: A music recommendation system based on music data grouping and user interests. In: Proceedings of the Tenth International Conference on Information and Knowledge Management, pp. 231–238. ACM (2001)
3. Ye, M., Yin, P., Lee, W.-C., Lee, D.-L.: Exploiting geographical influence for collaborative point-of-interest recommendation. In: Proceedings of the 34th International ACM SIGIR Conference on Research and Development in Information Retrieval, pp. 325–334. ACM (2011)
4. Yin, P., Luo, P., Lee, W.-C., Wang, M.: App recommendation: a contest between satisfaction and temptation. In: Proceedings of the Sixth ACM International Conference on Web Search and Data Mining, pp. 395–404. ACM (2013)
5. Linden, G., Smith, B., York, J.: Amazon. com recommendations: Item-to-item collaborative filtering. IEEE Internet Computing **7**(1), 76–80 (2003)
6. McMillan, C., Grechanik, M., Poshyvanyk, D.: Detecting similar software applications. In: 2012 34th International Conference on Software Engineering (ICSE), pp. 364–374. IEEE (2012)
7. Kawaguchi, S., Garg, P.K., Matsushita, M., Inoue, K.: Mudablue: An automatic categorization system for open source repositories. Journal of Systems and Software **79**(7), 939–953 (2006)
8. Jiang, H., Ma, H., Ren, Z., Zhang, J., Li, X.: What makes a good app description?. In: Proceedings of the 6th Asia-Pacific Symposium on Internetware on Internetware, pp. 45–53. ACM (2014)
9. Duan, C., Cleland-Huang, J., Mobasher, B.: A consensus based approach to constrained clustering of software requirements. In: Proceedings of the 17th ACM Conference on Information and Knowledge Management, pp. 1073–1082. ACM (2008)
10. Jiang, H., Ren, Z., Xuan, J., Wu, X.: Extracting elite pairwise constraints for clustering. Neurocomputing **99**, 124–133 (2013)
11. Dumitru, H., Gibiec, M., Hariri, N., Cleland-Huang, J., Mobasher, B., Castro-Herrera, C., Mirakhorli, M.: On-demand feature recommendations derived from mining public product descriptions, In: 2011 33rd International Conference on Software Engineering (ICSE), pp. 181–190. IEEE (2011)
12. Minelli, R., Lanza, M.: Software analytics for mobile applications-insights & lessons learned. In: 2013 17th European Conference on Software Maintenance and Reengineering (CSMR), pp. 144–153. IEEE (2013)
13. Ruiz, I.J.M., Nagappan, M., Adams, B., Hassan, A.E.: Understanding reuse in the android market. In: 2012 IEEE 20th International Conference on Program Comprehension (ICPC), pp. 113–122. IEEE (2012)
14. Linares-Vásquez, M., Bavota, G., Bernal-Cárdenas, C., Di Penta, M., Oliveto, R., Poshyvanyk, D.: Api change and fault proneness: a threat to the success of android apps. In: Proceedings of the 2013 9th Joint Meeting on Foundations of Software Engineering, pp. 477–487. ACM (2013)
15. Spertus, E., Sahami, M., Buyukkokten, O.: Evaluating similarity measures: a large-scale study in the orkut social network. In: Proceedings of the Eleventh ACM SIGKDD International Conference on Knowledge Discovery in Data Mining, pp. 678–684. ACM (2005)

The Regulation of Steam Pressure in a Drum Boiler by Neural Network and System Identification Technique

Karin Kandananond[✉]

Valaya Alongkorn Rajabhat University, Phra Nakhon Si Ayutthaya, Thailand
kandananond@hotmail.com

Abstract. The dynamic characteristic of a drum boiler is complex and this complication leads to the difficulty in controlling the output of the system, i.e., steam pressure. Therefore, this study attempts to investigate the application of two model predictive methods, artificial neural network (ANN) and system identification, in order to assess the performance of each method. According to the system, the inputs are feed water flow rate and applied heat while the output is the steam pressure. The ANN method used is based on a training algorithm, Levenberg-Marquardt back propagation. On the other hand, the optimal model of system identification method is the output error (OE). The performance measurement is compared by considering the mean squared error (MSE) after fitting the simulated prediction from each model to the observation. The results show that ANN slightly outperforms the system identification technique. Moreover, another finding is that ANN method is capable of identifying the outlier among the observations so it is robust to the disturbances.

Keywords: Drum boiler · Artificial neural network · Steam pressure · System identification

1 Introduction

Drum boiler is a standard type of boiler which has the water tube inside. It is operated when the water is fed through the tube into the unit and the certain amount of heat is supplied. After the steam is generated, the water level will decrease. As a result, the level of water in the tube must be maintained at a constant level. Moreover, the inputs of drum boiler, feed water rate and supplied heat, should be controlled while the pressure of steam is the output of the boiler. However, the complex control scheme is required in order to derive the relationship between inputs and output. Therefore, the feed water rate and heat must be adjusted efficiently to achieve the desired level of pressure. Anyway, it is difficult for the operators on the shop floor, sometimes without the strong background in the heat transfer area, to understand this complex relationship and to regulate the steam pressure. As a result, the simplification of the relationship by using the empirical model might be another alternative to pave the way for the successful operation of a drum boiler. Therefore, two approaches, data mining and system identification technique, are selected and benchmarked to construct empirical models for regulating the steam pressure.

© Springer International Publishing Switzerland 2015
M. Ali et al. (Eds.): IEA/AIE 2015, LNAI 9101, pp. 425–434, 2015.
DOI: 10.1007/978-3-319-19066-2_41

One of the data mining methods, artificial neural network, is popular for many years since the first introduction. It is a powerful method for identifying the model for explaining the relationship between inputs and outputs of the system. On the other hand, the system identification method is another approach also widely used to develop a mathematical model for determining the dynamic behavior of the system. The major difference between these two methods is the mechanism and complexity, i.e., the amount of variables which are taken into the consideration. While the data mining method considers only the paired output and input data of the system, the system identification technique also includes the historical data (time series) of the system noise in the model as well. As a result, since the complexity and conditions of each method are different, it is interesting to characterize the performance of each method to predict the output of a black box system. The system used in this case study used is a drum boiler which is considered as a black box with three variables, steam pressure as an output and two inputs, i.e., feed water flow rate and applied heat.

2   Literature Review

The dynamic mechanism regarding the heat exchange is always a challenging issue for practitioners to characterize the relationship between inputs and outputs. As a result, a number of methods are deployed to determine the complicated relationship. The first chosen method is the direct characterization of the dynamic system while the alternative is the application of data mining method, e.g., artificial neural network (ANN). Another method is the utilization of system identification method to predict the output of the system.

According to the literature, Astrom and Bell [1] deployed a nonlinear dynamic model for explaining the complex dynamic system of a drum boiler. Similarly, the system identification of steam pressure from a fire-tube boiler by a second order linear model, autoregressive moving average with exogenous inputs (ARMAX), was carried on by Rodriguez Vasquez et al. [2]. Artificial neural network (ANN) is also widely used to determine the output of complex dynamical system. Vasickaninova et al. [3] utilized ANN to calculate the optimal inputs to the heat exchanger. The result was compared with the traditional PID control. Yu and Li [4] also deployed ANN to identify the on-line nonlinear system. Subudhi and Jena [5] used Levenberg-Marquardt algorithm associated with the differential evolution and opposition based differential evolution to identify the nonlinear system. The back propagation neural network was utilized by Wu and Tsai [6] to identify the output of speaker system.

Xia [7] has compared the performance of neural network and system identification method to construct the model for magnetorheological (MR) damper. For neural network, the multi-layer perceptron (MLP) architecture is used with Levenberg-Marquardt training algorithm while the autoregressive with exogenous inputs (ARX) is the model utilized for the system identification. The ANN based system identification and model predictive control are designed for controlling the interface level of a flotation column by Mohanty [8].

Other methods used for modelling the system include fuzzy logic and wavelet based models. For example, dynamic fuzzy model is utilized by Habbi, Zelmat and Bouamama [9] to explain the complex dynamic of drum boiler based on the model of Astrom and Bell. Adaleesan et al. [10] have deployed Wiener type Lagurre-wavelet network model to construct a model for controlling a bioreactor.

Therefore, according to the literature, there is more than one method focusing on the exploration of the relationship between inputs and outputs of the dynamic system of a boiler. Two methods, data mining and system identification, are selected to be characterized and the study is based on determining the best performed method to predict the steam pressure of a drum boiler. The results will lead to the findings which empower the operators to understand the predicting capability of each method. Consequently, they can choose the appropriate method and it will equip them with the potential to control the output of a drum boiler efficiently.

3   Method

The dynamic system representing the relationship between inputs and output in this study is based on two methods, system identification and neural network.

3.1   System Identification

System identification method is the utilization of different statistical methods to construct the mathematical model of a dynamic system by focusing on inputs and outputs. The general equation of dynamic model is the discrete linear time-invariant, where output ($y(t)$) and input ($u(t)$) are shown as follows:

$$y(t) = G(q)u(t) + H(q)e(t) \tag{1}$$

where G(q) is the relationship between input and output and it is represented in the form of the following transfer function or parameter of dynamic system as illustrated in (2). The pole of dynamic system is determined in F(q).

$$G(q) = B(q)/F(q) \tag{2}$$

On the contrary, H(q) is the relationship between noise and output or parameters of noise system as shown in (3) where D(q) determines the pole of system noise.

$$H(q) = C(q)/D(q) \tag{3}$$

where B(q), F(q), C(q) and D(q) are polynomials with

$$B(q) = b_1 q^{-1} + \cdots + b_n q^{-n} \tag{4}$$

$$F(q) = f_1 q^{-1} + \cdots + f_n q^{-n} \tag{5}$$

$$C(q) = c_1 q^{-1} + \cdots + c_n q^{-n} \tag{6}$$

$$D(q) = d_1 q^{-1} + \cdots + d_n q^{-n} \tag{7}$$

q^{-n} is the shift operator so $q^{-n}u(t)$ is the output signal at time $t = t-n$ or $u(t-n)$. According to (1)-(7), the resultant equation is the Box-Jenkins (BJ) model as shown in (8).

$$y(t) = \left[\frac{B(q)}{F(q)}\right] u(t) + \left[\frac{C(q)}{D(q)}\right] e(t). \tag{8}$$

BJ model has the capability to estimate both dynamic and noise parameters. Therefore, the construction of a model is based on the measurement noise, not the input noise. As a result, the structure of BJ model provides the high flexibility for modeling noise. Moreover, the above model can be extended into different subclasses of how to model dynamic and noise parameters. Among these models is the output error (OE) model which focuses on the estimation of dynamic model only while a noise model is set at the constant level and equal to 1. The OE model is shown in (9).

$$y(t) = \left[\frac{B(q)}{F(q)}\right] u(t) + e(t) \tag{9}$$

3.2 Artificial Neural Network (ANN)

The objective of ANN models is to explore the relationship between input variables and output variables. Basically, the neural architecture consists of three or more layers, i.e., input layer, output layer and hidden layer as shown in Fig. 1. The function of this network can be described as follows:

$$y_j = f(\Sigma_i w_{ij} x_{ij}) \tag{10}$$

where y_j is the output of node j, f (.) is the transfer function, w_{ij} the connection weight between node j and node i in the lower layer and x_{ij} is the input signal from the node i in the lower layer to node j.

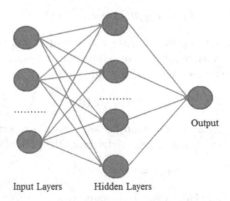

Output

Input Layers   Hidden Layers

Fig. 1. ANN Structure

One of the algorithms used in ANN is the Levenberg–Marquardt algorithm or LMA. The main purpose of this algorithm is to solve the non-linear problems based on the least square method. When y_i and x_i are the inputs and outputs, β is the regression parameter and f(x, β) is the curve fitting function, the LMA is based on the idea that the sum of the squares of the deviations S(β) as shown in (11) is minimized.

$$S(\beta) = \sum_{i=1}^{n}[y - f(x_i, \beta)]^2 \qquad (11)$$

4   Manufacturing Process

One of the most important manufacturing units in factories or power plants is the drum boiler. The inputs of a drum boiler are feed water flow rate (kg/sec) and heat (kJ) as concluded in Table 1 while the controlled output is pressue (kPa). The drum boiler used as a pilot for this case study is located in a power plant and the maximum working pressure is 1050 kPa. The drum boiler diagram is shown in Fig. 2. To conduct the empirical study, the inputs and output are sampled every minute for 474 minutes and they are shown graphically in Fig. 3, 4 and 5 consecutively.

Table 1. Ouput and inputs

Type of variable	Variable	Unit
Output	Pressure	kPa
Input	Feed water flow rate	kg/sec
Input	Applied heat	kJ

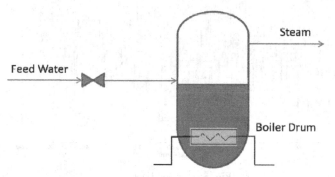

Fig. 2. The Schematic Diagram of A Drum Boiler

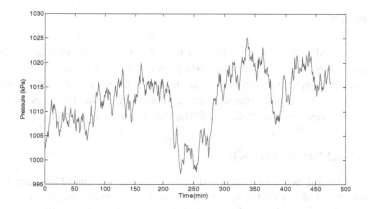

Fig. 3. Output Pressure

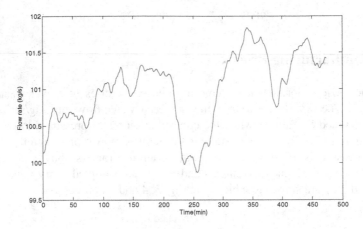

Fig. 4. Feed Water Flow Rate

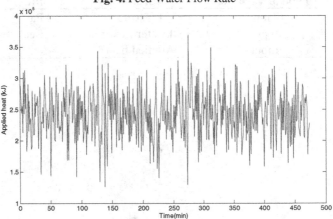

Fig. 5. Applied Heat

5   Results

The data achieved from the observations of a drum boiler is analyzed using neural network and system identification toolbox provided by MATLAB software. The predicting errors after applying the fitting model to the observations of each method are calculated and used as a performance index. It is assumed that the forecasting method leading to the lowest error is preferred to the one whose error is high.

5.1   Artificial Neural Network

The artificial neural network is applied to fit the observations. There are two inputs (water flow rate and applied heat) while there is an output (pressure). The training algorithm used is Levenberg-Marquardt back propagation while the mean squared

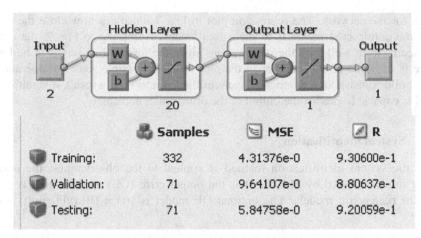

Fig. 6. The Summary of ANN Model

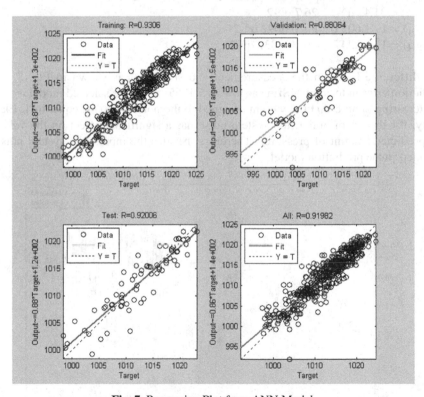

Fig. 7. Regression Plot from ANN Model

error (MSE) is as low as 4.3137 for the training stage, 9.64107 for the validation stage and 5.84758 for the testing stage. The whole data set is divided for different purposes (70% for training, 15% for validation and 15% for testing). The summary of ANN model used in this study is illustrated in Fig. 6 and it shows that there are 20 hidden layers for the network. The regression plot in Fig. 7 illustrates how close the output from the constructed model is to the actual values. According to Fig. 7, the overall plot seems to fit well with the observations since the formed line seems to be linear. There is only one obvious outlier in the plot which is possible since the operation of drum boiler consists of a number of uncontrollable factors. As a result, it signifies that ANN is capable to predict the output of the drum boiler accurately.

5.2   System Identification

After the system identification method is applied to the observations, the optimal fitting model selected by MATLAB is the output error (OE) model which is a subclass of parametric models. The optimal OE model is $y(t) = [B(q)/F(q)]u(t) + e(t)$, where

$B_1(q) = 3.447\text{e-}005\ q^{\wedge}\text{-}1 - 3.073\text{e-}005\ q^{\wedge}\text{-}2$
$B_2(q) = 31.41\ q^{\wedge}\text{-}1 - 26.7\ q^{\wedge}\text{-}2$
$F_1(q) = 1 - 1.603\ q^{\wedge}\text{-}1 + 0.6092\ q^{\wedge}\text{-}2$
$F_2(q) = 1 - 0.2247\ q^{\wedge}\text{-}1 - 0.2332\ q^{\wedge}\text{-}2.$

The difference between the measurement and simulation is shown in Fig. 8. The prediction error in term of MSE is as low as 5.1265. Since the selected model is OE, it is interesting to note that the weight assigned to the noise or $e(t)$ is equal to 1. Elaborately, for this drum boiler, the system noise has a significant effect on the value of the predicted amount of pressure. Therefore, besides the inputs, the system noise is included in the prediction model.

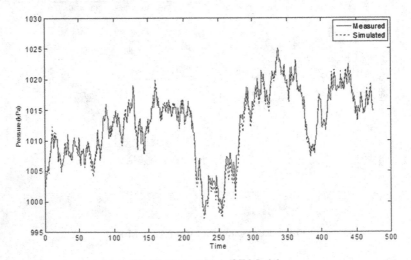

Fig. 8. Output from OE Model

6   Conclusions

The empirical study is conducted on a drum boiler in a power plant. The observed inputs of the boiler are water feed rate and applied heat while the objective is to regulate the output pressure. Two approaches, ANN and system identification, are utilized to construct empirical models for predicting the pressure of the boiler. The observed data is sampled every minute for 474 minutes. The performance of these two methods is based on the predicting errors. According to the results from ANN model, if only the training set of data is considered, ANN seems to slightly outperform the OE model. However, the MSE of validation stage is quite high since there is an outlier among the output data. This fining implies that ANN is capable of identifying the outlier embedded in the observations. In conclusion, the MSE from the overall period is higher than the one from the training stage.

7   Discussions

When the predicting errors are considered, the performance of both methods is not significantly different. However, the advantage of ANN method is its robustness to the outliers in the process. Although the outliers might be identified in the validating stage and not included in the training stage, the overall MSE of ANN is slightly higher than the one of OE model which is based on the whole data set. Another point that the practitioners should keep in mind for selecting the appropriate method is the complexity of each method. In term of parsimony, OE might be preferred to ANN because OE has a standardized model with a fixed number of variables which is straightforward and simple to understand. Last but not least, the results indicate that the inclusion of the system noise in the fitting model of OE does not lead to a significant improvement to the predicting performance.

References

1. Astrom, K.J., Bell, R.D.: Drum-Boiler Dynamic. Automatica **36**, 363–378 (2000)
2. Rodriquez Vasquez, J.R., Rivas Perez, R., Sotomayor Moriano, J., Peran Gonzalez, J.R.: System Identification of Steam Pressure in A Fire-tube Boiler. Comput. Chem. Eng. **9**, 2839–2848 (2008)
3. Vasickaninova, A., Bakosova, M., Meszaros, A., Klemes, J.J.: Neural Network Predictive Control of A Heat Exchange. Appl. Therm. Eng. **31**, 2094–2100 (2011)
4. Yu, W., Li, X.: Some New Results on System Identification with Dynamic Neural Networks. IEEE T. Neural Networks **12**, 412–417 (2001)
5. Subudhi, B., Jena, D.: A Differential Evolution Based Neural Network Approach to Nonlinear System Identification. Appl. Soft Comput. **11**, 861–871 (2011)
6. Wu, J.-D., Tsai, Y.-J.: Speaker Identification System Using Empirical Mode Decomposition and An Artificial Neural Network. Expert Syst. Appl. **38**, 6112–6117 (2011)

7. Xia, P.-Q.: An Inverse Model of MR Damper Using Optimal Neural Network and System Identification. Journal of Sound and Vibration **266**, 1009–1023 (2003)
8. Mohanty, S.: Artificial Neural Network Based System Identification and Model Predictive Control of A Floatation Column. J. Process Contr. **19**, 991–999 (2009)
9. Habbi, H., Zelmat, M., Bouamama, B.O.: A Dynamic Fuzzy Model for A Drum-Boiler-Turbine System. Automatica **39**, 1213–1219 (2003)
10. Adaleesan, P., Miglan, N., Sharna, R., Saha, P.: Nonlinear System Identification Using Wiener Type Laguerre-Wavelet Network Model. Chem. Eng. Sci. **63**, 3932–3941 (2008)

Pattern Recognition using the TTOCONROT

César A. Astudillo[1][(✉)] and B. John Oommen[2]

[1] Department of Computer Science, Universidad de Talca,
Km 1. Camino a los Niches, Curicó, Chile
castudillo@utalca.cl
[2] School of Computer Science, Carleton University,
Ottawa K1S 5B6, Canada
oommen@scs.carleton.ca

Abstract. We present a method that employs a tree-based Neural Network (NN) for performing classification. The novel mechanism, apart from incorporating the information provided by unlabeled and labeled instances, re-arranges the nodes of the tree as per the laws of Adaptive Data Structures (ADSs). Particularly, we investigate the Pattern Recognition (PR) capabilities of the Tree-Based Topology-Oriented SOM (TTOSOM) when Conditional Rotations (CONROT) [8] are incorporated into the learning scheme. The learning methodology inherits all the properties of the TTOSOM-based classifier designed in [4]. However, we now augment it with the property that frequently accessed nodes are moved closer to the root of the tree.

Our experimental results show that on average, the classification capabilities of our proposed strategy are reasonably comparable to those obtained by some of the state-of-the-art classification schemes that only use labeled instances during the training phase. The experiments also show that improved levels of accuracy can be obtained by imposing trees with a larger number of nodes.

Keywords: Tree-based SOMs · TTOSOM · CONROT · Pattern recognition

1 Introduction

In a previous work [1,6], we presented a clustering algorithm that combined the philosophies defined by the tree-structured families of Self Organizing Maps (SOMs) and the field of Adaptive Data Structures (ADSs). The pioneering manner in which we perceived clustering, attempted to generate asymptotically

---

C.A. Astudillo—Assistant Professor. IEEE Member. The work of this author is partially supported by the FONDECYT grant 11121350, Chile.

B.J. Oommen—*Chancellor's Professor; Fellow: IEEE* and *Fellow: IAPR*. This author is also an *Adjunct Professor* with the University of Agder in Grimstad, Norway. The work of this author was partially supported by NSERC, the Natural Sciences and Engineering Research Council of Canada.

M. Ali et al. (Eds.): IEA/AIE 2015, LNAI 9101, pp. 435–444, 2015.
DOI: 10.1007/978-3-319-19066-2_42

optimal trees based on the access probabilities of the neurons[1]. In this paper we design a classifier based on these principles, and show the effect of the classification accuracies obtained on different real-world domains.

To report the contribution of this paper in the context of the work done in [4], we mention that in [4], we designed a classifier based solely on the Tree-Based Topology-Oriented SOM (TTOSOM) algorithm, and showed that it was able to learn the decision boundaries based on labeled and unlabeled samples simultaneously. This so-called "semi-supervised" learning classifier utilized the strategy presented by Zhu [12], which identified clusters by using a possibly large number of unlabeled samples, and subsequently, associating each neuron with a label by utilizing a small number of labeled instances. We showed that this approach, indeed, can produce accuracies that are reasonably comparable to the ones achieved by state-of-the-art classifiers.

A natural extension to our investigation is to develop a study analogous to the one performed in [4], but now considering the effect of the rotations in the tree. In this sense, our proposed methodology consists of deriving a classifier similar to the TTOSOM-based classifier, but using the TTOCONROT [6] as a foundation.

The remainder of the paper is organized as follows: The next section describes the necessary background regarding the SOM. Section 3 explains how the TTO-CONROT is used to perform classification. Subsequently, Section 3 focuses on the experimental results, and finally, Section 4 contains the conclusions.

1.1   Literature Review

The Self Organizing Maps (SOMs) is a family of Neural Networks (NN) suitable for visualization and data clustering. The model uses a network of neurons arranged in a grid which are trained using a concept called *competitive learning*. In each training step, a new instance (also called input vector) is presented to the network and the most similar neuron is declared as the winner or *best matching unit* (BMU). To achieve this, each neuron is associated to a weight vector that possess the same dimensionality as the input vectors and a dissimilarity function (such as the Euclidean distance) is used to compare them. The SOM introduces the concept of neighborhood function, that identify a subset of neurons in the vicinity of the BMU. A central process of the training mechanism of the SOM is the migration phase, in which the BMU and its neighboring neurons are moved closer to the input vector. This movement is controlled by the so-called update rule:

$$\mathbf{w}_i(t+1) = \mathbf{w}_i(t) + \phi_{ci}(t)(\mathbf{x}(t) - \mathbf{w}_i(t)) \tag{1}$$

where $\mathbf{x(t)}$ is a d-dimensional vector that represents the input vector at time t, c is the index of the BMU, $\mathbf{w}_i$ is the weight vector associated to the i-th neuron and ϕ_{ci} is the neighborhood function.

---

[1] A paper that reported the preliminary results of the TTOCONROT won the best paper award in a well-known AI conference [2].

As a result of the training process the weight vectors of the neurons absorb the properties of the original data distribution and its topological structure.

There are hundreds of papers reporting the applicability of the SOM in almost all branches of engineering (if not all) [7]. However, in spite of all these benefits, the SOM has known handicaps, some of which are discussed in [5,7]. As a result, numerous variants of SOM has been designed in an effort to render the topology more flexible or to accelerate the learning process.

One of this variants is the Tree-based Topology Oriented SOM (TTOSOM) [3], which uses a neural tree instead of a grid. The SOM trains a user-defined tree in a similar manner compared to the SOM but using a neighborhood function defined over the tree which produces a completely different mapping. The TTOSOM has shown holographic properties and also reduces to the 1D SOM, when the tree is a "linear" sequence of neurons.

2   The TTOCONROT-Based Classifier

In [4], the authors designed a classifier based on the TTOSOM algorithm that was able to learn from labeled and unlabeled samples. This so-called "semi-supervised" learning classifier utilized the strategy presented by Zhu [12], and consisted of clustering the instances using a "massive" number of unlabeled samples, and subsequently, identifying the label of each neuron based on a possibly small number of labeled instances. The authors of [4] showed that this approach can, indeed, produce accuracies that are comparable to the ones achieved by state-of-the-art classifiers.

Our goal in this paper is to devise a classifier analogous to the one presented in [4], but this time based on the foundation of the TTOCONROT instead of the TTOSOM. In order to clarify the way in which this classifier is built, we will first summarize the main properties of the above-mentioned clustering technique.

The reader will recall that in [6], we had merged the fields of SOMs and ADSs. The adaptive nature of the strategy presented, namely the Tree-Based Topology-Oriented Topology Using Conditional Rotations (TTOCONROT), is unique because adaptation is perceived in two forms: The migration of the code-book vectors in the feature space is a consequence of the SOM update rule, and the rearrangement of the neurons *within* the tree is a result of the ADS-related rotations. This reorganization can be perceived to be both automatic and adaptive, such that on convergence, the DS tends towards an optimal configuration with a minimum average access time. In most cases, the most probable element will be positioned at the root (head) of the tree (DS), while the rest of the tree is recursively positioned in the same manner.

The TTOCONROT [6] is a further enhancement of the TTOSOM [3] which considers how the underlying tree itself can be rendered dynamic and adaptively transformed. To do this, we presented a method by which a SOM with an underlying BST structure can be adaptively re-structured using Conditional Rotations [8]. These rotations on the nodes of the tree are local, can be done in constant time, and performed so as to decrease the WPL of the entire tree. In [6], we also introduced the concept referred to as *Neural Promotion*, where neurons gain prominence in the NN as their significances increase. The advantages of such a scheme is that the leaned tree learns the topological peculiarities of the stochastic data distribution, and at the same time recursively positions neurons accessed more often close to the root. As a result, the TTOCONROT, converges in such a manner that the neurons are ultimately placed in the input space so as to represent its stochastic distribution, and additionally, the neighborhood properties of the neurons suit the best BST that represents the data.

Even though, the advantages of the CONROT algorithm are explained in [6], the proposed architecture allows the inclusion of alternative restructuring modules other than the CONROT. Potential candidates which can be used to perform the adaptation are the Splay and the Mehlhorn's D-Tree algorithms, among others [8].

Analogously to the classifier devised in [4], our aim is to design a classifier that works in two stages. First of all, the data distribution and its structure is learned in an unsupervised manner using the TTOCONROT scheme. In a second phase, we utilize some labeled samples to categorize the decision regions that have been previously created.

The TTOCONROT-based classifier uses the cluster-then-label paradigm [11], leading to an algorithm similar to the one described in [4], with the difference that the TTOCONROT is used as the unsupervised learning algorithm.

3   Experimental Results

3.1   Experimental Setup

In order to verify the capabilities of the TTOCONROT for classifying items belonging to the real-world domain, and for making the results comparable to the ones obtained by the TTOSOM-based classifier, we have chosen the same datasets described in [4]. These six datasets are Iris, Wisconsin Diagnostic Breast Cancer (WDBC), Wine, Yeast, Wine Quality and Glass datasets, all publicly available from the UCI Machine Learning repository [9]. For an explanation of each of these datasets, we refer the reader to Section 4.3 included in [4].

Analogous to the experiments performed in [4], the classifiers considered in this comparison include five supervised classifiers, namely, Bayes Networks (BN), Naive Bayes (NB), C4.5, k-Nearest Neighbors (KNN) and Learning Vector Quantization 1 (LVQ1), and three "semi-supervised" classifiers, namely the TTOCONROT, the TTOSOM and the SOM. The sampling method utilized to measure the accuracy was the stratified 10-fold cross validation.

3.2   Comparison to Other Classifiers

We started our experimental analysis by comparing the accuracies of the TTO-CONROT with the rest of the classifiers mentioned above, using the parameter settings specified in [4]. The results obtained are presented in Table 1, which shows the accuracy of the classifiers obtained for the various datasets.

For example, Table 1 shows that the TTOCONROT classifier, using *only* 15 neurons, accurately predicts, with an accuracy of 96.07%, the correct label of the instances belonging the *wine* dataset, which is outperformed only by the BN and the NB schemes. On the other hand, the SOM classifies correctly the same dataset with an *accuracy* of only 67.98%!

Table 1. General classification results of the TTOCONROT and other methods investigated, reported in terms of the accuracy

Dataset	ROT15	TTO15	BN	NB	C4.5	KNN	LVQ1	SOM
iris	94.00	92.00	92.67	96.00	96.00	95.33	96.00	84.67
wdbc	93.32	92.09	95.08	93.15	93.15	96.66	92.09	90.51
glass	53.74	52.34	71.96	49.07	67.76	67.76	61.22	63.08
wine	96.07	95.51	98.88	97.19	93.82	94.94	74.16	67.98
yeast	51.08	51.82	56.74	57.61	55.86	54.78	24.33	46.16
winequality	53.60	53.41	57.72	55.03	62.91	57.79	44.15	49.59

We have also compared the TTOCONROT-based classifier with other VQ-based methods. One such strategy that belongs to the supervised family is the LVQ1, while the SOM, the TTOSOM and the TTOCONROT primarily learn the distributions using an unsupervised learning paradigm. All four classifiers used the same values for their parameters (the ones specified in [4]). In addition, the LVQ1 and the SOM used 128 neurons, and the results shown for the TTO-SOM and the TTOCONROT include *only* 15 neurons, respectively. As per our results, the TTOCONROT, using only a small percentage of the neurons used in the SOM and LVQ1 (almost 10%), outperforms their recognition capabilities in almost *all* six datasets[2]. The differences with respect to the TTOSOM are more subtle and are analyzed in a subsequent section.

Similar to the TTOSOM, we observe that the TTOCONROT offers accuracies comparable to the ones obtained by certain supervised classifiers. For instance, the results are similar to the one obtained using the KNN. However, as stated in [4], even though the KNN is internally used for labeling the neurons, this is done only once, and applies to only a small subset of the neurons, which represent a small fraction of the total number of instances, i.e., those which are involved in the computations of the KNN every time a query is performed.

Another advantage that the scheme presents is its "semi-supervised" nature, that allows it to associate the neurons with a class label using only a minimum number of tagged instances. In cases when these samples are scarce (but when the unlabeled samples are abundant), it has been shown that other schemes that belong to the same "semi-supervised" family, yield competitive results as pointed out in [10] and as our results in [4] demonstrate.

---

[2] The table show the results for the case when we have only 15 neurons. But if the number of neurons is increased to 127, the accuracy is superior in all the VQ-based algorithms.

3.3 Effect of the Number of Neurons

We now consider the effect of varying the number of neurons involved in the TTOCONROT tree. To test this, we trained the TTOCONROT with the configuration presented in [4], and steadily augmented the size of the respective tree. Analogous to the experimental settings used in [4], we permitted the starting value for the radius to be twice the depth of the tree, so as to ensure that all the neurons are initially considered as part as the BoA.

Table 2 shows the accuracies obtained by the TTOCONROT, where in the respective column, the specific tree size is systematically increased. The respective graphical curves are illustrated in Figure 1. To cite an example, Table 2 shows that the TTOCONROT classifier, using 127 neurons, predicts with an accuracy of 55.60%, the actual category of the instances belonging to the *winequality* dataset. The experiments use an analogous parameter configuration in which we systematically increase the size of a full binary tree with depths ranging from 3 to 12. Observe that the tree was restricted to be binary, even though it could have been of an arbitrary size. The reason for this was the TTO-CONROT nature which is constrained by a BST structure[3]. Trees with small size were used to test the capabilities of the classifiers with a very condensed representation of the feature space. On the other hand, trees with a larger size were utilized so as to observe the effect of adding artificial data points which, in some cases was even greater than the number of sample points themselves. These artificial points attempted to preserve the original properties of the feature space.

Table 2. The accuracy of the TTOCONROT as the number of neurons increases

Dataset ↓ Neurons →	7	15	31	63	127	255	511	1023
glass	51.87	53.74	63.08	67.76	68.22	66.36	66.36	67.29
iris	92.00	94.00	92.67	94.67	93.33	94.67	94.00	94.00
wdbc	88.23	93.32	94.55	95.96	94.55	95.08	96.13	95.43
wine	91.01	96.07	97.19	94.38	97.19	97.19	96.07	95.51
winequality	53.85	53.60	53.28	54.72	55.60	56.85	56.66	58.79
yeast	50.74	51.08	53.23	55.73	55.32	50.27	51.21	52.29

3.4 Difference of Classifying with and Without Conditional Rotations

We have also investigated the effect of the rotations and studied how the accuracies vary as the number of neurons is increased.

The results for the *wine* dataset when the number of neurons is increased are shown in Table 3. In the table, each row presents the accuracies obtained

---

[3] We are currently investigating the generalization of the CONROT, which will allow rotations on trees with an arbitrary number of children per node.

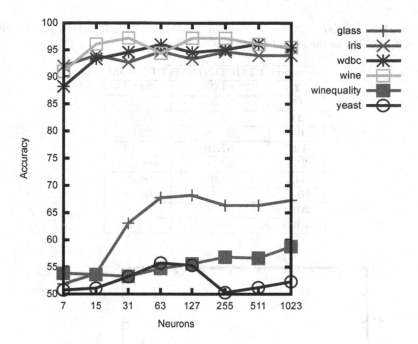

Fig. 1. The accuracies for the different datasets as obtained by using the TTOCONROT-based classifier and an increasing number of neurons.

by using a specific tree size, and each column indicates the accuracies obtained by the TTOCONROT and the TTOSOM, respectively. In order to verify if one strategy performs better than the other, we have computed the average accuracy and ranking indices. As per our observations, both algorithms possess a similar pattern. The maximum accuracy obtained in both cases is 97.19%. However this "peak" is reached by the TTOSOM only once (when using 127 neurons), while the TTOCONROT-based classifier achieve this in 3 instances, i.e., when using 31, 127 and 255 neurons respectively. In this regard, it is worth mentioning that based on this results, only the BN, which belongs to the "supervised" family, could outperform this accuracy, obtaining in that case, 98.88%, i.e., only a fraction better than the results obtained by our proposed methods. It is remarkable that the TTOCONROT was able to provide almost the highest accuracy possible, in comparison to the state-of-the-art classifiers included in our study, using only a limited number of 31 neurons. The fact that this result can be replicated by using a larger tree, further demonstrates the consistency of the method. From our perspective, this evidence suggests that the user does not need to know *a priori* the exact number of neurons required to train the tree effectively.

Another dataset that we have considered belongs to the same problem domain, i.e., the *winequality* dataset. However, the latter presents a harder classification problem, in which the state-of-the-art supervised classifiers provide accuracy rates which are roughly between 50% and 60%. Figure 2 illustrates the differences in

Table 3. *Wine* dataset – Accuracy rate in % obtained by using the TTOCONROT and the TTOSOM as the size of the tree is increased

Neurons	TTOCONROT	TTOSOM
7	91.01	94.94
15	96.07	95.51
31	97.19	95.51
63	94.38	96.07
127	97.19	97.19
255	97.19	96.63
511	96.07	96.07
1023	95.51	96.63
2047	96.07	96.63
4095	96.07	96.07

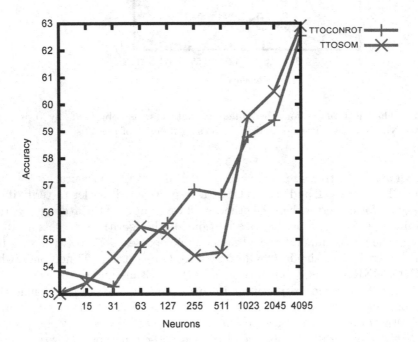

Fig. 2. The *winequality* dataset is learned using the TTOCONROT-based classifier and the TTOSOM-based classifier. In each case the number of neurons is increased systematically.

accuracy obtained by the TTOCONROT and the TTOSOM as the size of the tree is increased.

Observe that in both cases, the classifiers have a tendency to increase their recognition capabilities as the number of neurons is increased. However, we observe that the TTOCONROT presents an almost monotonic non-decreasing behavior. From this behavior we believe that when solving practical problems, it is worth training the classifier with trees that possess even more neurons than the number of training samples. From our experiments, we infer that it is possible to improve the accuracy rates, if additional computational power, time and/or space are available. We believe that this occurs because the TTOCONROT tree also effectively covers those regions where no samples lie, and this is used by the classifier to accurately predict the class labels of those regions when labeled instances become available.

4 Conclusions

This paper has presented the design and experimental analysis of a Patter Recognition (PR) scheme based on the Tree-Based Topology Oriented SOM using Conditional Rotations (TTOCONROT). The approach utilizes the tree-based neural network to learn the distribution of *all* the samples available, regardless of the fact that their labels are known, and then utilize a set of labeled samples (expected to be scarce), to categorize the regions of the feature space. In particular, the proposed scheme constrains the neural tree as per the laws of the field of Adaptive Data Structures (ADS).

Our experiments demonstrated that, the TTOCONROT-based classifier is able to sometimes outperform state-of-the-art classifiers that use the supervised learning paradigm, i.e., those which are unable to learn from unlabeled samples. This concurs with the results of other researchers who observe that under certain scenarios, semi-supervised schemes like the one presented in this paper, can lead to performance levels that are comparable to the ones obtained by true supervised methods [10]. Particularly, in most of our experiments, trees whose sizes are only a small fraction of the cardinality of the dataset, are sufficient to obtain accuracies comparable to the ones provided by the best supervised classifiers.

Additionally, we have performed a meticulous analysis to identify the advantages of incorporating the neural rotations provided by the TTOCONROT. To do this, we compared the results with the TTOSOM-based classifier (presented in [6]), using analogous parameter settings and using different tree sizes. Our results showed that regardless of the inclusion of the rotations, competitive accuracy rates can be obtained. Moreover, our experiments also suggest that in certain cases, the rotations lead to accuracy rates that increase in a smoother manner, in comparison to the ones obtained by the TTOSOM-based classifier, as more neurons are incorporated in the tree.

References

1. Astudillo, C.A., Oommen, J.B.: A Novel Self Organizing Map Which Utilizes Imposed Tree-Based Topologies. In: Kurzynski, M., Wozniak, M. (eds.) Computer Recognition Systems 3. AISC, vol. 57, pp. 169–178. Springer, Heidelberg (2009)

2. Astudillo, C.A., Oommen, B.J.: On using adaptive binary search trees to enhance self organizing maps. In: Nicholson, A., Li, X. (eds.) 22nd Australasian Joint Conference on Artificial Intelligence (AI 2009), pp. 199–209 (2009)
3. Astudillo, C.A., Oommen, B.J.: Imposing tree-based topologies onto self organizing maps. Information Sciences 181(18), 3798–3815 (2011)
4. Astudillo, C.A., Oommen, B.J.: On achieving semi-supervised pattern recognition by utilizing tree-based SOMs. Pattern Recognition 46(1), 293–304 (2013)
5. Astudillo, C.A., Oommen, B.J.: Fast BMU Search in SOMs Using Random Hyperplane Trees. In: Pham, D.-N., Park, S.-B. (eds.) PRICAI 2014. LNCS, vol. 8862, pp. 39–51. Springer, Heidelberg (2014)
6. Astudillo, C.A., Oommen, B.J.: Self-organizing maps whose topologies can be learned with adaptive binary search trees using conditional rotations. Pattern Recognition 47(1), 96–113 (2014)
7. Astudillo, C.A., Oommen, B.J.: Topology-oriented self-organizing maps: A survey. Pattern Analysis and Applications 17(2), 1–26 (2014)
8. Cheetham, R.P., Oommen, B.J., Ng, D.T.H.: Adaptive structuring of binary search trees using conditional rotations. IEEE Trans. on Knowl. and Data Eng. 5(4), 695–704 (1993)
9. Frank, A., Asuncion, A.: UCI machine learning repository (2010). http://archive.ics.uci.edu/ml
10. Gabrys, B., Petrakieva, L.: Combining labelled and unlabelled data in the design of pattern classification systems. International Journal of Approximate Reasoning 35(3), 251–273 (2004). Integration of Methods and Hybrid Systems
11. Zhu, X.: Semi-supervised learning literature survey. Technical Report 1530, Computer Sciences, University of Wisconsin-Madison (2005)
12. Zhu, X., Goldberg, A.B.: Introduction to Semi-Supervised Learning. Morgan & Claypool Publishers, January 2009

Classification

Approximate is Enough: Distance-Based Validation for Geospatial Classification

Yangping Li and Tianming Hu[(⊠)]

Dongguan University of Technology, Dongguan, China
tmhu@ieee.org

Abstract. In geospatial classification, the validation criterion functions should incorporate both error rate based classification accuracy and distance based spatial accuracy. However, due to the difference in subject and scale between the two accuracies, it is not trivial to combine them in a reasonable way. To circumvent this difficulty, we develop approximate functions for spatial accuracy that preserve distance ranking instead of distance values. The resultant criterion functions not only take less computation cost, but also get more commensurate with classification accuracy. Finally, the approximation power of the proposed criterion functions are validated on real-world datasets.

1 Introduction

Geospatial data distinguish themselves from other data in that associated with each object, the attributes under consideration include not only non-spatial normal attributes which also exist in other databases, but also spatial attributes which are often unique or emphasized in spatial database and describe the object's spatial information such as location and shape. Independent and identical distribution, a fundamental assumption often made in data sampling, is no longer valid with geospatial data [1]. Instead, in the geospatial sense, everything is related to everything else, but nearby things are more related than distant things. This property is also referred to as the first law of geography. For example, Figures 1a and 1b show a large data set for 1980 US presidential election results covering 3107 counties [2]. In the original regression problem, income, home ownership and population with college degrees are used to predict voting rate. One can observe clear spatial trend in northwest-southeast direction, along which the voting rate decreases. This roughly coincides with the decreasing degree of education. Also, the data show positive spatial correlation.

Geospatial contextual classification [3,4] is probably the sub-area in spatial data mining that receives most attention. It differs from other classification in that spatial context, in addition to the local attributes, must be taken into account, in both the classification model construction phase (training) and decision phase (test). In this paper, we focus on a crucial component in the

---

T. Hu — This work was supported by UTPG(2013-246) and STPG.

M. Ali et al. (Eds.): IEA/AIE 2015, LNAI 9101, pp. 447–456, 2015.
DOI: 10.1007/978-3-319-19066-2_43

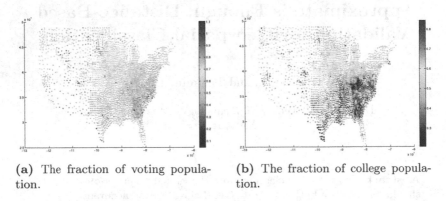

(a) The fraction of voting population.

(b) The fraction of college population.

Fig. 1. The election data

test phase, the validation criterion functions, which are used to evaluate the estimated class labeling against the true class labeling. Appropriate validation criteria for geospatial data should capture both classification and spatial accuracies. However, traditional classification accuracies, such as classification rate, discard spatial information. One straightforward means of disambiguating the definition of a good multiobjective solution is to assign the accuracies different weights before combining them together. For instance, given classification accuracy C and spatial accuracy S, the weighted scheme gives $(1 - \alpha)C + \alpha S$, where α is the relative significance coefficient, determined by the user according to his/her preference. However, since C is usually a classification rate-like measure with unit of percentage and S is a distance measure with unit of length, direct combination is cumbersome and can be even meaningless, due to their totally different subjects and scales. Besides, spatial accuracy computation is usually expensive, which involves nearest neighbor search and distance evaluation. To that end, in this paper, we propose new criterion functions that, instead of directly evaluating near neighbor distance values, approximate distance rankings using contiguity matrix. Finally, their effectiveness is validated on real-world databases.

Overview. The rest of the paper is organized as follows. Section 2 introduces problem background and related work. Section 3 presents new criterion functions that approximate distance-based ranking. Section 4 discusses the ideal properties that a validation criterion function must have. Empirical results are reported in Section 5. Finally, Section 6 concludes this paper.

2   Related Work

Geospatial classification problem can be informally described as follows. Given a spatial framework of n sites, where each site s_i has a class label $z_i \in \{0, 1\}$ (we focus on binary classification in this paper) and a vector of explanatory attributes $\mathbf{x}_i$. We need to construct a classification model to predict the class

Table 1. Notations in the criterion functions

notation	description
a_1	number of actual class 1 sites
a_0	number of actual class 0 sites, $n = a_1 + a_0$
p_1	number of predicted class 1 sites
p_0	number of predicted class 0 sites, $n = p_1 + p_0$
$\mathbf{a_1}$	$\mathbf{a_1} \equiv \mathbf{z}$, actual vector for class 1, $a_1 = \mathbf{a}_1^T \mathbf{a}_1$
$\mathbf{a_0}$	$\mathbf{a_0} \equiv \mathbf{1} - \mathbf{z}$, actual vector for class 0, $a_0 = \mathbf{a}_0^T \mathbf{a}_0$
$\mathbf{p_1}$	$\mathbf{p_1} \equiv \hat{\mathbf{z}}$, predicted vector for class 1, $p_1 = \mathbf{p}_1^T \mathbf{p}_1$
$\mathbf{p_1}$	$\mathbf{p_0} \equiv \mathbf{1} - \hat{\mathbf{z}}$, predicted vector for class 0, $p_0 = \mathbf{p}_0^T \mathbf{p}_0$

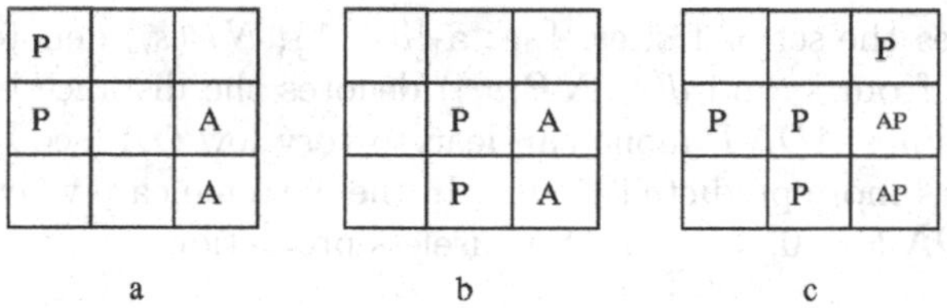

Fig. 2. CA vs ADNP. "A" denotes actual (class 1) site. "P" denotes predicted (class 1) site.

label with explanatory attributes. Unlike conventional classification, the class label of each site is not only determined by its own attributes, but also affected by its neighbors. The neighbor relationship is given by $N \subseteq S \times S$, i.e., sites s_i and s_j are neighbors iff $(s_i, s_j) \in N$. Often N is encoded in a contiguity matrix W for which $W(i,j) > 0$ if sites s_i and s_j are neighbors, $W(i,j) = 0$ otherwise.

2.1   Accuracies of Interest

To ease the following discussion, we first introduce the relevant notations in Table 1, where $\mathbf{1}$ is a n-D vector of 1's.

Traditional validation criterion functions for geospatial contextual classification include likelihood and posterior. Although they have well-established theoretic justification, they are not always satisfactory in practice. In classification, the most popular criterion is probably the classification accuracy, CA. As defined in Equation (1), it simply computes the fraction of data that is correctly classified.

$$CA \equiv \frac{\mathbf{p}_1^T \mathbf{a}_1 + \mathbf{p}_0^T \mathbf{a}_0}{n} \tag{1}$$

However, hypotheses yielding a higher likelihood do not always yield a higher classification rate. For location prediction, it could be worse, for spatial accuracy has to be considered and CA alone is not enough. See Figure 2 for an example, where CA cannot distinguish the two predicted class labelings in Figure 2a and b. Suppose we are interested in the sites of class 1, e.g., the location of gold mine. Domain scientist may prefer Figure 2b where predicted 1 locations are near actual 1 locations. In the case of gold mine, it is a reasonable expectation that even if we cannot accurately predict the real locations of gold mine, our predicted ones are not far away from them.

Along this line, as defined in Equation (2), Shekhar et al. [5] proposed a new measure, Average Distance to Nearest Predicted (class 1) location from actual (class 1) location ($ANDP$)

$$ADNP \equiv \sum_{s_i \in S_1} \frac{1}{a_1} d(s_i, NP(s_i)) \tag{2}$$

where S_1 denotes the set of 1 sites, $\{s_i : a_1[i] = 1\}$, $NP(s_i)$ denotes the nearest predicted 1 site from s_i and $d(s_i, NP(s_i))$ denotes the distance between them. However, sometimes $ADNP$ alone can lead to very low CA (see Figure 2c) and alway encourages more predicted 1 sites. In the extreme case where all sites are predicted 1, $ADNP = 0$, a perfect but useless prediction.

2.2   Multiobjective Optimization

Since a good criterion function for geospatial classification should consider both classification accuracy and spatial accuracy, which is a typical two-objective optimization problem, below we brief three representative approaches to multi-objective optimization [6].

The first approach, probably the most popular by far, is the weighted formula, i.e., transforming multiobjective optimization into a single-objective problem by assigning a numerical weight to each objective and then combining the values of the weighted criteria into a single value by either adding or multiplying all the weighted criteria. While simple and easy to use, the weight assignment is subjective and appropriate combination of multiple criteria is not trivial. The second approach makes use of lexicographic order to assign different priorities to different objectives, and then focuses on optimizing the objectives in their order of priority. Hence, when comparing candidate solutions, if one candidate is significantly better than the others by more than the tolerance threshold with respect to the highest-priority objective, it is chosen. Although the lexicographic approach recognizes the non-commensurability of different quality criteria, it introduces a new set of ad-hoc parameters, the tolerance thresholds for the criteria, which is not trivial to specify in a principled manner. The third approach employs Pareto dominance and directly use a multi-objective algorithm to solve the original multi-objective problem. Since the Pareto approach returns a set of non-dominated solutions, rather than just a single solution as in a single-objective algorithm, it explores a considerably wider area of the search space and is more complex than its single-objective counterpart.

3   Approximate Distance Ranking Functions

To take both classification accuracy and spatial accuracy into account, the most straightforward way is to utilize the weighted scheme [5]:

$$M0 \equiv (1 - \alpha)CA + \alpha ADNP' \tag{3}$$

where α is a relative significance coefficient, $ADNP' = \exp(-ADNP)$ is normalized to $[0, 1]$.

However, while the normalization step is necessary, it is hard to select the appropriate normalization functions [7]. For instance, it is difficult to justify the use of $\exp(x)$ over $\frac{x-min}{max-min}$, which may be domain-dependent or even dataset-dependent. Even if $ADNP$ is appropriately normalized to $[0, 1]$, it is questionable if the normalized $ADNP$ can be linearly combined with CA, which is essentially a percentage. Besides, the computation of $ADNP$ is not trivial, which involves nearest neighbor search and distance evaluation. In the worse case, both time and space complexities are $O(n^2)$.

Further examination of $M0$ reveals that $ADNP$ plays a role of regularization. That is, in practice, when two estimated class labelings do not differ much in the values of classification accuracy, we need $ADNP$ to rank them. Indeed, even if we make wrong predictions of gold sites, we hope they are not far from the true ones. Moreover, it is also observed that if sites are evenly distributed and there are predicted 1 sites in the neighborhood of every misclassified 1 site, we can approximate the rank very well by leveraging the contiguity matrix W. Since what we really need is the relative ranking by $ADNP$ rather than the raw distance values, in the following, we develop a series of approximate distance ranking functions with W that do not require finding nearest neighbors or computing distance. The new criteria of W are functions of fraction of site size, which are more commensurate with the classification rate. Last but not least, it should be noted that since W is a sparse matrix for which we only need to store those non-zero elements, a lot of storage space is also saved.

For ease of expression, we first define a new vector $\mathbf{q} \equiv \mathbf{a}_1 \wedge \mathbf{p}_0$, where $\mathbf{a}_1$ and $\mathbf{p}_0$ are treated as boolean vectors and $\wedge$ is bit-AND operator. That is, $\mathbf{q}[i] = 1$ iff s_i is 1 but predicted 0. Below we present the three functions together with their respective assumptions.

3.1 $M1$

If sites are evenly distributed and there are predicted 1 sites in the neighborhood of every misclassified 1 site, the size of such sites is usually proportional to $ADNP$. In this case, we can count those misclassified 1 sites instead of computing the distance for them. This gives $M1$ defined in Equation (4).

$$M1 \equiv (1 - \alpha)CA + \alpha(1 - \frac{\mathbf{q}^T \mathbf{q}}{a_1}) \tag{4}$$

3.2   *M2*

During search for the nearest prediction s_j for the misclassified site s_i of 1, naturally we hope this search could terminate as early as possible from the starting point s_i. Specifically, it would be good if s_j is a neighbor of s_i in terms of $N(s_i)$. Moreover, it would be even better if many sites in $N(s_i)$ are predicted 1, i.e., i-th component in $W\mathbf{p_1}$ is close to 1. On the other hand, we prefer the number of such s_i be small. This results in $\mathbf{q}^T(\mathbf{1} - W\mathbf{p_1})$, which is preferred small. Finally, we note $(\mathbf{1} - W\mathbf{p_1}) = W\mathbf{p_0}$. Combining all of the above gives $M2$ in Equation (5).

$$M2 \equiv (1 - \alpha)CA + \alpha(1 - \frac{1}{a_1}\mathbf{q}^T W \mathbf{p_0}) \tag{5}$$

3.3   *M3*

$M3$ is defined in Equation (6).

$$M3 \equiv (1 - \alpha)CA + \alpha(1 - \frac{1}{a_1}\mathbf{q}^T W^* \mathbf{p_0}) \tag{6}$$

The binary matrix W^* is defined as

$$W^*(i,j) \equiv \begin{cases} 1 & j = \text{argmax}_j W(i,j), \\ 0 & \text{otherwise.} \end{cases} \tag{7}$$

where we assume closer neighbors have a larger value in W. For each s_i, we only allow the element in W^* corresponding to its nearest neighbor s_j to be 1 and 0 elsewhere. Consider s_i for which $\mathbf{q}[i] = 1$. In search for nearest predicted 1 neighbor s_j, we hope $W^*(i,j) = 1$, i.e., s_j is just the nearest neighbor of s_i. This is a much stronger requirement than that in $M2$.

4   Ideal Properties

To compare the various criterion functions for geospatial classification validation, we propose a series of ideal properties that a criterion function must satisfy. In Equation (3), the user-defined significance coefficient α is still confusing. In the ideal properties, we will give α a more meaningful interpretation.

In the proposed ideal properties, α is determined by the user and is typically small, e.g., 0.05. In a way, it plays the similar role as the tolerance threshold in the lexicographic approach to multiobjective optimization. It provides a meaningful threshold, based on which we decide which measure, CA or $ADNP$, should be emphasized. If the absolute difference in CA is greater than α, the hypothesis with larger CA is favored, because a much larger CA often means a larger $ADNP$. If the absolute difference in CA is less than α, we ignore the slight difference in CA and favor the hypothesis with greater $ADNP$, i.e., spatial accuracy matters. Let $\mathbf{y}, \hat{\mathbf{y}}_1, \hat{\mathbf{y}}_2$ denote the actual class labeling and two predicted

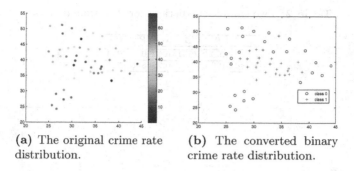

(a) The original crime rate distribution.

(b) The converted binary crime rate distribution.

Fig. 3. The crime dataset

class labelings, respectively. Let $CA_i \equiv CA(\mathbf{y}, \hat{\mathbf{y}}_i)$, $ADNP_i \equiv ADNP(\mathbf{y}, \hat{\mathbf{y}}_i)$, where $i = 1, 2$. In detail, an ideal criterion function M combining CA and $ADNP$ should satisfy the following properties.

1. If $|CA_1 - CA_2| > \alpha$, the estimation with larger CA is favored.
2. If $|CA_1 - CA_2| \leq \alpha$, the estimation with larger $ADNP$ is favored.

We can see $M0$ defined in Equation (3) does not meet property 2. Suppose $\alpha = 0.05$, $C_1 - C_2 = 0.04$, and $ADNP_1 < ADNP_2$. At this time, for $M_1 < M_2$, we need that $ADNP_2 - ADNP_1 \approx 1$. But it is nearly impossible, considering the small difference in CA.

5 Experimental Evaluation

In this section, we first introduce the experimental setting. Then we use two real-world datasets to compare the criterion functions.

5.1 Experimental Setting

Spatial Autoregression Model(SAR) is employed as the classification model in this paper [8]. Because the output from SAR, $\hat{\mathbf{y}}$, is estimated probability $P(Z_i = 1)$, we try various cutoff probability (threshold) θ, sampled evenly at a constant interval 0.01 in $[0, 1]$. With a total of 101 θs, we transform $\hat{\mathbf{y}}$ to class labeling $\hat{\mathbf{z}}$, i.e., $\hat{\mathbf{z}}[i] = 1$ if $\hat{\mathbf{y}}[i] > \theta$, $\hat{\mathbf{z}}[i] = 0$ otherwise. As θ increases, more sites with moderate $\hat{\mathbf{y}}[i]$ are classified 0 and only those with high $\hat{\mathbf{y}}[i]$ remain predicted 1. To emphasize the impact of spatial accuracy $ADNP$, we set $\alpha = 0.2$ for all criterion functions containing α as the coefficient.

To compare the validation criterion functions for geospatial classification, the most direct way is to visualize the results, e.g., plotting their curves together for the common set of class labelings. We can also compare them based on the pairwise ranking. In detail, given the true class labeling and a set of estimated class labelings, all possible pairs of estimates $(\hat{\mathbf{z}}_i, \hat{\mathbf{z}}_j)$ are taken and it is determined,

Table 2. Two sample predictions for the crime dataset

	θ	CA	$ADNP$	$M0$
Pred1	0.3	0.7551	0.2210	0.7644
Pred2	0.5	0.8571	0.2848	0.8361

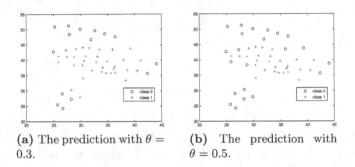

(a) The prediction with $\theta = 0.3$.

(b) The prediction with $\theta = 0.5$.

Fig. 4. Two predictions for the crime dataset

whether they are treated in the same manner by two criterion functions M_a and M_b. Then we compute Rand index and Pearson coefficient [9], for which a larger value indicates a larger degree of similarity.

Among the two datasets for evaluation, one is the election dataset introduced earlier in Section 1. The other dataset, as shown in Figure 3a, records crime statistics in 49 neighborhoods in Columbus Ohio, USA. In the original regression problem of the crime dataset, household income and housing values are treated as explanatory attributes to predict crime rates [2]. For binary classification, the original continuous target attribute y is transformed to binary z as $z = 1$ if $y > \text{avg}(y)$, $z = 0$ otherwise. The binary crime rate is illustrated in Figure 3b, which exhibits significant spatial characteristics.

5.2   Empirical Results

Sample Predictions. To give readers some flavor of predictions from the SAR model, we illustrate two predictions with cutoff threshold $\theta = 0.3$ and 0.5 in Figures 4a and 4b, respectively. Their accuracies are reported in Table 2. Although Prediction Pred1 leads to a lower $ADNP$ than Pred2, its CA is significantly worse than Pred2. This shows again the risk of considering $ADNP$ alone. Fortunately, the low accuracy of Pred1 is successfully detected by $M0$, which takes into consideration both classification accuracy and spatial accuracy.

Comparison with $M0$. Comparison of $M0$ and its approximate functions is illustrated in Figure 5. In the crime dataset, one can see that all of the approximate functions, except $M3$, follow $M0$ (and therefore $M0$'s ranking) quite well. It shows that we can use approximate distance measures to save computation

Table 3. Comparison with $M0$

crime	$M1$	$M2$	$M3$	election	$M1$	$M2$
Rand	0.9682	0.9241	0.8673	Rand	0.9958	0.9925
Pearson	0.9145	0.8562	0.7443	Pearson	0.9913	0.9844

Table 4. Comparison with ideal properties

crime	$M0$	$M1$	$M2$	$M3$	election	$M0$	$M1$	$M2$
Rand	0.8559	0.8037	0.7626	0.6884	Rand	0.6824	0.6865	0.6749
Pearson	0.6605	0.6076	0.5652	0.5116	Pearson	0.3606	0.3637	0.3550

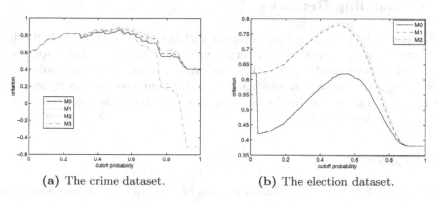

(a) The crime dataset.   (b) The election dataset.

Fig. 5. The validation criterion functions

entailed by $ADNP$. The reason why $M3$ deviates from $M0$ as cutoff probability θ gets large is that it is more unlikely that for any misclassified 1 site, its nearest neighbor remains in the set of predicted 1 sites most of time, as such a set gets smaller and smaller.

As for the election dataset, it is noticed that $M0$ decreases sharply around $\theta = 0.04$, the same place as $ADNP$ did, as a consequence of direct combination of transformed $ADNP$. On the other hand, except around $\theta = 0.04$, both $M1$ and $M2$ approximate the trend of $M0$ in most places. The reason is that around $\theta = 0.04$, many 1 sites far away from other 1 sites (contribute much to $ADNP$) are misclassified when $\theta > 0.04$, but they are correctly classified when $\theta < 0.04$. This results in a big increase in $ADNP$ and consequently in $M0$. However, their number is small compared to the total number of 1 sites, which means they are unable to have an equally great impact on $M1$ and $M2$ which only considers size of sites. The reason why we did not include $M3$ is that the non-zero elements in each row of contiguity matrix is the same, resulting from the large size of the dataset. So there is no maximum element in each row, as computed in Equation 7.

As we discussed previously, what we care most is the capability of these approximate functions to reflect the pairwise ranking of $M0$, rather than the raw distance values of $ADNP$. Table 3 reports their performance in terms of Rand index and Pearson coefficient. Apparently, all measures indicate a high resemblance.

Comparison with Ideal Properties. Comparison with ideal properties is given in Table 4. Although $M0$ employs the weighted scheme to explicitly incorporate $ADNP$ and thus appears to satisfy the ideal properties (also specified in $ADNP$) best, it is comparable to $M1$ in terms of both Rand index and Pearson coefficient. Again, this confirms the usefulness and advantage of our proposed criterion functions to approximate ranking in $ADNP$.

6 Concluding Remarks

In this paper, by leveraging the contiguity matrix that encodes the neighborhood relationship, we proposed new criterion validation functions that approximate the pairwise ranking of $ADNP$. They not only circumvent the difficulty in the direct weighted scheme, but also lead to smaller time and space complexity. For the future work, we plan to devise specialized optimization algorithms for the proposed validation functions.

References

1. Chun, Y., Griffith, D.: Spatial Statistics and Geostatistics: Theory and Applications for Geographic Information Science and Technology. SAGE (2013)
2. Pace, R.K., Barry, R.: Quick computation of spatial autoregressive estimators. Geographical Analysis **29**, 232–247 (1997)
3. Hu, T., Xiong, H., Gong, X., Sung, S.Y.: ANEMI: An adaptive neighborhood expectation-maximization algorithm with spatial augmented initialization. In: Proceedings of the Pacific-Asia Conference on Knowledge Discovery and Data Mining, pp. 160–171 (2008)
4. Liu, X., Chen, F., Lu, C.: Robust prediction and outlier detection for spatial datasets. In: Proceedings of the IEEE 12th International Conference on Data Mining, pp. 469–478 (2012)
5. Shekhar, S., Schrater, P., Raju, W.R., Wu, W., Chawla, S.: Spatial contextual classification and prediction models for mining geospatial data. IEEE Transactions on Multimedia **4**(2), 174–188 (2002)
6. Coello, C.A.: An updated survey of ga-based multiobjective optimization techniques. ACM Computing Surveys **32**(2), 109–143 (2000)
7. Hu, T., Sung, S.Y., Sun, J., Ai, X., Ng, P.A.: A linear transform scheme for building weighted scoring rules. Intelligent Data Analysis **16**(3), 383–407 (2012)
8. Anselin, L.: Spatial regression. In: The Sage handbook of spatial analysis. Sage Thousand Oaks, CA, pp. 255–275 (2009)
9. Hubert, L.J., Arabie, P.: Comparing partitions. Journal of Classification **2**, 63–76 (1985)

Efficient Classification of Binary Data Stream with Concept Drifting Using Conjunction Rule Based Boolean Classifier

Yiou Xiao[⊠], Kishan G. Mehrotra, and Chilukuri K. Mohan

EECS Department, Syracuse University, Syracuse, NY, USA
{yixiao,mehrotra,mohan}@syr.edu

Abstract. We propose a conjunction rule based classification technique that has good classification performance, is simple, automatically identifies important attributes, and is extremely fast. Due to these properties the classifier is most suitable for "big"/streaming data. Empirical study, using multiple datasets, shows that time complexity, compared with other classifiers, is faster by several factors, especially for large number of attributes without sacrificing performance.

Keywords: Boolean classifier · Conjunction rule · Data stream · Concept drifting

1 Introduction

Supervised learning is one of the most significant data mining problems. Classifiers based on different models have been proposed to tackle the problem. Logic based algorithms and decision trees [17], and their variations summarized in [13], are popular because of their simplicity and easy interpretation. Other classification models include Artificial Neural Network (ANN) [20] based on non-linear perceptron; Bayesian network [8] that focuses on building a probability model; k-Nearest Neighbor [5] that uses proximity of data points to determine their classes; and SVM [18] which employs kernel-based approach. Other models, such as Particle Swarm Optimization, are used in domain specific classification problems, e. g., image classification [15].

In recent years, problems related to "Big Data", on-line system, and data streams, have been studied. The related classification problems are difficult due to enormous size of data and limited resources. One important question in data stream mining problems is: how to incrementally update the classifier? The problem has been addressed by several researchers including Domingos et al. [6] who have adopted Hoeffding Tree model to induce decision tree (VFDT) for extremely large data or data stream. In [12], the authors use grid of different resolutions and k-Nearest Neighbor algorithm to efficiently update the classifier; [19] uses the weighted ensemble classifier to handle the concept drifting in data streams, [14] uses an efficient feature hashing method to classify textual data streams and

© Springer International Publishing Switzerland 2015
M. Ali et al. (Eds.): IEA/AIE 2015, LNAI 9101, pp. 457–467, 2015.
DOI: 10.1007/978-3-319-19066-2_44

Incremental Linear Discriminant Analysis (ILDA) is proposed as an extension for Linear Discriminant Analysis in [16].

A Boolean function is defined as $f : B^k \rightarrow B$ where $B = \{1, 0\}$. Boolean functions are used to model state transitions in deterministic systems or relationships between Boolean vectors. In this paper, we propose application of Boolean functions in supervised 2-class classification problems, restricting our study to Boolean functions of conjunction form:

$$B_1 \wedge B_2 \wedge \ldots \neg B_m \implies (C = 1) \tag{1}$$

where C is the predicted label and B_i's are boolean attributes. We focus on this specific formalism of conjunction rule so that the learning process only focuses on the "positively" labeled data points.

This paper is organized as follows: In sections 2, we introduce the (Boolean Classifier based on Conjunction rule); in Section 3 we compare its performance with several well known classifiers such as CART [3] for static data. In Section 4 we apply the algorithm to 'big' streaming data and data where generating process may change over time. Section 5 contains conclusion and possible extensions.

2   Boolean Classifier Based on Conjunction Rule

This classifier uses binary training data $D = (B_{N \times M} : C_{N \times 1})$ where N is the number of training data objects, M is the number of attributes and where all B_{ij}'s and $C_i \in \{0, 1\}, 1 \leq i \leq N, 1 \leq j \leq M$.

One important component of classification involves attribute selection, i. e., determination of which attributes should be included in the final evaluation of the classifier. In BCC it is performed using two properties of an attribute, the *attribute importance score* and its *significance score*. Attribute importance score determines usefulness of the attribute for the classification task, its sign determines whether it appears with a negation, and significance score is used to penalize a low variation attribute.

We calculate the contingency table by counting the number of 1s and 0s in the training dataset in one pass of the dataset. Table 1 illustrates the contingency table for attribute j. The learning phase uses these M contingency tables only. In Equation 1 we focus on $(C = 1)$, therefore we are only interested in the second row of the contingency tables.

Table 1. Contingency table for attribute j and C in training data. Here $n_{k\ell}^{(j)} =$ the number of data points that satisfy $(B_{ij} = k, C_i = \ell)$; N_0, N_1 are number of data points of 0-class and 1-class respectively.

		Attribute j		Total
		0	1	
Class	0	$n_{00}^{(j)}$	$n_{01}^{(j)}$	N_0
	1	$n_{10}^{(j)}$	$n_{11}^{(j)}$	N_1

For attribute j we calculate the following two ratios:[1]

$$p_j^+ = \frac{n_{11}^{(j)}}{N_1} \quad \text{and} \quad p_j^- = \frac{n_{10}^{(j)}}{N_1}$$

and the difference, $s_j = p_j^+ - p_j^-$, is used as a measure of importance of attribute j. If $p_j^+ \approx p_j^-$ then attribute j is considered "weak" because it is equally favorable to $C = 1$ and $C = 0$ and should be omitted in Equation 1. A user specified threshold λ is used to prune weak attributes

$$\begin{cases} \text{If } |s_j| \geq \lambda \text{ then attribute } j \text{ is important,} \\ \text{If } -\lambda < s_j < \lambda \text{ then attribute } j \text{ is weak.} \end{cases}$$

The sign of s_j determines the role of the attribute; if $s_j < 0$, then the attribute appears as a negative literal, positive otherwise.

We employ the background distributions of an attribute to evaluate its significance score. In an extreme case if an attribute is 0 (or 1) for all data points, then it should be ignored even though its importance score is high. This notation can be relaxed to include cases when attribute's values are *mostly* 0 (or *mostly* 1). Towards this, the background ratios:

$$r_j^+ = \frac{n_{01}^{(j)} + n_{11}^{(j)}}{N} \quad \text{and} \quad r_j^- = \frac{n_{00}^{(j)} + n_{10}^{(j)}}{N}. \tag{3}$$

are calculated and two binomial random variables are defined as:

$$X_j^+ \sim \text{Binomial}(N_1, r_j^+) \quad \text{and} \quad X_j^- \sim \text{Binomial}(N_0, r_j^-) \tag{4}$$

which provide two significance scores[2] $\mathcal{P}_j^+ = P(X_j^+ \geq n_{11})$ and $\mathcal{P}_j^- = P(X_j^- \geq n_{10})$ If $\mathcal{P}_j^+$ is small, then the attribute is significant, and if $\mathcal{P}_j^-$ is small, then the attribute is significant as a negative literal.

2.1 The Boolean Classifier

The set of attributes that should be included and whether an attribute appears as a positive or negative literal is determined as described in Algorithm 1. In other words $\{I^+, I^-\}$ completely describes the BCC. Moreover, the set of all M

---

[1] Note that these two ratios are actually special cases of Jaccard indexes [10] between subsets of the training data:

$$p_j^+ = \mathcal{J}(B^+, C^+) = \frac{|B^+ \cap C^+|}{|B^+ \cup C^+|} \quad \text{and} \quad p_j^- = \mathcal{J}(B^-, C^+) = \frac{|B^- \cap C^+|}{|B^- \cup C^+|} \tag{2}$$

where $B^+ = \{i \mid B_{ij} = 1 \wedge C_i = 1\}$, $B^- = \{i \mid B_{ij} = 0 \wedge C_i = 1\}$, $C^+ = \{i \mid C_i = 1\}$ is the set of class-1 points, and as a result, $B^+ \subseteq C^+$ and $B^- \subseteq C^+$.

[2] In the implementation, these probabilities are approximated by using Gaussian distribution; this approximation is reasonable because generally N_0 and N_1 are large.

Algorithm 1. Using contingency table to obtain conjunction rule

---

procedure GETRULE((CT, λ, τ))
    Initially, let $I^+ = \emptyset$ and $I^- = \emptyset$
    for $1 \le j \le M$ **do**
        calculate $s_j, \mathcal{P}_j^+$ and $\mathcal{P}_j^-$ based on CT
        if $s_j < -\lambda$ and $\mathcal{P}_j^- \le \tau$ **then**
            $I^- = I^- \cup j$,
        else if $s_j > \lambda$ and $\mathcal{P}_j^+ \le \tau$ **then**
            $I^+ = I^+ \cup j$
        end if
    end for
    return $[I^+, I^-]$
end procedure

---

contingency tables, CT, acts as the *sufficient statistic* for Boolean function rule inference.

Thus, the corresponding rule is:

$$f(b_1, b_2, \ldots, b_M) \triangleq (\wedge_{j \in I+} b_j) \wedge (\wedge_{k \in I-} \neg b_k) \tag{5}$$

2.2 Data Preprocessing

The Boolean classifier can be applied only to datasets with binary attributes. However in real-world applications attributes are categorical, real or integer values. With appropriate transformations, these datasets can be modified so that BCC can be implemented. These data preprocessing steps are described below.

Categorical attributes. We represent a categorical attribute using k binary attributes $B_1, B_2, \ldots, B_k$, where k is the number of different categories (possible values) for the attribute.

Real-valued attributes. For real-valued attributes, we adopt a simple thresholding method:

$$B_j = \begin{cases} 1 & \text{if } x_j > \mu_j \\ 0 & \text{otherwise} \end{cases}$$

where μ_j is the mean of the jth random variable (median can also be used).

2.3 Why Boolean Classifier Should Have Good Performance?

The proposed algorithm and the naive Bayes share some similarities. In the naive Bayes all attributes are assumed to be conditionally independent, given the class. In spite of the fact that this assumption is often invalid it's performance is very good [22, Ch 9]. In the proposed algorithm, the joint probabilities of Boolean attributes and the class variables are independent. Therefore, as in naive Bayes, BCC has good performance.

In BCC we also implement the concept of *quality of selected jth attribute* using the binomial distributions described above. Thus, irrelevant attributes are excluded from the final conjunction rule. In contrast naive Bayes estimates the probability of class conditioned on every attribute.

3   Performance of BCC on Static Datasets

To test the performance of BCC we used the following four datasets, obtained from UCI-repository. The details of adopted datasets are summarized in Table 2

Table 2. Datasets Description: 1.Wisconsin Breast Cancer dataset [21] (WBC). 2.Wisconsin Diagnostic Breast Cancer [21] (WDBC). 3. Adult Income Datset [11] (AID). 4. MADELON artificial dataset [9] (MAD)

Dataset	M	Attribute Type	$|D_{\text{training}}|$	$|D_{\text{test}}|$	N_1/N
WBC	9	Integer,$[1,10]$	420	279	35%
WDBC	30	Real	340	229	37%
AID	14	6 Real,8 Categorical	32561	16281	24%
MAD	500	Integer,$[1,1000]$	2000	600	50%

Table 3. The comparison of Boolean classifier and C4.5 with respect to F1 score and accuracy. In all experiments, the importance threshold λ is set to 0.7 and $\tau = 0.05$.

Methods	F1 score				Accuracy			
	WDB	WDBC	AID	MAD	WDB	WDBC	AID	MAD
CART	0.94	0.91	0.53	0.76	0.91	0.88	0.79	0.77
C4.5	0.96	0.92	0.28	0.76	0.93	0.94	0.77	0.77
Random Forest	0.95	0.92	0	0.50	0.94	0.87	0.76	0.67
BCC	0.95	0.90	0.58	0.67	0.94	0.86	0.83	0.69

As seen in Table 3, BCC performs competitively with other algorithms, but the main advantage is that BCC requires one pass of the training set to calculate the contingency table, therefore its time complexity is $O(NM)$. On the other hand, time complexity of C4.5 is $O(MN \log N)$. Clearly, BCC algorithm is very efficient especially when applied to dataset with large number of features. For example, for the MAD dataset, the training times are 0.0384 seconds for BCC and 3 minutes and 2 seconds for CART.

4   Incremental Boolean Classifier Based on Conjunction Rule (IBCC)

Encouraged by competitive efficiency of BCC, we focus on the incremental BCC classification algorithm for streaming data.

A data stream is defined as an ordered sequence of data points that can be read in a short period of time. Examples of data streams include communication logs, transaction data flow, and censoring video stream [20]. Since the preprocessing steps are easy to implement, we assume that the streaming data consists of data points with binary attributes only. We assume that the streaming data comes in chunks of $\Delta \geq 1$ observations per round. In all experiments described in this section, Δ is set to be 1; which also represents the worst case scenario due to the reason that the classifier has to be updated for each new data point.

Incremental Update of Contingency Tables. BCC depends only on M contingency tables, CT. In order to use BCC for streaming data, the contingency table of attribute j can be easily updated to:

$$CT_{k\ell}^{(j)} \leftarrow CT_{k\ell}^{(j)} + n_{k\ell}^{(j)}, \text{ for } k, \ell = 0, 1 \text{ and } j = 1, 2, \ldots, M \tag{6}$$

where $CT_{k\ell}^{(j)}$'s are the elements of contingency table for the old data and $n_{k\ell}^{(j)}$ are the elements of the contingency table for the new streamed data.

4.1   Performance Comparison Between IBCC and ILDA

We compare IBCC with ILDA [16]. Two real-life datasets and 9 synthetic datasets are used in this comparison.

I - Iris Dataset: We use this well known data, created by Fisher [2,7], for illustrative purposes only. Iris dataset contains four real-valued attributes and three classes: Setosa,Versicolor and Virginica. Since IBCC is a two-class classifier, we construct 2 two-class datasets as described below.

1. **The VV dataset** contains all 100 data points belonging to *Versicolor* and *Virginica. It is well known that these two classes are not linearly separable.* Class *Versicolor* is labeled 1 and class *Setosa* is labeled as 0.
2. **The SV dataset** contains all 100 data points belonging to *Versicolor* and *Setosa. These two classes are linearly separable.* Class *Virginica* is labeled 1 and class *Versicolor* as 0.

Finally, since IBCC can only take binary values, all four attributes were transformed to binary using the mean value of each attribute whereas ILDA uses the original real-valued attributes.

The experiment is performed as follows. We randomly chose 75 data points to be used as streaming data stream and 25 data points for testing. First four observations of the training set are used to build the classifiers, IBCC and IDLA.

The performance of these two classifiers is tested using the 25 data points that were saved for testing. Next, we stream fifth data point of the training set, build the classifier and test their performance on the test dataset, and continue in this manner with one more streaming data point at a time. That is, after a new data point in the training data stream is analyzed by ILDA and IBCC, the updated classifiers are immediately evaluated using 25 testing data points. In Figure 1, we illustrate the performance of two methods for one randomly partitioned dataset of VV dataset. In the top of Figure 1 the data is displayed when 33, 66, and 100% of the training data point were acceptable to the two algorithms, and in the bottom part we plot the accuracy at each stage of the process. We note that the accuracy of IBCC is superior than of ILDA.

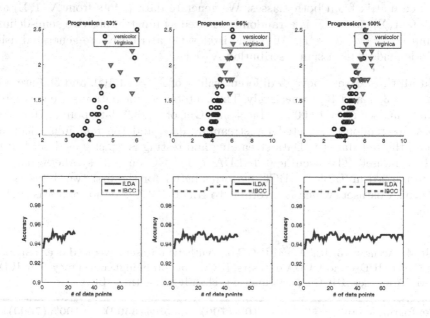

Fig. 1. Illustration of data stream and performance comparison: Progression figures (top) illustrate the positions of accumulated data points on attribute 3 and 4; whereas performance figures (bottom) show the accuracy of ILDA (solid line) IBCC (broken line) at each iteration. This only shows one of the 10 reputations in our experiments.

Because the order of streaming data points may affect performance of the classifiers, we repeat the above experiment 10 times and evaluate the average accuracy. For the SV dataset, IBCC performs as good as ILDA after 10% of data points arrived; and IBCC performs 5% better than ILDA for the VV dataset.

II - Synthetic Datasets We generated 3 types of synthetic datasets with different number of attributes:

1. **Independent Uniformly Distributed Attributes (IUDA)**: Each generated data point x contains M i.i.d. $x_j \sim U(0,1)$. A random subset $I \subset \{1,2,3,\ldots,M\}$ of attributes is selected as the "relevant attribute" set. Data points in one randomly determined I-dimensional corner (e.g. $x_{I_1} \leq \theta, x_{I_2} > \theta, \ldots, x_{I_k} \leq \theta$) of the subspace are labeled as 1, other data points are labeled as 0.

2. **Dependent Uniformly Distributed Attributes (DUDA)**: This synthetic data is created in a way similar to IUDA. However, some linear dependency among "relevant attributes" is introduced[3]. If necessary, the dependent variable is normalized to $[0,1]$.

3. **Dependent Multivariate Gaussian Attributes (DMGA)**: LDA assumes that data points are Gaussian random variables with the same covariance matrix from both classes. We generate data points from $\mathcal{N}(1,\Sigma)$ and $\mathcal{N}(-1,\Sigma)$, where Σ is a randomly created symmetric positive semi-definite matrix of size $|I| \times |I|$. All other (irrelevant) attributes are generated using independent Gaussian distribution $\mathcal{N}(0,1)$.

In all three cases above, 3 different values of $M = 10, 100$, and 500 are used with $|I| = 3, 5$, and 10, respectively. Thus, a total of nine datasets are generated. In each dataset $N = 10,000$ and $\theta = 0.8$. Out of 10,000 data points, 7000 data points are randomly selected for streaming data and the remaining 3000 are used as the test dataset. Data streaming and testing is exactly as described for the Iris dataset. The accuracy of ILDA and IBCC on nine synthetic datasets is summarized in Table 4. IBCC performs better for the first two datasets and for DMGA dataset, which is favorable to ILDA, IBCC's performance is slightly worse.

Table 4. Accuracy of ILDA and IBCC on synthetic datasets averaged over ten trials. Note that in IUDA and DUDA datasets, IBCC can obtain higher accuracy than ILDA. In DMGA dataset, IBCC's performance is slightly worse than ILDA.

Performance at Synthetic Datasets	10%(700)		50%(3500)		100%(7000)	
	IBCC	ILDA	IBCC	ILDA	IBCC	ILDA
IUDA ($M = 10, |I'| = 3$)	1.00	0.892	1.00	0.891	1.00	0.891
IUDA ($M = 100, |I'| = 5$)	1.00	0.904	1.00	0.931	1.00	0.931
IUDA ($M = 500, |I'| = 10$)	1.00	0.916	1.00	0.949	1.00	0.955
DUDA ($M = 10, |I'| = 3$)	1.00	0.901	1.00	0.908	1.00	0.908
DUDA ($M = 100, |I'| = 5$)	1.00	0.897	1.00	0.854	1.00	0.854
DUDA ($M = 500, |I'| = 10$)	1.00	0.92	1.00	0.981	1.00	0.981
DMGA ($M = 10, |I'| = 3$)	0.94	1.00	0.915	1.00	0.932	1.00
DMGA ($M = 100, |I'| = 5$)	0.897	1.00	0.943	1.00	0.97	1.00
DMGA ($M = 500, |I'| = 10$)	0.782	1.00	0.845	1.00	0.949	1.00

---

[3] For example, $x_{I_2} = a x_{I_1} + b + \epsilon$ is linearly dependent on the first attribute x_{I_1}. As before, $x_{I_1} \sim U(0,1)$. Here $\epsilon \sim \mathcal{N}(0,1)$.

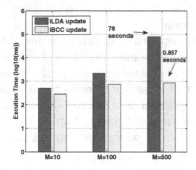 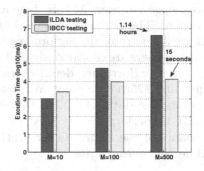

(a) Execution time of incremental **(b)** Execution time of classification
update

Fig. 2. Comparison of computational efficiencies of ILDA (dark bars) and IBCC (light bars). The running time is scaled in $\log_{10}$ of milliseconds. When $M = 500$, total time is 78 seconds for updating ILDA and 1 second for updating IBCC (left chart); and the total execution time is 4112 seconds for testing ILDA and 15 seconds for testing IBCC (right chart);

The execution time for ILDA and IBCC are presented in Figure 2a and the testing costs in Figure 2b[4]. The time complexity of IBCC is far superior to that of ILDA.

4.2   Concept Drifting in Data Stream

Concept drifting is loosely defined as change in data generating mechanism [1, 4, 6, 20]. For example, in network intrusion detection problem, methods of attack might change but they are still considered as class 1 (threat); or in natural disaster monitoring systems, flood, typhoon or hurricane may have different characteristics in attribute space yet they are classified as a "hazard".

To simulate concept drifting we assume that the data generation mechanism changes at random times; we assume that the inter generation time follows uniform distribution ($U[T_{\min}, T_{\max}]$) where $T_{\min}$ and $T_{\max}$ are preassigned integers. Starting from the first model, which we denote by $\{I_1^+, I_1^-\}$, we generate streaming data T_1 times and then the model switches to $\{I_2^+, I_2^-\}$ for T_2 streaming data points before it switches to the third model and the process continues in this manner. The process of attribute generation is essentially the same for IUDA as described in the previous section.

In the concept-drifting context, we keep records of confusion matrix and accuracy measure = (True positive + True Negative)/ Number of Observations. If accuracy continues to go down for L consecutive streaming data observations,

---

[4] For this experiment, we used Matlab code running on a Ubuntu 14.04 machine with Intel i-7 CPU and 8 Gb RAM.

then we conclude that the concept (data generation mechanism) has changed; L is a preassigned integer. At this stage a new classifier is learnt, using the past L observations. The process continues in this manner.

Figure 3 illustrates the performance of the algorithm. In this experiment four distinct classifiers are continuously updated using the incoming data chunks until accumulated performance deteriorates. In this synthetic dataset, the concept changed after 125th data point. We observe that classifier 1 begins to misclassify data points (see dotted line). At this stage classifier 2 is learned and it can be seen that its performance is very good (see the solid line) . Figure 3 depicts performance of the algorithm for synthetic data in which the concept changes four different times.

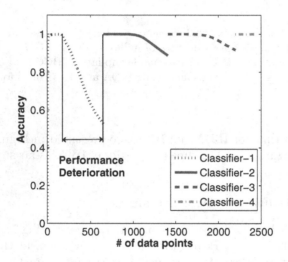

Fig. 3. Illustration of classifier performance and reevaluation for data in which the data generation process four times

5   Conclusion

In this paper, we proposed BCC which in essence is an attribute-wise voting method. We used contingency tables as the sufficient statistics for classifier learning; compressing the data required for further evaluation from $O(NM)$ to $O(M)$ Since the update process of contingency tables is nothing but counting 1s and 0s in each attribute, BCC is easy to extend for streaming data. Time complexity of IBCC is very low. Given that BCC only focus on "class-1" objects, the decision of which class is "1" is crucial to competitive accuracy performance. Limitation of this approach is that conjunction rule model works only for two-class classification problems.

References

1. Babcock, B., Babu, S., Datar, M., Motwani, R., Widom, J.: Models and issues in data stream systems. In: Proceedings of the Twenty-First ACM SIGMOD-SIGACT-SIGART Symposium on Principles of Database Systems, pp. 1–16. ACM (2002)
2. Bache, K., Lichman, M.: UCI machine learning repository (2013)
3. Breiman, L.: Bagging predictors. Machine learning **24**(2), 123–140 (1996)
4. Chen, Y., Tu, L.: Density-based clustering for real-time stream data. In: Proceedings of the 13th ACM SIGKDD International Conference on Knowledge Discovery and Data Mining, pp. 133–142. ACM (2007)
5. Cover, T., Hart, P.: Nearest neighbor pattern classification. IEEE Transactions on Information Theory **13**(1), 21–27 (1967)
6. Domingos, P., Hulten, G.: Mining high-speed data streams. In: Proceedings of the Sixth ACM SIGKDD International Conference on Knowledge Discovery and Data Mining, pp. 71–80. ACM (2000)
7. Fisher, R.A.: The use of multiple measurements in taxonomic problems. Annals of eugenics **7**(2), 179–188 (1936)
8. Friedman, N., Geiger, D., Goldszmidt, M.: Bayesian network classifiers. Machine learning **29**(2–3), 131–163 (1997)
9. Guyon, I.: Design of experiments of the nips 2003 variable selection benchmark. In: NIPS 2003 Workshop on Feature Extraction and Feature Selection (2003)
10. Jaccard, P.: The distribution of the flora in the alpine zone.1. New Phytologist **11**(2), 37–50 (1912)
11. Kohavi, R.: Scaling up the accuracy of naive-bayes classifiers: A decision-tree hybrid. In: KDD, pp. 202–207 (1996)
12. Law, Y.-N., Zaniolo, C.: An adaptive nearest neighbor classification algorithm for data streams. In: Jorge, A.M., Torgo, L., Brazdil, P.B., Camacho, R., Gama, J. (eds.) PKDD 2005. LNCS (LNAI), vol. 3721, pp. 108–120. Springer, Heidelberg (2005)
13. Murthy, S.K.: Automatic construction of decision trees from data: A multi-disciplinary survey. Data mining and knowledge discovery **2**(4), 345–389 (1998)
14. anculef, R., Flaounas, I., Cristianini, N.: Efficient classification of multi-labeled text streams by clashing. Expert Systems with Applications **41**(11), 5431–5450 (2014)
15. Omran, M., Salman, A., Engelbrecht, A.P.: Image classification using particle swarm optimization. In: Proceedings of the 4th Asia-Pacific Conference on Simulated Evolution and Learning, vol. 1, pp. 18–22, Singapore (2002)
16. Pang, S., Ozawa, S., Kasabov, N.: Incremental linear discriminant analysis for classification of data streams. Trans. Sys. Man Cyber. Part B **35**(5), 905–914 (2005)
17. Quinlan, J.R.: Induction of decision trees. Mach. Learn. **1**(1), 81–106 (1986)
18. Suykens, J.A., Vandewalle, J.: Least squares support vector machine classifiers. Neural processing letters **9**(3), 293–300 (1999)
19. Wang, H., Fan, W., Yu, P.S., Han, J.: Mining concept-drifting data streams using ensemble classifiers. In: Proceedings of the Ninth ACM SIGKDD International Conference on Knowledge Discovery and Data Mining, pp. 226–235. ACM (2003)
20. Wang, S.-C.: Artificial neural network. In: Interdisciplinary Computing in Java Programming, pp. 81–100. Springer, Heidelberg (2003)
21. Wolberg, W.H., Mangasarian, O.L.: Multisurface method of pattern separation for medical diagnosis applied to breast cytology. Proceedings of the national academy of sciences **87**(23), 9193–9196 (1990)
22. Wu, X., Kumar, V.: The top ten algorithms in data mining. CRC Press (2010)

Semi-supervised SVM-based Feature Selection for Cancer Classification using Microarray Gene Expression Data

Jun Chin Ang, Habibollah Haron(✉), and Haza Nuzly Abdull Hamed

Department of Computer Science, Faculty of Computing,
Universiti Teknologi Malaysia, Skudai, Johor, Malaysia
jun_chin2001@yahoo.com.sg, jcang2@live.utm.my,
{habib,haza}@utm.my

Abstract. Gene expression data always suffer from the high dimensionality issue, therefore feature selection becomes a fundamental tool in the analysis of cancer classification. Basically, the data can be collected easily without providing the label information, which is quite useful in improving the accuracy of the classification. Label information usually difficult to obtain as the labelling processes are tedious, costly and error prone. Previous studies of gene selection are mostly dedicated to supervised and unsupervised approaches. Support vector machine (SVM) is a common supervised technique to address gene selection and cancer classification problems. Hence, this paper aims to propose a semi-supervised SVM-based feature selection (S^3VM-FS), which simultaneously exploit the knowledge from unlabelled and labelled data. Experimental results on the gene expression data of lung cancer show that S^3VM-FS achieves the higher accuracy yet requires shorter processing time compares with the well-known supervised method, SVM-based recursive feature elimination (SVM-RFE) and the improved method, S^3VM-RFE.

Keywords: Support vector machines · Semi-supervised · Feature selection · Cancer · Gene expression

1 Introduction

During the past decade, the exponential growth of technologies simultaneously led to the high-throughput growth of harvested data with respect to the ultra-high dimensionality and sample size. These voluminous data occupy large storage space and cause difficulties in handling. Therefore, data mining and machine learning techniques were put in place to consolidate the pattern recognition and knowledge discovery process. These techniques are extensively used in the field of bioinformatics [1], especially for gene expression analysis. However, microarray gene expression data usually contain more than ten thousands of features and relatively limited number of samples. This curse of dimensionality is a major obstacle in machine learning and data mining, hence feature selection and dimensionality reduction always an active research topic in bioinformatics.

© Springer International Publishing Switzerland 2015
M. Ali et al. (Eds.): IEA/AIE 2015, LNAI 9101, pp. 468–477, 2015.
DOI: 10.1007/978-3-319-19066-2_45

Feature selection falls into three categories that is unsupervised [2–4], supervised [5–7] and semi-supervised feature selection [8–11], where it respectively uses fully unlabeled, fully labeled and partially labeled data in the learning algorithm. However, in real applications especially for microarray gene expression analysis, labeled data are given with external knowledge, ultimately expensive and scarce [11,12]. This fact aggravates the risk of over-fitting in the supervised learning process. Moreover, gene expression data that include high dimensionality features versus limited number of samples have deteriorated the performance of supervised methods over the small training set. In order to tackle this problem, the semi-supervised feature selections are broadly evolved [8–11], which exploits the knowledge of marginal distributions from unlabeled samples and simultaneously utilizes the label information provided by labeled samples. Consequently, it offers an opportunity to produce a more reasonable estimation of the incoming samples especially when the training labels are rare.

This paper attempts to propose a semi-supervised feature selection model based on the Support Vector Machine (SVM). The supervised SVM classifier is well known in bioinformatics due to its high accuracy, and effectiveness in handling high dimensional data such as gene expression [13]. SVM has been found suitable to feature selection, other than their superior applicability to classification [14]. However, majority of the studies are only involved in the supervised learning paradigm. Semi-supervision is only included in two of the studies, which are [15] and [10]. The authors of [15] proposed a transductive-based SVM (TSVM) feature selection method while the authors of [10] suggested a partial supervised feature selection with regularized linear models. The study as shown in [15] is similar to this paper. However, this paper is more focusing on the inductive instead of transductive manner, whereby the test set is independently from the learning process.

The next section of this paper discusses the previous works of semi-supervised feature selection and SVM as feature selection techniques. Section 3 presents the SVM-based recursive feature elimination (SVM-RFE) and the proposed semi-supervised SVM-based feature selection method. The experimental settings and results are presented in Section 4 and 5 respectively while conclusion is demonstrated in the last section.

2   Methods

This paper proposed two semi-supervised feature selection methods, one is the semi-supervised SVM-based RFE (S^3VM-RFE), the improvement of supervised SVM-RFE, and another proposed method is the modification model of S^3VM-RFE, which is known as semi-supervised SVM-based feature selection (S^3VM-FS).

2.1   SVM-Based Recursive Feature Elimination (SVM-RFE)

SVM-RFE is generating the features ranking using backward feature elimination, whereby the features are arranged in decreasing order of their weights in hyperplane, and the lowest ranked features are removed. The remaining features will recursively train with SVM until all the features are ranked on the list. It is a simple and competent method to handle binary classification problems.

The feature elimination is based on the SVM ranking criteria, a squared coefficient w_i^2 of the weight vector w. The i-th feature with highest ranking score $c_i = (w_i)^2$ are the most informative, contrary to the feature with smallest ranking score will be discarded. According to Optimal Brain Damage (OBD) algorithm [16], $c_i = (w_i)^2$ is chosen as the principle of ranking. The features eliminated by this criterion have the smallest change in the objective function, $J = \frac{1}{2}\|w\|^2$. The approximates change in criterion J can be expanded in second order Taylor series:

$$\Delta J(i) = \frac{\delta J}{\delta w_i}\Delta w_i + \frac{\delta^2 J}{\delta w_i^2}(\Delta w_i)^2 \tag{1}$$

At the optimum, first derivative can be neglected by using $J = \frac{1}{2}\|w\|^2$, and the Equation 1 becomes $\Delta J(i) = (\Delta w_i)^2$. Let say the i-th feature is discarded by setting the corresponding weight to 0, thus $\Delta w_i = w_i$. Therefore, discarding the feature with smallest w_i^2 will generate the least increase of J, which simultaneously improves the generalization performance.

2.2   Semi-Supervised SVM-Based RFE (S³VM-RFE)

Semi-supervised SVM-based RFE (S³VM-RFE) is developed as the improvement of the supervised SVM-RFE, which integrated the unsupervised algorithm into the original program. The improvements have been done in S³VM-RFE are:

1. Involved labeled and unlabeled data in training process. The labeled training samples are randomly selected among the training set. The rest of labeled training samples are invoked as unlabeled set.
2. Labeled data are trained with supervised SVM [20] and unlabeled data are trained with unsupervised one-class SVM algorithm [17].
3. The feature weight vector produced from supervised SVM (w^l) and unsupervised SVM (w^u) are summed up as the final ranking criterion (W).

The procedure of semi-supervised SVM-based RFE (S³VM-RFE) are shown in Algorithm 1.

2.3   Semi-Supervised SVM-Based Feature Selection (S³VM-FS)

The proposed method is known as semi-supervised SVM-based feature selection (S³VM-FS), which is a modification method from S³VM-RFE. In the proposed method, the step 6 till step 9 stated in the S³VM-RFE method are skipped. After computing the final ranking scores, the modified method will sort the feature list based on the ranking scores. Thus, the dataset is updated by eliminating the lowest ranked feature based on the pre-defined number of feature. The modified procedure is indicated in Figure 1, the flowchart of S³VM-FS with the nested cross validation (CV) classification model.

Algorithm 1. S^{3}VM-RFE (X_0, y):

Input

      Training examples, $X_0 = [x_1, x_2, ?, x_l, x_{(l+1)}, ?, x_u]^T$

      Class labels, $y = [y_1, y_2, ?, y_l, y_{(l+1)}, ?, y_u]^T$,

          where $[y_1, ?y_l] \in -1, +1$ and $[y_{(l+1)}, ?, y_u] = 0$

Initialize

      Subset of surviving features, $s = [1, 2, ?, n]$

      Feature ranked list, $r = [\,]$

      Repeat until $s = [\,]$

Start

1. Restrict training examples to good feature indices

$$labelx = X_0(:, s);$$
$$unlabelledx = X_0(:, s);$$

2. Train the classifier

$$\propto_{label} = \text{SVM-train } (X, y); \quad \%\text{Supervised SVM}$$
$$\propto_{unlabelled} = \text{SVM-train } (X); \quad \%\text{Unsupervised one-class SVM}$$

3. Compute the weight vector of dimension length(s), for all i

$$w_i^l = \sum_k \propto_{(label_k)} y_k x_k$$
$$w_i^u = \sum_k \propto_{(unlabelled_k)} y_k x_k$$

4. Determine the sum of w_i^l and w_i^u,

$$W_i = w_i^l + w_i^u;$$

5. Compute the ranking criteria, $c_i = (W_i)^2$;

6. Find the feature with smallest ranking criterion, $f = argmin(c)$;

7. Update feature ranked list, $r = [s(f), r]$;

8. Eliminate the feature with smallest ranking criterion

$$s = s(1 : f - 1, f + 1 : length(s));$$

9. Repeat until all the features are ranked.

Output

Ranked feature set r.

---

3   Experimental Setup

A series of experiments are conducted to examine the effectiveness and efficiency of the proposed semi-supervised SVM-based feature selection method. The experiments are performed on a laptop with Intel Core i5 @ 1.70GHz, and 11.9 GB of RAM.

This study first apply the model to microarray data from [18] who reported 181 lung cancer tissue samples with 31 malignant pleural mesothelioma (MPM) and 150 adenocarcinoma (AD). The gene expression intensities are obtained from Affymetrix high-density oligonucleotide microarrays containing probes for 1,626 genes[18].

As the supervised method is widely applied in the gene expression analysis, hence the available data set has been completely labelled. However, for a semi-supervised learning method, the input data should consist of a numbers of n labelled samples, and a number of u unlabelled samples. The labelled data are expressed in the form $(X_n, Y_n)=(X_1, y_1),\cdots,(X_n, y_n)$, $y \in +1, -1$, where X

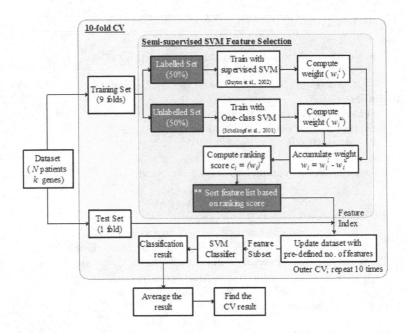

Fig. 1. Flowchart for S^{3}VM-FS with the nested CV Classification model

denotes the feature and Y denotes the class label. Whereas, the unlabelled data is expressed in the form $(X_u, y_u) = (X_{(n+1)}, y_u), \ldots, (X_{(n+u)}, y_u)$, where $y_u = 0$ assumed it as unlabelled. A complete data set can be denotes as $X \rightarrow Y = (X_n, y_n) \bigcup X_u$. In addition, each feature X contains several number of k genes, that is expressed in $X = x^1, \ldots, x^k$.

In the input data design, the X-value is arranged in columns, which represented as each of the gene expression, where the number of features (genes) depends on the collected data set. The last column indicates the labelled of the individual sample. However, each row refers to each patient sample in term of labelled and unlabelled samples. Figure 2 shows input data designs with sample-by-gene matrix form. For example, x_n^k indicates that the feature on n-th samples with k-th genes.

In order to bring the data set into a notionally common scale, a rescaling scheme has to be introduced. In this paper, normalized z-score scale is used as the rescaling scheme. The rescaling scheme is a common pre-processing step in clustering and classification as it often can improve the accuracy of the results [19]. Given X to be the sample-by-gene matrix, the following defines the normalized scales used here:

$$x_{ij} \leftarrow \frac{x_{ij} - \mu}{\sigma} \tag{2}$$

	Gene$_1$	Gene$_2$			Gene$_k$	Class
Labelled Sample$_1$ $\Rightarrow$	x_1^1	x_1^2	...	...	x_1^k	y_1
.	.	.			.	.
.	.	.			.	.
.	.	.			.	.
Labelled Sample$_n$ $\Rightarrow$	x_n^1	x_n^2	...	...	x_n^k	y_n
Unlabelled Sample$_1$ $\Rightarrow$	x_{n+1}^1	x_{n+1}^2	...	...	x_{n+1}^k	y_u
.	.	.			.	.
.	.	.			.	.
Unlabeled Sample$_u$ $\Rightarrow$	x_{n+u}^1	x_{n+u}^2	...	...	x_{n+u}^k	y_u

Fig. 2. Input data designs with Sample-by-gene Matrix Form

where x_{ij} is entry (i, j) of X, μ denote the mean of X, and σ is the standard deviation of X. By inspection it is clear that the normalized scale will bring all entries of the data matrix to the range of [-1,1].

The normalised data are divided into ten sets by 10-fold CV, where nine sets for the training process and one set for testing purpose. Between the training set, fifty percent of the samples are randomly selected and trained with the presence of class labels and another fifty percent of samples are trained without the presence of class labels. The CV procedure is repeated ten times and the *Accuracy* is expressed in term of % accuracy calculated as the average classification accuracy on testing samples as described in Equation 3.

$$Accuracy\ (\%) = \frac{1}{k}\sum_{i=1}^{k}\frac{\#\text{correctly predicted data}}{\#\text{total testing data}} \times 100 \qquad (3)$$

where k is the number of k-folds CV, # correctly predicted data denotes as the number of samples that is correctly predicted, and # total testing data is the total number of testing samples.

The SVM consists of few parameters called hyper-parameters that affect the performance of an SVM. The hyper-parameters include the soft margin constant (C), and parameters in kernel function, which are the degree of a polynomial kernel, or width of a Gaussian kernel (γ). The parameter C can be viewed as a way to control over-fitting. The larger the value of C, the more the error is penalized. The parameter C setting in the experiments is in the range of $[2^{-2}, 2^{-1}, \ldots, 2^9, 2^{10}]$. Kernel parameters also have a significant effect on the decision boundary. The degree of the polynomial kernel and the width parameter of the Gaussian kernel control the flexibility of the resulting SVM in fitting the data. The lowest degree polynomial is the linear kernel, while for the non-linearly separable features, a higher degree polynomial is needed for greater curvature. This is consistent with the parameter γ where smaller γ resulting smooth decision boundary. As γ increased the locality of the support vector expansion increases, leading to the greater curvature of the decision boundary. However, large value of γ and degree polynomial may lead to over-fitting. The parameter γ used in polynomial, and sigmoid kernel function is set in the range of $[2^{-5}, 2^{-4}, \ldots, 2^3, 2^4]$.

4   Result and Discussion

In order to demonstrate the performance of the proposed method, comparison of classification accuracy and computational processing time between supervised SVM-RFE, S^3VM-RFE, S^3VM-FS are presented. Table 1 report the 10-folds CV classification accuracy results based on the pre-defined number of features with different types of kernel function used in the experiments.

Table 1 show that supervised SVM-RFE is performed slightly better than another two methods by using the linear kernel, where it able to achieve 85.64 percent of accuracy with only 10 features in classification. However, by using sigmoid kernel, the proposed method, S^3VM-FS is able to achieve even higher accuracy (86.16 percent) but with more features (100 features) involved in classification.

The proposed method, S^3VM-FS is the best among the three feature selection methods in average. Generally, the means of classification accuracy achieved by S^3VM-FS with linear, polynomial and sigmoid kernel (respectively 83.11, 83.95 and 80.41) are higher than the means of the accuracy of SVM-RFE (respectively 80.58, 78.58, and 74.20) and S^3VM-RFE (respectively 80.45, 81.96, and 79.58). Moreover, S^3VM-FS with polynomial kernel attained highest means of accuracy results when compared with linear and sigmoid kernel.

The proposed semi-supervised method, S^3VM-RFE gives better results than the supervised method, SVM-RFE, because by using both unlabeled and labeled

Table 1. Comparison of classification accuracy

Feature Selection	10-folds CV classification accuracy (%)										Means
No. of feature	10	20	40	80	100	140	180	220	260	300	
Linear Kernel											
SVM-RFE	**85.64**	79.59	80.17	79.56	80.67	82.30	81.22	77.42	79.03	80.17	*80.58*
S^3VM-RFE	79.00	78.45	81.78	76.78	**83.45**	82.89	82.89	81.22	79.56	78.45	*80.45*
S^3VM-FS	83.42	82.89	82.33	82.89	**84.00**	82.89	84.00	81.78	83.42	83.45	***83.11***
Polynomial Kernel											
SVM-RFE	78.45	79.61	**80.73**	78.47	78.97	80.08	75.11	76.90	76.81	80.70	*78.58*
S^3VM-RFE	81.22	82.33	83.42	**83.45**	82.89	83.45	80.76	79.00	81.81	81.22	*81.96*
S^3VM-FS	83.45	84.53	82.92	83.45	83.45	84.56	**85.11**	84.00	84.00	84.00	***83.95***
Sigmoid Kernel											
SVM-RFE	70.14	71.84	77.89	78.97	**80.11**	65.78	79.00	75.67	67.95	74.61	*74.20*
S^3VM-RFE	75.76	74.67	80.11	81.22	79.00	81.78	80.11	**82.33**	81.22	79.56	*79.58*
S^3VM-FS	80.67	78.01	80.05	81.81	**86.16**	83.42	80.67	80.17	77.39	75.73	***80.41***

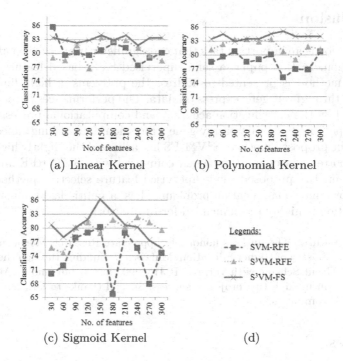

(a) Linear Kernel   (b) Polynomial Kernel

(c) Sigmoid Kernel   (d)

Fig. 3. Comparison of Classification Accuracies in Lung Cancer Dataset

data, S^3VM-RFE improves the generalization ability of a given supervised task. Figure 3 shows the comparison of classification accuracy between SVM-RFE, S^3VM-RFE, and S^3VM-FS with different kernel function. Table 2 reports the computational processing time of the best classification accuracy achieved by using different kernel functions in three methods.

As showed in Table 2, SVM-RFE and S^3VM-RFE require more processing time to complete the iteration compared with the proposed method. The shortest processing time of S^3VM-FS to achieve the highest accuracy is 19.35 seconds, which is nearly fifty times faster than method SVM-RFE, and S^3VM-RFE.

Table 2. Comparison of computational processing time

Types of kernel	Processing Time (seconds)		
	SVM-RFE	S^3VM-RFE	S3VM-FS
Linear	1097.58	1033.34	30.10
Polynomial	1432.42	1430.21	73.84
Sigmoid	462.86	1109.37	19.35

5   Conclusion

Feature selection is an important task that can significantly affect the performance of classification. In this paper, a semi-supervised SVM-based feature selection (S^3VM-FS) method is presented to address the problems of high dimensionality and high throughput gene expression data. The performances of the methods are evaluated by the classification accuracy and computational processing time. Based on experiments on microarray gene expression data of lung cancer, it was found that the proposed method, S^3VM-FS has achieved the slightly higher accuracy with shorter processing time when compared with SVM-RFE and S^3VM-RFE. However, the proposed semi-supervised feature selection method is only applicable for binary classification problem. Thus, a multi-class semi-supervised feature selection is highly recommended for future work.

Acknowledgments. The authors honorably appreciate Research Management Center, UTM and Ministry of Higher Education, Malaysia for funding through the Exploratory Research Grant Scheme with Vot. No. R.J130000.7828.4L146, and UTM Zamalah for the support in making this project a success. Special thanks to Dr. Lim Seng Poh for his valuable comments.

References

1. Saeys, Y., Inza, I., Larraaga, P.: A review of feature selection techniques in Bioinformatics. Bioinformatics. **23**, 2507–2517 (2007)
2. Loscalzo, S., Yu, L., Ding, C.: Consensus group stable feature selection. Proceedings of the 15th ACM SIGKDD International Conference on Knowledge Discovery and Data Mining, pp. 567–576. ACM, New York, NY, USA (2009)
3. Witten, D.M., Tibshirani, R.: A Framework for Feature Selection in Clustering. Journal of the American Statistical Association **105**, 713–726 (2010)
4. Xu, R., Damelin, S., Nadler, B., Wunsch II, D.C.: Clustering of high-dimensional gene expression data with feature filtering methods and diffusion maps. Artificial intelligence in medicine **48**, 91–98 (2010)
5. Du, W., Sun, Y., Wang, Y., Cao, Z., Zhang, C., Liang, Y.: A novel multistage feature selection method for microarray expression data analysis. International journal of data mining and bioinformatics **7**, 58–77 (2013)
6. Gaafar, M.A., Yousri, N.A., Ismail, M.A.: A novel ensemble selection method for cancer diagnosis using microarray datasets. IEEE 12th International Conference on BioInformatics and BioEngineering, BIBE 2012. pp. 368–373 (2012)
7. Liang, Y., Liu, C., Luan, X.-Z., Leung, K.-S., Chan, T.-M., Xu, Z.-B., Zhang, H.: Sparse logistic regression with a L1/2 penalty for gene selection in cancer classification. BMC Bioinformatics **14**, 198 (2013)
8. Barkia, H., Elghazel, H., Aussem, A.: Semi-supervised Feature Importance Evaluation with Ensemble Learning. In: 2011 IEEE 11th International Conference on Data Mining (ICDM), pp. 31–40 (2011)
9. Benabdeslem, K., Hindawi, M.: Efficient Semi-supervised Feature Selection: Constraint, Relevance and Redundancy. IEEE Transactions on Knowledge and Data Engineering 1, (2013)

10. Helleputte, T., Dupont, P.: Partially supervised feature selection with regularized linear models. In: Proceedings of the 26th Annual International Conference on Machine Learning. pp. 409–416. ACM, New York, NY, USA (2009)
11. Kalakech, M., Biela, P., Macaire, L., Hamad, D.: Constraint scores for semi-supervised feature selection: A comparative study. Pattern Recognition Letters **32**, 656–665 (2011)
12. Zhao, Z., Liu, H.: Semi-supervised Feature Selection via Spectral Analysis. SDM, 641–646 (2007)
13. Kotsiantis, S.B.: Supervised Machine Learning: A Review of Classification Techniques. Proceedings of the 2007 Conference on Emerging Artificial Intelligence Applications in Computer Engineering: Real Word AI Systems with Applications in eHealth, HCI, Information Retrieval and Pervasive Technologies. pp. 3–24. IOS Press, Amsterdam, The Netherlands (2007)
14. Zhili, W.: Kernel Based Learning Methods for Pattern and Feature Analysis (2004)
15. Wu, Z., Li, C.: Feature Selection with Transductive Support Vector Machines. In: Guyon, I., Nikravesh, M., Gunn, S., Zadeh, L.A. (eds.) Feature Extraction. Studies in Fuzziness and Soft Computing, vol. 207, pp. 325–341. Springer, Heidelberg (2006)
16. LeCun, Y., Denker, J.S., Solla, S.A.: Optimal Brain Damage. Advances in Neural Information Processing Systems. pp. 598–605. Morgan Kaufmann (1990)
17. Scholkopf, B., Platt, J.C., Shawe-Taylor, J., Smola, A.J., Williamson, R.C.: Estimating the Support of a High-Dimensional Distribution. Neural computation 13, 1443–1471 (2001)
18. Gordon, G.J., Jensen, R.V., Hsiao, L.-L., Gullans, S.R., Blumenstock, J.E., Ramaswamy, S., Richards, W.G., Sugarbaker, D.J., Bueno, R.: Translation of Microarray Data into Clinically Relevant Cancer Diagnostic Tests Using Gene Expression Ratios in Lung Cancer and Mesothelioma. Cancer research **62**, 4963–4967 (2002). http://algorithmics.molgen.mpg.de/Static/Supplements/CompCancer/datasets.htm
19. Chowdary, D., Lathrop, J., Skelton, J., Curtin, K., Briggs, T., Zhang, Y., Yu, J., Wang, Y., Mazumder, A.: Prognostic Gene Expression Signatures Can Be Measured in Tissues Collected in RNAlater Preservative. The journal of molecular diagnostics **8**, 31–39 (2006)
20. Guyon, I., Weston, J., Barnhill, S., Vapnik, V.: Gene Selection for Cancer Classification using Support Vector Machines. Machine learning **46**, 389–422 (2002)

Opponent Classification in Robot Soccer

Asma S. Larik and Sajjad Haider[✉]

Artificial Intelligence Lab, Faculty of Computer Science,
Institute of Business Administration, Garden Road, Karachi 74400, Pakistan
{asma.sanam,sajjad.haider}@khi.iba.edu.pk

Abstract. The paper presents an approach to perform post-hoc analysis of RoboCup Soccer Simulation 3D teams via log files of their matches and to learn a
model to classify them not only as being strong, medium or weak but also
through their game playing styles such as frequent kickers, frequent dribblers,
heavy/lean attackers, etc. The learned model can then be used to further cluster
teams to predict game style of similar opponents. We have applied the presented approach to 22 teams from RoboCup 2011 in a fully automated fashion
and the results show the validity of our approach.

Keywords: RoboCup soccer · Opponent modeling · Behavior analysis · Classification · Machine learning

1 Introduction

Opponent modeling[1], [2] focuses on the analysis and interpretation of adversary's
actions. In real world adversarial environments, humans usually build an internal
model/representation of their opponents and use it to understand the nature of the
opponent being faced. This aids in informed decisions and effective counter actions
against different opponents. The modeling, however, is not as easy when it comes to
autonomous multi-agents environments [3][4], where agents are capable of doing
decentralized decision making. In such situations, the opponent model consists of a
collective set of tactical behaviors, termed as strategy, that are exhibited by agents in
a dynamic and uncertain environment. For instance, during a RoboCup Soccer [5]
match, each agent/robot can take up specific role and perform certain set of actions.
The synergetic effect of these individual actions would emerge as a team strategy and
would reveal the quality of the team.

This paper presents a novel framework that analyzes key agent behaviors in Robo-
Cup Soccer Simulation 3D[6] using a set of pre-defined rules and classifies a team
based upon these extracted team quality features. Team quality features were identified by a group of experts and extraction of these features is automated via a parser
application. These features include team players positioning, frequency of players in
attacking region, ball possession, average velocity of team players, frequency of kicks
executed by teams, shots attempted towards goal etc. Once a team is classified upon
its strength parameters then grouping of teams that play similar is done via clustering
that would aid in counter strategy formulation against similar opponents. The major

© Springer International Publishing Switzerland 2015
M. Ali et al. (Eds.): IEA/AIE 2015, LNAI 9101, pp. 478–487, 2015.
DOI: 10.1007/978-3-319-19066-2_46

benefit of this approach is that if we are able to judge our opponent's strategies we can apply specific rules to deal with them. For example, certain teams make excessive use of kicking while many others have the tendency to dribble the ball to maintain possession of it. If the opponent has strong dribbling skill then our best strategy would be to block the path of the opponent and put most of our players to this job. On the other hand, if the opponent team has a tendency to frequently kick the ball then we would like to keep most of our players around our goal area in order to save the goal (especially if the ball is in our half).

Opponent modeling has been extensively researched in the domain of RoboCup Soccer Simulation 2D[6] due to the simplicity of actions such as Kick, Turn and Dash as well as their availability in the log files. In the Simulation 3D league, however, the task is not as easy as perceiving behavior is quite difficult due to the unavailability of labeled actions. The Simulation 3D log file only contains positional data of each and every joint of a player in the form of 4 X 4 transformation matrices. Several matrices have to be multiplied to compute the position and orientation of the player in the field. In addition, one has to use this computed players and ball positioning data to predict the behavior perform by agents. The approach presented in this paper has following major phases: Data Extraction Phase, Behavior Definition Phase, Team Classification Phase and Team Clustering Phase. To support data extraction as mentioned previously, a parser application[7] has been developed that performs offline analysis of the Simulation 3D log files and extracts positional data for home team players, opponent players and ball for every time cycle. The application also derives speed of the ball/agents and their velocities. It is important to note that there is no other way to identify behavior as the log files do not give us any indication of the behavior performed by a soccer agent. Subsequently, the behavior definition phase identifies key behaviors, such as kick, dribble and shots that resulted in goal from log files using pre-defined rules. In the third phase classification of teams as being Strong, Medium and Weak is done on the basis of ranking metrics published by Champion team of RoboCup 2011 namely UT Austin Villa [8]. In future we will perform same multi-class classification task using teams participating in RoboCup 2012, 2013 and 2014 respectively with teams reaching the quarter finals being classified as Strong, Others that reach Round two as Medium and the remaining as Weak teams. Lastly for a particular class further clustering can give us an insight into similar game styles adopted by certain group of teams. As clustering is a black box in this phase we analyze teams that are being placed into a certain cluster and try to investigate their key strengths such as frequent kickers/dribblers, heavy/lean attackers, etc. Thus, if two teams attack with more number of attackers then they will fall into single cluster at runtime and then we can activate the cluster specific strategy of blocking goal path to minimize the chances of goal. The rest of the paper is organized as follows: Section 2 provides review of the efforts made in the domain of opponent modeling specifically within the context of RoboCup Soccer Simulation 2D. Section 3 describes the proposed approach in detail, while Section 4 discusses the design of experiments and results. Finally, Section 5 concludes the paper and provides future research directions.

2   Related Work

S. Pourmehr and C. Dadkhah [9] have written a comprehensive literature review of various techniques applied in the domain of 2D. L. Agapito et al. [10] present OMBO, an opponent modeling approach based on observations, which purely focuses on the skills of goal keeper and attacker. They first build a classifier to label opponent actions, next they place a dummy player in the field just for the purpose of recording opponent actions and lastly they predict actions of opponent based on training dataset. T. Warden [11] uses dynamic scenes analysis and geometrical flow of scenes for action / event recognition. Opponent modeling efforts have also been made in the context of coaching competition within the Simulation 2D league where the focus was to learn normal based patterns of team plays and predict the team strategy. The coach that recognizes more patterns wins. The most notable of these efforts are presented in [12]- [13]. For analysis of team performance F.Almeida et al. [14] proposes a set play based approach for automatic goal plans extractionin the context of RoboCup Soccer Simulation 2D. In the domain of RoboCup Small Size League opponent strategy identification has also been explored by C.Erdogan and M.Veloso[15]. They use log files data for extraction of spatial as well as temporal attributes and create geometrical trajectory curves. Next agglomerative hierarchical clustering is performed for trajectory matching and corresponding team behavior is learnt incrementally. Finally online opponent strategy is predicted via clustering. K.Yausui et al. [16] propose a dissimilarity function that classifies opponent strategies based upon cluster analysis.

3   Proposed Approach

The proposed architecture comprises of the phases depicted in Fig 1. It takes a log file generated by server as input, extracts features that form the individual and team actions, devises rules that create a mapping between these attributes and behavior, classifies teams as strong, medium and weak and lastly performs clustering for counter strategy prediction.

3.1   Data Extraction Phase

In the Simulation 3D league, the server continuously records a log file during a match. The messages recorded in the log are in the form of S-Expressions which essentially are grammatical expressions written in Lisp family of programming languages. Log file records messages such as environment information, game state, scene graph header and scene graph contents. The scene graph is a structure that arranges logical and spatial representation of a graphical scene. The scene graph is a tree with base nodes, transformation nodes, geometry nodes, static mesh nodes, light nodes, etc. A header expression that is sent initially contains information regarding field length, field width, field height, goal width, goal depth, goal height, border size, free kick distance, wait before kickoff time, radius of the agent, ball radius, ball mass, play mode (such as goal kick, play on, side kick, etc.), time, score, SLT (single linear transform), light nodes, TRF (transformation matrices), etc. A new header is sent whenever the scene changes, such as, loading of a player, removal of a player from the field, etc.

In order to analyze the log data we have developed a parser that traverses each and every S-Expression and identify nodes for ball and players. A detailed description of the formal extraction process is described in [7]. The parser application performs the following tasks:

- Extraction of node information from the header
- Computation of players' and ball's position
- Extraction of updated nodes from the timestamp information

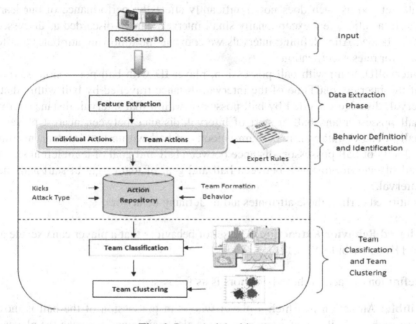

Fig. 1. Proposed Architecture

3.2   Behavior Definition and Identification Phase

Behavior Definition
A behavior at a particular instance is an action performed by player. Behaviors are predetermined and uniform amongst all teams however its execution depends upon the ability of the team exhibiting the action. In context of this paper we are primarily focusing on the behavior of an agent who is in possession of the ball. A ball possessor is an agent maintaining minimum distance with the ball.

Interval Definition
We also need to define the time interval in which the behavior persisted as the actions are not atomic in nature and may last around 10 to 20 cycles with one cycle

corresponding to 20 milliseconds. If we define a static window size, say 20 cycles, then it may not be correct since a dribbling event by a player may last 50 cycles or a single action may finish in a shorter time. Thus we have defined a dynamic window/interval. An interval comprises of the time cycles during which there is no change in ball possession. A single behavior is executed during the interval. It is worth mentioning that there are bound to be situations which will not be properly covered using the above two definitions. For instance, the ball might be lying in the field and none of the player is actually possessing it, the ball might be moving from the nearby player who is actually not in possession of the ball, two player might be struggling for the ball, a possessing player might be moving away from the ball as per a new game plan, etc. During our initial (but thorough) experiments, however, we found out that these situations are a tiny portion of the data set and as such does not significantly affect the performance of our learned models. In addition, few exceptionally small intervals can be discarded as discussed in Section 4 below. After defining intervals we compute the following attributes that form the basis for rules specification:

> IntervalID, team with ball possession, player ID with ball possession, startTime of the interval, endTime of the interval, distance travelled by ball within that interval, distance travelled by ball possessor within that interval, distance between ball possessor and ball at start of interval, distance between nearest player and ball at end of interval, maximum velocity of ball within that interval, maximum velocity of ball possessor, distance between ball and goal of team left at start and end of interval, distance between ball and goal of team right at start and end of interval.

As mentioned earlier, these attributes aid in defining behaviors.

Rule-based Behavior Extraction. The set of behaviors that a player can execute are:
> B = {Kick, Dribble, ShotToGoal, Clear}

The definition of each of these behaviors is as follows:

1. **Dribble:** An action in which a player who is in possession of the ball is moving along with the ball. An average distance is also maintained between the player and the ball.
2. **Kick:** An action in which at a certain timestamp the ball is very near (less than 1 meter) from the ball possessing player and in the next cycle the ball is far from that player and the velocity of the ball also increases drastically.
3. **Shot to Goal:** An action in which the distance between ball and goal was less than 0.5 m at the start of interval and greater at the end of interval. We are considering both goal post with goal left and goal right. In our approach we only consider a shot that resulted in goal.
4. **Clear:** Any undefined action is classified as Clear. Similarly time stamp during which the ball goes out of bounds is also termed as Clear.

The rules for behavior identification are listed in Table 1. It must be mentioned that these rules are expert defined and the threshold values of parameters (α, β, λ, μ, γ, δ, ρ and π) can be tweaked further based on experts' opinion.

Table 1. Rules for behavior definition

```
// Kick:
if ((DistTravelledByBall > α and VelocityBall> β and
  ((DistBWPlayerAndBallAtStart< λ and
   (DistBWPlayerAndBallAtEnd> μ)
   or
   (DistTravelledByBall> ɤ and
    DistTravelledByPlayer<π ) )
then 'Kick'


// Dribble:
Elseif ((abs(DistTravelByBall-DistTravelbyPlayer)< δ)
  or
  ((DistBWPlayerAndBallAtStart-
   DistBWPlayerAndBallAtEnd)< δ)
then 'Dribble'

//Shot ToGoal :
elseif(DistOfBallFromGoalLeftAtStart< ρ     or
   DistOfBallFromGoalLeftAtEnd< ρ )
then 'ShotGoalLeft'
elseif
 (DistOfBallFromGoalRightAtStart< ρ or
  DistOfBallFromGoalRightAtEnd<ρ )
then 'ShotGoalRight'

// Clear:
Else 'Clear'
```

3.3   Team Classification Phase

In this phase we utilize the derived attributes and behaviors for analyzing different type of teams. Our approach consists on using the built dataset to create automatic learners capable of correctly classifying the opponent team. For this, we have decided to perform experiments using two classifiers namely Naive Bayes and Neural Networks and to finally choose the one with highest prediction accuracy. Artificial Neural Networks[17] are systems made up of connected nodes capable of approximating nonlinear functions of their weighted inputs. We have chosen feed forward multilayer perceptron architecture for NN with one hidden layer and sigmoid activation function to perform the task. On the other hand, Naive Bayes classifier[17] is a simple probabilistic classifier based on Bayes' Theorem with a strong independence assumption. In spite of its naïve design and over-simplified assumptions, it has worked quite well in many complex, real-world situations. We applied the supervised feature selection algorithm available in WEKA[18] and then selected the following set of features that gave promising results : average players in home regions, average number of players in opponent region, minimum distance maintained with the ball, ball possession, average velocity of players, maximum velocity of players, average number of kicks, average dribble, frequency of successful shots to goal, number of attackers, opponent type etc. The reason for using the above mentioned attributes is that a thorough game analysis reveals that Strong teams usually have a high velocity of players, maintain ball possession approximately 70% of the time, attack with more than two attackers, kick ball frequently or either dribble often to create more goal scoring opportunities. The problem is modeled as a multi-class classification task utilizing Strong, Medium and Weak classes. As mentioned earlier, we have used ranking published by UT Austin Villa team [8]. They reported an average goal difference over 100 games of all teams that participated in World RoboCup 2011 against UT Austin (the champion team of 2011 World RoboCup Soccer 3D). We made bins of the goal difference and created team categories. Thus teams like UTAustinVilla, Apollo3D, BoldHearts, Robocanes, Cit3D, and FCPortugal3D are classified as Strong teams while Magma, OxBlue, Kylinsky, DreamWing3D, Seuredsun, Karachi Koalas, Beestanbul, Nexus3D, Hfutengine3D, Futk3D and NaoTH are classified as Medium teams. Finally, the remaining teams such as Nomofc, Kaveh/rail, Bahia3D, L3msim and Farzanegan belong to the Weak team's category.

3.4   Team Clustering Phase

Clustering is an unsupervised learning technique that segments data items into different groups. Once teams are classified, the next task is to further cluster them based on their playing style. For instance, few teams are strong due to their frequent kicks, while the others dribble the ball really well. Similarly, few teams attack with more attackers while others with a single attacker. Thus, it is extremely important to have a good understanding of the opponent playing style in order to devise a good counter strategy. Instead of specifying strategy for each and every team, in this paper we suggest clustering teams with similar style of play in one group. This allows us to have one strategy against all the teams that play with a similar style instead of having a team specific strategy.

4   Design of Experiment and Results

For the experiments, we have used the log files of 25 matches of World RoboCup 2011. As discussed earlier, this data is used as a benchmark since rating of teams is available. Further investigation regarding current data is in progress. The parser application executes a complete log file containing 1500 instances within 10 seconds and produces the attributes discussed in Section 3.1 above. The attributes were then passed to the next phase for interval creation and behavior identification. These intervals signified the events that were of interest during the match. Further pre-processing of intervals was performed and those containing less than five cycles with each cycle corresponding to 20ms were discarded. Rules were applied on these intervals with the parameters value set to α = 2.5 m, β= 0.5 m/s^2, λ=0.5 m/ s^2, μ=0.5 m/ s^2 ɤ=3, δ=2, π=0.5 and ρ=1. In this paper, we have set these parameters after analyzing several matches. In future, however, we aim to apply evolutionary algorithms to learn these parameters. From the log file of a single match, two records, one for team left and one for team right, were obtained and stored separately in database. The procedure was repeated for all the 25 matches to obtain 50 records that were later used during the classification phase. The teams were labeled as Strong, Medium and Weak as mentioned before.

4.1   Classification Results Analysis

For classification we used two classifiers namely Naïve Bayes Classifier and Multi-layer Perceptron. Table 2 shows the prediction accuracy for each class.

Table 2. Prediction Accuracy of each classifier per class

Class	Naive Bayes (%)	Neural Networks (%)
Strong	88.23	82.35
Medium	64.7	65
Weak	94.11	84.21

As evident from the table, the class Strong and Weak have a high prediction accuracy. The class Medium has less accuracy because the boundaries between Strong and Medium are less crisp. It is possible that in some matches Medium teams may outperform Strong teams or employ strategy that is similar to those exhibited by Strong teams.

4.2   Clustering Analysis

After various experiments cluster size was fixed to two and results for k-means is shown in Table 3.

Table 3. Clustering Results via k-means

Class	Cluster 0	Cluster 1
Strong	Cit3D, Robocanes	Apollo3D, UTAustinVilla
Medium	Kylinsky, FUTK, Magma	Dreamwing, NaoTH, KK

The clustering results provide us further insight into the data. For instance, CIT3D and Robocanes are in the same cluster as their kick frequency is higher in contrast to Apollo3D and UTAustinVilla who dribble the ball well and maintain high ball possession. Similarly, among Medium teams, Kylinsky, FUTK and Magma have more aggressive attack while Dreamwing and NaoTH play with more defensive team formation. Our future task is to develop cluster-specific strategy to optimize our performance.

5   Conclusion

The paper presented an approach for classification of opponent teams based upon their spatial and temporal attributes as well as their behavior. The behaviors pertain to the skills of the agents such as Kick, Dribble, Pass, and Intercept. A parser application was developed and used to extract positional data. This data was then divided into temporal intervals based upon specific behaviors and expert-specified rules. The technique presented in this paper is first of its kind since it creates an opponent model based on 3D soccer simulation log files. From a single half of the match consisting of 1500 records, we obtain around 100-120 intervals and from these intervals we obtain two records one for each team. The approach has been tested on 22 teams that participated in World RoboCup 2011 and it provides promising results. In the future, we wish to extend this approach and try to devise a mechanism that would recommend a counter strategy based upon team classification and playing style.

References

1. Borghetti, B.J.: Opponent Modeling in Interesting Adversarial Environments. ProQuest (2008)
2. Schadd, F., Bakkes, E., Spronck, P.: Opponent modeling in real-time strategy games. In: Proceedings of the GAME-ON 2007, pp. 61–68 (2007)
3. Bjarnason, R.V., Peterson, T.S.: Multi-agent learning via implicit opponent modeling. In: Proceedings of the 2002 Congress on Evolutionary Computation, 2002. CEC '02, vol. 2, pp. 1534–1539 (2002)
4. Julia Frolova, "Review of Multi-Agent Systems and Applications," 29-Mar-2005. [Online]. http://jasss.soc.surrey.ac.uk/8/2/reviews/frolova.html. [accessed: 29-Mar-2013]
5. Kitano, H., Tambe, M., Stone, P., Veloso, M., Coradeschi, S., Osawa, E., Matsubara, H., Noda, I., Asada, M.: The RoboCup synthetic agent challenge 97. In: RoboCup-97: Robot Soccer World Cup I, vol. 1395, H. Kitano, Ed. Berlin, Heidelberg: Springer Berlin Heidelberg, pp. 62–73 (1998)
6. "Robocup official website (www.robocup.org)."
7. Larik, A.S., Haider, S.: Rule-based behavior prediction of opponent agents using robocup 3D soccer simulation league logfiles. In: Iliadis, L., Maglogiannis, I., Papadopoulos, H. (eds.) Artificial Intelligence Applications and Innovations. IFIP AICT, vol. 381, pp. 285–295. Springer, Heidelberg (2012)
8. MacAlpine, P., Collins, N., Lopez-Mobilia, A., Stone, P.: UT Austin Villa: RoboCup 2012 3D simulation league champion. In: Chen, X., Stone, P., Sucar, L.E., van der Zant, T. (eds.) RoboCup 2012. LNCS, vol. 7500, pp. 77–88. Springer, Heidelberg (2013)
9. Pourmehr, S., Dadkhah, C.: An overview on opponent modeling in RoboCup soccer simulation 2D. In: Röfer, T., Mayer, N., Savage, J., Saranlı, U. (eds.) RoboCup 2011. LNCS, vol. 7416, pp. 402–414. Springer, Heidelberg (2012)
10. Ledezma, A., Aler, R., Sanchis, A., Borrajo, D.: OMBO: An opponent modeling approach. AI Commun. 22121–35 IOS Press (2009)
11. Warden, T., Visser, U.: Real-time spatio-temporal analysis of dynamic scenes. Knowl. Inf. Syst. 32(2), 243–279 (2011)
12. Riley, P.: Coaching: Learning and Using Environment and Agent Models for Advice. Carnegie Mellon University, (2005)

13. Nardi, D.: The UT Austin Villa 2003 Champion Simulator Coach: A Machine Learning Approach
14. Almeida, F., Abreu, P.H., Lau, N., Reis, L.P.: An automatic approach to extract goal plans from soccer simulated matches. Soft Comput. **17**(5), 835–848 (2012)
15. Erdogan, C., Veloso, M.: Action selection via learning behavior patterns in multi robot domains. In: Proceedings of the Twenty-Second International Joint Conference on Artificial Intelligence - Volume Volume One, Barcelona, Catalonia, Spain, pp. 192–197 (2011)
16. Yasui, K., Kobayashi, K., Murakami, K., Naruse, T.: Analyzing and learning an opponent's strategies in the robocup small size league. In: Behnke, S., Veloso, M., Visser, A., Xiong, R. (eds.) RoboCup 2013. LNCS, vol. 8371, pp. 159–170. Springer, Heidelberg (2014)
17. Russell, S., Norvig, P.: Artificial Intelligence: A Modern Approach, 3rd edn. Prentice Hall, Upper Saddle River (2009)
18. "Weka Software." [Online]. Available: http://www.cs.waikato.ac.nz/ml/weka/

Improving Document Classification
Using Fine-Grained Weights

Soo-Hwan Song[1] and Chang-Hwan Lee[2]([✉])

[1] Department of Computer Science, KAIST, 291 Daehak-ro,
Yuseoung-Gu, Deajeon, Korea
dramanet30@kaist.ac.kr
[2] Department of Information and Communications, DongGuk University, 3-26
Pil-Dong, Jung-Gu, Seoul, Korea
chlee@dgu.ac.kr

Abstract. In this paper document classification methods using multi-nomial naïve Bayes are improved in a number of ways. We use the value weighting method, a new fine-grained weighting method, to calculate the weights of the feature values. Our experiments show that the proposed approach outperforms other state-of-the-art methods.

Keywords: Multinomial naïve Bayes · Value weighting

1 Introduction

Correctly identifying documents into a particular category is still a challenging task because of large and vast amount of features in a dataset. In this paper we explore the multi-label text classification problem. To improve the performance of multi-label text classification tasks, we investigate the reasons behind the poor performance of multinomial naïve Bayes (MNBs). In text classification problem, multinomial naive Bayes (MNB) algorithm has been most commonly used. MNB classifier is an efficient and reliable text classifier, and many researchers usually regard it as the standard naive Bayes text classifier. However, their performances are not as good as some other learning methods such as support vector machines and boosting. In this paper, we firstly investigate the reasons behind the poor performance of MNB. Then to enhance the performance of MNB method, we propose a new paradigm of weighting method, called value weighing method, for MNB learning. Furthermore we compare the performance of the proposed model, the value weighting method, to that of other state-of-the-art multi-label classifiers.

2 Related Work

Text classifiers based on naive Bayes have been studied extensively in the literature. Especially there have been many researches for using MNB model in text classification.

© Springer International Publishing Switzerland 2015
M. Ali et al. (Eds.): IEA/AIE 2015, LNAI 9101, pp. 488–492, 2015.
DOI: 10.1007/978-3-319-19066-2_47

McCallum and Nigam [1] compares classification performance between multi-variate Bernoulli model and multinomial model. They show that the multi-variate Bernoulli model performs well with small vocabulary sizes, but the multinomial model usually performs even better at larger vocabulary sizes.

Rennie et al. [2] introduced Complement Naive Bayes (CNB) for the skewed training data. CNB estimates parameters using data from all classes except the currently estimated class. Furthermore they demonstrated that MNB can achieve better accuracy by adopting a TFIDF representation, traditionally used in Information Retrieval.

3   Improving the Performance of Multinomial naïve Bayes

In this paper, we assume that documents are generated according to a multinomial event model. Thus a document is represented as a vector $\mathbf{x} = (f_1, \ldots, f_{|V|})$ of word counts where $|V|$ is the vocabulary size and each f_t indicates how often t-th word W_t occurs in $\mathbf{x}$. Given model parameters $p(W_t|c)$ and $p(c)$, assuming independence of the words, the most likely class value c for a document $\mathbf{x}$ is computed as

$$c^*_{MNB}(\mathbf{x}) = argmax_c p(c) \prod_{t=1}^{|V|} p(W_t|c)^{f_t}$$

where $p(W_t|c)$ is the conditional probability that a word W_t may happen in a document $\mathbf{x}$ given the class value c and $p(c)$ is the prior probability that a document with class label c may happen in the document collections.

MNB provides reasonable prediction performance and easy to implement. But it has some unrealistic assumptions that affect overall performance of classification. The first is that all features are equally important in MNB learning. Because this assumption is rarely true in real-world applications, the predictions estimated by MNB are sometimes poor. The second is that the probabilities for word occurrence are independent of document length. Because of this assumption, the parameters are clearly dominated by the word counts that come from long documents. The performance of MNB can be improved by mitigating these assumptions. The following section describes these issues in detail.

3.1   Fine-Grained Weighting Method in Text Classification

Since the assumption that all features are equally important hardly holds true in real world applications, there have been some attempts to relax this assumption in naïve Bayesian algorithm. Feature weighting in the naïve Bayesian approach is one method for easing the independence assumption, and it assigns a continuous value weight to each feature [4]. MNB classification is a special form of feature-weighted naïve Bayesian learning. The MNB classification with feature weighting is represented as follows.

$$c^*_{MNB-FW}(\mathbf{x}) = \arg \max_c \; p(c) \prod_{t=1}^{|V|} p(W_t|c)^{w_t} \qquad (1)$$

where $p(W_{tj}|c)$ is the conditional probability that a word W_t may happen in a document **x** given the class value c, and $p(c)$ is the prior probability that a document with class label c may occur within the document collection. In feature-weighted naïve Bayes, each word W_t has its own weight w_t.

In traditional MNB, the frequency f_t of each word W_t plays the role of the significance of the word. Therefore, the basic assumption in MNB is that when a certain word appears frequently in a document, the importance of the word grows in proportion to its occurrence. Each word is given a weight, which is the frequency of the word in the document.

Kim et al. [5] proposed a feature weighting scheme using information gain. Information gain for a word given a class, which becomes the weight of the word, is calculated as follows:

$$w_t = f_t \sum_c \sum_{W_i \in \{W_t, \overline{W}_t\}} p(c, W_i) log \frac{p(c, W_i)}{p(c)p(W_i)}$$

where $p(c, W_i)$ is the number of documents with the word W_i and class label c divided by the total number of documents, and $p(W_i)$ is the number of documents with the word W_i divided by the total number of documents, respectively.

In this paper, we have thought of a new method in which weights are assigned in a more fine-grained manner where we treat each occurrence of a word differently in terms of its importance. In order to implement fine-grained weighting, we first discretize the term frequencies of each word. The discretization task converts a continuous term frequency f_t to a categorical *word frequency bin* a_{tj}, which represents the j-th discretized value of the term frequency. Instead of assigning a weight to each word feature(e.g. MNB), we assign a weight to each word frequency bin.

In order to implement the fine-grained weighting, we first discretize the term frequencies of each word. The discretization task converts a continuous term frequency f_t to a categorical *word frequency bin* a_{tj} , which represents the j-th discretized value of the term frequency. In other words, instead of assigning a weight to each word feature(e.g. MNB), we assign a weight to each word frequency bin. After that, the weights of these word frequency bin are automatically calculated using training data.

We label this method as *value weighting* method. Unlike with the current feature weighting methods, the value weighting method calculates a weight for each word frequency bin. The value weighting method in MNB can be defined as follows.

$$c^*_{MNB-VW}(\mathbf{x}) = \arg \max_c p(c) \prod_{a_{tj} \in \mathbf{x}} p(a_{tj}|c)^{w_{tj}} \qquad (2)$$

where w_{tj} represents the weight of word frequency bin a_{tj}. You can easily see that each word frequency bin is assigned a different weight.

3.2   Calculating Value Weights of Word Frequency Bins

This section describes the method for calculating weights of frequency bins. We use an information-theoretic method for assigning weights to each word frequency bin. The basic assumption of the value weighting method is that when a certain word frequency bin is observed, it gives a certain amount of information to the target word feature. The more information a word frequency bin provides to the target class, the more important the bin becomes.

In this paper, the amount of information that a certain word frequency bin contains is defined as the difference between a priori probability distribution and a posteriori distribution of class label under the word frequency bin. We employ a Kullback-Leibler (KL) measure [3] in order to calculate the weight of word frequency bin. The KL measure for a word frequency bin a_{tj} is defined as

$$KL(C|a_{tj}) = \sum_c p(c|a_{tj}) \log \frac{p(c|a_{tj})}{p(c)} \tag{3}$$

The formula $KL(C|a_{tj})$ calculates the average mutual information between the events c and a_{tj} with the expectation taken with respect to a posteriori probability distribution of C. It can be used as a proper weighting function, so the value weight for a word frequency bin a_{tj} is defined as $w_{tj} = \frac{1}{Z_t} KL(C|a_{tj})$ like

$$w_{tj} = \frac{1}{Z_t} \sum_c p(c|a_{tj}) log \frac{p(c|a_{tj})}{p(c)}$$

where Z_t is a normalization constant given as $Z_t = \frac{1}{|a_t|} \sum_{j|t} w_{tj}$, and $|a_t|$ represents the number of word frequency bins for word feature t.

4   Experimental Evaluation

In order to measure the effects of the value weighting method, we compare the performance of the value weighting method to MNB. We used four text classification benchmark datasets [6] [7] to conduct our empirical text classification study. All these datasets have been widely used in text classification, and are publicly available. Table 1 provides a brief description of each dataset.

"New3" (TunedIT 2012) dataset contains a collection of news stories and "Ohsumed" (TunedIT 2012) is a dataset of medical articles. "Amazon" (Frank and Asuncion 2010) dataset is derived from the customer reviews in Amazon Commerce Website for authorship identification. "CNAE-9" (Frank and Asuncion 2010) is a dataset of free text business descriptions of Brazilian companies. The continuous features in datasets were discretized using equal distance method with 5 bins.

The continuous features in the datasets were discretized using the equal distance method with five bins. In this experiment, single-label classification was conducted, so the value weighting method (MNB-VW) was applied. The proposed MNB-VW is compared to regular naïve Bayes (NB) and multinomial naïve Bayes (MNB).

Table 1 provides a brief description of each dataset and results of the accuracies of these methods. The numbers with bold letters indicate the best accuracy among NB, MNB, and MNB-VW. MNB-VW shows the best performance

Table 1. Description of the single-label data sets and their accuracies

Dataset	#(Data)	#(Label)	#(Feature)	NB	MNB	MNB-VW
New3	3204	6	13196	0.8620	0.8846	**0.9104**
Ohsumed	1003	10	3183	0.9402	**0.9860**	0.9571
Amazon	1500	50	10000	0.9760	0.9826	**0.9893**
CNAE-9	1080	9	857	0.9602	0.9783	**0.9787**

in three cases, and it always shows better performance than the NB method. These results clearly indicate that assigning weights to each word frequency bin could improve the performance of the classification task of naïve Bayesian for document classification.

5 Conclusions

In the paper, the multinomial naïve Bayes algorithm for document classification is improved in a number of ways. The value weighting method, a new, fine-grained weighting method is proposed for multinomial naïve Bayesian learning. Unlike traditional weighting methods, it assigns a different weight to each feature value. The experimental results show that the value weighting is successful and show better performance in most cases than other existing algorithms. As a result, this work suggests that we could improve the performance of multinomial naïve Bayes in text classification by using value weighting approach.

References

1. McCallum, A., Nigam, K.: A comparison of event models for Naive Bayes text classification. In: Proc. AAAI-1998 Workshop on Learning for Text Categorization. AAAI Press, pp. 41–48 (1998)
2. Rennie, J., Shih, L., Teevan, J., Karger, D.: Tackling the poor assumptions of naive bayes text classifiers. In: Proceedings of the 20th International Conference on Machine Learning (ICML), pp. 616–623 (2003)
3. Kullback, S., Leibler, R.A.: On Information and Sufficiency. The Annals of Mathematical Statistics, 79–86 (1951)
4. Lee, C.H., Gutierrez, F., Dou., D.: Calculating Feature Weights in Naive Bayes with Kullback-Leibler Measure. In: 11th IEEE International Conference on Data Mining, pp. 1146–1151. IEEE Press, Vancouver (2011)
5. Kim, S., Han, K., Rim, H., Myaeng, S.: Some effective techniques for naive bayes text classification. IEEE Transactions on Knowledge and Data Engineering **18**(11), 1457–1466 (2006)
6. Tunedit Data Mining Blog. http://tunedit.org/
7. UCI Machine Learning Repository. http://archive.ics.uci.edu/ml/

Diagnosis of Epilepsy in Patients
Based on the Classification of EEG Signals
Using Fast Fourier Transform

Thao Nguyen Thieu and Hyung-Jeong Yang(✉)

Department of Electronics and Computer Engineer, Chonnam National University,
77 Yongbong-ro, Buk-gu, Gwangju 500-757, Korea
thieuthaonguyen@gmail.com, hjyang@jnu.ac.kr

Abstract. The brain signals of human or animal is recorded from many sensors placed on the scalp, called EEG signals. Based on this signal, many brain diseases which occur in human and animal is simply found and prevented. A popular brain disease is epileptic seizure. Nowadays, many scientists use the different methods to recognize abnormal activities of the brain functionality, thence diagnosis of epilepsy is easier. In this paper, we propose a way to detect seizure in human. Fast Fourier transform is used to convert the EEGs signals into the simpler form, remove some noises and get better features.

Keywords: EEG · Fast fourier transform · Neural network · Epilepsy · Seizure

1 Introduction

Electroencephalography (EEG) is developed as the method to support the diagnosis and treatment of the abnormal symptoms in human. EEG is sampled from many electrode placed in the scalps to record of electrical activity along the scalp. An EEG signals is a measurement of currents that flow during synaptic excitation of the dendrites of many pyramidal neurons in the cerebral cortex [1]. EEG signals is always used to diagnose many brain disorders. Commonly, the signals of the brain of a diseased man and a healthy man are different, it is discriminated based on the difference of the brain waves, especially the frequency ranges. These frequency bands from low to high frequencies respectively are called alpha, theta, beta, delta and gamma [1].

EEG signals is usually applied into the healthy fields to detect the diseases concerned brain, especially epileptic seizures. Developing the human EEG signals is important application in the field of epilepsy. The detection of epileptic seizures using different techniques has been successful, particularly for the detection of adult seizures. Epilepsy is a disorder of the normal brain function, making some differences in activities of brain. It is many seizures occur frequently, repeated spontaneous, unpredictable and uncontrollable. According to statistics, epileptic seizures affects about 1% of the population [2]

Nowadays, the scientists are constantly trying to find more effective methods to control this diseases based on the human EEGs. EEG remains the most useful way for

© Springer International Publishing Switzerland 2015
M. Ali et al. (Eds.): IEA/AIE 2015, LNAI 9101, pp. 493–500, 2015.
DOI: 10.1007/978-3-319-19066-2_48

the study of epilepsy [1]. In order for a responsive neurostimulation device to success-fully stop seizures, a seizure must be detected and electrical stimulation applied as early as possible. The method proposed in this paper will be used for detecting sei-zures in human and to reduce the diseased ability in human.

Fourier transform is a linear operator, it is useful for analyzing nonstationary sig-nals by expressing frequency as a rate of change in phase, so that the frequency can vary with time domain [3]. The benefits of the signal processing is clearly confirmed. It can also be applied in many different types with special effects, especially in scien-tific fields, not only in a subject. With the increasing development, its methods and application ability attracted a lot of engineers, physicists and mathematicians re-searched into [4].

After using fast Fourier transform (FFT) to remove phase information and taking frequency domain, Neural Network is applied to classify the seizure data and non-seizure data. Naïve Bayes and Logistic Regression are also investigated for compari-son.

2   Proposed Method

2.1   Fast Fourier Transform

Fourier transform was named the French mathematician, Joseph Fourier, is an integral transforms used to develop a function based on the basic sine function [5]. Fourier transform has many scientific applications, such as in physics, arithmetic, signal processing, probability, statistics, oceanography, geometry and many other areas. In signal processing and related industries, the Fourier transform is often thought as the switching signal into component amplitude and frequency. The widespread applica-tion of FT derived from the useful properties of this transformation. It usually uses for non-periodic signals. Typically, Fourier transform is attached to the continuous Fouri-er transform and also called Continuous-time Fourier transform (CTFT).

Given a signal $x(t)$, the Fourier transform $X(\omega)$ of $x(t)$ is defined to be the fre-quency function

$$X(\omega) = \int_{-\infty}^{\infty} x(t) \, e^{-j\omega t} dt, \qquad -\infty < \omega < \infty \qquad (1)$$

where ω is the continuous frequency variable.

In mathematics, the Discrete Fourier transform (DFT) is used in Fourier analysis for the discrete-time systems, it is also called Discrete time Fourier transform (DTFT). The input of this method is a finite sequence of real numbers or complex numbers, it is a good ideal for handling information on the computer. In particular, this method is widely used in signal processing to analyze the frequencies contained in a signal, to discard the phase information in this signal and to do as convolution operations. Fast Fourier transform (FFT) is an efficient algorithm to compute the DFT.

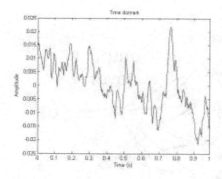 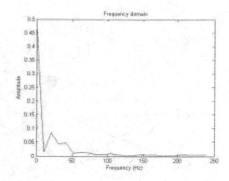

Fig. 1. The time domain and frequency domain of a seizure EEG segment using FFT

$$X(\Omega) = \sum_{n=-\infty}^{\infty} x[n] \, e^{-jn\Omega} \tag{2}$$

Ω in this case is called D-T frequency of the signals.

Based on the foundation of DFT, FFT is proposed as a good algorithm to analyze the signal data. There are many FFT methods. In this paper, we use the FFT function in Matlab function. Figure 1 shows the plot of the first second of the seizure data and its graph in frequency domain after applying FFT to convert.

2.2   Classification : Neural Network (NN)

Neural Network (NN), or Artificial neural network (ANN), was proposed to try to simulate intelligence activities of human. Since its inception, the neuron network has rapidly developed in many fields of identification, classification, noise reduction, predictions…[5];

ANN are regeneration of the functions of the human nervous system with a multitude of neural associated with each other through network. Like as human brains, ANN are learned by experiences, save these experience and use it in appropriated situations.

NN is the organization of computing units and the types of connections permitted. It consists of three components: Input layer, Hidden layer and Output layer. In particular, the hidden layer consists of neurons, receiving the input data from the previous neuron layers and convert the input data to the next layer based on computing the values of the network parameters (weights) [6][7]. A Neural network model can have a lot of hidden layers.

Each output element of the ANN is computed as follows:

$$o = f\left(\left(\sum_{i=1}^{N} x_i w_i\right) - t\right) \tag{3}$$

This models computes the weight sum of its input based on the weight w_i , compares with threshold t and passes the results to a non-linear function f.

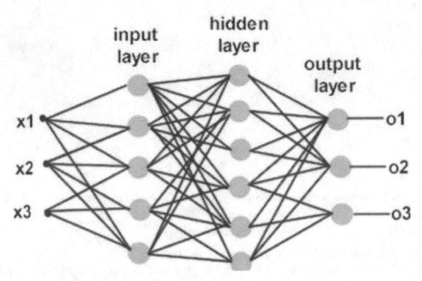

Fig. 2. An example of the multilayer network

3   Experimental Results and Discussion

3.1   Data Acquisition

The dataset is recorded from eight American patients with temporal and extra-temporal lobe epilepsy undergoing evaluation for epilepsy surgery. The EEG recordings are from depth electrodes implanted along anterior-posterior axis of hippo-campus, and from subdural electrode grids in various locations. Data sampling rates vary from 500 Hz to 5,000 Hz. The number of channels used for each patients is also not same.

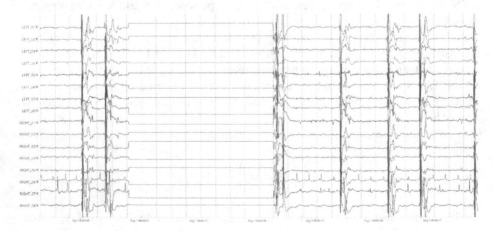

Fig. 3. An example of the original data

The data is separated into many segments based on the recorded time. Each segment corresponds with data signal part sampled in one second. The data signals used in this paper are arranged sequentially, the seizure segments are first arranged, followed by the non-seizure segments. Figure 3 presents the data signal of a subject sampled at 500Hz with 16 channels in the first 30 seconds. It is plotted by the International Epilepsy Electrophysiology Portal's tool [17].

Tables 1 and 2 respectively show the number of channels of eight patients and the quantity of seizure and non-seizure segments in each patient. The number of non-seizure segments is always more than the seizure segment.

Table 1. The number of electrodes placed in the scalp of each patient and the data's sampling rates

	Channels
Patient 1	68
Patient 2	16
Patient 3	55
Patient 4	72
Patient 5	64
Patient 6	30
Patient 7	36
Patient 8	16

Table 2. The number of electrodes placed in the scalp of each patient and the data's sampling rates

	Seizure segments	Non-seizure segments
Patient 1	70	104
Patient 2	151	2990
Patient 3	327	714
Patient 4	20	190
Patient 5	135	2610
Patient 6	225	2772
Patient 7	282	3239
Patient 8	180	1710

This dataset is gotten freely at the International Epilepsy Electrophysiology portal [17] and was developed by the University of Pennsylvania and the Mayo Clinic.

3.2 Experiment

For data preprocessing, Fast Fourier transform is applied to each segment, taking frequency magnitudes in the range 1-47Hz and removing noise information.

Actually, the signal of this data is a discrete variable type. Therefore, applying FFT into this dataset is necessary and judicious. Fast Fourier transform is applied to each segment of whole data in turn. Normalizing them after converting is quite necessary.

In data mining, selecting good features or reducing the number of feature is also a problem to solve. The good features will give better results and they must be fitting to a classifier. To get the features for classification, the original data must be transformed from data space into another space, the feature space. The simplest way to find the features of this dataset is to calculate its eigenvalues. The eigenvalues is computed from the covariance matrix which received from the normalized matrix. We used these eigenvalues as the new features to become the inputs of Neural Network model. We also present the work process in figure 4.

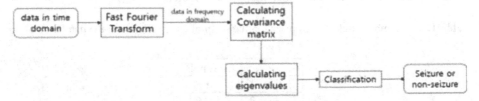

Fig. 4. The flowchart of our process

After classification, the accuracy is calculated by following formula [19]:

$$ACC = \frac{\#PredictedLabels == \#TestLabels}{\#Total}$$

In this paper, we got a haft of seizure segments and about 75 percent of non-seizure segments to train while the others are to use as a test part. Afterward, we used Neural Network for classification. We also used Naïve Bayes and Logistic Regression for comparison. The results are presented in table 3 for three classification methods. The best way to detect seizure is to use Neural Network to classify, the average accuracy in this case is more than 88 percent. Some of subjects get the high accuracies, while the others get very low accuracies. As the first subject, it always receives bad results because its signal is not clear, retains many noise. Furthermore, the number of recorded samples is also less, not enough to classify exactly.

Table 3. The accuracy after classifying seizure and non-seizure segments of 8 patients

Subjects	1	2	3	4	5	6	7	8
Neural Network (%)	78.79	98.02	80.03	91.73	82.46	95.03	88.76	89.8
Naïve Bayes (%)	74.37	97.7	83.37	88.79	76.78	91.61	91.01	87.16
Logistic Regression (%)	69.03	97.3	83.38	90.46	77.87	92.49	91.99	89.47

4   Conclusion

Using Fast Fourier Transform, the signal will be converted from time domain into another domain, frequency domain. In this domain, the data signal become simpler to analyze and handle. EEG signals usually contain many noise because of the recorded condition, recorded environment, including the patients. Thus, application of FFT to remove some noises gets better results. The detection seizure or non-seizure of a patient or a canine become easier.

Furthermore, the combination FFT and NN makes the classification process more advantageously. The seizure diagnose sometimes incorrect or in accurate because of several issues such as the complexity of the input data, the value of options as deviation, error threshold, the number of neurons per layer not selected correctly

Acknowledgment. This research was supported by Basic Science Research Program through the National Research Foundation of Korea(NRF) funded by the Ministry of Education, Science and Technology(2014-054530) and was also supported by Basic Science Research Program through the National Research Foundation of Korea(NRF) funded by the Ministry of Education, Science and Technology(2014-046797).

References

1. Sanei, S., Chambers, J.A.: EEG signal Processing. Centre of Digital Signal Processing, UK (2007)
2. World Health Organization. http://www.who.int/mediacentre/factsheets/fs999/en/
3. Kamen, E.W., Heck, B.S.: Fundamentals of Signals and Systems using the web and mat lab, 3rd edn. Prentice-Hall, Inc. (2000)
4. Quiroga, R.Q.: Quantitative analysis of EEG signals: Time-frequency methods and Chaos theory, Institute of Physiology - Medical University Lubeck and Institute of Signal Processing - Medical University Lubeck (1998)
5. Wikipedia. http://en.wikipedia.org/wiki
6. Johnson, R.A., Wichern, D.W.: Applied Multivariate Statistical Analysis, Pearson International Edition, the Sixth Edition
7. Christos Stergiou and Dimitrios Siganos: Neural Network. http://www.doc.ic.ac.uk/~nd/surprise_96/journal/vol4/cs11/report.html
8. Aleksander, I., Morton, H.: An introduction to neural computing, 2nd edition
9. Lachaux, J.P., Rudrauf, D., Kahane, P.: Intracranial EEG and human brain mapping. Journal of Physiology – Paris **97**, 613–628 (2003)
10. Shoeb, A., Guttag, J.: Application of Machine Learning To Epilepsy Seizure. Massachussettes Institute of Technology, 02139 (2010)
11. Kharbouch, A.A.: Automatic Detection of Epileptic Seizure Onset and Termination using Intracranial EEG. Massachussettes Institute of technology, June 2012
12. Iasemidis, L.D.: Epileptic seizure prediction and control. IEEE Trans. Biomed. Eng. **5**, 549–558 (2003)
13. Bruns, A.: Fourier-, Hilbert- and wavelet-based signal analysis: are they really different approaches? J. Neurosci. Meth. **137**, 321–332 (2004)
14. Adeli, H., Zhou, Z., Dadmehr, N.: Analysis of EEG records in an epileptic patient using wavelet transform. Journal of Neuroscience Methods **123**, 69–87 (2003)
15. Meenakshi, Dr. R.K Singh, Prof. A.K Singh: Frequency Analysis of Healthy & Epileptic Seizure in EEG using Fast Fourier Transform, International Journal of Engineering Research and General Science **2**(4), ISSN 2091 -2730, June-July 2014

16. Subasi, A., Ercelebi, E.: Classification of EEG signals using neural network and logistic regression. Computer Methods and Programs in Biomedicine **78**, 87–99 (2005)
17. The International Epilepsy Electrophysiology portal. https://www.ieeg.org/
18. Sasse, H.G., Duffy, A.P.: Numerical Noise Reduction in the Fourier Transform Component of Feature Selective Validation. Progress in Electromagnetics research symposium Proceedings, Morocco, March 2011
19. Nguyen, H.Q., Yang, H.J., Thieu, T.N.: Feature Extraction from Covariance by Using Kernel Method for Classifying Polysomnographys Data. 9th ICUIMC 2015, Bali, Indonesia, January 2015

Unsupervised Learning

A New Similarity Measure by Combining Formal Concept Analysis and Clustering for Case-Based Reasoning

Mohsen Asghari[✉] and Somayeh Alizadeh

Industrial Engineering Department, Khaje Nasir Toosi University, Tehran, Iran
mohsen.asghari@gmail.com, S_alizadeh@kntu.ac.ir

Abstract. This paper represents a new similarity measure by combining clustering, and Formal concept analysis (FCA). The novelties of this research are calculating the importance of features by clustering methods, and utilizing FCA to improve the accuracy of retrieving similar cases. Also, using the FCA helps us, manage the case-base structure. Finally, after performing several experiments on the UCI datasets with cross validation, our new similarity measure improve the accuracy of classification CBR significantly when compare to the other six measures proposed before.

Keywords: Formal concept analysis · Case based reasoning · Clustering · Similarity measure

1 Introduction

A Case-Based Reasoning (CBR) approach relies on the concept that similar problems have similar solutions [19]. Facing a new problem, a cased-based system retrieves similar cases stored in a case base and adapts them to fit the problem prepared. Also, case representation in a knowledge-based system and drawing implication between the cases in CBR system can be done by Formal Concept Analysis [20].

Cases in CBR system consist of many features that define the cases, and the importance of each feature also is considerable; therefore, it is significant to find out a solution to calculate the importance of each feature. In addition to formal concept analysis, Clustering methods are helpful in this research. These methods analysis the data based on their statistical behavior. Clustering methods are useful for multivariate statistical analysis and unsupervised classification analysis [21, 22]. We believe that clustering can help us to find out the importance of features.

In this paper CBR, Clustering techniques and FCA are utilized together to improve the retrieval process, which is usually regarded as the most important step in CBR cycle. In essence, a good assessing similarity between cases is a key to success in this approach. In the meantime, the retrieval process and knowledge base system construction must be designed to accord with each other. Hence, a new similarity measure based on the feature weighting and supporting retrieval process from FCA is proposed.

© Springer International Publishing Switzerland 2015
M. Ali et al. (Eds.): IEA/AIE 2015, LNAI 9101, pp. 503–513, 2015.
DOI: 10.1007/978-3-319-19066-2_49

1.1   FCA – Formal Concept Analysis

Formal concept analysis (FCA), invented by Radulf Wille [2], provides a conceptual framework for structuring, analyzing and visualizing data, in order to make them more understandable [1]. The aim and meaning of formal concept analysis as a mathematical theory of concept and concept hierarchies is to support rational communication of human by mathematically developing appropriate conceptual structure, which can be logically activated [2].

A formal context is a triple of sets (G, M, I), where G is called a set of objects; M is a set of attributes, and $I \subseteq G \times M$. The notation $g I m$ indicates that $(g, m) \in I$ and denotes the fact that the object g possesses the attribute m. The set of all formal concepts of (G, M, I) can be ordered in this way denoted by $\mathfrak{B}(G, M, I)$ and is called the concept lattice of the formal concept (G, M, I). Concept lattice can be considered as a new structure for knowledge-based system in CBR. To extract knowledge for solving the new problem, the Basic theorem on concept lattice [15] is used. This theorem provides implication among attributes that are used to identify solution in our work.

1.2   Case-Based Reasoning and FCA

CBR is quite different from any other artificial intelligence (AI) problem solving methodology in that it searches for the most specific case or set of cases reused. Most other AI problem solvers, such as Bayesian networks, artificial neural networks, or decision trees, have a known tendency to overgeneralize, thus they cannot shape their behavior on the most specific instance in their training set. Therefore, it can be expected that CBR can achieve excellent accuracy provided that it capitalizes on a significant and broad-spanning memory of cases. One of the main assets of CBR is its eagerness to learn. Learning in CBR can be as simple as memorizing a new case or can entail refining the memory organization or meta-learning schemes.

The reasoning cycle in CBR system involves [19]: (1)accepting a new problem description (a case without solution and feedback); (2) retrieving relevant cases from a case-base (solve problem with similar input); (3) adapting retrieve cases to fit the problem at hand and producing the solution for it; (4) evaluating the solution[4].

The structure of knowledge-based system that directly supports four steps above, and similarity measure that retrieve the relevant cases to propose a new solution are two important tasks in CBR systems. The first task that is related to the structure of knowledge-based system in CBR has a great effect on the performance of CBR. Formal concept analysis (FCA) can elicit knowledge embedded in previous cases to solve new problems. Implication drawn from FCA can suggest solutions from dependency inside the knowledge based system. Thus, we apply FCA to build knowledge-based system for CBR. The second task is to measure how relevant is a case to guarantee retrieving only highly relevant cases. Considering the CBR cycle, one can say that the more similar the cases are the less adaption is necessary, and consequently, the proposed solution may be more probably correct. In many processes, it's better to retrieve fewer cases, or none, than to retrieve less useful cases that would result in a poor solution. Therefore these are our motivations to propose a new model to improve the quality of CBR system.

1.3   Case Based-Reasoning and Clustering

Case selection and retrieval is usually regarded as the most important step within the case-based reasoning. This problem has been studied in both supervised and unsupervised framework by many researches and practitioners, and hundreds of different algorithms and various approaches have been developed to retrieve similar cases from the case-base [26]. In the process of case matching and retrieval, the searching space in the entire case base, not only makes the task costly and inefficient, but also sometimes leads to poor performance. To address such a problem, many classifications and clustering methods are applied before selecting the most similar case or cases. After the cases are partitioned into several sub-clusters, the task of case matching and retrieval boils down to matching the new case with one of the several sub clusters, and finally, the desired number of similar cases can be obtained.[25] Thus, various classification/clustering algorithms, such as K-means, C-means, and Hierarchical clustering, play an important role in this process.

Clustering is a type of multivariate statistical analysis also known as cluster analysis, or unsupervised classification analysis. The main objective of clustering is to find similarities between samples, and then group similar samples together to assist in understanding relationships that might exist among them. More formally, the object of clustering is to group data points into subsets (clusters) in such a way that the objects within each cluster are more related to one another than are objects within another subset. Thus, the notion of similarity is essential to clustering [24].

In 2008 Tzung-Pei Hong and Yan-Liang Liou [8] used Feature Clustering with Case base reasoning. Their proposed method followed two advantages:

1. Indexing and representing case by representative attributes may reduce the complexity of case retrieval and matchmaking.
2. Selecting feature by attribute clustering preserves the relation among features.

All these advantages make the CBR framework more flexible.

As we saw in section 1.2, FCA can be used as case representation and similarity measure. This novelty is proposed by Tadrat and others [4]. Although the result of their work shows that they have 100% accuracy on hierarchical data, their solution did not work well on non hierarchical data. To cover the hierarchical and non-hierarchical data structure we believed that Clustering methods could be helpful.

2   Similarity, Case-Based Reasoning and FCA

Case-based reasoning (CBR) is a relatively recent problem solving technique that is attracting increasing attention [5]; therefore, various researches have been developed on CBR. The work by Schank and Abelson in 1977 is considered to be the original work on Case-based reasoning. They proposed that our knowledge about situation is recorded as scripts that allow us to set up expectation and perform inference [5]. Afterward; Aomadt and plaza 1994 have described a CBR typically as a cyclical process comprising the four REs: 1. RETREVE 2.REUSE 3.REVISE 3.RETAIN. Sankar K.Pal discussed the use of fuzzy classification and clustering of cases. He explained that the process of case matching or in other words retrieving, that is the first step of CBR based on Aomadt model, encounters vast search area problem, which makes it costly and inefficient, also sometimes leads to poor performance. Therefore, to address such a problem, he proposed that using classification and clustering should be applied before selecting the relevant case or cases. He states that two solutions named Intracluster similarity and Intercluster similarity [24].

Our paper combines the methods of clustering, that defines the frequency of features based on the frequency of each attribute on individual clusters, and methods of FCA, which are utilized to optimize of knowledge representation.

We are going to define a formula to optimize the similarity measure in CBR retrieval. These methods are applied to CBR system separately, Tzung and Yan 2008 utilized feature clustering to CBR system [8]. Belen & Pedro 2001; pattaraintakorn et al 2008; Tadra, Boonjing & Pattaraintakorn 2008, 2012[6, 7, 4]) used FCA for problem solving CBR system. We proposed a knowledge base model based on FCA and Clustering methods. Nevertheless, optimal similarity measure was computed separately based on the clustering methods. It will be more efficient to compute directly from FCA knowledge based system.

Several researchers have developed similarity measure to retrieve formal concept. Lengenink (2001) [11] defined similarity measures to find similar and relevant concepts: Local similarity and Global similarity as follows:

$$S_l\big((A,B),(C,D)\big) = \frac{1}{2}\left(\frac{|A \cap C|}{|A \cup C|} + \frac{|B \cap D|}{|B \cup D|}\right) \tag{1}$$

$$S_g\big((A,B),(C,D)\big) = \frac{1}{2}\left(\frac{|A \cap C|}{|G|} + \frac{|B \cap D|}{|M|}\right) \tag{2}$$

Saquer and Deugan (2001) proposed similarity measure relying on similarity between approximated sets. Suppose that A is a set of objects and B is a set of features. The objective is to find a formal concept C as the extent of C similar to A and the intent of C similar to B, as possible. Therefore, they define a similarity measure $F_c(A,B)$ that indicates how well the formal concept C approximates the pair (A, B) as follows:

$$F_c(A,B) = \frac{1}{2}\left(\frac{|A \cap Extent(C)|}{|A \cup Extent(C)|} + \frac{|B \cap Intent(C)|}{|B \cup Intent(C)|}\right) \tag{3}$$

The expression $|A \cap Extent(C)|/|A \cup Extent(C)|$ indicates how similar A is to Extents(C) and the expression $|B \cap Intent(C)|/|B \cup Intent(C)|$ indicates how similar B is to Intent(C). It is also easy to see that the range of $F_c(A,B)$ is interval (0,1).[12] Formica (2006,2008) proposed a similarity measure for FCA based on the similarity of the intents (the set of attributes) which has been addressed according to information content approach, rather than similarity graph, whose defining requires human domain expertise [1,9]. Dau, Ducrou, and Ekludan (2008) combined the local similarity and global similarity measure to develop a new measure [10]. Alqadah (2010) proposed three similarity measures based on existing set-based measures [13]. For any two sets of intention in formal concept x and y, Jaccard index (SJac), Sorenese coefficient (SSor) and symmetric difference (Sxor) [13].

Tadra, Boonjing & Pattaraintakorn (2012) proposed a framework that constructs knowledge base system in CBR, based on rough sets and FCA. Also they computed a similarity measure separately by employing vector model idea. Their concept similarity measure is defined as [4].

$$sim(C_p, C_N) = \frac{1}{2}\left(\frac{\sum_{u \in I_n \cap I_p}\left(\log\frac{N}{Fa_u}\right)^2}{\left[\sum_{i \in I_n}\left(\log\frac{N}{Fa_i}\right)^2 \sum_{j \in I_p}\left(\log\frac{N}{Fa_j}\right)^2\right]^{1/2}} + \frac{\sum_{v \in E_n \cap E_p}\left(\log\frac{N}{Fc_v}\right)^2}{\left[\sum_{k \in E_n}\left(\log\frac{N}{Fc_k}\right)^2 \sum_{l \in E_p}\left(\log\frac{N}{Fc_l}\right)^2\right]^{1/2}}\right) \quad (4)$$

Where N is a total of formal concepts, Fa_u, Fa_j and Fa_i are frequency of attributes u, j and i respectively $\{u, i, j\} \in M$ and Fc_k, Fc_l and Fc_v are frequencies of v, k cases and l, respectively $\{v, k, l\} \in G$.

3   Proposed Similarity Measure in CBR

Through all studies done in this research, we catch two points. First, weights are selected based on user's requirement or by matching user's query and previous cases are binary relation form. Alternatively the weight of attribute is determined directly from data. Second, in the entire studies researcher tried to compare the concepts with each other in spite of the fact that the main parts of the cases are the attributes, which define the problem. Thus, we proposed a similarity measure based on the vector model and feature weighting which is computed by clustering methods.

Case retrieval in concept lattice can be done by two distinct methods: lattice traversal and similarity measure. [4] The objective is to use similarity measure that exploit the vector model that is a classical model of information retrieval [15] and feature weighting to invent new similarity measure that is described below:

$$Sim(C_P, C_N) = \frac{\sum_{u \in I_n \cap I_p}\left[\left(\log\frac{N}{Fa_u}\right)^2 * Wa_u\right]}{\left[\sum_{i \in I_n}\left[\left(\log\frac{N}{Fa_i}\right)^2 * Wa_i\right]\sum_{j \in I_p}\left[\left(\log\frac{N}{Fa_j}\right)^2 * Wa_j\right]\right]^{1/2}} \quad (5)$$

Old Cases are stored in case-base as sets of concepts, so a new case must be compare with old concepts. Cn is a new case and Cp is an old concept that stored in case base. A new case defined as $Cn := (I_n)$ and I_n is a set of new case descriptions provided by the user. The old concept defined as $Cp := (E_p, I_p)$ and E_p is a set of previous cases that have similar problem description and solution while I_p comprises of all concept description and solution shared by all those cases. N is a total number of concepts. Fa_u, Fa_j and Fa_i are a frequency of attribute u, j and I. Wa_u, Wa_i and Wa_j are weights of attributes that calculated based on the frequency of attributes in each cluster.

3.1   Frequency Assignment for Features

After clustering the data, by one of the clustering methods such as K-means, we calculated the frequency of attribute in each cluster by the formula that is defined as follow:

$$\left(Wa_y\right)_{cluster} = \frac{(F_{data_y})_{cluster}}{I_{cluster}} \tag{6}$$

Where F_{data_y} is a frequency of attribute y in a cluster and I is a total number of attributes on that cluster.

The weight of each attribute in the case-base is defined as follows:

$$Wa_y = \sum_{i=1}^{cluster} \left(Wa_y\right)_i \tag{7}$$

4   Problem Solving in CBR by Proposed Similarity Measure

In this section, we examine the new similarity measure, in order to enhance the retrieval process of CBR. Car Data[1] is selected to make the procedure of proposed similarity measure more transparent. This dataset has evaluated the cars by six attributes: Buying, Maint, doors, Persons, Lug_boot, safety.

Table 1. description of the data set

Field name	Values
Buying	Vhigh, high, med, low
Maint	Vhigh, high, med, low.
Doors	2, 3, 4, 5more.
Persons	2, 4, more.
Lug_boot	Small, med, big.
Safety	Low, med, high.

4.1   Pre Processing

Because all datasets are mixed-type: consist of both discrete and continues variables. And many datasets are incomplete: contain Null values. The datasets must be normalized in order to utilize by FCA. In this paper combination of data-mining and FCA concepts are used together. Although data-mining algorithm can utilize the ordinary type of data such as discrete or continues variable, the FCA algorithm such as In-Close algorithm (The explanation of this algorithm can be found in the paper written by Simon Andro[17]) use the Boolean matrix to extract the hidden concepts from a context. Therefore, the data must be converted to a Boolean matrix.

---

[1] Standard UCI dataset.

4.2   Clustering and FCA Model

In some cases, dataset have been divided to specific classes that can be used as cluster label for each record in dataset. However, Clustering techniques apply when there is no class to be predicted, but the instances are to be divided into natural groups [16].

Selected dataset involves classification of four types of cars: Vgood, good, acc, Unacc. Each class label defined as a cluster-name that can be used in our algorithm. Table 2 shows details of six records of 1728 number of instances.

Table 2. 10 records of data set that we choose for an example

No.	Buying	Maint	Doors	Persons	Lug_boot	Safty	Cluster Name	Class Lable
1	high	Vhigh	2	2	big	Low	cluster-1	Unacc
2	high	Med	4	4	big	Med	cluster-2	Acc
3	med	High	3	more	med	Low	cluster-1	unacc
4	Low	Vhigh	4	more	med	Med	cluster-2	Acc
5	Low	Low	4	4	med	High	cluster-4	Vgood
6	Low	Low	4	more	med	Med	cluster-3	good

Next step is to generate the hidden Concepts in the dataset. We selected In-Close algorithm in order to generate the concepts. This algorithm uses incremental closure and matrix searching by computing all formal concepts in a formal context [17]. But, before using that algorithm, the dataset must be changed in to a Boolean-matrix format. For example Buying is one of the attributes that shows the buying status with three values: low, med, high. To change the format of this attribute we have used three new attributes: Buying-low, Buying-med, Buying-high. Each new attributes just have two values: true – false. The other attributes that they did not have Boolean format must be changed. After that, concepts will be generated by In-Close algorithm.

Each generated concepts have sets of intents (attributes) and extents (cases).Next step is to calculate the Weights of intents based on two parameters, first frequency of intent in each cluster, second frequency of intent in the whole dataset.

For example we compute the weight of attribute named "safety-mid". Respectively, $y = safety - med$ and the number of intents which describe cluster_1 is equal to $15 (I_{cluster\_1} = 15)$. The set of intents for cluster_1 are as follows:

$\{class - unacc, \quad safety - low, \quad safety - mid, \quad lugboot - big,$
$ugboot - med, \quad person - 2, \quad person - mid,$
$door - 2, \quad door - 3, \quad maint - vhigh, \quad maint - high,$
$maint - med, \quad buying - high, \quad buying - mid,$
$buying - low\}$

Cluster_2 is equal to 12 ($I_{cluster\_2} = 12$), cluster_3 is equal to 10($I_{cluster\_3} = 10$) and cluster_4 is equal to 10($I_{cluster\_4} = 10$).

Frequency of y in each cluster is equal to$(F_{data_y})_{cluster\_1} = 1$,$(F_{data_y})_{cluster\_2} = 3$, $(F_{data_y})_{cluster\_3} = 0$ and $(F_{data_y})_{cluster\_4} = 2$

Based on the equation that we defined in section 3.1 then we calculated the weight of y in each cluster $(Wa_y)_{cluster\_1} = 1/15$, $(Wa_y)_{cluster\_2} = 3/12$, $(Wa_y)_{cluster\_3} = 0/10$ and $(Wa_y)_{cluster\_4} = 2/10$, consequently the weight of this attribute is equal to 0.05 (Wa_y=0.05).

Table 3. frequency and weight of some intents

Intent	Frequency in Concept	Weight
Safety- low	3	0.033
Lug boot –big	9	0.025
Person-2	3	0.033
Door-2	3	0.033
maint -v high	3	0.021
buying – high	3	0.021
safety –med	18	0.05
person -4	6	0.025
door -4	21	0.025
maint – med	6	0.025
Lug boot – med	19	0.025
person –med	14	0.05
door -3	3	0.025
maint –high	3	0.021
buying – med	6	0.025
safety –high	3	0.05
maint –low	6	0.05
Buying- low	13	0.025

At this point we encounter a new case (see below):

I_N = {*buying-high, door-2, person-2, lugboot-med, safety-med, maint-high*}

The aim of this method of CBR is to retrieve the similar cases by similarity measure defined in section 3. Table 4 shows the similarity value between new case and the concepts that generated by our similarity measurement in order to find the relevant cases with the new case. The number of concepts that are utilized in equation was 47, which is generated by In-Close Algorithm.

$$\sum_{i \in I_n} \left[\left(\log \frac{N}{Fa_i} \right)^2 * Wa_i \right]$$

$$= \left(\log \frac{47}{3} \right)^2 * 0.02 + \left(\log \frac{47}{3} \right)^2 * 0.033 + \left(\log \frac{47}{3} \right)^2 * 0.033$$

$$+ \left(\log \frac{47}{19} \right)^2 * 0.025 + \left(\log \frac{47}{18} \right)^2 * 0.05 + \left(\log \frac{47}{3} \right)^2 * 0.02$$

$$= 0.167.$$

As seen, a formal concept in table 4 is retrieved with *similarity equation*. Selected concept shows the new case is similar to {Case 1, Case 8} and the attributes of that concept is { **class_unacc,person_2,door_2,lugboot_big** }. Thus, a solution of new unseen case, I_n, is class **class_unacc**.

Table 4. concept similarity measure obtained from figure 2

Concepts:	$\sum_{u \in I_n \cap I_p}\left[\left(\log \frac{N}{Fa_u}\right)^2 * Wa_u\right]$	$\sum_{j \in I_p}\left[\left(\log \frac{N}{Fa_j}\right)^2 * Wa_j\right]$	simi- larity
[2,5,6,7,8,10] > [saftey_med]	0.009	0.0087	0.227
⋮	⋮	⋮	⋮
[1,8]>[class_unacc,person_2,door_2,lugboot_big]	0.095	0.148	0.605
[1]>[saftey_low,maint_vhigh,buying_high,class_u nacc,person_2,door_2,lugboot_bing]	0.125	0.2551	0.604
⋮	⋮	⋮	⋮
[5,9,10] > [maint_low]	0	0.03995703	0

5   Evaluation of the Method

We used two bench mark data set from UCI repository: Balance scaling data set, and Hayes-Roth.   The types of these data sets are non-hierarchical. At the end we run the cross validation to calculate the accuracy of the methods. Therefore Fig3-4 shows the comparison of obtained accuracy for two data sets. Unlike the proposed similarity measure by Terada et.al we improve the accuracy of classification and problem solving in non-hierarchical data sets.

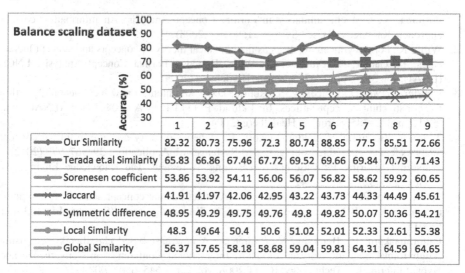

Fig. 1. Comparison of our similarity and the other one for Balance scaling dataset

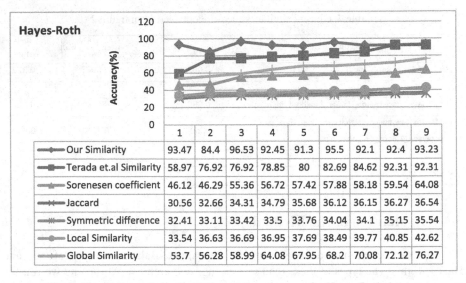

Hayes-Roth	1	2	3	4	5	6	7	8	9
Our Similarity	93.47	84.4	96.53	92.45	91.3	95.5	92.1	92.4	93.23
Terada et.al Similarity	58.97	76.92	76.92	78.85	80	82.69	84.62	92.31	92.31
Sorenesen coefficient	46.12	46.29	55.36	56.72	57.42	57.88	58.18	59.54	64.08
Jaccard	30.56	32.66	34.31	34.79	35.68	36.12	36.15	36.27	36.54
Symmetric difference	32.41	33.11	33.42	33.5	33.76	34.04	34.1	35.15	35.54
Local Similarity	33.54	36.63	36.69	36.95	37.69	38.49	39.77	40.85	42.62
Global Similarity	53.7	56.28	58.99	64.08	67.95	68.2	70.08	72.12	76.27

Fig. 2. Comparison of our similarity and the other one for Hayes-Roth data set

References

1. Formica, A.: Concept similarity in Formal Concept Analysis: An information content approach. Knowledge-Based System **21**, 80–87 (2008)
2. Wille, R.: Formal concept analysis as mathematical theory of concepts and concept hierarchies. In: Ganter, B., Stumme, G., Wille, R. (eds.) Formal Concept Analysis. LNCS (LNAI), vol. 3626, pp. 1–33. Springer, Heidelberg (2005)
3. van der Merwe, D., Obiedkov, S., Kourie, D.G.: Addintent: a new incremental algorithm for constructing concept lattices. In: Eklund, P. (ed.) ICFCA 2004. LNCS (LNAI), vol. 2961, pp. 372–385. Springer, Heidelberg (2004)
4. Tadrat, J., Boonjing, V., Pattaraintakorn, P.: A new similarity measure in formal concept analysis for case-based reasoning. Expert Systems with Applications **39**, 967–972 (2012)
5. Watson, I., Marira, F.: Case-based reasoning: A review. The Knowledge Engineering Review **9**(4), 327–354 (1994)
6. Dõâaz-agudo, B., Gonzaâlez-calero, P.A.: Abstract Formal concept analysis as a support technique for CBR. International Journal in Knowledge-based Systems **14**, 163–171 (2001)
7. Pattaraintakorn, P., Boonjing, V., Tadrat, J.: A new case-based classifier system using rough formal concept analysis. In: Third International Conference on Convergence and Hybrid Information Technology, ICCIT 2008, vol. 2, pp. 645–650 (2008)
8. Hong, T.-P., Liou Y.-L.: Case-based reasoning with feature clustering. Paper Presented at the 7th IEEE International Conference on Cognitive Informatics, ICCI 2008, Stanford, CA, August 14–16 2008
9. Formica, A.: Ontology-based concept similarity in Formal Concept Analysis. Information Sciences **176**(18), 2624–2641 (2006)
10. Eklund, P., Ducrou, J.: Navigation and Annotation with Formal Concept Analysis. Lecture Notes in Computer Science **5465**, 118–121 (2009)

11. Dau, F., Ducrou, J., Eklund, P.: Concept similarity and related categories in search sleuth. In: Eklund, P., Haemmerlé, O. (eds.) ICCS 2008. LNCS (LNAI), vol. 5113, pp. 255–268. Springer, Heidelberg (2008)
12. Saquer, J., Deogun, J.S.: Concept approximations based on rough sets and similarity measures. International Journal of Applied Mathematics and Computer Science **11**(3), 655–674 (2001)
13. Alqadah, F., Bhatnagar, R.: Similarity measures in formal concept analysis. Journal Annals of Mathematics and Artificial Intelligence **61**(3), 245–256 (2011)
14. Ganter, B., Wille, R.: Formal concept analysis: Mathematical foundation. Springer (1999)
15. Baeza-Yates, R., Ribeiro-Neto, B.: Modern Information Retrieval Addison Wesley (1999)
16. Witten, I.H., Frank, E., Hall, M.A.: Data Mining Practical Machine Learning Tools and Techniques, third edn. Morgan Kaufmann Publishers is an imprint of Elsevier (2011)
17. Andrews, S.: In-Close, a Fast Algorithm for Computing Formal Concepts. The Seventeenth International Conference on Conceptual Structures (2009)
18. Bohanec, M., Zupan, B.: Car Evaluation Data Set UCI machine learning repository (1997). http://archive.ics.uci.edu/ml/datasets/Car+Evaluation
19. Leake, D.B.: Case-Based Reasoning: experiences, lessons and future directions. AAAI Press (2000)
20. Tadrat, J., Boonjing, V., Pattaraintakorn, P. Building classification rules for case based classifier using fuzzy sets and formal concept analysis. In: Proceedings of the Fifth International Conference on Soft Computing as Transdisciplinary Science and Technology (IEEE/ACM CSTS08), Paris, France, pp. 13–18 (2008)
21. Hartuv, E., Shamir, R.: A clustering algorithm based on graph connectivity. Information Processing Letters **76**(4–6), 175–181 (2000)
22. Marzouk, Y.M., Ghoniem, A.F.: K-means clustering for optimal partitioning and dynamic load balancing of parallel hierarchical N-body simulations. Journal of Computational Physics **207**(2), 493–528 (2005)
23. Bichindaritz, I., Montani, S.: Advances in case-based reasoning in the health sciences. Artificial Intelligence in Medicine **51**(2), 75–79 (2011)
24. Sollenborn, M.: Clustering and Case-Based Reasoning for User Stereotypes. Mälardalen University (2004)
25. Wattuya, P., Nuawan, S., Xiaoyi, J.: Using soft Case-Based Reasoning in Model order Selection for Image Segmentation Ensemble, The 26 Annual conference of Japanese Society for Artificial Intelligent (2012)
26. Sankar, K., Simon, C.: Foundation of soft case-Based Reasoning (2004)

Ensemble Algorithms for Unsupervised Anomaly Detection

Zhiruo Zhao$^{(\boxtimes)}$, Kishan G. Mehrotra, and Chilukuri K. Mohan

Syracuse University, Syracuse, NY, USA
{zzhao11,mehrotra,mohan}@syr.edu

Abstract. Many anomaly detection algorithms have been proposed in recent years, including density-based and rank-based algorithms. In this paper, we propose ensemble methods to improve the performance of these individual algorithms. We evaluate approaches that use score and rank aggregation for these algorithms. We also consider sequential methods in which one detection method is followed by the other. We use several datasets to evaluate the performance of the proposed ensemble methods. Our results show that sequential methods significantly improve the ability to detect anomalous data points.

Keywords: Anomaly detection · Ensemble method · Density-based anomaly detection · Rank-based anomaly detection

1 Introduction

Anomaly detection methods are used in a wide range of applications. We consider unsupervised detection because in most realistic scenarios training data sets are not available. Supervised anomaly detection is equivalent to the classification problem with the caveat that the training set is often extremely small and ensemble methods such as simple or weighted averaging, boosting, and bagging can be applied to improve the performance of individual classifiers. Oza and Tumer [8] provide a survey of these techniques.

A particular algorithm may be well-suited to the properties of one data set and be successful in detecting anomalous observations of the particular application domain, but may fail to work with other datasets whose characteristics do not agree with the first dataset. The impact of such mismatch between an algorithm and an application can be alleviated by using ensemble methods where a variety of algorithms are pooled before a final decision is made. Mathematically, ensemble methods help by addressing the classical bias-variance dilemma, at least in the classification context.

Recently, a few researchers have studied ensemble methods such as anomaly detection and distributed intrusion detection in Mobile Ad-Hoc Networks (see Cabrera *et al.* [3]). Others have studied supervised anomaly detection using random forests and distance-based outlier partitioning (see Shoemaker and Hall [12]). In the semi-supervised case, one possible approach is to convert the problem to

© Springer International Publishing Switzerland 2015
M. Ali et al. (Eds.): IEA/AIE 2015, LNAI 9101, pp. 514–525, 2015.
DOI: 10.1007/978-3-319-19066-2_50

a supervised anomaly detection by exploiting strong relationships between features; Tenenboim-Chekina et al. [15] and Noto et al. [7] provide two different approaches to accomplish this goal. Pevny et al. [11] proposes anomaly detector processing for a continuous stream of data, which requires high acquisition rate, limited resources for storing the data, and low training and classification complexity. The key idea is to use bagging on multiple weak detectors, each implemented as a one-dimensional histogram. In these approaches the word 'ensemble' has a different connotation.

Aggarwal et. al. [1] has categorized the ensemble methods for outlier detection into two types: independent and sequential, and outlined the framework for both. Their paper argued that sequential methods have not been quite explored in unsupervised outlier ensemble area as a general meta-algorithms. In our paper, we have explored both the independent and sequential ensemble methods. Our proposed method is an application of a two-phase sequential ensemble method, where we use one algorithm to determine the subset of dataset which contains all the anomalies, and focus on that part in the second phase. We evaluate the performance for recently proposed two families of outlier detection algorithms: density-based and rank-based outlier detection. Zimek et. al. [17] have argued that drawing multiple subsamples from the dataset and combine the output as the final outlier detection is more efficient than detection on the whole dataset. In this paper, we use this procedure that uses a two-phase sequential method by eliminating the non-anomalies(noise) at the first phase. We achieve better performance both in quality and efficiency. Using correlation coefficient as a similarity measure we select the best pair of algorithms to use in the two-phase sequential method. We observe that the sequential methods outperform aggregation methods that combine the results of different anomaly detection algorithms, on several well-known data sets.

In Section 2 we briefly describe six anomaly detection algorithms; three based on local density and the other on ranks. In Section 3 we consider ensemble methods; all use the algorithms described in Section 2; however, the proposed ensemble methods are generally applicable. Section 4 contains description of datasets used to compare the performance of the algorithms and results. In Section 5, we provide a procedure that finds the best pair of algorithms for sequential ensemble method. Section 6 presents concluding remarks.

2   Density-Based and Rank-Based Outlier Detection

In this section we briefly describe six recent density and rank based algorithms that we intend to apply in ensemble methods. The common theme among the three density-based algorithms (LOF, COF, and INFLO) is that outlierness-score is assigned based on density comparison of the object with its k nearest neighbors; an object is considered an outlier if its outlierness-score is greater than a pre-defined threshold. The other three algorithms are based on the notion of *rank*, and use the concept that if k nearest neighbors of an object consider the object as their close neighbor then it is less likely to be an outlier.

2.1   Density-Based Algorithms

These algorithms assume a symmetric distance function, with $d(p,q) = d(q,p)$, where $p, q \in D$ (the data set). An important notion for various algorithms is $N_k(p)$, i.e., the kth nearest neighborhood (k-NN) of a point $p \in D$ [4]. The *reverse nearest neighborhood* of an object p, $R_k(p)$, is defined to be the set of points q such that p is among the k nearest neighbors of q, i.e.,

$$R_k(p) = \{q : p \in N_k(q)\}$$

Note that $N_k(p)$ has at least k objects but $R_k(p)$ may be empty, because p may not be in any of the k-NN sets of any of its k nearest neighbors.

LOF (Local Outlier Factor) LOF [2], a density-based outlier detection algorithm, captures the degree of outlierness of an object, which is essentially the average of the ratio of the local reachability distance of an object with the average distance of the object's k nearest neighbors $N_k(p)$. The reachability distance of a point q from p, is either the distance of p from q or the distance of the kth neighbor of p, whichever is larger. LOF > 1 indicates that the object is potentially an outlier, whereas if LOF is ≤ 1 then the object's local density is as large as the average density of its neighbors, i.e., it is a non-outlier.

COF (Connectivity-based Outlier Factor) The COF [14] of a point p is defined as the ratio of p's *average-chaining distance* with the average of average-chaining distances of its k nearest neighbors. To define the concept of average chaining distance, Tang *et al.* [14] define the concept of Set Based Nearest Path (SBN-path) which represents the order in which nearest neighbors of p are successively obtained; define Set-based trail (SBT) as an ordered collection of $k-1$ edges associated with a given SBN path; the ith edge e_i connects the $(i+1)$th point p_{i+1} in SBN-path to one of the nearest earlier points in the path; each such edge is assigned a weight proportional to the order in which it is added to SBT set; and finally the average-chaining distance of p is the weighted sum of the lengths of the edges. COF was designed to work better than LOF in data sets with sparse neighborhoods (such as a straight line), but its computation cost is larger than LOF.

INFLO (INFLuential measure of Outlierness by symmetric relationship) INFLO [6] is a variation of LOF based on the reverse nearest neighbor relationship. Influential outlierness of a point p is the ratio of the density in the neighborhood of p with average density of objects in $R_k(p)$. By using the reverse neighborhood, INFLO enhances its ability to identify the outliers in more complex situation, but its performance is poor if p's neighborhood includes data objects from groups of different densities.

2.2   Rank-Based Algorithms

RBDA (Rank Based Detection Algorithm) As in INFLO, RBDA [5] also exploits the reverse neighborhood concept. But in RBDA, outliers are detected on mutual closeness of a data point and its neighbors using ranks instead of distance. If $q \in N_k(p)$ and in turn $p \in N_k(q)$, then p and q are mutually close; i.e., with respect to each other p and q are not anomalous data points. However, if $q \in N_k(p)$ but not $p \in N_k(q)$, then with respect to q, p is an anomalous point. If p is an anomalous point with respect to most of its neighbors, then p should be declared to be an anomaly. The rank $r_q(p) = \ell$ if p is ℓth nearest neighbor of q in $N_k(q)$. Outlierness of p is defined as

$$O_k(p) = \frac{\sum_{q \in N_k(p)} r_q(p)}{|N_k(p)|}. \tag{1}$$

RADA (Rank with Averaged Distance Algorithm) RADA [4] adjusts the rank of p by average distance from its k nearest neighbors. Measure of outlierness is

$$W_k(p) = O_k(p) \times \frac{\sum_{q \in N_k(p)} d(p,q)}{|N_k(p)|}.$$

ODMR (Outlier Detection using Modified-Ranks) For a point near a dense and large cluster all of the k nearest neighbors of p may find their neighbors in a close vicinity and p may not be their neighbor; a point near a dense and large cluster may be declared an anomaly, although it may not be so. ODMR [4] modifies the rank of an observation by assigning a weight to overcome the cluster density effect. In ODMR all clusters (including isolated points viewed as a cluster of size 1) are assigned weight 1, i.e., all $|C|$ observations of the cluster C are assigned equal weights $= 1/|C|$. The "modified-rank" of p with respect to q is defined as the sum of weights associated with all observations within the circle of radius $d(q,p)$ centered at q, and the measure of outlierness is given by Equation (1) with $r_q(p)$ replaced by the modified rank of p.

3   The Ensemble Methods

Ensemble methods for anomaly detection have been categorized into two groups [1]: the independent ensemble and the sequential ensemble approaches. In independent ensemble approach, the results from executions of different algorithms are combined in order to obtain a more robust model. In sequential ensemble approaches, a set of algorithms is applied sequentially. We describe two independent ensemble methods in Section 3.1 and a sequential method in Section 3.2. The descriptions below assume that the set of six algorithms in Section 2 are to be used; this set can be changed to include other algorithms for anomaly detection, using the same approaches described in this section.

3.1   Averaging and Other Methods to Combine the Anomalousness

Each algorithm, described in the previous section, provides an anomaly score to objects in D; objects that receive higher score are considered more anomalous. Equivalently these scores can be sorted in decreasing order and thus assign ranks (object that receives the highest score is ranked 1, and so on). Scores as well as ranks can be used in such ensemble methods.

The Averaging Method The averaging method combines anomaly scores, first normalized to be between 0 and 1. Let $\alpha_i(p)$ be the normalized anomaly score of $p \in D$, according to algorithm i. Then the normalized score, averaged over all six algorithms, is obtained as follows:

$$\alpha(p) = \frac{1}{6} \sum_{i=1}^{6} \alpha_i(p).$$

The rank $r(p)$ of p, based on the average scores, is:

$$r(p) = |D| - |\{q | \alpha(q) < \alpha(p)\}|.$$

That is, the observation with the highest anomaly score is ranked 1 (i.e., is most anomalous observation in the datasets, the second ranked observation has the second largest anomaly score, and so on).

The Min-Rank Method Let the anomalous rank of p, assigned by algorithm i, be given by:

$$r_i(p) = |D| - |\{q | \alpha_i(q) < \alpha_i(p)\}|.$$

As mentioned earlier, a smaller rank implies that the observation is more anomalous. The Min-rank method assigns

$$\text{rank}(p) = \min_{1 \le i \le 6} r_i(p).$$

Thus, if object p is found to be most anomalous by at least one algorithm then the Min-rank method also declares it to be the most anomalous object. If all six algorithms give substantially different results, six different points may have the same rank.

3.2   Sequential Application of Two algorithms

In this section, we use the output of one algorithm on the original dataset D, extract a subset of D, and use another algorithm on the extracted dataset. Our argument is that similar algorithms are likely to make similar errors; by sequentially applying two substantially different algorithms, the second algorithm may 'correct' the previous one's errors and thus provide better results. As in the previous section, for each object in the dataset the ranks are calculated using the anomaly scores. In the sequential method, we run the first algorithm on the whole dataset, and obtain the ranks of all observations. Next we consider the dataset D_α obtained by retaining the top $\alpha \times 100\%$ of D, suspected to contain all anomalies. The second anomaly detection algorithm calculates the anomaly scores of all objects in D with reference to D_α. In the following discussion, this algorithm is referred as Sequential-1, whose pseudo-code is given in Algorithm 1.

Algorithm 1. Sequential-1 Algorithm

Input: dataset D, detection algorithm A and B
Output: a list of ranks associated with each of the objects in D

1. ScoreList$_A(D)$ = Apply algorithm A on D.
2. RankList$_A(D)$ = Assign each object in D a rank sorted by ScoreList$_A(D)$ in decreasing order.
3. D_α = Identify the top $\alpha \times 100\%$ objects in RankList$_A(D)$.
4. ScoreList$_{Final}(D)$ = For each object $o \in D$, apply algorithm B and get an anomaly score calculated with reference to D_α.
5. RankList$_{Final}(D)$ = Assign each object in D a rank sorted by ScoreList$_{Final}(D)$ in decreasing order.

Zimek *et al.* [17] have argued that anomaly detection performance on a subsample of the dataset could be better than detection performance on the whole dataset. They select a simple random sample from the dataset D; evaluate anomaly score of each object in D with reference to the data in the sample; repeat the above experiment multiple times and report that the average score gives much better performance. In our second sequential approach, we take a subsample from D_α and evaluate anomaly score of each $p \in D$ using another algorithm with respect to the subsample. As in Zimek *et al.* [17], we repeat this multiple times and evaluate the average score, used in the final anomaly ranking. This algorithm is referred as Sequential-2.

4   Experiments and Results

In this section first we give a brief description of datasets used in evaluating the performance of proposed methods, next we described evaluation metrics, followed by the results.

4.1   The Datasets

1. **Packed Executables dataset**[1]   Executable packing is the most common technique used by computer virus writers to obfuscate malicious code and evade detection by anti-virus software. Originally, this dataset was collected

---

[1] http://roberto.perdisci.com/projects/cpexe

from the Malfease Project [10] to classify the non-packed executables from packed executables so that only packed executables could be sent to universal unpacker. In our experiments, we select 1000 packed executables as normal observations, and insert 8 non-packed executables as anomalies. All 8 features are used in experiments.

2. **KDD 99 dataset**[2]   KDD 99 data-set is constructed from DARPA intrusion dataset evaluation program. This dataset has been widely used in both intrusion detection and anomaly detection area. There are 4 main attack categories in KDD 99 dataset. In our experiments, we select 1000 normal connections from test dataset and insert 14 attack-connections as anomalies. All 41 features are used in experiments.

3. **Wisconsin Dataset**[3] The Wisconsin diagnostic breast cancer dataset contains 699 instances and has 9 attributes. There are many duplicate instances and instances with missing attribute values. After removing all duplicate and instances with missing attribute values, 236 instances labeled as benign class and 236 instances as malignant were left. In our experiments the final dataset consisted 213 benign instances and 10 malignant instances.

4.2   Evaluation Metrics

In this paper, we use three evaluation metrics: AUC, Rank Power and number of false positives when detection rate =100%.

Area Under Curve (AUC) The Receiver Operating Characteristic (ROC) curve is widely used in to measure the performance of a classifier [9], and plots the true positive rate vs. the false positive rate. The area under the ROC curve, AUC, is often used to measure the performance of an algorithm.

Rank Power (RP) Rank Power, proposed by Tang *et al.* [13], evaluates the ratio of the number of known anomalies and anomalies returned by an algorithm, along with their rankings. Precisely,

$$\text{RankPower} \ (m) = \frac{m_t \times (m_t + 1)}{2 \times \Sigma_{i=1}^{m_t} R_i}$$

where m_t is the number of true outliers among top m objects returned by the algorithm and R_i is the rank of the ith true outlier.

Number of False Positives when Detection Rate=100% (FP-100) When we evaluate performance of an algorithm we know which observations in D are anomalous and we want to capture all of them with as few false positive as possible. For example, if the dataset has 10 known anomalies and the rank of the 10th

---

[2] http://kdd.ics.uci.edu/databases/kddcup99/kddcup99.html
[3] http://archive.ics.uci.edu/ml/machine-learning-databases/breast-cancer-
    wisconsin/

anomaly is 17 by algorithm 'A' and 20 by algorithm 'B', then clearly algorithm 'A' is better because for 100% detection rate there are only 7 false positives. We use this FP-100 metric to evaluate the performance of an algorithm.

4.3   Ensemble versus Individual Algorithms

Ensemble methods were compared using all six individual algorithms, and results are summarized in Table 1. To implement Sequential-1 we choose $\alpha = 0.20$ both for COF and LOF followed by RADA. First six columns show the performance of single algorithms, the last four columns show the performance of the ensemble methods. For example, for Wisconsin dataset, $k = 2$, if we choose COF as the first and only algorithm, the associated FP-100 is 113, whereas, FP-100 value for COF followed by RADA is 9. Performances of the Average and Min-rank methods are mixed for FP-100, slightly better as measured by AUC and RankPower; sequential-1 has far superior performance than any single application algorithm in all cases. Performance of Sequential-1 was exhaustively compared with all six individual algorithms for the Wisconsin data (table not shown). Purpose of the exhaustive comparison was to determine if Sequential-1 has good performance only for the selected pairs of algorithms or in general; and it was observed that the results are consistently better for the Sequential-1. For example, it was observed that if we apply COF alone, then FP-100 = 113, AUC = 0.573, and RankPower = 0.049; whereas COF followed by INFLO has FP-100 = 26, AUC = 0.979, and RankPower = 0.455. We show the performance using different k value on dataset Wisconsin in Table 1. Results were not sensitive to the value of k.

Table 1. Performance Comparison of Sequential-1 and other ensemble methods, including single algorithms, for all datasets

DataSet		Single						Ensemble			
	$k = 2$	LOF	INFLO	COF	RBDA	RADA	ODMR	Ave	Min	COF-RADA	LOF-RADA
	FP100	143	123	113	180	94	63	110	131	**9**	12
	AUC	0.612	0.642	0.575	0.513	0.837	0.850	0.719	0.683	**0.996**	0.987
	RP	0.052	0.056	0.049	0.042	0.109	0.121	0.070	0.062	**0.776**	0.592
	$k = 4$	LOF	INFLO	COF	RBDA	RADA	ODMR	Ave	Min	COF-RADA	LOF-RADA
Wis-	FP100	175	168	127	158	62	51	101	114	**11**	12
consin	AUC	0.698	0.715	0.530	0.691	0.889	0.912	0.782	0.777	**0.995**	0.992
	RP	0.063	0.067	0.045	0.063	0.156	0.188	0.088	0.085	**0.738**	0.672
	$k = 10$	LOF	INFLO	COF	RBDA	RADA	ODMR	Ave	Min	COF-RADA	LOF-RADA
	FP100	98	100	164	90	42	28	75	76	**11**	11
	AUC	0.801	0.731	0.795	0.732	0.867	0.916	0.816	0.781	**0.994**	**0.994**
	RP	0.076	0.094	0.037	0.117	0.239	0.300	0.111	0.130	**0.726**	0.714
	$k = 2$	LOF	RBDA	RADA	INFLO	COF	ODMR	Ave	Min	COF-RADA	LOF-RADA
PEC	FP100	953	667	81	495	720	654	526	179	28	14
	AUC	0.525	0.685	0.968	0.705	0.549	0.690	0.720	0.934	0.994	**0.997**
	RP	0.008	0.012	0.108	0.014	0.009	0.013	0.014	0.055	0.371	**0.507**
	$k = 2$	LOF	RBDA	RADA	INFLO	COF	ODMR	Ave	Min	COF-RADA	LOF-RADA
KDD	FP100	934	849	75	972	692	741	587	180	18	**10**
	AUC	0.832	0.853	0.984	0.801	0.803	0.875	0.881	0.961	0.993	**0.995**
	RP	0.031	0.034	0.255	0.026	0.028	0.040	0.043	0.122	0.441	**0.549**

4.4   Sub-sampling Approach (Sequential-2 Method)

To evaluate if $p \in D$ is an anomaly Zimek et. al. [17] have proposed to repeatedly
(typically 25 times) draw a sample of size $100 \times \theta\%$ from the original dataset D.
We select a sample of size $100 \times \theta\%$ from D_α and, as in their approach, compute
the anomaly score for each object with respect to the sub-sampled data. Zimek
et. al. have argued that the performance will be good when random sampling
is performed 25 times or more. Our experiment indicates that it is sufficient to
use as few as 5 subsamples from D_α. We use $k = 2$ and LOF to provide initial
rankings and evaluate the performance for multiple values of $\alpha = 0.2, 0.3, 0.4,$
$0.5, 0.6, 0.7, 0.8, 0.9, 1.0$ to measure the effect of this parameter[4]. For comparison,
we run each experiment 100 times; Figure 1 represents the corresponding boxplots for all three metrics. Results for sub-sampling approach are summarized in
Table 2. To evaluate the effect of k, these experiments were repeated for $k = 5$
and 10. These results are presented in Table 3 for PEC dataset. Clearly, the
performance of subsamples from D_α is far better than simple random subsample
from D, especially when α is less than 0.5.

Subsamples drawn from D_α give better performance, so, we recommend that
the Zimek et. al. approach of random sub-sampling should be applied to selected
subset of the entire dataset. In fact our result show that drawing multiple subsamples from D_α is not advantageous over selection of entire D_α.

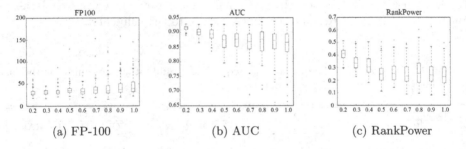

(a) FP-100   (b) AUC   (c) RankPower

Fig. 1. Performance of subsampling approach when samples are drawn from D_α for
different values of α, for Wisconsin dataset

Table 2. Performance of Sequential-2 Algorithm; $k = 2$. LOF is used in anomaly
detection.

	Wisconsin		PEC		KDD(1)	
	$D_{0.20}$	$D_{1.00}$	$D_{0.20}$	$D_{1.00}$	$D_{0.20}$	$D_{1.00}$
FP-100	**30.73**	46.98	**131.054**	390.122	**10.62**	19.37
AUC	**0.888**	0.855	**0.930**	0.728	**0.979**	0.957
RankPower	**0.409**	0.243	**0.064**	0.031	**0.359**	0.265

---

[4] When α=1.0, our approach is equivalent to Zimek's random selection from the whole
dataset.

5 Selection of Pair of Algorithms for Sequential Application

It has been argued that diversity of algorithms is very important [1, 16] when combining different anomaly detection algorithms. Since correlation coefficient provides a measure of similarity, we employ it towards this goal. We calculate the Pearson correlation coefficient between each pair of algorithms using the associated score vectors. Table 4 summarizes the average correlation between every pair of algorithms over nine different datasets. Although all correlation coefficients are large, the smallest average value corresponds to COF and RADA. In other words, we expect that, in general, COF followed by RADA (or in reverse order) should give good performance. Largest average correlations are observed between three algorithms of rank-family, i.e., between RBDA, RADA,

Table 3. Performance of Sequential-2 Algorithm on PEC dataset for $k = 5, 10$. LOF is used in anomaly detection.

	$k = 5$			$k = 10$		
	$D_{0.20}$	$D_{1.00}$	p-value	$D_{0.20}$	$D_{1.00}$	p-value
FP-100	**150.5**	218.7	0.04	**37.48**	155.94	≤ 0.001
AUC	**0.946**	0.915	0.02	**0.976**	0.894	≤ 0.001
RankPower	**0.075**	0.048	0.02	**0.272**	0.078	≤ 0.001

Table 4. Average correlation between pair of algorithms over all datasets

	LOF	INFLO	COF	RBDA	RADA	ODMR
LOF	1.000	0.910	0.622	0.666	0.559	0.635
INFLO	0.910	1.000	0.566	0.636	0.536	0.582
COF	0.622	0.566	1.000	0.482	**0.390**	0.446
RBDA	0.666	0.636	0.482	1.000	0.874	0.943
RADA	0.559	0.536	0.390	0.874	1.000	0.860
ODMR	0.635	0.582	0.446	0.943	0.860	1.000

Table 5. Performance over different pair of algorithms

	FP100		AUC		RankPower	
DataSet	(C, R)	(R, R)	(C, R)	(R, R)	(C, R)	(R, R)
PEC	**15**	111	**0.996**	0.988	**0.486**	0.194
Wisconsin	**13**	17	**0.993**	0.989	**0.692**	0.600
KDD	29	**27**	0.991	0.991	0.377	**0.390**
Iris(1)	**0**	34	**1.000**	0.816	**1.000**	0.121
Iris(2)	**0**	**0**	**1.000**	**1.000**	**1.000**	**1.000**
Iono(1)	**36**	76	**0.973**	0.805	**0.410**	0.104
Iono(2)	**12**	**12**	**0.989**	0.979	**0.250**	0.176
Eeg	34	**18**	0.983	**0.987**	0.156	**0.233**
Segment	68	**57**	0.980	**0.985**	0.153	**0.189**

(C, R) represnts the pair (COF, RADA) and (R, R) represents (RBDA, RADA)
Bold items represent superior performance.

and ODMR; implying that selecting two algorithms among these will not perform as well as if one is chosen out of these three and the other from the distance based algorithm.

We note that the pair (COF, RADA) has the least average correlation and therefore compare the performance of sequential application of these two algorithms with sequential application of the pair (RBDA, RADA). Table 5 summarizes the results.

We observe that, using AUC metric, pair (COF, RADA) performs better than the pair (RBDA, RADA) for seven out of nine datasets. For FP-100 and RankPower, the pair (COF, RADA) performs better than (RBDA, RADA) only for six out of nine sets, but, in general, for the remaining four datasets the performance is comparable.

6   Conclusions

In this paper, we used three metrics to evaluate the performance of several ensemble methods and noted that the average and the Min-rank methods provide incremental improvement. This is due to the reason that one anomaly detection algorithm is not able to detect variety of anomalies present in a dataset. For this reason we argued that a sequential method, where the second algorithm builds upon the findings of the first algorithm should perform better. Our empirical study suggests that either LOF or COF followed by RADA achieve very good performance in most cases. Furthermore, deviating from Zimek *et. al.* finding, we observe that anomaly detection using a subset of the given dataset, that is suspected to contain all anomalies, is better both in terms of quality of results and time complexity. Finally, we provided a procedure of using correlation coefficient to find the best pair to use in sequential ensemble method.

References

1. Aggarwal, C.C.: Outlier Ensembles (position paper)
2. Breunig, M.M., Kriegel, H.-P., Ng, R.T., Sander, J.: Lof: identifying density-based local outliers. In: ACM Sigmod Record, vol. 29, pp. 93–104. ACM (2000)
3. Cabrera, J.B., Gutiérrez, C., Mehra, R.K.: Ensemble methods for anomaly detection and distributed intrusion detection in mobile ad-hoc networks. Information Fusion 9(1), 96–119 (2008)
4. Huang, H., Mehrotra, K., Mohan, C.K.: Algorithms for detecting outliers via clustering and ranks. In: Jiang, H., Ding, W., Ali, M., Wu, X. (eds.) IEA/AIE 2012. LNCS, vol. 7345, pp. 20–29. Springer, Heidelberg (2012)
5. Huang, H., Mehrotra, K., Mohan, C.K.: Rank-based outlier detection. Journal of Statistical Computation and Simulation 83(3), 518–531 (2013)
6. Jin, W., Tung, A.K.H., Han, J., Wang, W.: Ranking outliers using symmetric neighborhood relationship. In: Ng, W.-K., Kitsuregawa, M., Li, J., Chang, K. (eds.) PAKDD 2006. LNCS (LNAI), vol. 3918, pp. 577–593. Springer, Heidelberg (2006)
7. Noto, K., Brodley, C., Slonim, D.: Anomaly detection using an ensemble of feature models. In: 2010 IEEE 10th International Conference on Data Mining (ICDM), pp. 953–958. IEEE (2010)

8. Oza, N.C., Tumer, K.: Classifier ensembles: Select real-world applications. Information Fusion **9**(1), 4–20 (2008)

9. Pang-Ning, T., Steinbach, M., Kumar, V., et al.: Introduction to data mining. In: Library of Congress (2006)

10. Perdisci, R., Lanzi, A., Lee, W.: Classification of packed executables for accurate computer virus detection. Pattern Recognition Letters **29**(14), 1941–1946 (2008)

11. Pevný, T., Fridrich, J.: Novelty detection in blind steganalysis. In: Proceedings of the 10th ACM Workshop on Multimedia and Security, pp. 167–176. ACM (2008)

12. Shoemaker, L., Hall, L.O.: Anomaly detection using ensembles. In: Sansone, C., Kittler, J., Roli, F. (eds.) MCS 2011. LNCS, vol. 6713, pp. 6–15. Springer, Heidelberg (2011)

13. Tang, J., Chen, Z., Fu, A.W., Cheung, D.W.: Capabilities of outlier detection schemes in large datasets, framework and methodologies. Knowledge and Information Systems **11**(1), 45–84 (2007)

14. Tang, J., Chen, Z., Fu, A.W., Cheung, D.W.: Enhancing effectiveness of outlier detections for low density patterns. In: Chen, M.-S., Yu, P.S., Liu, B. (eds.) PAKDD 2002. LNCS (LNAI), vol. 2336, pp. 535–548. Springer, Heidelberg (2002)

15. Tenenboim-Chekina, L., Rokach, L., Shapira, B.: Ensemble of feature chains for anomaly detection. In: Zhou, Z.-H., Roli, F., Kittler, J. (eds.) MCS 2013. LNCS, vol. 7872, pp. 295–306. Springer, Heidelberg (2013)

16. Zimek, A., Campello, R.J., Sander, J.: Ensembles for unsupervised outlier detection: challenges and research questions a position paper. ACM SIGKDD Explorations Newsletter **15**(1), 11–22 (2014)

17. Zimek, A., Gaudet, M., Campello, R.J., Sander, J.: Subsampling for efficient and effective unsupervised outlier detection ensembles. In: Proceedings of the 19th ACM SIGKDD International Conference on Knowledge Discovery and Data Mining, pp. 428–436. ACM (2013)

Comparison of Adjusted Methods for Selecting Useful Unlabeled Data for Semi-Supervised Learning Algorithms

Thanh-Binh Le and Sang-Woon Kim$^{(\boxtimes)}$

Department of Computer Engineering, Myongji University, Yongin 449-728, Korea
{binhlt,kimsw}@mju.ac.kr

Abstract. This paper presents a comparison of the methods of selecting a small amount useful unlabeled data to improve the classification accuracy of semi-supervised learning (SSL) algorithms. In particular, three selection approaches, namely, the simply adjusted approach based on an uncertainty level, the normalized-and-adjusted approach, and the entropy based adjusted approach, are considered and compared empirically. The experimental results, which are obtained from synthetic and real-life benchmark data using semi-supervised support vector machines (S3VMs), demonstrate that the entropy based approach works slightly better than the other ones in terms of the classification accuracy.

Keywords: Machine learning · Semi-Supervised Learning · Semi-supervised support vector machines

1 Introduction

In semi-supervised learning (SSL) [4,13] algorithms, both a limited number of labeled data ($L = \{(x_i, y_i)\}_{i=1}^{n_l}$, where $x_i \in \mathbb{R}^d$, and $y_i \in \{+1, -1\}$) and a multitude of unlabeled data ($U = \{x_i\}_{i=1}^{n_u}$) are utilized to learn a classification model. However, it is also well known that the utilization of U data is not always helpful for SSL algorithms. Therefore, in order to select (and utilize) a small amount of useful data ($U_s = \{x_i\}_{i=1}^{n_s}$) from U data, various approaches have been proposed in the literature, including the procedures that are used in self-training [12] and co-training [2], confidence-based approaches [8,9], and other approaches used in active learning (AL) algorithms [6,11]. However, in AL, selected instances are useful when they are labeled, thus it is required to query their true class label from a human annotator. From this point of view, the approaches for SSL and AL algorithms are different.

Meanwhile, in the confidence-based approaches, the confidence values are calculated for each $x_i \in U$. Choosing top confident examples will make sure that the helpful examples are included in the training data. Thus, in order to select a small amount of useful U_s data, various selection criteria have been proposed in the literature. One criterion, for example, is based on the prediction

© Springer International Publishing Switzerland 2015
M. Ali et al. (Eds.): IEA/AIE 2015, LNAI 9101, pp. 526–535, 2015.
DOI: 10.1007/978-3-319-19066-2_51

by a component classifier and the similarity between pairwise training examples. Since the criterion is only concerned with the distance information among the examples, however, sometimes it doesn't work appropriately, particularly when the U examples are near the boundary. In order to address this concern, in [8], a method of training semi-supervised support vector machines (S3VMs) using a selection criterion has been investigated; this method is a modified version of that used in SemiBoost [9].

In particular, in SemiBoost, the confidence value of $x_i \in U$ is computed using two quantities, named p_i and q_i, which are measured using the pairwise similarity between x_i and other U and L examples. p_i and q_i can be used to guide the selection at each iteration using differences in their values ($|p_i - q_i|$) as well as to predict the pseudo class labels using a signum function ($sign(p_i - q_i)$). Therefore, the difference in values between p_i and q_i (i.e., $p_i - q_i$) should be measured first and can be computed as: $X_i^+ - X_i^- + X_i^u$ (the details of this computation will be explained later in the present paper). Here, the three terms of X_i^+, X_i^-, and X_i^u denote respectively three measurements obtained with the examples of the positive class of L (L^+), the negative class of L (L^-), and U data. Thus, when x_i is between L^+ and L^-, both X_i^+ and X_i^- have similar values and, in turn, the difference depends on X_i^u only. From this observation, it can be noted that L data don't have an influence in selecting x_i as well as predicting its label; consequently, the confidence value obtained might be inappropriate for selecting useful U_s.

In order to address this issue, adjusted criteria, including the criterion in [8], that minimize the errors in estimating the confidence value are considered and compared in this paper. That is, the computation of $p_i - q_i$ can be adjusted by taking a balance among X_i^+, X_i^-, and X_i^u. This adjustment can be achieved with reducing the impact of X_i^u using certain *uncertainty* levels of x_i, which are reflecting the incompleteness in measuring the confidence value. Therefore, for the original criterion, the confidence values are computed using the quantities of p_i and q_i only, whereas, for the adjusted criteria, they are measured using the uncertainty levels as well as the p_i and q_i quantities.

The remainder of the paper is organized as follows. In Section 2, the confidence-based approaches are briefly explained. In Section 3, adjusted criteria for selecting a small amount of useful U_s to improve S3VMs through utilizing the uncertainty level are presented. In Section 4, the experimental setup and the results obtained are presented. In Section 5, concluding remarks are described.

2   Selection Strategies

The goal of SemiBoost, which is a boosting framework for SSL, is to iteratively improve the performance of a supervised learning algorithm ($\mathcal{A}$) by regarding it as a black box, using U and pairwise similarity. In order to follow the boosting idea, SemiBoost optimizes performance through minimizing an objective loss function defined as Proposition 2 in [9], where $h_i(= h(x_i))$ is the base classifier learned by $\mathcal{A}$ at the iteration, α is the weight for combining h_i's, and

$$
\begin{aligned}
p_i &= \sum_{j=1}^{n_l} S_{i,j}^{ul} e^{-2H_i} \delta(y_j, 1) + \frac{C}{2} \sum_{j=1}^{n_u} S_{i,j}^{uu} e^{H_j - H_i}, \\
q_i &= \sum_{j=1}^{n_l} S_{i,j}^{ul} e^{2H_i} \delta(y_j, -1) + \frac{C}{2} \sum_{j=1}^{n_u} S_{i,j}^{uu} e^{H_i - H_j}.
\end{aligned}
\tag{1}
$$

Here, $H_i(= H(x_i))$ denotes the final combined classifier and S denotes the pairwise similarity. For all x_i and x_j of the training set, for example, S can be computed as follows: $S(i,j) = exp(-\|x_i - x_j\|_2^2/\sigma^2)$, where σ is the scale parameter controlling the spread of the function. In addition, S^{lu} (and S^{uu}) denotes the $n_l \times n_u$ (and $n_u \times n_u$) submatrix of S. Also, S^{ul} and S^{ll} can be defined correspondingly; C $(= |L|/|U| = n_l/n_u)$ is introduced to weight the importance between L and U; and $\delta(a,b) = 1$ when $a = b$ and 0 otherwise.

Using these settings, in [9], p_i and q_i have been used to guide the selection of U data at each iteration using the confidence measurement $|p_i - q_i|$, as well as to assign the pseudo class label $sign(p_i - q_i)$.

From (1), the difference in values between p_i and q_i can be measured as:

$$
\begin{aligned}
p_i - q_i = {} & \left(e^{-2H_i} \sum_{x_j \in L^+} S_{i,j}^{ul} \right) - \left(e^{2H_i} \sum_{x_j \in L^-} S_{i,j}^{ul} \right) \\
& + \left(\frac{C}{2} \sum_{x_j \in U} S_{i,j}^{uu} (e^{H_j - H_i} - e^{H_i - H_j}) \right).
\end{aligned}
\tag{2}
$$

Here, by substituting $X_i^+ \equiv e^{-2H_i} \sum_{x_j \in L^+} S_{i,j}^{ul}$ and $X_i^- \equiv e^{2H_i} \sum_{x_j \in L^-} S_{i,j}^{ul}$ in the first two summations of (2), the expression can be simplified to:

$$
p_i - q_i = X_i^+ - X_i^- + X_i^u,
\tag{3}
$$

where $X_i^u \equiv \left(\frac{C}{2} \sum_{x_j \in U} S_{i,j}^{uu} (e^{H_j - H_i} - e^{H_i - H_j}) \right)$.

From this, it can be seen that if the difference of X_i^+ and X_i^- is nearly zero, then x_i could remain on the boundary of the classifier and, consequently, classification of x_i is a complicated problem.

3   Proposed Method

In order to overcome the problem, the selection/prediction criterion based on p_i and q_i is adjusted using a posteriori probability and a learning algorithm is proposed continuously.

3.1   Adjusted Criteria

As mentioned previously, using p_i and q_i can lead to incorrect decisions in the selection and labeling steps; this is particularly common when the summation of the similarity measurement from $x_i \in U$ to $x_j \in L$ is too weak, as follows:

$$X_i^+ - X_i^- \ll X_i^u, \tag{4}$$

or

$$X_i^+ \approx X_i^-. \tag{5}$$

In this case, the confident measurement is formulated as

$$p_i - q_i \simeq X_i^u. \tag{6}$$

From (6), it can be observed that the confidence level of a certain example $x_i \in U$ is measured using the distance between x_i and $x_j \in U$, while excluding L. As a consequence, the measurement is determined using U only and, therefore, sometimes it does not function as a criterion for selecting strong examples. In order to avoid this, the confidence level of (3) can be adjusted as follows:

$$|\rho(x_i)| = \left| X_i^+ - X_i^- + X_i^u - (1 - P_i) \right|, \tag{7}$$

where $P_i(= P(x_i))$ denotes the class posterior probability of instance of x_i (i.e. a *certainty* level); meanwhile $(1 - P_i)$ denotes an *uncertainty* level reflecting the incompleteness in measuring the confidence value. That is, in computation of the confidence level of x_i, in particular, when $X_i^+ \approx X_i^-$, the confidence is adjusted by considering the uncertainty level.

In [8], for example, the quantity of P_i is computed as follows:

$$P_i = p_E(x_i), \tag{8}$$

where $p_E(x_i)(= 1 - \bar{p}_E(x))$ denotes the class-conditional probability, which is estimated from L data under the assumption of the Gaussian distribution. However, sometimes, the value computed is relatively too large, compared to the values of the other three terms. From this observation, in this paper the quantity of P_i can be measured as follows:

$$P_i = sign(f_\theta(x_i)) \left(\widetilde{p}_E(x_i) - \widetilde{\bar{p}}_E(x) \right)^2, \tag{9}$$

where $f_\theta(x_i)$ denotes the predicted label ($\{+1, -1\}$) of a decision function f (e.g. SVM) for an example x_i, and $\theta = (w, b)$ denotes the parameters of the function; $\widetilde{p}_E(x_i)$ denotes the normalization of $p_E(x_i)$.

Also, P_i can be measured based on the entropy for 2-class problems as follows:

$$P_i = - \sum_{j=1}^{2} p(c_j|x_i) log p(c_j|x_i), \tag{10}$$

where $p(c_j|x_i)$ denotes the class posterior probability of instance x_i belonging to the j-th class c_j.

Methods of computing the class-conditional probability ($p_E(x_i)$) and using it are omitted here, but can be found in the related literature, including [3, 10].

3.2 Proposed Algorithm

The proposed learning algorithm begins with predicting the labels of U using a supervised SVM classifier trained with L only. After initializing the related parameters, e.g. the kernel function and its related conditions, the confidence levels of U data are first calculated using the criteria in (2), (8), (9), or (10). Next, the confidence levels are sorted in descending order to select strong examples. The sampling function described can be formulated as follows.

Algorithm 1. Sampling

Input: labeled data (L) and unlabeled data (U).
Output: selected unlabeled data (U_s).
Procedure: perform the following steps to select U_s from U.

1. For each example of $x_i \in U$, compute the confidence levels $\{|p_i - q_i|\}_{i=1}^{n_u}$ (and $\{|\rho(x_i)|\}_{i=1}^{n_u}$) using the criterion in (2), (8), (9), or (10).
2. After sorting the confidence levels in descending order, choose a small portion from the top of the unlabeled data (e.g. 10% top) as U_s, according to their levels.
3. Update the estimated label for any selected example x_i using $sign(\rho(x_i))$.

End Algorithm

After selecting a few examples ranked with the highest confidence levels, combining them with L creates a training set for the base classifier (h). The selection and training steps are repeated while verifying the training error rates of the classifier. The repeated regression leads to an improved classification process and, in turn, provides better prediction of the labels over iterations. Consequently, the best training set, which is composed of L and U_s examples, constitutes the final classifier for the problem in hand. In particular, for the t-th iteration, the classification performance of h^t is measured using validation data (V) as well as U data through a weighted cost function. As in [5,7], the weighted cost function (J) can be defined for any scalar decreasing function c (and its derivative c') as follows:

$$J^{(t)} = \sum_{i=1}^{|V|+|U|} w_i^{(t)} y_i h_i^{(t+1)}, \tag{11}$$

where

$$w_i^{(t)} = \frac{c'(y_i h_i^{(t)})}{\sum_{j=1}^{|V|+|U|} c'(y_j h_j^{(t)})}. \tag{12}$$

Based on this brief explanation, the proposed algorithm for improving the S3VM using the modified criterion is summarized as follows, where the labeled and unlabeled data (L and U), cardinality of U_s, number of iterations (e.g. $t_1 = 100$), and type of kernel function and its related constants (i.e. C and C^*), are given as input parameters. As outputs, the labels of all data and the classifier model are obtained:

Algorithm 2. Proposed Learning Algorithm

Input: labeled training data (L), labeled validation data (V), unlabeled data (U), and number of iteration (IT).

Output: the final combined classifier (H_f).

Initialization: select $U_s^{(0)}$ from U through a supervised classifier (e.g., SVM) trained with L; set the parameters, e.g. C and C^*, and kernel function (Φ); train the first combined classifier (H_f) with $L \cup U_s^{(0)}$; compute the training error ($\epsilon(H_f)$) using L only; $J(H_f) = -\infty$; and initialize weight vectors of all $x_i \in \{V \cup U\}$ with $w_i^{(0)} = \frac{1}{n_v + n_u}$, where $n_v = |V|$ and $n_u = |U|$.

Procedure: repeat the following steps while increasing t from 1 to IT in increments of 1.

1. Choose $U_s^{(t)}$ from U using the sampling function (i.e., Algorithm 1), where the previously trained S3VM is invoked.
2. Train a new classifier (SVM), $h^{(t)}$, using both L and $U_s^{(t)}$, and obtain the training error ($\epsilon(h^{(t)})$) with L.
3. Compute $J(h^{(t)})$ using the weighted cost function in (11) for the $L \cup U$ examples.
4. If $J \leq 0$ then break; else go on;
5. If $J(h^{(t)}) \geq J(H_f)$ then keep $h^{(t)}$ to the combined classifier, i.e $H_f \leftarrow H_f + h^{(t)}$ and $J(H_f) \leftarrow J(h^{(t)})$.
6. After updating the weight $w^{(t)}$ go to (1).

End Algorithm

---

4   Experimental Results

In order to illustrate the functioning of the modified criterion, the original and modified criteria are compared using a synthetic data first and then experimented using real-life data.

4.1   Experiment # 1 (Synthetic Data)

First, a two-dimensional, two-class synthetic data set was generated using MAT-LAB codes: [0.1*randn(5,2)+1; 0.1*randn(5,2)+5] for L data and [0.5*randn(100,2)+2.6; 0.5*randn(100,2)+3.4] for U data. Second, confidence values were computed for all $x_i \in U$ using the four criteria in (2), (8), (9), and (10). Hereafter, they are named SB, $\rho 1$, $\rho 2$, and $\rho 3$, respectively. Finally, top 30% the cardinality of U was selected as a U_s. Fig. 1 presents a comparison of the selection results obtained with the original criterion and the three adjusted criteria for the synthetic data.

From Fig. 1 (a), it should be observed that SB has a difficulty when selecting the useful U_s from U, especially from U data being on the margin. This is clearly demonstrated by different colors of the 'o' and '◇' symbols. Meanwhile, from Fig. 1 (b) and (c), using the adjusted criteria of $\rho 1$ and $\rho 2$ leads to choosing the examples most closely to the labeled examples and being far away from the boundary. However, from Fig. 1 (d), it should also be observed that $\rho 3$ is much more sensitive to changes in the confidence value than $\rho 1$ and $\rho 2$ (refer to the difference in the regions of the boundary in Fig. 1 (b), (c), and (d)).

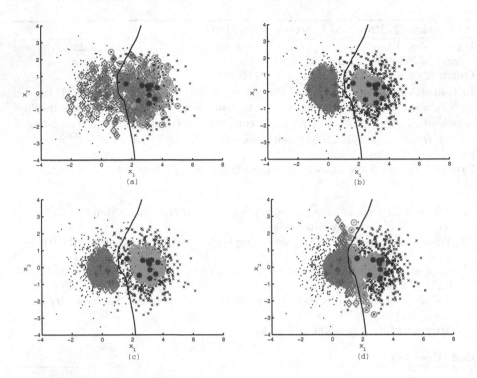

Fig. 1. Comparison of the selections based on (a) the original criterion (SB) and (b) - (d) the three adjusted ones ($\rho1$, $\rho2$, and $\rho3$) for the synthetic data. Here, the objects in the positive and negative classes are respectively denoted by '+' (in red) and '*' (in blue) symbols, while the unlabeled objects of the two classes are indicated using '·' and '×' symbols, respectively. In addition, the selected objects from the two classes are marked with '◇' (in pink) and 'o' (in green) symbols. Also, the boundary lines, with which the two classes are divided, are obtained using the corresponding learning algorithms.

4.2   Experiment # 2 (Real-life Data)

In order to further investigate the run-time characteristics obtained using the three selection criteria, the experiment was repeated for real-life data sets: the SSL-Book benchmark data [4] [1] and UCI data cited from UCI Machine Learning Repository [1] [2], which are summarized in Table 1.

First, each dataset was divided into four subsets as: the labeled training subset, the labeled validation subset, the test subset, and the unlabeled subset at a ratio of 20%: 10%: 20%: 50% [3]; second, a subset (U_s) was selected from U

---

[1] http://www.kyb.tuebingen.mpg.de/ssl-book/

[2] http://archive.ics.uci.edu/ml.

[3] In this experiment, as was done in the related references [8], [9], each dataset was randomly divided into the four subsets under the condition, $|L| \ll |U|$. However, the entire problem of investigating the impacts of having different ratios remains open.

Table 1. Characteristics of the data

Data types	Dataset names	# of features	# of classes	# of objects [# class1 , # class2]
SSL-Book	Digit1	241	2	1500 [734, 766]
	BCI	117	2	400 [200, 200]
	USPS	241	2	1500 [1200, 300]
	g241n	241	2	1500 [748, 752]
UCI	Heart	13	2	297 [160, 137]
	Breast	9	2	683 [444, 239]
	Ionosphere	34	2	351 [225, 126]
	Diabetes	8	2	768 [500, 268]

after computing the confidence values using the four criteria (i.e., SB, $\rho1$, $\rho2$, and $\rho3$) and, in turn, their pseudo-labels were predicted; finally, using the enlarged training data (i.e., $L \cup U_s$), a S3VM classifier was trained first and then evaluated using the test data. The training and evaluation procedures were repeated *fifty* times and the classification error rates were computed by averaging the obtained results. Fig. 2 presents a comparison of the mean error rates obtained from the experiment.

Observations obtained from the figure are as follows. First, when comparing SB to $\rho1$ (as well as $\rho2$ and $\rho3$) in Fig. 2, the latter criterion is generally superior

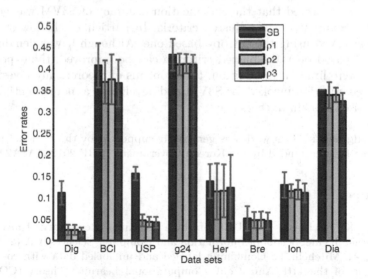

Fig. 2. Comparison of the classification error rates (mean values and standard deviations) obtained using four *S3VM* classifiers for the SSL-Book and UCI datasets. Here, each classifier was trained with $L \cup U_s$'s (top 10% cardinality of $|U|$). Also, as was done in Fig. 1, four U_s's were selected from U using SB, $\rho1$, $\rho2$, and $\rho3$ criteria, respectively. In addition, the datasets are represented with three letter acronym.

to the former in terms of classification accuracies. For all of the datasets, the three adjusted criteria work better than the original. Second, when comparing the error rates among $\rho1$, $\rho2$, and $\rho3$, $\rho3$, albeit not always, is superior to the others. In particular, for Digit1, BCI, USPS, Breast, Ionosphere, and Diabetes data, the lowest rates were achieved using $\rho2$, while, for g24ln data, both the error rates of $\rho1$ and $\rho3$ are the same. However, for Heart data, the lowest rate was achieved using $\rho1$.

In summary, it is not easy to decide which kind of significant adjusted criterion is more suitable for selecting useful unlabeled data for SSL algorithms. However, as a matter of comparison, it is clear that with regard to the classification accuracies, the entropy based criterion is *marginally* better than the others for some SSL-Book and UCI data.

5 Conclusions

In order to improve the classification performance of semi-supervise learning algorithms, in this paper, an empirical study on comparing the adjusted criteria for selecting a small amount of useful unlabeled data for semi-supervised support vector machine (S3VM) algorithms was performed. Four selection criteria, including the original and the three adjusted criteria of the simply adjusted, the normalized-and-adjusted, and the entropy based adjusted, were considered and evaluated with publicly available benchmark datasets. The experimental results obtained demonstrated that the classification accuracy of S3VM was generally improved by employing the adjusted criteria. In particular, the lowest accuracies were achieved using the entropy based one. Although it was demonstrated that S3VM based on the adjusted criterion can be improved, the experiments performed were limited. In addition, the problems of theoretically investigating the adjusted criterion invoked for S3VM and developing a more scientific model for comparison remain unchallenged.

Acknowledgments. This work was generously supported by the National Research Foundation of Korea funded by the Korean Government (NRF-2012R1A1A2041661).

References

1. Bache, K., Lichman, M.: UCI Machine Learning Repository. University of California, Irvine, School of Information and Computer Sciences, CA (2013)
2. Blum, A., Mitchell, T.: Combining labeled and unlabeled data with co-training. In: Proc. of the 11th Ann. Conf. Computational Learning Theory (COLT 98), Madison, WI, pp. 92–100 (1998)
3. Chang, C.-C., Lin, C.-J.: LIBSVM: a library for support vector machines. ACM Trans. Intelligent Systems and Technology **2**(3), 1–27 (2011). http://www.csie.ntu.edu.tw/~cjlin/libsvm
4. Chapelle, O., Schölkopf, B., Zien, A.: Semi-Supervised Learning. The MIT Press, MA (2006)

5. d'Alché-Buc, F., Grandvalet, Y., Ambroise, C.: Semi-supervised marginboost. Advances in Neural Information Processing Systems (NIPS), pp. 553–560. The MIT Press, London (2002)
6. Dagan, I., Engelson, S. P.: Committee-based sampling for training probabilistic classifiers. In: Proc. of the 12th Int'l Conf. on Machine Learning (ICML 1995), pp. 150–157. Morgan Kaufmann, Tahoe City, CA (1995)
7. Le, T.-B., Kim, S.-W.: On incrementally using a small portion of strong unlabeled data for semi-supervised learning algorithms. Pattern Recognition Letters **41**, 53–64 (2014)
8. Le, T. -B., Kim, S. -W.: On selecting helpful unlabeled data for improving semi-supervised support vector machines. In: Proc. of the 3rd Int'l Conf. on Pattern Recognition Applications and Methods (ICPRAM 2014), Angers, France, pp. 48–59 (2014)
9. Mallapragada, P.K., Jain, A.K., Liu, Y.: SemiBoost: boosting for semi-supervised learning. IEEE Trans. Pattern Anal. and Machine Intell. **31**(11), 2000–2014 (2009)
10. Platt, J.C.: Probabilistic outputs for support vector machines and comparison to regularized likelihood methods. In: Smola, A., Bartlett, P., Schölkopf, B., Schuurmans, D. (eds.) Advances in Large Margin Classifiers. The MIT Press, Cambridge (2000)
11. Reitmaier, T., Sick, B.: Let us know your decision: Pool-based active training of a generative classifier with the selection strategy 4DS. Information Sciences **230**, 106–131 (2013)
12. Yarowsky, D.: Unsupervised word sense disambiguation rivaling supervised methods. In: Proc. of the 33rd annual meeting on Association for Computational Linguistics (ACL1995), Cambridge, MA, 189–196 (1995)
13. Zhu, X.: Semi-Supervised Learning Literature Survey. Technical Report 1530, Dept. of Computer Sciences, University of Wisconsin at Madison, MA (2006)

A Novel Clustering Algorithm Based on a Non-parametric "Anti-Bayesian" Paradigm

Hugo Lewi Hammer[1]([✉]), Anis Yazidi[1], and B. John Oommen[2]

[1] Department of Computer Science, Oslo and Akershus University
College of Applied Sciences, Oslo, Norway
hugo.hammer@hioa.no
[2] School of Computer Science, Carleton University, Ottawa, Canada

Abstract. The problem of clustering, or unsupervised classification, has been solved by a myriad of techniques, all of which depend, either directly or implicitly, on the Bayesian principle of optimal classification. To be more specific, within a Bayesian paradigm, if one is to compare the testing sample with only *a single* point in the feature space from each class, the *optimal* Bayesian strategy would be to achieve this based on the distance from the corresponding means or *central* points in the respective distributions. When this principle is applied in clustering, one would assign an unassigned sample into the cluster whose mean is the closest, and this can be done in either a bottom-up or a top-down manner. This paper pioneers a clustering achieved in an "Anti-Bayesian" manner, and is based on the breakthrough classification paradigm pioneered by Oommen *et al*. The latter relies on a radically different approach for classifying data points based on the non-central *quantiles* of the distributions. Surprisingly and counter-intuitively, this turns out to work equally or close-to-equally well to an optimal supervised Bayesian scheme, which thus begs the natural extension to the unexplored arena of clustering. Our algorithm can be seen as the Anti-Bayesian counter-part of the well-known k-means algorithm (The fundamental Anti-Bayesian paradigm need not just be used to the k-means principle. Rather, we hypothesize that it can be adapted to any of the scores of techniques that is *indirectly* based on the Bayesian paradigm.), where we assign points to clusters using quantiles rather than the clusters' centroids. Extensive experimentation (This paper contains the *prima facie* results of experiments done on one and two-dimensional data. The extensions to multi-dimensional data are not included in the interest of space, and would use the corresponding multi-dimensional Anti-Naïve-Bayes classification rules given in [1].) demonstrates that our Anti-Bayesian clustering converges fast and with precision results competitive to a k-means clustering.

John Oommen—Chancellor's Professor; *Fellow: IEEE and Fellow: IAPR*. This author is also an *Adjunct Professor* with the University of Agder in Grimstad, Norway.

M. Ali et al. (Eds.): IEA/AIE 2015, LNAI 9101, pp. 536–545, 2015.
DOI: 10.1007/978-3-319-19066-2_52

1 Introduction

Clustering is a key task in data analysis [2] [3]. There exist a range of different clustering methods that vary in the understanding of what a cluster is. For instance, density models, such as OPTICS [4] and DBSCAN [5], find the most dense regions in the space. As opposed to this, in hierarchical clustering [6] [7], the aim is to arrange the data points in an underlying hierarchy. A third group of clustering algorithms constitute the so-called "centroid" methods where each cluster is represented by a single point. The most prominent example of a scheme within this family is the k-means clustering algorithm where a centroid is represented by the mean value of the points in the cluster. The central strategy motivating *these* clustering schemes involves classifying data points to the different clusters based on the distances to the means (or centroids) of the clusters.

In this paper we introduce a novel alternative to the k-means clustering algorithm. We shall demonstrate that we can obtain excellent clustering performance by operating in a diametrically opposite way, i.e., a so-called "Anti-Bayesian manner. Indeed, we shall show the completely counter-intuitive result that by working with a few points distant from the "mean (centroid), one can obtain remarkable clustering performances. While the clustering algorithm in this paper follows the steps of a typical k-means clustering algorithm, it assigns the data points to the already-formed clusters using completely different criteria – by invoking the concepts of Anti-Bayesian Pattern Recognition (PR). Rather, unlike the k-means clustering strategies which rely on centroid-based criteria, our paradigm advocates the association of points to clusters based on *quantiles distant from the cluster means* [8] [1] [9], which is a concept that is unreported in the literature. Indeed, it is actually both un-intuitive and non-obvious.

We consider the non-parametric clustering problem where the distribution of each cluster is unknown. This is in contrast to the work in [8] [9] reported in the area of classification where the distributions are known, and more in the directions of [1], where a non-parametric case is considered. In [1] the quantiles are estimated by assuming that the data points are sampled from Gaussian distributions. In this paper we work with a totally distribution-free model which is more natural when the aim is to achieve flexible and robust clustering.

In this paper, in the interest of space and brevity, we merely propose the Anti-Bayesian clustering strategy for one and two-dimensional data. This is because the aim of the paper is to introduce the paradigm in a *prima facie* setting. Although the extensions to multi-dimensional data are still open, they would not be too complicated – they would use the corresponding multi-dimensional Anti-Naïve-Bayes classification rules given in [1].

2 Anti-Bayesian Clustering

Let $x_1, x_2, \ldots, x_n$ represent n points in $\mathbb{R}^p$. In this paper we present an algorithm, which is based on the Anti-Bayesian classification framework [8] [1] [9], to cluster these points into k clusters. The present algorithm follows the same steps as a

k-means clustering algorithm. To motivate this discussion, we start this section with a short review of the classical k-means algorithm. Then, in Section 2.2 we briefly review the Anti-Bayesian classification framework which forms the basis for the Anti-Bayesian clustering algorithm presented in Section 2.3. Before we proceed, we re-iterate our earlier comment: Since the fundamental Anti-Bayesian paradigm need not just be used to the k-means principle, our paradigm can be adapted to any of the scores of techniques that are *inherently* Bayesian.

2.1 k-means Clustering

Let $m_1, m_2, \ldots, m_k \in \mathbb{R}^p$ represent the respective centroids of each of the k clusters currently described, $C_1, \ldots, C_k$. The algorithm starts by associating an initial value to each centroid. A simple approach is to assign each of centroids to some of the points $x_1, x_2, \ldots, x_n$. The algorithm then consists of two steps:

1. **Assignment:** Assign each point x_i to the nearest[1] centroid.
2. **Update:** When all points are assigned to a cluster, update the centroid value of each cluster as the average of all the points in the cluster:

$$m_j = \frac{1}{|C_j|} \sum_{i=1}^{|C_j|} x_{j(i)}, \quad j = 1, \ldots, k,$$

   where $|C_j|$ is the number of points in cluster C_j and $j(1), j(2), \ldots, j(|C_j|)$ represent the points assigned to cluster C_j.
3. **Loop:** Repeat the above two steps until no points switch their clusters.

2.2   Anti-Bayesian Classification

The following classification method is based on the "Anti-Bayesian" methodology described and proven in [8] [1] [9]. To explain how it works, we start with the uni-dimensional case.

Let $x_1, x_2, \ldots, x_{n_1}$ and $y_1, y_2, \ldots, y_{n_2}$ be random samples from some unknown probability distributions $f_X(x)$ and $f_Y(y)$, respectively. Our task is to classify a new point z to see whether it is a sample from $f_X(x)$ or $f_Y(y)$. If we assume that the variances of $f_X(x)$ and $f_Y(y)$ are equal, the optimal classification strategy is to assign z to $f_X(x)$ or $f_Y(y)$ if z is closer to the means of X or Y respectively. This, in turn, assigns z to $f_X(x)$ if the average of the samples $x_1, x_2, \ldots, x_{n_1}$ is closer to z than the average of $y_1, y_2, \ldots, y_{n_2}$. It is otherwise assigned to $f_Y(y)$. The reader should observe this is precisely how points are assigned to clusters in the k-means paradigm.

In the Anti-Bayesian classification approach, classification is achieved based on quantile-based comparisons rather than comparisons with regard to the mean. To render this formal, we denote the quantiles as follows: $q_p^X = P(X > p)$.

---

[1] It should be mentioned that the concept of "nearest" can be based on the specific metric being used, for example, a simple Euclidean metric.

Although, in practice, the quantiles have to be, estimated (or learned), for ease of clarification, in the descriptions below, we assume that the quantiles are known. We also assume that $q_{1-p}^X < \text{Median}(X)$ so that q_p^X is always greater than q_{1-p}^X. In such a case, the Anti-Bayesian classification method operates as follows:

1. Determine which of the distributions $f_X(x)$ or $f_Y(y)$ is to the left by using the quantiles of the distributions. We have three possible cases:

   Case 1: If $q_p^X < q_p^Y$ and $q_{1-p}^X < q_{1-p}^Y \implies f_X(x)$ is to the left of $f_Y(y)$.
   Case 2: If $q_p^X > q_p^Y$ and $q_{1-p}^X > q_{1-p}^Y \implies f_Y(y)$ is to the left of $f_X(x)$.
   Case 3: Else, we determine their relative positions by comparing the averages of the quantiles as follows:
   If $\frac{q_p^X + q_{1-p}^X}{2} < \frac{q_p^Y + q_{1-p}^Y}{2} \implies f_X(x)$ is to the left of $f_Y(y)$.
   Else[2] $f_Y(y)$ is to the left of $f_X(x)$.

   Figure 1 depicts the above three cases. We see that for Cases 1 and 2, $f_X(x)$ and $f_Y(y)$ is the distribution to the left, respectively. In the bottom figure (Case 3), the decision is not that obvious because the classes are highly overlapping.

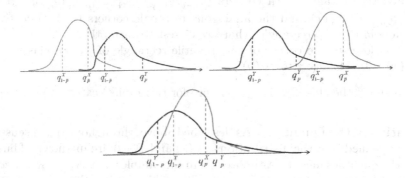

Fig. 1. This figure depicts Cases 1, 2 and 3 – arranged from left to right and from top to bottom respectively

2. Once the relative positions of the distributions are determined, the classification rule must now be specified. For simplicity, we describe this just for Case 1 since the rules for the "mirrored" cases are analogous. The Anti-Bayesian rule classifies using the *right* quantile of the left distribution and the *left* quantile of the right distribution. If $B = \frac{q_p^X + q_{1-p}^Y}{2}$, we classify as follows:
   If $z < B$, classify z to $f_X(x)$.
   Else, classify z to $f_Y(y)$.
   This approach works even when the distributions overlap such that q_{1-p}^Y is to the left of q_p^X as shown in Figure 2.

The theoretical motivation for this algorithm is given in [8] and [9] and not repeated here.

---

[2] This case occurs rarely in practice except when the classes are highly overlapping, in which case the PR problem is often meaningless.

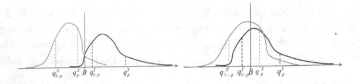

Fig. 2. The left panel shows the standard situation under Case 1, while the right panel shows a situation when q^Y_{1-p} is to the left of q^X_p

We now go over to the case where the points are two dimensional. Let $x_{11}, \ldots, x_{n_1 1}$ and $x_{12}, \ldots, x_{n_1 2}$ be $2n_1$ independent samples from $f_X(x)$ and define the points (x_{i1}, x_{i2}), $i = 1, 2, \ldots n_1$. Similarly, define the points (y_{i1}, y_{i2}), $i = 1, 2, \ldots n_2$ based on samples from $f_Y(y)$. Again we want to classify a point z to $f_X(x)$ or $f_Y(y)$. The classification is done as per the ideas in [1] and [9]. It is a natural generalization of the one dimensional case above and follows two steps.

1. Define the rectangle with corners (q^X_{1-p}, q^X_{1-p}), (q^X_{1-p}, q^X_p), (q^X_p, q^X_{1-p}) and (q^X_p, q^X_p) for $f_X(x)$, and the analogous rectangle corners for Y. Locate the corners in the two rectangles that are closest to each other.
2. If z is closer to the corner of the quantile rectangle of $f_X(x)$, classify z to $f_X(x)$. Else classify z to $f_Y(y)$.

Figure 3 shows the classification procedure for two typical cases.

Estimation of the Quantiles. As described above, throughout this discussion, we have assumed that the distributions $f_X(x)$ and $f_Y(y)$ are unknown. Thus, in reality, the quantiles must be estimated from the samples. Let $x_{(1)}, x_{(2)}, \ldots, x_{(n_1)}$ be n_1 samples from $f_X(x)$ sorted by value. To achieve this estimation, we adapt the method recommended in [10] where the cumulative distribution function of $f_X(x)$ is estimated with a linear interpolation (spline) through the points $(x_{(i)}, p_i)$, $i = 1, 2, \ldots, n_1$, and where p_i is defined as:

$$p_i = \frac{i - 1/3}{n_1 + 1/3}, \ i = 1, 2, \ldots, n_1.$$

The estimate of q^X_p can then be easily read (or rather, inferred) from this curve.

2.3   Clustering Based on Anti-Bayesian Classification

We now present a clustering algorithm that uses that Anti-Bayesian classification methodology presented in Section 2.2 combined with the assignment/update steps in the k-means clustering algorithm. Suppose that we have a set of points $x_1, x_2, \ldots, x_n$ that we want to group into k clusters denoted by $C_1, C_2, \ldots, C_k$. Let q^j_{1-p} and q^j_p represent the $(1-p)$-valued and p-valued quantiles of the data in cluster C_j. The algorithm starts by associating values to the quantiles of all

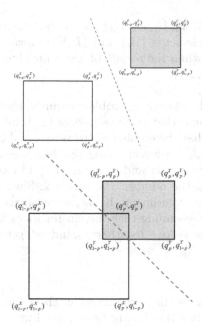

Fig. 3. Classification in the two dimensional scenario for two typical cases. The gray dashed line shows the border of the discriminant regions when z is classified to $f_X(x)$ or to $f_Y(y)$.

the clusters (q_{1-p}^j, q_p^j), $j = 1, 2, \ldots, k$. One way to achieve this is by randomly selecting $2k$ points and by associating the quantiles to the values of these points. To prevent an excessive initial overlapping of the clusters, it is natural to first sort the points before associating them to the quantiles. Similar to the k-means scheme, the present algorithm now consists of two steps:

1. **Assignment:** Assign each point to a cluster. For each point, x_i, we do the following by repeated classifications using the methodology in Section 2.2. We start to determine if x_i is most likely to belong to C_1 or C_2. Assume that x_i is most likely to belong to C_2. We then say that C_2 is the current candidate cluster for x_i. Next we do an evaluation between C_2 and C_3 and repeat this for all the remaining clusters $C_4, \ldots, C_k$. In the last step we do an evaluation with the current candidate cluster from the previous evaluations, say C_a, and C_k. If x_i is more likely to belong to C_a, we assign x_i to this cluster, else we assign x_i to C_k.
2. **Update:** When all the points are assigned to clusters, we estimate the quantiles of all the clusters, i.e., (q_{1-p}^j, q_p^j), $j = 1, 2, \ldots, k$, using the estimator presented at the end of Section 2.2.

2.4   Evaluation of Clustering Performance

The question of how to measure the performance of a clustering algorithm is far from being obvious or trivial. Consider the following simple example. Suppose that six points $\{A_1, A_2, A_3, B_1, B_2, B_3\}$ are to be clustered into two clusters, with the goal that the elements $\{A_i\}$ and $\{B_i\}$ are located in the same cluster. Consider now the case when the results of a specific clustering algorithm are:

Cluster C_1: $\{A_1, B_2, B_3\}$,

Cluster C_2: $\{A_2, A_3, B_1\}$.

If the requirements of the clustering problem required that all the elements $\{A_i\}$ are to be in cluster C_1, and that all the elements $\{B_i\}$ are to be in cluster C_2, we could immediately see that the number of errors incurred by the above clustering is 4. On the other hand, if the clustering problem merely stipulated that the $\{A_i\}$'s were to be in one cluster and that the $\{B_i\}$'s in another (irrespective of whether it is C_1 or C_2), the number of errors is 2. Since the latter is the more meaningful issue, in the experiments below, we evaluate the performance of a clustering algorithm by assuming that the cluster index is irrelevant as long as the elements that should belong together do, indeed, get clustered together.

3   Experiments

In this section we compare the performance of the k-means and Anti-Bayesian clustering algorithms when the data to be clustered was generated from a range of different distributions. For all the experiments we considered $k = 3$ clusters. The parameters of the distributions are shown in Table 1.

Table 1. Distributions used in the experiments. The second column shows the parameters of the distributions of the leftmost cluster. The third and the fourth columns show how much the distribution is shifted to the right for cluster two and three for the one and two dimensional experiments.

Distribution	Parameters Used	Shift: 1D Case	Shift 2D Case
Normal	$\mu = 0,\ \sigma = 1$	3, 6	2, 4
Uniform	$a = 0,\ b = 1$	9/10, 9/5	4/3, 8/3
Laplace	$\mu = 0,\ \lambda = 1/4$	4/5, 8/5	2/3, 4/3
Beta (symmetric)	$a = 2,\ b = 2$	5/8, 5/4	5/11, 10/11
Beta (asymmetric)	$a = 3,\ b = 1$	0.65, 1.29	0.42, 0.84
Gamma	$a = 3,\ s = 3$	2.13, 4.27	1.67, 3.33

A plot of all the distributions in the one dimensional experiment are shown in Figure 4. The distributions are the same for the two dimensional experiment, except that the distributions are less separated (shifted) from each other.

We considered two cases: In the first case, we generated $N = 10$ independent samples from each distribution, i.e. a total of 30 samples, and in the second where we generated $N = 1,000$ independent samples from each distribution. To evaluate the performance of the clustering algorithms, we first generated the synthetic data, performed the clustering and then measured the portion of samples that were classified to wrong cluster. This procedure was repeated a large amount of times to minimize the Monte Carlo error. In all the cases, we set $p = 1/3$ in the computation of the quantiles for the Anti-Bayesian clustering.

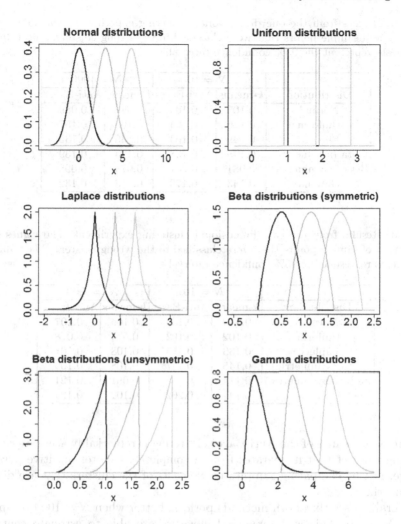

Fig. 4. Plot of all the distributions presented in Table 1 for the one dimensional case

Tables 2 and 3 show the results for the one and two dimensional cases respectively. It is appropriate to mention that similar to the k-means algorithm, in some rare cases, the Anti-Bayesian algorithm did not converge, but cycled between a few different configurations. In such cases, we terminated it after 100 iterations and reported the final clusters. We also computed the 95% confidence intervals for the portions in Tables 2 and 3 and the width of these intervals were ≈ 0.002, implying that the Monte Carlo error was almost completely removed. These were thus not included in the tables.

There are many factors that distinguish the performances of the k-means and Anti-Bayesian clustering algorithms.

1. The k-means scheme needs to estimate the means (centroids) of the clusters, while the Anti-Bayes scheme estimates the quantiles from the samples.

Table 2. Results from the one dimensional clustering experiment. The values show the portion of sample points that were classified to the wrong clusters. The values in parentheses represent the 95% confidence interval.

Distribution	$N = 10$		$N = 1000$	
	k-means	Anti-Bayes	k-means	Anti-Bayes
Normal	0.105	0.105	0.090	0.089
Uniform	0.116	0.104	0.106	0.157
Laplace	0.149	0.163	0.136	0.135
Beta (symmetric)	0.145	0.138	0.125	0.139
Beta (asymmetric)	0.081	0.087	0.074	0.098
Gamma	0.143	0.170	0.113	0.132

Table 3. Results from the two dimensional clustering experiment. The values show the portion of sample points that were classified to the wrong clusters. The values in parentheses represent the 95% confidence interval.

Distribution	$N = 10$		$N = 1000$	
	k-means	Anti-Bayes	k-means	Anti-Bayes
Normal	0.140	0.145	0.106	0.107
Uniform	0.102	0.102	0.074	0.078
Laplace	0.139	0.156	0.108	0.108
Beta (symmetric)	0.145	0.147	0.108	0.109
Beta (unsymmetric)	0.114	0.168	0.081	0.121
Gamma	0.141	0.202	0.102	0.124

2. The performance of the Anti-Bayes clustering is remarkably accurate considering the fact that it operates from a completely counter-intuitive perspective, i.e., by comparing samples to elements of the clusters that are distant from the means.

3. Overall we see that both methods perform better when $N = 1000$ compared to $N = 10$. This is as expected since we are able to estimate centroids and quantiles with better precision. For $N = 10$, the k-means performs marginally better than Anti-Bayes for the symmetric distributions, but there are exceptions. For the one dimensional case, the Anti-Bayes performs the best for the Uniform and the Beta (symmetric) distributions.

4. When the distributions are asymmetric, it is known that the Anti-Bayes classification does not perform as well as the Bayes' bound [1] [9]. For these asymmetric distributions, the differences between the k-means and Anti-Bayes schemes is larger, but these difference are, really, quite small.

5. As mentioned in the introduction, in this paper, we have merely concentrated on the Anti-Bayesian clustering strategy for one and two-dimensional data. The extensions to multi-dimensional data are still open, but they would not be too complicated. Indeed, they would use the corresponding multi-dimensional *Anti-Naïve-Bayes* classification rules given in [1].

4   Conclusions

In this paper we have demonstrated how the "Anti-Bayesian" pattern classification framework formulated earlier in [8] [1] [9] can be used to build an efficient clustering algorithm competitive with the k-means clustering algorithm. The algorithm documents impressive clustering performance for a range of different distributions.

The k-means algorithm associates points to the clusters relying on the mean values (centroids) of the clusters which is known to be the optimal Bayesian bound. In contrast, the clustering algorithm presented in this paper associates points to clusters using quantiles distant from the cluster mean. Intuitively, one would thus expect a poor clustering performance and even serious convergence problems for the algorithm. But as demonstrated in [8] [1] [9], impressive classification precision can be achieved for the Anti-Bayesian approach which, thus, lays the foundation for the efficient clustering algorithm presented here.

The problems that are open are many. First of all, it would be interesting to demonstrate the power of such a strategy for multi-dimensional and real-life data. But more interestingly, we believe that the fundamental "Anti-Bayesian" paradigm can be applied to *any* clustering technique (apart from the k-means) that is inherently dependent on a Bayesian philosophy. We thus believe that this paper opens the doors to a host of unresolved problems.

References

1. Thomas, A., Oommen, B.J.: Order statistics-based parametric classification for multi-dimensional distributions. Pattern Recognition **46**(12), 3472–3482 (2013)
2. Jain, A.K., Dubes, R.C.: Algorithms for Clustering Dats. Prentice Hall, Englewood Cliffs (1988)
3. Xu, R., Wunsch II, D.: Survey of clustering algorithms. Trans. Neur. Netw. **16**(3), 645–678 (2005)
4. Ankerst, M., Breunig, M.M., Peter Kriegel, H., Sander, J.: Optics: ordering points to identify the clustering structure, pp. 49–60. ACM Press (1999)
5. Ester, M., Peter Kriegel, H., Sander, J., Xu, X.: A density-based algorithm for discovering clusters in large spatial databases with noise, pp. 226–231. AAAI Press (1996)
6. Murtagh, F., Contreras, P.: Methods of hierarchical clustering. CoRR abs/1105.0121 (2011)
7. Sibson, R.: SLINK: An optimally efficient algorithm for the single-link cluster method. The Computer Journal **16**(1), 30–34 (1973)
8. Thomas, A., Oommen, B.J.: The fundamental theory of optimal "anti-bayesian" parametric pattern classification using order statistics criteria. Pattern Recognition **46**(1), 376–388 (2013)
9. Oommen, B.J., Thomas, A.: Anti-Bayesian parametric pattern classification using order statistics criteria for some members of the exponential family. Pattern Recognition **47**(1), 40–55 (2014)
10. Hyndman, R.J., Fan, Y.: Sample quantiles in statistical packages. American Statistician **50**, 361–365 (1996)

Vision, Image and Text Processing

Research on an Intelligent Liquid Lamp Inspector Based on Machine Vision

Lasheng Yu$^{(\boxtimes)}$, Yanying Zeng, and Zhaoke Huang

School of Information Science and Engineering,
Central South University, Changsha 410083, China
yulasheng@csu.edu.cn, 13539823@qq.com, huangzhaoke@126.com

Abstract. The traditional oral liquid light inspect method has many defects and deficiencies. In this paper, an automatic oral liquid light inspection machine is designed to implement the detection of visible particles in oral liquid. Initially, the original images are captured, the background noise is removed using the image technique, and then the target image with visual particles and air bubble noise are acquired. Secondly, an adaptive threshold processing algorithm is used to extract moving target. After that, the position of the biggest object in two different images is compared to remove the overlap section. Finally, the real-time detection system to determine whether the biggest object is visual particle is implemented. A large number of experimental results show that the method proposed in this paper is more accurate and quicker than traditional detection method.

Keywords: Image processing and Vision · Oral liquid inspection · Intelligent control · Difference and energy accumulation · Adaptive threshold processing

1   Introduction

Oral liquid has been widely used in health protection due to its remarkable curative effect. Because of the shortage of traditional craft and environment, bottle quality, packing procedures, collisions, filtration and filling, different kinds of foreign substances such as rubber scraps, glass scraps, aluminum scraps, chemical fibers, hair and other visible particles may appear in oral liquid [1]. It may bring to serious consequences when these kinds of particles are taken by people. "Chinese Pharmacopoeia" has constraints about the diameter and length of insoluble visible particles which should be less than 50 μm [2].

According to the statistic taken by China Pharmaceutical Industry Association (CPIA), 99.6% pharmaceutical factories in China use traditional method to detect visible foreign particles by workers [3] which is totally based on their inspection experience [4]. Though this method is easy, many disadvantages exist in this process such as

Lasheng Yu—This paper has been supported by the cooperation project in industry, education and research of Guangdong province and Ministry of Education of P.R.China (Granted number:2011B090400316).

© Springer International Publishing Switzerland 2015
M. Ali et al. (Eds.): IEA/AIE 2015, LNAI 9101, pp. 549–558, 2015.
DOI: 10.1007/978-3-319-19066-2_53

the low inspection efficiency and high omission rates because of the repeated work while leading to worker's tiredness. Nowadays, the research on the detection of foreign particle in oral liquid based on the machine vision is rare [5,6], some   inspection technologies or equipment maily focus on ampoule and cannot get satisfactory detection results. Therefore, the research and development of online automatic intelligent particle inspection technology for China's oral liquid is significant both theoretically and practically [7,8].

2   System Structure

Learning from the traditional detection process [9], the machine vision based light detection system simulates the processing of manual inspection, in order to get an accurate image samples, the placement of Oral liquid bottle has been changed from vertical to horizontal direction. The structure of this system has been showed in Fig 1. Firstly, the oral liquid bottle is imported into the bottle horizontal channel by a conveyor one by one, the rotation axis rotate the oral liquid bottle quickly, and then make foreign substances in oral bottle suspending and moving, once it arrives at the detection field, it stops the bottle at once and send it to the testing station. At this moment, the bottle has already stopped, however, the solution in the bottle is still moving. Eight successive images have been captured by cameras and then these images are sent into the computer for further processing. The foreign substances could be found out by the position of particles in successive multi-frame images because of the relative movement of particles in the solution. Finally, computer sent commands to a robot on the basis of the results of image recognition, the robot would kick the disqualified bottle according to the order of the bottle and the bottle would be taken into the defective area, otherwise it would be taken into the authentic area.

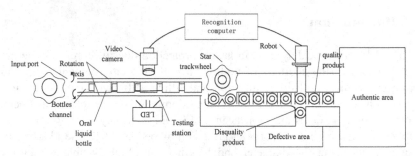

Fig. 1. System structure

3   Key Algorithms of Foreign Substances Detection

Fingerprints, dust, Scratches, graduations, can be seen on the surface of the oral bottles, even some foreign particles or bubbles, can be found in the solution, which make detection more difficult. The foreign substances lack structural information as distinct characteristics, especially they are small and usually occupy only several or dozens of pixels in a single image, and therefore it is hard to distinguish them from noise. and disturbances all depending on their gray values.

In order to solve problems, based on special foreign substances detection algorithms in 4 and 5, two sequence image difference algorithm[10] based on spatiotemporal continuity is proposed to remove existing static disturbances in the background, and then the cumulative energy of the differential image is computed. An adaptive threshold processing is used for image binary segmentation. The detection flowchart is shown in Fig 2.

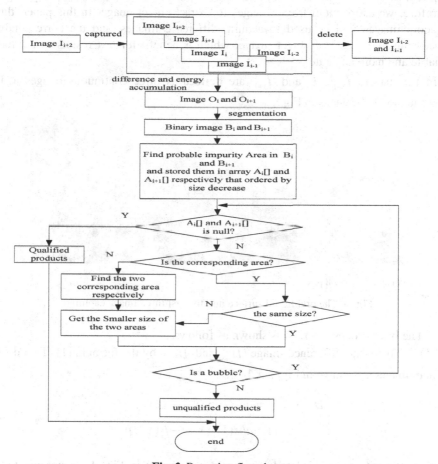

Fig. 2. Detection flowchart

3.1 Difference Algorithm

Two consecutive images are subtracted to remove the static background and isolate the moving objects in the difference image, then the foreign particles in the difference image got more likely to be identified through morphological processing, after treatment, the energy of two successive difference images accumulated, finally, an object image without static background can be obtained.

According to the quantity theory, in the condition of relative station between the camera and bottles, if the difference between two consecutive frames is not zero, this gray value is caused by noise and foreign substances , however the difference image contain less information about foreign substances. And the time interval T of capture image is a key aspect for the detection, if the value of T is too small, it's difficult to find foreign particles as the result of small of movement, but if the value of T is too large, it may affect the capture of moving targets because the foreign particles has stopped. In the other words, the value of T will affect the capture of moving targets, therefore, we adopt multi-frame images to extract object image. In this paper, three adjacent images will be used to acquire difference images, and the corresponding processing to extract the object image, which is benefit for test results to get more reliable and more accurate.

In this paper, I_{i-1}, I_i and I_{i+1} are defined as three continuous images at the time interval T shown as in Fig3.

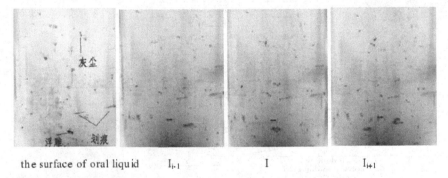

the surface of oral liquid I_{i-1} I I_{i+1}

Fig. 3. The oral liquid surface and the first three image captured

The basis processing flow is shown as follows:

1)   To get the difference image $D_{i,i-1}$ and $D_{i+1,i}$ by the formula (1). Two difference images are shown in Fig. 4 a, b.

$$\begin{cases} D_{i,i-1}(x,y) = abs(I_i(x,y) - I_{i-1}(x,y)) \\ D_{i+1,i}(x,y) = abs(I_{i+1}(x,y) - I_i(x,y)) \end{cases} \tag{1}$$

2)   To get an accumulated image $C_i(x,y)$ shown in Fig. 4,c according to the formula (2). The image $C_i(x,y)$ contains all the movement information of three consecutive frames.

$$C_i(x,y) = D_{i+1,i}(x,y) + D_{i,i-1}(x,y) \tag{2}$$

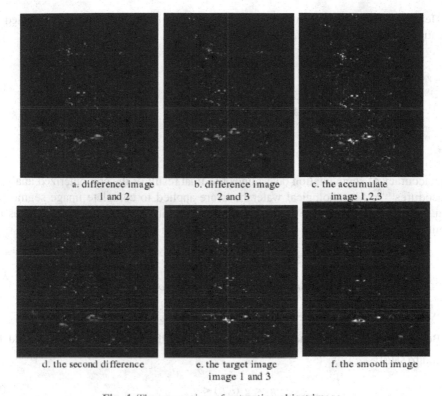

a. difference image 1 and 2	b. difference image 2 and 3	c. the accumulate image 1,2,3
d. the second difference	e. the target image image 1 and 3	f. the smooth image

Fig. 4. The processing of extracting object image

3) In order to extract the target image of the image i, the image i+1 subtract image i-1 to get the second difference image $D_{i+1,i-1}$, as show in Fig. 4 ,d .

$$D_{i+1,i-1} = abs(I_{i+1}(x, y) - I_{i-1}(x, y))$$ (3)

4) To get the target image in Fig.4,e by the formula (4).

$$O_i(x, y) = abs(C_i(x, y) - D_{i+1,i-1}(x, y))$$ (4)

5) A Gaussian Filter as equation (5)is adopted to smooth the target image.

$$G(x, y) = re^{-(x^2+y^2)/2\sigma^2}$$ (5)

Where r is standardized coefficient, it's defined as the equation (6),

$$r = \frac{1}{\sum_x \sum_y G(x, y)}$$ (6)

Here 3*3 Gaussian templates is used to convolute with the target image, expressed as equation (7), to get smooth image that is shown in Fig. 4, f.

$$Smooth_i = O_i(x, y) * G(x, y)$$

$$= \frac{1}{MN} \sum_{m=0}^{M-1} \sum_{n=0}^{N-1} O_i(m, n) G(x - m, y - n) \tag{7}$$

3.2   Image Segmentation

The accuracy of segmentation determines the final result of the computerized analysis procedures[9,10]. Morphological watersheds are applied to complete image segmentation. To get maximum edge information, the gradient image is generally used as an input image. The gradient of image $smooth_i$ is given by

$$g_i(x, y) = grad(Smooth_i(x, y))$$

$$= sqrt((\partial Smooth_i(x, y) / \partial x)^2 + (\partial Smooth_i(x, y) / \partial y)^2) \tag{8}$$

Where $grad(\bullet)$ is a gradient function, and $sqrt(\bullet)$ is a square root operator.

Because of noise and other local irregularities of the gradient, $g_i(x, y)$ should be forced to satisfy

$$g_s < g_i(l, k) < g_e \tag{9}$$

Where g_s and g_e are intensity value defined accordance with pre-knowledge that could remove useless information and avoid image oversegmentation.

The image segmentation algorithm is described as follow:

Let $S_1^i, S_2^i, ...,$ S_R^i be sets of the coordinates of points which in the local minima of image $g_i(x, y)$, and let $W_i[h]$ denote as a set of the coordinates (l, k) which satisfies the condition $g_i(l, k) < g_s$. That is,

$$W_i[h] = \{(l, k) \mid g_i(l, k) < g_s\} \tag{10}$$

where h is an integral variable in the range of $[g_s + 1, g_e + 1]$, and $W_i[h]$ is the set of coordinates of points in image $g_i(x, y)$, located below the planar of $g_i(x, y) = h$, and marked as black points, however other points in the image marks white.

$C^i(S_j^i)$ is a set of the coordinates of points in the catchment basin associating with local region minimum S_j^i, and $C_n^i(S_j^i)$ denotes as a set of coordinates of points in the catchment basin combining with S_j^i at the stage n. $C_n^i(S_j^i)$ can denote as

$$C_n^i(S_j^i) = C^i(S_j^i) \cap W_i[h] \tag{11}$$

Let $C^i[n]$ be the union set of $C_n^i(S_j^i)$,

$$C^i[n] = union(C_n^i(S_j^i)) = \bigcup_{j=1} C_n^i(S_j^i)) \tag{12}$$

Then let $C^i[g_e + 1]$ be the union of all catchment basins:

$$C^i[g_e + 1] = \bigcup_{j=1}^{R} C^i(S_j^i) \tag{13}$$

Based on equation (10) and (11), it's easy to find that each connected component of $C^i[n-1]$, $C^i[g_e + 1] = W_i[g_e + 1]$ as an initial condition to find the watershed lines. Let ST denote as the connected components of $W_i[h]$, to all $st \in ST[h]$ there exists the followings to obtain $C^i[n]$ from $C^i[n-1]$:

When st satisfies $st \cap C^i[n-1] = \varnothing$, let st to be incorporated into $C^i[n-1]$ to create $C^i[n]$. When $st \cap C^i[n-1]$ comprise one connected component of $C^i[n-1]$, let st to be combined into $C^i[n-1]$ to create $C^i[n]$. When $st \cap C^i[n-1]$ contains more than one connected components of $C^i[n-1]$, a dam must be built within st to avoid overflow between the catchment basins.

Gradient image $g_t(x, y)$ though this processing could get white water lines which are considered as segmentation lines in this image, show in Fig. 5. After filling the region within the segmentation line, the interesting areas in image is shown in Fig. 6.

Fig. 5. The gradient image after watershed segmentation

Fig. 6. After filling region with segmentation line

3.3   Foreign Substances Identification

After interesting objects in oral liquid have been selected from image, the most important step is to determine whether the impurities exist in these interesting objects. The key algorithm   is described as follow:

1) Let B_i and B_{i+1} be binary images that denote the segmented image i and $i+1$ after filling region with segmentation line respectively. A_i and A_{i+1} denote as a array stored with interesting areas in image B_i and B_{i+1} respectively.

2) Sorting the array A_i and A_{i+1} from largest to smallest and removing the interesting areas which are smaller than the standard size of impurities in oral liquid. Let S_i and S_{i+1} be a set of the probable area of B_i and B_{i+1} respectively given by

$$\begin{cases} S_i = \{A_i^j > STD \mid j <= length(A_i)\} \\ S_{i+1} = \{A_{i+1}^k > STD \mid k <= length(A_{i+1})\} \end{cases} \tag{14}$$

Where STD is a constant defined by oral manufacturers.

3) Let Cen_i and Cen_{i+1} denote as a vector stored with the coordinates of the central point of interesting areas in S_i and S_{i+1} respectively. Due to every particle in oral liquid all have a maximum speed and minimum speed when they moving. In this case, it's easy to determine the movement range of particles at the interval capture time T in two consecutive images. Let V_{max} and V_{min} denote as maximum speed and minimum speed separately. The range of movement is expressed by

$$\text{range} = \{v_{max}T \leq d \leq v_{min}T \mid d \in R\} \tag{15}$$

4) Let S_i^j denote as the element j in S_i, S_i^1 is the biggest element of S_i. Cen_i^j denotes as the coordinate of the central points of area S_i^j. The distance between S_i^j of S_i and S_{i+1}^k of S_{i+1} is given by

$$D = sqrt((Cen_i^j(1) - Cen_{i+1}^k(1))^2 + (Cen_i^j(2) - Cen_{i+1}^k(2))^2) \tag{16}$$

where $Cen_i^j(1)$ and $Cen_i^j(2)$ is x coordinate and y coordinate of the central point S_i^j , and $j \in [1, length(S_i)]$ $k \in [1, length(S_{i+1})]$. If the distanced D satisfies $D \in range$, then S_i^j and S_{i+1}^k are considered to be corresponding areas.

5) Compare the size of S_i^j and S_{i+1}^k to determine whether exist the phenomenon of overlap, for instance particles, noises and bubbles. If they are far away from each other, that's mean that the existence of overlapping appeared in movement. Then the smaller

area of S_i^j and S_{i+1}^k should be selected and go to step 6. Else the phenomenon of overlap may not exist, then go to step 7 to determine whether S_i^j is a bubble or not.

6) let A_{min}^j be the smaller size of S_i^j and S_{i+1}^k given by

$$A_{min}^j = \min(S_i^j, S_{i+1}^j) \tag{17}$$

7) The bubbles generated through the formation of oral liquid movement would be a ellipse in images, and satisfies

$$\frac{(x - Cen_i^k(1))^2}{a^2} + \frac{(y - Cen_i^k(2))^2}{b^2} = 1 \tag{18}$$

Where (x, y) is the coordinate of edge of area S_i^j, a and b is a constant. If the edge coordinate of S_i^j satisfy this equation, then S_i^j can be considered to be a bubble area. Otherwise S_i^k is impurity area, that's mean this oral liquid is not qualified.

4   Experiments and Results

In the system, an industrial camera TDC1300 with high-precision industrial lenses and side light is used. The system has four channels. Each channel can detect 60 bottles per minute, that is to say the system can detect 240 bottles per minute, which is much quicker than traditional detection method. And it can also record the detected number, unqualified number, the total number. All the parameters can be modified according to the result to make detecting precision higher. This system is tested in the worksite, 500 bottles high concentration oral solution and 500 bottles low concentration oral solution are used to check the possible exist particles such as color-dot, glass scraps, aluminum scraps, hair, short fiber and so on.

From the result of table 1, it can be seen that this system has a low default detection rate and low leak detection rate about the detection of color-dot, glass scraps, aluminum scraps, hair, short fiber, the rate of low concentration oral solution is higher than 92.8%. But for the high concentration oral solution, some of the glass scraps may adhere the bottle cliff, it would be treated as impurity of bottle cliff and cannot be detected, the rate is lower than 85%.

5   Conclusion

This paper describes a real-time visual based automatic intelligent inspection system for foreign substances in oral solution which has been applied in some pharmacy factories in China which largely weakened man-made factors in the light inspection, making the detection results more stable and accurate, and improving the efficiency of work and reducing labor intensity. Experiments and field test show that the system has good robustness, reliability, efficiency and low cost production. Future work will focus on promoting the use of this equipment and optimize the recognition algorithm, and then design higher precision servo driving system to reduce the influence brought by the vibration deriving from the mechanical system. More suitable illumination styles and different image processing algorithms should be explored with the current inspection platform.

Table 1. The result of visible impurity

kind of impurity	low concentration		high concentration	
	default detection (bottle)	leak detection (bottle)	default detection (bottle)	leak detection (bottle)
color-dot	4	3	3	8
glass scraps	5	6	5	30
aluminum-scraps	2	3	3	2
hair	1	2	2	3
short fiber	2	1	1	3
other	3	4	4	6

References

1. Guo, B., Xie, D., Cheng, J.: Research on the algorithm of visual particles inspection based on machine vision[C]. In: Proceedings of the world Congress on Intelligent Control and Automation, WCICA 2010, vol. 9, pp. 5395-5398 (2010)
2. Zhang, Z., Chen, Z., et al.: Surface roughness vision measurement in different ambient light conditions[C]. In: 15th Int. Conf. Mechatronics and Machine Vision in Practice, Auckland (New Zealand), pp. 2-8, December 2008
3. Zhen, Y.: Chinese Pharmacopoeia[S]. Chemical Industry Press, Beijing (2000)
4. Ji, G., Yao-nan, W., Bo-wen, Z., et al.: Intelligent foreign particle inspection machine for injection liquid examination based on modified pulse-coupled neural networks[J]. Sensors 9, 3386–3404 (2009)
5. Gamage, P., Xie, S.: A real-time vision system for defect inspection in cast extrusion manufacturing process. The International Journal of Advanced Manufacturing Technology 40, 144–156 (2008)
6. Graves, M., Batchelor, B.: Machine vision for the inspection of natural products [J]. New York, NY, USA, Spring Verlag. pp. 35-86, (2008). ISBN:1-85233-525-4
7. Han, L.-W., Xu, D.: Statistic Learning-based Defect Detection for Twill Fabrics[J]. International Journal of Automation and Computing 7(1), 86–94 (2010)
8. Patel, Krishna Kumar, Kar, A., Jha, S.N., Khan, M.A.: Machine vision system: a tool for quality inspection of food and agricultural products [J]. Journal of Food Science and Technology 49(2), 123–141 (2012)
9. Huan-jun, L.I.U.: Research on intelligent methods to inspect impurity in liquid [J]. Modern Manufacturing Engineering. 2011(8), 100–105 (2011)
10. Bo-wen, Z.H.O.U., Yao-nan, W.A.N.G., et al.: A Machine-Vision-Based Intelligent Inspection System for Pharmaceutical Injections [J]. Robot. 31(2), 53–60 (2009)

Event Extraction for Gene Regulation Network Using Syntactic and Semantic Approaches

Wen Juan Hou(✉) and Bamfa Ceesay

Department of Computer Science and Information, National Taiwan Normal University,
No.88, Section 4, Ting-Chou Road, Taipei 116, Taiwan, Republic of China
{emilyhou,bmfceesay}@csie.ntnu.edu.tw

Abstract. Gene Regulation Network (GRN) is a graphical representation of the relationship between molecular mechanisms and cellular behavior in system biology. This paper examines the extraction of GRN from biological literatures using text mining techniques. The study proposes two independent methods first, a syntactic method and a semantic method in text mining, to extract biological events from the unstructured text. The paper presents the performance of the two methods and then experiments with the combined strategy to construct a gene regulation network from texts. The results show that the graph-based approach obtains a better result on event extraction and produces a much better regulation network than the semantic analysis method. The combination of the two approaches has yet a much slightly better result than that with the individual approach. This exhilarates us to find more future directions in the biological event extraction research.

Keywords: Biological event extraction · Gene regulation network · Graph-based approach · Semantic approach · Slot error rate

1 Introduction

Given a biological literature, a Gene Regulation Network (GRN) can be used as a graphical description of the interactions of molecular events and cellular entities to visualize their regulatory relationship. In text mining, representing biological literatures with a regulatory network requires the extraction of biological events, relations, and entities. To illustrate this, consider the biological literature:

FAS ligation induces the activation of the src family kinase of jun kinase.

This text mentions about the relationship between two biological entities, *FAS* and *jun kinase*. This paper examines how to represent such a statement graphically to show relationship between entities and events mentioned. For this study topic, we develop a system with two independent components that can be used to generate gene regulation networks from biological literatures. In text mining, semantic and syntactic analyses are significant techniques to analyze texts. The question is that which of the two techniques is better in practice. How can we take advantage of the two practices to have better performance? We will explore the issues in the paper.

There are several pioneering works in the development of event extraction systems for biological literatures. The related works are introduced as follows. Björne *et al.* [1]

© Springer International Publishing Switzerland 2015
M. Ali et al. (Eds.): IEA/AIE 2015, LNAI 9101, pp. 559–570, 2015.
DOI: 10.1007/978-3-319-19066-2_54

proposed an approach to extracting the complex biological events using the syntactic dependency graph. They made an extensive use of the dependency analysis of sentences to derive some features. They enhanced their event extraction by using semantic and syntactic features of tokens. They finally used multiclass support vector machine as described by Tsochantaridis *et al.* [2]. Huang *at el.* [3] suggested an automatic extraction of information about molecular interactions in biological pathways from texts based on ontology and semantic processing. They developed a framework using the ontology inference and semantic processing techniques to extract biological knowledge from a large scale of literature in NCBI PubMed [4]. The system integrated various thesauri of WordNet [5] , MesH [6] and Gene Ontology (GO) [7]. In addition to these novel approaches, this paper exploits the both syntactic feature and semantic feature using graph-based feature sets and semantic analysis respectively. The combination of features from the two perspectives (syntactic and semantic) outperforms the individual results for event extraction in the biological domain.

The rest of this paper is organized as follows. Section 2 presents the overview of our system architecture. We describe syntactic and semantic methods in details in Sections 3 and 4. The experimental results are shown and discussed in Section 5. Section 6 introduces a combined strategy for improving the performance. Finally, we express our main conclusions and the future research directions.

2   Architecture Overview

2.1   Overall Architecture

Figure 1 shows the overall architecture of our method for the biological event extraction and gene regulation network. For a given sentence, the first processing step is parsing it using the Stanford full parser [8, 9]. In the event extraction stage of using graph-based feature sets as shown in Part A, we make use of some key statistical dependency features to detect and classify our event triggers and edges. In the semantic processing phase, the study analyzes the semantic relationships between the detected triggers and event triggers. Event triggers with improper arguments or with no arguments will be pruned. The method in Part A will be described in Section 3.

In our semantic annotation based on natural language processing method, we focus on using syntactic and semantic features to extract events and event arguments from the parsed sentences [10]. The method corresponding to the Part B of Figure 1 will be described in Section 4. In the result of extracted events and event arguments, we combine the results from Part A and Part B via a combined strategy. The detailed steps will be explained in Section 6.

2.2   Experimental Data

The experimental data come from the BioNLP-2013 Shared Task's GRN task [11]. The purpose of the GRN task is to use information extraction techniques to retrieve all the genetic interaction of reference networks at least one occurrence per interaction independent of where it is mentioned in the given literature. The results are then represented as a GRN network. The evaluation of the system is based on its capacity to reconstruct a reference network. The experimental data consists of three parts:

training data, development data, and the test data. The file formats contain the text file, the entity annotation file and the event annotation file. The text files in the training and development data have both entity and event annotation files. However, the test data only have entity annotation files. Therefore, the task in the shared task is simplified to find the event annotations for the test data and to generate a GRN network.

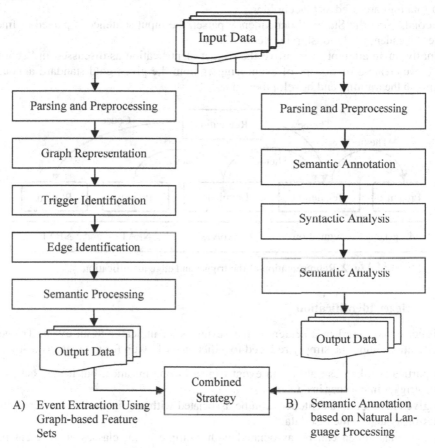

Fig. 1. The main system architecture

3 Syntactic Event Extraction Using Graph-Based Feature Sets

This approach presents the use of a syntactic practice in text mining. The method employs a pipeline of three main processing steps which are as follows: (1) trigger identification, (2) edge identification, and (3) semantic processing.

The final result is a graphical representation of the target extraction. The nodes in the graph represent named entities or events. The edges are instances of possible relations that exist between any given nodes. As an illustration, consider the expression:

IL-4 gene regulation involves NFAT1 and NFAT2.

This expression can be represented as in Figure 2 where square rectangles represent events nodes and round rectangles represent entity nodes. The syntactic approach includes three basic steps which are stated as follows.

1. First, the input sentence is tokenized and all tokens that are named entities are replaced with placeholders. The aim is to prevent our parser from segmenting multiple word named entities. The annotated IDs in the given gold standard entity annotations are used as placeholder.
2. Second, using the Stanford dependency parser, the input sentence is parsed to find the dependency relationship between tokens.
3. Finally, in an attempt to improve the trigger identification as discussed in Section 3.1, we create a dictionary of event triggers from the given gold standard annotations in the training and development data.

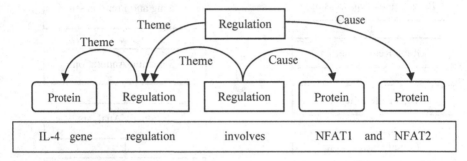

Fig. 2. Representation of the trigger and edge identification

3.1   Trigger Identification

A trigger is any word in a sentence that severs as an indicator to an event. Trigger identification cannot be simply reduced to a dictionary lookup for the reasons below:

- A particular token can act as an event trigger in one instance, but it may act as a non-trigger in another instance.
- A given input data (a sentence) can be associated with multiple event classes in the dictionary and in the test data.
- A particular token can be associated with multiple event classes at different instances in a given input.

In light of these challenges, we approach trigger identification with a multi-class Support Vector Machine classifier [12]. This allows us to assign multiple classes to a token. Tokens that cannot be classified into any event class are assigned a negative class and are considered as non-triggers. A combined class is assigned to a trigger that belongs to a multiple class. For instance, *"transcription/regulation"* denotes a combined class for a token that is classified as a transcription event class at one instance and a regulation event class at another instance. This implies that our trigger identification produces a single event node for each identified trigger as shown in Figure 2.
Training the Classifier. In this work, several features of gold standard triggers given in the training data are used to train the classifier. Table 1 gives a list of feature sets use to train the classifier for trigger identification. Let's explain each type as follows.

Token Features. Token features define the properties of the candidate token and generally test for the state (i.e., presence or absence) of the token in known trigger collections in the training data. They are testing for token capitalization, testing for punctuation and numeric characters, testing for stemmed tokens and token frequency, as shown in the row "Token" of Table 1.

Table 1. Features of trigger candidate for trigger identification

Type	Features
Token	An upper case form, in our trigger dictionary, including a numeric character, an entity, after "_", a stemmed word, token frequency
Candidate in the parser output	Token features of word with dependencies from candidate bigrams of dependencies Bigram and trigram of words (base form + POS)
Shortest path	Shortest path features between candidate and their closest entity: vertex walk, edge walk, n-gram of dependencies (n=2, 3), n-gram of words (base form + POS) representing governor-dependent relationship (n= 1, 2, 3) Length of the path between candidates and entities

Candidates in the Parse Tree. The goal of this feature set is to determine the relationship of tokens in the parser output and in the input data. It uses two token features *(Dictionary and Frequency)* as mentioned in Table 2. In addition, the following N-gram features *Bigram dependency* and *Bigram and trigram of words* are defined as follows.

$$BD = BD\text{-}sum / Number\ of\ tokens \qquad (1)$$

$$BT = BT\text{-}sum / Number\ of\ tokens \qquad (2)$$

where BD-sum is the summation of all bigrams of candidate tokens and their dependency tokens; and BT-sum is the summation of all bigrams of token in the parser result and trigrams of token in the input data.

Table 2. Candidate in parser output for "IL-4 gene regulation involves NFAT1 and NFAT2"

Token	Candidate in parser output			
	Token features		Bigram dependency	Bigram and trigram of words
	Dictionary	Frequency		
regulation	1	1	0.356	0.524
involves	1	1	0.271	0.445

Shortest Path. This phase aims to identify relationship between candidate token and their most nearest token in the dependency tree using the shortest path distance. A depth of three starting from the candidate token node and a set of features are used as illustrated in Table 3. The study uses the typed-dependency representation to represent dependencies as simple tuples. For example: *reln(gov, dep)* represents a relation tuple where *reln* is the dependency relation, *gov* is the governor word and *dep* is the dependent word. Table 3 shows the shortest path features for "*IL-4 gene regulation involves NFAT1 and NFAT2*." In Table 3, the vertex walk is the sequence of nodes from the candidate node and their substructures. In Figure 2, the vertex walk from the candidate node regulation has a score of four (three *theme* and one *cause*). Similarly, the edge walk of the shortest path features accounts for the *theme* and *cause* pair attributed to each candidate node. In Figure 2, the candidate node *regulation* has three pairs attributed to it, whereas the candidate node *involve* has a single pair attributed to it. The N-gram of words are the sum of bigram and trigram of the candidate token and its dependency words in the dependency tree and the token in the input sentence normalized by the token size as shown in Equations 1 and 2 respectively.

Table 3. Shortest path feature for "IL-4 gene regulation involves NFAT1 and NFAT2"

Token	Shortest path features			
	Vertex walk	Edge walk	N-gram of dependency	N-gram of words
regulation	4	3	0.782	0.510
involves	2	1	0.453	0.431

3.2 Edge Identification

The goal of edge identification is to find event arguments with respect to the identified triggers. Any potential edge is defined to be from an event node to another event node (nested events) or from an event node to an entity node. An edge can be related to multiple event arguments or linking nested events. In edge identification, edges are predicted independently to prevent classification of one edge affecting another. We use SVM multiclass for edge classification. Using the dependency relationship in the parse tree, the goal of edge identification is simplified to generate features of potential event arguments (edges) represented as the shortest undirected path in the dependency graph. Features used in the edge identification are shown below in Table 4.

Table 4. Features for candidates of edge identification

Type	Features
Terminal node	An upper case form, in the trigger dictionary, including a numeric character, an entity, after "_", a stemmed word, token frequency
Three words around the edge pair	N-gram (n=1, 2, 3, 4) of words
Shortest path	• Shortest path features listed in Table 1 between terminal nodes • Shortest path features listed in Table 1 between the argument trigger and the closest entity

N-grams are defined in respect to the potential edge. For each token and its two flanking dependencies, the trigram is calculated and similarly for dependencies and their flanking tokens. Bigrams are defined for consecutive tokens in the dependency to determine their governor dependent relationship. Edges are classified into *Theme, Cause* or *Negative*. *Theme* indicates the target of the event that is either an entity or an event in the case of nested events. *Cause* indicates the agent of an event occurrence. In cases where no *Theme* or *Cause* is predicted, a *Negative* is assigned.

3.3 Post-processing

The graph obtained from Sections 3.1 and 3.2 can still contain unwanted nodes. Event nodes with improper combination of arguments (edges) or event nodes with no argument can be presented in the graph. For solving this problem, we use pruning techniques to remove event nodes with no argument, and represent event nodes with the same class and the same trigger by a single node. Certain event nodes can only have one *Theme* argument and one *Cause* argument. For such cases, a heuristic rule is used to generate a new event for theme-cause combination. Figure 3 gives an example.

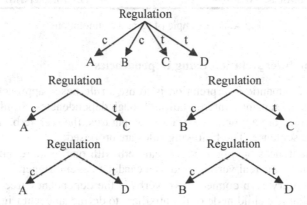

Fig. 3. An illustration of node duplication, where c=cause, t=theme

4 Semantic Approach

Besides using syntactic analysis of text mining, an attempt in this section is to seek answers to the following questions:

- What are results by using semantic analysis of text mining?
- How will the result differ from the result in Section 3?
- How can the result be improved utilizing syntactic and semantic analysis?

The proposed semantic approach uses three layers of natural language processing (NLP) method to extract events. They are explained in the following subsections.

4.1 Semantic Annotation

Figure 4 gives the example of using semantic annotations. This work uses four different thesauri to assign semantic tags to noun phrases of subjects and objects as described in the work of Huang *et al.* [3]. The thesauri are MeSH, GO, WordNet and a stopword list [13]. In Figure 4, the semantic codes <ME.D> indicates descriptors in MeSH; <SW.P> and <SW.N> indicate prepositions and nouns in the stopword list respectively; <WN.N> indicates the noun in WordNet; and <GO.COM> indicates the cellar component in GO.

Input sentence (from PMID-10075739-S9):	
We show that GerE binds to two sites that span the -35 region of the cotD promoter.	

Results of semantic annotation:	
<SW.N>We</SW.N>	<WN.N>two</WN.N>
<ME.D>show</ME.D>	<WN.N>sites</WN.N>
<SW.P>that</SW.P>	<ME.D>span</ME.D>
<GO.COM>GerE</GO.COM>	<ME.D>region</ME.D>
<ME.D>binds</ME.D>	<SW.P>of</SW.P>
<SW.P>to</SW.P>	<SW.P>the</SW.P>
<ME.D>sites</ME.D>	<GO.COM>CotD</GO.COM>
<SW.P>that</SW.P>	<ME.D>promoter</ME.D>

Fig. 4. An example of semantic annotations

4.2   Semantic Interpretation Using Dependencies

The objective of semantic interpretation is to use a rule-based approach to providing clues to subsequent semantic interpretation through dependencies. Similar to the work of [3], several rules are applied to extract noun phrases, the real verb, the subject and the object of the sentence. The following rules are proposed:

- For simple sentences having a single real verb with one or more auxiliaries in the dependency tree, the real verb is used as a candidate event trigger.
- When an auxiliary verb connects two verbs in the dependency tree, find the node with a verb that is a child node of the auxiliary to define an event trigger.
- In instances where the root verb is an intransitive verb that has a clause of a sentence after it; the verb in the clause defines the event or event trigger.

For every identified verb, its subject and the object is determined from the input sentence. A verb will be defined as an event if it has the related subject and the object in the sentence and the subject or the object is a named entity.

In some situations, we have to deal with compound sentences that have clauses connected by conjunctions, and the relationship between these clauses is determined by the conjunction. We assume that dependent clauses connected by dependent markers such as *"after," "although," "as," "because," "if," "since," "until," "when," "whether"* and *"while"* may have causal relationships between them. Clauses joined by the coordinating conjunctions such as *"and," "or," "but," "so"* and *"yet,"* or joined by independent markers such as *"also," "consequently," "however," "moreover"* and *"therefore"* are assumed to be independent. These sentences are separated into two clauses and traversed in the dependency tree for extracting the real verbs. The real verb is our identified trigger and we can query our ontology to assign it to a category.

4.3   Semantic Analysis

This phase uses two basic patterns to identify event triggers: triggers in noun forms and triggers in verb forms. To facilitate pattern identification, the following processing steps are adopted.

- All the gold standard annotated triggers given in the training and development data are extracted and categorized according to event types.

- For candidate triggers in the test data, determine its frequency in the extracted annotated trigger.
- For n classes, the classifier uses n classes versus all other classes by setting n classes as positive and the rest as negative, where n ranges from 1 to the total number of event types. Then determine the distance from the hyper line for each instance.

For a candidate trigger, its frequency in the annotated event trigger and its closest distance to the hyper line for a given class is used to determine the confident score. Finally we use pattern analysis to strongly determine events. The objective here is to extract event triggers from noun phrases (NPs). To achieve this, the candidate trigger is checked if it is a noun. In case of a noun trigger, a noun phrase joined to the candidate trigger with at least one named entity is determined. Two forms are considered.

- **Form 1.** Consider cases where NP is followed by a prepositional phrase (PP) tag.
- **Form 2.** Consider cases when the candidate trigger is the head of the NP.

   Form 1: NP → NN NN

   Form 2:

   NP → NP PP,   NP → DT NN,

   NP → NN CC NN   PP → IN NP

where NP means a noun phrase, NN means a noun, PP stands for a prepositional phrase, DT is a determiner, CC means a conjunction, and IN is a preposition. For the simple or regulatory event type, if the NP does not have a PP tag, we extract all proteins on the left of the trigger from the NP and form event pairs *<trigger, argument>*. Otherwise, we extract all proteins on the right of the trigger from the NP and form event pairs *<trigger, argument>*. To extract events from a verb phrase (VP), the trigger is checked if it is a verb, then find a VP, which is a direct parent of this trigger from the parse tree, and find a sister NP immediately preceding this VP. Next, we extract the candidate events as in the NP case mentioned above.

5   Experimental Results and Discussion

This work uses the BioNLP-2013 shared task test evaluation service to test the experimental result [11] and the result is compared to those of the shared task's participants. The objective of this comparison is to outline the performance of our system, especially the combined strategy among systems that uses the single perspective of events for event extraction. The evaluation service gives recall, precision and F-score and in addition, it also gives the total number of predictions, matches, deletions and Slot Error Rate (SER) after comparison with the gold standard network for the test data. For evaluation metrics, recall is the number of correctly aligned slots divided by the number of all correct slots. Precision is the number of correctly aligned slots divided by the number of all returned aligned slots. F-score considers both precision and recall rates. For the test corpus, the scores of participating teams are ranked according to the SER of their results [14]. Equations (3) to (6) define key evaluation terms.

$$\text{Recall (R)} = TP/\,(TP + FN) \tag{3}$$

$$\text{Precision (P)} = TP/\,(TP + FP) \tag{4}$$

$$\text{F-score} = 2PR/\,(P + R) \tag{5}$$

SER = (total number of slot errors) / (total number of slots in the reference)   (6)

where *TP* is the number of true positives, *FN* is the number of false negatives, and *FP* is the number of false positives. The total number of slot error is the number of slots in the reference that do not align to any slot in the hypothesis [14]. Lower SER values are preferred in the experiment. The experimental results for individual components are shown in Tables 5 and 6.

Table 5. Result from event extraction using graph-based feature sets

Substitutions	Deletions	Insertions	Recall	Precision	F-score	SER
8	62	9	0.2045	0.5143	0.2927	0.8977

Table 6. Result of semantic annotation based on NLP

Substitutions	Deletions	Insertions	Recall	Precision	F-score	SER
9	61	10	0.2045	0.4865	0.2880	0.9091

In the tables, the columns are defined as:

- "Substitutions" (incorrect slots) is the number of slots in our system that are aligned to slots in the reference and are scored as incorrect.
- "Deletions" (missing slots or false rejections) is the number of slots in the reference that do not align to any slot in our system.
- "Insertions" (spurious slots or false acceptances) represent the number of slots in our system that are not aligned to any slot in the reference.

Comparing with the results in Tables 5 and 6, the semantic approach has lower F-score, deletion and precision. However, this result has higher substitution, insertion, and SER score. Using the SER score as the ranking criterion, it can be concluded that our syntactic approach has slightly better performance than the semantic one.

6   Combined Strategy

In this phase, we want to further improve the SER score using the result of individual syntactic and semantic methods with proposing the combined strategy. This is achieved using the stacking and filtering technique. For a given input sentence, it is processed independently by the two approaches mentioned in Sections 3 and 4. The results from the two approaches are stacked together to form a larger collection of extracted events. For the filtering process, the collection of extracted events and arguments is processed as follows:

- For an event with similar arguments that appear from the two components (i.e., syntactic and semantic ones), the approach retains a single case.
- In the case where the similar events extracted from the two components but with different arguments, we keep both cases of the events.

The final evaluation score for combining the syntactic and semantic approaches is shown in Table 7 below. It can be understood that our combined result is dominated by the approach of event extraction using the graph-based feature sets. However, the result of individual components of our system is very much similar. Our semantic annotation approach has a poor score in the number of deletions causing a higher SER.

Table 7. System score with the combined strategy

Substitutions	Deletions	Insertions	Recall	Precision	F-score	SER
9	59	10	0.2273	0.5128	0.3150	0.8864

7   Conclusion

In this paper, we implemented the syntactic and semantic approaches individually and with the combined one, in order to extract biological events from the unstructured texts. We form a gene regulation network for the GRN shared task. The paper uses the test evaluation system provided by the BioNLP-2013 shared task community to test the results. Comparing our results with participants in the BioNLP shared task 2013 as shown in Table 8, our approach achieves better SER score than certain participants. It also shows we have some space for further improvement.

Our future work will focus on the other statistical approaches in computational linguistics. Because mathematical modeling has played a significant role in many research areas, we believe it can be applied to extraction of biological events from any given biological structure. The statistical natural language processing includes several quantitative approaches such as probabilistic modeling, information theory, and linear algebra. We believe a careful study of these approaches can be of great interest in the area of biological event extraction.

Table 8. Results of participants in BioNLP shared task 2013

Participant	Rank	SER	Recall	Precision	F1
University of Ljubljana	1	0.73	0.34	0.68	0.45
K.U.Leuven	2	0.83	0.23	0.50	0.31
TEES-2.1	3	0.86	0.23	0.54	0.32
IRISA-TexMex	4	0.91	0.41	0.40	0.40

Acknowledgements. Research of this paper was partially supported by Ministry of Science and Technology, Taiwan, under the contract MOST 103-2221-E-003-014.

References

1. Björne, J., Heimonen, J., Ginter, F., Airola, A., Pahikkala, T., Salakoski, T.: Extracting Contextualized Complex Biological Events with Rich Graph-based Feature Sets. Comput. Intell. **27**(4), 541–557 (2011)
2. Tsochantaridis, I., Hofmann, T., Joachims, T., Altun, Y.: Support vector machine learning for interdependent and structured output spaces. In: Proceedings of the 21st International Conference on Machine Learning (2004)
3. Huang, Y.T., Yeh, H.Y., Cheng, S.W., Tu, C.C., Kuo, C.L., Soo, V.W.: Automatic extraction of information about the molecular interactions in biological pathways from texts based on ontology and semantic processing. In: Proceedings of IEEE International Conference on Systems, Man and Cybernetics (SMC 2006), vol. 5, pp. 3679–3684. IEEE Press (2006)

4. Roberts, R.J.: PubMed Central: The GenBank of the Published Literature. Proceedings of the National Academy of Sciences **98**(2), 381–382 (2001)
5. Miller, G.A.: WordNet: a Lexical Database for English. Commu. ACM **38**(11), 39–41 (1995)
6. MeSH. http://www.ncbi.nlm.nih.gov/mesh
7. Ashburner, M., et al.: Gene Ontology: Tool for the unification of biology. Nature genetics **25**(1), 25–29 (2000)
8. McClosky, D., Surdeanu, M., Manning, D.C.: Event extraction as dependency parsing. In: Proceedings of the 49th Annual Meeting of the Association for Computational Linguistics, pp. 1626–1635 (2011)
9. Standford Full Parser. http://nlp.stanford.edu/software/lex-parser.shtml
10. Bui, Q.C., Sloot, M.A.P.: Extracting biological events from text using simple syntactic patterns. In: Proceedings of BioNLP Shared Task 2011 Workshop, pp. 143–146 (2011)
11. Nédellec, C., Bossy, R., Kim, J.D., Kim, J.J., Ohta, T., Pyysalo, S., Zweigenbaum, P.: Overview of BioNLP shared task 2013. In: Proceedings of the BioNLP Shared Task 2013 Workshop, pp. 1–7 (2013)
12. Crammer, K., Singer, Y.: On the Algorithmic Implementation of Multiclass Kernel-based Vector Machines. J Mach Learn Res **2**, 265–292 (2002)
13. Stopword list. http://jmlr.org/papers/volume5/lewis04a/a11-smart-stop-list/english.stop
14. Makhoul, J., Kubala, F., Schwartz, R., Weischedel, R.: Performance measures for information extraction. In: Proceedings of DARPA Broadcast News Workshop, pp. 249–252 (1999)

Video Stitching System of Heterogeneous Car Video Recorders

Teng-Hui Tseng[1] and Chun-Ming Tsai[2(✉)]

[1] Department of Communication Engineering, Oriental Institute of Technology,
New Taipei City 220, Taiwan, Republic of China
alex@mail.oit.edu.tw
[2] Department of Computer Science, University of Taipei,
No. 1, Ai-Kuo W. Road, Taipei 100, Taiwan
cmtsai2009@gmail.com

Abstract. Many heterogeneous car video recorders are sold in the market. In order to record the ultra-wide-angle road scenes, at least two heterogeneous recorders are used. These heterogeneous recorders have different viewing angles, different resolutions, and different lens sensors. Because of different hardware and software of the heterogeneous recorders, the captured videos are heterogeneous, which is very challenging for video stitching research. The traditional image stitching system includes color correction, feature detection, feature descriptor, feature matching, and video stitching. When a traditional image stitching system is used to process the images captured by homogeneous cameras, the results are reasonably good. However, when a traditional image stitching system is used to process the images captured by heterogeneous cameras, the results are not so good. Furthermore, applying a traditional image stitching system to process videos captured by heterogeneous car video recorders is time-consuming. This paper presents a study that tested multiple existing methods to evaluate their capability when used to stitch heterogeneous images to allow a driver to see an ultra-wide-angle driving view without blind spots and to record these images. Experimental results show that some methods used in this study can be used for this purpose, but they have significant error rates and are time-consuming.

Keywords: Video stitching · Heterogeneous recorder · Ultra-wide-angle road scene · Big view · Car video recorder

1   Introduction

A blind spot in a vehicle is an area around the vehicle that cannot be directly observed by the driver while at the controls [1]. Blind spots may occur in the front of the vehicle because the a-pillar, side-view mirror, and/or interior rear-view mirror block the driver's view of the road ahead [1]. Good driver visibility is essential to safety, so reducing the blind spot and enhancing the driver visibility is very important for car makers and for road safety.

© Springer International Publishing Switzerland 2015
M. Ali et al. (Eds.): IEA/AIE 2015, LNAI 9101, pp. 571–580, 2015.
DOI: 10.1007/978-3-319-19066-2_55

Blind spots, in particular, the front-end blind spots, can easily create traffic collision in roundabouts, intersections, and road crossings when a vehicle collides with another vehicle, pedestrian, animal, road debris, or other stationary obstruction, such as a tree or utility pole that the driver cannot see because of a blind spot [2].

Traffic collisions may result in injury, death, vehicle damage, and property damage [2]. If the driver is at fault, he may be forced to pay compensation, or he may be charged with a driving violation. However, how to determine whether the driver is at fault in a car accident? Policemen usually consult witnesses about how the accident occurred. However, eye-witness accounts are often inconsistent and unreliable, and in any case often there are no witnesses, so it is very difficult to determine who is at fault in a car accident. To address this problem, many car owners install car video recorders to record a video of the road ahead, both to prevent traffic accidents, and to reduce disputes if there is an accident.

Usually, the record resolution of car video recorders is 1080p with a wide angle lens of 120 degrees. However, when a wide angle lens is used, many blind spots in the driver's front view still exist. For example, a driver may be unable to see a pedestrian or a vehicle approaching when negotiating a turn to the left or the right because of the thick a-pillar on either side of the windscreen. In principle, the blind spot can be eliminated by a camera on the side pillar and able to provide a plus pillar display [3] that can recognize the full scene outside of the car and allow the driver to proceed cautiously avoiding any mishap.

Thus, a car video recorder can, in theory, provide the driver an ultra-wide-angle driving view free of front-view blind spots and record that ultra-wide-angle view of traffic on the road ahead. That view would reduce accidents and when an accident occurs, provide actual evidence to establish the truth and reduce disputes.

Many heterogeneous car video recorders are sold in the market. In order to record the ultra-wide-angle road scenes, at least two heterogeneous recorders are used and the two images "stitched" together to produce an ultra-wide-angle view. These heterogeneous recorders, however, have different viewing angles, different resolutions, and different lens sensors. Because of different hardware and software of the heterogeneous recorders, the captured videos are heterogeneous and very challenging for video stitching research.

The traditional image stitching system includes color correction, feature detection, feature descriptor, feature matching, and image stitching. In this paper, the traditional image stitching method is modified to process the videos captured by two heterogeneous cameras. The modified method includes motion blur detection, color correction, feature detection, feature descriptor, feature matching, and image stitching methods. The advantages and the disadvantages of the modified method will be found by conducting the experiments. In the future, the disadvantages will be solved.

2   Motion Blurred Detection

Motion blur occurs because the car is moving, objects in the scene are rapidly moving, the road surfaces are uneven, and/or the car video recorders are shaking. These blurred frames cannot be seen clearly. Thus, the blurred frame problem should be solved for the system to produce usable images. In Tsai's papers [4-5], a motion blur detection algorithm has been proposed to detect the motion blur of moving business cards held in a blind person's hands. Herein, this algorithm will be extended to detect the motion blur of the videos captured by the two heterogeneous car video recorders. To detect whether the video frame is motion blur the following detection criterion one is satisfied:

$$UM < T_{UM},$$ (1)

where UM and T_{UM} represent the uniformity measure and the threshold value for uniformity measure, respectively.

The uniformity measure, first stated by Levine and Nazif [6], is primarily used as a criterion for measuring the quality of image segmentation. In [7], the uniformity of a feature over a region is defined as being inversely proportional to the variance of the values of that feature, evaluated at every pixel belonging to that region, with an appropriate weighting factor. Ng and Lee [7] proved that the optimality of uniformity measure is basically equivalent to the criterion measure proposed by Otsu [8]. Herein, the uniformity measure (UM) adopted from Ng and Lee [7] is defined as:

$$UM = 1 - \frac{\sigma_w^2}{C},$$ (2)

where σ_w^2 denotes the within-class variance of the given threshold value and C is the normalization factor which limits the maximum value of UM to 1. Then, to optimize the uniformity measure for a given threshold value, Eq. (2) can be maximized, producing:

$$t_u^* = \underset{0<t<D}{Max}\left[1 - \frac{\sigma_w^2}{C}\right],$$ (3)

Fig. 1. The frame #15 in the video captured by the left car video recorder (1920×1080)

Fig. 2. The frame #15 in the video captured by the right car video recorder (1280×720)

where t_u^* represents the threshold value at which the uniformity measure is optimal.

Parameters t and D are the different value and the maximum different value, respectively, in the different image.

Uniformity measure is based upon the region property. If the differencing image for two successive frames is motion blur, its uniformity measure value is smaller.

3   Color Correction

Because the captured videos are captured by two heterogeneous car video recorders, the two captured videos have different luminance and color. So that the resulting panorama looks coherent and natural, the color correction problem should be solved.

Many color image enhancement algorithms [9-12] have been proposed to enhance the color images. Herein, the algorithm in [12] is extended to enhance the captured videos by the two heterogeneous car video recorders. This enhancement method has low computational cost due to its simplicity and efficiency and can be processed in a real-time system. In paper [12], an adaptive local power law transformation method with low computational complexity and high performance is proposed to enhance color images. Furthermore, in order to reduce the computational time, a fast method is

Fig. 3. The matching result from using SIFT detector, SURF descriptor, and BF Matcher

Fig. 4. The matching result from using SIFT detector, SURF descriptor, and FLANN Matcher

proposed to compute the local mean from the neighboring pixels to adapt the exponential value in the power-law transformation for enhancing the color images. This enhancement method has low computational cost and can be processed in a real-time system due to its simplicity and efficiency.

4   Features Finding and Matching

SIFT keypoint detector and descriptor [13] has proven successful in many applications. However, it imposes a large computational complexity. SURF [14] has been demonstrated to achieve robustness and speed. ORB [15] has efficiently computed orientations based on the intensity centroid moment. BRISK [16] searches maxima in a 3D scale-space and achieves comparable quality of matching at much less computation time. RANSAC [17-18] is a general parameter estimation approach designed to cope with a large proportion of outliers in the input data. It is a resampling technique that generates candidate solutions by using the minimum number of data points required to estimate the underlying model parameters. FLANN [19-21] is an excellent software package developed by Muja and Lowe that includes implementations of forests of randomized kd-trees (KDT) and hierarchical k-means (HKM). This method also includes a mechanism for automatically selecting and tuning KDT and HKM algorithms for a given data set.

Fig. 5. The matching result from using SURF detector, SURF descriptor, and BF Matcher

Fig. 6. The matching result from using SURF detector, SURF descriptor, and FLANN Matcher

In this study, SIFT [13], SURF [14], ORB [15], and BRISK [16] methods are used to detect and describe the features from the selected frames. The RANdom Sample Consensus (RANSAC) [17-18] is used to remove the outlier correspondences from the combined feature correspondences. Finally, the FLANN [19-21] method is used to be the matching method to match the left and the right features descriptors of the left and the right car video recorders, respectively.

5   Experimental Results

This study system was implemented in Microsoft Visual C++ 2012 on an Intel(R) Core(TM) i7-3667U CPU @ 2.00GHz Notebook, carried out on two video clips. These two video clips are captured by two heterogeneous car video recorders to capture a real scene. The left car video recorder is DOD F900HD and the right car video recorder is Trywin DTN-3DX. The camera angles for these two recorders are 100 and 120, respectively. Their resolutions are 1920×1080 and 1280×720, respectively.

Figures 1 and 2 are the frame #15 in the videos captured by the left and the right car video recorders, respectively. In Fig. 1, the depth of the captured frame is near. But in Fig. 2, the depth of the captured frame is far. Furthermore, the color in these two frames is different. Figure 1 is brighter than Fig. 2 because these two car video recorders captured the scene with different exposures.

Fig. 7. The matching result from using ORB detector, ORB descriptor, and BF Matcher

Fig. 8. The matching result from using ORB detector, ORB descriptor, and FLANN Matcher

Herein, the open source given by the computer vision library OpenCV [22] is used to stitch the two video images together. The feature detectors SIFT [13], SURF [14], ORB [15], and BRISK [16] are applied for the videos captured by the mentioned two heterogeneous car video recorders. As Table 1 shows, the SIFT feature detector detects the most keypoints, while the ORB feature detector detects the fewest keypoints.

Table 1. The key points detected by SIFT, SURF, ORB, and BRISK

Feature detectors	Key points in Fig. 1	Key points in Fig. 2
SIFT	6961	15486
SURF	3797	3365
BRISK	1327	2461
ORB	500	500

Fig. 9. The matching result from using BRISK detector, BRISK descriptor, and BF Matcher

Furthermore, four feature detectors, three feature descriptors, and two matching methods are combined and are applied for Figs. 1 and 2 to obtain the execution times and matching results. Figures 3-10 show the matching results of these eight different combinations. The matching results from using SIFT detector, SURF descriptor, and Brute Force Matcher are shown in Fig. 3. These matching lines are almost all erroneous. The matching results from using SIFT detector, SURF descriptor, and FLANN Matcher are shown in Fig. 4. These matching lines are almost erroneous too.

Fig. 10. The matching result from using BRISK detector, BRISK descriptor, and FLANN Matcher

The matching results from using SURF detector, SURF descriptor, and Brute Force Matcher are shown in Fig. 5. Here, one matching line between the zebra crossings is correct, as is one matching line between the direction marks. The other matching lines are within error. The matching results from using SURF detector, SURF descriptor, and FLANN Matcher are shown in Fig. 6. One matching line between the zebra crossings is correct. The other matching lines are incorrect. The matching results from using ORB detector, ORB descriptor, and Brute Force Matcher are shown in Fig. 7. Those from using ORB detector, ORB descriptor, and FLANN Matcher are shown in Fig. 8; those from using BRISK detector, BRISK descriptor, and Brute Force Matcher are shown in Fig. 9; and those from using BRISK detector, BRISK descriptor, and FLANN Matcher are shown in Fig. 10. The matching lines in Figs. 7 to 10 are all incorrect.

Table 2 shows the execution times of eight different combinations. As shown in Table 2, the execution time of the combination: ORB feature detector + ORB feature descriptor + BF matcher is least. However, this combination of the matching result is worst as shown in Fig. 7. The matching lines are all incorrect.

Fig. 11. The example resulting from using OpenCV's stitcher: Cylindrical warper

For real-time video stitching system, the above-mentioned feature detectors, feature descriptors, and matching methods will be improved to produce correct matching lines. That is, the first step is to find the correct or good corresponding pixel points between the two planes [23]. The second step is using RANSAC to remove outliers. The third step is to compute the homography by the inliers. Last, when the homography is calculated, the target image can be warped in order to fit the plane of the reference image. The images must be padded to have the same size, and the seams for the stitching must be found in the overlapping regions.

Table 2. Four feature detectors, three feature descriptors, and two matching methods are combined to study their performance

Feature detector: times	Descriptor extractor: times	Matcher: times	Total times
SIFT: 11.90s	SURF: 3.05s	BF: 6.47s	21.42s
SIFT: 12.03s	SURF: 3.40s	FLANN: 5.44s	20.87s
SURF: 1.46s	SURF: 3.43s	BF: 0.77s	5.66s
SURF: 1.46s	SURF: 3.42s	FLANN: 0.78s	5.67s
ORB: 0.21s	ORB: 0.22s	BF: 0.01s	0.44s
ORB: 0.23s	ORB: 0.23s	FLANN: 0.15s	0.61s
BRISK: 0.85s	BRISK: 0.58s	BF: 0.20s	1.63s
BRISK: 0.84s	BRISK: 0.58s	FLANN:0.48s	1.90s

We also tried the open source solutions given by computer vision library OpenCV [22] which are based on the automatic panoramic image stitcher by Brown et al. [24], i.e., the Stitcher class using cylindrical wrapper is used. Example of the respective panorama image is given in Figure 11. The resulting panorama image resolution and per image execution time are 3097×1212 and 9.05 seconds, respectively. As this figure show, this implementation can provide an acceptable stitching image. However, its execution time is too longer to be used for real-time stitching.

6   Conclusions

A system for heterogeneous car video recorders is presented to stitch two heterogeneous videos. This system includes motion blur detection, color correction, feature detection, feature descriptor, feature matching, and image stitching methods. The experimental results show that the execution time is still time-consuming. In the future, this study will speed the feature detection, feature descriptor, and image stitching methods to obtain the complete real-time video stitching system for heterogeneous car video recorders.

Acknowledgments. The authors would like to express their gratitude to Dr. Jeffrey Lee and Walter Slocombe, who assisted editing the English language for this article.

References

1. Blind spot. http://en.wikipedia.org/wiki/Blind_spot_(vehicle)
2. Traffic collision. http://en.wikipedia.org/wiki/Traffic_collision
3. A Plus Pillar concept gives eyes to a driver's blind spot. http://www.damngeeky.com/2012/10/30/6751/a-plus-pillar-concept-gives-eyes-to-a-drivers-blind-spot.html
4. Tsai, C.M.: Non-motion blur detection for helping blind persons to "see" business cards. ICMLC **2012**, 1901–1906 (2012)
5. Tsai, C.M.: Text detection in moving business cards for helping visually impaired persons using a wearable camera, IJWMIP, **12:01** (2014) 1450010-1–1450010-16
6. Levine, M.D., Nazif, A.M.: Dynamic measurement of computer generated image segmentations. IEEE Trans. on PAMI **7**(2), 155–164 (1985)
7. Ng, W.S., Lee, C.K.: Comment on using the uniformity measure for performance measure in image segmentation. IEEE Trans. on PAMI **18**(9), 933–934 (1996)
8. Otsu, N.: A threshold selection method from gray-level histogram. IEEE Trans. SMC **9**, 62–66 (1979)
9. Tsai, C.M., Yeh, Z.M.: Contrast enhancement by automatic and parameter-free piecewise linear transformation for color images. IEEE Trans. on CE **54**(2), 213–219 (2008)
10. Tsai, C.M., Yeh, Z.M.: Contrast compensation by fuzzy classification and image illumination analysis for back-lit and front-lit color face images. IEEE Trans. on CE **56**(3), 1570–1578 (2010)
11. Tsai, C.M., Yeh, Z.M., Wang, Y.F.: Decision tree-based contrast enhancement for various color images. MVA **22**(1), 21–37 (2011)
12. Tsai, C.M.: Adaptive local power-law transformation for color image enhancement. Appl. Math. Inf. Sci. **7**(5), 2019–2026 (2013)
13. Lowe, D.G.: Distinctive image features from scale-invariant keypoints. Int. J. Comput. Vis. **60**(2), 91–110 (2004)
14. Bay, H., Ess, A., Tuytelaars, T., Gool, L.V.: Speeded-Up Robust Features (SURF). CVIU **110**(3), 346–359 (2008)
15. Rublee, E., Rabaud, V., Konolige, K., Bradski, G.: ORB: An efficient alternative to SIFT or SURF. ICCV **2011**, 2564–2571 (2011)
16. Leutenegger, S., Chli, M., Siegwart, R.Y.: BRISK: Binary Robust Invariant Scalable Keypoints. ICCV **2011**, 2548–2555 (2011)
17. Fischler, M.A., Bolles, R.C.: Random sample consensus: a paradigm for model fitting with applications to image analysis and automated cartography. Commun. of the ACM **24**(6), 381–395 (1981)
18. Choi, S., Kim, T., Yu, W.: Robust video stabilization to outlier motion using adaptive ransac. Proceedings of IEEE/RSJ ICIRS **2009**, 1897–1902 (2009)
19. Muja, M., Lowe, D.G.: Fast approximate nearest neighbors with automatic algorithm configuration. In: International Conference on Computer Vision Theory and Application (VISAPP 2009), pp. 331–340 (2009)
20. Muja, M., Lowe, D.G.: Fast matching of binary features. In: Proceedings of the 2012 9th Conference on Computer and Robot Vision, pp. 404–410 (2012)
21. Muja, M., Lowe, D.G.: Scalable nearest neighbor algorithms for high dimensional data. IEEE Trans. on PAMI **36**(11), 2227–2240 (2014)
22. OpenCV (Open source computer vision). http://opencv.org/
23. Hartley, R.I. and Zisserman, A.: Multiple View Geometry in Computer Vision, second edition. Cambridge University Press (2004)
24. Brown, M., Lowe, D.G.: Automatic panoramic image stitching using invariant features. Int. J. Comput. Vis. **74**(1), 59–73 (2007)

A Self-adaptive Genetic Algorithm
for the Word Sense Disambiguation Problem

Wojdan Alsaeedan and Mohamed El Bachir Menai[⊠]

Department of Computer Science, College of Computer and Information Sciences,
King Saud University, P.O. Box 51178, Riyadh 11543, Saudi Arabia
wojdan@ccis.imamu.edu.sa, menai@ksu.edu.sa

Abstract. Genetic algorithms (GAs) have widely been investigated to
solve hard optimization problems, including the word sense disambiguation (WSD). This problem asks to determine which sense of a polyse-
mous word is used in a given context. The performance of a GA may
drastically vary with the description of its genetic operators and selection methods, as well as the tuning of its parameters. In this paper, we
present a self-adaptive GA for the WSD problem with an automated
tuning of its crossover and mutation probabilities. The experimental
results obtained on standard corpora (Senseval-2 (Task#1), SensEval-
3 (Task#1), SemEval-2007 (Task#7)) show that the proposed algorithm
significantly outperformed a GA with standard genetic operators in terms
of precision and recall.

1 Introduction

Natural languages are known for their ambiguity. This means that words can
have different meanings/senses depending on the context in which they occur.
Such words are called homographs. They are spelled the same way but have
different meanings and sometimes different pronunciations. For instance, the
English word *bear* can be a verb (to support or carry) or a noun (the animal)
according to the context. Ambiguity can be present at many other levels, such
as syntactic, semantic, constituent boundary, and anaphoric ambiguity.

Word sense disambiguation (WSD) is critically important in natural language
processing (NLP). It is necessary for many NLP applications, such as machine
translation, information retrieval, information extraction, part of speech tagging,
and text categorization. WSD asks to determine which sense of a word is used in
a particular context. It is an AI-complete (Artificial Intelligence-complete) problem [12] that is analogous to NP-complete problems in complexity theory. It has
been tackled using several approaches, including knowledge-based approaches
and machine learning-based approaches. Genetic algorithms (GA) have also been
investigated for solving the WSD problem in Spanish [5], in English [28] and in
Arabic [13].

Genetic algorithms have successfully been applied in various optimization
fields. However, their performance is influenced by the type of genetic operators, the selection and initial population methods and parameters. In this paper

© Springer International Publishing Switzerland 2015
M. Ali et al. (Eds.): IEA/AIE 2015, LNAI 9101, pp. 581–590, 2015.
DOI: 10.1007/978-3-319-19066-2_56

we present a self-adaptive GA (SAGA) with an automated tuning of the probabilities of its crossover and mutation operators, apply it to the word sense disambiguation problem, and evaluate its performance on standard English corpora. In the rest of this paper, section 2 presents some related work describing how to improve a GA. Section 3 presents SAGA for WSD. Section 4 reports and discusses the experimental results. Section 5 concludes this paper.

2   How to Improve a Genetic Algorithm?

The results of several studies applying GAs in different fields suggest that there is no best GA, but each GA can be better than others, depending on the application domain. Particularly, an appropriate choice of genetic operators and their parameter settings can increase significantly the performance of the GA.

Syswerda [22] showed that the uniform crossover operator is more efficient when compared with a 2-point crossover. De Jong and Spears [4] examined the interaction role of crossover operators and population size. They empirically showed that the uniform crossover is an effective operator for problems in which there are constraints on the size of the population. This study suggests that more disruptive crossover operators such as uniform or n-point $(n \gg 2)$ crossovers are likely to yield better results with small populations, whereas less disruptive crossover operators (e.g., 2-point) are more likely to work better with larger populations. Liepins and Vose [11] anticipated that the characterization of the crossover operator will lead to a better understanding of its role in the GA, and will assist in determining how different crossover operators influence GA performance. Wu and Chow [27] compared the single-point, 2-point, 3-point and 4-point crossover operators in a GA for the discrete optimization of trusses. Their results suggest that the 2-point, 3-point and 4-point crossovers should be favored over the single-point crossover. Jenkin's study [8] suggests that multiparent in partial string crossover technique yields to a faster progress to the optimum than a single-point crossover. Vrajitoru [25] introduced a new crossover operator, called dissociated crossover, closely related to the 2-point crossover. It was implemented in a GA used in an information retrieval system to answer users' queries. This algorithm significantly outperformed a GA with standard genetic operators. Hasançebi and Erbatur [6] compared several crossover techniques, including single-point, 2-point, multi-point (involving more than two cut points), variable to variable, and uniform crossover operators in structural design fields. Their study shows that n-point crossover $(n \geq 2)$ yields better solution and faster progress to the optimum. Park et al. [16] proposed a GA using a stochastic crossover and an artificial initial population scheme. They successfully applied it to the least-cost generation planning in electric utilities. The improved GA gave better solutions than the conventional simple GA.

In a more recent study, Pick et al. [17] compared 10 crossover operators used in binary-coded GAs: single-point crossover, 2-point crossover, half-uniform crossover, uniform crossover, shuffle crossover, segmented crossover, reduced surrogate crossover, non-geometric crossover, orthogonal array crossover, and hybrid Taguchi crossover. They found that the most successful ones appear to be the uniform, single-point, and reduced surrogate cross-over. The orthogonal array and hybrid Taguchi crossovers were the worst performing ones. However, they do not recommend to take this as a general rule because those operators may behave differently on other benchmark problems. Kaya [9] proposed two new crossover operators: sequential crossover and random mixed crossover. The random mixed crossover yielded higher fitness values than some existing crossover operators when used in a GA applied to a deep beam problem and a concrete mix design problem.

Other studies underlined the positive impact of dynamically varying the crossover and mutation ratios. Vasconcelos et al. [24] examined three versions of the GA: simple GA, steady state GA, and replacement GA. They experimentally demonstrated that all these GAs were effective when they were used with their best genetic operations and values of parameters. Their study underlines the positive impact of the space reduction and the global elitism. Hong et al. [7] proposed a dynamic GA (DGA) using different crossover and mutation operators to generate the population of the next generation based on the performance of the current population. The crossover and mutation ratios also were adjusted dynamically. Their results show that DGA performed better than GAs with single-point crossover and single mutation operators.

Other techniques were investigated to enhance the performance of GAs, such as using efficient data structures to avoid recalculating fitness values and incorporating a local search to exploit better the search landscape. Approximately, a third of the time in a GA may be spent testing individuals that have already been tested [3]. Cooper and Hinde [3] showed that the efficiency of a GA can be improved by using an intelligent fitness that has both short and long term memories to reduce the number of duplicates and repeated fitness tests. Povinelli and Feng [18] used a hash table in a GA to save calculated fitness values of the most recently evaluated chromosomes. They reported a significant drop in computation time by over 5% and a reduction in performance variation.

Wan and Birch [26] investigated the performance of several GAs in which directional local searches are embedded. They demonstrated that both their computational efficiency and accuracy can be improved over the traditional GA. Someya and Yamamura [21] presented a GA with search area adaptation (GSA) that adapts to the landscape of the solution space by controlling the tradeoff between global and local searches. A generation of GSA consists of two phases: a crossover phase and a mutation phase. The crossover phase consists of a selection for reproduction, crossover, and selection for survival. The mutation phase consists of a selection for reproduction, mutation, and a selection for survival. They applied it to the floorplan design problem. Toğan and Daloğlu [23] showed that incorporating member grouping strategies to a GA, applied to design optimization of truss structures, may enhance its convergence and performance. They reported better performance results of the improved GA than those obtained with conventional mathematical programming methods.

3   Self-adaptive Genetic Algorithm for WSD

We propose SAGA based on the following observations:

- Our survey of some existing works on GAs shows that the uniform, single-point, and 2-point crossover operators generally outperformed the multi-point, orthogonal array, and hybrid Taguchi crossover operators. Furthermore, a decrease in the number of cut points may encourage exploitation, but discourages exploration.
- At the beginning of a search process, more exploration is needed to find new solutions in promising regions, which may be refined to improve their fitness at its end.
- More exploration is needed when the search process is in a low fitness region, while more exploitation is needed when it is in a high fitness region.
- High fitness solutions should be protected, while subaverage fitness solutions should be disrupted.

In SAGA, uniform crossover and mutation operators are used with high probabilities to promote exploration, while single-point crossover and mutation operators are used with low probabilities to promote exploitation. The probabilities of crossover and mutation operators are dynamically adapted depending on the largest fitness of the mating parents f', and the largest fitness f_{max} and the mean fitness $\bar{f}$ in the current population. The crossover probability P_c is defined by:

$$P_c = \begin{cases} \dfrac{f_{max} - f'}{f_{max} - \bar{f}} & \text{if } f_{max} \neq \bar{f}, \\ 0 & \text{otherwise.} \end{cases} \tag{1}$$

Let f be the fitness of an individual. The mutation probability P_m within this individual is defined by:

$$P_m = \begin{cases} \dfrac{0.5(f_{max} - f)}{f_{max} - \bar{f}} & \text{if } f_{max} \neq \bar{f}, \\ 0 & \text{otherwise.} \end{cases} \tag{2}$$

Note if $f_{max} = \bar{f}$, then $P_m = P_c = 0$ to protect high fitness solution and transfer it undisrupted to the next generation.

The main steps of SAGA are outlined by Algorithm 1. To encourage exploration, the uniform crossover and mutation operators are applied with the probabilities $P_c = 1$ and $P_m = 0.5$, respectively.

The WSD problem is solved by SAGA using the following formulation:

- An individual is represented by an integer string of a fixed length, where each gene encodes a word sense. The initial population is randomly generated: the value of each gene in an individual is randomly selected among the senses (indexes) retrieved for each word. Parents are selected by a tournament selection method. The population for next generation is selected using a generational or elitist survivor schemes.

Algorithm 1. SAGA

---

input : $Population_{size}$, $Problem_{size}$
output: S_{best}

$Population \leftarrow$ Initialize($Population_{size}$, $Problem_{size}$);
Evaluate($Population$);
$S_{best} \leftarrow$ BestSolution($Population$);
while *not exit criterion* **do**
    $Parents \leftarrow$ SelectParents ($Population$);
    $Children \leftarrow \phi$;
    for $Parent_1, Parent_2 \in Parents$ **do**
        if $f' \geq \bar{f}$ **then**
            Calculate P_c using (1);
            $(Child_1, Child_2) \leftarrow$ SinglePointCrossover ($Parent_1, Parent_2, P_c$);
            Calculate P_m using (2);
            $Children \leftarrow$ SinglePointMutation ($Child_1, P_m$);
            $Children \leftarrow$ SinglePointMutation ($Child_2, P_m$);
        else
            $P_c = 1$;
            $(Child_1, Child_2) \leftarrow$ UniformCrossover ($Parent_1, Parent_2, P_c$);
            $P_m = 0.5$;
            $Children \leftarrow$ UniformMutation ($Child_1, P_m$);
            $Children \leftarrow$ UniformMutation ($Child_2, P_m$);
        end
    end
    Evaluate($Children$);
    $S_{best} \leftarrow$ BestSolution($Children$);
    $Population \leftarrow$ SurvivorSelection ($Population, Children$);
end
Return(S_{best})

- Three measures of semantic relatedness are used to measure the fitness of an individual:
    1. The *Lesk* relatedness measure [10] calculates the sense which leads to the highest overlap between the glosses of two or more words. Given two target words w_1 and w_2, and their respective senses $Senses(w_1)$ and $Senses(w_2)$, for each pair of senses $S_1 \in Senses(w_1)$ and $S_2 \in Senses(w_2)$, the *Lesk* relatedness measure is defined by:

$$score_{Lesk}(S_1, S_2) = |gloss(S_1) \cap gloss(S_2)| \qquad (3)$$

    where $gloss(S_i)$ represents the bag of words corresponding to the definitions of the sense S_i.
    2. The semantic relatedness measure proposed by Banerjee and Pedersen [1] is an extension of the *Lesk* measure. The glosses of the words are extended to include those of other words to which they are related according to a given hierarchy. Given two target words w_1 and w_2 and

their respective senses $Senses(w_1)$ and $Senses(w_2)$, for each pair of senses $S_1 \in Senses(w_1)$ and $S_2 \in Senses(w_2)$, the Banerjee and Pedersen relatedness measure $Relate_{B\&P}(S_1, S_2)$ is defined by:

$$Relate_{B\&P}(S_1, S_2) = \sum_{S':S_1 \xrightarrow{rel} S'} \sum_{S'':S_2 \xrightarrow{rel} S''} |gloss(S') \cap gloss(S'')| \quad (4)$$

where the sense S' is either S_1 itself or has a relation rel with S_1, and the sense S'' is either S_2 itself or has a relation rel with S_2.

3. The ontology-based measure introduced by Sánchez et al. [19] uses taxomonical features to measure the semantic distance between words. Given two concepts c_1 and c_2, their semantic distance $Dist_{S\&B\&I\&V}(c_1, c_2)$ is defined by:

$$Dist_{S\&B\&I\&V}(c_1, c_2) =$$
$$\log \left(1 + \frac{|\phi(c_1) \setminus \phi(c_2)| + |\phi(c_2) \setminus \phi(c_1)|}{|\phi(c_1) \setminus \phi(c_2)| + |\phi(c_2) \setminus \phi(c_1)| + |\phi(c_1) \cap \phi(c_2)|} \right),$$
$$|\phi(c_1) \setminus \phi(c_2)| + |\phi(c_2) \setminus \phi(c_1)| + |\phi(c_1) \cap \phi(c_2)| \neq 0 \quad (5)$$

where $\phi(c_1)$ and $\phi(c_2)$ are features of the concepts c_1 and c_2, respectively, and $\phi(a) \setminus \phi(b) = \phi(a) - \phi(b) = \{c \in C | c \in \phi(a) \wedge c \notin \phi(b)\}$.

The fitness function is based on the similarity of the senses, instead of their distance. The distance $Dist_{S\&B\&I\&V}(c_1, c_2)$ between the concepts c_1 and c_2 can be considered to define their similarity by determining the intersection of their features, rather. We introduce FWN a similarity measure based on this method.

4   Experimental Evaluation

We present the results of experiments conducted with SGA (GA with standard genetic operators), $SAGA_{Gener}$ (Generational scheme), and $SAGA_{Elitist}$ (Elitist scheme) on standard corpora, including SensEval-2 (Task#1: English all words (2001)) [15], SensEval-3 (Task#1: English all words (2004)) [20], and SemEval-2007 (Task#7: Coarse-grained English all-words (2007)) [14]. Table 1 summarizes the characteristics of these corpora. Table 2 presents the experimental settings of the three algorithms SGA, $SAGA_{Gener}$, and $SAGA_{Elitist}$. Each of these algorithms was evaluated using different fitness functions based on the three semantic relatedness measures $Lesk$, $Relate_{B\&P}$ and FWN. We used the precision and recall as performance evaluation criteria. Let TP, TN, FP, and FN be the true positives, true negatives, false positives, and false negatives, respectively. The precision is given by: $P(\%) = (TP/(TP + FP)) \times 100, TP + FP \neq 0$. The recall is given by: $R(\%) = (TP/(TP + FN)) \times 100, TP + FN \neq 0$.

Tables 3, 4 and 5 show the mean precision $\overline{P}(\%)$ and mean recall $\overline{R}(\%)$ of the three variants of the algorithms SGA, $SAGA_{Gener}$, and $SAGA_{Elitist}$ on SensEval-2, SensEval-3 and SemEval-2007, respectively.

Table 1. Corpora used for the experiments

Corpus	Task	Number of Data sets	Number of words	Source of glosses and lexical relations
SensEval-2	Task#1: English all words (2001) (Fine-grained)	3	2473	WordNet 3.0
SensEval-3	Task#1: English all words (2004) (Fine-grained)	3	2081	WordNet 3.0
SemEval-2007	Task#7: Coarse-grained English all-words (2007)	5	2269	WordNet 2.0

Table 2. Experimental settings of the algorithms

	SGA	SAGA$_{Gener}$	SAGA$_{Elitist}$
Initial population population size	Random generation 50	Random generation 50	Random generation 50
Parent selection	Tournament selection ($k = 20$)	Tournament selection ($k = 20$)	Tournament selection ($k = 20$)
Crossover	Single-point ($P_c = 0.75$)	Adaptive	Adaptive
Mutation	Uniform ($P_m = 0.15$)	Adaptive	Adaptive
Survivor selection	Generational	Generational	Elitist (best 5)
Termination condition	100 generations	100 generations	100 generations

The three variants of SAGA$_{Gener}$ and SAGA$_{Elitist}$ outperformed the related variants of SGA on all the data sets at both fined-grained and coarse-grained levels. SAGA$_{Elitist}$ with $Relate_{B\&P}$ as relatedness measure gave the best precision on SensEval-2 (49.82%), SensEval-3 (43.95%) and SemEval-2007 (75.51%). The results show that its precision improvement over the precision of SGA on SensEval-02, SensEval-03 and SemEval-2007 is 28.93%, 51.08% and 14.63%, respectively. SAGA$_{Elitist}$ with FWN as relatedness measure gave the best recall on SensEval-2 (61.28%), SensEval-3 (56.23%) and SemEval-2007 (85.07%). The results show that its recall improvement over the recall of SGA on SensEval-02, SensEval-03 and SemEval-2007 is 18.23%, 28.96% and 5.02%, respectively.

We compared the results of SAGA$_{Elitist}$ to an unsupervised WSD method called UoR-SSI [2] considered as the best performing method on SemEval-2007 (Task#7) (precision=83.21%; recall=83.21%). The performance of SAGA$_{Elitist}$ with $Relate_{B\&P}$ ($\overline{P}(\%) = 75.51, \overline{R}(\%) = 78.77$) is close to that of UoR-SSI. SAGA$_{Elitist}$ with FWN gave a better recall ($\overline{R}(\%) = 85.07$), but a worse precision ($\overline{P}(\%) = 64.52$).

Table 3. Results of performance comparison of the GAs on SensEval-2 (Task#1: English all words (2001)). The best results are highlighted.

Algorithm	Fitness	Lesk		Relate$_{B\&P}$		FWN	
		$\overline{P}$(%)	$\overline{R}$(%)	$\overline{P}$(%)	$\overline{R}$(%)	$\overline{P}$(%)	$\overline{R}$(%)
SGA		38.21	53.81	38.64	54.79	36.07	51.83
SAGA$_{Gener}$		41.28	57.92	48.38	51.29	38.33	59.86
SAGA$_{Elitist}$		41.67	60.43	49.82	53.27	38.96	61.28

Table 4. Results of performance comparison of the GAs on SensEval-3 (Task#1: English all words (2004)). The best results are highlighted.

Algorithm	Fitness	Lesk		Relate$_{B\&P}$		FWN	
		$\overline{P}$(%)	$\overline{R}$(%)	$\overline{P}$(%)	$\overline{R}$(%)	$\overline{P}$(%)	$\overline{R}$
SGA		28.41	45.41	29.09	47.37	27.44	43.60
SAGA$_{Gener}$		29.97	48.74	42.03	46.73	29.69	54.22
SAGA$_{Elitist}$		30.36	53.34	43.95	48.59	29.93	56.23

5   Conclusion and Future Work

This paper proposed a self-adaptive GA for the WSD problem. We evaluated six variants of SAGA using two different survivor selection methods and three different fitness functions (semantic relatedness measures) with a corpora including SensEval-2 (Task#1), SensEval-3 (Task#1) and SemEval-2007 (Task#7). A significant improvement of both precision and recall was achieved by all the variants of SAGA in comparison to a GA with standard operators. The elitist variants of SAGA were the best performing algorithms. They demonstrated a precision and recall approaching the top performing methods on SemEval-2007 (Task#7).

The future work includes evaluation of our algorithms on more data sets, such as the fine-grained SemEval-2007 WSD (Task#17), and its comparison to other related methods. We also plan to investigate the impact of parent selection methods and other semantic relatedness measures on the performance of SAGA.

Table 5. Results of performance comparison of the GAs on SemEval-2007 (Task#7: Coarse-grained English all-words (2007)). The best results are highlighted.

Algorithm	Fitness	Lesk		Relate$_{B\&P}$		FWN	
		$\overline{P}$(%)	$\overline{R}$(%)	$\overline{P}$(%)	$\overline{R}$(%)	$\overline{P}$(%)	$\overline{R}$
SGA		64.46	76.95	65.87	78.93	61.48	81.00
SAGA$_{Gener}$		66.49	80.18	74.34	77.36	64.42	84.10
SAGA$_{Elitist}$		67.18	83.18	75.51	78.77	64.52	85.07

References

1. Banerjee, S., Pedersen, T.: Extended gloss overlaps as a measure of semantic relatedness. In: Proceedings of the 18th International Joint Conference on Artificial Intelligence, pp. 805–810. Morgan Kaufmann Publishers Inc., San Francisco (2003)
2. Chen, P., Ding, W., Bowes, C., Brown, D.: A fully unsupervised word sense disambiguation method using dependency knowledge. In: Proceedings of Human Language Technologies: The 2009 Annual Conference of the North American Chapter of the Association for Computational Linguistics, NAACL 2009, pp. 28–36 (2009)
3. Cooper, J., Hinde, C.: Improving genetic algorithms' efficiency using intelligent fitness functions. In: Chung, P.W.H., Hinde, C.J., Ali, M. (eds.) IEA/AIE 2003. LNCS, vol. 2718, pp. 636–643. Springer, Heidelberg (2003)
4. De Jong, K.A., Spears, W.M.: An analysis of the interacting roles of population size and crossover in genetic algorithms. In: Schwefel, H.-P., Männer, R. (eds.) PPSN. LNCS, vol. 496, pp. 38–47. Springer, Heidelberg (1991)
5. Gelbukh, A., Sidorov, G., Han, S.Y.: Evolutionary approach to natural language word sense disambiguation through global coherence optimization. WSEAS Transactions on Communications 1, 11–19 (2003)
6. Hasançebi, O., Erbatur, F.: Evaluation of crossover techniques in genetic algorithm based optimum structural design. Computers & Structures 78(1–3), 435–448 (2000)
7. Hong, T.P., Wang, H.S., Lin, W.Y., Lee, W.Y.: Evolution of appropriate crossover and mutation operators in a genetic process. Applied Intelligence 16(1), 7–17 (2001)
8. Jenkins, W.: On the application of natural algorithms to structural design optimization. Engineering Structures 19(4), 302–308 (1997). Structural Optimization
9. Kaya, M.: The effects of two new crossover operators on genetic algorithm performance. Applied Soft Computing 11(1), 881–890 (2011)
10. Lesk, M.: Automatic sense disambiguation using machine readable dictionaries: how to tell a pine cone from an ice cream cone. In: Proceedings of the 5th Annual International Conference on Systems Documentation, SIGDOC 1986, pp. 24–26. ACM, New York (1986)
11. Liepins, G.E., Vose, M.D.: Characterizing crossover in genetic algorithms. Annals of Mathematics and Artificial Intelligence 5(1), 27–34 (1992)
12. Mallery, J.C.: Thinking about foreign policy: Finding an appropriate role for artificial intelligence computers. PhD thesis, MIT Political Science Department, Cambridge, MA, USA (1988)
13. Menai, M.E.B.: Word sense disambiguation using evolutionary algorithms-Application to Arabic language. Computers in Human Behavior 41, 92–103 (2014)
14. Navigli, R., Litkowski, K.C., Hargraves, O.: Semeval-2007 task 07: coarse-grained english all-words task. In: Proceedings of the 4th International Workshop on Semantic Evaluations, SemEval 2007, pp. 30–35. ACL (2007)
15. Palmer, M., Fellbaum, C., Cotton, S., Delfs, L., Dang, H.T.: English tasks: all-words and verb lexical sample. In: The Proceedings of the Second International Workshop on Evaluating Word Sense Disambiguation Systems, SENSEVAL 2001, pp. 21–24. ACL (2001)
16. Park, J.B., Park, Y.M., Won, J.R., Lee, K.: An improved genetic algorithm for generation expansion planning. IEEE Transactions on Power Systems 15(3), 916–922 (2000)

17. Picek, S., Golub, M., Jakobovic, D.: Evaluation of Crossover Operator Performance in Genetic Algorithms with Binary Representation. In: Huang, D.-S., Gan, Y., Premaratne, P., Han, K. (eds.) ICIC 2011. LNCS, vol. 6840, pp. 223–230. Springer, Heidelberg (2012)
18. Povinelli, R.J., Feng, X.: Improving genetic algorithms performance by hashing fitness values. In: Proceedings of the International Conference on Artificial Neural Networks in Engineering, pp. 399–404 (1999)
19. Sánchez, D., Batet, M., Isern, D., Valls, A.: Ontology-based semantic similarity: A new feature-based approach. Expert Syst. Appl. **39**(9), 7718–7728 (2012)
20. Snyder, B., Palmer, M.: The english all-words task. In: The Proceedings of the Third International Workshop on the Evaluation of Systems for the Semantic Analysis of Text, July 21–26. ACL (2004)
21. Someya, H., Yamamura, M.: A geneti algorithm without parameters tuning and its appliation on the oorplan design problem. In: Banzhaf, W., Daida, J., Eiben, A.E., Garzon, M.H., Honavar, V., Jakiela, M., Smith, R.E. (eds.) Proceedings of the Geneti and Evolutionary Computation Conferene (GECCO 1999), vol. 1, pp. 620–627. Morgan Kaufmann, Orland (1999)
22. Syswerda, G.: Uniform crossover in genetic algorithms. In: Proceedings of the 3rd International Conference on Genetic Algorithms, pp. 2–9. Morgan Kaufmann Publishers Inc., San Francisco (1989)
23. Toğan, V., Daloğlu, A.T.: An improved genetic algorithm with initial population strategy and self-adaptive member grouping. Computers & Structures **86**(11–12), 1204–1218 (2008)
24. Vasconcelos, J., Ramirez, J., Takahashi, R., Saldanha, R.: Improvements in genetic algorithms. IEEE Transactions on Magnetics **37**(5), 3414–3417 (2001)
25. Vrajitoru, D.: Crossover improvement for the genetic algorithm in information retrieval. Inf. Process. Manage. **34**(4), 405–415 (1998)
26. Wan, W., Birch, J.B.: An improved hybrid genetic algorithm with a new local search procedure. Journal of Applied Mathematics **2013**(Article ID 103591) (2013)
27. Wu, S.J., Chow, P.T.: Steady-state genetic algorithms for discrete optimization of trusses. Computers & Structures **56**(6), 979–991 (1995)
28. Zhang, C., Zhou, Y., Martin, T.: Geneti word sense disambiguation algorithm. In: Proeedings of the 2008 Seond International Symposium on Intelligent Information Tehnology Appliation, IITA 2008, vol. 01, pp. 123–127. IEEEComputer Soiety, Washington, DC (2008)

Improving Noisy T1-Weighted MRI Spatial Fuzzy Segmentation Based on a Hybrid of Stationary Wavelet Thresholding and Filtering Preprocess

Papangkorn Pidchayathanakorn and Siriporn Supratid[✉]

School of Information Technology, Rangsit University, Pathumthani, Thailand
papangkorn.p@gmail.com, siri_sup1@hotmail.com

Abstract. This paper proposes an improved spatial fuzzy segmentation of a noisy image, based on a hybrid of stationary wavelet thresholding and filtering preprocess. The proposed methods aim to improve the segmentation by reducing the effect of additive noise during preprocess. Noise filtering as well as wavelet thresholding are carried out in each stationary wavelet subbands. Thus, noise distributed across any subband coefficients can be examined. This would lead to image denoising improvement. Afterwards, fuzzy c-means incorporated with spatial information (sFCM) is utilized for segmenting the denoised image. The denoising preprocess and segmentation measurements rely on peak signal-to-noise ratio (PSNR) and Xie-Beni (XB) validity index respectively. T1-weighted MRI is tested with salt-and-pepper and Gaussian additive noise. Based on experimental results, the proposed hybrid methods improve the segmentation more efficiently than comparative traditional denoising methods.

Keywords: Stationary wavelet transform · Noise filtering · T1-weighted MRI · Fuzzy c-means

1 Introduction

The primary goal of brain magnetic resonance image (MRI) is to partition a given brain image into six different regions, representing anatomical structures. The accurate representation of gray matter (GM), white matter (WM), cerebrospinal fluid, fat, bone, and air [3] provides a way to identify brain disorders such as Alzheimer disease, dementia or schizophrenia. By this particular reason, brain MRI segmentation is especially significant. Fuzzy c-means (FCM) [1] is a method that allows one piece of data to belong to two or more clusters. The FCM has been widely used in the segmentation of brain MRI [2] because it can preserve most information from the original image. However, most of medical images possess intrinsic structure, that is pixels in the immediate neighborhood usually have similar features, and the pixels with similar features have high probability of belonging to the same cluster. Such spatial relationship of neighboring pixels is an important issue for medical image segmentation. The conventional FCM does not consider such relationship; as a result, it is very sensitive to the intensity inhomogeneity problem, emerged from improper image acquisition. Fuzzy c-means clustering with spatial information (sFCM) [3] was developed for

© Springer International Publishing Switzerland 2015
M. Ali et al. (Eds.): IEA/AIE 2015, LNAI 9101, pp. 591–600, 2015.
DOI: 10.1007/978-3-319-19066-2_57

dealing with intensity non-uniformity during clustering iterations, where suitable cluster centroids as well as membership degrees are discovered.

Nevertheless, most of medical images are inevitably corrupted by noise during acquisition and transmission. Such types of noise can severely degrade the segmentation. Filtering image before segmentation process usually promotes the segmentation quality. Median filter has been established as a reliable method to remove the salt and pepper noise without harming MRI edge details [4]. Gaussian smoothing is applied in [5] for reducing noise effect and enabling FCM to create a more homogeneous clustering in brain MRI segmentation. Wiener filter was employed for quantum Gaussian noise reduction in digital mammography, where the necessary parameters are adaptively estimated directly from the noisy image [6]. Nevertheless, those traditional denoising approaches and many more are prone to over-smooth edges and image detail. To relieve this drawback, discrete wavelet transform (DWT) [7] was developed as efficiently useful tool for noise removal [8,9]; while most of image details could still remain. DWT decomposes an image into different subband images; then, reduces the noise in those subbands. Nevertheless, a critical disadvantage of DWT is that it is not a shift-invariant transform due to down- sampling in each subband. This causes information loss in the respective subbands. The stationary wavelet transform (SWT) [10] was designed to overcome the lack of such shift-invariance, regarding DWT. SWT wavelet coefficients contain many redundant information which helps distinguishing the noise from meaningful, clean data. It also reports in [11,12] that SWT works more effectively with Bayes shrinkage [13] in image denoising than some other powerful shrinkages methods such as VisuShrink and universal   threshold.

This paper proposes an improved spatial fuzzy segmentation of a noisy image, based on a hybrid of stationary wavelet thresholding and filtering preprocess. The proposed method, namely SWMsFCM and SWWsFCM enhance the segmentation quality by respectively reducing the effect of salt-and-pepper and Gaussian additive noise during preprocess. Then, sFCM is performed for spatially segmenting the denoised image. The preprocessing and segmentation evaluations sequentially rely on peak signal-to-noise ratio (PSNR) and Xie-Beni (XB) validity index. The T1-weighted brain MRI [14] is tested with salt-and-pepper as well as Gaussian additive noise. The rest of this paper is organized as follows. Section 2 reviews filtering and SWT for image preprocessing. sFCM is described in section 3. Experimental comparisons are delineated in section 4 and 5. The overall conclusion is drawn in the last section.

2   Filtering and SWT for Image Preprocessing

Median, Gaussian, and Wiener filters [15,16] can be referred as effective low-pass filters. Median filter is a non-linear digital filtering technique. It is most appropriate for removing salt and pepper and impulse noise. On the other side, Gaussian filter utilizes a linear smoothing technique that applies weights specified by the probability density function of a bivariate Gaussian with a particular variance. The Gaussian smoothing filter is suitable for removing noise drawn from a normal distribution; at the same time it blurs the detail information. Wiener filter is an optimum linear approach, which minimizes the mean square error for images degraded by additive noise and blurring. However, such traditional filters still have a tendency to lower the integrity of image detail.

To cope with this difficulty, discrete wavelet transform (DWT) [7] employs mathematical functions to separate noise from an image; while maintain a large part of the desired information. DWT has become a useful computational tool for a variety of signal and image processing, including noise removal. However, DWT still suffers from the lack of shift invariance. This means that small shifts in the input signal can cause major variations in the distribution of energy between coefficients at different levels; then, it may cause some errors in reconstruction. This problem can be solved by eliminating the down sampling steps after filtration at each level in stationary wavelet transform (SWT) [10]. Similar to DWT, SWT requires three operational steps for image denoising, which refer to decomposition, threshold estimation and shrinkage, and reconstruction.

2.1 Decomposition

Suppose an input image is decomposed by two-level SWT into different subbands. In Fig. 1(a), LH_1, HL_1 and HH_1 denote the first level high frequency detailed information along the horizontal, vertical and diagonal subbands, respectively; whereas LL_1 represents the low frequency estimation which consists of low resolution residual. The LL_1 subband is further split at the next level of decomposition to achieve LL_2, LH_2, HL_2 and HH_2 Fig. 1(b) displays two-level decomposition of T1-weighted MRI.

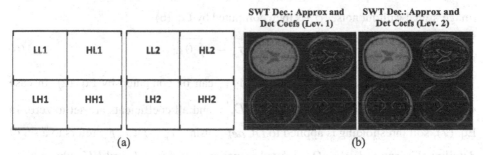

(a)   (b)

Fig. 1. Decomposition using SWT (a) Two-level decomposition structure (b) Two-level decomposition of T1-weighted MRI

2.2 Threshold Estimation and Shrinkage for Denoising

Noise can be separated from the detail subbands of an image through threshold estimation and shrinkage. The model of the image, corrupted by additive noise is basically shown in the following form:

$$\mathbf{X} = \mathbf{S} + \mathbf{E} \qquad (1)$$

where $\mathbf{X}$, $\mathbf{S}$ and $\mathbf{E}$ are matrices of size $M \times N$. $\mathbf{X}$ refers to a noisy image, degraded by noise $\mathbf{E}$; $\mathbf{S}$ denotes an image without noise. Here, Bayes shrink [13] is used for noise reduction by minimizing Bayesian risk. The Bayes threshold, t_B is defined as Eq. (2)

$$t_B = \frac{\sigma_e^2}{\sigma_s}$$

(2)

where σ_s^2 represents a variance of S. In other words, it is a signal variance without noise. σ_e^2 is a variance of noise E. σ_e is estimated using median estimator, as shown in Eq. (3)

$$\sigma_e = \frac{median\left(|cD_1|\right)}{0.6745}$$

(3)

cD_1, corresponding to HH_1 and having the largest spectrum represents the detail wavelet coefficients at the first level. 0.6745 is a normalization factor. Since the noise E and the image signal without noise S are independent, then Eq. (4) can be stated.

$$\sigma_x^2 = \sigma_s^2 + \sigma_e^2$$

(4)

where σ_x^2 is a variance of the noisy image X that can be calculated as Eq. (5)

$$\sigma_x^2 = \frac{1}{(MN)} \sum_{\forall m,n} x^2(m,n)$$

(5)

$x(m,n)$ denotes a pixel in image X, $m = 1,..,M$ and $n = 1,...,N$. The variance of the image signal without noise σ_s^2 is then computed by Eq. (6)

$$\sigma_s = \sqrt{\max(\sigma_x^2 - \sigma_e^2, 0)}$$

(6)

With σ_e^2 and σ_s, the Bayes threshold t_B can be computed by Eq. (2). In case where σ_s is 0, then t_B is set to $\max(|cD_1|)$; and all coefficients are set to zero. In Eq. (7), soft thresholding is applied to $DCoef_j$, using t_B. $DCoef_j$ refers to a set of detail coefficients cH_j, cV_j, cD_j, related respectively to LH_j, HL_j and HH_j subbands.

$$DCoef_j' = \begin{cases} \text{sgn}(DCoef_j)(DCoef_j - t_B), & if\ |DCoef_j| > t_B \\ 0, & otherwise \end{cases}$$

(7)

At the end of shrinkage, one would achieve a new set of modified wavelet detail coefficients, $DCoef_j'$; it consists of cH_j', cV_j', cD_j' consecutively associated within LH_j', HL_j' and HH_j' subbands.

2.3   Reconstruction

After denoising, the reconstructed image $\hat{\mathbf{S}}$ is computed based on the approximate and modified detail coefficients. The low-pass and high-pass filters at each level are up sampled by putting zeroes between each filter's coefficients of the previous level.

3   Spatial Fuzzy C-means (sFCM)

Fuzzy c-means (FCM) [1] has been successfully applied to image segmentation [2]. The success of segmentation is evaluated by FCM cost function. This cost function, defined in Eq. (8) iteratively minimizes the distance between pixels $\hat{s}(m,n)$ in image $\hat{\mathbf{S}}$ and cluster center v_c

$$J = \sum_{\forall m,n} \sum_{\forall c} u_{\hat{s}(m,n),c}^{k} \left\| \hat{s}(m,n) - v_c \right\|^2 \tag{8}$$

where $u_{\hat{s}(m,n),c}$ represents membership degree of $\hat{s}(m,n)$ in cluster c. k specifies the degree of fuzziness. $u_{\hat{s}(m,n),c'}$ and cluster center $v_{c'}$ are iteratively updated by Eq. (9) and Eq. (10).

$$u_{\hat{s}(m,n),c'} = \sum_{\forall \varrho} \left(\frac{\left\| \hat{s}(m,n) - v_{c'} \right\|}{\left\| \hat{s}(m,n) - v_c \right\|} \right)^{-\frac{2}{k-1}} \tag{9}$$

$$v_{c'} = \frac{\sum_{\forall m,n} u_{\hat{s}(m,n),c'}^{k} \hat{s}(m,n)}{\sum_{\forall m,n} u_{\hat{s}(m,n),c'}^{k}} \tag{10}$$

However, one of the important characteristics of a medical image is that neighboring pixels are highly correlated. There is a high probability for them to belong to the same cluster. This significantly spatial relationship is not recognized in a standard FCM algorithm. Spatial fuzzy c-means (sFCM) [3] exploits the spatial information during clustering iterations.   A spatial function is defined as Eq. (11).

$$h_{\hat{s}(m,n),c'} = \sum_{k \in NB(\hat{s}(m,n))} u_{k,c'} \tag{11}$$

Neighborhood of $\hat{s}(m,n)$, $NB(\hat{s}(m,n))$ represents a square 3x3 window. The spatial function is incorporated into membership function as follows:

$$u'_{\hat{s}(m,n),c'} = \frac{u_{\hat{s}(m,n),c}^{p} h_{\hat{s}(m,n),c'}^{q}}{\sum_{\forall c} u_{\hat{s}(m,n),c}^{p} h_{\hat{s}(m,n),c}^{q}} \tag{12}$$

p and *q* are parameters to control the relative importance of both functions. In a homogenous region, the spatial functions simply fortify the original membership, and the clustering result remains unchanged. In a noisy area, this formula reduces the weighting of a noisy cluster by the labels of its neighboring pixels.

4   An Improved Spatial Fuzzy Image Segmentation Method, Based on a Hybrid of SWT and Filtering Preprocessing

As aforementioned, median filter has been remarkably effective for removing salt-and-pepper noise; whereas Wiener technique has been widely used for Gaussain noise reduction. Therefore, the proposed hybrid method, SWMsFCM and SWWsFCM sequentially utilizes a combination of SWT with median filter (SWM) and a fusion of SWT using Bayes shrinkage and Wiener filter (SWW) to reduce the effect of additive noise during segmentation preprocess. SWM and SWW are employed for salt-and-pepper and Gaussian noise removal respectively, as depicted in Fig. 2 (a) and (b).Noise filtering is executed on all individual SWT subbands in the whole decomposition levels. Thus, noise distributed across all subband coefficients can be examined.

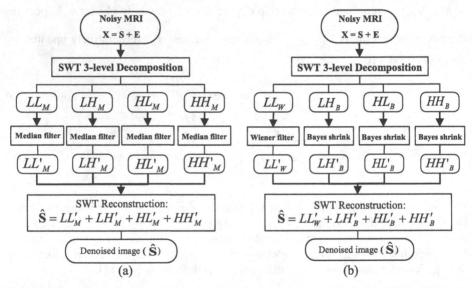

Fig. 2. Over all mechanism of the proposed denoising preprocesses (a) Using SWM for salt-and-pepper noise removal (b) Using SWW for Gaussian noise removal

With respect to SWM and SWW denoising preprocesses, 3-level decomposition of SWT is executed to generate different subbands LL_m, LH_m, HL_m and HH_m as well as LL_W, LH_B, HL_B and HH_B consecutively. Fig. 2(a) illustrates the SWM removes salt-and-pepper additive noise in each subband using median filter. Then, denoised image $\hat{S}$ is reconstructed by the denoised subbands LL LL'_m, LH'_m, HL'_m and HH'_m.With respect to SWW shown in Fig. 2(b), noise associated within all levels of approximate subband LL_W is eliminated by the Wiener filter. Denoised approximate subband LL'_W is then obtained. On the other side, Bayes shrinkage is employed to manage the noise hidden in detail subband coefficients cH_B, cV_B and cD_B, respectively corresponding to LH_B, HL_B and HH_B. Unlike in the traditional Bayes shrinkage, Bayes threshold t_B in this work takes into account the entire coefficients in every level. It is shown in Eq. (13) that the standard deviation of *the* noise σ_e is calculated using a global median, where the whole coefficients can be determined.

$$\sigma_e = \frac{median\ (|cA_W, cH_B, cV_B, cD_B|)}{0.6745} \tag{13}$$

Thereafter, soft shrinkage is carried out for each one of cH'_B, cV'_B and cD'_B, related to LH'_B, HL'_B and HH'_B. Such denoised detail subbands and the denoised approximate LL'_W are all reconstructed to produce the restored image $\hat{S}$. Afterwards, the segmentation would be implemented on the denoised image $\hat{S}$ based upon sFCM.

5   Experimental Results and Discussion

A *T1-weighted* MRI [14] is tested with 0.02 noise-density of salt-and-pepper as well as 0.01-variance Gaussian additive noise. 10-fold cross-validation is performed to evaluate the denoising preprocess and segmentation. The measurements rely on peak signal-to-noise ratio (PSNR) and Xie-Beni (XB) validity index [17], respectively shown in Eq. (14) and Eq. (15)

$$PSNR = 10 \log_{10}\left(\frac{255^2}{MSE}\right) \tag{14}$$

where 255 is the maximum value of the pixels, presented in an image. MSE represents the mean square error between original and denoised image.

$$XB = \frac{\sum_{\forall m,n} \sum_{\forall c} \mu^k_{\hat{s}(m,n),c} \|\hat{s}(m,n) - v_c\|^2}{(M \times N) \times (\min_{c' \neq c} \|v_{c'} - v_c\|^2)} \tag{15}$$

The numerator of Eq. (15) repeats the FCM cost function, defined in Eq. (8). Parameters used in the comparative denoising methods are declared in Table 1. Such parameters are chosen, relying on experimental comparison over several effective basis functions.

Table 1. Parameters used in the denoising methods for each type of additive noise

Additive noise	Denoising methods	Parameters used
Salt-and-pepper	Gaussian filter	$\sigma = 0.5$ with window size = [3 3]
	Wiener filter	window size = [5 5], noise power=0.01
	SWT	'db10' with 1 and 3 decomposition levels
	Median filter	window size =[3 3]
	SWM	'rbio3.1' with 1 and 3 decomposition levels
Gaussian	Gaussian filter	$\sigma = 1.3$ with window size = [3 3]
	Wiener filter	window size = [3 3], noise power=0.03
	Median filter	window size =[3 3]
	SWT	'bior1.5' with 1 and 3 decomposition levels
	SWW	'bior5.5' with 1 and 3 decomposition levels

Derived boxplots in Fig. 3(a) indicate that exceptional PSNR values are yielded by median filter, followed by 3-level SWM (3L-SWM) for dealing with salt-and-pepper noise. However, Fig. 3(b) shows that 1-level SWMsFCM contributes the best segmentation quality in terms of minimum median and variance of XB values. Although median filter with sFCM (MsFCM) and 3-level SWMsFCM also achieve competetive XB results, both of them generate higher degree of variances than 1-level SWMsFCM does. With respect to Gaussian noise removal, Fig. 4(a) shows 3-level SWT and 1-level SWW consecutively produce 19.75% and 10.2% higher PSNR than Wiener filter alone , based on median values of derived boxplots; while the lowest PSNR is resulted by 3-level SWW. On the contrary, Fig. 4(b) points the best segmentation performance, yielded by 3-level SWWsFCM. Such a method along with 3-level SWTsFCM and 1-level SWWsFCM respectively provide 65.87 %, 41.21 % and 25.92 % lower XB degrees than Wiener with sFCM (WsFCM). Athough the 3-level SWWsFCM generates rather high XB variance, there exists none of outliers; whereas the 3-level SWTsFCM and 1-level SWWsFCM provide many outliers. Based on Fig. 4(a) and (b), one would see that the denoising method yielding low PSNR possibly provides high segmentation performance afterwards. This mean the balance between inhomogeneity problem reduction and image detail can probably occur in a low PSNR image.

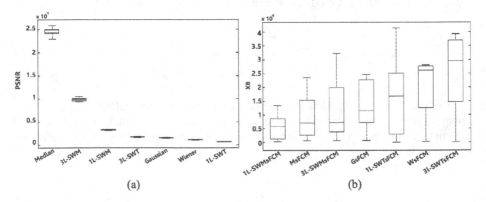

(a)   (b)

Fig. 3. Results of salt-and-pepper noise removal (a) PSNR (b) XB degrees

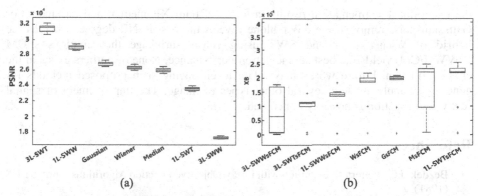

Fig. 4. Results of Gaussian noise removal (a) PSNR (b) XB degrees

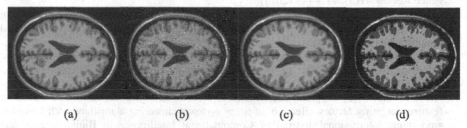

Fig. 5. Visual results of T1-weighted MRI denoising and segmentation. (a) Original, (b) Original with salt-and-pepper noise, (c) Using Median, (d) Using 1L-SWMsFCM

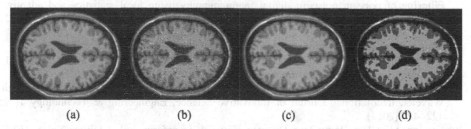

Fig. 6. Visual results of T1-weighted MRI denoising and segmentation. (a) Original, (b) Original with Gaussian noise, (c) Using 3L-SWT, (d) Using 3L-SWWsFCM

Fig. 5(c) and 6(c) represent the best denoised images with highest values of PSNR; while Fig. 5(d) and 6(d) refer to the best segmented images with lowest XB degrees.

6   Conclusions

This paper points that using a hybrid of SWT and a suitable filter during preprocess can effectively improve spatial fuzzy segmentation of a noisy T1-weighted MRI. Properly selected basis wavelet function, level of decomposition and other related parameters are also essential. Here, noise distributed across any subband coefficients in each level is globally investigated. Based on 10-fold cross-validation, median filter indicates the outstanding PSNR results for salt-and-pepper noise removal. However, the hybrid of SWT and median filter, along with spatial FCM (SWMsFCM) provides the exceptional segmentation results in terms of both XB median and variance. For Gaussian noise removal, the SWT alone reveals the best PSNR degree; whereas the hybrid of Wiener filter and SWT using Bayes shrinkage that employs sFCM (SWWsFCM) yields the best segmentation performance. None of outliers exists in the cross-validation. Future works may relate to enhancing such proposed methods for generically applying with several other types of image. The improvement of spatial fuzzy segmentation is possibly concerned.

References

1. Bezdek, J.C.: Pattern Recognition with Fuzzy Objective Function Algorithms. Springer US (1981)
2. Balafar, M.A.: Fuzzy C-mean based brain MRI segmentation algorithms. Artificial Intelligence Review **41**(3), 441–449 (2014)
3. Chuanga, K.S., Tzenga, H.L., Chena, S., Wua, J., Chenc, T.J.: Fuzzy c-means Clustering with Spatial Information for Image Segmentation. Computerized Medical Imaging and Graphics **30**(1), 9–15 (2006)
4. Babu, K.R., Sunitha, K.V.N.: Image de-noising and enhancement for salt and pepper noise using improved median filter-morphological operations. In: Das, V.V., Stephen, J. (eds.) CNC 2012. LNICST, vol. 108, pp. 7–14. Springer, Heidelberg (2012)
5. Xiao, K., Ho, S.H., Bargiela, A.: Automatic brain MRI segmentation scheme based on feature weighting factors selection on fuzzy c-means clustering algorithms with Gaussian smoothing. International Journal of Computational Intelligence in Bioinformatics and Systems Biology **1**(3), 316–331 (2010)
6. Vieira, M.A.C., Bakic, P.R., Maidment, A.D.A., Schiabel, H., Mascarenhas, N.D.A.: Filtering of poisson noise in digital mammography using local statistics and adaptive wiener filter. In: Maidment, A.D.A., Bakic, P.R., Gavenonis, S. (eds.) IWDM 2012. LNCS, vol. 7361, pp. 268–275. Springer, Heidelberg (2012)
7. Mallat, S.: A wavelet tour of signal processing. Academic Press, New York (1998)
8. Om, H., Biswas, M.: An Improved Image Denoising Method Based on Wavelet Thresholding. Journal of Signal and Information Processing **3**, 109–116 (2012)
9. SavajiP, S., AroraP, P.: Denoising of MRI Images using Thresholding Techniques through Wavelet. International Journal of Innovative Science, Engineering & Technology **1**(7), 422–427 (2014)
10. Fowler, J.E.: The Redundant Discrete Wavelet Transform and Additive Noise. IEEE Signal Processing Letters **12**(9), 629–632 (2005)
11. Sudha, S., Suresh, G.R., Sukanesh, R.: Comparative Study on Speckle Noise Suppression Techniques for Ultrasound Images. International Journal of Engineering and Technology **1**(1), 57–62 (2009)
12. Ruikar, S., Doye, D.D.: Image denoising using wavelet transform. In: International Conference on Mechanical and Electrical Technology, pp. 509–515. IEEE (2010)
13. Chang, S.G., Yu, B., Vetterli, M.: Adaptive Wavelet Thresholding for Image Denoising and Compression. IEEE Transactions of Image Processing **9**(9), 1532–1546 (2000)
14. BrainWeb Simulated Brain Database. http://brainweb.bic.mni.mcgill.ca/brainweb/
15. Jain, R., Rangachar, K., Schunck, B.G.: Machine Vision. McGraw-Hill, New York (1995)
16. Vaseghi, S.V.: Advanced digital signal processing and noise reduction. John Wiley & Sons, New York (2000)
17. Xie, X.L., Beni, G.: A Validity Measure for Fuzzy Clustering. IEEE Transactions on Pattern Analysis and Machine Intelligence **13**(8), 841–847 (1991)

Smoke Detection for Autonomous Vehicles using Laser Range Finder and Camera

Alexander Filonenko, Danilo Cáceres Hernández, Van-Dung Hoang, and Kang-Hyun Jo[✉]

Intelligent Systems Laboratory, Graduate School of Electrical Engineering,
University of Ulsan, Ulsan 680-749, Korea
{alexander,danilo,hvzung}@islab.ulsan.ac.kr, acejo@ulsan.ac.kr

Abstract. This paper describes the smoke detection for autonomous vehicles using sensor fusion. The main difference from another algorithms is the ability to perform detection during ego movement. Laser data were used to shrink the region of interest suchwise decreasing processing time of the whole algorithm. Color and shape characteristics of smoke are used to detect possible smoke clouds which then refined by removing small objects and by filling holes. Sky region and lane markings are removed by checking edge density of the region. Other rigid objects are expelled by the boundary roughness feature. Finally, the shape descriptor was utilized to compare smoke regions in frame sequence to delete static objects.

Keywords: Smoke detection · Autonomous vehicle · Laser range finder

1 Introduction

The problem discussed in this paper is detection of smoke if one appeared above the road surface for reasons like car accident or forest fire. This task is not trivial for a computer when is it not possible to build a background model of the scene. It is just such a case if the observer is moving. There are many kinds of sensors used to solve problems which arise when a car should move autonomously. In this work the combination of lidar distance data and camera smoke detection techniques was used. Authors consider scenarios where the robot is moving along roads outside big cities with a few number of unique objects, which can be used for tracking and feature detection, or this kind of objects may not appear at all. This problem was initially the one of the tasks of the autonomous vehicle competition organized by Hyundai Motor Company.

2 Camera and LRF Calibration

This section briefly describes the calibration method of a camera and LRF. More details and evaluation can be found in [3]. The difficulty of extracting the relative position and direction of a camera and LRF is that laser beams are not visible by a camera when projected on a calibration pattern. The system presented in Fig.1

© Springer International Publishing Switzerland 2015
M. Ali et al. (Eds.): IEA/AIE 2015, LNAI 9101, pp. 601–610, 2015.
DOI: 10.1007/978-3-319-19066-2_58

Fig. 1. Autonomous vehicle system

consists of multiple cameras and two LRFs. For the smoke detection purposes it is enough to utilize bottom LRF and the right camera.

Calibration pattern shown in Fig.2(a) consists of two adjacent right-angled triangles ABC and ADC. The lower one is a triangular hole which makes it easier to estimate the position of laser beams on the pattern and to detect corresponding features between image and laser beam. Four black rectangles are used to detect corners on the image. The pattern is placed in the field of view of a camera while letting laser beams cross the triangles and form the line with points E, F, and G. The point F is directly estimated based on laser beams. Since the distance from the center of LRF to the points E, F, and G and angles between them are known, the magnitudes of line segments EF, FG, and EG can be calculated by the law of cosines. Taking into account that AFG and CFE are similar triangles, the distance AF can be found:

$$AF = \frac{AC * FG}{FE + FG} \tag{1}$$

The angle θ is given by:

$$\theta = \begin{cases} asin\left(\frac{GH}{GE}\right) \\ 180° - \left(\frac{GH}{GE}\right) \end{cases} \tag{2}$$

AF and CE are given by the law of cosines:

$$AG = \sqrt{AF^2 + FG^2 - 2 * AF * FG * cos\left(\pi - \widehat{DAC} - \theta\right)} \tag{3}$$

$$CE = \sqrt{CF^2 + EF^2 - 2 * CF * EF * cos\left(\pi - \widehat{ACB} - \theta\right)} \tag{4}$$

The next step is the estimation of world points in camera coordinates based on perspective-n-point (PnP) method. In Fig.2(b) p_i and p_j are the norm vectors oriented from the camera center O_C to the points P_i and P_j in the real world. p_i and pi_j are estimated based on camera image [4]. Points should follow constraints $P_i = d_i p_i$ and $P_j = d_j p_j$, where d_i and d_j are scalars. d_i, d_j, and distance between P_i and P_j are incorporated into the law of cosines:

$$d_{ij}^2 = d_i^2 + d_j^2 - 2d_i d_j cos(a_{ij}) \tag{5}$$

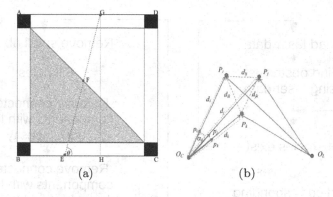

Fig. 2. (a) Calibration pattern; (b) Model of recovering real world points using three perspective points

where distance d_{ij} is given by LRF and angle a_{ij} should be calculated by using pair of reflection rays from the center of camera to the world points. By combining equations for 3 points in the real world P_i, P_j, and P_k, camera-point distances can be found by solving the equation system:

$$\begin{cases} d_i^2 + d_j^2 - 2d_id_j\cos(a_{ij}) - d_{ij}^2 = 0 \\ d_i^2 + d_k^2 - 2d_id_k\cos(a_{ik}) - d_{ik}^2 = 0 \\ d_k^2 + d_j^2 - 2d_kd_j\cos(a_{kj}) - d_{kj}^2 = 0 \end{cases} \qquad (6)$$

Finally, rotation and translation of the camera according to the LRF is defined by directly following method described in [5].

3 Smoke Detection

This section describes the algorithm for smoke detection above road surface which is an extension of the previous work [6]. The outline of the algorithm is shown in Fig.3. Unlike [6], using the fact that dense smoke reflects laser beams to the LRF, image processing is performed only if LRF detects any object. This approach resulted in not appearing most of false detected regions which had to be filtered out in [6] leading to significantly lower average CPU load.

3.1 Obstacle Detection

2D LRF is used to detect obstacles in front of an autonomous vehicle. A single laser scan returns a set of points $L_i = \{x_i, y_i, r_i | i = 1, ..., n\}$, where x_i and y_i are coordinates determined by the reflected position of the ith laser beam with distance r_i [7]. To make sure that reflected points are not noise, but some real object, the nearby points should be grouped into clusters. It is not possible to predefine the number of possible clusters. To deal with this problem, the

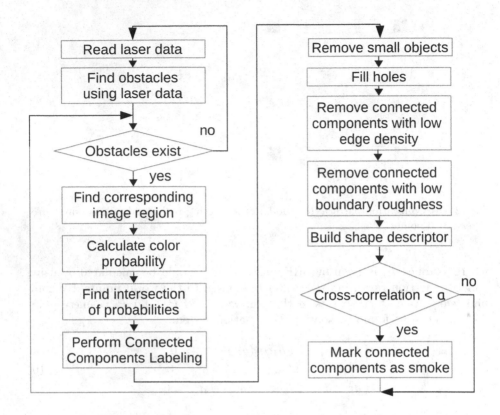

Fig. 3. Smoke detection method outline

DBSCAN method is used [8] which depends on mostly two parameters such as maximum radius of the neighbourhood (R) and the minimum number of points (Pm) belonging to the same cluster. One of the features of LRF is that density of beams is inversely proportional to the distance to an object. It is assumed that the distance between consecutive points approximately equals to the arc length (L) for small angular resolution (ψ) of LRF.

$$L = \frac{r\psi\pi}{180} \qquad (7)$$

$$R = nL \qquad (8)$$

where n is the coefficient which should be chosen according to a distance. At close distances smaller n may help to distinguish objects placed close to each other. For longer distances higher n allows to deal with beams shifted or lost by noise. Tests have shown that n=2 for the first case and n=4 for the second performs well. Pm should also be varied. For example, the number of laser beams pointing at 40 cm object will be 5 at 20 meters and only 3 at 40 meters away from LRF. Fig.4 shows the clusters of smoke regions detected by LRF. Distance is defined in

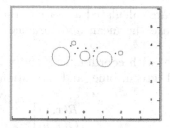

Fig. 4. Smoke regions detected by LRF

Fig. 5. Example of training database members

meters with respect to the LRF position. Smoke clusters are depicted by circles with the center coinciding with the centroid of the cluster and with radius which equals to the distance to the farthest member of the cluster. Dots show laser beams excluded from cluster since they were shifted because of non-constant density of the cloud of smoke. But they were considered during the process of building the region of interest for the image processing part by reason of their proximity to detected clusters. If the horizontal distance between two clusters or between a cluster and a point is smaller that 0.3 meters, they are united to the same bigger cluster. After the clusters and nearby points are defined, their position is mapped to the image coordinates. The image is then divided to parts containing objects. Left and right borders of each object coincide with leftmost and rightmost position of the corresponding cluster. The bottom border is selected by removing everything situated closer than the cluster. The top border is set at horizon.

3.2 Image Processing

In most scenarios autonomous cars should be moving. That is why existing smoke detection algorithms based on background subtraction cannot be applied. Thus, image processing part should take into account features of the smoke itself comparing to its possible surrounding. To gather the key features of the smoke the training database was gathered for different lightning conditions (Fig.5).

For all the pixels in the database probability density functions (PDF) of a normal distribution are built for:

1. Red (R), green (G), and blue (B) channels of the RGB color space;
2. Saturation (S) channel of the HSV color space;

Mean and variance values are calculated for each of the PDFs mentioned above. For example, μ_S and σ_S are the mean and variance values for the saturation channel respectively.

Each pixel in the image with coordinates x and y can be represented as an array E with 4 values: red, green, blue, and saturation (the 4-channel image is formed).

$$E(x,y) = \begin{bmatrix} R(x,y) \\ G(x,y) \\ B(x,y) \\ S(x,y) \end{bmatrix} \tag{9}$$

$$C_i(x,y) = e^{\dfrac{(E_i - \mu_i)^2}{-2(\sigma_i)^2}} \tag{10}$$

where i is one of the four channels of the pixel $E(x,y)$. $C_i(x,y)$ is the normalized color probability which shows the probability that the current pixel belongs to a smoke region according to the ith channel distribution. After applying (10) to all the channels of E, the color probability image will be formed.

$$C_P(x,y) = \begin{bmatrix} C_R(x,y) \\ C_G(x,y) \\ C_B(x,y) \\ C_S(x,y) \end{bmatrix} \tag{11}$$

It is not convenient to work with all 4 channels simultaneously. To simplify calculations of the following steps the intersection of all channels of C_P was found:

$$I_P(x,y) = C_R(x,y) \cdot C_G(x,y) \cdot C_B(x,y) \cdot C_S(x,y) \tag{12}$$

For the further analysis of the image those pixels should be united into groups by connected components labeling (CCL) method which assumes that the input image is binary. For binarization the threshold λ should be decided heuristically. A good alternative to the predefined threshold is the dynamic one introduced in [9].

$$B(x,y) = \begin{cases} 1, & \text{if } I_P > \lambda \\ 0, & \text{otherwise} \end{cases} \tag{13}$$

Very small possible smoke regions mostly appear due to noise. These small blobs (connected regions) can be deleted from the image. In this research objects with number of pixels less than 0.1% of the image area are deleted. Sometimes noise or low density smoke in form of holes occur in the blobs considered as potential smoke regions. Those pixels should be converted to the value of the surrounding blob [10].

Based on tests with videos in different scenarios, it became clear that the sky region and the light color road are often considered as smoke regions due to their similar color characteristics. Also it was noticed that smoke regions usually contain many edges whilst sky regions and roads are almost empty. The edge density filter can deal with this problem:

$$D_E = \frac{A_e}{A_b} \tag{14}$$

where A_e is the number of edge pixels within the blob and A_b is the total number of pixels in the blob. The following rule applied then:

$$I(m,n) = 0, \text{if } D_E < \xi \tag{15}$$

where I is the image obtained by applying all the previous steps of the algorithm. (m,n) are the coordinates of pixels of the currently considered blob. ξ is the threshold value. In tests the best value of ξ was 3%.

In some rare cases cars or banners may appear with similar color characteristics with smoke. Usually they have the shape similar to the convex hull while the smoke regions are very random in their shape. Another filter called boundary roughness can distinguish real smoke regions from that objects [11]:

$$R_B = \frac{P_b}{P_{CH_b}} \tag{16}$$

where P_b is the perimeter of the blob and P_{CH_b} is the perimeter of the convex hull of the same blob. Too "ideal" blobs are deleted according to the rule:

$$I(m,n) = 0, \text{if } R_B < \beta \tag{17}$$

where β is the threshold. Tests have shown that β=1.1 is high enough.

Without any background model it is very difficult to know which pixels belong to the static and dynamic objects. Withal static object do not change their shape when time goes. Based on this feature the blobs shape descriptor (S) is introduced. This shape descriptor represents an array of Euclidean distances from the centroid of the blob to the edge pixels along directions from 0 to 359 degrees with a step of 1 degree. Thereby, shape of each blob can be described by 360 members array. In the next step cross-correlations (CC) between (S) of the the current blob and nearby blobs in the previous frame are evaluated. If distance to the nearest blob with CC < α is closer than 10% of the image size, then this blob is marked as smoke one. Otherwise, it is just added to the comparison list of the next frame. If there are no regions of interest in the current frame, the program loop returns to the reading new data from LRF.

4   Experimental Results

All the experiments were conducted on the following computer: Intel i7 870 CPU, 8 GB DDR3 RAM. LRF data processing speed exceeds physical capabilities of

Table 1. Detection performance

Precision	Recall	F-measure
0.9032	0.9180	0.9106

Table 2. LRF processing time (ms)

Process:	Reading Data	Preprocessing	DBSCAN	Total
Time:	0.9122	0.0508	10.0136	10.4227

the current SICK laser scanner (see Table 2). Processing of 640x512 image gives from 4 to 5 fps as shown in Table 3 according to number and size of objects detected in the scene. During day time under moderate brightness all smoke scenes were detected correctly excluding the moments when smoke occupied the whole image. In that moments the LRF became blind. Smoke parts with low density were not discovered by LRF (left side of Fig.6(s)). Also, image processing part was not able to detect the total volume of smoke (Fig.6(u)). Thereby, the algorithm can tell whether there is a smoke region on front of car, but exact dimensions of it will remain unknown. Sometimes automatic exposure correction of the camera was not working right making most of the image white which it made impossible to use image processing. Conversely, when wet objects appeared in the scene, LRF failed to detect them causing the image processing algorithm to be never started. Sometimes reflection of smoke appeared on the car surface and became false positive when smoke is close enough to make ROI cover the front part of the car. Also, worn lane markings were introduced to the smoke region and were considered as false positives. The overall performance of the algorithm is represented in Table 1.

Table 3. Image processing time (ms)

Process	Time (small)	Medium	Large
Preprocessing	17	18	17
Color Probability	138	141	139
Connected Components	2	9	15
Refining	5	4	4
Edge Density	4	10	16
Boundary Roughness	3	18	32
Matching	5	25	39
Total	178	235	278

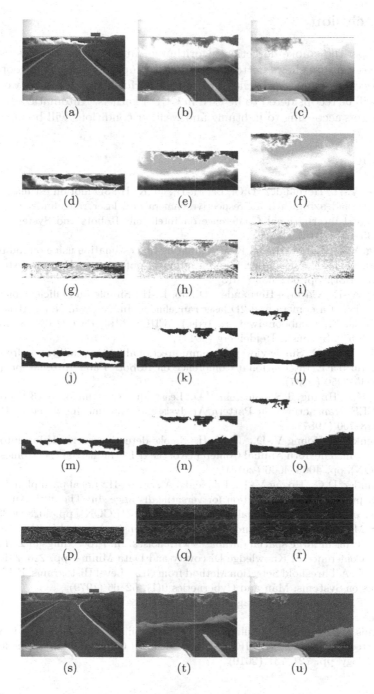

Fig. 6. Image processing steps example for small, medium, and large smoke blobs: (a)(b)(c) Original image; (d)(e)(f) Color probability of red channel; (g)(h)(i) Saturation probability; (j)(k)(l) Binarized intersection of all probabilities; (m)(n)(o) Remove small blobs and fill holes; (p)(q)(r) Edges; (s)(t)(u) Result.

5   Conclusion

In this paper the smoke detection algorithm for autonomous vehicles was improved. Sensor fusion shows better performance comparing to their uncooperative work. Processing time approached to real-time performance. In future work some of steps will be reconsidered to be used in CUDA kernels. Automatic adjustment of parameters according to lightning and weather conditions will be introduced.

References

1. Hoang, V.-D., Hernández, D.C., Le, M.-H., Jo, K.-H.: 3D motion estimation based on pitch and azimuth from respective camera and laser rangefinder sensing. In: IEEE/RSJ International Conference on Intelligent Robots and Systems (IROS), pp. 735–740 (2013)
2. Hoang, V.-D., Le, M.-H., Jo, K.-H: Planar motion estimation using omnidirectional camera and laser rangefinder. In: International Conference on Human System Interactions (HSI), pp. 632–636 (2013)
3. Hoang, V.-D., Cáceres Hernández, D., Jo, K.-H.: Simple and efficient method for calibration of a camera and 2D laser rangefinder. In: Nguyen, N.T., Attachoo, B., Trawiński, B., Somboonviwat, K. (eds.) ACIIDS 2014, Part I. LNCS, vol. 8397, pp. 561–570. Springer, Heidelberg (2014)
4. Mei, C., Rives, P.: Single view point omnidirectional camera calibration from planar grids. In: IEEE International Conference on Robotics and Automation (ICRA), pp. 3945–3950 (2007)
5. Arun, K.S., Huang, T.S., Blostein, S.D.: Least-squares fitting of two 3-D point sets. In: IEEE Transactions on Pattern Analysis and Machine Intelligence (PAMI-9), pp. 698–700 (1987)
6. Filonenko A., Hoang V.-D., Jo, K.-H.: Smoke detection on roads for autonomous vehicles. In: The 40th Annual Conference of the IEEE Industrial Electronics Society (IECON), pp. 4063–4066 (2014)
7. Hernández, D.C., Hoang V.-D., Filonenko A., Jo, K.-H.: Local path planning strategy: a practical implementation for versatile distance. In: The 40th Annual Conference of the IEEE Industrial Electronics Society (IECON), pp. 4028–4033 (2014)
8. Ester, M., Kriegel, H.-P., Sander, J., Xu, X.: A density-based algorithm for discovering cluster in large spatial databases with noises. In: Proceedings of 2nd International Conference of Knowledge Discovery and Data Mining, pp. 226–231 (1996)
9. Otsu, N.: A Threshold Selection Method from Gray-Level Histograms. IEEE Transactions on Systems, Man and Cybernetics 9(1), 62–66 (1979)
10. Soille, P.: Morphological image analysis: principles and applications, pp. 173–174. Springer-Verlag (1999)
11. Vinicius P., Borges K., Izquierdo E.: A probabilistic approach for vision-based fire detection in videos. In: IEEE Transactions on Circuits and Systems for Video Technology, pp. 721–731 (2010)

Tracking Failure Detection Using Time Reverse Distance Error for Human Tracking

Joko Hariyono, Van-Dung Hoang, and Kang-Hyun Jo[⊠]

Graduate School of Electrical Engineering, University of Ulsan, Ulsan 680–749, Korea
{joko,hvzung}@islab.ulsan.ac.kr, acejo@ulsan.ac.kr

Abstract. This paper proposes a tracking failure detection method based on time reverse error distance. Corner feature is used as interest point to perform the task. The algorithm consists of three stages. First, the corner feature is extracted from the image. Then, a tracker produces a trajectory by tracking the point in the previous frame to the current frame. Second, the location of point in the current frame is initialized as a reference point. The validated trajectory is obtained by tracking-reverse the reference point in the current frame to the previous frame. Third, both trajectories are compared with each other, if they are significantly different, the reversed point is considered as an incorrect. To evaluate the performance of this method, an object tracking method is performed. A set of points is initialized from a rectangular bounding box. These points are tracked and evaluated using the proposed tracking failure detection method. The correct tracked points are used to estimate bounding boxes in consecutive images. The performance results show that the proposed method is efficient for human detection and tracking in omnidirectional images.

Keywords: Tracking · Time reverse · Human detection · Omnidirectional camera

1 Introduction

Nowadays, one of the critical challenges in video surveillance systems is how to reliably detect human to perform high level tasks. In general, cameras are getting adopted because it is relatively economic than other sensors so that it is more appropriately used for indoor and outdoor surveillance system. The omnidirectional camera which provides a 360° horizontal field of view in a single image attracts wider attention. It is an important advantage in many application areas such as object detection and tracking based surveillance systems.

Over the last decade, the question of how to detect human in an image has been thoroughly investigated [1, 2, 5, 7]. Due to random influences, such as scene structure, variation of clothes, occlusion, the problem remains challenging and continues to attract research [14]. Simple and applicable methods also attract research in real-time mobile robot applications. Several methods for human detection have been actively developed. Gavrila et al. [1] employed hierarchical shape matching to find pedestrian candidates in images that obtained from a moving camera. Local descriptor is

© Springer International Publishing Switzerland 2015
M. Ali et al. (Eds.): IEA/AIE 2015, LNAI 9101, pp. 611–620, 2015.
DOI: 10.1007/978-3-319-19066-2_59

proposed for object recognition and image retrieval. Some authors presented methods for human detection using HOG and SVM, such as [2, 7, 8, 14]. HOG features are calculated by taking orientation histograms of edge intensity in a local region. HOG features are extracted from all locations of a dense grid on an image region. Thus, HOG features are fed to SVM to classify objects.

In computer vision task, point tracking is commonly used. Suppose a point location in time t, the goal is to estimate its location in time t + 1. In practice, tracking often faces with a problem where the points dramatically change appearance or disappear from the camera view. Under such conditions, tracking often results in failures. The method commonly used to detect tracking failures is to describe the tracked point by a surrounding patch which is compared from time t to t + 1 using sum-of-square differences [5, 12]. The differential error enables detection of failures caused by occlusion or rapid movements, but does not detect slowly drifting trajectories. The detection of a drift can be approached by defining an absolute error, such as a comparison between the current patch and affine warps of the initial appearance. This method is applicable only to planar targets. Recently, a general method for assessing the tracking performance was proposed [13]. The method was designed for particle filters with a static measurement model. Adaptation of their method to point tracking was not suggested.

In summary, the contribution of this paper is tracking failure detection is proposed using time reverse distance error measurement. Then, motion flow analysis is proposed using window tracking. Motion analysis is employed for detection and tracking moving object in omnidirectional images.

This work proposed a method to detect the tracking failure based on time reverse error distance. Corner feature is used as the interest point to perform the task. The algorithm consists of three stages. First, the corner point is extracted from the image. Then, a tracker produces a trajectory by tracking the point in the previous frame to the current frame. Second, the location of point in the current frame is initialized as a point reference. The validation trajectory is obtained by tracking-reverse the point reference in the current frame to the previous frame. Third, both trajectories are compared and if they different significantly, the point reverse is considered as incorrect. To evaluate the performance of this method, an object tracking method is proposed. It is used for human detection in omnidirectional images.

2   The Omnidirectional Camera System

This section presents the omnidirectional camera system which is used in this work. The camera consists of the perspective camera and the hyperboloid mirror [6, 9]. It captures an image reflecting from the mirror so that the image obtains reflective scene. Actually those images are very useful, because it shows in 360 degree field of view. However some pre-processing is needed. For this task, unwrapping into panoramic image is performed. It will take a little bit computational cost, however it is needed for easier to analyze pattern of motion.

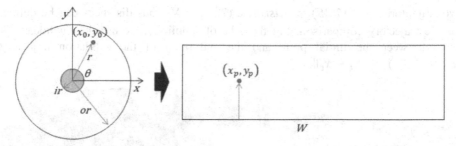

Fig. 1. A simple transformation method

A simple transformation is used for unwrapping from omnidirectional into pano-ramic images [3]. Fig. 1 shows the transformation model. The necessary parameters for this unwrapping are center and radius of the projected circles from both mirror borders, inner and outer. Given a point (x_0, y_0) in the omnidirectional image, (c_x, c_y) is the center point of the image, radius r and angle θ. (x_p, y_p) is a point, located in the panoramic image corresponding to (x_0, y_0). W is the width of panoramic image, i_r is the inner radius and o_r is the outer radius. The transformation of the output coordi-nates to coordinates of the captured image can be written as,

$$(x_0, y_0) = (r \sin \theta + c_x, \, r \cos \theta + c_y) \tag{1}$$
$$(x_p, y_p) = (W0 \, /2\pi, \, r - i_r) \tag{2}$$

Actually, the calculated pixels in the omnidirectional image will not corresponding exactly, one to one, to the pixels of projected image in the panoramic view. So, sub pixel anti-aliasing methods should be used. The bilinear interpolation method is used that may lead to aliasing in case the omnidirectional image is under-sampled.

3   Time Reverse Distance Error

Time reverse distance error is performed for detecting tracking failure. First, the cor-ner point is extracted from the image. Then, a tracker produces a trajectory by track-ing a point in the previous frame to the current frame. Second, the location of the point in the current frame is initialized as a point reference. The validation trajectory is obtained by tracking-reverse the point reference in the current frame to the previous frame. Third, both trajectories are compared and if they different significantly, the point reverse is considered as incorrect.

Suppose that $S = (I_t, I_{t+1}, \dots I_{t+i})$ is an image sequence and x_t be a point location in time t. Using an arbitrary tracker, the point x_t is tracked forward for i steps. The resulting trajectory is $T_f^i = (x_t, x_{t+1}, \dots x_{t+i})$, where f stands for *forward* and i indi-cates the length. Our goal is to estimate the error (reliability) of trajectory T_f^i given the image sequence S. For this purpose, the validation trajectory is first constructed. Point x_{t+i} is tracked *backward* up to the first frame and produces $T_b^i = (\hat{x}_t, \hat{x}_{t+1}, \dots \hat{x}_{t+i})$, where $\hat{x}_{t+i} = x_{t+i}$. The time reverse error is defined as the distance between these two trajectories: $e_{TR}(T_f^i|S) = distance\ (T_f^i, T_b^i)$. Various distances can be defined for the trajectory comparison. For the sake of simplicity, we use the Euclidean distance between the initial point and the end point of the validation trajectory, tance$(T_f^i, T_b^i) = \|x_t - \hat{x}_t\|$.

Fig. 2. Forward-backward tracking failure detection

Fig. 2 illustrates the method when a point is tracked between two consecutive images. The point no. 1 is visible in both images and the tracker is able to localize it correctly. Tracking this point forward or backward results in identical trajectories. On the other hand, the point no. 2 is not visible in the bottom image and the tracker localizes a different point. Track this point backward ends in a different location then the original one.

4   Human Detection and Tracking

4.1   Human Detection Method

Our motion feature is motivated by the fact that strong cues exist in the movements of different body parts when the human is walking. They include the motion of two legs, those between two parts of an arm, or those between a leg and an arm. They provide useful cues to identify the walking motion.

The sparse optical flow method is defined to segment out moving objects from static environment. Landmarks, such as building shape, door, stairs, announcement board etc. are used as our reference in the case of indoor application. Because our camera system is moving, so in order to calculate the camera ego-motion, interest points from landmarks are needed. Sparse optical flow is chosen, and corner features are used as interest points.

Corner features are used in this work as the interest point to perform the task for feature tracking. Using method [5], corner features are extracted on the image. Then, all the features are defined within patches (cells) which generate by divided an image using n x n cell windows. Two consecutive images are compared and tracked corresponding features from previous frame to the current frame. The feature which located in a group cell is used to find the motion distance of each pixel in a group of cells. The motion distance d in x and y − axis by of features cell in the previous frame

$g_{t-1}(i,j)$ is obtained by finding most similar cell $g_t(i,j)$ in the current frame, $g_{t-1}(i,j) = g_t(i + d_x(i,j), j + d_y(i,j))$. It represented as affine transformation of each group as $g_t(i,j) = Ag_{t-1}(i,j) + d(i,j)$. Where A is 2x2 transformation matrix and $d(i,j)$ is 2x1 translation vector. The camera ego-motion compensated frame difference I_d is calculated based on the tracked corresponding pixel groups using $I_d(i,j) = |g_{t-1}(i,j) - g_t(i,j)|$. where $I_d(i,j)$ is a pixel group located at (i,j) in the grid cell.

The affine transformation is applied. It is calculated by the least square method using three corresponding features in those two consecutive frames. Then, the affine parameter is applied on the all of window cells. Thus, each pixel in previous frame is corresponded, one to one, with the pixel in current frame. The regions which are difference from previously registered are obtained by subtracting frame difference on those corresponding pixel cells. Those regions are segmented from the motion of camera ego-motion. Fig. 3 shows two consecutive images 3(a) and 3(b), then, features extraction and grid cell windows are performed on the first image 3(c). When only frame difference is applied, the result is showed in 3(d), and frame difference with ego-motion compensation result is showed in 3(e).

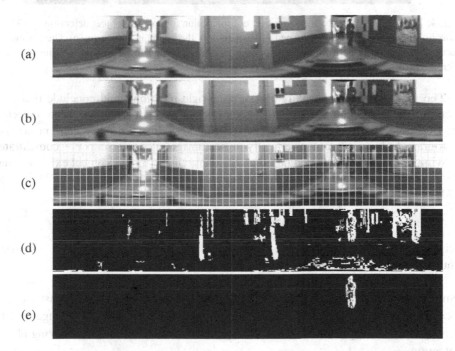

(a)

(b)

(c)

(d)

(e)

Fig. 3. Given two consecutive images (a) and (b), then performed features extraction and grid cell windows on first image (c) Frame difference result (d) Ego-motion compensation result (e)

The segmentation method is devised for accurately locates bounding boxes of the motion in the different image. Each pixel output from the previous step cannot show clearly as silhouette. It just gives information about motion region of moving objects. Due to those regions are obtained from subtracting pixel to pixel according to its corresponding, a hole appears caused by the intersection of two regions. So, those regions should be applied morphological process, opening and closing, to obtain smoothness region of moving objects.

(a)

(b)

(c)

Fig. 4. The region, result from ego-motion compensation for moving object detection (a). Then, morphological process are performed on that region, the result is shown in (b) The histogram vertical projection is performed with specific threshold, so that the region of moving object is localized (c).

The fact that humans usually appear in upright positions, and conclude that segmenting the scene into vertical strips is sufficient most of the time. Then, detected moving objects are represented by the position in width in x axis. Using projection histogram h_x by pixel voting vertically project image intensities into $x -$ coordinate. Moving object area is detected based on the constraint of moving object existence that the bins of histogram in moving object area must be higher than a threshold and the width of these bins should be higher than a threshold.

$$h_x(i \pm 10) > A \, max(h_x) \tag{3}$$

Where A is a control constant and the threshold of bin value is dependent on the maximum bin's value. Fig. 4 shows moving object localization process.

Adopting the region segmentation technique proposed in [4], the region is defined using boundary salience. It measures the horizontal difference of data density in the local neighborhood. The local maxima correspond to where maximal change in data density occurs. They are candidates for region boundaries of human in moving object detection.

4.2   Object Tracking

The main algorithm of our proposed tracking method is consisted of three stages. First, a set of points is initialized from a rectangular detected bounding box. Those points are located 4 pixel interval each other. Second, KLT [5] tracker is used to track which generates sparse motion flow between current frame and next frame. Then, those tracked points are evaluated using time reverse tracking failure detection. Threshold filter is performed to filter out the minimum failure points are detected. Based on our observation, threshold value 0.65 is selected to be the number of remaining points from the initial generation. Third, based on correct points are tracked, bounding box is updated. It is estimated using the existing distances of set points within the bounding box. Fig. 5 shows our tracking algorithm.

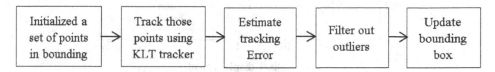

Fig. 5. Tracking algorithm

5   Experiment and Results

Our robot system moved in the corridor. The omnidirectional camera captured more than 6,000 sequent images with around 12,000 people. Our omnidirectional camera is a Blackfly BFLY-PGE-12A2C-CS from Point Grey Research Inc., which provides 1280×960 pixels resolution at 52 frames per second (fps). The System is evaluated using those sequential images. Proposed algorithm is programmed in MATLAB and executed on a Pentium 3.40 GHz, 64-bit operating system with 8 GB Random Access Memory.

The ability of time reverse error detection is evaluated on synthetic data. One hundred panoramic images are used and Gaussian noise added on those images. A set of points is initialized from a rectangular detected bounding box. Those points are located 4 pixel interval each other. KLT tracker is used to estimate displacements of points. Threshold value using 2 pixel errors, points are displaced closer than 2 pixels are classified as inliers. The results are compared to ground truth, and the performances are shown in Fig. 6. When the threshold value 1 is used, the recall is 94% and precision 97%.

For the human detection system, the original HOG, which proposed by Dalal and Triggs is implemented. The HOG features are extracted from 16×16 pixel regions on the candidate region image. Candidate images are obtained from motion segmentation results. The first, Sobel filter is performed to obtain the gradient orientations from each pixel in this local region. The local region is divided into small cells with cell size is 4×4 pixels. Histograms of gradients orientation with eight orientations are calculated from each of the local cells. Then, the total number of HOG features becomes $128 = 8 \times (4 \times 4)$ and they constitute a HOG feature vector. For the training process, the person INRIA datasets [2] are used. The total number of sample images is 3,000. From those images, 1,000 person images and 2,000 negative samples are used as training samples to determine the parameters of the linear SVM.

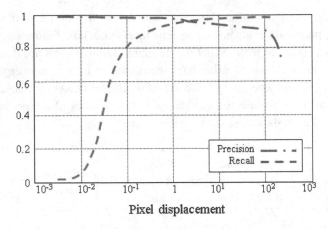

Fig. 6. Pixel displacement threshold in precision and recall

The original HOG [2] is implemented using sequential panoramic images. The recognition rate for the test of those dataset is 95.33% at 0.3 false positive rates. Then, combinations of the HOG and our proposed method, ego-motion compensation (EMC) are tested. The HOG feature vectors are extracted on the candidate image from motion segmentation results. Thus, the feature vectors were used as input of the linear SVM. The selected subsets are evaluated by cross validation. And also the recognition rates of the constructed classifier using test samples are evaluated. The relation between the detection rates and the number of false positive rate are shown in Fig. 7. The best recognition rate 97.25 % was obtained at 0.3 false positive rates for HOG + EMC, cell size 8 x 8, while original HOG obtain lowest. It means that higher detection rate with smaller false positives rate is obtained.

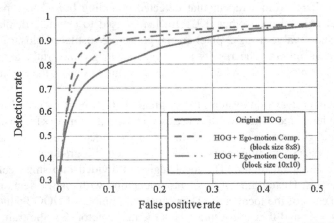

Fig. 7. Comparison result the proposed method and the original HOG

Detected human bounding box is used to perform our proposed object tracking. A set of points is initialized from a rectangular detected bounding box. Those points are located 4 pixel interval each other. KLT tracker is used to track which generates sparse motion flow between current frame and next frame. Then, those tracked points are evaluated using proposed tracking failure detection. Threshold filter is 0.65 as minimum correct points tracked to estimate the next bounding box. Our omnidirectional images dataset is used.

The objects were manually initialized in the first frame and tracked up to the end of the sequence. The trajectory was considered correct if the bounding box overlap with ground truth was larger than 50%. Performance was assessed as the maximal frame number up to which the tracker was correct. Table 1 shows the comparison of several object tracking approaches [10, 12, 11]. The best results obtained using our object tracking approach using time-reverse failure detection method.

Table 1. Comparison of performance several object tracking approaches

Method	The number of frame	The number of correct frame
Kalal [10]	6,000	4,731
Babenko [12]	6,000	3,956
Avidan [11]	6,000	3,177
Ours	6,000	4,982

6   Conclusion

This paper presents a tracking failure detection method based on the time reverse error distance. Corner feature is used as the interest point to perform the task. The algorithm consists of three stages. First, the corner point is extracted from the image. Then, a tracker produces a trajectory by tracking the point in the previous frame to the current frame. Second, location of point in the current frame is initialized as a point reference. The validated trajectory is obtained by tracking-reverse the point reference in the current frame to the previous frame. Third, both trajectories are compared and if they different significantly, the point reverse is considered as incorrect. To evaluate the performance of this method, an object tracking method is performed. A set of points is initialized from a rectangular bounding box. These points are tracked and evaluated using proposed tracking failure detection. The correct points tracked are used to estimate the consecutive bounding box. Then, it is used for human tracking in omnidirectional images. The detection rate 97.25 % was obtained at 0.3 false positive rate. Performance of object tracking was assessed in omnidirectional input images. The best results obtained using our object tracking approach using time-reverse failure detection method.

References

1. Gavrila, D.M., Munder, S.: Multi-cue Pedestrian Detection and Tracking from a Moving Vehicle. International Journal of Computer Vision **73**(1), 41–59 (2007)
2. Dalal, N., Triggs, B.: Histograms of Oriented Gradients for Human Detection. In: IEEE Conference on Computer Vision and Pattern Recognition (CVPR) (2005)
3. Hoang, V.D., Vavilin, A., Jo, K.H.: Fast Human Detection Based on Parallelogram Haar-Like Feature. In: The 38th Annual Conference of The IEEE Industrial Electronics Society, Montreal, pp. 4220–4225 (2012)
4. Hariyono, J., Hoang, V.-D., Jo, K.-H.: Motion Segmentation using Optical Flow for Pedestrian Detection from Moving Vehicle. In: Hwang, D., Jung, J.J., Nguyen, N.-T. (eds.) ICCCI 2014. LNCS, vol. 8733, pp. 204–213. Springer, Heidelberg (2014)
5. Tomasi, C., Kanade, T.: Detection and Tracking of Point Features. In: Proceedings of Fourteenth International Conference on Pattern Recognition, vol. 2, p. 1433 (1998)
6. Hariyono, J., Wahyono, Jo, K.H.: Accuracy Enhancement of Omnidirectional Camera Calibration for Structure from Motion. In: International Conference on Control, Automation and Systems, Korea (2013)
7. Hariyono, J., Hoang, V.D., Jo, K.H.: Moving Object Localization using Optical Flow for Pedestrian Detection from a Moving Vehicle. The Scientific World Journal (2014). http://dx.doi.org/10.1155/2014/196415
8. Hoang, V.D., Le, M.H., Jo, K.H.: Hybrid cascade boosting machine using variant scale blocks based HOG features for pedestrian detection. Neurocomputing **135**, 357–366 (2014)
9. Hariyono, J., Hoang, V.D., Jo, K.H.: Human Detection from Mobile Omnidirectional Camera Using Ego-motion Compensated. In: 6th Asian Conference on Intelligent Information and Database Systems, Bangkok (2014)
10. Kalal, Z., Matas, J., Mikolajczyk, K.: P-N Learning: Bootstrapping Binary Classifiers by Structural Constraints. In: CVPR (2010)
11. Avidan, S.: Ensemble Tracking. IEEE Trans. Pattern Analysis and Machine Intelligence **29**(2), 261–271 (2007)
12. Babenko, B., Yang, M.-H., Belongie, S.: Visual tracking with online multiple instance learning. In: IEEE Conference on Computer Vision and Pattern Recognition (CVPR) (2009)
13. Wu, H., Chellappa, R., Sankaranarayanan, A., Zhou, S.: Robust visual tracking using the time-reversibility constraint. In: International Conferences on Computer Vision (2007)
14. Enzweiler, M., Gavrila, D.M.: A Multilevel Mixture-of-Experts Framework for Pedestrian Classification. IEEE Transaction on Image Processing (2011)

Intelligent Systems Applications

Optimization of Trading Rules for the Spanish Stock Market by Genetic Programming

Sergio Luengo[1], Stephan Winkler[2], David F. Barrero[3],
and Bonifacio Castaño[1]($\boxtimes$)

[1] Department of Physics and Mathematics, University of Alcalá, Madrid, Spain
bonifacio.castano@uah.es
[2] Bioinformatics Research Group, University of Applied Sciences Upper Austria,
Hagenberg, Austria
[3] Department of Computer Science, University of Alcalá, Madrid, Spain

Abstract. This paper deals with the development of a method for generating input and output signals in the Spanish stock market. It is based on the application of set of simple trading rules optimized by genetic programming. To this aim we use the HeuristicLab software. To evaluate the performance of our method we make a comparison with other traditional methods such as Buy & Hold and Simple Moving Averages Crossover. We study three different market scenarios: bull market, bear market and sideways market. Empirical test series show that market global behavior has a great influence on the results of each method and that strategies based on genetic programming perform best in the sideways market.

Keywords: Stock exchange markets · Genetic programming · Optimization · Trading rules · Market tendencies

1 Introduction

The application of genetic programming has a long tradition in predicting the behavior of Stock markets [2] [10] [8]. Specifically it develops methods that implement strategies, based on technical analysis, to predict future values of shares and indexes [12]. The main objective is to maximize the benefits while minimizing the losses [4]. Most of these approaches use the basic daily data generated by the market: open, high, low, close and volume, and many of the indicators associated with them such as Moving Averages, RSI (Relative Strength Index), MACD (Moving Averages Convergence Divergence), Bollinger Bands and others [1]. The results of this techniques are all kind of predictive heuristic objects: neural networks, tree programs, multivariate regressions. All of them are able to predict future stock values or give input and output signals into the market.

In this article we show a method for generating investing strategies in the Spanish stock market. These strategies are based on technical analysis trading rules [5] that are coded as program trees and optimized by genetic programming [6]. The aim is to generate buy and sell signals on the financial market in order to

© Springer International Publishing Switzerland 2015
M. Ali et al. (Eds.): IEA/AIE 2015, LNAI 9101, pp. 623–634, 2015.
DOI: 10.1007/978-3-319-19066-2_60

maximize profit. We only consider long positions, that is we buy shares at a price in order to sell them later at a higher price and make a profit after deducting commissions. For all our work on genetic programming we use the HeuristicLab software (HL) [13].

We have developed a specific plug-in that we have integrated into the HL program. This plug-in works out according to the operation of the stock markets and their characteristics. In principle, this is a first application where simple inversion rules has been implemented in HL. Using this tool we experiment with genetically evolved strategies on different kinds of market. Then we compare the results with other classic investment approaches: such as the Buy & Hold and Moving Averages Crossover methods [5]. We choose these classic methods because they are widely known and used by many investors. This work is an attempt to see if genetic programming improves traditional strategies.

For this experiment we use quotes from 10 of the most representative shares in the Spanish market for the past 10 years. In that period we identify three types of different market developments: bull market (a long upward price movement where prices in the end of the period are higher than in the beginning), bear market (a long downward price movement where prices in the end are lower than in the beginning) and sideways market (prices in the end and the beginning are almost the same) [5]. We choose this market because it is well known for us.

This paper is structured as follows: first we describes how we obtain Spanish stock market data and the way we adapt them to our study. Then we divide up these data into three periods according to the market behavior: bull market, bear market and sideways (no trend) market. Afterwards, we give the details of the main outlines of our experiment. next we present the results obtained by traditional trading methods and we explain the details on how to apply genetic programming to the study of markets and presents the results and conclusions of this experiment. Finally we outline our future work.

2   Data Retrieval for Spanish Stock Market

We have obtained the whole data from http://finance.yahoo.com/ page. This web contains most of the Spanish stock market daily historical data since 2000. For our experiment we select a set of relevant stocks all of them listed on the IBEX35 index. One of the most efficient ways to download the data is employing the language **R** and its library **quantmod** [11]. After downloading, data need some corrections to suit the type of analysis we develop in this work.

Because we are going to compare the total revenue we can get using three different methods we must use the adjusted close prices of the shares. The adjusted price takes into account the dividends and splits of any share and consistently it reflects the gains and the losses of each value over a period. Thus, it is necessary to set also opening prices, maximum and minimum according to the adjusted closing prices. Another important question is to remove Spanish holidays or days without market negotiation from the data. The companies we have chosen are 10 of the most representative and for which reliable data can be obtained in a period of 10 years from January 2003 to January 2013. These are:

Table 1. Market periods

Type of market	Precalculus	Training	Test
Bull market	1/02/2003-31/01/2004	1/02/2004 - 31/01/2006	1/02/2006-31/01/2007
Bear market	1/05/2008-30/04/2009	1/05/2009 - 30/04/2011	1/05/2011-30/04/2012
Sideways market	1/03/2009-28/02/2010	1/03/2010 - 28/02/2012	1/03/2012-28/02/2013

- Telecommunication: TEF (Telefónica)
- Electricity: ELE (Endesa)
- Banks: POP (Popular), SAN (Santander) and BBVA (Bilbao Vizcaya)
- Oil industry: REP (Repsol)
- Construction: FCC (Fomento de Construcciones y Contratas) and SYV (Sacyr Vallehermoso)
- Insurance: MAP (Maphre)
- Engineering: IDR (Indra)

In addition, we collect the IBEX35 index (which summarizes the most capitalized companies in the Spanish stock market) data for that period. We use this index as the main reference when determining the different types of markets during the 10 years of the study. The chart for the IBEX35 index during this 10 years showing the three test periods of the experiment is shown in Fig. 1 (See also Table 1).

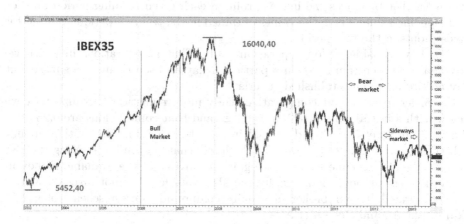

Fig. 1. Test periods in 10 years of the IBEX35 index

3   General Framework

The first task is to divide the 10 years period into three smaller intervals, four years long each (that sometimes overlap), according to the three main definitions of markets: bull, bear and sideways markets (See Table 1). Then, each one of these intervals is also divided into three different periods. First of all we have a initial one year time interval, that we call "precalculus" period, to set the indicators properly, then a two years interval for training and optimization of our trading systems (both Moving Averages Crossover and Genetic Programming) and finally a one year period for testing. This last test period is the one we are interested in because it corresponds to the period when the real investment, based on the optimized trading strategies developed in the training period, would be accomplish in the market. These test period are shown in Fig. 1. For each one of these types of markets we experiment with three different trading methods:

- Buy & Hold.
- Usual optimized Moving Averages Crossover.
- Genetically optimized trading rules.

In the Buy & Hold method the shares are purchased at the beginning of each test period and sold at the end. Obviously, the Buy & Hold method does not need a previous training period. In this case, we compute directly the results of the test period operation.

In the case of Moving Averages Crossover we employ the usual method of crossover between a long-term simple moving average and a short-term one. That is, a buy signal is generated when the short-term moving average crosses above the long-term one and a sell signal is triggered by a short-term moving average crossing below a long-term one [5]. To choose the most profitable set of moving averages we have used a comprehensive method to determine the number of days n, between $n = 2$ and $n = 100$, for each set of long-term and short-term moving averages, that produces the higher profit in each given training period and for each specific value. Then, the values obtained in the training period are used for investments in the test period.

We have considered long operations only. That is, operations in which we buy shares of a certain asset on a particular day and sell them on a subsequent day according to an established criteria.

In order to make our comparative study more realistic we assume that we begin with a initial capital of 10,000.00 € and that commissions are 0.25 % of the value of assets purchased or sold in each transaction. This initial amount will change according to the gains or the losses that we will have during the test period. We also assume that we are going to invest all the available money in every investment operation. Finally, we also consider that not invested money is remunerated by our broker in a daily compound interest scheme of a 5 % Equivalent Annual Return (EAR).

Purchases and sales are always performed at the opening price of the corresponding market day. The decision, to enter or exit the market, is made the previous day after the market closes because we need time to set all the values and indicators and study the market behavior. This determination is made equally in all market situations.

In all cases we assume that we will sell the assets held in our portfolio, if we have any, the last day of the period under experimentation. In that way, the comparison between the profit or the loss of each one of the different investment

Table 2. Revenues achieved using Buy & Hold and Moving Averages Crossover

Stock	BULL MARKET		BEAR MARKET		SIDEW. MARKET	
	B & H	Crossover	B & H	Crossover	B & H	Crossover
BBVA	11,752.11	10,987.94	6,162.99	8,086.08	7,727.44	9,213.98
ELE	16,816.34	10,607.02	5,716.28	6,465.14	7,864.00	9,553.28
FCC	15,144.01	13,410.30	5,940.65	10,142.98	8,500.47	7,621.48
IDR	11,835.28	11,508.52	5,285.66	10,048.83	6,959.87	6,976.49
MAP	64,448.24	61,623.04	8,126.66	8,159.3 0	10,654.52	9,370.14
REE	11,912.84	11,142.23	7,655.05	9,412.54	10,802.45	13,092.61
REP	11,767.40	11,207.21	6,226.20	10,013.45	12,910.32	11,393.10
SAN	12,989.41	12,635.34	6,135.66	8,582.98	7,697.23	8,392.15
SYV	22,180.83	17,153.47	1,631.06	9,104.55	5,193.04	7,594.95
TEF	13,593.50	11,933.79	6,557.32	7,144.64	8,564.87	8,175.43
Total	**192,439.96**	**172,208.86**	**59,437.53**	**87,160.49**	**86,874.21**	**91,383.61**

methods is more realistic. That is, we will compare the money the investor would have in his pocket after and before any market period.

In the following section we show the results of each type of market and the two types of classic strategies: Buy & Hold and Moving Averages Crossover.

4 Results for Classic Methods

In this section we present the results obtained by the methods of Buy & Hold and Moving Averages Crossover for the 10 chosen Spanish shares and the three types of market. The corresponding columns indicate the amount a person who invested 10,000.00 € at the beginning of the test period will have on hand at the end of this period.

In the Table 2 we present the results of the three test periods for each one of the 10 shares and for each market scenario.

We have to recall that the distinction between the three market periods has been made looking at the IBEX35 index. In this context, it may be possible that some shares could have a different behavior from the index evolution.

Viewing Table 2 it shows that in the bullish period we have gains with any of the two methods and in all shares. In this case of Buy & Hold the revenues are higher than in the case of the Moving Averages Crossover which is a typical characteristic of bullish periods. We would like to stand out the extraordinary result of MAP in the bull market test period. This is due to a 5 × 1 split that occurs during this period (investor shares are multiplied by 5). In the case of a bear market both methods suffer a lot of substantial losses over that period. For some of them, like SYV, the losses are very large. Nevertheless using Moving Averages Crossover the losses are not as large as they are by the Buy & Hold method. Sideways market shows that globally the best results are obtained with Moving Averages Crossover. The two methods lose but the losses are not as large as in the bear market period.

5   Optimization of Trading Rules using Genetic Programming

In this section we describe all the elements for the creation and manipulation of trading rules coded as program trees. These trees are handled and optimized through HL.

HL is a software environment for heuristic and evolutionary algorithms developed by the members of the Heuristic and Evolutionary Algorithm Laboratory (HEAL) at the University of Applied Sciences Upper Austria, Campus Hagenberg. This software also provides the option to integrate plug-ins and algorithms for specific problems according to the users' needs . HL can read data from a **.csv** file and the user can decide which columns are involved in the computation and which of them are used as a target.

For this work we have developed a plug-in that takes into account all the usual criteria used in trading rules applied to time series of securities. The goal we aim is to use genetic programming to obtain a program tree that encodes an investment strategy that tells us when we have to be in the market or out the market, i.e. we want a binary response.

Working one generation after another, the best individual of any generation will be the one that encodes the trading rule that makes us get the maximum benefit from a time series of stock values. This will be the fitness criterion for selecting individuals in a generation. This is a special case in genetic programming because there is no objective function that serves to evaluate the fitness. Besides, a tree encoding a trading rule has to be strongly typed [9], its value should be true or false (meaning: in or out the market and coded as: 1 or -1) and its grammar has to be quite strict. All of these constrains will also affects the crossover operations between the individuals of each generation [7]. The elements defining the experiment are described below.

5.1   Solution Encoding

In this section we show how we build program trees that codify trading rules. A very important aspect of these trees is that they compute a binary result. That is, for each day, whether we should be in or out of the market. For this reason the root node must represent a function with a boolean output. In addition, because there are no binary variables and functions involved, we need to take into account some restrictions in the tree representation and genetic operators in order to generate valid trees. With this aim in mind we select a type of GP algorithm named Strongly Typed Genetic Programming [9], which handles pretty well this kind of situations.

Table 3 contains the set of functions we used in non-terminal and terminal nodes. We distinguish four types of functions: logic, relational, arithmetic and specialized. The logic functions operate in the classic sense, relational operators (Greater than, lower than) return a boolean value TRUE or FALSE, codified as 1 or -1, and arithmetic functions perform the usual arithmetic operations.

Table 3. Functions and terminal nodes

Root node	Non-terminal nodes	Terminal nodes
AND	Addition	C
OR	Substraction	n
NOT	Multiplication	INT
Boolean values	Lag	TRUE, FALSE
Greater than	SMA	
Less than	Max	
	Min	

Table 4. Grammar definition

Parent node	Offspring nodes
Root	AND, NOT, OR, <, >
AND, OR, NOT	TRUE, FALSE, AND, NOT, OR, <, >
<, >	+, -, ×, Close, INT, Lag, Max, Min, SMA
+, -, ×	+, -, ×, Close, INT, Lag, Max, Min, SMA
Lag, Max, Min, SMA	C, n

The symbol C stands for the closing price of a share, $n \in [1, 100]$ for a certain number of days and INT is an integer constant. The values C and n are the arguments for the specialized functions that are specific to our trading domain: SMA(C,n) (Simple moving average value of C in the last n days), Max(C,n) (Maximum value of C in the last n days), Min(C,n) (Minimum value of C in the last n days) and Lag(C,n) (Value of C n days ago). These functions are codified specifically for this experiment because there are not included in HL. The Simple Moving Average (SMA) required some special attention because of some implementation details in HL. To this end, we include in the input file one column $Sum(j)$ that stores the sum of the closing values of all the days since the first one to day j.

On top of that, we have to take into account that Strongly Typed Genetic Programming requires a set of formal constrains to determine which types of nodes might be offspring of each type of node. Table 4 represents these restrictions. Basically, the root node only can represent logic or relationship functions and the offspring must follow the rules expressed in Table 4. Figure 2 shows an example of a valid tree with 15 nodes encoding the trading rule:

(Min(C,59) < Lag(C,76)) AND (SMA(C,55) >   Max(C,21))

5.2   Fitness Computation

We define the fitness of a program that codifies a trading rule as the profit it gets when that rule is put into practice in a certain period of time and with a time series of stock prices. The more return a rule gets the higher fitness the program has. In our case, the time series contains all the values we need in our

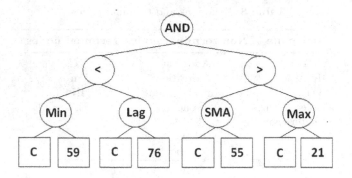

Fig. 2. Example of a valid tree

Table 5. Buying and selling rule from vector **R**

Yesterday	Today	Tomorrow
-1	1	Buy
-1	-1	Stay out
1	1	Stay in
1	-1	Sell

trading strategy and for all shares we have chosen. These series are the input data of our experiment and they are described in Sect. 5.4.

In order to compute the profit of each individual tree we develop two programs. The first one is the **Interpreter**. It goes along the time series and evaluates the tree for each day, from 1 to N, of the period under study. It returns a vector **R** of size N with the form $\{-1, 1\}^N$. This vector represents the recommended buy/sell strategy for that period. Then using this list we run a second program, named **Evaluator**, that simulates an investor that buys/sells shares according a fixed schema defined as described in Table 5. That is, at the end of today we look at the values **R(Today)** and **R(Yesterday)** and take the corresponding action pointed out in **Tomorrow**.

In addition, if we have some purchased shares on the last market day the rule indicates to sell them at the opening of the upcoming market day.

We always use all the money we have and buy as many shares as possible. We sell all the shares we have in any sell operation as well. We pay a 0,25 % buy and sell commission and get a 5% EAR, in a daily basis, for the ready cash. The money we have after the last day of any period and after the last sell is the fitness of the corresponding tree.

5.3 Selection, Genetic Operators and Population Initialization

The selection used in our experimentation is binary tournament [3]. We selected this method because after testing all the selection mechanisms implemented by HL we found that binary tournament gave the best performance.

Given the strongly typed representation, genetic operators must be selected with care to avoid the generation of unable trees. For this reason we selected Context Aware Crossover [7], which always generates a tree according to the defined grammar.

The initial population is generated using the Ramped Half&Half method, with a maximum size of 17 nodes and maximum depth value equal to six. These trees are created according to the grammar.

5.4 Input Data

Input data is composed by a collection of .csv files whose rows represent market days, one row for one day. These data are read at the beginning of the experiment and are the time series we use for computing the return of our trading strategies. A description of the file columns follows:

- **Date.** Date that identifies the market day.
- **Open.** Numerical field with the opening value adjusted for the given day. It is used as market entry value.
- **Close.** Numerical field with the closing value adjusted to the given day. This number summarizes the share market evolution.
- **Sum.** Numerical field that stores the sum of all the market closing values to date. It is used to compute the SMA of the closes for any number of days.
- **Days.** Numerical field that stores the number of natural days since the first day in the file. It is used to compute the daily ready cash return.

6   Experimental Results

Once the algorithm configuration is established, we proceed to carry out the experiment to test our approach based on GP running in HL. We use the set of parameters presented in Table 6 and take the same companies (see section 2), periods of time for training and test, initial 10,000.00 € investment, same commissions and same EAR. For each stock and market period we make 10 runs and compute the profit average and standard deviation.

The results of the experiment, in comparison with the outcome gotten by classic methods, are presented in Table 7, which summarizes the results for bull market, Table 8 for bear market and finally Table 9 for sideways market. In the following we discuss them in more detail.

We observe in Table 7 that the method Buy & Hold achieves the highest profit in bull market as it should be and the method based on GP is the worse in terms of average profit in this test. We want to emphasize the extraordinary results of MAP, they are due to a 5 × 1 split in the test period. Unfortunately GP method does not exploit this split in some experiments but it does in others. This means lower profit average and high standard deviation. This could be this way because MAP split is an isolated event and therefore handling this is quite difficult for any learning algorithm. If we remove MAP shares from the

Table 6. Tableau with the GP parameters setting

Parameter	Value
Population size	100
Generations	25
Selection	Binary Tournament
Crossover	Context Aware Crossover
Initial population	Rampled H & H (D=9, L=17)
Max tree size	Depth=9, length=50
Fitness	Maximum profit
Initial investment	10,000.00 €
Periods	Given in Table1

experiment we will get lower differences among GP and the rest of the methods and standard deviations is approximately 10% of average value.

In the case of the bear market (shown in Table 8) all the investment strategies are losers. The worse method is Buy & Hold, which is perfectly logical. The method that achieves less losses is Moving Averages Crossover. Please note that GP obtains better returns than Buy & Hold, which indicates that using GP might be advantageous in some circumstances. In this case standard deviations are a little higher than in bull market.

In the last case shown in Table 9, corresponding to the sideways market, we should stress that GP is the best method and the only one which gets some profit even paying commissions expenses. As in the bull market standard deviation is around 10%. This las result suggests the viability of trading methods based on GP and the interest of going further in its study to enhance the results.

7 Future Work

As we have shown GP might provide valid results in the context of investment in stock markets. We should take into account that the developed method is quite simple, and hence, it seems reasonable to hypothesize that with more elaborated approaches we can achieve better results, the literature points in this direction. In this sense, some of the future research lines are the following ones.

One potential way to improve the trading system is to include more market indicators, providing additional information. Some of the best known and used market indicators in classic trading are Moving Averages Convergence Divergence (MACD) and Relative Strength Index (RSI), that measure different aspect of a share evolution. Other market indicators that could be included are Rate of Change (ROC), Williams Accumulation / Distribution(WAD), William's indicator (WPR) or Bollinger's bands (BBands). It would be interesting to investigate which market indicators are more informative to understand the market.

It would be also interesting experimenting how the different GP parameters influences the results. Let's say: mutation rates, number of generations, offspring selection, population size, depth and size of trees, etc.

Table 7. Performance of all tested methods on bull market data

Share	Buy and Hold	Average Crossover	GP Average	GP Stan.Dev.
BBVA	11,752.11	10,987.94	10,948.19	390.36
ELE	16,816.34	10,607.02	12,252.24	1,221.18
FCC	15,144.01	13,410.30	12,192.15	1,422.80
IDR	11,835.28	11,508.52	10,796.81	1,263.17
MAP	64,448.24	61,623.04	20,259.78	21,220.14
REE	11,912.84	11,142.23	10,478.73	409.37
REP	11,767.40	11,207.21	10,248.31	519.08
SAN	12,989.41	12,635.34	11,937.02	1,052.62
SYV	22,180.83	17,153.47	14,193.69	3,930.27
TEF	13,593.50	11,933.79	11,303.45	1,198.69
TOTAL	192,439.95	172,208.87	124,610.37	

Table 8. Performance of all tested methods on bear market data

Share	Buy and Hold	Average Crossover	GP Average	GP Stan.Dev.
BBVA	6,162.99	8,086.08	7,348.46	950.91
ELE	5,716.28	6,465.14	7,012.75	1,678.85
FCC	5,940.65	10,142.98	6,998.00	1,703.81
IDR	5,285.66	10,048.83	6,905.40	1,728.34
MAP	8,126.66	8,159.30	9,233.05	2,226.90
REE	7,655.05	9,412.54	7,999.64	1,104.69
REP	6,226.20	10,013.45	6,862.32	929.86
SAN	6,135.66	8,582.98	8,765.49	2,138.84
SYV	1,631.06	9,104.55	5,687.35	2,282.78
TEF	6,557.32	7,144.64	7,749.30	523.74
TOTAL	59,437.53	87,160.51	74,561.77	

Table 9. Performance of all tested methods on sidewalks market data

Share	Buy and Hold	Average Crossover	GP Average	GP Stan.Dev.
BBVA	11,805.46	9,213.98	10,836.76	729.03
ELE	11,594.52	9,553.28	10,923.69	631.35
FCC	4,820.19	7,621.48	6,765.82	1,753.30
IDR	10,415.03	6,976.49	10,621.24	1,794.72
MAP	10,327.95	9,370.14	10,446.72	2,615.08
REE	11,765.92	13,092.61	11,293.81	1,898.45
REP	8,837.53	11,393.10	11,246.52	3,476.75
SAN	10,264.12	8,392.15	10,303.64	1,241.18
SYV	5,804.95	7,594.95	10,647.07	2,278.17
TEF	8,184.48	8,175.43	9,821.19	1,099.33
TOTAL	93,820.15	91,383.60	102,906.47	

An additional research line could be the fundamental analysis of the components of our interest, and in this way to gain a better control of the relevant changes in the stock market. Sometimes the market reacts to events such as news; we expect to detect changes through the study of patterns and, on the basis of this information, infer the future shares evolution.

Another important enhancement of this strategy might be the integration of a portfolio system to control the investment in different shares at the same time. The objective would be the automatic selection, by means of a tendencies classification system, of those shares with higher revaluation potential.

Acknowledgments. This work is supported by the Project of Castilla-La Mancha PEII-2014-015A (PEII11-0079-8929).

References

1. Bodas-Sagi, D.J., et al.: A parallel evolutionary algorithm for technical market indicators optimization. Nat. Comp. (2012)
2. Brabazon, A., et al.: An Introduction to Evolutionary Computation in Finance. IEEE Computational Intelligence Magazine **3**, 42–55 (2008)
3. Goldberg, D.E., Kalyanmoy, D.: A Comparative Analysis of Selection Schemes Used in Genetic Algorithms. Foundations of Genetic Algorithms, pp. 69–93 (1991)
4. Kaboudan, M.: GP forecasts of stock prices for profitable trading. In: Evolutionary Computation in Economics and Finance, Heidelberg, pp 359–382 (2002)
5. Kirkpatrick, C.D., Dahlquist, J.: Technical Analysis: The Complete Resource for Financial Market Technicians. Financial Times Press (2006)
6. Koza, J.R.: Genetic Programming: On the Programming of Computers by Means of Natural Selection. The MIT press (1992)
7. Majeed, H., Ryan, C.: Using context-aware crossover to improve the performance of GP permission and/or a fee. In: Proc. GECCO 2006, pp. 847–854 (2006)
8. Mallick, D., Lee, V.C., Ong, Y.S.: An empirical study of genetic programming generated trading rules in computerized stock trading service system. In: Procc. Int. Conf. on Serv. Syst. and Serv. Man., pp. 1–6 (2008)
9. Montana, D.J.: Strongly Typed Genetic Programming. Evo. Comp. **3**(2), 199–230 (1995)
10. Potvina, J.-Y., Sorianoa, P., Vallee, M.: Generating trading rules on the stock markets with genetic programming. Comp. & Oper. Res. **31**, 1033–1047 (2004)
11. Ryan, J.A., et al.: Package quantmod: Quantitative Financial Modeling Framework (2015). http://cran.r-project.org/web/packages/quantmod/quantmod.pdf
12. Summers, B., et al.: Back to the future: an empirical investigation into the validity of stock index models over time. Appl. Fina. Econ. **14**, 209–214 (2004)
13. Wagner, S., et al.: Architecture and design of the HeuristicLab optimization environment. In: Advanced Methods and Applications in Computational Intelligence. Topics in Intelligent Engineering and Informatics, pp. 197–261. Springer (2014)

Meta-Diagnosis for a Special Class of Cyber-Physical Systems: The Avionics Test Benches

Ronan Cossé[1,2]([✉]), Denis Berdjag[2], Sylvain Piechowiak[2],
David Duvivier[2], and Christian Gaurel[1]

[1] AIRBUS HELICOPTERS, Marseille International Airport,
13725 Marignane, France
ronan.cosse@airbus.com

[2] LAMIH UMR CNRS 8201, University of Valenciennes, 59313 Valenciennes, France

Abstract. Several researches have been made to determine the best system representation for process diagnosis. Many approaches were considered, such as heuristic searches, event driven models, structural models... As of today, there are no clear winners considering real applications on industrial systems.

In this article, based on a specific system model, a diagnostic approach for complex systems is discussed. The proposed diagnostic algorithm processes a multi-layered system model and converges to the right representation granularity, suitable to explain the fault and to identify the faulty subsystem or component. The method is adapted for a very specific application: the ATB of the NH90, a medium weight multi-role military helicopter. The problem is reformulated as a meta-diagnosis problem since the test bench may also host the fault. A detailed case study is presented for illustration.

Keywords: Model-based reasoning · Diagnosis · Avionics systems

1 Introduction

In order to monitor a complex system, it is required to provide an accurate representation of the behavior. Two classes of representation exist: data-based or model-based. Data-based diagnosis is powerful to empirically realize industrial-strength automation of failure diagnosis [10]. It has also been used with filtering and classification for software defect prediction [12]. Model Based Diagnosis (MBD) was proposed by de Kleer and Williams [14], [13]. Davis [7] defined the structure of diagnosis with hypothesis generating from symptoms. Those works have clearly defined the classical logical framework of MBD. De Kleer and Williams models the idea that a faulty state can be explained with mathematical theory and observations. In parallel, many approaches and techniques can be used in a diagnostic process. Venkatasubramanian has reviewed several methods about fault detection and diagnosis from qualitative models to quantitative models [19]. One weakness of MBD is when dealing with large systems as aircrafts, the size of the mathematical representation is overwhelming if the type of the representation is not selected

© Springer International Publishing Switzerland 2015
M. Ali et al. (Eds.): IEA/AIE 2015, LNAI 9101, pp. 635–644, 2015.
DOI: 10.1007/978-3-319-19066-2_61

carefully. Kuntz in [15] has made an application of MBD for avionics systems. He has decreased the complexity of large systems diagnosis with an automated diagnostic solution. Bayesian networks can be used for MBD taking advantage of the graphic probabilistic models to diagnose complex systems [8]. Another solution is to provide a multi-layered representation of the system, and then process each layer separately [17]. For large systems, the computation of subsystems is decomposing the set of all components into smaller sets that reduce the problem complexity. The idea of using functional or topological granularity for complex systems diagnosis has been studied by Ressencourt [17]. Several abstractions and hierarchies might be needed to describe a sufficiently accurate behavior of the fault: Chittaro used structural and behavioral knowledges for this task [6]. Bregon [5] has developed a framework for a distributed diagnosis based on the structural model decomposition. To limit complexity, submodels are extracted from a complete system model, based on some criteria, also explained in [18]. This computation can be automated to obtain the smallest partition with respect to specifications as done in [3]. In [15], submodels are generated with specification documents that describe the system.

In industry, every system is fully tested before coming into operation but the tests in real conditions are expensive. To overcome this difficulty, testbeds have been developed with the following purposes defined in [16]: test a plurality of hardware and embedded softwares configurations with the same testbed; run automatic tests and diagnosis; emulate the real environment; be an intermediary between the software development and integration on the aircraft; develop and test new methods and algorithms to deal with the above purposes. However, being a technological system also subject to failures, it is mandatory to guarantee a faultless behavior of the testbed itself. In [4], a similar testbed is used with prognostics to recreate the conditions of the real environment and evaluate the system under test degradations. Using a MBD representation, Belard has defined the meta-diagnostic approach in [2] to identify the inconsistencies in models, observations or the diagnostic algorithm. The Avionics Test Bench (ATB) diagnosis differs the Belard's one because we consider the representation of the ATB itself in order to be closer to the ATB specificities and issues. Giap in [11] has introduced an iterative process for the diagnosis to resolve the diagnosis problem with the available knowledge into a solvable problem. In this paper, a specific meta-diagnostic approach is presented to prevent failures of testbeds. For the reasons stated above, the diagnosis of the testbed is what we call meta-diagnosis. For an aerospace qualification process, the impacts can range from the simple loss of few hours to the qualification refusal of the aircraft. The diagnosis of modern airplanes avionics has also been studied in [1].

In this article, we define a model-based meta-diagnostic approach to supplement the ATB functionality. The model of the system is built following a lattice-like topology compatible with the meta-diagnosis iterative procedure. We present this topic in the second section. This theory has many benefits: first of all, it is general enough to master different types of ATB; secondly this representation can cope with large-dimension systems of systems such as ATB; at last, it is compatible with many diagnostic algorithms so it can be used to test different methods on the same system. They can be implemented in the monitoring tools to have quickly insights about the faults. The description of the testbed is given in the third section, providing an illustration of a complex system composed of a plurality of different components. Our contribution is the modeling and the diagnostic of the ATB using the proposed method. An illustration of the approach, based on a case study of the ATB is provided in the fourth section.

2   Diagnostic Framework

2.1   System Representation

A diagnosis framework is needed to enable the development of different diagnostic methods and algorithms. The objective of diagnosis is to give insights on occurring failures as soon as possible. Compare to [11], the decompositions are provided only thanks to the structural knowledge of the system.

A system is composed of several subsystems that interact together to achieve a global function. The definitions of subsystems is guided by the interaction of components that communicate to achieve a global function. So, depending on the type of the interaction between the subsystems, the system can be seen as a semi-lattice where the nodes are subsets of components of the system and each relation between nodes is an inclusion relation. As an example, we suppose that p_1 is associated with a functionality of the system with $P_1 = \{\sigma_1; \sigma_2\}$, $\sigma_1 = \{C_1\}$ and $\sigma_2 = \{C_2, C_3\}$. In the following paragraphs, we give formal definitions of a partition and operators to manipulate them.

Definition 1. *COMPS is the set of all the components of a system. A partition P is a set of subsystems: $P = \{\sigma_i | \sigma_i \cap \sigma_j = \emptyset, i \neq j \text{ and } \bigcup_{i=1}^{n} \sigma_i = COMPS\}$ where $\sigma_i \subseteq COMPS$. We note $\mathscr{P}$ the set of all partitions and Σ the set of all the subsystems.*

Definition 2. *The multiplication of two partitions is defined by: $\forall P, Q \in \mathscr{P}^2$, $\exists \sigma_{P \times Q} \in P \times Q \Leftrightarrow \exists \sigma_P \in P, \exists \sigma_Q \in Q, \sigma_{P \times Q} = \sigma_P \cap \sigma_Q$.*

2.2   Diagnostic Approach

Three basic diagnostic functions are defined to diagnose a system : *observation*, *faulty* and *check* functions. An observation is the evaluation of a component. The *obs* function is used to determine if a component is faulty or not. However, it is impossible to define how a unique component behaves regarding to a fault. So we need to define the *faulty* function of a subsystem. In our study, unobservable faults are not detectable. The behaviour of a faulty subsystem is also not sufficient to explain a fault. In fact, subsystems are interconnected making a granularity of the entire system. To describe the fault, it is necessary to introduce the verification function *check* of a partition and the following propositions.

Definition 3. *The obs function of a component c_i is defined by: $obs : COMPS \rightarrow \{0, 1\}$ s.a $obs(c) = 0$ if the component c is faulty and $obs(c) = 1$ if the component c is fault-free;*

Definition 4. *The faulty function of a subsystem* $\sigma_i \subseteq COMPS$ *is defined by:* $faulty : \Sigma \to \{0,1\}$ *s.a* $faulty(\sigma_i) = 0 \Leftrightarrow \forall c_i \in \sigma_i, obs(c_i) = 1$ *and* $faulty(\sigma_i) = 1 \Leftrightarrow \exists c_i \in \sigma_i, obs(c_i) = 0$.

Definition 5. *The check function of a partition p is defined by: check* $: \mathscr{P} \to \{0,1\}$ *s.a* $check(P) = 1 \Leftrightarrow \forall \sigma_i \in P, faulty(\sigma_i) = 0$ *and* $check(P) = 0 \Leftrightarrow \exists \sigma_i \in P, faulty(\sigma_i) = 1$.

2.3 Diagnostic Algorithm

Once the formalism about diagnosis of complex systems has been described, it is now necessary to introduce a diagnostic method whose aim is to solve the above problem. An iterative approach is very helpful in this case of distributed systems [11] since diagnosis can focus on subsystems and partitions. The results of the diagnosis is re-injected in the upper system to deduce informations. The algorithm is based on the two following propositions.

Proposition 1. $\forall P, Q \in \mathscr{P}^2, \exists \sigma_{P \times Q} \in P \times Q, \exists \sigma_P \in P, \exists \sigma_Q \in Q, \sigma_{P \times Q} = \sigma_P \cap \sigma_Q, faulty(\sigma_{P \times Q}) = 1 \Rightarrow faulty(\sigma_P) = 1 \wedge faulty(\sigma_Q) = 1$

Proposition 2. $\forall P, Q \in \mathscr{P}^2, check(P \times Q) = 0 \Rightarrow check(P) = 0 \wedge check(Q) = 0$.

The algorithm is composed of three procedures: *Initialization* initializes all variables, *CheckMultiplicationPartition* checks partitions and updates P^-, F_c sets and finally *CheckComponents* is used to check components.

Algorithm 1. *Initialization*

Input: $d = \{p_i\}$
Output: $\Delta(Diagnosis)$
Global variables: $F_c(faulty\ components), U_c(unfaulty\ components),$
$P^-(faulty\ partitions), P^+(unfaulty\ partitions), end$
Initialization(d)
$\Delta, F_c, U_c, P^+, P^- \leftarrow \{\}$; $end \leftarrow false$;
while $\neg end$ **do**
  | $CheckMultiplicationPartition(d)$;
  | $CheckComponents(F_c)$;

---

3   ATB Meta-Diagnosis

3.1   Avionics System

The avionics system of the NH90 is designed to support multiple hardware and software platforms for more than 12 national customers in over 20 different basic helicopter configurations. It can be found in details in [9]. The NH90 Avionics System consists of two major subsystems: the CORE System and the MISSION

Algorithm 2.
CheckMultiplicationPartition

Input: $d = \{p_i\}$

Outputs:

$F_c(faulty\ components)$,

$P^-(faulty\ partitions)$.

foreach $p_i \in \mathscr{P}, \exists (p_j, p_k) \in P^2$:

$p_i = p_j \times p_k$ **do**

> **if** $check(p_i) = 0$ **then**
>> $P^- \leftarrow P^- \cup \{p_i\}$
>> **foreach** $\sigma_i \in p_i$ **do**
>>> $\sigma_i \leftarrow \sigma_i \setminus U_c$
>>> **if** $\sigma_i = \{c_i\}$ **then**
>>>> $F_c \leftarrow F_c \cup \sigma_i$
>
> **if** $check(p_i) = 1$ **then**
>> $P^+ \leftarrow P^+ \cup \{p_i\}$
>> **foreach** $\sigma_i \in p_i$ **do**
>>> **if** $\sigma_i = \{c_i\}$ **then**
>>>> $U_c \leftarrow U_c \cup \sigma_i$

Algorithm 3. *CheckComponents*

Inputs: $F_c(faulty\ components)$

Outputs: $\Delta(Diagnosis)$

$F_c(faulty\ components)$,

$U_c(unfaulty\ components)$,

end

Initialization: $\sigma_+, \sigma_- \leftarrow I$;

foreach $c_i \in F_c$ **do**

> **if** $obs(c_i = 0)$ **then**
>> $\Delta \leftarrow \Delta \cup \{c_i\}$
>> *end* $\leftarrow$ *true*
>
> **else**
>> $F_c \leftarrow F_c \setminus \{c_i\}$
>> $U_c \leftarrow U_c \cup \{c_i\}$

System. A computer, also known as the bus controller, is managing each subsystem communications: the Core Management Computer (CMC) for the CORE System and the Mission Tactical Computer (MTC) for the MISSION System.

The main computers of the CORE System are the Display and Keyboard Unit (DKU) and the Multi-Functional Display (MFD). Each computer is connected to one or both subsystems via a multiplex data bus (MIL-STD-1553), point to point connections (ARINC429) and serial RS-485 lines. Redundant computers are used as backup. The avionics system $COMPS_{ATB}$ of the ATB is composed of 14 computers and the above connections: two CMC; two Plant Management Computer (PMC); five MFD; two DKU; two Inertial Reference System (IRS); one Radio Altimeter (RA); The avionics system under test $COMPS_{SUT}$ is a subsystem of $COMPS_{ATB}$. It is described on Figure 1.

$COMPS_{SUT} = \{CMC1, CMC2, PMC1, PMC2, MFD1, DKU1, IRS1, RA\}$.
The PMC is used to monitor the status of all the avionics computers. It displays the informations on the MFD. We define the performances partition p_{PERF} $=\{\sigma_{PERF}, \sigma_{\neg PERF}\}$ with: $\sigma_{PERF} = \{PMC1, PMC2, RA, IRS1, MFD1\}$, $\sigma_{\neg PERF} = \{CMC1, CMC2, DKU1\}$.

The test consists in the simulation of a high roll maneuver. Normally the RA should be deactivated above the value of 40 degrees. The procedure contains the following actions: engage the RA with the DKU; simulating a roll of 50 degrees; check that the function is deactivated on the DKU. Several messages are sent to achieve this functionality, see Table 2. The function under test consists in the

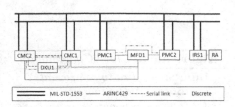

From	To	Information	Subsystems
$DKU1$	$CMC1$	Mode on	$\sigma_{Serial1}$
$CMC1$	$IRS1$	Mode on	σ_{MIL}
$IRS1$	RA	Mode on	$\sigma_{trackMode}$; σ_{ARINC}
RA	$IRS1$	Alert	$\sigma_{trackMode}$
$IRS1$	$CMC1$	Alert	σ_{MIL}
$CMC1$	$DKU1$	Alert	$\sigma_{Serial1}$

Fig. 1. Architecture of the avionics subsystem

Fig. 2. Messages

control of the RA, transmitting track mode power level, horizontal beam and alert. We define the RA track mode $p_{trackMode} = \{\sigma_{trackMode}, \sigma_{\neg trackMode}\}$ with: $\sigma_{trackMode} = \{RA, IRS1, MFD1\}$; $\sigma_{\neg trackMode} = \{CMC1, CMC2, DKU1, PMC1, PMC2\}$.

3.2 System Under Test (SUT) Decomposition

We define the subsystems σ_i and meta-subsystems p_i with regards to the connections of the avionics system of Figure 1, the serial communications:
$\sigma_{Serial1} = \{CMC1, CMC2, DKU1\}$, $\sigma_{Serial2} = \{PMC1, PMC2\}$, $\sigma_{\neg Serial} = \{MFD1, IRS1, RA\}$, $p_{Serial} = \{\sigma_{Serial1}; \sigma_{Serial2}; \sigma_{\neg Serial}\}$;
the ARINC communications:
$\sigma_{ARINC} = \{CMC1, CMC2, PMC1, PMC2, MFD1, IRS1, RA\}$,
$\sigma_{\neg ARINC} = \{DKU1\}$, $p_{ARINC} = \{\sigma_{ARINC}; \sigma_{\neg ARINC}\}$;
the MIL-STD-1553 communications:
$\sigma_{MIL} = \{CMC1, CMC2, PMC1, PMC2, IRS1\}$,
$\sigma_{\neg MIL} = \{MFD1, DKU1, RA\}$, $p_{MIL} = \{\sigma_{MIL}; \sigma_{\neg MIL}\}$.

We define the partitions into two categories: functional partitions and communication partitions. The functional blocks contains the subsystems that compute and send informations. The communication blocks contains the subsystems that relay these informations. In our example, the navigation functionality is tested. Functional partition sets are: $\{p_{trackMode}, p_{PERF}\}$, connection partitions sets are $\{p_{MIL}; p_{Serial}; p_{ARINC}\}$. We need to define additional partitions that can be checked with the *check* function on the system thanks to this representation: $p_{trackMode.MIL} = p_{trackMode} \times p_{MIL} = \{\{MFD1, RA\}$; $\{IRS1\}$; $\{CMC1, CMC2, PMC1, PMC2\}$; $\{DKU1\}\}$; $p_{trackMode.Serial} = \{\{CMC1, CMC2, DKU1\}$; $\{PMC1, PMC2\}$; $\{MFD1, IRS1, RA\}\}$; $p_{trackMode.ARINC} = \{\{MFD1, IRS1, RA\}$; $\{CMC1, CMC2, PMC1, PMC2\}$; $\{DKU1\}\}$.

The performance function can give insights about the fault. We compute the partitions with this function: $p_{PERF.MIL} = \{\{MFD1, RA\}$; $\{DKU1\}$; $\{CMC1, CMC2\}; \{PMC1, PMC2, IRS1\}\}$; $p_{PERF.Serial} = \{\{CMC1, CMC2, DKU1\}$; $\{PMC1, PMC2\}$; $\{MFD1, IRS1, RA\}\}$; $p_{PERF.ARINC} = \{\{PMC1, PMC2, MFD1, IRS1, RA\}$; $\{CMC1, CMC2\}$; $\{DKU1\}\}$.

3.3   Outlooks About the Decompositions

If the result of the algorithm is not sufficient to check the system, an other decomposition is required. We describe an iterative method to update the diagnostic result by providing new granularities of the system to upgrade the model. The subsystems are computed with the framework of the previous section. Given the components, the messages sent between them, and the protocol of these messages, we can obtain an overview of the system decomposition: p_{SUT} can be decomposed into $d_{protocol} = \{p_{SUT} \times p_{MIL}; p_{SUT} \times p_{Serial}; p_{SUT} \times p_{ARINC}\}$, see Figure 3.

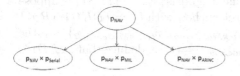

Fig. 3. Navigation function decomposition with $d_{protocol}$

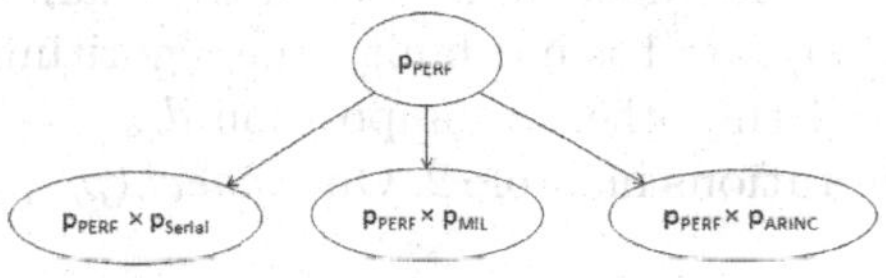

Fig. 4. Performance function decomposition with $d_{protocol}$

The decomposition is also applied for the performance function, see Figure 4. The previous test can also be described with the following partitions that describe the route of the informations "RA mode on" and "RA alert" on pilot side: $\sigma_{com_1} = \{\{DKU1, CMC1, IRS1, RA\}\}$; $\sigma_{\neg com_1} = \{\{MFD1, CMC2, PMC1, PMC2\}\}$; $p_{com_1} = \{\sigma_{com_1}, \sigma_{\neg com_1}\}$. The route of the informations "RA mode on" and "RA alert" on copilot side defines an other decomposition: $\sigma_{com_2} = \{\{CMC2, IRS1, RA, DKU1\}\}$; $\sigma_{\neg com_2} = \{\{MFD1, CMC1, PMC1, PMC2\}\}$; $p_{com_2} = \{\sigma_{com_2}, \sigma_{\neg com_2}\}$. We compute the partitions:

$p_{trackMode.com1} = \{\{RA, IRS1\}; \{MFD1\}; \{CMC1, DKU1\}; \{CMC2, PMC1, PMC2\}\}$;

$p_{trackMode.com2} = \{\{RA, IRS1\}; \{DKU1, CMC2\}; \{MFD1\}; \{CMC1, PMC1, PMC2\}\}$;

$p_{PERF.com1} = \{\{RA, IRS1\}; \{CMC1, DKU1\}; \{CMC2\}; \{MFD1, PMC1, PMC2\}\}$;

$p_{PERF.com2} = \{\{RA, IRS1\}; \{DKU1, CMC2\}; \{CMC1\}; \{MFD1, PMC1, PMC2\}\}$.

4   Meta-Diagnostic Approach

4.1   Application of the Meta-Diagnosis with Several Connection Partitions

We suppose that the navigation function does not behave normally. We describe the iterations of the algorithms with two decompositions. First, we have launched the meta-diagnostic algorithm with the decomposition:

$d_{protocol} = \{p_{trackMode.MIL}, p_{trackMode.ARINC}, p_{trackMode.SERIAL}\}$, see Tables 1.

Table 1. Iterations of $CheckMultiplicationPartition$ with $d_{protocol}$

p_i	$check(p_i)$	$P^+; P^-; U_c; F_c$
$p_{trackMode.ARINC}$	0	$\emptyset; \{p_{trackMode.ARINC}\}; \emptyset; \{DKU1\}$
$p_{trackMode.SERIAL}$	1	$\{p_{trackMode.SERIAL}\}; \{p_{trackMode.ARINC}\}; \emptyset; \{DKU1\}$
$p_{trackMode.MIL}$	0	$\{p_{trackMode.SERIAL}\};$ $\{p_{trackMode.ARINC}, p_{trackMode.MIL}\}; \emptyset; \{IRS1, DKU1\}$

The third step checks the components in F_c set that can be faulty: $DKU1$ and $IRS1$. If the components are faulty, this may explain the system behavior and the algorithm ends. In the example, the diagnosis is $\Delta = \{IRS1\}$. Suppose that $IRS1$ is not faulty, the algorithm is relaunched with $U_c = \{DKU1, IRS1\}$ and the other decomposition $d_{com} = \{p_{trackMode.com1}, p_{trackMode.com2}\}$. see the iterations in Table 2. Once $check(p_{trackMode.com2}) = 1$, we deduce that $MFD1$ is

Table 2. Iterations of $CheckMultiplicationPartition$ with d_{com}

p_i	$check(p_i)$	$P^+; P^-; U_c; F_c$
$p_{trackMode.com1}$	0	$\emptyset; \{p_{trackMode.com1}\}; \{DKU1, IRS1\}; \{RA, MFD1\}$
$p_{trackMode.com2}$	1	$\{p_{trackMode.com2}\}; \{p_{trackMode.com1}\};$ $\{DKU1, IRS1, MFD1\}; \{RA\}$

not faulty. At this step, the fault-free components are $\{DKU1, IRS1, MFD1\}$, and the diagnosis is $\{RA\}$. If the observation of the possible faulty component RA state is 0, then the algorithm ends. Else, an other decomposition must be used until the faulty components are found.

4.2 Application of the Meta-Diagnosis with Several Functional Partitions

We suppose that the navigation function and the performance function are not behave normally. Once $check(p_{PERF.com2}) = 1$, we deduce that $CMC1$ is not faulty. We now describe the iterations of the algorithms with the decomposition $d_{protocol}$ knowing the $CMC1$ is not faulty in Table 3. The third step checks the components of F_c set: $DKU1$ and $CMC2$.

At this state, since the reparation of $CMC2$ has fixed the system, we conclude that $CMC2$ has been faulty. The diagnosis is $\Delta = \{CMC2\}$. The algorithms are then implemented in a spy software of ARINC and MIL-STD-1553 buses and developed with C++ for effective diagnosis. Compare to traditional tools and methods about diagnosis evaluated in [19] and about iterative diagnosis [11], our approach aims to upgrade the diagnosis applications since it is integrated on an industrial systems. It is a complement to the work on avionics systems [15] since it takes the test system environment into account. It gives an other approach for the meta-diagnosis problem defined in [2] because we define

Table 3. Iterations of $CheckMultiplicationPartition$ with $d_{protocol}$

p_i	$check(p_i)$	$P^+; P^-; U_c; F_c$
$p_{PERF.ARINC}$	0	$\emptyset; \{p_{PERF.ARINC}\}; \{CMC1\}; \{DKU1, CMC2\}$
$p_{PERF.SERIAL}$	1	$\{p_{PERF.SERIAL}\}; \{p_{PERF.ARINC}\}; \{CMC1\};$ $\{DKU1, CMC2\}$
$p_{PERF.MIL}$	0	$\{p_{PERF.SERIAL}\}; \{p_{PERF.ARINC}, p_{PERF.MIL}\};$ $\{CMC1\}; \{DKU1, CMC2\}$

the system description with the lattice concept whereas Belard has extended the whole theory of MBD. Finally, it is quite similar to [18] since we perform a logical system decomposition but differences appear in the model using Analytical Redundancy Relations (ARR) in their work and lattice representations in our study to cope with the ATB specificities.

5   Conclusion

An approach for complex systems diagnosis is proposed, driven by the structural decomposition of embedded functions and component connections. A specific form of MBD is introduced and enhanced using the lattice-like modeling structure to cope with a special class of test system: the ATB. The variable granularity decomposition is used to obtain a variable size subsystem-based structure to overcome the initial problem complexity. Then a meta-diagnostic algorithm is used to find the faults. The proposed decomposition is based on avionics and connection/communication functions. If additional avionics functions are involved, it is easy to augment the diagnostic model using partitions addition. The case study shows results of this procedure on a real industrial system.

References

1. Balin, C.E., Stankunas, J.: Investigation of fault detection and analysis methods for central maintenance systems. Aviation technologies 1(1), January 2013
2. Belard, N., Pencole, Y., Combacau, M.: A theory of meta-diagnosis: reasoning about diagnostic systems. In: Proceedings of the Twenty-Second International Joint Conference on Artificial Intelligence. IJCAI 2011, pp. 731–737. Catalonia, Spain, Barcelona (2011)
3. Berdjag, D., Cocquempot, V., Christophe, C., Shumsky, A., Zhirabok, A.: Algebraic approach for model decomposition: Application for fault detection and isolation in discrete-event systems. International Journal of Applied Mathematics and Computer Science (AMCS) 21(1), 109–125 (2011)
4. Braden, D.R., Harvey, D.M.: Aligning component and system qualification testing through prognostics. Electronics System-Integration Technology Conference (ESTC) 2014, 1–6 (2014)
5. Bregon, A., Daigle, M., Roychoudhury, I., Biswas, G., Koutsoukos, X., Pulido, B.: Improving distributed through structural model decomposition. DX 2011 22nd International Workshop on Principles of Diagnosis (2011)

6. Chittaro, L., Ranon, R.: Hierarchical model-based diagnosis based on structural abstraction. Artificial Intelligence **155**(1–2), 147–182 (2004)
7. Davis, R., Hamscher, W.: Exploring artificial intelligence, pp. 297–346. Morgan Kaufmann Publishers Inc., San Francisco, CA, USA (1988)
8. Delcroix, V., Maalej, M.-A., Piechowiak, S.: Bayesian networks versus other probabilistic models for the multiple diagnosis of large devices. International Journal on Artificial Intelligence Tools **16**(3), 417–433 (2007)
9. Dordowsky, F., Bridges, R., Tschope, H.: Implementing a software product line for a complex avionics system. In: 2011 15th International Software Product Line Conference (SPLC), pp. 241–250, August 2011
10. Duan, S., Babu, S.: Empirical comparison of techniques for automated failure diagnosis. In: Proceedings of the Third Conference on Tackling Computer Systems Problems with Machine Learning Techniques, SysML 2008, pp. 2–2. USENIX Association Berkeley, CA, USA (2008)
11. Giap, Q.-H., Ploix, S., Flaus, J.-M.: Managing diagnosis processes with interactive decompositions. In: Iliadis, Maglogiann, Tsoumakasis, Vlahavas, Bramer, (eds.) Artificial Intelligence Applications and Innovations III. IFIP International Federation for Information Processing, vol. 296, pp. 407–415. Springer, US (2009)
12. Gray, D., Bowes, D., Davey, N., Sun, Y., Christianson, B.: Reflections on the NASA MDP data sets. IET Software **6**(6), 549–558 (2012)
13. de Kleer, J., Mackworth, A.K., Reiter, R.: Characterizing diagnoses and systems. Artificial Intelligence **56**(2–3), 197–222 (1992)
14. de Kleer, J., Williams, B.: Diagnosing multiple faults. Artificial Intelligence **32**(1), 97–130 (1987)
15. Kuntz, F., Gaudan, S., Sannino, C., Laurent, R., Griffault, A., Point, G.: Model-based diagnosis for avionics systems using minimal cuts. DX 2011 22nd International Workshop on Principles of Diagnosis (2011)
16. Poll, S., Patterson-hine, A., Camisa, J., Garcia, D., Hall, D., Lee, C., Mengshoel, O.J., Nishikawa, D., Ossenfort, J., Sweet, A., Yentus, S., Roychoudhury, I., Daigle, M., Biswas, G., Koutsoukos, X.: Advanced diagnostics and prognostics testbed. In: Proceedings of the 18th International Workshop on Principles of Diagnosis, pp. 178–185 (2007)
17. Ressencourt, H.: Hierarchical modelling and diagnosis for embedded systems. 17th International Workshop on Principles of Diagnosis DX 2006, pp. 235–242 (2006)
18. Roychoudhury, I., Daigle, M.J., Bregon, A., Pulido, B.: A structural model decomposition framework for systems health management. Big Sky, MT, United States, March 2013
19. Venkatasubramanian, V., Rengaswamy, R., Kavuri, S.N., Yin, K.: A review of process fault detection and diagnosis: Part III: Process history based methods. Computers & chemical engineering **27**(3), 327–346 (2003)

Finding Similar Time Series in Sales Transaction Data

Swee Chuan Tan[✉], Pei San Lau, and XiaoWei Yu

School of Business, SIM University, 535A Clementi Road, Singapore, Singapore
jamestansc@unisim.edu.sg, {jessenie,iewoaixuy}@gmail.com

Abstract. This paper studies the problem of finding similar time series of product sales in transactional data. We argue that finding such similar time series can lead to discovery of interesting and actionable business information such as previously unknown complementary products or substitutes, and hidden supply chain information. However, finding all possible pairs of n time series exhaustively results in $O(n^2)$ time complexity. To address this issue, we propose using *k-means* clustering method to create small clusters of similar time series, and those clusters with very small intra-cluster variability are used to find similar time series. Finally, we demonstrate the utility of our approach to derive interesting results from real-life data.

1   Introduction

The term market basket analysis in data mining [2, 5] involves analyzing and understanding products and their categories that are commonly purchased by a customer over a specific period of time, e.g., during a visit to a departmental store. This is an intriguing research topic because the insight derived from this kind of study can be very useful.

In the retailing context, identifying complementary products (e.g., coffee and sugar) or substitute products (e.g., sugar and artificial sweetener) can help a marketer to design better promotion campaigns to increase sales of related product lines. Sales managers can also use such information to formulate better pricing strategies that could lead to improved market share or profits. Supermarket store managers can use insight from a market basket analysis to improve store layout and design, so that commonly purchased items can be more easily accessed by shoppers.

Obviously, the notion of 'market basket' can also be extended into many other areas beyond the retailing space. For example, a pharmacy manager can use the information to organize storage of medicines so as to improve the efficiency of locating, retrieving, and even packaging of some medicinal drugs commonly used to treat a certain disease.

Despite all the potential benefits we can get from market basket analysis, the analytics task per se is not straight forward and a lot of research is still ongoing. In marketing science, the common approach is to use traditional statistical techniques to analyze customer purchases. Most of such studies aim to improve our general understanding of customer purchasing behaviors (e.g., see [8, 17]).

© Springer International Publishing Switzerland 2015
M. Ali et al. (Eds.): IEA/AIE 2015, LNAI 9101, pp. 645–654, 2015.
DOI: 10.1007/978-3-319-19066-2_62

In the area of data mining applications, the problem is more applied in nature because the analyst is usually asked to solve a particular business problem at hand. For example, given a sales transaction dataset, the problem is usually to draw insight from commonly purchased products so as to improve sales and operations in the near future. To this end, the association rule mining (ARM) approach proposed [1] two decades ago has been a very important topic frequently studied by data mining researchers.

In this paper, we are motivated to tackle the 'market basket analysis' problem using a time series clustering approach instead of the ARM approach. This is because the proposed approach uses time series of product sales as records of interest, in which quantity and time information are used in the analysis, yielding richer and more insightful results. Furthermore, time series patterns are generally more concise as compared to large number of rules produced by ARM. Indeed, previous research has shown that storing and analyzing data in time series format actually takes up less computer memory space, compared to the amount of memory space used to store binary format data in ARM [16].

Another advantage of finding similar time series of product sales is that, it is easy to study how each time series sales pattern is associated with a certain entity of interest (e.g., with a particular customer). The entity then provides a unique context in which potentially useful domain information can be used to interpret the time series patterns so as to produce meaningful insight. Here we show two example entities that could be associated with time series patterns.

The first type of entity is *product category*. We can study the time series of product sales patterns associated with product categories. If two products are in the *same product category* (e.g., tooth paste) and are having similar time series patterns of product sales, then they could potentially be *substitute* products. On the other hand, if two products are in *different product categories* (e.g., tooth paste and tooth brush) and are having similar time series patterns of product sales, then they could potentially be *complementary* products.

The second type of entity is *customer making purchases*. We can study the time series pattern of the sales of a product purchased by a particular customer. If there are two similar time series patterns of two different customers purchasing the same product, then the pattern suggests that these two customers may have similar demand on the same product. This implies that these two customers could be related, such as working on a common project.

The rest of this paper is organized as follow. Section 2 reviews related work. Section 3 presents the general proposed approach to find similar time series in sales transaction data. Section 4 illustrates how the approach is used for analyzing a sales transaction dataset containing more than 300 thousand records. We then present a case where real-life insights were discovered using our approach. We also show how similar time series can be discovered from the dataset while association rule mining fails to produce useful results. Finally, Section 5 concludes this paper by discussing some future research directions.

2   Related Work

The first association rule mining technique that has been proposed in the literature, known as the *Apriori* algorithm [1], is also a tool commonly used by data miners. Despite its popularity, Apriori has several weaknesses that are still not well addressed today.

The first weakness arises from the fact that the *Apriori* algorithm requires the purchase status of every product associated with a transaction to be recorded in binary format, e.g., Product_A_Purchased = {Yes, No}. Hence the mining problem is sometimes referred as the Boolean Association Rules problem [14]. This form of data representation gives rise to several issues. Firstly, the analysis cannot take quantity information into account. In fact, quantitative variables have to be discretized either during the preprocessing stage or during the rule mining process [14]. Apart from losing information, the need to use discretization will increase the complexity of the rule mining process because the user has to choose an appropriate discretization method and find the right parameter settings (e.g., the number of categories). Despite many attempts (e.g., [9, 11, 13, 18]), there is still no established approach to address this problem. Secondly, Boolean Association Rules problem does not consider the time dimension [12]. As such, the method cannot discover rules that may only be interesting in certain time window (e.g., Christmas period) but uninteresting in other times. To address this problem, some authors propose solving the temporal data mining problem [3], in which the main difficulty lies in finding the time window that contains interesting rules.

The second weakness of Apriori is that it tends to generate a huge number of rules. It is not uncommon to see Apriori (and in fact many other ARM methods) producing hundreds or thousands of rules from a dataset with only a few hundred transactions. This poses a serious problem to the data miner because there is a risk that some 'interesting' rules could have happened just by chance alone [19]. To the user, it is also difficult to find the few most interesting rules among many uninteresting ones. Currently researchers are looking at possible solutions such as rule reduction (e.g., [7]), concise rule generation (e.g., [10, 20]), rule visualizations (e.g., [4]), etc. Such research directions help move towards developing a more usable rule mining system, but as of now many business users are still facing difficulty in using existing systems.

In this paper, we explore an alternative approach that is easier to implement and use compared to association rule mining. This alternative market basket analysis approach involves time series clustering on real-world sales transaction data. We show that this approach can help alleviate some of the above-mentioned issues.

3   Proposed Approach to Find Similar Time Series

Here we describe the steps to prepare data for time series clustering and the procedure to find similar time series.

Given a dataset that contains a finite number of transaction records over a specified time period T. The time domain T can be represented as v time units, $T = \{0, 1, 2, \ldots, v-1\}$. When a transaction occurs at time i, the transaction records the quantity (q_i) of a product (p_i) being purchased at time i. Let $T = \{w_0, w_1, \ldots, w_{m-1}\}$ be made up of m

non-overlapping time windows. Then the jth time window w_j having d consecutive time units will have time units $w_j = \{j \times d, j \times d +1, ..., j \times d +d-1\}$. Once d, the window width is defined, we also know that $m = v/d$, and the total quantity of a product P ordered in time window w_j is $Q_j = \sum q_i \times I(p_i \in P)$, for all i that satisfy the constraint $j = \lfloor i/d \rfloor$. Note that $I(p_i \in P) = 1$ if $p_i \in P$; otherwise $I(p_i \in P) = 0$.

Eventually, we can represent the time series of a product P as a tuple $S := \{P, Q_0, Q_1, ..., Q_{m-1}\}$. Note that the quantities ordered for different products can be quite different, so each total ordered quantity value in a time window should be normalized to a range of $[0, 1]$. This is achieved using min-max normalization, $Q^*_j := (Q_j - Q_{min})/(Q_{max} - Q_{min})$, where $Q_{min} := min(Q_0, Q_1, ..., Q_{m-1})$, and $Q_{max} := max(Q_0, Q_1, ..., Q_{m-1})$. This gives an equal weight to all product sales time series and facilitates clustering of similar time series structures. Note that S is excluded if $Q_{max} = Q_{min}$.

Once the time series data is prepared, we use a four-step procedure to find similar time series.

- Step 1: Given x time series, decide the expected number of time series (y) to be included in each cluster.
- Step 2: Apply k-means using $k = \lfloor x/y \rfloor$, where k is the number of clusters.
- Step 3: For each cluster, sort each time series by its distance to the centroid.
- Step 4: Select all consecutive pairs of ordered time series in each cluster that has similarity greater than some predefined threshold.

Note that our main aim is very different from the most common intent of cluster analysis – to find most natural groupings. Instead, our aim here is to find many (small) groups of similar time series. Hence the choice of y is not critical in Step 1, and we recommend it to be a small number, say 5. Once y is determined, k can be computed and k-means [6] can be applied. In Steps 3 and 4, we use the centroid regions to find the most similar time series as these are areas where highly similar time series reside.

4   Finding Similar Time Series in Sales Transaction Data

We demonstrate the utility of the proposed approach using two cases. The first case involves analyzing a real-life sales transaction dataset that we have purchased for academic research purposes. We present selected time series clustering results derived from this dataset to demonstrate how the proposed method works. We shall call this dataset "Sales Transaction".

The second case involves analyzing a sales transaction dataset obtained from a global part supplier. The actual identity of this company cannot be disclosed due to confidentiality reasons. We shall call this company as "Company Z". Although we are not allowed to show the actual time series patterns of Company Z data, we are allowed to reveal some insights derived from the analysis.

Data Preparation of the Sales Transaction data: The Sales Transaction dataset contains 350 thousand transactional records of about 800 products purchased over a period of one year. For the purpose of time series clustering, we only require the Product Code, Sales Date, and Quantity Ordered of each transaction.

During the data preparation stage, the transaction records are sorted by Product Code and then Sales Date. The records are then aggregated by a two-week time window and the total quantity ordered for each product within each two-week period is computed. The result of aggregation is a table with Product Code and Window Period Identifier (an integer than ranges from 1 to 26), and Period Sales Quantity. This table is then transposed to have each Window Period Identifier representing a column capturing the total sales quantity of each product for the period. The resulting table ends up with about eight hundred records over a period of 26 time-points. Each record captures periodic sales quantities of a product sold over the one-year period.

Figure 1 gives an example of the data conversion from sales transaction format to time series format. Since the data spans over only 26 time points, the data size is small and we can apply clustering directly on the time series data.

Once the time series data is prepared, the sales quantities for each product are then normalized to a range of [0, 1] using the above-mentioned min-max normalisation. Finally, the dataset is imported into the IBM SPSS Modeler®, which is the software package used for this study.

Product Code	Period ID	Sales Qty
D23	1	73
D23	2	59
D23	...	...
D23	26	67
D24	1	15
D24	2	22
D24	...	...
D24	26	23
...	...	...

	Period ID			
Product Code	1	2	...	26
D23	73	59	...	67
D24	15	22	...	23
...	...	...	...	...

Fig. 1. Data conversion from sales transaction table format to time series format

Analysis of the Sales Transaction Data: In this project, there are about 800 products in the dataset; the expected number of time series in each cluster is set as 4. As a result, we use *k-means* clustering algorithm to generate 200 clusters. The following presents how interesting business information can be derived from the results.

Since most of the patterns found are similar, we present some typical example patterns to highlight the key insight. The first example pattern is presented in Figure 2, which shows a positive relationship between Product 281 (Product Category: HiFi) and Product 293 (Product Category: Telephony) found in Cluster 117. Notice that these two products are from different product categories. This result suggests the first possible insight that may be derived from this type of time series pattern: *Two products in different product categories could be complementary products if they have similar product sales time series.*

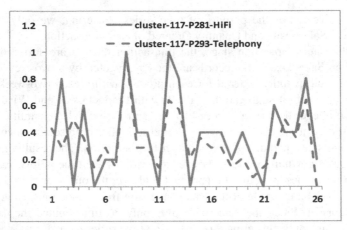

Fig. 2. Similar time series of products sales in different categories suggests opportunities for cross-selling

The second example pattern is presented in Figure 3, which shows a positive relationship between Products P408 and P613. Upon further investigation, it is found that these products all belong to the Hardware product category. In huge procurement systems, the actual relationships amoung thousands of products are usually hidden. Sometimes, previously unknown substitute products could be discovered this way. This result suggests the second possible insight that may be derived from this pattern: *Two products in the same product category could be substitutes products if they have similar product sales time series.*

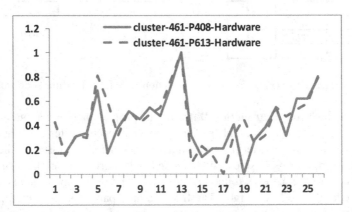

Fig. 3. Similar time series of products sales in the same category suggests opportunities for understanding the reason(s) for the demand

Sometimes, time series may be similar due to coincidence. It is important to validate every possible insight with the domain expert. The above examples also suggest that the product sales time series should be interpreted within a predefined context. In these examples, knowing whether two products are in the same or different product categories suggests that the products are substitutes or complementary, respectively. In the following, we present a case study where similar product time series, when interpreted in different contexts, result in different types of insights that are useful for decision making.

Case Study of Z Company: Here, we report a case study that has been conducted with a real company. Due to confidentiality reasons, we cannot reveal the name of the company, and shall call it **Company Z**. This company is a supplier of raw parts to many manufacturing companies around the world. At the time of the study, the company wanted to better understand their customers' needs in order to better manage their productions, stock control and customers. One of the initiatives towards this end was to analyze the product sales transaction data in order to discover interesting customer purchasing behaviors. In fact, some early results of Company Z have been presented in previous papers [15, 16]. Here we present more recent results derived from the same project.

The first type of discovery involves grouping the product sales time series according to products ordered by the *same customer*. In doing so, we examined a few hundreds of raw parts (e.g., nuts and screws) that have been ordered by buyers of *a single manufacturing plant*. Using the proposed method, we were able to find some groups of parts that exhibit similar product sales time series. Subsequently, the customer confirmed that a number of these parts with similar time series are indeed complementary parts for manufacturing certain consumer electronics products. Armed with such knowledge, we were then able to advise the sales team on pricing and sales strategies of these parts.

The second type of discovery involves grouping the product sales time series according to products ordered by *different customers*. Sometimes, this type of analysis can reveal previously unknown interactions among customers. In this case, we found that certain parts purchased by two different customers, namely Company B and Company C, did have very similar time series patterns. On further investigation, it was found that Companies B and C were actually supplying products to another customer, Company A. It is interesting to note that Companies A, B and C are all customers of Company Z (i.e., the raw part supplier). The supply chain involving companies A, B and C are illustrated in Figure 4.

Figure 4 shows that Company Z has been supplying a variety of products to three companies. Company A is a special product maker for automotive market. Companies B and C are competitors – both provide value-added services for raw materials used in white goods. This finding is counterintuitive because Company A is known to serve a market that is different from the market served by Companies B and C.

To ensure the validity of this finding, we talk to Company A and confirm that Companies B and C are indeed their qualified vendors for parts used in making certain products. Companies B and C are chosen due to their expertise in upgrading parts to be used in their products, and because of their factory locations, they are able to offer shorter lead-time for goods delivery.

This finding suggests that the engagement of these companies can now be more organised. Currently, Companies A, B and C are handled by three different sales staff as these companies are perceived as serving different markets. However, the clustering result reveals that these companies all serve the same market within the discovered supply chain. This is valuable information for sales and operations. For example, the company could now assign one sales staff (instead of three) to take care of all three companies. This is likely to result in a more productive and consistent engagement of these companies. Other issues such as (fairness in) pricing, sharing of sensitive information, and identififcation of cross-selling opportunities can now be beter managaed by one (and not three) sales staff.

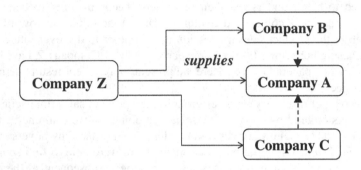

Fig. 4. Hidden supply chain information (the dotted lines) discovered from our analysis. Companies B and C are competitors supplying to Company A. All Companies A, B and C are customers of Company Z.

Applying Apriori on the Data: We apply the *Apriori* method on the Sales Transaction dataset using a minimum antecedent support threshold of 5% and minimum confidence threshold of 20%. This produces 55 rules with very low support and confidence values. Figure 5 shows a scatter plot of support and confidence percentage of these 55 rules. It is easy to see that the all rules have confidence lower than 25% and support lower than 4%. This suggests that association rule mining is unable to discover any interesting rules from the data.

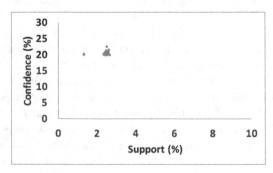

Fig. 5. The 55 rules generated from the Sales Transaction data set are all of low support and low confidence

5   Concluding Remarks

This paper shows that finding similar time series of sales data can help reveal interesting information about customers. We have shown that it is possible to discover complementary or substitute products that can lead to better formulation of sales strategies. Also, it is possible to discover hidden processes that are important information for implementing better customer engagement. Our discovery of hidden supply chain information is a case in point.

One question that arises from this study is that why the *Apriori* method is unable to produce any interesting rules from the given sales transaction data? One reason is because the combinations of individual products are not purchased often enough by individual customers. However, when we examine the time series pattern of the total sales quantity of each product over different time windows, we do find relationship between certain products that has similar time series patterns. One caveat against misusing such results is that we should always check that the purported relationship of any two products is not due to coincidence, and this can be easily verified with the domain expert.

References

1. Agrawal, R., Srikant, R.: Fast algorithms for mining association rules. In: Proceedings of the International Conference on Very Large Data Bases, pp. 487–499 (1994)
2. Charlet, L., Kumar, A.: Market Basket Analysis for a Supermarket based on Frequent Itemset Mining. International Journal of Computer Science Issues. **9**(5), 257–264 (2012)
3. Chen, X., Petrounias, I.: Discovering Temporal Association Rules: Algorithms, Language and System. In: Proceedings of the 16th International Conference on Data Engineering, pp. 306–306 (2000)
4. Hahsler, M., Chelluboina, S.: Visualizing association rules in hierarchical groups. Presented at the 42nd Symposium on the Interface: Statistical, Machine Learning, and Visualization Algorithms (Interface 2011). The Interface Foundation of North America, June 2011 (unpublished)
5. Kim, H.K., Kim, J.K., Chen, Q.Y.: A Product Network Analysis for Extending the Market Basket Analysis. Expert Systems with Applications **39**(8), 7403–7410 (2012)
6. MacQueen, J.B.: Some Methods for classification and Analysis of Multivariate Observations. In: Proceedings of 5th Berkeley Symposium on Mathematical Statistics and Probability 1, pp. 281–297. University of California Press (1967)
7. Mafruz, Z.A., David, T., Kate, S.: Redundant association rules reduction techniques. International Journal Business Intelligent Data Mining **2**(1), 29–63 (2007)
8. Manchanda, P., Ansari, A.: Gupta, S: The ``shopping basket'': A model for multicategory purchase incidence decisions. Marketing Science **18**(2), 95–114 (1999)
9. Minaei-Bidgoli, B., Barmaki, R., Nasiri, M.: Mining numerical association rules via multi-objective genetic algorithms. Information Sciences **233**, 15–24 (2013)
10. Palshikar, G., Kale, M., Apte, M.: Association Rules Mining Using Heavy Itemsets. Data & Knowledge Engineering **61**(1), 93–113 (2007)

11. Tong, Q., Yan, B., Zhou, Y.: Mining Quantitative Association Rules on Overlapped Intervals. In: Li, X., Wang, S., Dong, Z.Y. (eds.) ADMA 2005. LNCS (LNAI), vol. 3584, pp. 43–50. Springer, Heidelberg (2005)
12. Qin, L.X., Shi, Z.Z.: Efficiently Mining Association Rules from Time Series. International Journal of Information Technology. 12(4), 30–38 (2006)
13. Salleb-Aouissi, A., Vrain, C., Nortet, C., Kong, X., Rathod, V., Cassard, D.: QuantMiner for Mining Quantitative Association Rules. Journal of Machine Learning Research 14, 3153–3157 (2013)
14. Srikant, R., Agrawal, R.: Mining Quantitative Association Rules in Large Relational Tables. In: Proceedings of the 1996 ACM SIGMOD International Conference on Management of Data, pp. 1–12 (1996)
15. Lau, P.S.: Time Series Clustering for Market Basket Analysis. Analytics Project Report. SIM University (2013)
16. Tan, S.C, Lau, P.S.: Time Series Clustering: A Superior Alternative for Market Basket Analysis. In: Herawan, T., Deris, M.M., Abawajy, J. (eds.) Proceedings of the First International Conference on Advanced Data and Information Engineering. LNEE(LNCS), vol. 285, pp. 241–248. Springer, Singapore (2013)
17. Vindevogel, B., Poel, D., Wets, G.: Why promotion strategies based on market basket analysis do not work. Expert Systems with Applications 28(3), 583–590 (2005)
18. Webb, G.: Discovering associations with numeric variables. In: Proceedings of the Seventh ACM SIGKDD International Conference on Knowledge Discovery and Data Mining (KDD 2001), pp. 383–388. ACM, New York (2001)
19. Webb, G.: Discovering Significant Patterns. Machine Learning 68(1), 1–33 (2007)
20. Xu, Y., Li, Y., Shaw, G.: Reliable representations for association rules. Data & Knowledge Engineering 70(6), 555–575 (2011)

Prediction of Package Chip Quality
Using Fail Bit Count Data of the Probe Test

Jin Soo Park and Seoung Bum Kim[(⊠)]

Department of Industrial Management Engineering, Korea University, Seoul, Korea
{dmqmpjs,sbkiml}@korea.ac.kr

Abstract. The quality prediction of the semiconductor industry has been widely recognized as important and critical for quality improvement and productivity enhancement. The main objective of this paper is to establish a prediction methodology of semiconductor chip quality. Although various research has been conducted for predicting a yield, these studies predict a yield by lot-level and do not consider characteristics of the data. We demonstrate the effectiveness of the proposed procedure using a real data from a semiconductor manufacturing.

Keywords: Quality prediction · Probe test · Smote · Nonparametric variable selection

1   Introduction

The quality is directly connected to the competitiveness of the companies. High quality products improve the reliability and customer satisfaction. For this reason, many manufacturing companies are currently working on an effort to improve the quality [1]. In particular, semiconductor market is rapidly growing, and manufacturers are focusing on early development of new products, mass production improvement, and quality management to strengthen business competitiveness power. Quality control can be divided into prediction and follow-up service. The former is predicting the quality by using manufacturing process parameters and the product properties to prevent defects in advance. The latter is to correspond and action when the customer claims occur. To improve competitiveness of the company, quality prediction and pre-detection of defects is very important.

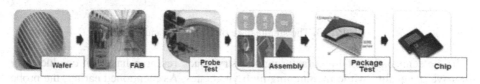

Fig. 1. The semiconductor manufacturing process

The semiconductor manufacturing usually consists of 200–300 process steps, and it takes three to four months to produce the final chips. As shown in Fig. 1, the semiconductor manufacturing processes can be generally divided into four basic

© Springer International Publishing Switzerland 2015
M. Ali et al. (Eds.): IEA/AIE 2015, LNAI 9101, pp. 655–664, 2015.
DOI: 10.1007/978-3-319-19066-2_63

processes: fabrication (FAB), probe test, assembly, and package test [2]. The FAB process forms hundreds to thousands of chips on the pure wafer by going through a process unit such as the photo and the etch process. Probe test, known as wafer test, provides key information about the performance of the wafer fabrication process. It involves testing of individual chips for their functionality based on different electrical probes. Assembly step separates the chips from a wafer and packs them to protect physical impacts from the outside. Packaged chips are sent for packaging test to determine the quality of chips by checking in harsh environments than users use [3].

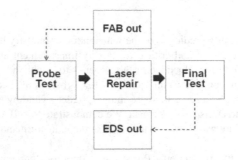

Fig. 2. Probe test process

Probe test is the first step in which the chip-level data are generated and thus an important process. Fig. 2 describes the probe test process in which the fail bit count (FBC) data are generated. Probe test is referred to as electric die sorting (EDS) process. As described in Fig. 3, the semiconductor memory chip is made up of the Giga cell, and hence defects can be present. Redundancy cells are a set of spare cells for repairing. The redundancy cells consist of spare rows and columns. Laser repair replaces the defect cells with the redundancy cells. Through this process, the yield can be significantly enhanced[4].

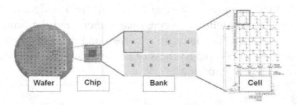

Fig. 3. The architecture of the DRAM chip

In the semiconductor manufacturing industry, much research has been conducted to improve the quality while maintaining high yield rates. As mentioned earlier, the probe test plays a significant role in the prediction of final chip quality because the probe test first generates the chip-level data. Through the chip quality prediction of the probe test step, high-quality chips and low-quality chips can be classified. Quality grading can lead to a dual package test and a dual manufacturing process. Through this, test time in the manufacturing process is reduced and the improvement in the yield can be achieved.

Thus, the studies using a probe test data were conducted because the probe test process is important. A semiconductor yield prediction using stepwise support vector machine [5], a package test yield prediction using wafer bin map [6] are conducted, however, these studies predicted yield by lot-level and focused on the accuracy of the overall model, not the sensitivity. Generally, the performance measure in classification is the accuracy, however, in the quality level, sensitivity should be taken into consideration. In particular, the semiconductor data have an imbalance problem. Data imbalance problem occurs when the number of high-quality chips are much larger than low-quality chips [7]. In this paper, we propose an efficient quality prediction methodology considering the characteristic of the FBC data from the probe test.

The rest of this paper is organized as follows. In Section 2, we illustrate the data imbalance problem. Section 3 presents the nonparametric variable selection. In Section 4, we present the quality prediction methodology with the FBC data from the probe test. In Section 5, we give some concluding remarks.

2   Data Imbalance Problems and Solutions

The FBC data from the probe test exhibit an imbalance problem. However, most classification algorithms are well trained under the assumption that the number of observations in each class is roughly equal [7]. In general, to deal with the problem caused by the imbalanced data, three methods have been previously proposed.

First, undersampling methods [8] address imbalance problems by sampling a small number of observations of the majority class. Not only can undersampling methods enhance the classification performance, but also reduce the computational costs since they samples a small number of observations from the majority class. However, applying undersampling methods has a possible drawback of biasing the distribution of the majority class. If the sampled observations from the majority class do not follow the original distribution, it may decrease in the classification performance. This possible disadvantage can be happened if the number of minority class observations is very small.

Second, oversampling methods [9, 10] solve the imbalance problems by copying observations from the minority class. In contrast with the undersampling methods, since oversampling methods contain all of the information on the original observations, it can accomplish a comparatively high performance. However, the computational costs training the classification models increase since the number of observations used in training is much larger than the number of the original observations.

The third methods, which are a combination of the two methods above, deal with the imbalance problems by sampling a small number of observations of the majority class and copying observations of the minority class [11]. The combination of these two methods is not always a good thing than using only undersampling methods. If we replicate the minority class, the decision region of the minority class becomes very specific, but does not spread into the majority class region, which makes the decision region as overfitted. One of the solutions to resolving this overfitting problem is the synthetic minority oversampling technique (SMOTE). SMOTE is a method that the minority class is oversampled by creating synthetic observations rather than by oversampling with replacement. For more details, please refer to Chawla (2002) [9].

3   Nonparametric Variable Selection

In the high-tech manufacturing process such as semiconductor process, a large amount of variables that are correlated with each other is generated. In this case, variable selection is critical. The main objective of variable selection is to identify a subset of variables that are most predictive of a given response variable. Variable selection is particularly of interest when the number of candidate explanatory variables is large, and when many redundant or irrelevant variables are thought to be present [12].

Therefore a dimensionality reduction process, which find the significant variables, is essential. In addition, since the FBC data from the probe test used in this paper have a number of variables, the dimensionality reduction technique can be necessary. In general, by performing the two-sample t-test [13] for each variable, the important variables will be selected. However the two-sample t-test is based on the parametric assumption. The FBC data from the probe test do not follow the normal distribution, as shown in Fig. 4. Therefore, we use a nonparametric variable selection technique. In this paper, we use the nonparametric resampling t-test among the various variable selection techniques [14].

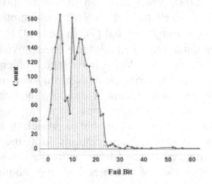

Fig. 4. The distribution of the fail bit count data

The nonparametric resampling t-test is how to find statistically significant variables in comparison to original t-statistics and resampling t-statistics without distinction of groups. Now we calculate the statistics of all the variables using following equation for each groups by employing a two-sample t-test.

$$t_i = \frac{\overline{X_{i,normal}} - \overline{X_{i,abnormal}}}{\sqrt{\frac{S^2_{i,normal}}{n_{normal}} + \frac{S^2_{i,abnormal}}{n_{abnormal}}}} \tag{1}$$

for i=1,2,...,29. i is the index of the predictor variables, $\overline{X_{i,normal}}$ and $S^2_{i,normal}$ are the sample mean and variance of normal groups. Likewise, $\overline{X_{i,abnormal}}$ and $S^2_{i,abnormal}$ are taken from abnormal groups. Next, we calculate p-values using a permutation method due to repeated measurements and thus we cannot clearly assume that each t_i follows a t-distribution. Under the assumption that there is no differential fail bit count level between the two classes (normal and abnormal groups), the t-statistic should have the same distribution regardless of how we make the permutation of fail bit count. Therefore, we can permute (shuffle) the two groups, and re-compute a set of t-statistics for each individual fail bit count feature based on the permuted dataset. If this procedure is repeated N times, we can obtain N sets of t-statistics as follows : $t_1^n, t_2^n, \ldots, t_{29}^n, n = 1, 2, \ldots, N$. The nonparametric p-value for i=29 and N=1,000 is obtained by

$$p_i = \sum_{n=1}^{1000} \frac{\#\{k : |t_k^n| \geq |t_i|, k = 1, 2, \ldots, 29\}}{29 \cdot 1000}. \tag{2}$$

Finally, we conduct a variable selection by applying the false discovery rate (FDR) [18] using these p-values. We can summarize the procedure of the variable selection based FDR as follows:

- Ordering the p-values in ascending $(p(1) \leq p(2) \leq \ldots \leq p(29))$
- Select a desired FDR level($= \alpha$) between 0 and 1 in this paper, we select 0.05
- Calculate the largest i denoted as w

$$w = \max\left[i : p(i) \leq \frac{i}{m} \frac{\alpha}{\delta}\right], \tag{3}$$

where m is the total number of variables (here m=29) and δ denotes the proportion of true null hypothesis. Many studies discuss the assignment of δ. In this paper, we use δ=1, the most conservative value.

- The threshold of the p-value is $p_{(w)}$, and declare the fail bit count feature t_i significant if and only if $p_i \leq p_{(w)}$.

4   The Quality Prediction Using the FBC Data from the Probe Test

4.1   Data Description and Performance Measure

The data used in this study are obtained from the real FBC data from the probe test in semiconductor manufacturing. The dataset contains 29 variables and 2,623 observations (2,000 high-quality chips, 623 low-quality chips). As shown in Table 1, the predictor variables, X are discrete count data, and the response variable, Y is binary, indicating whether the corresponding chip is high quality or low quality. Fig. 5 describes a three-dimensional principal component score plot [15] showing that high-quality observations and low-quality observations are overlapped with each other.

Table 1. The fail bit count data of probe test

Wafer	X1	X2	...	X28	X29	Y
1	9	21	...	1	0	0
2	8	16	...	0	0	0
⋮	⋮	⋮	...	⋮	⋮	⋮
2622	5	12	...	0	1	1
2623	3	6	...	2	0	1

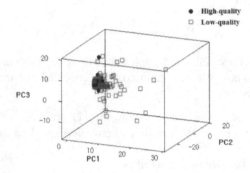

Fig. 5. A three-dimensional principal component score plot

We applied a 10-fold cross validation [15] to obtain reliable results. The performance of the proposed method is evaluated by sensitivity [16]. That can be calculated by the following equation:

$$\text{Sensitivity} = \frac{TP}{TP+FN}. \tag{4}$$

Table 2. Confusion matrix

		Predicted	
		Abnormal	Normal
Actual	Abnormal	TP	FN
	Normal	FP	TN

· TP (True Positive): the number of positive examples correctly predicted by the classification model.
· FN (False Negative): the number of positive examples wrongly predicted as negative by the classification model.
· FP (False Positive): the number of negative examples wrongly predicted as positive by the classification model.
· TN (True Negative): the number of negative examples correctly predicted by the classification model.

4.2   A Prediction Methodology of Package Chip Quality

In this study, we used the three steps to construct prediction models. Step 1 solves the imbalance problem using SMOTE. Step 2 identifies important variables by using the non-parametric techniques. Finally, Step 3 determines the relevant values of the parameters such that the sensitivity is maximized. We provide detailed descriptions of these steps as follows:

Step 1. Solving the Data Imbalance Problem

As mentioned in Chapter 2, the FBC data from the probe test have an imbalance problem. Because we are more interested in detecting low-quality chips of the minority class, the imbalance problem should be resolved [16]. In this paper, we used the SMOTE algorithm.

Fig. 6 shows the sensitivity values from various classification algorithms before and after applying the SMOTE. It can be clearly seen that the sensitivity is improved by using SMOTE.

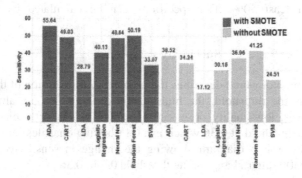

Fig. 6. The sensitivity with SMOTE technique

Step 2. Variable Selection

The number of predictor variables considered in this study is 29. Here we used a non-parametric approach to select important variables because the probability distribution of the data is unknown [17]. Especially, we performed the nonparametric resampling t-test [14] to obtain a set of p-values for each variable. We applied the false discovery rate (FDR) [18] using these p-values to select a significant variables.

As a result of variable selection, 13 of the 29 variables were selected. Looking at the non-selected variables, their correlation is high or they do not affect the outcome variable since a high-quality class and a low-quality class are almost equal.

Step 3. Adjusting the Parameters of the Model

As shown in Fig. 5, we recognized that high-quality observations and low-quality observations are overlapped with each other. In the quality prediction, the classification accuracy of the low-quality observations is very important. Therefore, we adjust the parameters of the model for increasing the sensitivity. In this study, we adjust the threshold of a logistic regression [15]. The results of the logistic regression come with probability values that belong to a particular category, as shown in the following equation.

$$p(x) = \frac{e^{\beta_0 + \beta_1 x}}{1 + e^{\beta_0 + \beta_1 x}} \tag{5}$$

Threshold is a criteria for determining whether observations belong to one category of the two categories. In general use of logistic regression, the threshold is set at 0.5 if the number of observations in each class is roughly equal. However, the FBC data have an imbalance problem. Therefore, if we use 0.5 as the threshold, the minority class can not be classified properly. Hence, we should find an appropriate threshold to maximize the sensitivity. Furthermore, the classification accuracy of the high-quality observations is important, thus, we adjust the threshold that the specificity is accomplished at least 50%. The specificity can be calculated by the following equation:

$$Specificity = \frac{TN}{FP + TN} \tag{6}$$

Fig. 7 is the box plot which describes the probability distribution of the two classes. If we use 0.5 as the threshold, the high-quality observations are almost classified properly. However, low-quality observations cannot be correctly classified. In this study, we set the threshold at 0.36 which is a Q1 (the first quartiles) of the low-quality observations. Fig. 8 is a diagram showing the change in sensitivity. Sensitivity is improved about 30% after changing the threshold 0.5 to 0.36.

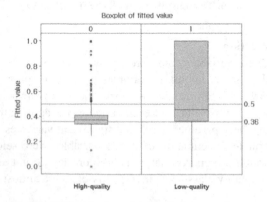

Fig. 7. Fitted value of the logistic regression

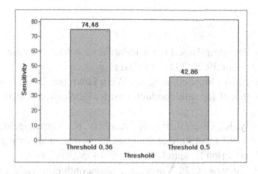

Fig. 8. The sensitivity difference between thresholds

4.3   Experiment Results

Table 3 shows the sensitivities of applying the SMOTE, the nonparametric variable selection and adjustment of the parameters of the model. The sensitivity was improved about 10% by the SMOTE technique. Besides, important variables are selected with eliminating redundant information, which eventually raise the sensitivity of the prediction model. Finally, by adjusting the threshold of the logistic regression, the sensitivity was improved from 42.86% to 74.48%. This results clearly demonsrate the effectiveness of the proposed procedure to predict the final chip quality based on FBC data obtained from the probe test.

Table 3. The result of experiment

	Sensitivity
Original data	30.18%
SMOTE	40.13%
SMOTE + Variable Selection	42.86%
SMOTE + Variable Selection + Adjusting the threshold	74.48%

5   Conclusions

The probe test is a critical step in the prediction of final chip quality. In this paper, we propose a quality prediction methodology using the FBC data obtained from the probe test. Most classification algorithms are well trained with balanced data. However, the FBC data from the probe test are highly imbalanced, and hence, we proposed to use the SMOTE algorithm to address imbalanced problems. In addition, since the FBC data from the probe test do not follow the normal distribution, we used a nonparametric variable selection technique to identify the important variables for prediction. Finally, by adjusting the parameters of the prediction model, the high sensitivity can be obtained.

References

1. Kang, J., Kim, S.: Bootstrap-Based Fault Identification Method. Journal of Korean Society for Quality Management **39**(2), 234–244 (2011)
2. Chen, H., Harrison, J.M., Mandelbaum, A., Wein Lawrence, M.: Empirical evaluation of a queueing network model for semiconductor wafer fabrication. Oper. Res. **36**(2), 202–215 (1988)
3. Kumar, N., Kennedy, K., Gildersleeve, K., Abelson, R., Mastrangelo, C.M., Montgomery, D.C.: A review of yield modelling techniques for semiconductor manufacturing. International Journal of Production Research **44**(23), 5019–5036 (2006)
4. Cunningham, S.P., Spanos, C.J., Voros, K.: Semiconductor yield improvement: Results and best practices. IEEE Transactions on Semiconductor Manufacturing **8**(2), 103–109 (1995)
5. An, D., Ko, H.H., Gulambar, T., Kim, J., Baek, J.G., Kim, S.S.: A semiconductor yields prediction using stepwise support vector machine. In: IEEE International Symposium on Assembly and Manufacturing, ISAM 2009, pp. 130–136. IEEE (2009)
6. Hsu, S.C., Chien, C.F.: Hybrid data mining approach for pattern extraction from wafer bin map to improve yield in semiconductor manufacturing. International Journal of Production Economics **107**(1), 88–103 (2007)
7. Kang, P., Cho, S.: EUS SVMs: ensemble of under-sampled SVMs for data imbalance problems. In: King, I., Wang, J., Chan, L.-W., Wang, D. (eds.) ICONIP 2006. LNCS, vol. 4232, pp. 837–846. Springer, Heidelberg (2006)
8. Kubat, M., Matwin, S.: Addressing the curse of imbalanced training sets: one-sided selection. In: Proceedings of the Fourteenth International Conference on Machine Learning, pp. 179–186 (1997)
9. Chawla, N.V., Hall, L., Kegelmeyer, W.: SMOTE: Synthetic Minority Oversampling Techniques. Journal of Artificial Intelligence Research **16**, 321–357 (2002)
10. Chawla, N.V., Lazarevic, A., Hall, L.O., Bowyer, K.W.: SMOTEBoost: improving prediction of the minority class in boosting. In: Lavrač, N., Gamberger, D., Todorovski, L., Blockeel, H. (eds.) PKDD 2003. LNCS (LNAI), vol. 2838, pp. 107–119. Springer, Heidelberg (2003)
11. Solberg, A., Solberg, R.: A Large-scale evaluation of features for automatic detection of oil spills in ers sar images. In: International Geoscience and Remote Sensing Symposium, pp. 1484–1486 (1996)
12. Shmueli, G., Patel, N.R., Bruce, P.C.: Data mining for business intelligence: concepts, techniques, and applications in Microsoft Office Excel with XLMiner. John Wiley & Sons (2007)
13. Walpole, R.E., Myers, R.H., Myers, S.L., Ye, K.: Probability and statistics for engineers and scientists, vol. 5. Macmillan, New York (1993)
14. Kim, S.B., Chen, V.C., Park, Y., Ziegler, T.R., Jones, D.P.: Controlling the False Discovery Rate for Feature Selection in High-resolution NMR Spectra. Statistical Analysis and Data Mining **1**(2), 57–66 (2008)
15. Hastie, T., Tibshirani, R., Friedman, J., Hastie, T., Friedman, J., Tibshirani, R.: The Elements of Statistical Learning, vol. 2(1). Springer, New York (2009)
16. Tan, P., Steinbach, M., Kumar, V.: Introduction to data mining, 1st edn. Addison Wesley (2006)
17. Conover, W.J.: Practical Nonparametric Statistics, 3rd edn. Wiley, New York (1999)
18. Benjamini, Y., Hochberg, Y.: Controlling the false discovery rate: a practical and powerful approach to multiple testing. Journal of the Royal Statistical Society. Series B (Methodological), 289–300 (1995)

Scene Understanding Based on Sound and Text Information for a Cooking Support Robot

Ryosuke Kojima[✉], Osamu Sugiyama, and Kazuhiro Nakadai

Graduate School of Information Science and Engineering,
Tokyo Institute of Technology, 2-12-1 Ookayama, Meguro-ku Tokyo, Japan
kojima@cyb.mei.titech.ac.jp

Abstract. We address noise-robust "auditory scene understanding" for a robot defined by extracting 6W (What, When, Where, Who, Why, hoW) information on the surrounding environment. Although such a robot has been studied in the field of robot audition, only the first four Ws except for "why" and "how" were in scope. Thus, this paper mainly focuses on extracting "how" information, in particular, on cooking scenes to realize a cooking support robot. In this case, "how" information is regarded as a cooking procedure, we construct sound-based cooking procedure recognition based on two models. One is a conventional statistical model, Gaussian Mixture Model (GMM), which is used for an acoustic model to recognize a cooking sound event such as stirring, cutting and so on. The other is a Hierarchical Hidden Markov Model (HHMM), which is used for a recipe model to recognize a sequence of cooking events, i.e., a cooking procedure. We constructed a prototype system for cooking recipe and procedure recognition. Preliminary results showed that the proposed GMM-HHMM based system outperformed a conventional GMM-HMM based system in terms of noise-robustness in cooking recipe recognition and our system was able to correct misrecognition of cooking sound events using recipe model in cooking procedure recognition.

Keywords: Scene understanding · Robot audition · HHMM

1 Introduction

Scene understanding is a key technology for a robot to support people since a robot should choose an appropriate action according to a scene constructed from sensory data. A difficulty in scene understanding is to extract useful information from various kinds of and a large amount of sensor data. We define "scene understanding" as extracting 6W (What, When, Where, Who, Why, and hoW) information and constructing a structure of the scene.

Scene understanding has been studied in the field of robot vision. However, it is difficult to achieve scene understanding only from visual information. Robot audition [11], thus, proposed to study audio-based scene understanding for a robot. Many researches to extract four of the 6W information have been reported such as sound source identification (What), voice activity detection (When),

© Springer International Publishing Switzerland 2015
M. Ali et al. (Eds.): IEA/AIE 2015, LNAI 9101, pp. 665–674, 2015.
DOI: 10.1007/978-3-319-19066-2_64

sound source localization (Where), and speaker recognition (Who). Although extracting "Why" and "How" information has been studied well in the domains of text mining and knowledge discovery [1,6], only limited research has been reported for real-world applications such as a robot.

This paper, therefore, addresses "How" information extraction for a cooking support robot equipped with a microphone as a first step. "How" information in a cooking scene can be regarded as a cooking procedure. For example, a cooking procedure for stir-fried vegetables is described by "First, cut vegetables. Second, mix vegetables. Finally stir-fry them." We hence tackle cooking procedure recognition for a robot.

To realize such cooking procedure recognition, we divided it into two steps. The first step is sound-based recipe recognition by identifying recipes from a sequence of cooking sounds. The second step is estimation of a cooking procedure represented by cooking actions and its objectives using the recipe obtained in the first step. To solve these steps, we have to consider three factors as follows:

1. Knowledge acquisition and representation for cooking,
2. Knowledge utilization, and
3. Noise robustness for sensor data.

The first factor obtains and maintains knowledge related to cooking. The knowledge is acquired utilizing web contents such as cooking recipe sites and cooking log sites; however these contents are usually written in natural language. Graph structure representing recipes has been already proposed as a manageable format of the recipe data [10,15]. We use a similar graph structure, called *flow graph*, as an internal manageable data structure for recipes. Furthermore, we convert it to a Hierarchical Hidden Markov Model (HHMM)[3], which is a probabilistic model to deal with sequential data. We use this HHMM as a model representing recipe, called *recipe model*. As the other models for recipe model, hierarchical Conditional Random Field (CRF)[9,14] and (semi-)Markov model can be considered. The former method requires an annotated sequential dataset for supervised learning but this paper deals with unannotated sequential data. The latter cannot represent cooking procedures that are represented by a hidden state of an HHMM.

The second factor is utilization of the cooking knowledge mentioned above to tackle the cooking procedure recognition. To solve the first step of cooking procedure recognition, that is, recipe recognition, we compute a recipe with the maximum likelihood based on the recipe models. The second step is extraction of a cooking procedure represented by a sequence of the event labels. We also compute a sequence of the labels with maximum likelihood in the estimated recipe using Viterbi algorithm for an HHMM [3] where each label corresponds to a state of HHMM.

The last factor is noise robustness. Many studies related to cooking recognition utilize visual information obtained from cameras equipped within a kitchen, its floor and/or human [4,13]. This approach cannot capture actions at dead camera angles or fine movements such as those of a fingertip. The sound information

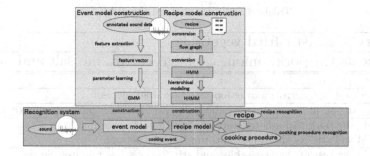

Fig. 1. Outline of our proposal system

compensates for these disadvantages since the radiated sound can be easily captured in a small area like a kitchen. On the other hand, noise-robustness is crucial since sound is always contaminated with noise like environmental sounds. This is solved using special cooking devices with microphones such as [16]. We feel that it is tough to prepare such devices, and it is reasonable to use conventional devices like a smart phone or a small robot equipped with a microphone. Our approach is utilizing a multimodal model consisting of a recipe and an event model. The event model is an acoustic model for cooking events such as "cut cabbage". We use a Gaussian Mixture Model (GMM) as an event model since the effectiveness of this model has been known in speech and sound recognition. Since the event model are affected by the environmental noise, the recognized events include misrecognition. We expect the recipe model corrects such misrecognition and our multimodal model is noise-robust.

2   Proposed System

Fig.1 shows the outline of our proposed system. It is divided into three parts; recipe model construction, event model construction, and recognition system. The recipe model construction uses an HHMM (explained in Section 2.2) converted from a flow graph (explained in Section 2.1) via an HMM. Through this process, our system obtains cooking knowledge from web sites and maintains it in the form of a recipe model. The event model construction trains parameters of GMMs from annotated sound data. The recognition system utilizes the cooking knowledge for cooking recipes and procedure recognition by unifying the recipe and event models (explained in Section 2.3). Since recipe models make Utilizing these two models is also expected to yield noise-robustness by. The following sections explain the three parts in detail.

2.1   Construction of Flow Graph

To construct a recipe model from recipes consisting of category, list of ingredients and cooking procedure like Table 1, recipes are converted to a *flow graph* which represents an abstracted recipe shown in Fig.2. It includes nodes and directed

Table 1. recipe: stir-fried vegetables

category	stir-fried vegetable
ingredients list	pork, onion, cabbage, carrot, oil, salt and pepper

procedure	
Step 1	Cut cabbage, carrot, and onion into bite sized pieces.
Step 2	Put the oil in a wok. Stir fry the garlic over low heat until it's slightly colored.
Step 3	Add the pork and stir-fry until cooked through, over low heat.
	Add and stir-fry the carrot and the salt and pepper.
Step 4	Add the rest of the vegetables and stir-fry for a few minutes. Serve.

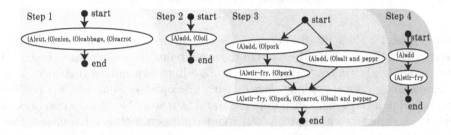

Fig. 2. Flow graph generated from recipe Table 1

edges. A node shows a cooking event consisting of a cooking action (e.g. cut, add and stir-fry) and its objectives (e.g. onion, cabbage and pork), and an edge shows the order of the two events.

The flow graph can be constructed from recipes using a dependency parser by assuming that the order of events is represented by a relationship between words. The dependency parser extracts words representing an action and the corresponding objectives, and constructs a node from the words. It also constructs a directed edge according to the dependency relationship between words representing cooking actions. The detailed procedure is shown as follows:

1. Perform dependency parsing for each sentence in recipes.
2. Using this result, a node consists of a word contained in the action list (action word) and a set of words depending on the action word and contained in the ingredient list (objective words) is built.
3. Connect the nodes using dependency relationships between their action words.

This procedure converts the recipe in Table 1 into the flow graph in Fig.2 where (A) and (O) indicates action and its objectives respectively. Note that "start" and "end" indicate special nodes representing start and terminal of each step, respectively. A node which has two or more parent nodes like the node in step 3 in Fig.3 denotes non-deterministic order of parent nodes. We call such a node a *joint node*.

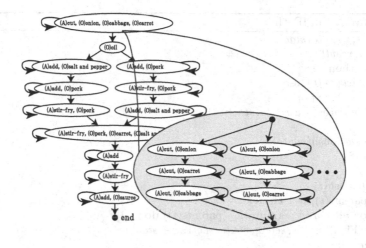

Fig. 3. Action HMM (left) and event HHMM (the whole graph)

2.2 Construction of Recipe Model

We converted a flow graph to a recipe model represented by an HHMM with two steps. First, we built a non-hierarchical HMM. Secondly, we constructed a hierarchical structure for an HHMM based on the HMM.

An HMM is a standard probabilistic model for sequential data, which includes several parameters regarding state transition and output probabilities. A flow graph is converted to an HMM through the following three steps.

1. Convert a joint node into state transitions in HMM (mentioned below).
2. Add a self-loop for every state to the HMM.
3. Add a transition from the end node to the start node in the next step for each step.

A joint node is converted to multiple state transition paths by combination of its ancestor sub-graphs by an enumeration algorithm shown in Algorithm 1[1]. This code contains two functions: parents(*node*), permutation(*set*). The parents (*node*) returns a set of parent nodes of the given *node*. The permutation(*set*) returns a set of all arrays generated by permutation of a given *set* of nodes. Using these functions, FindPath(*s, t, route, stack*) computes all possible paths from node *s* to node *t* using two stacks: *route* and *stack*. All searched paths are stored in the global variable *result*. All arrays in *result* are converted into state transitions. In this operation, the number of states increases exponentially as the number of joint nodes increases; however this did not cause any trouble in our experiments described in Section 3.2 using real recipes.

---

[1] For this algorithm, we referenced the topological sort algorithm [2] and the algorithms for counting topological orders [5].

Algorithm 1. FindPaths

Require: $s, t, route, stack$

 1: **globals** $result$
 2: **if** $s = t$ **then**
 3:     **if** $|stack| > 0$ **then**
 4:         $c \leftarrow stack.$pop
 5:         FindPaths($s, c, route, stack$)
 6:     **else**
 7:         $result.$push($route$)
 8:     **end if**
 9: **else**
10:     $route.$push(t)
11:     **if** $|$parents(t)$| > 1$ **then**
12:         **for all** seq in permutation(parents(t)) **do**
13:             FindPaths($s, seq[0], route, seq[1...] + stack$)
14:         **end for**
15:     **else**
16:         FindPaths($s,$ parents(t)$, route, stack$)
17:     **end if**
18: **end if**
19: **return** $result$

When this procedure is applied to the flow graph in Fig.2, we can obtain an HMM shown in Fig.3. We call it *"action HMM"* since each state of the HMM contains an action and a set of its objectives.

A state in an action HMM can be divided into fine-grained states. A fine-grained state represents an event consisting of a pair of an action and an objective selected from those in the action HMM. We, thus, defined an *event HHMM* which is an extension of an action HMM so that each state in action HMM has finer HMMs with fine-grained states. Permutations of these fine-grained states determine state transitions in the finer HMM. We show an example of event HHMMs in Fig.3.

2.3   Construction of Event Model and Proposal System

A recipe model provides us with the most likely cooking recipe and procedure from a given sequence of cooking events by computing likelihood with a recipe based on an HHMM. To integrate them with the recipe model, we also built a sound-based cooking event recognition using a GMM-based acoustic model trained with preprocessed cooking sound events and annotated event labels.

As preprocessing, sound activity detection is performed for the recorded data by zero-crossing and amplitude-based thresholding. For each frame of the sound obtained from sound activity detection, MFCC(Mel-Frequency Cepstrum Coefficients) is computed. A GMM is trained with MFCC vectors and the annotated event labels.

Table 2. Categories and the number of recipes

category	# of recipes	category	# of recipes
beef stew	32	ginger pork saute	35
macaroni au gratin	16	okonomi-yaki	26
plain omelet	16	udon	25
stir-fried vegetable	94		

In our system, a recipe is estimated with the following three steps: preprocess, recognizing cooking event, estimation of the recipe. Input sound is converted to MFCC vectors by the same way as making training data in the first step. In the next step, a cooking event is estimated from the vectors at every frame. In the last step, a recipe is obtained from a recipe model and the sequence of the estimated cooking events.

This system also can compute a cooking procedure with maximum likelihood. A cooking procedure is computed by Viterbi algorithm for an HHMM [3] in the estimated recipe.

3 Evaluation

To evaluate our system[2,3], we built three experimental settings. The first experiment is a preliminary experiment to evaluate an event model in order to know the performance of cooking event recognition. In the second experiment, we evaluate a recipe model to show the noise-robustness in terms of recipe estimation using artificial data. In the last experiment, we tackle cooking procedure recognition. For the second and third experiments, we used recipe data belonging to 7 categories shown in Table 2[4] from COOKPAD (http://cookpad.com/) where the recipes are written in Japanese.

3.1 Experiment 1: Evaluation of the Event Model

In this experiment, we tackled a task of cooking event recognition to evaluate an event model. We prepared 13 types of cooking sounds. Table 3 shows a confusion matrix of this task where the numbers of training data and validation data are 124,473 and 31,118, respectively. Note that the number of mixtures in GMM is 30, the sampling rate is 16kHz and the input feature is 130-dim MFCCs

---

[2] To implement this system, we use a Japanese morphological analyzer Mecab [8] and a Japanese dependency structure analyzer CaboCha [7] to build a recipe model. We also use a logic-based programming language intended for symbolic-statistical modeling PRISM [12] to implement HMM and HHMM.

[3] Parameters of HHMMs are usually learned from data. We however determine them in ad hoc since it is difficult to collect a sufficient number of recorded cooking sound. This is a future task.

[4] We selected these categories since a sufficient number of recipes was obtained in this site and variances in cooking procedures of the recipes were small.

Table 3. Confusion matrix in cooking event recognition : the columns and rows indicate results and answer sets respectively

		crush	cut									wash		bake
		egg	cabbage	green-pepper	onion	pork	green-onion	chinese-cabbage	cucu-mber	radish	carrot	chopping board	knife	vege-tables
crush	egg	**73**	7	4	11	4	14	7	13	23	7	0	0	0
cut	cabbage	0	**1025**	80	51	2	7	1	9	80	27	0	0	2
	green-pepper	1	128	**577**	14	32	0	0	7	15	27	0	0	32
	onion	0	4	7	**1513**	3	5	3	9	6	0	0	0	0
	pork	0	47	240	3	**394**	2	1	1	4	7	0	0	78
	green-onion	18	30	23	116	52	**870**	96	9	65	114	14	3	0
	chinese-cabbage	0	0	2	5	3	22	**227**	0	0	2	0	0	0
	cucumber	10	160	85	296	28	67	5	**826**	556	188	2	2	0
	radish	52	972	878	430	202	385	7	364	**6720**	994	5	3	0
	carrot	28	213	415	111	259	581	12	145	730	**1834**	1	1	0
wash	chopping board	3	7	4	8	6	37	47	3	10	20	**198**	3	0
	knife	0	1	0	9	10	21	13	4	48	14	6	**153**	0
bake	vegetables	1	25	55	4	53	5	0	0	0	0	5	0	**6500**

((12-dim MFCC + power)× 10 frames) where the frame size is 25 msec. In this experiment, the same kitchen and devices were used.

As a result, the accuracy was 67 % with 5-fold cross-validation. This result shows that sounds are a good clue for cooking event recognition. We however noticed that events with a low sound level such as "cut pork" and events having similar sounds with each other such as "cut carrot" and "cut radish" were difficult to be identified. To solve this, audio-visual multimodal recognition would be promising.

3.2   Experiment 2: Evaluation of Cooking Recipe Recognition

In this experiment, we conducted recipe recognition using synthetic data generated by event HHMM. The generated data can be regarded as ideal, that is, the accuracy when the cooking event recognition is 100%. Considering a more realistic case, we added noise to each event in an event sequence at a probability r. It means that the noise randomly replaces an event to another event, which corresponds to misrecognition of the cooking event. We defined *category accuracy Acc* as

$$Acc = \frac{N_t}{\# \text{ of generated data}}$$

where N_t is the number of estimated recipes containing the same category with the generated one. By controlling r, we observed the category accuracy.

Fig.4 shows a result of this experiment when ten sequences were sampled from one recipe and we used 10 recipes for each category. Fig.4 also shows a result when the action HMM is simply used for the recipe model.

From this result, the system using an event HHMM as a recipe model is superior to an action HMM by means of category accuracy for all noise rates. Let us elaborate this result. The accuracy of the event HHMM decreases rapidly when the noise rate is greater than 0.5; nevertheless, according to the result in

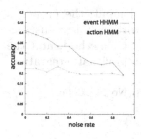

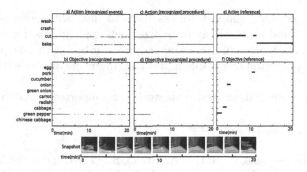

Fig. 4. Accuracy of recipe recognition under artificial noise

Fig. 5. Recognized events a), b), an extracted cooking procedure from a recipe model c), d), and reference e), f)

Section 3.1, the accuracy of cooking event recognition is 67%, which corresponds to a 0.33 noise rate. Hence we guess that this system works around 0.3 in this graph where the system performs relatively high accuracy. Furthermore, this result implies that improvement of the cooking event recognition brings us high category accuracy of the whole system.

3.3 Experiment 3: Cooking Procedure Recognition

In the third experiment, we extract a cooking procedure from cooking sounds in real cooking scenes where we actually cooked dishes belonging to a category : stir-fried vegetables using our system that is constructed in the same setting in Section 3.2. Here, we furthermore extract a cooking procedure on the most likely recipe. These results are shown in Fig.5. Note that this result is low accuracy compared to the experiment in Section 3.1, since this real scene includes a some other acion such as putting ingredients on the cutting board.

As we can see, the recognized events a)(inputs of the recipe model) contain the small segments and misrecognition of "bake" and "cut" but the recognized cooking process c) was smoothened. On the other hand, misrecognition of "onion" and "cabbage" remains in b) and d). We can guess that this misrecognition was caused by the misrecognition of event recognition. This graph shows progress of cooking whose information has the possibility to be used for the a support robot. For example, when a mismatch with a recipe is detected, a robot can point it out.

4 Conclusion

This paper presented cooking procedure recognition as an example of scene understanding, in particular, focusing on "how" information among 6Ws. We proposed a sound-based cooking recipe and procedure recognition system using a recipe model represented by an HHMM from web sites, and a cooking sound event model based on GMM. It showed high noise-robustness compared to a system with a non-hierarchical HMM model. Since real-world applications such as cooking support robots have to deal with noisy sound input, the proposed method is promising to realize such applications. Future work includes construction of an actual cooking support robot with the proposed method, and the exploration of a multimodal cooking event model with audio-visual integration.

Acknowledgments. This work was supported by KAKENHI-No.24220006.

References

1. Cimiano, P., Hotho, A., Staab, S.: Learning concept hierarchies from text corpora using formal concept analysis. J. Artif. Intell. Res. (JAIR) **24**, 305–339 (2005)
2. Uslar, M., Specht, M., Rohjans, S., Trefke, J., Gonzalez, J.M.V.: Introduction. In: Uslar, M., Specht, M., Rohjans, S., Trefke, J., Vasquez Gonzalez, J.M. (eds.) The Common Information Model CIM. POWSYS, vol. 2, pp. 3–48. Springer, Heidelberg (2012)
3. Fine, S., Singer, Y., Tishby, N.: The hierarchical hidden markov model: Analysis and applications. Machine Learning **32**(1), 41–62 (1998)
4. Hashimoto, A., Mori, N., et al.: Smart kitchen: A user centric cooking support system. Proc. of IPMU **8**, 848–854 (2008)
5. Inoue, Y., Minato, S.: An Efficient Method for Indexing All Topological Orders of a Directed Graph. In: Ahn, H.-K., Shin, C.-S. (eds.) ISAAC 2014. LNCS, vol. 8889, pp. 103–114. Springer, Heidelberg (2014)
6. Khoo, C.S., Chan, S., Niu, Y.: Extracting causal knowledge from a medical database using graphical patterns. In: Proc. of the 38th Annual Meeting on Association for Computational Linguistics, pp. 336–343. ACL (2000)
7. Kudo, T., Matsumoto, Y.: Japanese dependency analysis using cascaded chunking. In: Proc. of the 6th Conf. on Natural Language Learning, vol. 20, pp. 1–7 (2002)
8. Kudo, T., Yamamoto, K., Matsumoto, Y.: Applying conditional random fields to japanese morphological analysis. EMNLP **4**, 230–237 (2004)
9. Liao, L., Fox, D., Kautz, H.: Hierarchical conditional random fields for gps-based activity recognition. In: Robotics Research, pp. 487–506. Springer (2007)
10. Mori, S., Maeta, H., et al.: Flow graph corpus from recipe texts. In: Proc. of the 9th International Conf. on Language Resources and Evaluation (2014)
11. Nakadai, K., Lourens, T., et al.: Active audition for humanoid. In: AAAI/IAAI, pp. 832–839 (2000)
12. Sato, T., Kameya, Y.: Parameter learning of logic programs for symbolic-statistical modeling. J. of Artificial Intelligence Research **15**(1), 391–454 (2001)
13. Spriggs, E.H., De La Torre, F., Hebert, M.: Temporal segmentation and activity classification from first-person sensing. In: CVPR Workshops 2009. IEEE Computer Society Conference, pp. 17–24 (2009)
14. Truyen, T.T., Phung, D., et al.: Hierarchical semi-markov conditional random fields for recursive sequential data. In: Advances in Neural Information Processing Systems, pp. 1657–1664 (2009)
15. Yamakata, Y., Imahori, S., Sugiyama, Y., Mori, S., Tanaka, K.: Feature Extraction and Summarization of Recipes Using Flow Graph. In: Jatowt, A., Lim, E.-P., Ding, Y., Miura, A., Tezuka, T., Dias, G., Tanaka, K., Flanagin, A., Dai, B.T. (eds.) SocInfo 2013. LNCS, vol. 8238, pp. 241–254. Springer, Heidelberg (2013)
16. Yamakata, Y., Tsuchimoto, Y., et al.: Cooking ingredient recognition based on the load on a chopping board during cutting. In: 2011 IEEE International Symposium on Multimedia. pp. 381–386 (2011)

A Practical Approach to the Shopping Path Clustering

In-Chul Jung[1], M. Alex Syaekhoni[2], and Young S. Kwon[2](✉)

[1] LG CNS, Seoul, Korea
uhhaha@gmail.com
[2] Department of Industrial and Systems Engineering, Dongguk University, Seoul, Korea
{alexs,yskwon}@dongguk.edu

Abstract. This paper proposes a new clustering approach for customer shopping paths. The approach is based on the Apriori algorithm and LCS (Longest Common Subsequence) algorithms. We devised new similarity and performance measurements for the clustering. In this approach, we do not require data normalization for preprocessing, which leads to an easy and practical application and implementation of the proposed approach. The experiment results show that the proposed approach performs well compared with k-medoids clustering.

Keywords: Clustering · Shopping path · LCS (Longest Common Subsequence)

1 Introduction

Customer behavior pattern is information essential to store managers in order to make decisions and manage stores. To understand customer behavior, customer travel in stores has been tracked. Many technologies have been implemented well in stores to track customer behavior, such as CCTV cameras [4], customer invoice generation [6], and more. In addition, one of the technologies used for this particular purpose is RFID (radio frequency identification).

However, tracking shopper behavior using RFID in real time produces raw data that is difficult to understand. This raw data cannot be used directly because of the complexity of unstructured paths. Therefore, Data Mining approaches, such as the clustering method that groups customer shopping paths, has been utilized in an effort to make such data easier and more useful to interpret. It is our hope that some strategies for different customer groups can be applied by store managers using different treatments based on long-term profitability.

We use the statistical clustering method for this purpose. However, with this method, there are still difficulties computing distance, or similar measures, because every customer has a different shopping path, which means that every shopper has different measurements. With regard to this issue, Larson et al. [3] assumed that each path was recorded with 100% locations. On an RFID reader, each blink represents a percentage of the way through the path in a given store. However, the original information might be lost. Therefore, in order to avoid data loss, we propose a new approach for customer shopping path clustering using LCS (Longest Common Subsequence) [2] and the Apriori algorithm [1]. Furthermore, our clustering approach is easier to implement for

© Springer International Publishing Switzerland 2015
M. Ali et al. (Eds.): IEA/AIE 2015, LNAI 9101, pp. 675–682, 2015.
DOI: 10.1007/978-3-319-19066-2_65

shopping path data because it can cluster such data without normalizing the number of shopping path points (of locations visited by shoppers).

To compare the performance of our proposed method with existing methods, we conducted experiments using the simulated sequence data that represents the shopping path from customers' movements in a grocery store. The results show that our clustering method has better performance for shopping path data compared to the k-medoids clustering method.

2   Shopping Path Clustering

Shopping path clustering has been performed because we assume that shoppers who visit similar product shelves (locations) have similar shopping patterns in a store. This means that the longer the sequence of visited locations between similar shopper paths, the more similar are their shopping properties. In this case, clustering methods are commonly used. Cluster analysis is the task of grouping a set of objects in such a way that objects in the same group (cluster) are more similar to each other than to those in other clusters. However, traditional clustering using the Euclidean measure works only if two compared paths have the same length. This is why all paths have to be recorded in such a way that every path has the same length. However, important original information might be lost. In order to solve this particular problem using the Euclidean distance, we propose a new similarity measurement using LCS. Originally, the LCS problem is to find the longest common subsequence for all sequences in a set of sequences.

In this paper, we regard shopper moving paths as sequences of locations visited by shoppers in the store. We use the RFID system to detect shopper moving locations. That is, the longer the LCS between two paths, the more similar are the paths. Therefore, the degree of similarity among shopping paths can be represented by the LCS length among paths. For example, let us assume that customers 1, 2, 3, and 4 have shopping paths A→B→C→D→E→F, A→B→E→Z→F, A→B→C→F, and A→B→Y→Z→F, respectively. Then, the LCS of customers 1 and 2 is A-B-E-F (length is four), the LCS of customers 1 and 3 is A-B-C-F (length is four), and the LCS of customers 1 and 4 is A-B-F (length is three). The LCS of customers 1 and 2 has the same length as the LCS of customers 1 and 3. This implies that customer 1 has a more similar shopping path to customers 2 and 3, and not to customer 4: LCS (customer 1, customer 2) = LCS (customer 1, customer 3) > LCS (customer 1, customer 4).

In this case, we cannot actually determine the shopping path that is more similar between LCS (customer1, customer2) and LCS (customer1, customer3). For this problem, we develop a new similarity measure, *Sim_LCS*, which is modified to consider the travel distance of each customer precisely.

$$Sim\_LCS(x, y) = \frac{length\_of\_LCS(x, y)}{Length\_of\_x + Length\_of\_y} \tag{1}$$

When we apply the new similarity measure (Eq.1), *Sim_LCS* (customer 1, customer 2) is calculated to be 0.36, and 0.4 for *Sim_LCS* (customer 1, customer 3), as shown in Fig. 1. This implies that the shopping path of customer 1 is more similar to that of customer 3 than to that of customer 2: *Sim_LCS* (customer 1, customer 3) > *Sim_LCS* (customer 1, customer 2) > *Sim_LCS* (customer 1, customer 4).

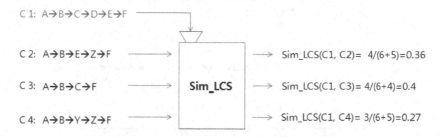

C 1: A→B→C→D→E→F

C 2: A→B→E→Z→F   →   → Sim_LCS(C1, C2)= 4/(6+5)=0.36

C 3: A→B→C→F   →   **Sim_LCS**   → Sim_LCS(C1, C3)= 4/(6+4)=0.4

C 4: A→B→Y→Z→F   →   → Sim_LCS(C1, C4)= 3/(6+5)=0.27

Fig. 1. Proposed similarity measure: *Sim_LCS*

In practice, our proposed approach consists of two steps. First, we determine the possible core sequences of shopper paths as initial clusters using the Apriori algorithm under the assumption that frequent item sets of shopping sequences represent the core sequences of store locations most visited by many shoppers in each cluster. Setting the frequent item sets for the shopping sequence to be the initial reference paths helps to form good clustering. In the second step, by utilizing the initial reference path obtained by the Apriori algorithm that consists of the most overlapped locations among customer paths in the store, our clustering approach attempts to find the shopping path most similar to the initial reference path, and sets it as the initial cluster center. Then, the algorithm continues to form the clustering by finding the shopping paths that are similar to the initial center using the value *Sim_LCS* between the initial center and the remaining shopping paths, until the user-defined threshold value for *Sim_LCS* is satisfied. Fig. 2 illustrates the pseudocode of the proposed clustering algorithm.

Furthermore, the proposed clustering approach works as explained in Fig. 3, which shows that the virtual map has 25 locations that can be visited. Location *1* is the entrance and *21* is the exit. Suppose we have the five shopping paths *C1* = <1 2 3 4 5 10 15 20 25 24 23 22 21>, *C2* = <1 2 3 8 7 6>, *C3* = <1 2 3 4 9 14 13 12 17 18 19 24 23 22 21>, *C4* = <1 2 7 11 16 21>, and *C5* = <1 6 11 16 21>.

Because {*1*}, {*21*} are trivial among the shopping sequences, given that all customers must visit the entrance and exit, we remove them, and thus the first frequent item set found by the Apriori algorithm is {3, 4, 22, 23, 24}. The frequent item set is the initial reference path used as the initial cluster. Then, the algorithm computes the *Sim_LCS* between the initial reference path and the remaining shopping paths. C1 has the largest *Sim_LCS* value at 0.16, which is selected as the cluster center. Subsequently, the algorithm computes *Sim_LCS* between C1 and the remaining shopping paths C2, C3, C4, and C5, and finds that C3 has the largest value, at 0.28, that satisfies the user-defined threshold value of 0.20, and places C3 into the same cluster group. Our algorithm then finds the second frequent item set from the remaining C2, C4, and C5 (not C1 and C3 because they are already grouped in the same cluster). As this process continues, C5 and C4 are assigned to one cluster. C2 is ignored because it does not satisfy the threshold. Finally, two clusters are formed, as shown in Fig. 3.

Function:	INITIAL_CENTER
Input:	Moving Path Trajectories (T), Support (S)
Output:	Initial Center of Moving Path (CT)
1	L = Apriori(T, S) // reference path
2	FOR (list in L) {
3	FP = max(L)
4	}
5	FOR (traj in T) {
6	CT=max(Sim_LCS (FP, traj))//selecting initial center
7	}
8	
9	RETURN CT

Function:	Path _CLUSTERING
Input:	Moving Path Trajectories T={x1, x2, ..., xn}, Support (S), Sim_
Output:	LCS Threshold (D)
	Clusters (Cn)
1	WHILE (CT = INITIAL_CENTER (T, S)) {
2	for (traj in T) {
3	If (Sim_LCS (traj, CT) > D) {
4	Cn <= traj // Insert Cluster group
5	}
6	RETURN Cn

Fig. 2. Pseudocode for proposed clustering approach

3   Experiments and Result

In this section, we compare our method with k-medoids. We conducted experiments using simulated data. For performance comparison, the *ADWC* (Average Distance Within Clusters), *ADBC* (Average Distance Between Clusters), and the Silhouette coefficient were used. ADWC is all possible distances between all objects in one cluster, whereas ADBC is all possible distances between all objects in one cluster, and all objects in another cluster [5]. The Silhouette coefficient is one of the popular approaches for measuring the average of all minimum values of ADWC, and the average of all maximum values of the ADBC.

3.1   Datasets (Simulated Path)

We simulated customer movement shopping paths on a virtual map. The process for creating the movement trajectories is by creating two initial paths that represent two major path trends on the map, and then, by creating new paths repeatedly, those paths change at some points and shapes, randomly. For example, let we assume that the image in Fig. 4 is a virtual store with areas ranging from area 1 (the entrance) to area 91 (the exit). The first movement data <1,2,3,13,23,...,30,...,93,92,91> represents the short pattern used as the benchmark for the customer with the shortest shopping path, and the second movement path <1,2,3,4,5,6,7,...,10,20,30,...,100,..92, 91> represents the longest shopping pattern. We created 200 new paths from these two paths.

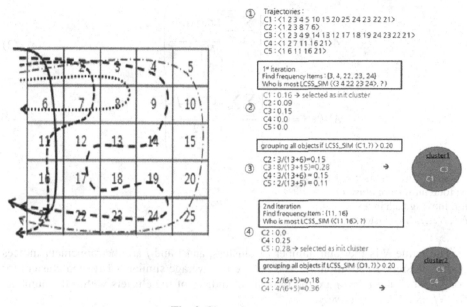

Fig. 3. Clustering process

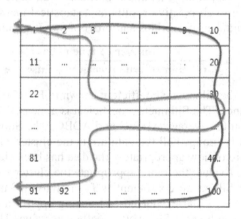

Fig. 4. Two basic shopping paths for simulation dataset

3.2   Measurement Definition

First, we defined *ADWC* and *ADBC* as the clustering performance measures. *ADWC* is an indicator that measures the distance of internal objects for each cluster, and ADBC is an indicator that measures the distance between clusters. This means that a low *ADWC* and a high *ADBC* are an excellent cluster result. *ADWC* and *ADBC* are calculated using Eqs. (2) and (3), respectively.

$$ADWC^* = 0.5 - \left(\frac{\sum_{i=o}^{N} \sum_{j=i+1}^{N} LCSS(i,j)}{\binom{N}{2}} \right) \tag{2}$$

$$ADBC^* = 0.5 - \left(\frac{\sum_{i=o}^{N1} \sum_{j=0}^{N2} LCSS(i,j)}{N1 \times N2} \right) \tag{3}$$

where 0.5: maximum similarity value,
N: total number of clusters
i, j: moving path indexes
$N1, N2,...,Nn$: cluster counts

The value N is the total number of clusters, and i and j are the movement indices within clusters. In addition, $N1$ and $N2$ are the average similarity between clusters for comparison, and i and j are the movement indices of the clusters within the numbers for $N1$ and $N2$, respectively.

Second, we used the precision measure, Eq. (4), against the clustering result. Precision is used as an evaluation benchmark, and this is an indicator that expresses the ratio of true positive to all positive results as a statistical method. We computed the similarity with the number of overlaps and order of moving points.

$$\text{precision} = \frac{\text{number of true positives}}{\text{number of true positives} + \text{false positives}} \tag{4}$$

Third, we used the Silhouette coefficient shown in Eq. (5). Tables 1 and 2 list the clustering performance. The Silhouette coefficient is a measure of how dissimilar two clusters are based on the average ADWC and ADBC. The Silhouette coefficient also represents how tightly grouped all the data are in the cluster. Overall, this coefficient determines a measure of how appropriately the data has been clustered. The higher the Silhouette coefficient value, the more appropriate the cluster has been produced. The range value for the Silhouette coefficient is from -1 to 1. Based on the experiment results (listed in Tables 1 and 2), our new clustering approach produces more appropriate clusters with the highest Silhouette coefficient value. This means that, in this particular data, LCS performs better than the k-medoids.

$$s(x) = \frac{ADBC - ADWC}{\max\{ADBC, ADWC\}} \tag{6}$$

3.3   Comparison Results

Table 1 lists the performance of the cluster results that depend on the changes of the *Sim_LCS* threshold for the proposed clustering algorithm. For our proposed approach (the LCS clustering algorithm), we changed the *Sim_LCS* threshold from 0.1 to 0.35.

Table 1 shows that LCS clustering performed accurately using 0.15 as the threshold, it achieved 0.804 as the highest Silhouette coefficient value, and 100% precision. In addition, our proposed method performs better compared with k-medoids clustering, which obtained 0.786 on the Silhouette coefficient, and has 0.99% precision, as indicated in Table 2.

Table 1. LCS clustering performance using simulated data

Threshold	Precision (%)	ADWC* (range: 0 – 0.5)	ADBC (range: 0 – 0.5)	Silhouette Coefficient (range: 0 – 1)
0.1	0.91	0.154	0.45	0.657
0.15	**100**	0.098	0.50	**0.804**
0.2	0.99	0.103	0.49	0.792
0.25	0.90	0.162	0.44	0.631
0.3	0.57	0.297	0.39	0.251
0.35	0.26	0.253	0.41	0.390

Table 2. k-medoids clustering performance using simulated data

k-cluster	Precision (%)	ADWC (Euclidean distance)	ADBC (Euclidean distance)	Silhouette Coefficient (range: 0 – 1)
2	**0.99**	7,836	36,685	**0.786**
3	0.77	7,575	27,571	0.725
5	0.82	5,509	22,930	0.760

In addition, our clustering approach can conduct direct clustering of the shopping path data without requiring normalized pre-processing, and it properly predefined the cluster number by selecting the proper initial value. Therefore, we determined that the proposed clustering algorithm might be easier to perform and more effectively used than k-medoids in terms of the Silhouette coefficient measurement.

4   Conclusion

We proposed a new clustering approach (called LCS clustering) for customer shopping that works directly on sequence data without normalizing the number of shopping path points (locations visited by shoppers in the store) as proposed by previous papers. This type of normalization can suffer from loss of original information. Therefore, in order to avoid data loss, we proposed a new approach that is easier for practical application and implementation. Furthermore, our experiment results showed that LCS clustering performs well compared to k-medoids clustering.

Using this approach, we expect to perform well and produce better results on real shopping path data. Furthermore, the information regarding groups of customer movement can be extremely useful for helping store managers make decisions, such as improving display layout, right-on-target promotions, etc.

Furthermore, we plan to record real shopping paths in grocery stores using RFID (radio frequency identification) technology, and apply LCS clustering to it. Then, we will analyze the trends of customer movement in the stores.

References

1. Agrawal, R., Srikant, R.: Fast Algorithms for Mining Association Rules. In: Proceedings of the VLDB Conference, Santiago, Chile. Expanded version available as IBM Research Report RJ9839 (1994)
2. Hirschberg, D.S.: Algorithms for the Longest Common Subsequence Problem. Journal of ACM **24**(4), 664–675 (1977)
3. Larson, J.S., Bradlow, E.T., Fader, P.S.: An Exploratory Look at Supermarket Shopping Paths. J. of Research in Marketing **22**, 359–414 (2005)
4. Newman, A.J., Yu, D.K.C., Oulton, D.P.: New Insights into Retail Space and Format Planning from Customer-tracking Data. Journal of Retailing and Customer Service **9**(5), 254–258 (2002)
5. Shmueli, G., Patel, N.R., Bruce, P.C.: Data Mining for Business Intelligence. John Wiley & Sons, Inc. (2007)
6. Pandit, A., Talreja, J., Agarwal, M., Prasad, D., Baheti, S., Khalsa, G.: Intelligent Recommender System using Shopper's Path and Purchase Analysis. In: Proceedings of International Conference on Computational Intelligence and Communication Networks, pp. 597–602 (2010)

iDianNao: Recommending Volunteer Opportunities to Older Adults

Wen-Chieh Fang$^{(\boxtimes)}$, Pei-Ching Yang, Meng-Che Hsieh,
and Jung-Hsien Chiang

Department of Computer Science and Information Engineering,
National Cheng Kung University, Tainan City, Taiwan
wcf123@ntu.edu.tw, {yang.peiching,kamei1120}@iir.csie.ncku.edu.tw,
jchiang@mail.ncku.edu.tw

Abstract. Older volunteers can provide valuable, needed services for communities and organizations. However, older people who wish to help often have difficulty finding the right volunteer opportunities. In this paper, we present an online system that exploits recommender system techniques to provide volunteer opportunities for older people. We rank the opportunities according to a weight sum of four scores, which are neighbor score, click score, time score, and area score.

For each opportunity, we first select the neighborhood of the user by considering preference similarity for the opportunity and the organization as well as the social interactions involved. We then compute the neighbor score as the ratio of neighbors choosing the target opportunity to all neighbors. The click score is calculated as the rate that an user clicks on the interested opportunity pages. The time and area scores indicate the availability of time and the ability to travel, respectively. For the particular volunteer opportunity recommendations, we further consider the coverage issue by adding a deadline factor when computing the opportunity score. At the end of the paper, we report the system implementation and provide a summary.

Keywords: Recommender system · Volunteerism · Social interaction

1 Introduction

The many benefits of volunteering in the case of older adults have been reported in many studies, and include increased well-being, more life satisfaction, better physical and psychological health, lower mortality risk, and so on [2]. Some surveys have reported that more and more older people have interest in volunteering [10] [14].

The study in [14] further reports that older people have a difficult time finding volunteer opportunities that interest them. It is reported that older users are considerably less likely than younger users to find an opportunity of interest. Many older non-volunteers do not volunteer because they have not found the right opportunity. Another article [6] also reports that many older adults

© Springer International Publishing Switzerland 2015
M. Ali et al. (Eds.): IEA/AIE 2015, LNAI 9101, pp. 683–691, 2015.
DOI: 10.1007/978-3-319-19066-2_66

encounter and highly value programs such as workforce transition centers and specially designed web sites that provide links to volunteer opportunities by connecting to existing portals. Not only do older people encounter difficulty, but also non-profit organizations still report difficulty finding the volunteers they need [14]. Although there are many web-sites that provide volunteer opportunities, such as *WorldWide Helpers* and *VolunteerMatch*, a web service specific to older people for volunteering is still unavailable.

Therefore, building a web-site or web service that can recommend appropriate volunteer opportunities fitting older people's needs and interests is a promising approach to the above-mentioned problem. In this paper, we propose a method that exploits the recommender system techniques to provide older users with volunteer opportunities.

The rest of this paper is organized as follows: In Section 2, we provide some background knowledge about recommender systems and related work relevant to our work. In Section 3, we present our recommender system for volunteering. In Section 5, we summarize this paper and outline some significant future research directions.

2   Related Work

Recommender systems have been used in a wide variety of applications, such as for recommending movies, books, music, and news. There are three main categories of recommendation approaches: content-based, collaborative filtering, and hybrid recommendation approaches [1] [13]. We refer readers who are interested in recommender systems to books such as [7] or [13].

Social recommender systems have obtained much attention in recent years. For example, the authors in [9] investigated the relationship between self-defined social connections and interest similarity in a collaborative tagging system *CiteU-Like*. Interest similarity is measured by the similarity of items and the meta-data they share and the tags they use. They found that traditional item-level similarity may be a less reliable way to find similar peers in social bookmarking systems. The users connected by social networks exhibit significantly higher similarity than non-connected users.

Morse et. al [12] developed an online system, doGooder, to foster volunteer communities that serve the homeless. doGooder connects volunteers with opportunities and develops social ties among volunteers and service organizations. Usability testing points out that doGooder successfully helps agencies to recruit, retain, and organize volunteers.

Kane and Klasnja [8] presented mobile social software intended to motivate users to volunteer and to help users find volunteer opportunities. In the study, the authors found social connections to be a strong motivation to volunteer. The authors interviewed nine volunteers about their volunteer work in order to investigate how technology might support and influence volunteering.

Although the systems built in [12] and [8] are used for volunteering, it is not clear whether recommendation system techniques are explicitly applied in these systems.

Chen et. al. [3] built a location-based volunteer matching system for volunteers and service organizations. The recommendation algorithm computes user similarity according to the information obtained from questionnaires. However, the details of the algorithm are unclear, and the method seems too basic.

3   Recommender Approach

In the recommender system, 'item' is the general term used to denote what the system recommends to users. Therefore, in this paper, 'item' denotes 'volunteer opportunity.' There are some particular constraints or settings in our volunteer opportunity recommender system. Each item needs a specific number of volunteers, and the item is withdrawn when sufficient volunteers are recruited. Each item has a deadline. We want to recommend as great of a percentage of all items as possible. We count the amounts such as the number of likes and the number of wall posts weekly.

Our system provides a top-N list of recommendation to users according to the item score for the items. We define an item score function be a weighted sum of four scores, which include a *neighbor score*, a *click score*, a *time score*, and an *area score*, for the item i as follows:

$$Score_i = \gamma_i(\lambda_n S_n + \lambda_c S_c + \lambda_t S_t + \lambda_a S_a), \tag{1}$$

where γ_i is the deadline parameter, and λ_n, λ_c, λ_t, and λ_a are the corresponding weights of the scores. $0 < \lambda_n, \lambda_c, \lambda_t, \lambda_a < 1$, and $\lambda_n + \lambda_c + \lambda_t + \lambda_a = 1$.

Our algorithm aims to make items be chosen as many times as possible before their *deadlines*. We set different values of γ_i to indicate whether sufficient users choose the item. For the item that no users choose, the closer the item is to the deadline, the larger the value of γ_i. We let γ_i roughly follow the function $\gamma_i(t) = 1 + exp(-\alpha t)$ such that the γ_i value increases when the current time reaches the time limit. α is the decay constant.

$$\gamma_i(\Delta t) \approx \begin{cases} 1 + exp(-\alpha \Delta t) & \text{if no users choose i} \\ 1 & \text{otherwise.} \end{cases}$$

Here α is the decay constant.

In the following sections, we describe each score function in detail.

3.1   Neighbor Score

According to [4], similarity leads to social interaction, and then social interaction leads to further similarity. In the real world, people with similar interests tend to be friends or are more likely to have social connections. This phenomenon is based on *the homophily principle*: similarity breeds connection [11] [15]. Therefore, we use both preference similarity and social interaction to find the user's nearest neighbors that can be used to predict the user's selection of volunteer opportunities.

User Profile. We define item features as the salient attributes of the item, such as teaching, general affairs, labor and so on. There are nine item features in this paper: accompanying, general affairs, teaching, delivering goods, guide, labor, tutor, activity, and other. The organizational features are defined as the categories or types of service organizations. We have seven organizational features: culture, benevolence, health, elderliness, children, gender, and other. We assign a binary value to each feature value of a volunteer opportunity in advance.

According to the questionnaire and the historical volunteering data, we represent each user as a 16-dimensional vector of the item features and the organizational features. In the case of a user u, its feature vector $\mathbf{x}_u = [x_1^s, x_2^s, \cdots, x_9^s, x_1^p, x_2^p, \cdots, x_7^p]$.

When new users sign up for our system, they are asked to fill out a questionnaire about their preferences related to the community services. They use binary values to represent their initial preferences for the item features and the organizational features. 1 indicates that the user has interest in the feature while 0 means the opposite. These values of all the features are used for the initial feature vector of a user.

When user u goes on our system, and N historical item data instances are recorded, we accumulate the values of the item features and the organizational features of all the N items and calculate the x_α^s and x_β^p of the item feature α and the organization feature β respectively, as follows:

$$x_\alpha^s = \begin{cases} \frac{n_\alpha^s + 1}{N+1} & \text{if initial value} = 1 \\ \frac{n_\alpha^s}{N+1} & \text{otherwise,} \end{cases}$$

$$x_\beta^p = \begin{cases} \frac{n_\beta^p + 1}{N+1} & \text{if initial value} = 1 \\ \frac{n_\beta^p}{N+1} & \text{otherwise.} \end{cases}$$

Here n_α^s and n_β^p are the sum of the item feature α values and organizational feature β values, respectively, from the N historical item data.

Since each user is represented as a feature vector, the user similarity between two users u and v can be computed using the traditional similarity computation, such as cosine similarity or Pearson correlation similarity. We denote this similarity as *preference similarity* S_{uv}^p.

Social Interaction. A number of people start a volunteer activity because a friend or family member is already involved in the activity [8] [12] [14]. Therefore we consider social connections in our system. We consider three social behaviors: follow, 'like' and wall post.

First we apply a social relationship similar to the *following* function in *Twitter* [5] or the *watching* function in *CiteULike* [9]. This social relation refers to linking to another user and receiving the linked user's information and posts after that. We call the user creating such a link the *follower*, and the linked user is known as the *followee*. Note that the relationship is unilateral and that it does not require mutual agreement to be connected or obtain information. All objects assembled by the followee are automatically shown to the follower as a part of the follower's 'watchlist'.

Let f_{uv} be a variable that denotes whether user u follows the news of user v so that the website content of user v directly appears on the personal website of user u. If users u and v follow each other, $f_{uv} = 2$. If user u follows user v, but user v does not follow user u, $f_{uv} = 1$. If there is no follower relationship between user u and v, $f_{uv} = 0$.

For 'like' behavior and wall post behavior, we define a numerical statistic that is intended to reflect the extent of these social behaviors. Given that b is a social behavior such as 'like' behavior or wall post behavior from user u, we define *social frequency* sf:

$$sf(b, v) = n_b^v, \tag{2}$$

where n_b^v is the number of times the social behavior b that user u applies on user v.

We define *inverse user frequency* iuf:

$$iuf(b, v, U) = \log \frac{|U|}{1 + |\{v' \in U : sf(b, v') \geq sf(b, v)\}|}, \tag{3}$$

where U is the user set, and where $\{v' \in U : sf(b, v') \geq sf(b, v)\}$ indicates the set of the users on whom u applies the social behavior b, for which the social frequency on v' is greater or equal to the social frequency on v. If u is a new user and does not engage in the social behavior b, this will lead to a division-by-zero. Therefore, we adjust the denominator to $1 + |\{v' \in U : sf(b, v') \geq sf(b, v)\}|$. The inverse user frequency is a measure of how much particularity there is of v for u. In other words, if the social frequency of u on v is larger than that of u on others, we reasonably consider that u and v have especially rich social interactions.

Then the like variable is calculated as follows:

$$l_{uv} = sf(l, v) \times iuf(l, v, U), \tag{4}$$

where l indicates the 'like' behavior.

Similarly, we calculate the wall post variable as follows:

$$p_{uv} = sf(w, v) \times iuf(w, v, U), \tag{5}$$

where w indicates the wall posting behavior.

We define *social similarity* S_{uv}^s between user u and v as the weighted sum of the above-mentioned three variables,

$$S_{uv}^s = \omega_f f_{uv} + \omega_l l_{uv} + \omega_p p_{uv}, \tag{6}$$

where ω_f, ω_l, and ω_p are the relative weights of importance of the variables f_{uv}, l_{uv}, and p_{uv}, respectively. $0 < \omega_f, \omega_l, \omega_p < 1$, and $\omega_f + \omega_l + \omega_p = 1$.

If a user clicks the 'like' button associated with an object, this indicates that the user is highly interested in the object or the content of another user. Therefore, frequent 'like' behavior means frequent positive social interactions.

Here 'positive' indicates that the behavior is highly associated with high interest similarity. Because the post content may contain negative issues, for example, an argument, a quarrel or a disagreement, frequent post behavior does not imply high interest similarity. That is, the impact of wall posting on social relationships depends heavily on the post content. However 'like' behavior is a more positive indicator for social connection than wall posting behavior. Therefore, we always set $\omega_l > \omega_p$.

Score Computation. We combine preference similarity with social similarity to obtain the *user similarity* of user u with respect to user v as follows:

$$S_{uv}^u = S_{uv}^p + S_{uv}^s. \tag{7}$$

We use user similarity to obtain the subset $U' \subseteq U$ of the K users most similar to user u.

Among the K users, if there are n_n users that select item i, we compute the neighbor score S_n:

$$S_n = \frac{n_n}{K}. \tag{8}$$

3.2   Click Score

The score S_c considers the relative weight of the number of times a click is made on the item i page by user u compared to the number of clicks made on all the item pages by user u.

$$S_c = \frac{n_c}{N_c}, \tag{9}$$

where n_c is the number of times a click is made on the item i page by user u. N_c is total number of clicks made on the item pages by user u.

3.3   Time Score

We simply divide each day into three time periods: morning, afternoon, and evening. Therefore there are 21 time periods in each week. We accumulate the total number of all time periods that the historical items occupy and denote the number as N_t. We can calculate the probability P_k of each time period k as the total number of the items for which the time periods include k divided by N_t. In other words,

$$P_k = \frac{|T|}{N_t}, \tag{10}$$

where T is the set of items for which time periods include k.

For an item i, if its occupied time periods form a set Γ, we compute the time score as follows:

$$S_t = \min\{P_k : k \in \Gamma\}, \tag{11}$$

The reason why we choose the minimum of probabilities is that user u is likely unavailable at the time periods that the historical items did not occupy. We reasonably infer that user u is still likely unavailable at these time periods if a new item occupies these time periods. However, each user is assumed to participate *fully* in a volunteer opportunity (item). If the user might be unavailable at a specific time period of the opportunity, we consider that he cannot take full part in this opportunity. Therefore the minimum of probabilities is chosen as the time score.

We use a simple example to illustrate the computation of the time score. For simplicity, we only consider the weekend. Given that a user u has 5 historical items (A, B, C, D, E), the time periods of these items are shown in Fig 1. The total number of the time periods N_t that these 5 items occupy is 10. Assuming that the new item i occupies time periods 2 and 3, we calculate their time probabilities using equation 10. Thus, the set of items for which time periods include 2 is $T = \{A, B\}$, and we obtain $P_2 = \frac{2}{10}$. Similarly, we obtain $P_3 = \frac{1}{10}$. Finally we calculate the time score $S_t = \min\{P_k : k \in \{2, 3\}\} = P_3 = \frac{1}{10}$.

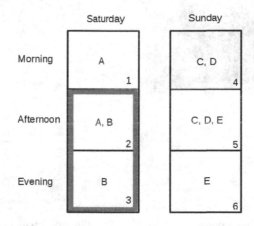

Fig. 1. An example for computation of time score

3.4   Area Score

If the item i is at area a, we compute the area score as follows:

$$S_a = \frac{n_a}{N},\tag{12}$$

where n_a is the number of the historical items at area a, and N is the total number of historical items.

4   System Implementation

We develop an online system, iDianNao, that is intended to motivate users to volunteer and to help users find volunteering opportunities. Fig. 2a shows the iDianNao homepage. Users can log in and get opportunity recommendations from our algorithm.

The system works as follows: The recommender system has a task that is executed weekly to pull volunteer opportunity data from multiple service organizations and then to pass it on to our server. Users can browse all the volunteer opportunities in our system. See Fig. 2b. The recommendation algorithm will collect the information related to users' social interactions and then upload it to our server. Whenever a user logs in to our system, the server will perform the opportunity recommendation and will also provide a ranking list of opportunity recommendations for him/her.

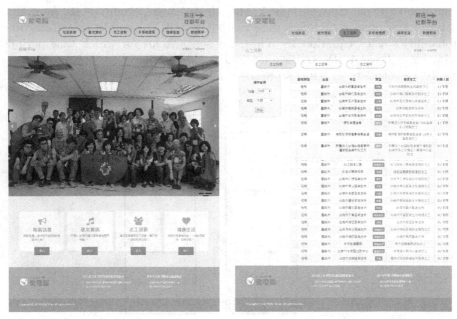

(a) The homepage   (b) The volunteer opportunity page

Fig. 2. The social network platform for older adults

5   Summary

In this paper, we present an online system, iDianNao, that helps connect individuals with volunteering opportunities. We describe the recommendation approach in detail and explain the score functions completely.

In the future, we plan to conduct usability testing and collect user data with several older participants and then analyze the data in order to evaluate our system. We want to investigate whether social interaction is a factor that influences neighborhood selection and whether the deadline factor impacts recommendation performance.

References

1. Adomavicius, G., Tuzhilin, A.: Toward the next generation of recommender systems: A survey of the state-of-the-art and possible extensions. IEEE Transactions on Knowledge and Data Engineering **17**(6), 734–749 (2005)
2. Brown, J., li Chen, S., LindaMefford, B.A., Callen, B.: PollyMcArthur: Becoming an older volunteer: A grounded theory study. Nursing Research and Practice, 1–8 (2011)
3. Chen, W.C., Cheng, Y.M., Sandnes, F.E., Lee, C.L.: Finding suitable candidates: The design of a mobile volunteering matching system. In: Proceedings of the 14th International Conference on Human-Computer Interaction, pp. 21–29 (2011)
4. Crandall, D., Cosley, D., Huttenlocher, D., Kleinberg, J., Suri, S.: Feedback effects between similarity and social influence in online communities. In: Proceedings of the 14th ACM SIGKDD International Conference on Knowledge Discovery and Data Mining (KDD 2008), pp. 160–168 (2008)
5. Hannon, J., Bennett, M., Smyth, B.: Recommending twitter users to follow using content and collaborative filtering approaches. In: Proceedings of the 4th ACM Conference on Recommender Systems (RecSys 2010), pp. 26–30 (2010)
6. Hoffman, L., Andrew, E.: Maximizing the potential of older adults: Benefits to state economies and individual well-being. NGA Center for Best Practices (2010). http://www.nga.org/files/live/sites/NGA/files/pdf/1004OLDERADULTS.PDF
7. Jannach, D., Zanker, M., Felfernig, A., Friedrich, G.: Recommender Systems: An Introduction. Cambridge University Press, 1st edn. (2010)
8. Kane, S.K., Klasnja, P.V.: Supporting volunteer activities with mobile social software. In: CHI 2009 Extended Abstracts on Human Factors in Computing Systems, pp. 4567–4572 (2009)
9. Lee, D.H., Brusilovsky, P.: Social networks and interest similarity: the case of citeulike. In: Proceedings of the 21st ACM Conference on Hypertext and Hypermedia (HT 2010), pp. 151–156 (2010)
10. Ma, X., Ono, A.: Determining factors in middle-aged and older persons' participation in volunteer activity and willingness to participate. Japan Labor Review **10**(4), 90–119 (2013)
11. McPherson, M., Smith-Lovin, L., Cook, J.M.: Birds of a feather: Homophily in social networks. Annual Review of Sociology, 415–444 (2001)
12. Morse, J., Cerretani, J., Halai, S., Laing, J., Perez, M.: dogooder: fostering volunteer communities to serve the homeless. In: CHI 2008 Extended Abstracts on Human Factors in Computing Systems, pp. 3843–3848 (2008)
13. Ricci, F., Rokach, L., Shapira, B., Kantor, P.B. (eds.): Recommender Systems Handbook. Springer (2011)
14. VolunteerMatch: Great expectations: Boomers and the future of volunteering (2007). http://cdn.volunteermatch.org/www/nonprofits/resources/greatexpectations/GreatExpectations_FullReport.pdf
15. Yang, S.H., Long, B., Smola, A., Sadagopan, N., Zheng, Z., Zha, H.: Like like alike: joint friendship and interest propagation in social networks. In: Proceedings of the 20th International Conference on World Wide Web (WWW 2011), pp. 537–546 (2011)

Particle Filter-Based Approach to Estimate Remaining Useful Life for Predictive Maintenance

Chunsheng Yang[1(✉)], Qingfeng Lou[2], Jie Liu[2], Hongyu Gou[1], and Yun Bai[3]

[1] National Research Council Canada, Ottawa, ON, Canada
Chunsheng.Yang@nrc.gc.ca
[2] Departments of Mechanical and Aerospace Engineering, Carleton University,
Ottawa, ON, Canada
[3] School of Computing, Engineering and Mathematics, University of Western Sydney,
South Penrith, NSW, Australia

Abstract. Estimation of remaining useful life (RUL) plays a vital role in performing predictive maintenance for complex systems today. However, it still remains a challenge. To address this issue, we propose a Particle filter (PF)-based method to estimate remaining useful life for predictive maintenance by employing PF technique to update the nonlinear predictive models for forecasting system states. In particular, we applied PF techniques to estimate remaining useful life by integrating data-driven modeling techniques in order to effectively perform predictive maintenance. After introducing the PF-based algorithm, the paper presents the implementation along with the experimental results through a case study of Auxiliary Power Unit (APU) starter prognostics. The results demonstrated that the developed method is useful for estimating RUL for predictive maintenance.

Keywords: Particle Filter (PF) · Remaining Useful Life (RUL) · Predictive Maintenance (PM) · Predictive model · APU starter prognostics

1 Introduction

Predictive maintenance (PM) is an emerging technology that recommends maintenance decisions based on the information collected through system or component monitoring (or system state estimation) and equipment failure prognostics (or system state forecasting). It demands a predictive model which is able to predict the failure before it occurs and estimate the remaining use life (RUL) of a component or subsystem, in which RUL estimation still remains as the least mature element in real-world applications [1]. Prognostics or RUL estimation entail the use of the current and previous system states (or observations). Reliable forecast information such as RUL estimation is useful and necessary to perform predictive maintenance in advance and provide an alarm before faults reach critical levels so as to prevent system performance degradation, malfunction, or even catastrophic failures [2].

In general, PM can be performed using either data-driven methods or physics-based approaches. Data-driven methods use pattern recognition and machine learning

© Her Majesty the Queen in Right of Canada 2015
M. Ali et al. (Eds.): IEA/AIE 2015, LNAI 9101, pp. 692–701, 2015.
DOI: 10.1007/978-3-319-19066-2_67

techniques to detect changes in system states [3, 4]. It relies on past patterns of the degradation of similar systems to project future system states; their forecasting accuracy depends on not only the quantity but also the quality of system history data, which could be a challenging task in many real applications [2, 5]. Another principal disadvantage of data-driven methods is that the prognostic reasoning process is usually opaque to users [6]; consequently, they sometimes are not suitable for some applications where forecast reasoning transparency is required. Physics-based approaches typically involve building models (or mathematical functions) to describe the physics of the system states and failure modes; they incorporate physical understanding of the system into the estimation of system state and/or RUL [7-9]. Physics-based approaches, however, may not be suitable for some applications where the physical parameters and fault modes may vary under different operation conditions [10]. On one hand, it is usually difficult to tune the derived models *in situ* to accommodate time-varying system dynamics. On the other hand, physics-based approaches cannot be used for complex systems whose internal state variables are inaccessible (or hard) to direct measurement using general sensors. In this case, inference has to be made from indirect measurements using techniques such as particle filtering (PF). Recently the PF-based approaches have been widely used for prognostic applications [11-15], in which the PF is employed to update the nonlinear prediction model and the identified model is applied for forecasting system states. It is proven that FP-based approach, as a Sequential Monte Carlo (SMC) statistic method [16, 17], is effective for addressing the issues described above. In this work, we apply the PF method to estimate RUL for Auxiliary Power Unit (APU) Starter prognostics. This paper presents the PF-based methods along with the experimental results from APU Starter prognostic application.

The rest of this paper is organized as follows. Section 2 briefly describes background of PF technique; Section 3 presents the algorithm and implementation of PF based method; Section 4 provides some experimental results from the case study. Section 5 discusses the results and future work. The final section concludes the paper.

2   Background of Particle Filter

In forecasting the system state, if internal state variables are inaccessible (or hard) to direct measurement using general sensors, inference has to be made from indirect measurements. Bayesian learning provides a rigorous framework for resolving this issue. Given a general discrete-time state estimation problem, the unobservable state vector $X_k \in R^n$ evolves according to the following system model

$$X_k = f(X_{k-1}) + w_k \tag{1}$$

where $f : R^n \to R^n$ is the system state transition function and $w_k \in R^n$ is a noise whose known distribution is independent of time. At each discrete time instant, an observation (or measurement) $Y_k \in R^p$ becomes available. This observation is related to the unobservable state vector via the observation equation

$$Y_k = h(X_k) + v_k \tag{2}$$

where $h: R^n \rightarrow R^p$ is the measurement function and $v_k \in R^p$ is another noise whose known distribution is independent of the system noise and time. The Bayesian learning approach to system state estimation is to recursively estimate the probability density function (*pdf*) of the unobservable state X_k based on a sequence of noisy measurements $Y_{1:k}$, $k = 1, ..., K$. Assume that X_k has an initial density $p(X_0)$ and the probability transition density is represented by $p(X_k | X_{k-1})$. The inference of the probability of the states X_k relies on the marginal filtering density $p(X_k | Y_{1:k})$. Suppose that the density $p(X_{k-1} | Y_{k-1})$ is available at step k-1. The prior density of the state at step k can then be estimated via the transition density $p(X_k | X_{k-1})$,

$$p(X_k | Y_{1:k-1}) = \int p(X_k | X_{k-1}) p(X_{k-1} | Y_{1:k-1}) \, dX_{k-1} \tag{3}$$

Correspondingly, the marginal filtering density is computed via the Bayes' theorem,

$$p(X_k | Y_{1:k}) = \frac{p(Y_k | X_k) p(X_k | Y_{1:k-1})}{p(Y_k | Y_{1:k-1})} \tag{4}$$

where the normalizing constant is determined by

$$p(Y_k | Y_{1:k-1}) = \int p(Y_k | X_k) p(X_k | Y_{1:k-1}) \, dX_k \tag{5}$$

Equations (3)-(5) constitute the formal solution to the Bayesian recursive state estimation problem. If the system is linear with Gaussian noise, the above method simplifies to the Kalman filter. For nonlinear/non-Gaussian systems, there are no closed-form solutions and thus numerical approximations are usually employed.

The PF or so-called sequential important sampling (SIS), is a technique for implementing the recursive Bayesian filtering via Monte Carlo simulations, whereby the posterior density function $p(X_k | Y_{1:k})$ is represented by a set of random samples (particles) $x_k^i (i = 1,2, ..., N)$ and their associated weights $w_k^i (i = 1,2, ..., N)$.

$$p(x_k | Y_{1:k}) \approx \sum_{i=1}^{N} w_k^i \delta(x_k - x_k^i), \quad \sum_{i=1}^{N} w_k^i = 1 \tag{6}$$

The w_k^i, normally known as importance weight, is the approximation of the probability density of the corresponding particle. In a nonlinear/non-Gaussian system where the state's distribution cannot be analytically described, the w_k^i of a dynamic set of particles can be recursively updated through Equation 8.

$$w_k^i \propto w_{k-1}^i \frac{p(y_k | x_k^i) p(x_k^i | x_{k-1}^i)}{q(x_k^i | x_{k-1}^i, y_k)} \tag{7}$$

where $q(x_k^i | x_{k-1}^i, y_k)$ is a proposal function called *importance density function*. There are various ways of estimating the importance density function. One common way is to select $q(x_k^i | x_{k-1}^i, y_k) = p(x_k^i | x_{k-1}^i)$ so that

$$w_k^i \propto w_{k-1}^i p(y_k | x_k^i) \tag{8}$$

3   PF Implementation: A Case Study of APU Prognostics

This section demonstrates the implementation of PF-based method for estimating remaining useful life through a case study, APU starter prognostics. First some background of APU and operational data are briefed, and then an algorithm of PF implementation is introduced.

3.1   APU and APU Data

The APU engines on commercial aircrafts are mostly used at the gates. They provide electrical power and air conditioning in the cabin prior to the starting of the main engines and also supply the compressed air required to start the main engines when the aircraft is ready to leave the gate. APU is highly reliable but they occasionally fail to start due to failures of components such as the Starter Motor. APU starter is one of the most crucial components of APU. During the starting process, the starter accelerates APU to a high rotational speed to provide sufficient air compression for self-sustaining operation. When the starter performance gradually degrades and its output power decreases, either the APU combustion temperature or the surge risk will increase significantly. These consequences will then greatly shorten the whole APU life and even result in an immediate thermal damage. Thus the APU starter degradation can result in unnecessary economic losses and impair the safety of airline operation. When Starter fails, additional equipment such as generators and compressors must be used to deliver the functionalities that are otherwise provided by the APU. The uses of such external devices incur significant costs and may even lead to a delay or a flight cancellation. Accordingly, airlines are very much interested in monitoring the health of the APU and improving the maintenance.

For this study, we considered the data produced by a fleet of over 100 commercial aircraft over a period of 10 years. Only ACARS (Aircraft Communications Addressing and Reporting System) APU starting reports were made available. The data consists of operational data (sensor data) and maintenance data. The maintenance data contains reports on the replacements of many components which contributed the different failure modes. Operational data are collected from sensors installed at strategic locations in the APU which collect data at various phases of operation (e.g., starting of the APU, enabling of the air-conditioning, and starting of the main engines). The collected data for each APU operation cycle, there are existing six variables related to APU performance: ambient air temperature (T_1), ambient air pressure (P_1), peak value of exhaust gas temperature in starting process (EGT_{peak}), rotational speed at the moment of EGT_{peak} occurrence (N_{peak}), time duration of starting process (t_{start}), exhaust gas temperature when air conditioning is enable after starting with 100% N (EGT_{stable}). There are 3 parameters related to starting cycles: APU serial number (S_n), cumulative count of APU operating hours (h_{op}), and cumulative count of starting cycles(cyc). In this work, in order to find out remaining useful life (cycle), we define a remaining useful cycle (RUC) as the difference of cyc_0 and cyc. cyc_0 is the cycle count when a failure happened and a repair was undertaken. When RUC is equal to zero (0), it means that APU failed and repair is needed. RUC will be used in PF prognostic implementation in the following.

3.2   APU Data Correction

The APU data collected in operation covers a wide range of ambient temperatures from -20° to 40° and ambient pressures relevant to the airport elevations from sea level to 3557ft. Since the ambient conditions have a significant impact on gas turbine engine performance, making the engine parameters comparable requires a correction from the actual ambient conditions to the sea level condition of *international standard atmosphere* (ISA). To improve the data quality, the data correction is performed based on the March number similarity from gas turbine engine theory. One main parameter, engine speed (N_{peak}) related APU performance is corrected using Equation 9.

$$\text{NP} = N_c = \frac{N_{peak}}{\Theta^{a_N}} \tag{9}$$

Here, *NP* stands for the APU rotational speed corresponding to EGT peak values after correction. N_{peak} is the data collected from aircraft under various environmental temperatures. The empirical exponent a_N is normally determined by running a calibrated thermodynamic computer model provided by engine manufacturers.

3.3   Implementation

This section presents an implementation of PF-based prognostics algorithm for APU starter. As we mentioned, one key parameter related to APU starter degradation is *NP* from data correction. In order to implement PF techniques for estimating RUL/RUC for APU Starter, we take *NP* as an example to demonstrate the implementation. In terms of the statistic result of *NP* parameter, the moving average $\mu_{X_{RUC}}$ and the moving standard deviation $\sigma_{X_{RUC}}$ are relatively stable in the normal phase, but decrease dramatically in the degraded phase. In the normal operation phase, *NP* measurements satisfy a stationary Gaussian $\mathcal{N}(\mu_{nor}, \sigma_{nor}^2)$. The starter is healthy in this phase, and this healthy state is indicated by the starter signal which is a relative constant value equivalent to μ_{nor}. Meanwhile, the noise signal is a stationary white noise with variance of σ_{nor}^2. In the degraded phase, *NP* measurements satisfy a non-stationary distribution that cannot be analytically described. The starter is experiencing degradation in this phase. Meanwhile, the noise signal is a non-stationary white noise with a variance that varies with the degradation level of starter.

Therefore, we can apply PF-based method to filter out the white noise and identify the degradation trend. To this end, we developed APU *NP* states estimation models. These models are as follows:

$$\overline{NP}_k: \quad x_{1_k} = x_{1_{k-1}} \left(\frac{x_{3_k}}{x_{3_{k-1}}} \right) \exp\left[x_{2_k} (RUC_k - RUC_{k-1}) \right] \tag{10}$$

$$\lambda_k: \qquad x_{2_k} = x_{2_{k-1}} + \omega_{2_k} \tag{11}$$

$$C_k: \qquad x_{3_k} = x_{3_{k-1}} + \omega_{3_k} \tag{12}$$

$$NP_k: \qquad y_k = x_{1_k} + v_k \tag{13}$$

where the subscript k represents the kth time step and RUC_k represents the starting cycle in this kth time step. There are three system states, $\overline{NP}$, λ, C, and one measurement, NP, in this system state model. These states and measurement are also denoted as x_1, x_2, x_3 and y respectively. ω_2 and ω_3 are independent Gaussian white noise processes, the v is approximate by the standard deviation of RUC in the collected dataset.

Table 1. The implementation algorithm

For $i = 1:N$ // Generate N particles from initial data

$$
\begin{cases}
x_{1_0}^{i+} = \mu_{nor} \\
x_{2_0}^{i} \sim \mathcal{N}\left(0, (\omega_{2_k})^2\right) \\
x_{3_0}^{i+} \sim \mathcal{N}\left(\mu_{nor}, (\omega_{3_k})^2\right)
\end{cases}
$$

End for {i}

For $k = 1$: size of RUC   // Obtain a priori particle

 For $i = 1:N$ // Compute importance weights w_i for each particle

$$
\begin{cases}
x_{2_k}^{i-} = x_{2_{k-1}}^{i+} + \omega_{2_k} \\
x_{3_k}^{i-} = x_{3_{k-1}}^{i+} + \omega_{3_k} \\
x_{1_k}^{i-} = x_{1_{k-1}}^{i+} \left(\dfrac{x_{3_k}^{i-}}{x_{3_k}^{i+}}\right) exp\left[x_{2_k}^{i-}(RUC_k - RUC_{k-1})\right]
\end{cases}
$$

$$w_i = P\left[(y_k)|x_{1_k}^i\right] \propto \frac{1}{(2\pi)^{\frac{1}{2}} v_k} exp\left(-\frac{v_k^2}{2}\right)$$

$$w_{reg_i} = \frac{w_i}{\sum_{i=1}^{N} w_i}$$

 End for {i}

 For $i = 1:N$ // Resample for posteriori particles $x_{j_k}^{i}$ ($j = 1,2,3$)

  Generate a random number of $r \sim U[0,1]$

  Find: $\sum_{i=1}^{j-1} w_{reg_i} < r$ & $\sum_{i=1}^{j} w_{reg_i} \geq r$

  Set $x_k^{i+} = x_k^{j-}$ with probability w_{reg_i}.

 End for {i}

$$
\begin{cases}
\hat{x}_{1_k} = E(x_{1_k}|y_k) \\
\hat{x}_{2_k} = E(x_{2_k}|y_k) \\
\hat{x}_{3_k} = E(x_{3_k}|y_k)
\end{cases}
$$

End for {k}

The first system state, $\overline{NP}$, represents the starter signal. As described in Equation 10, its value at time step k is determined from the system states at the previous time step. The second system state λ represents the starter degradation rate. It is located in the exponential part of Equation 10. Therefore, the starter degradation rate between two adjacent starting cycles is indicated by e^λ. The higher λ is, the faster a starter degrades along an exponential growth. When $\lambda = 0$, no degradation develops between two starting cycles. The third system state C represents a discrete change of the starter degradation between two adjacent starting cycles. During the PF iterations, the systems states are estimated in the framework of recursive Bayesian by constructing their conditional *pdf* based on the measurements. Consequently, APU starter prognostic is implemented by λ estimation. Once the measurement stops, both λ and C are fixed with their most recent values. Thus the future degradation trend is expressed as an exponential growth of e^λ. The PF-based method for estimating RUC is implemented as shown in Table 1 by using system state NP, using an MATLAB environment.

4   Experimental Results

By implementing PF technique for NP, we can use λ to perform APU starter prognostics. The idea is that λ is fixed at its most recent values updated by the available measurements. Then the future degradation trend is expressed as an exponential growth, e^λ, started from the latest $\overline{NP}$ estimations. The experiments were mainly conducted to learn the weight parameters for PF methods and to predict or estimate the NP using learned parameters. The triggering point for prediction is determined based on the statistic analysis given a failure mode.   Figure 1 and 2 show the PF results when the prognostics is triggered at 38% and 33% of NP_{peak} for NP prediction corresponding to RUC at -125 and -75 starting cycles prior to the failure or replacement, respectively. As we mentioned, $\overline{NP}$ (Y axis) is the rotational speed of APU when EGT reach its peak value during start. The unit is %. 100% means APU designed working rotational speed. In these figures, we use "negative" numbers to represent the remaining useful cycles to failure event. Zero (0) represents the timing of failure event.

From the results, the RUL estimation of APU starter can be easily performed by setting up a threshold for $\overline{NP}$. From Figure 1, $\overline{NP}$ threshold is set at 38%. In other words, when NP estimation from the learnt PF model reached 38% of NP_{peak} , it starts to use NP prediction to perform prognostics and the RUC corresponding to triggered EP will be used as onset point of RUC estimation. If the NP prediction is reducing to 38% of NP_{peak}, it means APU starter should be changed or replacement within 125 RUCs. In Figure 1, the "star dots" represents the measurements; the red points are estimation from PF model during learning phase; and black line is NP prediction from the learnt PF models.

Similarly, Figure 2 shows the result of prediction triggered at $NP = 33\%$ of NP_{peak} correspond RUC = -75. From that point, the trained PF model starts to predict the NP for APU Starter prognostics.

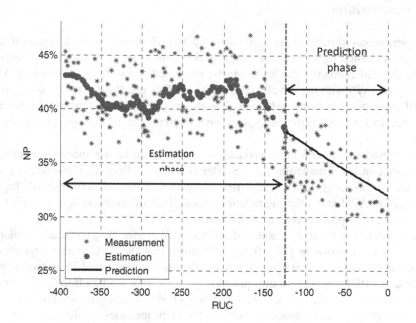

Fig. 1. PF prognostics result for NP (Triggered RUC= -125)

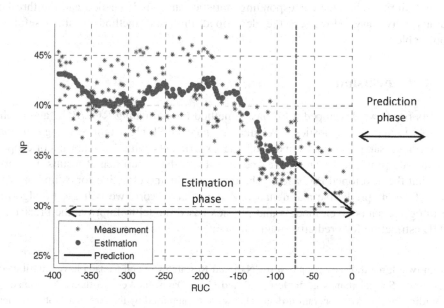

Fig. 2. PF prognostic result for NP (Triggered RUC= -75)

5   Discussions

The experimental results above demonstrated that PF-based method is useful and effective for estimating RUC for performing APU starter prognostics. If the threshold value is decided correctly, the RUC can be estimated precisely by monitoring APU engine speed (NP) with the developed techniques. This result is useful for developing onboard prognostic systems, which makes the prognostic decision transparent and simplified. In turn, it promotes application of prognostic techniques to real-world problems.

Since there is existing a large variance in the different failure models, the precise RUC prediction for a particular APU Starter is really challenged. However, our PF-based techniques suggested clearly that once the estimated $\overline{NP}$ is 38% higher than μ_{nor}, the APU starts the degradation phase. This information is also useful for helping make decision on predictive maintenance.

It is worth to note that the results of NP for APU starter prognostic are similar to EP, which was reported in [18]. It is also assume that the APU starter degradation follows a certain exponential growth pattern when we implemented PF-based method for estimating RUL for APU starter prognostics. This may not be effective for repetitive fluctuations of the starter degradation. In the future we should integrate data-driven prognostic techniques with PF-based prognostics to develop a hybrid framework for prognostics.

The results in this work only demonstrated one failure mode, "Inability to Start". The threshold value described above is determined only for this failure mode. For other failure modes, the corresponding statistics analysis is needed and the threshold values may vary. However, the developed PF-based method is still useful and applicable.

6   Conclusions

In this paper we developed a PF-based method for estimating RUL/RUC and applied it to APU Starter prognostics. We implemented the PF algorithm by using sequential importance sampling, and conducted the experiments with 10 years historic operational data provided by an airline operator. From the experimental results, it is obvious that the developed PF-based technique is useful and effective for estimating RUC in performing predictive maintenance. As our future work, we will use PF algorithm to integrate data-driven models and physics-based model to improve the precision of RUL estimation for predictive maintenance.

Acknowledgment. Many people at the National Research Council Canada have contributed to this work. Special thanks go to Jeff Bird and Craig Davison. We also thank Air Canada for providing us the APU operation data. This work is supported by the National Natural Science Foundation of China (Grant Nos. 61463031).

References

1. Jardine, A., Lin, D., Banjevic, D.: A review on machinery diagnostics and prognostics implementing condition-based maintenance. Mechanical Systems and Signal Processing **20**, 1483–1510 (2006)
2. Liu, J., Wang, W., Golnaraghi, F.: A multi-step predictor with a variable input pattern for system state forecasting. Mechanical Systems and Signal Processing **2315**, 86–99 (2009)
3. Gupta, S., Ray, A.: Real-time fatigue life estimation in mechanical structures. Measurement Science and Technology **18**, 1947–1957 (2007)
4. Yagiz, S., Gokceoglu, C., Sezer, E., Iplikci, S.: Application of two non-linear prediction tools to the estimation of tunnel boring machine performance. Engineering Applications of Artificial Intelligence **22**, 808–814 (2009)
5. Wang, W., Vrbanek, J.: An evolving fuzzy predictor for industrial applications. IEEE Transactions on Fuzzy Systems **16**, 1439–1449 (2008)
6. Tse, P., Atherton, D.: Prediction of machine deterioration using vibration based fault trends and recurrent neural networks. Journal of Vibration and Acoustics **121**, 355–362 (1999)
7. Adams, D.E.: Nonlinear damage models for diagnosis and prognosis in structural dynamic systems. Proc. SPIE **4733**, 180–191 (2002)
8. Luo, J., et al.: An interacting multiple model approach to model-based prognostics. System Security and Assurance **1**, 189–194 (2003)
9. Chelidze, D., Cusumano, J.P.: A dynamical systems approach to failure prognosis. Journal of Vibration and Acoustics **126**, 2–8 (2004)
10. Pecht, M., Jaai, R.: A prognostics and health management roadmap for information and electronics-rich systems. Microelectronics Reliability **50**, 317–323 (2010)
11. Saha, B., Goebel, K., Poll, S., Christophersen, J.: Prognostics methods for battery health monitoring using a Bayesian framework. IEEE Transactions on Instrumentation and Measurement **58**, 291–296 (2009)
12. Liu, J., Wang, W., Golnaraghi, F., Liu, K.: Wavelet spectrum analysis for bearing fault diagnostics. Measurement Science and Technology **19**, 1–9 (2008)
13. Liu, J., Wang, W., Golnaraghi, F.: An extended wavelet spectrum for baring fault diagnostics. IEEE Transactions on Instrumentation and Measurement **57**, 2801–2812 (2008)
14. Liu, J., Wang, W., Ma, F., Yang, Y.B., Yang, C.: A Data-Model-Fusion Prognostic Framework for Dynamic System State Forecasting. Engineering Applications of Artificial Intelligence **25**(4), 814–823 (2012)
15. García, C.M., Chalmers, J., Yang, C.: Particle filter based prognosis with application to auxiliary power unit. In: The Proceedings of the Intelligent Monitoring, Control and Security f Critical Infrastructure Systems, September 2012
16. Doucet, G.S.A.C.: On sequential Monte Carlo sampling methods for Bayesian filtering. Statistics and Computing, pp. 197–208 (2000)
17. Arulampalam, M., Maskell, S., Gordon, N., Clapp, T.: A tutorial on particle filters for online nonlinear/non-gaussian bayesian tracking. Trans. Sig. Proc. **50**(2), 174–188 (2002). http://dx.doi.org/10.1109/78.978374
18. Yang, C., Lou, Q., Liu, J., Yang, Y., Bai, Y.: Particle filter-based method for prognostics with application to auxiliary power unit. In: Ali, M., Pan, J.-S., Chen, S.-M., Horng, M.-F. (eds.) IEA/AIE 2014, Part I. LNCS, vol. 8481, pp. 198–207. Springer, Heidelberg (2014)

Combining Constrained CP-Nets and Quantitative Preferences for Online Shopping

Bandar Mohammed, Malek Mouhoub$^{(\boxtimes)}$, and Eisa Alanazi

Department of Computer Science,
University of Regina, Regina, SK, Canada
{mohammeb,mouhoubm,alanazie}@cs.uregina.ca

Abstract. Constraints and preferences coexist in a wide variety of real world applications. In a previous work we have proposed a preference-based online shopping system that handles both constraints as well as preferences where these latter can be in a qualitative or a quantitative form. Given online shoppers' requirements and preferences, the proposed system provides a set of suggested products meeting the users' needs and desires. This is an improvement to the current shopping websites where the clients are restricted to choose among a set of alternatives and not necessarily those meeting their needs and satisfaction.

For a better management of constraints and preferences, we extend in this paper the well known constrained CP-Net model to quantitative constraints and integrate it into our system. This extended constrained CP-Net takes a set of constraints and preferences expressing user's requirements and desires, and returns a set of outcomes provided in the form of list of suggestions. This latter list is sorted according to user's preferences. An experimental evaluation has been conducted in order to assess the time efficiency of the proposed model to return the list of suggestions to the user. The results show that the response time is acceptable when the number of attributes is of manageable size.

Keywords: Preferences · CP-nets · Constrained CP-nets · Online shopping

1 Introduction

People are prone to making decisions in day to day activities. When faced with many alternatives, people tend to distinguish alternatives based on their preferences. Therefore, preferences is a key aspect in decision making. The Last decade showed a growing number of online shopping systems and e-commerce applications. This stems from the fact that the number of online shoppers and buyers over the web has increased drastically. This brings many challenges when it comes to online shopping systems. One of the main challenges of such systems is to be able to learn and reason about buyers' interests and preferences. Most existing systems gather preference information without engaging the user.

© Springer International Publishing Switzerland 2015
M. Ali et al. (Eds.): IEA/AIE 2015, LNAI 9101, pp. 702–711, 2015.
DOI: 10.1007/978-3-319-19066-2_68

In principle, most of such systems keep track of the user history and reason about the user preference passively without engaging her with the process. Needless to say, such techniques are prone to inaccurate and missing preference information. In addition, users usually pose some restrictions over the set of available alternatives (or products) and require the system to satisfy them. For instance, a user looking for online tickets may be interested only in those with a specific airline. This would make itineraries with other airlines irrelevant when looking for the result. Therefore, online buyer's preferences and constraints usually overlap and coexist together.

We are interested in online shopping system with combinatorial nature. Such systems have been intensively studied in the literature [1–5]. However, many of these existing systems assume features aggregation to be additive and do not allow *dependencies* among the features. In a previous work [6,7], we have proposed a system that allows these dependencies and tackles two well known classes of preferences: qualitative and quantitative. In qualitative preferences, the buyer expresses her preferences comparatively over the feature values. For instance, a user may be interested in 8GB of RAM more than 4GB for her laptop. This induces that the RAM with 8GB is more preferred to the 4GB one. In contrary, quantitative preferences assign numeric values to the feature values. Qualitative preferences are known to be more intuitive and easier to elicit from the user. At the same time, unlike quantitative ones, qualitative preferences are computationally problematic when it comes to deciding which alternative is preferred to another. As a result, we have also introduced an approximation method to approximate qualitative information into quantitative ones. This allowed our system to process both types of preferences together and being able to response faster than solving the qualitative information alone.

For a better management of constraints and preferences, we extend in this paper the well known constrained CP-net graphical model [8] to quantitative constraints and integrate it into our system. This extended constrained CP-Net first takes a set of constraints and qualitative preferences and returns a list of Pareto optimal outcomes (outcomes that are not dominated by any other outcome). A total order is then enforced using a lexicographic method. Finally quantitative preferences, represented through utility functions, are combined to this total order in order to return a list of suggestions to the user. In order to assess the time performance of this process, an experimental study has been conducted on the case study of online shopping. The results demonstrate the efficiency of our extended model for a manageable size of attributes. To our best knowledge this is the first time constrained CP-nets have been used together with quantitative preferences in online shopping.

The rest of the paper is organized as follows. The following section describes the necessary background information. The related work is then presented in the section 3. Section 4 presents our extended constrained CP-Net model as well as its related algorithms. The online shopping system integrating the proposed model is described in section 5. In section 6, the experimental evaluation is reported. Finally, section 7 summarizes our research contribution and foreseeable future work.

2   Background

2.1   Conditional Preference Networks (CP-Nets)

A CP-net is a graphical model to compactly represent user conditional preferences [9]. A CP-net with respect to a set of variables V is a directed acyclic graph (DAG) $\mathcal{G}$ with vertex set V. For every variable $V_i \in V$, the user chooses a (possibly empty) set $Pa(V_i) \subseteq V \backslash \{V_i\}$ representing the direct predecessors of V_i. Every vertex V_i is associated with a conditional preference table $cpt(V_i)$ showing, for every assignment of $Pa(V_i)$, a total order over the values of V_i.

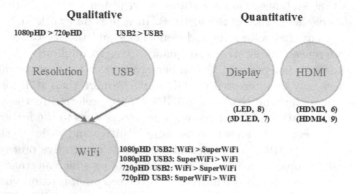

Fig. 1. Representation of qualitative and quantitative preferences

Example 1 (CPNet). The CP-net in Figure 1 defines a set of three variables {Resolution,USB,Wifi} each associated with its preference table. For variable Resolution, the user unconditionally prefers 1080HD to 720HD. However, for Wifi, the user preferences are conditional upon the values assigned to both Resolution and USB.

The CP-net defines a dominance relation over the set of outcomes $\mathcal{O}$ such that for any two outcomes $o_i, o_j \in \mathcal{O}$, o_i dominates o_j if and only if there is a sequence of worsening flips from o_i to o_j. A worsening flip is a change in the variable value to a less preferred one where all other things being equal (ceteris paribus).

Example 2 (Dominance Relation). Recall our running example in Figure 1, let $o_1 = $(1080HD,USB2,Wifi) and $o_2 = $ (1080HD,USB2,SuperWifi). Then going from o_1 to o_2 is a worsening flip to the variable Wifi. Moreover, o_1 dominates (720HD,USB3,Wifi) because there is a sequence of improving flips from the former to the latter. In particular, the sequence is (1080HD,USB2,Wifi),(720HD, USB2,Wifi),(720,USB3,Wifi).

2.2   Constrained CP-nets

Preferences usually take place in a constrained environment. Thus, solving CP-nets with the presence of constraints is an important step towards applying such models into real world applications. A Constrained CP-net [8] is a tuple $(\mathcal{N}, \mathcal{C})$ where $\mathcal{N}$ is a CP-net and $\mathcal{C}$ is a set of constraints restricting the values that some of the CP-net variables can take. Solving a Constrained CP-net consists of finding one or more feasible outcomes (called Pareto optimal solutions) that are not dominated by any other feasible outcome.

Example 3 (Pareto Set).
Consider the CP-net in Figure 1 and assume (1080HD,USB2) is infeasible. Then the best outcome of this CP-net (1080HD,USB2,Wifi) is no longer valid according to the above constraint.
Instead we have (1080HD,USB3,SuperWifi) and (720HD,USB2,Wifi) as the Pareto set.

In [8] a method, called Search-CP, has been proposed to solve the constrained CP-net problem. Search-CP is a backtrack search algorithm that assigns the variables to values from their domains, in a topological order. Following the semantics of CP-nets, the first solution generated by Search-CP is guaranteed to be Pareto optimal solution. However, when looking for more than one solution, we need to perform a dominance testing with earlier solutions every time a feasible solution is found. In order to improve its time performance in practice, Search-CP has been updated with constraint propagation and the most constrained variable ordering heuristic [10].

3   Managing Constraints and Preferences

3.1   Problem Statement

We assume a multivalued domain over a set of n variables $V = \{V_1, V_2, \ldots, V_n\}$. Each variable V_i is associated with a set of possible values $dom(V_i) = \{v_1^i, v_2^i, \ldots, v_{m_i}^i\}$. The set of variables V is the result of combining two disjoint sets $V = V_{Qual} \cup V_{Quan}$. V_{Qual} is the set of variables that represent the CP-net while V_{Quan} is the set of variables where each variable $V_i \in V_{Quan}$ has a utility function $\mu(V_i)$ associated to it. A utility function for variable V_i is a mapping from $dom(V_i)$ to $[1, 10]$. We also consider the existence of a set of constraints $\mathcal{C}$ over the variables in V. Therefore, A solution (i.e. mapping from V to its domain) is feasible if and only if it satisfies all the constraints in $\mathcal{C}$. Thus, the problem can be viewed as a triple $\varphi = \langle \mathcal{N}, \mathcal{U}, \mathcal{C} \rangle$ where:

- $\mathcal{N}$ is a CP-net over a subset of variables $V_{Qual} \subset V$,
- $\mathcal{U}$ is a set of utility functions over $V_{Quan} \subset V$,
- and $\mathcal{C}$ is a set of constraints over V.

Algorithm 1. Finding Pareto Set

Input: CP-net $\mathcal{N}$ and global constraint $\mathcal{C}$
Output: The Pareto Set $\mathcal{S}$ (initially empty)
1. Let $>$ be topological ordering over $\mathcal{N}$
2. Let T be a stack.
3. push $\{\}$ into T.
4. while(T is not empty)
5. $n \leftarrow$ pop first element
6.   if(n is complete assignment)
7.   if(n is not dominated by any element in $\mathcal{S}$)
8.   add n to $\mathcal{S}$
9.   else:
10.   for every value v_i of the next variable V_i according to $>$
11.   if($n \cup (V_i, v_i)$ is consistent with $\mathcal{C}$)
12.   push $n \cup (V_i, v_i)$ into T
13. return $\mathcal{S}$

Given φ, we are interested in finding the set of solutions that are feasible according to $\mathcal{C}$ and best according to both $\mathcal{N}$ and $\mathcal{U}$. In other words, our goal is find the Pareto set of the problem. A feasible solution is Pareto if its not dominated by any other feasible solution. The Pareto set is the set of all Pareto solutions.

3.2 Solving Method

We solve the problem by finding the Pareto set of the qualitative (i.e. CP-net) part first. Then, we approximate the set lexicographically. In order to find the set of Pareto outcomes, we use the algorithm described in [10]. It should be noted that this latter algorithm finds one Pareto but it is straightforward to make it look for all the Pareto set. When looking for the Pareto set, we need to perform dominance testing every time we encounter a solution. Algorithm 1 describes the general steps in order to find the Pareto set. At each step, we check whether the current assignment (n) is a complete assignment or not (line 6). If it is a solution, we perform a dominance testing with the current solutions found so far (line 7). Otherwise, we move to the next variable. When the search ends, $\mathcal{S}$ contains all the Pareto solutions of the problem.

Once we have the Pareto set from Algorithm 1, we use Algorithm 3 in order to obtain a total order over $\mathcal{S}$ following the lexicographic ordering. We do this by posing a total order over the set of variables V (consistent with the CP-net $\mathcal{N}$). Afterwards, for every two Pareto solutions, we break the ties by the first different value they have. The result is a total order over the Pareto set $s_1 > s_2 > \cdots > s_{|\mathcal{S}|}$. Thus, we mapped every Pareto solution to a number $i \in [1, \ldots, |\mathcal{S}|]$ representing the position of that solution in the lexicographic order $>$.

Algorithm 2. PrefC Method - Combining Qualitative and Quantitative together

---

Input: f^1 and f^2 mapping functions, constraints $\mathcal{C}$
Output: A set of ranked feasible solutions $\mathcal{S}$
$\mathcal{S} \leftarrow \emptyset$
while $(\mathcal{O}_{Qual} \neq \emptyset)$
  $o_1 \leftarrow \min(f^1(\mathcal{O}_{Qual}))$
  $copy \leftarrow Q_{Quan}$
  while$(copy \neq \emptyset)$
    $o_2 \leftarrow \min(f^2(copy))$
    if$(o_1 \times o_2$ consistent with $\mathcal{C})$
      add $o_1 \times o_2$ to $\mathcal{S}$
    remove o_2 from $copy$
  remove o_1 from $\mathcal{O}_{Qual}$
return $\mathcal{S}$

For the quantitative part, every outcome is defined over V_{Quan}. The utility of an outcome $o \in \mathcal{O}_{Quan}$ is defined as follows:

$$\mu(o) = \sum_{i=1}^{|V_{Quan}|} \mu_i(o[i]) \tag{1}$$

where $\mu_i(x)$ is the utility value when $V_i = x$. Therefore, the set of outcomes over V (denoted as $\mathcal{O}$) is the product of $\mathcal{O}_{Qual} \times \mathcal{O}_{Quan}$. Let $f^1 : \mathcal{O}_{Qual} \rightarrow N$ and $f^2 : \mathcal{O}_{Quan} \rightarrow N$ be the two mapping functions for qualitative and quantitative respectively. Then the utility of an outcome $o \in \mathcal{O}$ is defined as follows:

$$u(o) = f^1(o_1) + f^2(o_2) \tag{2}$$

where o_1 and o_2 correspond to the outcome o projected to V_{Qual} and V_{Quan} respectively. Once we are done using Algorithm 1 to get the Pareto set and Algorithm 3 in order to have a total order to break Pareto Set Lexicographically, we use Algorithm 2 PrefC method in order to combine qualitative and quantitative preferences together. More precisely, we combine all the tuples with their weight/cost for qualitative and quantitative preferences and apply PrefC method which consists of computing the sum of the quantitative and qualitative weights while satisfying the budget global constraints.

4   Case Study: Online Shopping System

The extended constrained CP-net is integrated into our online shopping system that is presented, in this section, via the problem of selecting products according to a set of constraints and preferences. Here the user has the capability to express her budget constraint as well as her preferences either in a quantitative (assigning numeric values to the attributes features) or in a qualitative manner (conditional of unconditional ranking of attributes features). Figure 2 shows a screenshot of our system Graphic User Interface (GUI).

Algorithm 3. Breaking Pareto Set Lexicographically

Input: Pareto Solutions $\mathcal{S}$

Output: Lexicographic ordering $>_{lex}$ over $\mathcal{S}$

1. Let $>_{lex} \leftarrow \{\}$
2. For every two elements $s_i, s_j \in \mathcal{S}$
3.   For $k = 1$ to $|V|$
4.   if $V_k(s_i) \neq V_k(s_j)$
5.   if $V_k(s_i)$ is better than $V_k(s_j)$
6.   add (s_i, s_j) to $>_{lex}$
7.   else
8.   add (s_j, s_i) to $>_{lex}$
9. Return $>_{lex}$

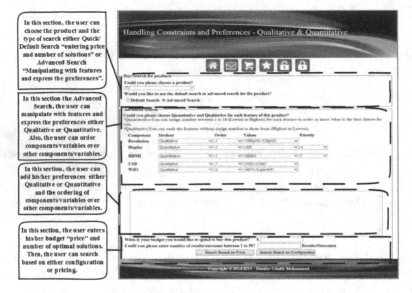

Fig. 2. Graphic User Interface of the Online Shopping System

Our online shopping system has been developed using the following.

1. Programming environment: Microsoft Visual Studio 2013 (ASP.NET - C Sharp).
2. Programming languages: Visual C Sharp 2012/2013 (ASP.NET MVC 5 Web Application), HTML and JavaScript.
3. Operating system: Web Hosting under Windows 7 and supporting .NET Runtime Version (ASP.NET 4.5) and PHP Runtime Version (PHP 5.4).
4. Database: SQL Server Database 2012.

As it is shown in Figure 3, there are three layers of our online shopping system: presentation layer, business layer and database layer. The presentation layer corresponds to the GUI presented in figure 2 and takes care of the user constraint and preference elicitation phase. As well, the presentation layer offers two types of search: default and advanced where the advanced search option corresponds to exploring the entire search space. The business layer represents the solving techniques we presented in section 3. The database layer takes care of the information storage and retrieval. We have reserved a domain and Deluxe Web Hosting on Godaddy for the online system.

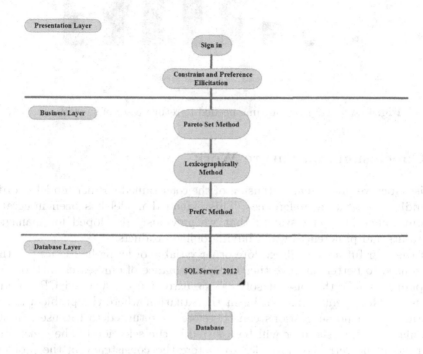

Fig. 3. Software Architecture

5   Experimentation

In order to evaluate the solving methods we presented in section 3 we have conducted several experiments on instances of the application we presented in the previous section.

The experiments have been c on a local machine with 64-bit Operating System, Intel Core i7 with a 2.4GHz CPU and 12.0 GB of RAM. The goal here is to determine the average time needed to return a list of suggestions.

Figure 4 reports this running time in milliseconds when varying the number of attributes from 3 to 10.

As we can see from the chart, while the running time is fast growing it is still under a second even when the number of attributes is equal to 10.

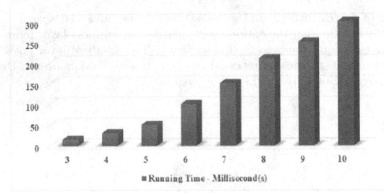

Fig. 4. Average running time needed to return a set of suggestions

6   Conclusion and Future Work

In this paper we propose an extension of the constrained CP-net model in order to handle quantitative preferences. This extended model has been integrated into our online shopping system that we previously developed for managing constraints and preferences when buying online products.

In the near future we will explore other variable ordering heuristics [11] that can be used to better improve the time performance of our search method. We also plan to study the case of solving the extended constrained CP-net in a dynamic environment. This can be in the situation where the problem is over constrained and no suggestions can therefore be returned to the user. In this particular situation the user will be assisted in the selection of the constraints that need to be retracted in order to restore the consistency of the problem. Another scenario is when there is a large number of Pareto optimal solutions to return to the user. In this case, the user needs to constrain the problem even more and eventually adds more preferences in order to obtain a manageable number of options. To handle both the addition and retraction of constraints, we will integrate the dynamic arc consistency algorithm proposed in [12] into the constraint propagation of our search method.

Acknowledgments. Bandar Ghalib Mohammed and Eisa Alanazi are sponsored by the Ministry of Higher Education, Saudi Arabia.

References

1. Pu, P., Faltings, B.: Enriching buyers' experiences: the smartclient approach. In: Proceedings of the SIGCHI Conference on Human Factors in Computing Systems, pp. 289–296. ACM (2000)
2. Stolze, M., Ströbel, M.: Recommending as personalized teaching: towards credible needs-based ecommerce recommender systems, designing personalized user experiences in ecommerce (2004)

3. Myers, K., Berry, P., Blythe, J., Conley, K., Gervasio, M., McGuinness, D.L., Morley, D., Pfeffer, A., Pollack, M., Tambe, M.: An intelligent personal assistant for task and time management. AI Magazine **28**, 47 (2007)

4. Sadaoui, S., Shil, S.K.: Constraint and qualitative preference specification in multi-attribute reverse auctions. In: Ali, M., Pan, J.-S., Chen, S.-M., Horng, M.-F. (eds.) IEA/AIE 2014, Part II. LNCS, vol. 8482, pp. 497–506. Springer, Heidelberg (2014)

5. Shil, S.K., Mouhoub, M., Sadaoui, S.: An approach to solve winner determination in combinatorial reverse auctions using genetic algorithms. In: Proceedings of the 15th Annual Conference Companion on Genetic and Evolutionary Computation, pp. 75–76. ACM (2013)

6. Alanazi, E., Mouhoub, M., Mohammed, B.: A preference-aware interactive system for online shopping. Computer and Information Science **5**, 33–42 (2012)

7. Mohammed, B., Mouhoub, M., Alanazi, E., Sadaoui, S.: Data mining techniques and preference learning in recommender systems. Computer and Information Science **6**, 88–102 (2013)

8. Boutilier, C., Brafman, R.I., Domshlak, C., Hoos, H.H., Poole, D.: Preference-based constrained optimization with cp-nets. Computational Intelligence **20**, 137–157 (2004)

9. Boutilier, C., Brafman, R.I., Domshlak, C., Hoos, H.H., Poole, D.: Cp-nets: A tool for representing and reasoning with conditional ceteris paribus preference statements. J. Artif. Intell. Res. (JAIR) **21**, 135–191 (2004)

10. Alanazi, E., Mouhoub, M.: Variable ordering and constraint propagation for constrained CP-nets. In: Ali, M., Pan, J.-S., Chen, S.-M., Horng, M.-F. (eds.) IEA/AIE 2014, Part II. LNCS, vol. 8482, pp. 22–31. Springer, Heidelberg (2014)

11. Mouhoub, M., Jashmi, B.J.: Heuristic techniques for variable and value ordering in csps. In: Krasnogor, N., Lanzi, P.L. (eds.) GECCO, pp. 457–464. ACM (2011)

12. Mouhoub, M.: Arc consistency for dynamic CSPs. In: Palade, V., Howlett, R.J., Jain, L. (eds.) KES 2003. LNCS, vol. 2773, pp. 393–400. Springer, Heidelberg (2003)

Focused Diagnosis for Failing Software Tests

Birgit Hofer, Seema Jehan$^{(\boxtimes)}$, Ingo Pill, and Franz Wotawa

Graz University of Technology, Graz, Austria
{bhofer,sjehan,ipill,wotawa}@ist.tugraz.at

Abstract. Ranging from firmware to cloud services, software is an essential part of almost any imaginable system, or at the very least assists us in their design or maintenance. The sheer complexity and sophisticated concepts of today's software products thus demand for solutions that assist us in assuring their quality. We aim to contribute in this direction by proposing a fault localization approach for failing test cases that draws on model-based diagnosis techniques from the AI community and focuses the search on dynamic executions. With this focus, we offer the scalability needed to consider also designs like service oriented architectures (SOAs). Furthermore, we opt for a flexible approach that allows a user to refine the reasoning by annotating our basic structure of a control flow graph with further information, e.g., for black box components. First experiments with standard software examples, as well as examples taken from the SOA domain show promising results.

Keywords: Model-based diagnosis · Software debugging · Constraint satisfaction problem · Control flow graph

1 Introduction

Software helps to meet challenges and offers flexible and cost-effective solutions - ranging from firmware that capitalizes a hardware's resources, via developer tools, to cloud solutions for complex, scalable demands. Ensuring a complex software product's quality is, however, a cumbersome task. For potentially huge designs, like service oriented architectures (SOAs) it becomes even more demanding (see [8]). Aside controllability and observability issues, we might have only partial knowledge about some system parts, e.g. invoked web services or 3rd party libraries. For SOAs we envisioned in [14] a corresponding, combined diagnosis and testing workflow that targets their BPEL processes (i.e., the descriptions of their business logic). Our first sketch at tackling diagnostic issues appeared at the DX Workshop [4]. For this paper, we rid ourselves of this focus and propose a diagnosis approach for failing test cases that is not aimed at some particular software architecture. Our underlying reasoning model is that of a software's control flow graph (CFG) that can be easily annotated with further knowledge about individual parts (like black boxes).

We draw on model-based diagnosis techniques [6,13] for localizing possible root cause variants explaining a failed test case. Most model-based diagnosis

© Springer International Publishing Switzerland 2015
M. Ali et al. (Eds.): IEA/AIE 2015, LNAI 9101, pp. 712–721, 2015.
DOI: 10.1007/978-3-319-19066-2_69

approaches use a model of the correct system for their reasoning. However, we debug the faulty system under scrutiny (that is, in this case our knowledge about it as aggregated in the CFG). Furthermore, while usually the system's entire model is considered for such a diagnostic endeavor, similar to dynamic slicing techniques [7,17], we focus our search to the scope of individual executions, i.e., those CFG parts executed by the given test case. In Section 3, we show how to implement such an approach. Similar to Wotawa et al. [16] , we compute the diagnoses directly from a CFG's constraint representation, isolating sets of constraints possibly responsible for the issue at hand. Our first experiments as reported in Section 4 showed promising results for programs taken from the software landscape and service-oriented architectures.

For illustration purposes, we use the computation of a circle's or square's area and circumference. For the input {type=circle,length=2}, the parameter *length* defines the circle's diameter, and the program's output is {area=3.14, circumference=3.14}. Obviously incorrect, this is due to a bug in line 6 s.t. the circumference is computed as $radius*pi (= 3.14)$ instead of $2*radius*\pi (= 6.28)$.

Example 1. Compute area and circumference of a circle or a square.

```
1    geometricFeatures(type, length){
2       pi − 3.14;
3       if(type=="circle"){
4          radius = length/2;
5          area = radius * radius * pi;
6          circumference = radius * pi;  //FAULT
7       else
8          area = length * length;
9          circumference = 4 * length;
10      endif
11   }
```

2   Preliminaries

The central model for our reasoning is the following generic control flow graph. Please note that we assume a program to be deterministic and currently do not consider parallel computations but sequential programs only.

Definition 1. *A* Control Flow Graph *G is defined by a tuple $G = \{V, E, v_0, F, S, \gamma_C(v \in V), \gamma_A(v \in V)\}$, where V is a finite set of vertices representing statements, $E \subseteq V \times V$ is a finite set of directed edges connecting the statements (edge $e = (v_1, v_2) \in E$ connects v_1 to v_2), $v_0 \in V$ is the start vertex, $F \subseteq V$ is the set of leaf vertices (with no outgoing edges), $S \subseteq V$ is the set of branching vertices (with more than one outgoing edge), and the functions $\gamma_C(v)$ and $\gamma_A(v)$ map vertices $v \in V$ to a statement's conditions and assignments respectively.*

The conversion of a sequential program into a CFG is straightforward, where further information can be easily added to $\gamma_C(v)$ and $\gamma_A(v)$. With G capturing the control flow structure of a program, a path π in G defines a valid scenario.

Definition 2. *A finite Path π of length n in a CFG $G = \{V, E, v_0, F, S, \gamma_C, \gamma_T\}$ is a finite sequence $\pi = \pi_1\pi_2...\pi_n$ such that we have that (1) for any $0 < i \leq n$, π_i is in V, (2) $\pi_1 = v_0$, (3) for any $0 < i < n$, the edge $e = (\pi_i, \pi_{i+1})$ is in E, and (4) $\pi_n \in F$. The length of some path π is denoted by $|\pi|$, where we use $f(\pi)$ to refer to the last vertex in sequence π. As we reason about finite computations only, per definition a path is finite. For such a path π, we define the set of* Path Constraints $C(\pi) = \bigcup_{(0 < i \leq n)} \gamma_C(\pi_i)$*, where variables are replaced by indexed variables in order to implement a static single assignment form. A path is feasible if its corresponding path constraints are satisfiable.*

In the path constraints $C(\pi)$ we collect our knowledge about π for reasoning. The static single assignment form (SSA) [1] implemented for $C(\pi)$ means that we use indexed variables s.t. whenever a variable is assigned a value, the index is incremented for further referrals along the path. This concept of "temporal" variable instances clocked by assignments ensures that every variable along a path is defined once only, but can be referenced as often as needed. For our example, $C(\pi)$ would be $\{\pi_0 == 3.14, (type_0 == "circle") == true, radius_0 == length_0/2, area_0 == radius_0 * radius_0 * pi_0, circumference_0 == radius_0 * pi_0\}$.

For our focused reasoning, we establish a connection between a control flow graph G and an actual program execution Π via our definition of a trace.

Definition 3. *The trace T in G for a program execution Π is defined as the path in G allowed for the input part of Π.*

Assuming a program to be deterministic and the inputs to be complete, the input values define a single specific path to be taken by the program. For our example computation, this means taking the then-branch. For our target scenario of a failing test case, we extract the relevant inputs from the very test case itself.

Definition 4. *A* test case *is a tuple $\tau = (I, O)$ where I (sometimes we will write also $I(\tau)$) is the set of value assignments to input variables, and O is the set of expected values for the output variables. A test case fails iff the observed output O' deviates from O, s.t. $O' \neq O$.*

Our diagnostic reasoning for localizing the fault(s) implements the basic concept of consistency-oriented model-based diagnosis as defined in [3, 6, 13]: Given a system's set of components $COMP$, a system description SD defining the correct behavior $\neg AB(c_i) \Rightarrow NominalBehavior(c_i)$ of components $c_i \in COMP$, and observations OBS about the system's behavior, the system is considered to be faulty iff $SD \cup OBS \cup \{\neg AB(c_i)|c_i \in COMP\}$ is unsatisfiable. The predicates $AB(c_i)$ represent the "health" state of components c_i: If $AB(c_i)$ is false, then c_i behaves as expected. Otherwise c_i is faulty and we assume to know nothing about it. While a minterm in the health predicates defines a specific system state, a diagnosis Δ is a subset-minimal set of faulty components that explains the observed faulty behavior OBS. $COMP$ is the set of individual constraints in $C(\pi)$ and SD is a correspondingly tailored constraint encoding of $C(\pi)$.

Definition 5. *A* diagnosis *for (SD, COMP, OBS) is a subset-minimal set $\Delta \subseteq COMP$ such that $SD \cup OBS \cup \{\neg AB(c_i)|c_i \in COMP \setminus \Delta\}$ is satisfiable.*

3 Localizing the Root Causes for Failed Software Tests

Based on a program's CFG, our aim was to isolate those program parts that could explain the deviation from expected behavior for the failed test case. With today's complex software designs, an important motivation was to focus the search for optimizing resources. In detail, and in contrast to a full temporal system model like the one we used for our work on LTL specification diagnosis [12], we thus focus on a specific path in the system's control structure, similar to the idea in [5]. This way, SD does not describe the program's entire CFG, but focuses on the trace T (a path) in the CFG that is exercised for the failed test case τ. In particular, we use the path constraints $C(T)$, and by encapsulating $C(T)$'s individual constraints c with corresponding health predicates $AB(c)$, we aim at identifying those constraints c whose incorrectness could be responsible for T's deviation from expected behavior (as defined by the output part O of τ).

Faults in branching conditions (or related variables) might cause a trace T to deviate from the correct path (in the repaired program). With our focused system model SD, this would mean that the "corrected" path as defined by a diagnosis Δ, let us call it π^{corr}, could leave SD at some branching vertex. For such a scenario, we have to (a) disable subsequent (in terms of temporal evolvement along T) constraints in $C(T)$, and (b) make the user aware of this. This is motivated by the fact that SD does then describe π^{corr} only up to this point, but afterwards describes an alternative route in the CFG. To address this, we divide the trace T into segments separated by the branching vertices, and introduce variables $intrace_i$ for all these segments. The variables' values define whether an individual segment is part of π^{corr} as defined by Δ.

Definition 6. *A segmented path (π, S) in a CFG G is a tuple s.t. π is a path in G, and S is the set of branching vertices s_i in π that divide the path into enumerated segments as follows. The branching vertices s_i are numbered according to their distance from π_0, the enumeration starting with 1. The first segment, numbered 0, starts at π_0 and ends with s_1. Starting with segment 1, a segment i starts right after s_i, and ends with either s_{i+1} or the path π's final vertex $f(\pi)$.*

Our example computation's trace T has two segments. The first one contains the constraints $\{pi_0 == 3.14, (type_0 == "circle") == true\}$, the second one contains the constraints $\{radius_0 == length_0/2, area_0 == radius_0 * radius_0 * pi_0, circumference_0 == radius_0 * pi_0\}$. Now let us formalize our specific diagnosis problem and introduce our corresponding technical implementation.

Definition 7. *A control flow graph diagnosis problem is a tuple (G, τ), where G is a CFG, and T is a trace from a failing test case τ.*

This describes the scenario where we have a program G that we find to be incorrect by observing unexpected behavior. For our automated reasoning, we define the following constraint encoding to be used with a corresponding solver.

Definition 8. *A constraint satisfaction encoding $CSP(G, \tau)$ for a control flow graph diagnosis problem (G, τ) is defined as follows:*

1. let (T, S) be a segmented path for the trace T for τ in G
2. let the set of variables V contain all the variables in $C(T)$, as well as Boolean variables $intrace_i$ for all of (T, S)'s segments as of Definition 6.
3. let $C'(T)$ be the path constraints $C(T)$ altered s.t. for any individual constraint $c \in C(T)$ of segment i, we add a predicate AB_c and do as follows:
   (a) if c is not a branching constraint from some $s \in S$, then c gets replaced by $\neg intrace_i \vee AB_c \vee c$.
   (b) if c is the branching constraint from $s_i \in S$, then c gets replaced by the following set of constraints: $\neg intrace_i \rightarrow \neg intrace_{i+1}$, $AB_c \rightarrow \neg intrace_{i+1}$ and $\neg intrace_i \vee AB_c \vee (c \leftrightarrow \circ intrace_{i+1})$ - with $\circ$ being the expected polarity of $intrace_{i+1}$ when constraint c is satisfied.
4. then let $CSP(G, \tau)$ be the combined constraints of $C'(\pi)$ and τ as well as the constraint $intrace_0$.

Example 2. The CSP for our example computation is as follows:

$\neg intrace_0 \vee AB_1 \vee pi_0 == 3.14$
$\neg intrace_0 \rightarrow \neg intrace_1$
$AB_2 \rightarrow \neg intrace_1$
$\neg intrace_0 \vee AB_2 \vee (type_0 == \text{"circle"}) \leftrightarrow intrace_1$
$\neg intrace_1 \vee AB_3 \vee radius_0 == length_0/2$
$\neg intrace_1 \vee AB_4 \vee area_0 == radius_0 * radius_0 * pi_0$
$\neg intrace_1 \vee AB_5 \vee circumference_0 == radius_0 * pi_0$
$type_0 == \text{"circle"}$
$length_0 == 2$
$area_0 == 3.14$
$circumference_0 == 6.28$
$intrace_0$

We do not use a conflict-based diagnosis engine like the computation suggested by Reiter [13]. Similar to the idea presented in [16] for Java programs, we formulate a constraint satisfaction problem adopting the CSP as of Definition 8.

The algorithm in Fig. 1 shows how we derive diagnoses directly via computing satisfying assignments that are limited in the amount of active health predicates. This is achieved via the constraint added in line 4 of the algorithm. A single query to the solver then returns all the solutions of size i, where a diagnosis (solution) is the set of active health predicates of a satisfying assignment. In the loop (lines 3 to 8), for each instance, we increase i (starting with one and stopping at a given bound) and save (in line 6) all the diagnoses of cardinality i. Furthermore, in line 7, we add for each new diagnosis Δ a blocking clause to the constraint model, which is basically a logic OR over all negated health predicates in Δ. This is needed to block all supersets of Δ for satisfying assignments, so that the reported solutions are indeed subset-minimal (see Def. 5). After reaching the bound for i, we report all diagnoses.

Deploying this algorithm on our running example's CSP as of Example 2 results in the computation of the two single fault diagnoses AB_2 and AB_5, where the latter catches the fault described in the introduction.

```
1: procedure GETDIAGNOSES(M, n)
2:     Let DS be {}
3:     for  i = 1 to n do
```

4:
$$M' = M \cup \left\{ \left(\sum_{j=1}^{|\pi|} AB_j \right) == i \right\}$$

```
5:         D = Solve(M')
6:         DS = DS ∪ D
```

7:
$$M = M \cup \left(\bigcup_{\Delta_j \in D} \left(\bigvee_{AB_k \in \Delta_j} (\neg AB_k) \right) \right)$$

```
8:     end for
9:     return DS
10: end procedure
```

Fig. 1. GetDiagnoses Algorithm

Please note that newer developments such as [10], and our experiments in [11] showed that direct approaches such as the one we used for this paper can offer superior advantage compared to conflict-based computations. The setup used in this paper is referred to as $Direct\text{-}MS_{CS}$ in the comparison found in [11].

For practical purposes, we propose that the user should be made aware of the different segments' being part (or being excluded) of the scenario described by a diagnosis. This is vital data when considering a diagnosis in practice, so that we suggest to report for a diagnosis Δ also the corresponding minterm in the intrace variables, that is, the evaluation of all the corresponding *intrace* variables.

Definition 9. *An* extended diagnosis *for a control flow graph diagnosis problem* (G, τ) *as of Def. 7 is a tuple* $(\Delta, T_S, INTRACE)$, *where* Δ *is a diagnosis (see Def. 5) in the predicates* AB_i *for the CSP as of Def. 8 derived from* (G, τ), T_S *refers to the segmented path/trace derived for* $CSP(G, \tau)$, *and INTRACE is the corresponding minterm in the intrace variables (cf Def. 8) describing which segments are in the path* π^{corr} *depicted by* Δ.

Considering our encoding, the restriction to deterministic programs is apparent in that we do not directly accommodate non-determinism in the program for scenarios where for one and the same input we could have different traces. However, for such scenarios one can apparently implement an exhaustive approach by considering all the options in a loop, and presenting the diagnoses grouped by the different choices.

4   First Empirical Experiments

For our first experiments, we implemented a compiler that converts a SOAs business logic programs (BPEL processes) into our control flow graph format (see [4]), and considered also standard examples from the testing and debugging community. As constraint solver, we used MINION [2] that supports logic and arithmetic operations on Booleans and integers out-of-the-box. All the experiments were carried out on a MacBook Pro (Late 2011) with a 2.4 GHz Intel Core i5, 4 GB 1333 MHz DDR3, running OS X 10.7.2.

For this paper, we report on our experience with six examples. Most of them come from the SOA domain. The Loan example is a typical SOA process which approves, delays, or rejects loan requests based on the amount requested and the client's history. The Body Mass Index (Bmi) example decides on given height and weight if the result should be "underweight", "healthy", "overweight", "obese", or "very obese". The Triangle (Tri) example determines depending on the input the type of triangle e.g. equilateral, isosceles, or scalene. The Geo example is the running example discussed in this paper. Tcas is a SOA variant of the well-known traffic collision avoidance system used in many software engineering examples[1]. The calculator (Calc) example demonstrates additions, subtractions, divisions and multiplications as well as control structures like if-then-else and while. For all the examples, we derived faulty versions containing a single or two faults each. The corresponding files can be downloaded from the following link[2] .

Table 1 shows our results when searching for single and double fault diagnoses. Depending on the injected fault and the specific scenario, there are not always single and double fault diagnoses. Column F indicates the number of faults injected for a specific sample. The total number of inputs and outputs to a program are given in column I/O. Column S indicates the number of statements in the diagnosis problem's execution trace. The number of constraints and variables derived for the trace are reported in the columns C and V. For single fault ($|\Delta| = 1$) and double fault diagnoses ($|\Delta| = 2$) we report in the corresponding columns on the number of solutions found, as well as the average time in milliseconds over three computations. The approach's diagnostic performance for the single fault scenarios is indicated in column R by means of the reduction in statements that have to be considered by using the following formula :

$$Reduction = 1 - \frac{|\text{single fault diagnoses}|}{|\text{executed statements}|}$$

In other words, this means how many statements out of the executed ones we could eliminate from consideration. For double fault scenarios, we need to establish a more appropriate metric and thus do not report corresponding numbers.

Our approach allows for small computation times for the considered examples. While most times were in the range of mere milliseconds, for TCAS it took up to a few seconds to consider also double fault diagnoses. In most cases we also saw an attractive reduction in the number of statements that would have to be checked for correctness. On average, we achieved a reduction of 39.07 % for our experiments. For most of the examples, the reduction ranged from 22 % to 50 %. For TCAS, the reduction got as high as 89.1 %. The specific structure of the Calc example did not allow for any reduction.

Summarizing, we saw promising results for these standard examples, but further experiments are needed in order to determine the approach's attractiveness for larger programs. As discussed in the next section, first steps towards such

---

[1] see also http://sir.unl.edu/

[2] http://a4s.ist.tugraz.at/downloads/ieaaie15.zip

Table 1. Results for programs with single and double faults

Program	F	I/O	S	C	V	$|\Delta| = 1$		$|\Delta| = 2$		R(%)
						#	$T_{avg}(ms)$	#	$T_{avg}(ms)$	
Loan-V1	1	2/1	9	17	12	8	19.03	0	218.97	11.0
Loan-V2	1	2/1	9	17	12	8	15.52	0	14.15	11.0
Loan-V3	2	2/1	8	15	12	3	18.84	4	15.82	
Loan-V4	2	2/1	8	15	12	3	17.85	4	15.99	
Loan-V5	2	2/1	8	15	12	3	15.37	0	12.54	
Loan-V6	2	2/1	8	15	12	3	15.81	0	12.89	
Bmi-V1	1	2/1	8	15	11	4	10.93	0	3.86	50.0
Bmi-V2	1	2/1	9	18	13	7	12.50	0	4.56	22.2
Bmi-V3	1	2/1	8	15	11	6	12.33	0	4.21	25.0
Bmi-V4	1	2/1	8	15	11	6	12.07	0	4.08	25.0
Bmi-V5	2	2/1	8	15	11	4	8.95	0	4.08	
Bmi-V6	2	2/1	8	15	11	4	8.73	0	3.93	
Bmi-V7	2	2/1	8	15	11	4	9.29	0	3.87	
Bmi-V8	2	2/1	8	15	11	4	8.88	0	3.72	
Tri-V1	1	3/1	5	12	10	3	4.31	0	3.28	40.0
Tri-V2	1	3/1	5	12	10	3	4.33	0	3.15	40.0
Tri-V3	1	3/1	5	12	10	3	4.30	0	3.37	40.0
Tri-V4	1	3/1	10	29	21	10	4.53	0	3.36	0.0
Tri-V5	2	3/1	5	12	10	2	4.52	1	3.80	
Tri-V6	2	3/1	5	12	10	2	4.53	1	3.77	
Tri-V7	2	3/1	5	12	10	2	3.97	1	3.17	
Tri-V8	2	3/1	5	12	10	2	4.22	1	3.21	
Geo-V1	1	2/2	5	13	10	2	5.52	2	6.48	60.0
Geo-V2	1	2/2	5	13	10	2	7.21	2	6.51	60.0
Geo-V3	2	2/2	5	13	10	2	9.47	1	5.46	
Gco-V4	2	2/2	5	13	10	3	9.70	0	4.16	
Geo-V5	2	2/2	5	13	10	1	8.21	5	12.15	
Geo-V6	2	2/2	5	15	10	1	7.74	2	9.34	
Tcas-V1	1	12/1	74	98	71	14	597.86	15	6200.84	81.0
Tcas-V2	1	12/1	74	98	71	18	602.45	2	4402.64	75.6
Tcas-V3	1	12/1	74	98	71	8	527.77	12	6374.44	89.1
Tcas-V4	1	12/1	74	98	71	18	653.73	9	4891.03	75.6
Tcas-V5	1	12/1	75	103	73	18	619.85	9	4217.57	76.0
Tcas-V15	2	12/1	75	96	77	7	413.66	13	3632.64	
Tcas-V40	2	12/1	73	90	73	13	613.24	9	2135.63	
Calc-V1	1	3/1	12	27	22	12	2.87	0	2.38	0.0
Calc-V2	1	3/1	8	22	19	8	2.79	0	1.78	0.0
Calc-V3	1	3/1	13	35	27	13	19.68	0	5.33	0.0
Calc-V4	2	3/1	12	28	18	12	2.32	0	2.36	
Calc-V5	2	3/1	15	39	27	15	3.13	0	3.60	

a study are the development of a modeling guide in order to support an efficient use of a solver, and we also need appropriate metrics for comparing the information delivered by our diagnoses with techniques like slicing.

5   Discussion and Future Work

In this paper we propose a focused approach at the functional diagnosis of software programs for failing test cases. Implementing a consistency-oriented model-based diagnosis approach, we compute corresponding diagnoses directly from a constraint satisfaction problem that encodes the program's control flow graph that can also be annotated with auxiliary data (e.g. for black box parts). Aiming to accommodate the sheer stunning complexity of today's programs, we focus our search on a single execution's trace in the control flow graph, instead of considering a full-fledged temporal model like an automaton. The diagnoses then tell which constraints involved in this trace could resolve the issue, or where a correct program might leave the trace.

Mayer and Stumptner [9] give an overview of related work on model-based software debugging (MBSD). Wotawa et al. [16] suggest an approach that also relies on constraint solving. Our work differs in one major aspect: instead of the source code, we rely on a concrete program execution. Thus, our approach could not eliminate as many diagnoses as [16], but has lower computational complexity since non-executed branches are abstracted. Wotawa [15] presented a lightweight MBSD approach that combines dynamic slicing with hitting-set computation. This approach captures which components participated in the computation of an erroneous value. In contrast, our approach captures the values of variables computed in the single components. Another difference concerns the diagnosis computation: while [15] computes the diagnoses via a hitting-set algorithm, we obtain the diagnoses directly from the constraint solver. The approach presented in [5] also relies on a concrete execution instead of the whole program. However, our work differs from theirs in particular in the used encoding: our new approach explicitly considers the possibility of changing the execution trace due to a fault.

First experiments with standard SOA-, testing-, and debugging examples showed that our focused approach allows for attractive runtimes. Future work will contain a more detailed experimental study, taking into account also the effects on the diagnostic performance of weakening and strengthening the auxiliary data. An important step for such a study will be to develop a modeling guideline that allows users to write models that are easily solvable and minimize consumed ressources (i.e. in terms of memory). While we currently focus on a weak fault model in that SD does not contain any information on faulty behavior, implementing strong fault models (encoding optional behavioral modes) is up to future work. In the context of reliability issues (and other non-functional properties of SOAs or cloud applications) moving from persistent faults to intermittent faults could be of interest. Extensions regarding concurrent and non-deterministic programs are further interesting topics. Additionally, a comparison of the approach with slicing techniques will be of interest.

Acknowledgments. The research leading to these results has received funding from the Austrian Science Fund (FWF) under project references P23313-N23 and P22959-N23.

References

1. Brandis, M.M., Mössenböck, H.: Single-pass generation of static assignment form for structured languages. ACM TOPLAS **16**(6), 1684–1698 (1994)
2. Gent, I.P., Jefferson, C., Miguel, I.: Minion: A fast scalable constraint solver. In: Proceedings of ECAI 2006, Riva del Garda, pp. 98–102. IOS Press (2006)
3. Greiner, R., Smith, B.A., Wilkerson, R.W.: A correction to the algorithm in Reiter's theory of diagnosis. Artificial Intelligence **41**(1), 79–88 (1989)
4. Hofer, B., Jehan, S., Pill, I., Wotawa, F.: Functional diagnosis of a SOA's BPEL processes. In: International Workshop on Principles of Diagnosis (DX) (2014)
5. Hofer, B., Wotawa, F.: Combining slicing and constraint solving for better debugging: The conbas approach. Adv. Software Engineering 2012 (2012)
6. de Kleer, J., Williams, B.C.: Diagnosing multiple faults. Artificial Intelligence **32**(1), 97–130 (1987)
7. Korel, B., Rilling, J.: Dynamic program slicing methods. Information & Software Technology **40**(11–12), 647–659 (1998)
8. Mayer, W., Friedrich, G., Stumptner, M.: On computing correct processes and repairs sing partial behavioral models. In: ECAI. Frontiers in Artificial Intelligence and Applications, vol. 242, pp. 582–587. IOS Press (2012)
9. Mayer, W., Stumptner, M.: Model-based debugging - state of the art and future challenges. Electron. Notes Theor. Comput. Sci. **174**(4), 61–82 (2007). http://dx.doi.org/10.1016/j.entcs.2006.12.030
10. Metodi, A., Stern, R., Kalech, M., Codish, M.: Compiling model-based diagnosis to Boolean satisfaction. In: 26th AAAI Conference on Artificial Intelligence, pp. 793–799 (2012)
11. Nica, I., Pill, I., Quaritsch, T., Wotawa, F.: The route to success - a performance comparison of diagnosis algorithms. In: International Joint Conference on Artificial Intelligence, pp. 1039–1045 (2013)
12. Pill, I., Quaritsch, T.: Behavioral diagnosis of LTL specifications at operator level. In: International Conference on Artificial Intelligence, pp. 1053–1059 (2013)
13. Reiter, R.: A theory of diagnosis from first principles. Artif. Intelligence **32**(1), 57–95 (1987)
14. Wotawa, F., Schulz, M., Pill, I., Jehan, S., Leitner, P., Hummer, W., Schulte, S., Hoenisch, P., Dustdar, S.: Fifty shades of grey in SOA testing. In: 9th Workshop on Advances in Model Based Testing; 2013 IEEE Sixth International Conference on Software Testing, Verification and Validation Workshops (ICSTW), pp. 154–157 (2013)
15. Wotawa, F.: Fault localization based on dynamic slicing and hitting-set computation. In: The 10th International Conference on Quality Software (QSIC). IEEE (2010)
16. Wotawa, F., Nica, M., Moraru, I.: Automated debugging based on a constraint model of the program and a test case. The journal of logic and algebraic programming **81**(4), 390–407 (2012)
17. Zhang, X., He, H., Gupta, N., Gupta, R.: Experimental evaluation of using dynamic slices for fault localization. In: Sixth International Symposium on Automated & Analysis-Driven Debugging (AADEBUG), pp. 33–42 (2005)

Smart Home Energy Conservation
Based on Context Awareness Technology

Dai Bin[1,2], Rung-Ching Chen[2(✉)], and Kun-Lin Chen[1,2]

[1] College of Computer and Information Engineering, Xiamen University of Technology,
Xiamen, China
bdai@outlook.com, 405740324@qq.com
[2] Department of Information Management, Chaoyang University of Technology,
Taichung, Taiwan
crching@cyut.edu.tw

Abstract. Significance of smart home is to provide the greatest degree of home comfort and convenience, while reducing energy consumption and expenses. Considering the scenario of intelligent analysis and real-time response, context awareness may be having the greatest feature and utilization. With the development of concept for smart home, more detailed needs and methods have emerged like smart home response mode, embedded processing mode, high-speed response mechanisms, energy optimization, cloud data analysis and processing. Focus on automatic processing, automatic response to situational changes, combined with the recommended mode of optimize energy consumption, for automatically respond to the surrounding context change, we developed an embedded smart home framework, it can also reduce home energy consumption. This framework uses efficient cloud data storage and analysis. To verify the actual performance of the framework, we designed an embedded control system and the experimental results show that framework is able to meet the purpose of the design and can be applied to smart home scenario, which can play a role in reducing energy consumption and can enhance the comfort of home.

Keywords: Smart home · Context awareness · Cloud computing · Energy consumption

1 Introduction

The future we used to dream is gradually turned into reality, the continuous development of smart technologies, including smart sensors, mobile Internet, Ubiquitous Computing, middleware and agent technologies, machine learning lead us into "smart world". As one of important patterns of ubiquitous computing, Smart home closely related to our daily life and played a special and important position.

As an associated application of ubiquitous computing, smart home is the key to provide services for human needs, humane care, home security, energy consumption, health care and etc. It is also able to autonomously acquire and applies knowledge

© Springer International Publishing Switzerland 2015
M. Ali et al. (Eds.): IEA/AIE 2015, LNAI 9101, pp. 722–731, 2015.
DOI: 10.1007/978-3-319-19066-2_70

about its inhabitants and their surroundings, and adapts to the inhabitants' behavior or preferences with the ultimate goal to improve their experience. [1]

Cloud computing has recently emerged as a new paradigm of ubiquitous computing that becomes significant for both private and public sectors[2]. The ultimate principle of cloud computing is that the computing is "in the cloud". It can quickly adapt to changing load owning to its powerful computing capabilities and massive storage capacity [3]. It deals with accessing applications or storing data in the "Cloud" via the Internet or a network and using associated services.

For automatic processing of the change of environmental parameters, we present an adaptive embedded framework which can automatically change work parameters according to the environmental context. This framework will adjust the work status of home appliances according the data collecting from the senses setting in the home. It also can adjust environmental parameters follows the activities of the residents. This framework has been put into practice and been verified to be efficient on improving human care and energy consumption. For considering the convenience of data processing and analysis, cloud computing and storage are adopted.

This paper is organized as follows. A schema of combined smart home, cloud computing and embedded control system is displayed in section 2. In the section 3, the overall structure and module components will be introduced. Section 4 will display the actual verification system and functions. We will discuss the performance of system and functionality in the section 5. Finally, we will conclude the paper in section 6.

2   Related Works

Type of context awareness services can be classified by contextual information sources and context service model, Contextual information[5, 6], in accordance with the classification of Schmidt, Beigl & Gellersen and other scholars, can be divided into User context as well as Physical context, So-called User context associated with the user, such as age, sex, emotions, habits, preferences, abilities, and refers to the Physical context such as time, location, temperature, light, the environment facilities. In accordance with the definition of David Kotz, the context service mode can be divided into Active context and Passive context, active service mode refers to the system will change behavior patterns in accordance with the received context information, such as: In the meeting, the phone will be pre-entered in accordance with the time on the meeting calendar, will switch to mute the ringtone or reject. Passive service mode refers to the system will change the presentation mode in accordance with the received context information, for example: the online ticketing system will recommend related movies according to the past preferences of customers.

Smart home is one of the most important realizations of context awareness service and pervasive Ubiquitous Computing. In the smart home environment, changes may occur at any time, smart home systems need to collect different signals from various types of sensors and analysis, processing and to judge the data for performing related tasks automatically based on the system settings. In this framework, infrared sensors will be used to detect human activities, light sensors to provide the light intensity data and temperature sensors provide temperature data. Meanwhile, to improve system response speed and enhance the stability, embedded control systems will be adapted and this approach also will be flexible for expense consideration.

Greatly improved stability and popularization of cloud computing has made the combination of smart home and cloud computing already the trend of the times. Cloud computing technology can provide good scalability for the framework. Context data collected by the smart home processing in the cloud, which not only can better release the pressure of local computing, while the cloud data storage and analysis can also be an excellent way to enhance the meaning and value of contextual data. Furthermore, from the cloud platform we can also get more information support and service sources. In this architecture, context data upload to the cloud through the Internet, energy consumption can be analyzed by consumption mode, and optimization scheme can be recommended according to the rules and strategy. For instance: according to different electricity consumption price of in different period or different amount consumption, combination recommendation will be given. Moreover, in the cloud data can also be used for the whole network retrieval, data combination, data mining and other field analysis.

3   Framework Design

3.1   Framework Overview

With the intelligent digital equipment in smart home continues to increase and update, how to adapt to this elastic demand, system expansibility is good or not is an important factor to consider. So, entirely depends on the local processing ability to adapt to the needs is not suitable, so the use of cloud computing is a reasonable solution.

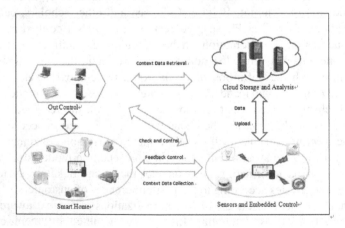

Fig. 1. Structure of Framework

As mentioned before, the embedded system has the advantages of low energy consumption, stable operation and low cost. Meanwhile, cloud computing can alleviate local computing pressure, transform the increasing system pressure into the cloud computing processing, and cloud-based computing need not to consider the problem of processing capacity[4]. Therefore, the scenario changes based intelligent environmental parameters and adjust its processing capacity into the clouds in the architecture design is an important consideration. The structure of the system framework is shown in Figure 1.

3.2   Description of Framework Function

The framework mainly consists of four parts: Context awareness data collecting embedded central control system, cloud computing and analysis, external control and interaction. Specifically, using various types of sensors for activities and the surrounding environment data collection and analysis, such as temperature and humidity sensors, smoke sensors, light intensity sensors, infrared sensors etc. At the same time, in order to simplify the realization of application level and cost considerations, using embedded system as the central control center. According to the feedback of sensor data, embedded control center can quickly adjust the work status of intelligent appliance.

In addition to automatic adjustment function of control center, but also from the external control devices can view, monitor, control, analyze the state of home appliances via Bluetooth, WIFI or cable network etc. To alleviate the pressure from local embedded computing control center to cloud, using cloud computing to solve the retrieval of data, analysis and other tasks is a better choice. Moreover, by adopting this structure, we can complete the data retrieval and analysis requirements at any time. If accompanied by Data Mining function, it can complete the preference analysis, energy consumption of recommended scheme and other functions.

For example: when the fire occurs, smoke and temperature sensors will monitor the abnormal value, the abnormal value will be sent to the control center, the control center will take the preset action, such as the fire emergency system will be activated, notice the resident by SMS and inform fire department at the same time, and uploads the data for cloud storage.

4   System Architecture Level and Analysis

For the actual operation test of the architecture, we designed a simple application system to the realization of this architecture function and analyzes its advantages and disadvantages. The architecture consists of the system is shown in Figure 2.

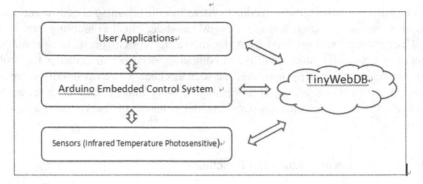

Fig. 2. System architecture level

4.1   Data Collection Using Sensors

Temperature sensors are used in smart hone for monitoring the ambient temperature, when the ambient temperature is lower or higher than the set value (E.g. above 30 degrees Celsius), the control center will use air-conditioning equipment for temperature control, the temperature will be adjusted to comfortable degree (e.g. near 26 degrees Celsius). Infrared sensors are used for tracking indoor activities, when the time meet the set value, light intensity can be automatically adjusted according to the collected data from light sensors. When the surrounding light is high, the light intensity will be reduced; when residents leave the room for a long time, lights will automatically turn off. In this way, the system can play a role in reducing energy consumption.

4.2   Embedded Control Center

In order to verify the work status of the framework, we use Arduino Platform as an embedded control central. Arduino is an open-source, single-chip microcomputer, which uses the Atmel AVR microcontroller, based on the open source software and hardware platform, built on open source simple I/O interface board, and has development environment using a similar Java, C language [7].

Arduino can use Arduino language and Macromedia Flash, Processing, Max/MSP, Pure Data and Supercollider, combined with electronic components, such as switches or sensors or other control device, LED, stepper motor or other output device to make interactive works. The Arduino can also operate independently as an interface can communicate with software [8].

As a reliable and low-cost embedded system is, Arduino can be simply connected with sensors such as infrared, ultrasound, thermistor, photoresist or, servo motors. Arduino also can also support various interactive programs, such as Adobe Flash, Max / MSP, VVVV, Pure Data, C, Processing ... etc.

For example: HC-06 Bluetooth module used in Arduino for the connection to external control centers, the photosensitive sensor is used for sensing the surrounding environment light intensity, when the light intensity changes, serial data output will give the order to adjust the brightness of the light so as to achieve the purpose of reducing electricity consumption. Meanwhile, the infrared sensor can also be used for indoor activity tracking, if there is no activity data for a long time, the control center will automatically turn off lights and appliances, which can also reduce energy consumption and safe use of electricity.

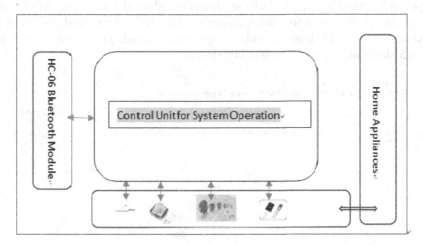

Fig. 3. Architecture of smart home control center

4.3   Client Application Design

For verifying the performance of the framework, we designed an Android application to communicate with the control center. Use Bluetooth to connect to the control center, the user can use the application program to control and adjust the operating status of home appliances, and also can implement device control, data retrieval, energy consumption analysis, schema optimization [9]. Figure 4 shows some of screens of the application program.

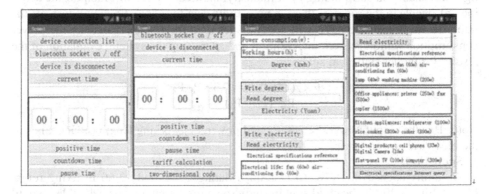

Fig. 4. Interface of client application

4.4   Cloud Data Storage and Analysis

In the data storage part, we use the TinyWebDB database of Google APP Engine, as a small cloud database system, TinyWebDB database fully consistent with the requirements of scene. Upload performing data acquisition and control center of the sensor data into the TinyWebDB database for data retrieval and analysis. The system uses the powerful calculating ability of the clouds to reduce the local load of operation. Upload data from sensors to control center for data retrieval and analysis, it can take advantage of powerful cloud computing capabilities, reducing local computation load. Figure 5 shows the Google TinyWebDB system.

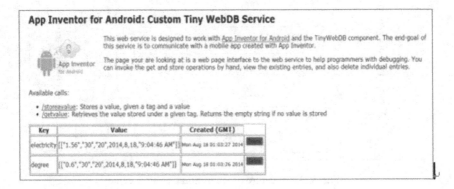

Fig. 5. Google TinyWebDB

5   Discussion

To verify the functionality of the architecture and practical operation, we consider the use the system described above to count the consumption of electricity in a family scene. For example, the residents come home at six o'clock every afternoon, go to sleep at 12:00 PM. During this period the light sensor will adjust the brightness of light depending on surrounding light intensity; infrared sensors will monitor the activities of people in the room, if there is no activity data for a long time, will turn off the lights to save electricity. We will set statistics time for a week, calculate the power consumption without system intervention and after using the system, evaluation of the results for further statistical analysis.

$$P = \begin{cases} 20 * t + x\,, (t \geq 0, x = 0.5 * P_e, P < P_e) \\ P_e,\ (P = P_e) \end{cases} \tag{1}$$

Note: P is the output power, P_e is the rated power, t is working length (hour), x is half the electrical power rating.

Scenario 1:

A family opened the 80W LED Lamps during 19:00 and 23:00 in living room. Meanwhile, at 20:00, the child stay in study room for learning, open a 40W lamp until to 22:00. Calculate the difference of electricity consumption of without the use of intelligent home control system and using the intelligent home control system, the statistical data analysis chart as Figure 6 , Table 1 and Figure 7, Table2.

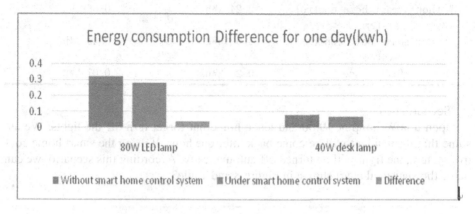

Fig. 6. Energy consumption difference for one day

Table 1. Energy consumption difference for one day

Electric type	80W LED lamp	40W desk lamp
Without smart home control system	0. 32 kWh	0. 08 kWh
Under smart home control system	0. 28 kWh	0. 07 kWh
Difference	0. 04 kWh	0. 01 kWh

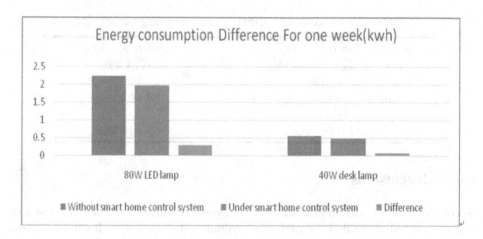

Fig. 7. Energy consumption difference for one week

Table 2. Energy consumption difference for one week

Electric type	80W LED lamp	40W desk lamp
Without smart home control system	2.24 kWh	0.56 kWh
Under smart home control system	1.96 kWh	0.49 kWh
Difference	0.28 kWh	0.07 kWh

Scenario two:

Open a 40W lamp at 19:00 and leave home but forget turn off the lights. We assume that the resident will be come back after one hour. If using the smart home control system, the light will be turned off automatically. According this scenario, we can make the statistical chart shown in Figure 8 and Table3.

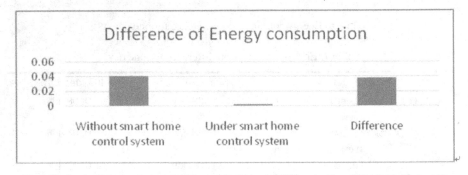

Fig. 8. Energy consumption difference for forgetting turn off light

Table 3. Energy consumption difference for forgetting turn off light

Electric type	40W desk lamp
Without smart home control system	0.04 kWh
Under smart home control system	0.002 kWh
Difference	0.038 kWh

6   Conclusions

Through the above system practical work situation examination, the smart home control system has showed significant effect from the statistical data. It can save about 12% of total consumption, because this system just use a few sensors and cannot provide accurate control, if it might improve, the further energy consumption expense will drop more.

Acknowledgement. This research was supported by the Ministry of Science and Technology, Taiwan, R.O.C., under contract number MOST103-2632-E-324-001-MY3 and MOST 103-2221-E-324 -028.

References

1. Nirmalya, R., Abhishek, R., Sajal, K.D.: Context-aware resource management in multi-inhabitant smart homes: a nash h-learning based approach. In: Proceedings of the Fourth Annual IEEE International Conference on Pervasive Computing and Communications (PERCOM 2006) 0-7695-2518-0/06 (2006)
2. Mina, D., Milan, P., Marco, N., Ilaria, B.: A home health care system in the cloud addressing security and privacy challenges. In: IEEE 4th International Conference on Cloud Computing (2011)
3. Yu, H.Z., Jian, Z., Wei, H.Z.: Discussion of a smart house solution basing cloud computing. In: International Conference on Communications and Intelligence Information Security (2010)
4. Xiaojing, Y., Junwei, H.: A framework for cloud-based smart home. In: International Conference on Computer Science and Network Technology (2011)
5. Schilit, B., Adams, N., Want, R.: Context-aware computing applications. In: IEEE Workshop on Mobile Computing Systems and Applications (WMCSA 1994), pp. 89–101 (1994)
6. Schilit, B.N., Theimer, M.M.: Disseminating Active Map Information to Mobile Hosts. IEEE Network **8**(5), 22–32 (1994)
7. http://arduino.cc/en/Guide/Introduction
8. http://arduino.cc/en/Main/ArduinoBoardUno
9. http://ai2.appinventor.mit.edu/
10. Cook, D.J., Das, S.K.: Smart Environments: Technology, Protocols and Applications. John Wiley (November 2004)

Logic Simulation with Jess for a Car Maintenance e-Training System

Gil-Sik Park, Dae-Sung Park, and Juntae Kim$^{(\boxtimes)}$

Department of Computer Science, Dongguk University, Seoul, Korea
{himdrura,basic,jkim}@dongguk.edu

Abstract. The research on self-directed learning has been accelerated by integrating education and IT technology. The process of self-directed learning in e-learning applications such as Car Maintenance Training is very difficult and complicated. Previous studies on car maintenance training applications provided simple training scenarios with predetermined action sequences. However, trainees must be able to perform various maintenance operations himself and experience various situations for self-directed learning. To provide such functionality, it is necessary to obtain an accurate response for various operations of trainees, but it requires complicated calculations with respect to varieties in the electrical and mechanical processes of a car. In this paper, we develop a logic simulation agent using JESS inference engine in which self-directed learning is achieved by capturing the behavior of trainees and simulating car operations without complicated physical simulations in car maintenance training.

Keywords: E-training · Car maintenance · Logic simulation · JESS

1 Introduction

With the development and spread of computer and internet, place of education is expanded from off-line to on-line. It is caused by the possibility that the educational service of the off-line education can also be provided on-line. E-learning enables interactive communication and activity in education, in which teaching-learning is not restricted by time or space utilizing IT technologies.

However, in e-learning, it is difficult for the teacher to directly concern the trainee's concentration and attitude, and for the student to participate in class spontaneously. According to this, e-learning has developed to e-training, in which trainee performs technical training in a virtual environment just as in the industrial site unlike the lectures with video[4, 7]. E-training not only can draw trainee's interest, but also has lots of advantages to solve safety problem of high risk equipment, cost problem, problem of space, etc.

It is essential to implement simulator to interpret the behavior of trainee and show the corresponding result in the virtual environment for this e-training. However, previous e-training systems only considered the result of the trainee's behavior which were predefined, and cannot properly respond to their various unexpected behavior[6].

© Springer International Publishing Switzerland 2015
M. Ali et al. (Eds.): IEA/AIE 2015, LNAI 9101, pp. 732–741, 2015.
DOI: 10.1007/978-3-319-19066-2_71

There are difficulties in time and economy to realize all the proper responses to trainee's various behavior. Therefore, there was no other choice but to develop e-training system by limiting the range of trainee's behavior and training processes. Since these training systems cannot perform simulation for the trainee's unexpected various behavior, trainee's learning effect is also restricted. Also, the training system along with a scenario made within a template cannot provide self-initiated learning, since the sense of reality and interest decrease.

This research presents an intelligent logic simulator that can respond not only to the predefined behavior in training scenario, but also to the trainee's unexpected various behavior in the car maintenance training education. Trainee's various behavior, which are not included in training scenario, can be perceived and corresponding results are made through reasoning. For this, JESS(Java Expert System Shell) reasoning engine was used. In JESS reasoning engine with the help of rules which are saved in its knowledge base we can consider various trainee's behavior in the form of new facts. These facts are again used to generate another facts consecutively by the rules. Therefore, the simulator based on JESS reasoning model can respond to the exceptional situations and various behavior.

Chapter 2 of this paper shows the related research, and chapter 3 shows the training system design. Chapter 4 shows the implementation of the system, and chapter 5 concludes the paper with future research direction.

2 Related Work

2.1 e-Learning and e-Training

E-learning means a method to promote and enhance learning, as all virtual action or process which is used to acquire technique or knowledge realized in a virtual space by electronic means[12-14]. E-learning is now becoming indispensable, and is the alternative of future education in the expectation to increase the efficiency and effectiveness of education. However, nevertheless the rapid development of e-learning, the reason why the preference of face to face education still appears is because there's a teacher in the center, who promotes learning.

E-training utilizes information and communication technology and various devices to acquire ability necessary for work and improve it. It is a learning model to enable practice based education. E-training can raise the sense of reality and interest, since it can show the result along with trainee's behavior by providing virtual environment. Furthermore, it is a self-initiated learning style of education. It has many advantages such as solving the safety problem of high risk equipment, cost problem, problem of space and place, etc. In the virtual environment for e-training, implementation of simulator that can show the result responding to the trainee's behavior is essential.

In addition to e-training, there are many types of learning utilizing computers: computer base education, which is the system to control the learning process using computers, adaptive e-learning that makes effective decision on the trainee's learning process, u-learning, the learning model which collects learner's situation information and provides diagnosis and service along with this[8, 9, 11, 15, 16].

Recently, as the advanced research related to car maintenance, there is a case to make a program by utilizing Visual Basic language with the environment to perform car maintenance in e-learning, and operate by establishing knowledge base by utilizing Microsoft Access program[1]. Car maintenance e-learning program suggested in this research can perform diagnosis in respond to trainee's behavior which the developer realized, however, it is difficult to diagnose the trainee's behavior which the developer did not realized. Therefore, there is a disadvantage of having difficulty in self-initiated learning, since it can only be performed according to the supervisor's preplanned scenarios.

The e-learning systems, which were developed previously, is difficult to respond to trainee's exceptional behavior one by one, which was not realized by the program developer. Since it cannot realize all of trainee's various behavior, it limits trainee's behavior, learning process, simulation range, etc. Therefore, the existing e-learning systems have restriction to perform self-initiated learning .

2.2   JESS Reasoning Engine

JESS is the abbreviation of Java Expert System Shell, which is a rule base reasoning engine, as well as knowledge base system development environment, which are operated in JAVA platform. It was developed by the U.S. Sandia National Laboratories in 1995 as a reasoning engine [10]. It succeeds the concept of CLIPS rule-based expert systems, and is simple to define rules, while it is the language similar to LISP.

Since JESS is operated in JAVA platform, various APIs of JAVA, e.g. networking, graphic, database, etc. can be utilized. This means that JESS can be applied in various environments of JAVA base such as Java Program, Java Applet, JSP (Java Server Page), etc. Also, the program developed with JAVA has the advantage that Java Virtual Machine interprets byte code and can be executed in various platforms. Therefore, JESS, also, which is operated in JAVA platform, has the advantage that it can be operated in various platforms which support JAVA. Also, it has an advantage that it doesn't require compile, and the knowledge base can be established by anyone who knows the grammar of composing rule without modification of program. Rule base reasoning engine as JESS uses facts in the working memory, generate a new fact by using matching rules in the knowledge base, then save the fact in working memory.

Existing researches utilizing JESS were mostly about implementing expert systems.Recently, there are researches which apply the ontology techniques to car system. AlHamad et al. [2] suggested the method of personalization of learning information based on the students' profile in learning management system in e-learning environment by using JESS. This research defined the problem to comprehend the students' level with template, and designed to proceed a learning scenario which is recommended along with the number of correct answer and wrong answer for the students' questions.

Other researches including various expert systems such as learning process recommendation system using JESS, and context aware system, combined reasoning with ontology, etc. are also being conducted[3, 5].

This paper presents the method how to express the complicated and huge car system and maintenance training process, and how to compose the rule by using JESS. And this paper also presents the design and implementation of the intelligent agent which can infer next states and conditions of the car through reasoning on the current condition of the car and the trainee behavior.

3   System Architecture

3.1   Car Maintenance Training

Car maintenance training is a process that a trainee finds out the cause along with the failure situation of a car, which was set beforehand by supervisor and take a proper step accordingly. E-training system, suggested for this, realized the simulation for the basic activities related with car operation, such as starting a car, changing gear, etc. It also realizes general activities of a car and the simulation related with maintenance activities defined in training process, the various maintenance tools such as multimeter, scan tool, etc. When we classify car maintenance training process, it can be divided into engine and car body, and again subdivides the phenomenon in engine, it can be divided into starting defect, engine defect, starting break, output acceleration performance defect, and fuel efficiency defect. Fig. 1 is a hierarchy diagram to classify the inspection process of car starting defect case. In Fig. 1, in case of car starting defect, maintenance process becomes different along with the availability of engine rotation. In case when engine rotation is not available, maintenance process again can be more subdivided along with the operation availability of solenoid. First, when the supervisor sets malfunction in order to perform inspection by selecting one of the various malfunction inspection scenarios, trainee can solve the malfunction problem set by the supervisor by performing each inspection process.

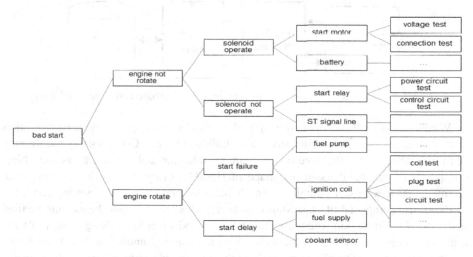

Fig. 1. A part of car maintenance training process hierarchy

For example, if the supervisor sets malfunction with battery defect, trainee can confirm that the starting is unavailable while starting the process, so that the trainee checks the rotation of start motor, also, checks solenoid operation availability when start motor does not rotates. If solenoid works properly, the trainee checks the lighting of the warning light and indicating light, if it is not turned on, the trainee will replace battery by checking voltage value is not normal after measuring voltage value by using multimeter to inspect battery.

3.2   Composition of Car Maintenance Training System

The e-training system is composed of supervisor (tutor), trainee, training expert, the knowledge base with the rule related to car maintenance training process, reasoning engine (JESS), and training simulator.

The supervisor enters the scenario of car maintenance training beforehand, and the trainee operates and inspects the car along with the given situation. Training simulator receives the entered car maintenance operation of the trainee and delivers to reasoning engine, and reasoning engine (JESS) performs reasoning with the delivered action. It delivers the advice on the wrong behavior to the trainee along with reasoning result of trainee's behavior, or enables the supervisor to concretely comprehend process performance information of trainee by delivering trainee's training record to the supervisor. Training process is composed of the training goal set by the car maintenance expert and training scenarios along with it. Training scenario composed by the expert is defined as the reasoning rule of trainee's action. Simulation agent updates training situation along with the domain knowledge related to the training by monitoring trainee's action. Fig. 2 shows the structure of the system.

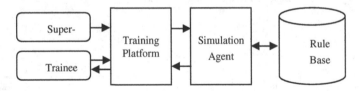

Fig. 2. The system structure of car maintenance training based on logic simulation

When the supervisor sets cause of malfunction in training platform, trainee takes various training behavior while performing training process. Consequently the training platform delivers the information of trainee's behavior and condition, as the object composed of the members in the shape of HashMap<Key, Value> to the simulation agent. In other words, it contains trainee's behavior and supervisor's setting condition in the object composed of HashMap members, and when calls out the execute method in JESS API, Simulation Agent comes into action and saves in working memory by the reasoning engine which reads the value from the object. Simulation Agent saves reasoning result also in HashMap, not only in the working memory, when the state change occurs in variable while reasoning. Simulation Agent performs reasoning by applying fact and rule saved in Knowledge base along with the Value(trainee's behavior) for the of HashMap Key(variable which saves trainee's behavior or supervisor's malfunction setting). Finally, simulation result is suggested, while reasoning result is delivered to training platform by being converted into the object composed with HashMap shape.

3.3   The Design of Training Simulator

In car maintenance training simulator, supervisor should be able to set the initial value of the car condition or the malfunction cause regarding car maintenance training. Trainee should be able to perform the process to operate and inspect the car along with the phenomenon on the car without knowing the malfunction cause. Intelligent agent should be designed to perform reasoning the current condition of the car with the condition set by the supervisor and the trainee's behavior of operating the car by exchanging data with knowledge base composed of JESS Rule. In order to build a simulator, firstly it is important to set the realization scope of the relevant field, also, a work to establish the expert knowledge as knowledge base.

For the car maintenance training, the car, which is the object of maintenance, should be prepared for simulation, not only the realization of car maintenance process, and maintenance tool should be arranged. Also, reasoning should be performed based on the electro-mechanical principles.

Maintenance process along with the phenomenon on the car can be divided into engine and car body. Among them, engine inspection process can be divided into 5 stages, 1. starting defect, 2. engine defect, 3. starting break, 4. output acceleration performance defect, and 5. fuel efficiency defect.

Fig. 3 shows the relation between this training process and simulator reasoning rule. JESS rule has the form of IF-THEN shape, IF phrase writes the cause of an event, and THEN phrase writes the result along with the cause.

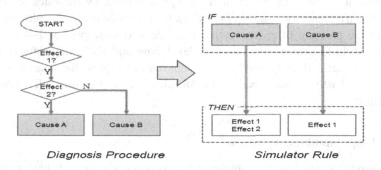

Fig. 3. The relationship between a diagnosis procedure and simulator rules

Therefore, the phenomena in the upper node in the training process flow chart should be composed in THEN phrase, and the cause and inspection process in lower node should be composed in IF phrase. Eventually, rule is composed in the shape as if training process flow chart is reversed, and Fig. 3 shows this general relation.

When expressing these rules and the variables designed beforehand with JESS rule, it becomes as Fig. 4. In Fig. 4, the parts or car condition which supervisor sets, which are SP variable, trainee's behavior, which is TR variable get in IF phrase, and current state of simulator gets in LS variable, also, the state of simulator which appears as the result, or the engine state, measurement result of inspection tool, etc. are expressed in THEN phrase.

```
(defrule rule_name
<SP : car problem setting> <TR : trainee's behavior>
<LS : current state of the simulator> <LS : current state of the car>…
=>
<LS : new state of the car> <LS : new state of the simulator>…)
```

Fig. 4. Logic simulation rule in JESS form

4   Implementation and Result

4.1   JESS Rules

The rule of reasoning car behavior in this research can be divided according to each function. Define_facts.clp file defines the template to save fact before composing the rule for reasoning. Also, it defined slot, fact, and other necessary parts for reasoning. In tool_rule.clp (61 rules), rules related to measurement tools such as scan tool, multimeter, charger, gage bar, and other   necessary parts for car maintenance are defined. Measurement results that are created through reasoning by the rules are defined in tool_rule.clp and electric_rule.clp (47 rules). In compression.clp (35 rules) file, rules to measure the pressure of engine cylinder are defined in respect of 1~4 cylinders. In etc_system.clp (17 rules) file, rules of developing and abolishing throttle valve and idle, etc. are defined. In igcoil.clp (16 rules), rules to express the state of ignition coil are defined. Lastly, in malfunction.clp (58 rules), rules of engine defect phenomenon are defined.

4.2   Logic Simulator Program

Logic simulator program was implemented to check simulator behavior, which was realized with Java, as Fig. 5. Simulator screen is divided into the left and the right. The taps for the supervisor and trainee to set the variable are on the left panel, and the result after JESS engine finishes reasoning on the entered value set on the left panel. For example, when the supervisor set malfunction on the left panel for training, trainee takes action for maintenance, and the result of reasoning on this entered value can be checked on the right panel. Right panel is divided into upper and lower part, reasoning result of entered value is printed out on the upper panel, and on the lower panel, internal state variables are printed out. Internal state variable exists as the abstract variable, which even cannot be actually checked by the trainee. It is composed to check the intermediate state how the simulator logically operates.

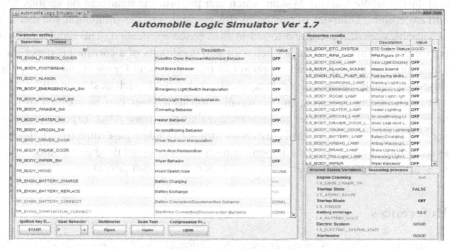

Fig. 5. The logic simulator program

In order to check the correctness of reasoning result by the simulator, the trainee's maintenance process regarding starting defect phenomenon is tested. To start car simulator, foot brake variable was set as TRUE. In the car applied with smart key, it is converted to ON and ACC state which supplies power when pushing smart key without stepping on the brake. Actually, in the simulator, when pushing start key without stepping on foot brake, it can be confirmed that only power is supplied without starting. When pushing start key after stepping on the brake, it can be confirmed that car is started with engine cranking sound

Fig. 6 shows the screen of performing simulation on a virtual car maintenance training system developed by Techvill Co.,Ltd. The logic simulation system model presented in this paper can be utilized in e-training system along with virtual reality and behavior recognition technologies.

Fig. 6. The car maintenance training system using the logic simulation

5   Conclusion

This paper presents design and implementation of a logic simulator using JESS, and its application to car maintenance training systems. While researches on the small size expert system establishment using JESS were conducted, there has been no research on simulating whole training system which can perform the treatment on various actions and exceptional behavior of trainee, as this research. For the simulation, the hierarchical variables are designed, and tested in Java based application. The simulation system performs reasoning on trainee's behavior and conditions of a car. In this paper the implemented logic simulator was applied to a training system for car maintenance, but the proposed simulation model can be applied to other fields such as ship or aviation.

For the future research, the personalization techniques can be adapted to this research to establish a better maintenance training system. Also, expert system tools should be developed to easily establish rule-knowledge base directly. If the expert system tool is developed car experts can correct and maintain the simulator, and add new function without the help of system expert or program developer.

References

1. Adsavakulchai, S., Ngamdumrongkiat, N., Chuchirdkiatskul, E.: E-Learning for Car Faulty Diagnosis. International Journal of Information and Communication Technology Research (IJICT) **1**(1), 20–26 (2011)
2. AlHamad, A.Q., Al-Omari, F., Yaacob, N: Applying JESS rules to personalized learning management system (LMS) using./ online quizzes. In: Interactive Collaborative Learning (ICL), pp. 1–4 (2012)
3. AlHamad, A.Q., Akour, M.A., Al-Omari, F.: Personalizing student's profiles using JESS in a learning management system (LMS). In: International Conference on Interative Mobile and Computer Aided Learning (IMCL), pp. 109–112. IEEE (2012)
4. van Dam, A., Laidlaw, D.H., Simplson, R.M.: Experiments in Immersive Virtual Reality for Scientific Visualization. Computers & Graphics **26**, 535–555 (2002)
5. Nambiar, A.N., Dutta, A.K.: Expert system for student advising using JESS. In: International Conference on Educational and Information Technology (ICEIT), vol. 1, pp. 312–315 (2010)
6. Shelton, B.E.: How Augmented Reality Helps Students Learn Dynamic Spatial Relationships. Unpublished doctorial dissertation, University of Washington (2003)
7. Dede, C.: The evolution of constructivist learning environments: Immersion in distributed virtual worlds. Educational Technology (1995)
8. en.wikipedia.org/wiki/elearning
9. Mikic Fonte, F.A., Burguillo, J.C., Nistal, M.L.: An intelligent tutoring module controlled by BDI agents for an e-learning platform. Expert Systems with Applications **39**(8), 7546–7554 (2012)
10. http://www.jessrules.com
11. McLuhan, M.: Understanding Media: The Extensions of Man (2011). MIT Press, Cambridge (1994)
12. Wang, M.: Integrating organizational, social, and individual perspectives in Web 2.0-based workplace e-learning. Information Systems Frontiers, Springer **13**, 191–205 (2011)

13. Houari, N., Far, B.H.: Application of intelligent agent technology for knowledge management integration. In: International Conference on Cognitive Informatics, pp. 240–249. IEEE (2004)
14. Yao, Y.-H., Trappey, A.J.C., Ku, C.C., Lin, G.Y.P., Tsai, J.P., Ho, P.-S., Hung, B.: Develop an intelligent equipment maintenance system using cyber-enabled JESS technology. In: International Conference on Mechatronics, pp. 927–932. IEEE (2005)
15. Kwon, S., Lee, J.: A Development Study of Learner Profile Model for u-Learning. Hanyang University Institute of Educational Technology, pp. 1–26 (2010)
16. Kim, J., Kim, B., Jung, H., Choi, E.: Ontology Reasoning in Intelligent U-learning for adaptation of learning contents. Korean Society for Internet Information 9(2), 267–270 (2008)

Dynamic Scoring-Based Sequence Alignment for Process Diagnostics

Eren Esgin[1(✉)] and Pinar Karagoz[2]

[1] Informatics Institute, Middle East Technical University, 068000 Ankara, Turkey
e124619@metu.edu.tr
[2] Computer Engineering Department, Middle East Technical University, 068000,
Ankara, Turkey
karagoz@ceng.metu.edu.tr

Abstract. Even though *process-aware information systems* are intensively utilized in the organizations, traditional process management paradigms majorly concentrate on the design and configuration phases. Instead of starting with a process design, *process mining* attempts to discover interesting patterns from process enactment namely event logs and extract business processes by distilling these event logs as knowledge base. One of the challenging issues in process mining domain is *process diagnostics*, which is complex and sometimes infeasible, especially when dealing with real-time, flexible and unstructured processes. In this aspect *sequence alignment* is applicable to find out common subsequences of activities in event logs that are found to recur within the process instances emphasizing some domain significance. In this study, we focus on a hybrid quantitative approach for performing process diagnostics, i.e. comparing the similarity among process models based on *dominant behavior* concept, *confidence metric* and *Needleman-Wunsch algorithm with dynamic pay-off matrix*.

Keywords: Process mining · Sequence alignment · Process diagnostics · Needleman-Wunsch algorithm with dynamic pay-off matrix · Confidence metric

1 Introduction

The market dynamics due to *globalization* are so rapid, instable and costly that only the most flexible forms of organizations will be able to interpret the trends of change to become more competitive. Unfortunately process design phase in this enterprise transformation is influenced by personal perceptions, e.g. business blueprints and reference process models are often normative in the sense that they reflect what *should* be done rather than describing the actual case [3]. Instead of manually designing the business processes, it is anticipated to reverse the design-centric approach by a more objective and automated way of modeling to collect the information related to that business process and discover the underlying process structure from the low-level information history [4]. In this aspect, *process mining* is the paradigm to distill

© Springer International Publishing Switzerland 2015
M. Ali et al. (Eds.): IEA/AIE 2015, LNAI 9101, pp. 742–752, 2015.
DOI: 10.1007/978-3-319-19066-2_72

significant patterns from the *event logs* to discover the business process model automatically [2, 4, 5, 6, 7, 8].

The most important information to improve an existing process design is *where* process participants deviate from the intended process definition. In business process management (BPM) this challenging issue is known as *process diagnostics*, i.e. encompassing process performance analysis, anomaly detection, diagnosis and inspection of interesting patterns [12]. Various hybrid methods at distinct domains, e.g. *sequence alignment* in bioinformatics, are employed to handle these issues.

In this work, we propose a hybrid quantitative approach that exploits sequence alignment for *delta analysis*, i.e. comparing the actual process, represented by a process model constructed through process mining, with *prescriptive* reference process model [8]. The approach initially derives *consensus activity sequence* that synthesizes the major *dominant behavior*. At this step, the method proposed in [11] is utilized for deriving the backbone sequence for the underlying business process. As the second phase, (sub)optimal alignment among derived consensus activity sequences is built up using *Needleman-Wunsch algorithm* with a *dynamic pay-off matrix*. Hence both structural and behavioral characteristics of business processes, which are inherited from *consensus activity sequence* and *confidence values*, are taken into consideration in similarity measurement. The main contribution of this paper lies in the second phase.

This paper is organized as follows: Section 2 includes literature review. Section 3 introduces the design of the proposed quantitative approach in measuring the similarity between the process models due to process diagnostics paradigm. Section 4 emphasizes experimental analysis and finally Section 5 summaries the concluding remarks.

2   Related Work

The equivalence of process models are usually subject to *verification* such as trace equivalence. This stream is based on a comparison of the sets of completed event logs as in [12]. In this work, Aalst does not wish to discover a graphical process model but rather aims to use such logs to check for deviation from a prescribed business process model, since he assumes that the activities of a mined process model are usually on a low level of abstraction.

Because of the existence of spaghetti processes, in [18] it is proposed to implement a new approach based on *trace alignment*. Trace alignment can be used in a preprocessing phase where the event logs are investigated or filtered and in later phases where detailed questions need to be answered. *Progressive alignment approach* for trace alignment is adopted such that; the basic idea of this progressive alignment is to iteratively construct a succession of pair-wise alignments. The selection of traces for alignment is based on their similarity at each iteration.

Unlike to prior trace clustering methods, which are fundamentally based on firing sequences (event logs), it is attempted to construct an abstract representation of the behavior captured by a process model in [19]. This abstraction is named as *casual footprints*. This causal footprint notion has some advantages such as; it includes information about the order of the activities beyond direct succession. The look-ahead and look-back links look at indirect dependencies and therefore it handles structural and behavioral aspects.

Esgin and Senkul [17] propose a distance metric, which is built on the vector model from information retrieval (IR) and an abstraction of process behavior as *process triple*. Process triple is a set, which covers transaction existence and interactions (successor/predecessors of each transaction) among activities. This metric takes into structural and behavioral perspectives into account.

In [9], Esgin and Karagoz aim to demonstrate that *process mining* can benefit from sequence mining techniques, which are strengthened with *standard Needleman-Wunsch algorithm* to quantify the similarities and discrepancies. Unlike to [17], proposed approach evaluates just consensus activity sequences by avoiding the requirement for well-structured process models. In [21], INDEL scores are determined according to the interactions among activities. These interactions are figured out by *confidence metric* and *case-based* INDEL scores are calculated at each iteration by Equation (5). On the other hand, default MATCH/MISMATCH values are set to *confidence FTC threshold value*.

Cook and Wolf [6] present an approach for delta analysis in the context of software processes. They use Artificial Intelligence (AI) algorithms to compare process models and compute the level of correspondence. Additionally they assume that there exists any difference at abstraction level of event logs and discovered process model.

In [20], Aalst aims to handle delta analysis paradigm with a quite distinct approach such that; *frequency profile* for each activity, which is a partial function indicating how many times certain transitions fired, is constructed, because enterprise resource planning (ERP) systems only logs the fact that a specific transaction has been executed without referring the corresponding case. Then an *integer programming* (IP) is formulated to check whether the modeled behavior and the observed behavior match.

Actually, *behavioral semantics* can lead to performance problems due to large sets of traces and state explosion. In [14], an approximation on behavioral semantics is given by the concept of causal footprint. In this study, instead of computing the similarity between each pair of process models, these models are represented as a point at Euclidian distance space. Hence the underlying problem is reduced to the nearest-neighbor problem.

A second way of defining behavioral semantic is monitoring the states in which the process model can be. This idea is realized by Nejati et al. [15] by taking the state-space of two process models and check if they can simulate one another. By counting such states, we can measure how dissimilar two process models are.

Another perspective in delta analysis is the *graph theory*, which is a useful means to analyze the process definitions. Especially, Bunke and Shearer in [13] have shown that with generic graphs, the maximum common sub-graph (MCS) is equivalent to computing the graph edit-distance emphasized in [16]. This MCS is the baseline to measure the common activities and transitions of workflow processes in [1].

3   Proposed Approach

Unlike prior studies [10, 11, 17], the main idea of this work is to perform a discrepancy analysis with respect to *consensus activity sequences* containing *dominant behavior* (i.e. common subsequence of activities in event log that are found to recur within a process instance or across process instances with some domain significance). Thus the adaptation of *sequence alignment* to process mining has created a new perspective to delta analysis; similarities are handled by analyzing just consensus activity sequences (thereby avoiding the requirement for well-structured process models).

In the previous studies, the most of the equivalence notions concentrate on atomic *true-false* answer. In reality there will seldom be a perfect fit. Hence we are interested in the *degree of similarity*. In order to do so, *dynamic scoring-based Needleman-Wunsch algorithm* is applied to quantify the similarities and deviations.

As the basic difference from the work in [21], the equations introduced in order to handle MATCH/MISMATCH scoring is modified such that while MATCH/MISMATCH default values are set to *confidence FTC threshold* in [21], these scores are dynamically determined with respect to interactions among current prefixes, their predecessors and successors. In the case of mismatch, *opportunity cost* ($oppCost(T_k^x, T_m^y, k)$ formulated by Equation (3)) is calculated for replacement of current prefixes. On the other hand, a *normalized* MATCH score is calculated with respect to *confidence FTC threshold* for match case.

3.1   Initialization

The starting point of the proposed approach is the creation of a so-called FROMTOCHART table by retrieving the activity labels from the *event logs*. Event logs consist of four major attributes: *activity, caseID, originator* and *timestamp*. For populating the table, event logs are arranged by process instances (e.g. *caseID*) and ordered by timestamp in ascending order. Then predecessor (*P*) and successor (*S*) are parsed for each transition in activity streams (traces) and the current score at $(P,S)^{th}$ element at FROMTOCHART table is incremented stepwise [9].

Afterwards accumulated scores at FROMTOCHART table are evaluated in order to prune down the weak scores. Majorly there are two evaluation metrics introduced in [10]: *confidence for from-to chart* (*confidence FTC*) and *support for from-to chart* (*support FTC*).

3.2   Derivation of Consensus Activity Sequences

This operation aims to find out the *consensus activity sequence* with the *minimum total moment value* at FROMTOCHART table by adapting *Genetic Algorithms* (GA) [11]. The coarse-grained GA stages are implemented as follows:

Initial population can be generated with or without a *schema*, which is a pattern of gene values modeled by a string of binary characters in the domain alphabet. The likelihood of finding even better solutions is increased by passing on more characteristics of this good schema to the next generation. Hence a *score top-list*, which is a data structure holding *top–n scores* at the FROMTOCHART table, is instantiated in order to build up the schema. The length of the score top–list, i.e. *n*, is formulated as $|\texttt{activityTypeDomain}|^2 \times 0.2$. Afterwards, a linear search with *O(n)* complexity is performed at this score top-list to populate the schema. As a result, a *non-intermittent schema* with the maximum length of $|\texttt{activityTypeDomain}|/3$ is constructed.

As far as GA are concerned, it is better to have higher fitness scores to provide more opportunities especially in selection stage. Therefore *moment function* introduced in [11] is used as the denominator of the fitness function to search for the individual with minimum moment.

$$f(z) = \frac{\displaystyle\sum_{i=1}^{N}\sum_{j=1}^{N} score_{ij}}{\displaystyle\sum_{i \in chromosome\, Z}\sum_{j \in chromosome\, Z} score_{ij} \times |j-i| \times p} \tag{1}$$

In Equation (1), *score*$_{ij}$ indicates total interactions from activity *i* to activity *j*, |*j – i*| is the distance from activity *i* to activity *j* at the generated sequence, and *p* is the backtracking penalty point assigned to each entry below the diagonal. Backtracking penalty point is doubled to enforce the model towards a linear arrangement.

As the selection method, *roulette wheel selection* is implemented. Roulette wheel selection is a kind of random selection type where the individual has a probability of $f(z)/\sum f(z)$ to be selected as a parent to mate.

In higher order domain alphabets, mutation and crossover framework may cause problems with chromosomes legality, e.g. multiple copies of a given activity type may occur at the offspring. Therefore in [11], we propose an alternative mutation scheme that automatically swaps the duplicate activity type with a randomly selected unobserved value. Hence a uniform chromosome that satisfies the chromosome legality is reproduced.

As a termination condition, if at least 95% of the individuals in the last population are in the *convergence band* (i.e. interval which holds upper limits for fitness score), no more new population is generated. In order to avoid premature convergence, convergence ratio has to be set appropriately. At the last iteration, individual with the highest fitness score implies the *consensus activity sequence* containing *dominant behavior*, which is the most common (likely) and typical behavior obtained on the basis of event logs. This essential sequence constitutes the backbone of the underlying business process. The transition from raw event logs to consensus activity sequence is represented in Figure 1.

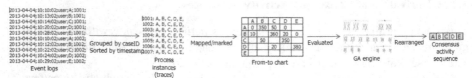

Fig. 1. Transition from Event Logs to Consensus Activity Sequence

3.3 Alignment of Consensus Activity Sequences

The fundamental idea of *Needleman-Wunsch algorithm* is to build up a global optimal alignment using previous solutions for optimal alignments of smaller consensus activity sequences. Let T_1 and T_2 be two *consensus activity sequences* generated at rearrangement step. A matrix F indexed by i and j, is constructed where the value $F(i,j)$ is the score of the best alignment between the prefix T^i_1 of T_1 and the prefix T^i_2 of T_2. $F(i,j)$ is constructed recursively by initializing $F(0,0)=0$ and then proceeding to fill the matrix from top left to bottom right. It is possible to calculate $F(i,j)$ according to neighboring values, $F(i-1,j)$, $F(i-1,j-1)$ and $F(i,j-1)$. There are *three possible ways* that the best score $F(i,j)$ of an alignment up to sub-sequences T^i_1 and T^i_2 can be obtained:

$$F(i,j)=\max\begin{cases} F(i-1,j-1)+M\left(T^i_1,T^j_2\right) \\ F(i-1,j)+I_2\left(T^j_2,T^i_1,T^{j+1}_2\right) \\ F(i,j-1)+I_1\left(T^i_1,T^j_2,T^{i+1}_1\right) \end{cases} \tag{2}$$

$M\left(T^i_1,T^j_2\right)$, $I_1\left(T^i_1,T^j_2,T^{i+1}_1\right)$ and $I_2\left(T^j_2,T^i_1,T^{j+1}_2\right)$ parameters at Equation (2) stand for MATCH/MISMATCH and INDEL (i.e. insertion or deletion) multipliers determined at *dynamic pay-off matrix*.

$$oppCost(T^x_k,T^y_m,k)=\frac{confFTC_k\left(T^{x-1}_k,T^x_k\right)\times confFTC_k\left(T^x_k,T^{x+1}_k\right)}{confFTC_k\left(T^{x-1}_k,T^y_m\right)\times confFTC_k\left(T^y_m,T^{x+1}_k\right)} \tag{3}$$

$$M\left(T^i_1,T^j_2\right)=\begin{cases} \text{if } T^i_1<>T^j_2 & avg(-\log_2(oppCost(T^i_1,T^j_2,1)),-\log_2(oppCost(T^j_2,T^i_1,2)))-confThr \\ \text{if } T^i_1=T^j_2 & -\log_2\left(\dfrac{confThr^2}{confFTC_1\left(T^i_1,T^{i+1}_1\right)\times confFTC_2\left(T^j_2,T^{j+1}_2\right)}\right)+confThr \end{cases} \tag{4}$$

While MATCH/MISMATCH default values are set to *confidence FTC threshold* (*confThr*) in [21], these scores are determined in a case-based (dynamic) manner with respect to interactions among current prefixes, T^i_1 and T^i_2, their predecessors, T^{i-1}_1 and T^{i-1}_2 and successors, T^{i+1}_1 and T^{i+1}_2. These interactions are figured out by *confidence values* ($confFTC_x(A,B)$). In the case of mismatch ($T^i_1<>T^i_2$), the *opportunity cost* ($oppCost(T^x_k,T^y_m,k)$ in Equation (3)) is calculated for replacement of current prefixes (e.g. replacement of T^i_1 with T^i_2 or vice-versa). On the other hand, a *normalized* MATCH score is calculated with respect to *confidence for from-to chart threshold* for match use-case ($T^i_1=T^i_2$) by Equation (4). Since predecessor and successor of current prefixes are subjected at *opportunity cost*, this metric resembles 3–gram model.

$$I_x(P,I,S)=-\log_2\left(\frac{confFTC_x(P,S)^2}{confFTC_x(P,I)\times confFTC_x(I,S)}\right) \tag{5}$$

Unlike to [9], INDEL scores are determined in a *dynamic manner* such that *confidence values* ($confFTC_x(A,B)$) belonging to predecessor (*P*), inserted (*I*) and successor (*S*) activity pairs are retrieved from FROMTOCHART table and the relative cost of inserting the activity *I* between activities *P* and *S* at T_x activity sequence is calculated by Equation (5). Hence *case-based* INDEL scores are generated per each element at matrix *F* for activity sequences T_1 and T_2.

The value at the *bottom right cell* of the matrix, $F(|T_1|,|T_2|)$, is the similarity score, i.e. $sim(T_1,T_2)$, for the alignment of activity sequences T_1 and T_2. This score consists of *structural* and *behavioral* components. In order to find out the optimal alignment, we must *backtrack* the path of choices from Equation (2) that led to this best score, i.e. we move from the current cell *(i,j)* to one of the neighboring cells from which the value *F(i,j)* is derived. At backtracking step, pair of symbols is added to alignment as follows:

   a. T^i_1 and T^j_2 if the move is to *(i-1,j-1)*. The value change at $sim(T_1,T_2)$ due to this move is assigned to *structural similarity*.
   b. T^i_1 and the gap symbol – if the move is to *(i-1,j)* or – and T^j_2 if the move is to *(i,j-1)*. The value change at $sim(T_1,T_2)$ due to this move is assigned to *behavioral similarity*.

Backtracking is terminated at the starting point (0,0).

4   Experimental Results

The validation of the underlying similarity measurement approach is based on the comparison of the similarity measurements of the proposed approach with the intuitive judgments of SAP consultants specialized on various modules. Basically, the intuitive judgments are collected by a questionnaire, which consists of a reference and five candidate process models designated for travel management business process. This process consists of 9 distinct activities: travel request create (A), travel request display (B), advance payment (C), travel request confirmation (D), travel request cancelation (E), travel completion (F), expense payment (G), transfer to accounting (H) and last check (I). SAP consultants rank the candidate process models according to the proposed similarities with reference model. Then these rankings are converted to *1-0 Likert chart*.

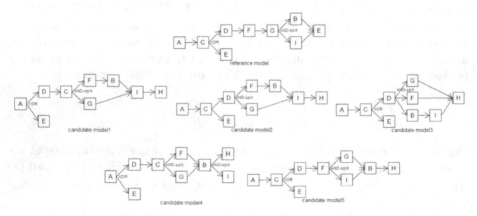

Fig. 2. Reference and Candidate Process Models

In parallel to intuitive judgments, 1000 process instances are generated for each process model by event log generator (ELG) application. This application generates event logs according to *Petri-net firing rule* such that; as the starting point, the initiator of the process (*I*) is fired and the successors of the initiator are tokenized. Then one of the tokenized activities is fired and the successors of this fired activity are tokenized. This tokenization-firing loop is continued up to a sink-typed activity (*S*) is fired. In addition to this Petri–net firing rule, a *priority factor* is assigned for each activity (e.g. travel request canceled activity seldom occurs). *Surprise effect* and *noise factor* are also taken into consideration at ELG application.

In order to eliminate inductive bias at verification phase, rearrangement and evaluation parameters introduced in [11] (e.g. confidence/support FTC thresholds, backtracking penalty point, population size, number of elite individuals, crossover and mutation probabilities) are configured in 30 consecutive sets. Afterwards, 30 distinct alignment runs are performed with respect to these underlying process configurations. According to reference process model alignment and similarity scores $sim(T_1, T_2)$, candidate process models are ranked and then these rankings are converted to 1-0 *Likert chart*.

Similarity measurements of the proposed approach (*DSBNW, dynamic scoring-based Needleman-Wunsch algorithm*) and intuitive judgments (*IJ*) obtained from SAP consultants are compared by dependent t-test, since generated event logs are dependent to reference and candidate process models, which are also the baseline for intuitive judgments. According to the t-value (-0.0117 versus $t_{0.05,48}$), the null hypothesis, H_0, which states that there is no clear distinction between similarity measurement of the proposed approach and intuitive judgments obtained from SAP consultants, is accepted. This result implies that, proposed approach appropriately reflects the perceptions of knowledge engineers (i.e. *tacit process variant assumption*). The result of t-test (α=0.05 and CI=95%) is given at Table 1(a).

Another comparison is performed with the *prior approach* (*CANW, context-aware Needleman-Wunsch algorithm*) introduced in [21]. As in the previous analysis, consensus activity sequences that belonged to reference and candidate process models are aligned by prior approach (*CANW*), while MATCH and MISMATCH default values are all set to *confidence FTC threshold value* of the current alignment run. According to reference process model alignment and similarity scores, candidate process models are ranked and then these rankings are converted to 1-0 Likert chart.

Then comparison of these two similarity metrics is performed by dependent t-test. According to the t-value (0.005 versus $t_{0.05,48}$), the null hypothesis, H_0, which states that there is no clear distinction between similarity measurements of these two approaches, is accepted. Actually there is a basic parallelism between *DSBNW* and *CANW* such that; they both avoid the requirement for well-structured process models and confidence values reflect the *behavioral* perspective of process models, while consensus activity sequence alignment majorly reflects *structural* perspective. The result of t-test (α=0.05 and CI=95%) is given at Table 1(b).

Although the previous t-test concludes that there is no significant distinction in similarity measurements of proposed approach (*DSBNW*) and prior approach (*CANW*), it is aimed to measure the relative distance of these two candidate approaches to intuitive judgments (*IJ*). Therefore, a third t-test is performed to compare similarity measurements of the prior approach (*CANW*) and intuitive judgments (*IJ*) obtained from SAP consultants. The difference between t-test values (-0.0117 versus -0.0154) given at Table 1(a) and 1(c) seemingly highlights the mechanism such that proposed approach (*DSBNW*) is *relatively more successful in reflecting structural and tacit features that are dominating intuitive judgments*. Additionally the cost functions formulized by Equations (3), (4) and (5) are more *robust* such that substitution of uncorrelated, negatively-correlated activities or insertion/deletion of activities not confirming to business rules are penalized heavily, while similar activities are encouraged to be substituted or inserted at lower costs. As a result, more sensitive similarity measurements are obtained by *DSBNW* approach. The result of underlying t-test (α=0.05 and CI=95%) is given at Table 1(c).

Table 1. t-tests for Similarity Measurement Comparison

(t-test(a) dynamic scoring-based NW vs. intuitive judgments, t-test(b) dynamic scoring-based NW vs. confidence-aware NW and t-test(c) confidence-aware NW vs. intuitive judgments)

	t–test(a)		t–test(b)		t–test(c)	
	DSBNW	IJ	DSBNW	CANW	CANW	IJ
Mean	3.5840	3.6000	3.5840	3.5720	3.5720	3.6000
Variance	27.4964	19.3067	27.4964	35.5016	35.5016	19.3067
Observations	25	25	25	25	25	25
Pooled Variance	23.4015		31.4990		27.4041	
Hypothesized Mean Difference	0.0000		0.0000		0.0000	
df	48		48		48	
t Stat	-0.0117		0.0050		-0.0154	
P(T<=t) one-tail	0.4954		0.4980		0.4979	
t Critical one-tail	1.6772		1.6772		1.6772	
P(T<=t) two-tail	0.9907		0.9960		0.9957	
t Critical two-tail	2.0106		2.0106		2.0106	

5   Conclusion

In this paper, we demonstrated that *process mining* can benefit from sequence alignment techniques, which are frequently used in *bioinformatics*. Unlike prior studies [10, 11, 17], the main idea is to realize discrepancy analysis with respect to consensus activity sequences containing *dominant behavior* (i.e. typical behavior obtained on the basis of event logs) and *confidence values* that are retrieved from *from-to chart*. Thus the adaptation of *sequence alignment* to process mining has highlighted a new perspective to delta analysis; deviations and violations are uncovered by analyzing just consensus activity sequences (thereby avoiding the requirement for well-defined process models).

While most equivalence notions concentrate on an atomic true-false answer, we are interested in the *degree of similarity*. In order to do so, *Needleman-Wunsch algorithm with a dynamic pay-off matrix* (i.e. MATCH/MISMATCH and INDEL scoring based on *confidence values*) is applied to quantify the similarities and discrepancies. Actually this similarity measurement takes *structural* and *behavioral* perspectives into account such that; consensus activity sequences reflect the structural perspective by a one-dimensional data structure, while confidence values pinpoint the interactions among activities. Additionally, Equations (3), (4) and (5) provide a robust cost function to handle substitution and insertion/deletion operations at alignment phase. According to experimental analysis, *dynamic scoring-based Needleman-Wunsch algorithm* successfully simulates the human assessment model and the results are consistent with the prior dissimilarity metric in [21].

References

1. Huang, K., Zhou, Z., Han, Y., Li, G., Wang, J.: An algorithm for calculating process similarity to cluster open-source process designs. In: Jin, H., Pan, Y., Xiao, N., Sun, J. (eds.) GCC 2004. LNCS, vol. 3252, pp. 107–114. Springer, Heidelberg (2004)
2. van der Aalst, W., Gunther, C., Recker, J., Reichert, M.: Using process mining to analyze and improve process flexibility. In: Proceeding of International Conference on Advanced Information Systems Engineering, pp. 168–177 (2006)
3. van der Aalst, W., Dongen, B.F., Herbs, J., Schimm, G., Weijters, T.A.J.M.M.: Workflow Mining: A Survey of Issues and Approaches. Data & Knowledge Engineering **47**(2), 237–267 (2003)
4. Gunther, C., van der Aalst, W.: Process mining in case handling systems. In: Proceeding of Multikonferenz Wirtschaftsinformatik (2006)
5. Agrawal, R., Gunopulos, D., Leymann, F.: Mining process models from workflow logs. In: Schek, H.-J., Saltor, F., Ramos, I., Alonso, G. (eds.) EDBT 1998. LNCS, vol. 1377, pp. 467–483. Springer, Heidelberg (1998)
6. Cook, J.E., Wolf, A.L.: Discovering Models of Software Processes from Event-Based Data. ACM TOSEM **7**(3), 215–249 (1996)
7. Weijters, T.A.J.M.M., van der Aalst, W.: Rediscovering Workflow Models from Event-Based Data Using Little Thumb. ICAE **10**(2), 151–162 (2003)
8. van der Aalst, W., Weijters, T.A.J.M.M., Maruster, L.: Workflow Mining: Discovering Process Models from Event Logs. Transaction on Knowledge and Data Engineering **16**(9), 1128–1142 (2004)
9. Esgin, E., Karagoz, P.: Sequence alignment adaptation for process diagnostics and delta analysis. In: Pan, J.-S., Polycarpou, M.M., Woźniak, M., de Carvalho, A.C., Quintián, H., Corchado, E. (eds.) HAIS 2013. LNCS, vol. 8073, pp. 191–201. Springer, Heidelberg (2013)
10. Esgin, E., Karagoz, P.: Hybrid approach to process mining: finding immediate successors of a process by using from-to chart. In: ICMLA, pp. 664–668 (2009)
11. Esgin, E., Senkul, P., Cimenbicer, C.: A hybrid approach for process mining: using from-to chart arranged by genetic algorithms. In: Graña Romay, M., Corchado, E., Garcia Sebastian, M. (eds.) HAIS 2010, Part I. LNCS, vol. 6076, pp. 178–186. Springer, Heidelberg (2010)
12. van der Aalst, W.: Business Alignment: Using Process Mining as a Tool for Delta Analysis and Conformance Testing. Requirements Engineering **10**, 198–211 (2005)
13. Bunke, H., Shearer, K.: A Graph Distance Metric Based on the Maximal Common Subgraph. Pattern Recognition Letters **19**, 255–259 (1998)
14. van Dongen, B.F., Dijkman, R., Mendling, J.: Measuring similarity between business process models. In: Bellahsène, Z., Léonard, M. (eds.) CAiSE 2008. LNCS, vol. 5074, pp. 450–464. Springer, Heidelberg (2008)
15. Nejati, S., Sabetzadeh, M., Chechik, M., Easterbrook, S., Zave, P.: Matching and merging of statecharts specifications. In: Proceeding of ICSE, pp. 54–63 (2007)
16. Zhang, K., Shasha, D.E.: Simple Fast Algorithms for the Editing Distance between Trees and Related Problems. SIAM Journal of Computing **18**(6), 1245–1262 (1989)
17. Esgin, E., Senkul, P.: Delta analysis: a hybrid quantitative approach for measuring discrepancies between business process models. In: Corchado, E., Kurzyński, M., Woźniak, M. (eds.) HAIS 2011, Part I. LNCS, vol. 6678, pp. 296–304. Springer, Heidelberg (2011)
18. Jagadeesh Chandra Bose, R.P., van der Aalst, W.: Trace alignment in process mining: opportunities for process diagnostics. In: Hull, R., Mendling, J., Tai, S. (eds.) BPM 2010. LNCS, vol. 6336, pp. 227–242. Springer, Heidelberg (2010)

19. Mendling, J., Dongen, B.F., van der Aalst, W.: On the degree of behavioral similarity between business processes. In: CEUR Workshop Proceedings, vol. 303, pp. 39–58 (2007)
20. van der Aalst, W.: Matching Observed Behavior and Modeled Behavior: An Approached Based on Petri Nets and Integer Programming. Decision Support Systems **42**(3), 1843–1859 (2006)
21. Esgin, E., Karagoz, P.: Confidence-aware sequence alignment for process diagnostics. In: SITIS, pp. 990–997 (2013)

Author Index

Printed in the United States
By Bookmasters

Moonis Ali · Young Sig Kwon
Chang-Hwan Lee · Juntae Kim
Yongdai Kim (Eds.)

Current Approaches in Applied Artificial Intelligence

28th International Conference
on Industrial, Engineering and Other Applications
of Applied Intelligent Systems, IEA/AIE 2015
Seoul, South Korea, June 10–12, 2015
Proceedings

 Springer

Editors
Moonis Ali
Texas State University
San Marcos, Texas
USA

Young Sig Kwon
Dongguk University
Seoul
Korea, Republic of (South Korea)

Chang-Hwan Lee
Dongguk University
Seoul
Korea, Republic of (South Korea)

Juntae Kim
Dongguk University
Seoul
Korea, Republic of (South Korea)

Yongdai Kim
Seoul National University
Seoul
Korea, Republic of (South Korea)

ISSN 0302-9743   ISSN 1611-3349   (electronic)
Lecture Notes in Artificial Intelligence
ISBN 978-3-319-19065-5   ISBN 978-3-319-19066-2   (eBook)
DOI 10.1007/978-3-319-19066-2

Library of Congress Control Number: 2015938754

LNCS Sublibrary: SL7 – Artificial Intelligence

Printed on acid-free paper

Springer International Publishing AG Switzerland is part of Springer Science+Business Media
(www.springer.com)

Preface

The International Society of Applied Intelligence (ISAI), through its annual IEA/AIE conferences, provides a forum for the international scientific and industrial community in the field of applied artificial intelligence to interactively participate in developing intelligent systems that are needed to solve the 21st century's ever-growing problems in almost every field.

The 28th International Conference on Industrial, Engineering & Other Applications of Applied Intelligent Systems (IEA/AIE 2015), held at COEX, Seoul, Korea from June 10 to 12, 2015, followed the IEA/AIE tradition of providing an international scientific forum for researchers in the diverse field of applied artificial intelligence.

IEA/AIE 2015 accepted 72 papers for inclusion in these proceedings out of many papers submitted from 36 different countries. Each paper was reviewed by at least two members of Program Committee, thereby facilitating the selection of high-quality papers. The papers in the proceedings cover a wide range of topics in applied artificial intelligence including reasoning, robotics, cognitive modeling, machine learning, pattern recognition, optimization, text mining, social network analysis, and evolutionary algorithms. These proceedings cover both the theory and applications of applied intelligent systems. Together, these papers highlight new trends and frontiers of applied artificial intelligence and show how new research could lead to innovative applications of considerable practical significance.

It was a great pleasure for us to organize this event. However, we could not have done it without the valuable help of many colleagues around the world. First we would like to thank the Organizing Committee members and Program Committee members for their extremely hard work and the timely return of their comprehensive reports. Without them, it would have been impossible to make decisions and to produce such high-quality proceedings on time. Second, we would like to acknowledge the contributions of all the authors of the papers submitted to the conference.

We also would like to thank our main sponsor, ISAI. The following cooperating organizations deserve our gratitude: the Association for the Advancement of Artificial Intelligence (AAAI), the Association for Computing Machinery (ACM/SIGART), the Austrian Association for Artificial Intelligence (GAI), the Catalan Association for Artificial Intelligence (ACIA), the International Neural Network Society (INNS), the Italian Association for Artificial Intelligence (AI*IA), the Japanese Society for Artificial Intelligence (JSAI), Korea Business Intelligence and Data Mining Society, Korean Information Processing Society (KIPS)., the Lithuanian Computer Society - Artificial Intelligence Section (LIKS-AIS), the Spanish Society for Artificial Intelligence (AEPIA), the Society for the Study of Artificial Intelligence and the Simulation of Behaviour (AISB), the Taiwanese Association for Artificial Intelligence (TAAI), the Taiwanese

Association for Consumer Electronics (TACE), and Texas State University – San Marcos. We would also like to thank the keynote speakers who shared their vision on applications of intelligent systems.

April 2015   Moonis Ali
                                                                                            Young Sig Kwon
                                                                                            Chang-Hwan Lee
                                                                                                  Juntae Kim
                                                                                              Yongdai Kim